AF595352

Cellular Oxidative Stress

Cellular Oxidative Stress

Editors

Silvia Dossena
Angela Marino

MDPI • Basel • Beijing • Wuhan • Barcelona • Belgrade • Manchester • Tokyo • Cluj • Tianjin

Editors
Silvia Dossena
Institute of Pharmacology and Toxicology
Paracelsus Medical University
Salzburg
Austria

Angela Marino
Department of Chemical, Biological, Pharmaceutical and Environmental Sciences
University of Messina
Messina
Italy

Editorial Office
MDPI
St. Alban-Anlage 66
4052 Basel, Switzerland

This is a reprint of articles from the Special Issue published online in the open access journal *Antioxidants* (ISSN 2076-3921) (available at: www.mdpi.com/journal/antioxidants/special_issues/Cellula_Oxidative_Stress).

For citation purposes, cite each article independently as indicated on the article page online and as indicated below:

LastName, A.A.; LastName, B.B.; LastName, C.C. Article Title. *Journal Name* **Year**, *Volume Number*, Page Range.

ISBN 978-3-0365-1644-8 (Hbk)
ISBN 978-3-0365-1643-1 (PDF)

© 2021 by the authors. Articles in this book are Open Access and distributed under the Creative Commons Attribution (CC BY) license, which allows users to download, copy and build upon published articles, as long as the author and publisher are properly credited, which ensures maximum dissemination and a wider impact of our publications.
The book as a whole is distributed by MDPI under the terms and conditions of the Creative Commons license CC BY-NC-ND.

Contents

About the Editors

Silvia Dossena

Silvia Dossena is senior scientist and Vice Chair of the Institute of Pharmacology and Toxicology of Paracelsus Medical University of Salzburg (Austria).

She obtained a Master of Biology from the University of Milan (Italy) and a PhD in Physiology from the same university. Later, she moved to Austria, where she obtained a venia docendi in pharmacology and toxicology.

The research activity of Dr. Dossena deals with the functional and molecular characterization of membrane ion transport systems in non-excitable cells and their role in human physiology and pathology.

Her current projects are related to the study of ion transport dysfunction in genetic and oxidative stress-related diseases, with a focus on ion channels and the anion exchanger pendrin (SLC26A4).

Angela Marino

Angela Marino is Associate Professor at the Department of Chemical, Biological, Pharmaceutical, and Environmental Sciences of the University of Messina, Italy.

The general interests of Prof. Marino deal with General Physiology, Cell Physiology, and Toxicology. Moreover, she teaches Human Physiology and Nutrition Physiology for studies in biology, medicine, and surgery.

Her scientific interests are related to membrane transport systems and cell volume regulation in invertebrates' cells, cultured cells, and erythrocytes; development of biological tests, such as hemolytic and cytolytic assays; and the impact of oxidative stress on cell homeostasis. Her current projects are related to the study of the anion exchanger Band 3 protein as a tool to monitor erythrocytes homeostasis and the impact of oxidative stress on Band 3 protein efficiency.

Preface to "Cellular Oxidative Stress"

Oxidative stress is a phenomenon that results from any imbalance between the production of reactive oxygen species (ROS) in cells and tissues and the ability of a biological system to detoxify these products. ROS are highly reactive molecules that consist of radical and non-radical species formed by the partial reduction of oxygen. These can be generated endogenously as a by-product of the aerobic metabolism or arise from interactions of biological macromolecules with xenobiotics or environmental stressors. Oxidative stress on a cellular level can lead to irreversible damage to nucleic acids, lipids, and proteins and activate specific intracellular signaling pathways, all of which may contribute to disease. Erythrocytes, neurons, epithelial, and immune cells are particularly exposed to oxidative stress, owing to their particular physiological function or metabolic activity. Diseases and pathophysiological conditions that have been unequivocally linked to oxidative stress include cancer, diabetes, inflammation, neurodegeneration, and aging.

This book includes 26 scientific articles, of which 17 are original research papers and 9 are reviews, unraveling the molecular mechanisms and targets of cellular oxidative stress and the potential beneficial effect of antioxidants.

As the Guest Editors, we would like to acknowledge all authors for their valuable contributions and the reviewers for their constructive remarks. A special thank you goes to the publishing team of the journal *Antioxidants* for the professional assistance in the completion of this Special Issue.

Silvia Dossena, Angela Marino

Editors

 MDPI

Editorial

Cellular Oxidative Stress

Silvia Dossena [1] **and Angela Marino** [2,*]

[1] Institute of Pharmacology and Toxicology, Paracelsus Medical University, Strubergasse 21, 5020 Salzburg, Austria; silvia.dossena@pmu.ac.at

[2] Department of Chemical, Biological, Pharmaceutical and Environmental Sciences, University of Messina, Viale F. Stagno D'Alcontres 31, 98166 Messina, Italy

* Correspondence: angela.marino@unime.it

Citation: Dossena, S.; Marino, A. Cellular Oxidative Stress. *Antioxidants* **2021**, *10*, 399. https://doi.org/10.3390/antiox10030399

Received: 2 March 2021
Accepted: 4 March 2021
Published: 6 March 2021

Publisher's Note: MDPI stays neutral with regard to jurisdictional claims in published maps and institutional affiliations.

Copyright: © 2021 by the authors. Licensee MDPI, Basel, Switzerland. This article is an open access article distributed under the terms and conditions of the Creative Commons Attribution (CC BY) license (https://creativecommons.org/licenses/by/4.0/).

Oxidative stress on a cellular level affects the function of tissues and organs and eventually of the whole body. The Special Issue "Cellular Oxidative Stress" has been conceived to collect and contribute to the dissemination of novel findings unraveling the impact of oxidative stress on cells, their subcellular components, and biological macromolecules. Great emphasis is placed on the effects of oxidative stress on erythrocytes and neurons and oxidative-stress-related conditions, including cardiovascular diseases, cancer, aging, and inflammation. In this context, the potential beneficial effects of natural and synthetic antioxidants have also been considered. Here, we offer an overview of the content of this Special Issue, which collects 17 original articles and nine reviews.

Both nucleated and anucleated cells have widely been used to study the effects of oxidative stress and antioxidants on a cellular level. Specifically, erythrocytes are continuously exposed to circulating oxidant molecules; oxidative stress associated to hyperglycemia represents an additional threat to cell homeostasis. Morabito et al. showed that the function and the expression of Band 3 protein (B3p), one of the most peculiar erythrocyte proteins, are affected in erythrocytes from diabetic subjects, and these effects were linked to the production of glycated hemoglobin (Hb) as well as oxidative stress. As oxidative stress, but not glycated Hb, was observed following the exposure of erythrocytes to increased glucose concentrations in vitro, these authors suggested that oxidative stress, rather than glycated Hb, is the key factor leading to early detrimental changes in poorly controlled hyperglycemia [1].

B3p has also been the focus of a study evaluating the effect of d-Galactose (d-Gal) on the anion exchanger activity. Despite d-Gal being known to induce oxidative stress, exposure of erythrocytes to relatively low (0.1–10 mM) d-Gal concentrations led to a reduced anion exchange capability independent from oxidative stress, but rather linked to glycated Hb production. This study sheds light on the early effects of excessive d-Gal on membrane transport systems and possible complications related to undiagnosed galactosemia [2].

Extracellular vesicles (EVs) are continuously produced in human blood from different cell types including erythrocytes, and their formation is triggered by various factors, including exposure to reactive oxygen species (ROS) and aging-associated oxidative damage. Sudnitsyna et al. showed that, following oxidative stress, erythrocytes produce EVs containing hemoglobin oxidized to hemichromes. Oxidative stress led to caspase-3 activation and B3p clustering in cells and EVs, events that are normally linked to eryptosis and removal of senescent erythrocytes from the blood circulation. Based on these findings, the authors suggested that erythrocytes might eliminate damaged hemoglobin by vesiculation as a protective mechanism to prolong their lifespan during oxidative stress [3].

In erythrocytes infected by *Plasmodium falciparum*, oxidative stress induces the production of hemichromes, which contain partially denatured hemoglobin with reactive iron. In turn, oxidative stress generated by hemichromes contributes to the activation of artemisin, a component of the standard treatment for *P. falciparum* infection. Tsamesidis et al. showed that Syk kinase inhibitors increase oxidative stress in parasitized erythrocytes by inhibiting the release of hemichromes and synergize with artemisin in producing a toxic effect against the parasite. These authors suggested that Syk kinase inhibitors might represent a useful strategy to increase the efficacy of artemisin-based combination therapies, especially in resistant strains [4].

Neurons are particularly vulnerable to oxidative damage; it is, therefore, not surprising that ROS play a fundamental role in the exacerbation and progression of neurodegeneration. Non-coding RNAs respond to oxidative stress with changes in their cellular abundance and, in turn, regulate gene expression networks involved in ROS homeostasis and buffering. Gámez-Valero and co-authors summarized the current knowledge on the role of non-coding RNAs in Alzheimer's disease, Parkinson's disease, Huntington's disease, and amyotrophic lateral sclerosis and suggested that these biomolecules might represent novel therapeutic targets in neurodegenerative conditions [5].

Oxidative stress in the central nervous system plays a fundamental role not only in neurodegeneration, but also in epilepsy. In *status epilepticus*, oxidative stress and neuroinflammation may lead to mitochondrial dysfunction, structural damage, and cell death. Lin et al. summarized the current knowledge supporting the use of antioxidants in epilepsy from animal studies as well as clinical trials [6].

Traumatic brain injury (TBI) is a major cause of death among youth in industrialized societies. The review "Management of Traumatic Brain Injury: From Present to Future" collects the current knowledge on TBI pathophysiology and examines existing and potential new therapeutic strategies, including the use of antioxidants, in the management of inflammatory events and behavioral deficits associated with TBI [7].

Barancik and co-authors reviewed the molecular mechanisms of the antioxidant and anti-inflammatory activities of molecular hydrogen in the cardiovascular and central nervous system and dissected the corresponding effects on intracellular signaling pathways, gene expression, and autophagy. Molecular hydrogen exerts a dual action in protecting cells against oxidative stress by directly reacting with ROS and inducing the cellular antioxidant machinery. These authors suggested that the therapeutic potential of molecular hydrogen in pathological conditions including ischemia-reperfusion injury, brain edema, and neurodegeneration deserves to be further investigated [8].

Oxidative stress plays a crucial role in cardiac and vascular abnormalities in different types of cardiovascular diseases. Han et al. summarized the current knowledge on how oxidative stress could contribute to the pathogenesis of one of the most serious complication of preeclampsia (PE), i.e., a prothrombotic state that can lead to ischemic heart disease, stroke, and venous thromboembolism, especially in those with pre-existing medical conditions [9]. Studying the impact of oxidative stress on hemostatic actions will help identify key targets for prediction, prophylaxis, and treatment of PE.

Cancer initiation and progression have been associated with oxidative-stress-enhanced DNA damage. In turn, the high metabolism of cancer cells is associated with an increase in ROS production. Consequently, cancer cells develop resistance to oxidative stress as one of their major adaptive changes. Peroni et al. reported the resistance of the human astrocytoma ADF cells to oxidative stress, with specific regard to the enzymatic activities involved in 4-hydroxynonenal (HNE) removal. ADF cells counteracted oxidative stress conditions better than normal cells, thus confirming the redox adaptation demonstrated for several cancer cells [10]. The clarification of this aspect may indicate new enzymatic targets to be inhibited to antagonize astrocytoma cells survival.

Gaikwad and Srivastava highlighted that ROS play a dual role in cancer; below certain threshold values, ROS act as signaling molecules leading to activation of oncogenic pathways. However, high levels of ROS exhibit an anti-tumoral effect leading to cellular death through various programmed cell death pathways including apoptosis, autophagy, ferroptosis, pyroptosis, and anoikis. Phytochemicals can stimulate ROS accumulation beyond the threshold value and could, therefore, be considered as part of a therapeutic strategy to promote programmed cell death in cancer [11].

Antineoplastic agents induce oxidative stress in cancer as well as in normal cells. Accordingly, ROS production and consequent impairment of mitochondrial function substantially contribute to cisplatin nephrotoxicity. The scaffolding protein Na^+/H^+ exchanger regulatory factor 1 (NHERF1) plays a fundamental role in the maintenance of a proper kidney ion transport and metabolism. Bushau-Sprinkle and co-authors showed that NHERF1 loss imposes a shift in cell metabolism towards the pentose phosphate pathway, which may sensitize kidney cells to the oxidative stress caused by cisplatin. These findings identify novel potential biomarkers and therapeutic targets in the detection and prevention of cisplatin-induced nephrotoxicity [12].

Oxidative stress plays a fundamental role in aging. In this regard, age-related macular degeneration (AMD) has been considered by Ulanczyk et al. Significant alterations in the levels of several crucial antioxidant enzymes in red blood cells and platelets from patients with AMD have been observed, indicating that the equilibrium of the endogenous antioxidant system is disrupted in AMD. A diet rich in green vegetables, fish, and omega-3-rich oils, together with physical exercise, might delay disease progression and help retain a better visual function in patients with AMD [13].

Elevated oxidative stress represents a striking aggravating factor of retinal dystrophies, including age-related macular degeneration (AMD) and retinitis pigmentosa (RP). Donato and co-authors analyzed genes differentially expressed and differentially alternatively spliced in human retinal pigment epithelium cells exposed to the oxidant agent N-retinylidene-N-retinylethanolamine (A2E) or left untreated. This study identifies novel pathways involved in the oxidative-stress-induced etiopathogenesis of retinal dystrophies [14].

The elderly are most likely the main population to be treated by stem cell therapy. Stem cells may be negatively influenced by oxidative stress in our body; therefore, the restoration of the cellular functions of stem cells injured by oxidative stress is crucial. In this context, rescue of adipose-derived stem cells (ADSCs) from diseased or aged patients and the modulation of their activity by substance P is anticipated to increase the therapeutic potential of stem cell transplantation [15].

Aging is also characterized by reduced immune response, a process known as immunosenescence. In particular, the immunological memory, which is a typical feature of the adaptive immune system, is impaired in the elderly. Oxidative stress was shown to negatively affect the maintenance of immunological memory in old age and promote the onset of a pro-inflammatory environment in the bone marrow [16]. Meryk et al. showed that the antioxidants vitamin C and N-acetylcysteine decreased oxidative stress as well as the expression of pro-inflammatory molecules in the bone marrow and spleen of aged mice, boosted the production of new memory T cells, and activated dendritic antigen-presenting cells. These findings suggest that the generation and maintenance of memory T cells in old age may be improved by targeting oxidative stress [17].

Continued oxidative stress can cause chronic inflammation. Cigarette smoke, which contains high concentrations of oxidants, promotes chronic inflammation in the airways, and is involved in the pathogenesis of chronic rhinosinusitis (CRS). In CRS, excessive production of ROS not only affects the inflammatory response but also leads to tissue remodeling. Park et al. demonstrated the role of the ROS/PI3K/Akt and NF-κB signaling pathways in mediating cigarette smoke extract-stimulated MMP-2/TIMP-2 imbalance in nasal fibroblasts, which might contribute to tissue remodeling in CRS [18].

Lazado et al. focused on the use of oxidative chemical compounds in several husbandry practices, to understand how chemical stressors impact fish physiology, especially at mucosal surfaces. Their study revealed the interplay between oxidative stress and the nasal microenvironment of Atlantic salmon at the molecular and cellular level. The results offer insights into the oxidative stress responses at the nasal olfactory mucosa, which is considered the most ancient arm of the mucosal immune system in vertebrates [19]. From a practical perspective, data suggest that some of the oxidant chemicals commonly used in aquaculture could trigger oxidative stress that, if chronic, may be detrimental to fish health.

"An Update of Palmitoylethanolamide and Luteolin Effects in Preclinical and Clinical Studies of Neuroinflammatory Events" focuses on neuroinflammation. The review analyzes the key role of N-palmitoylethanolamine that, though lacking in antioxidant effects, exhibits powerful neuroprotective and anti-inflammatory properties and its co-ultramicronization with the flavonoid luteolin, which makes it more effective than the molecule alone [20].

Vascular permeability is increased in inflammation. Freitas et al. studied the impact of free radical generation on microvascular permeability. These authors showed that bradykinin and histamine utilize different signaling pathways to increase microvascular permeability, i.e., ROS and nitric oxide, respectively, and cytokines may potentiate the bradykinin effect. Importantly, animals treated with simvastatin did not display potentiation of bradykinin-induced microvascular permeability by cytokines. This finding underpins the anti-inflammatory properties of statins, which may be directly associated with their cellular antioxidant activities [21].

Acute pancreatitis is an inflammatory process of the pancreatic tissue that may lead to liver injury, and obesity is a risk factor for the development of hepatic complications in the context of acute pancreatitis. Rius-Pérez and collaborators showed that pancreatitis leads to marked induction of the transcriptional co-activator PGC-1α in the mouse liver, with consequent protection from nitrosative stress. Obesity caused PGC-1α deficiency and enhanced nitrosative stress in the liver during pancreatitis. These findings underscore a novel protective role of PGC-1α in preventing nitrosative stress in the liver during the development of acute pancreatitis [22].

Epigenetic mechanisms regulate the expression of genes involved in inflammation. Phenolic compounds, which are very well known for their antioxidant properties, have the ability to modulate gene expression through the regulation of epigenetic mechanisms, including DNA methylation, histone modification, and miRNA expression. Ciz and co-authors reviewed these aspects and suggested that targeting epigenetics by dietary polyphenolic compounds may represent an attractive strategy in the prevention and treatment of inflammatory conditions [23].

The use of natural antioxidants and nutraceuticals in the context of oxidative-stress-related diseases are the subject of intense investigation. Among natural antioxidants with beneficial effects against inflammation, cashew (*Anacardium occidentale* L.) nuts improve the endogenous antioxidant activity and limit pro-inflammatory cytokines release [24]. This study increases the knowledge on the role of foods in the modulation of oxidative stress and inflammation in an in vivo experimental model, thus opening the way to new investigations on the benefits of a balanced diet in ensuring optimal health.

Rizzo reviewed the published clinical trials on the possible antioxidant effect of soy, soy foods, and soy bioactive substances. This author denoted that, despite the generally accepted beneficial role of soy on human health, its antioxidant properties have not been unequivocally demonstrated and deserve to be further investigated [25].

A diet enriched in legumes is fundamental to positively modulate redox balance, lipid metabolism, and insulin sensitivity. Centrone and co-authors studied the nutraceutical features of two genetically and phenotypically distinct chickpea cultivars, i.e., MG_13 and PI358934. Both chickpea accessions showed significant antioxidant ability; however, only MG_13 reduced the lipid over-accumulation in steatotic rat hepatoma cells and in the liver of mice fed on a high-fat diet. These findings underscore the importance of studies characterizing the distinct effects and precise composition of nutraceuticals [26].

Although widely recognized as beneficial for human health, antioxidants may exhibit toxic pro-oxidant effects following reaction with reactive chemical species. Giordano and co-authors showed that Trolox (6-hydroxy-2,5,7,8-tetramethylchroman-2-carboxylic acid), a hydrophilic analog of vitamin E well known for its strong antioxidant activity, develops pro-oxidant properties in HeLa cells when used at concentrations higher than 20 μM and leads to apoptotic cell death. This study denotes that the antioxidant activity of bioactive compounds may have a concentration threshold that must be carefully considered in the context of an antioxidant therapy [27].

Funding: This research received no external funding.

Conflicts of Interest: The authors declare no conflict of interest.

References

1. Morabito, R.; Remigante, A.; Spinelli, S.; Vitale, G.; Trichilo, V.; Loddo, S.; Marino, A. High Glucose Concentrations Affect Band 3 Protein in Human Erythrocytes. *Antioxidants* **2020**, *9*, 365. [CrossRef]
2. Remigante, A.; Morabito, R.; Spinelli, S.; Trichilo, V.; Loddo, S.; Sarikas, A.; Dossena, S.; Marino, A. d-Galactose Decreases Anion Exchange Capability through Band 3 Protein in Human Erythrocytes. *Antioxidants* **2020**, *9*, 689. [CrossRef] [PubMed]
3. Sudnitsyna, J.; Skverchinskaya, E.; Dobrylko, I.; Nikitina, E.; Gambaryan, S.; Mindukshev, I. Microvesicle Formation Induced by Oxidative Stress in Human Erythrocytes. *Antioxidants* **2020**, *9*, 929. [CrossRef] [PubMed]
4. Tsamesidis, I.; Reybier, K.; Marchetti, G.; Pau, M.C.; Virdis, P.; Fozza, C.; Nepveu, F.; Low, P.S.; Turrini, F.M.; Pantaleo, A. Syk Kinase Inhibitors Synergize with Artemisinins by Enhancing Oxidative Stress in *Plasmodium falciparum*-Parasitized Erythrocytes. *Antioxidants* **2020**, *9*, 753. [CrossRef]
5. Gamez-Valero, A.; Guisado-Corcoll, A.; Herrero-Lorenzo, M.; Solaguren-Beascoa, M.; Marti, E. Non-Coding RNAs as Sensors of Oxidative Stress in Neurodegenerative Diseases. *Antioxidants* **2020**, *9*, 1095. [CrossRef]
6. Lin, T.K.; Chen, S.D.; Lin, K.J.; Chuang, Y.C. Seizure-Induced Oxidative Stress in *Status Epilepticus*: Is Antioxidant Beneficial? *Antioxidants* **2020**, *9*, 1029. [CrossRef]
7. Crupi, R.; Cordaro, M.; Cuzzocrea, S.; Impellizzeri, D. Management of Traumatic Brain Injury: From Present to Future. *Antioxidants* **2020**, *9*, 297. [CrossRef]
8. Barancik, M.; Kura, B.; LeBaron, T.W.; Bolli, R.; Buday, J.; Slezak, J. Molecular and Cellular Mechanisms Associated with Effects of Molecular Hydrogen in Cardiovascular and Central Nervous Systems. *Antioxidants* **2020**, *9*, 1281. [CrossRef]
9. Han, C.; Huang, P.; Lyu, M.; Dong, J. Oxidative Stress and Preeclampsia-Associated Prothrombotic State. *Antioxidants* **2020**, *9*, 1139. [CrossRef] [PubMed]
10. Peroni, E.; Scali, V.; Balestri, F.; Cappiello, M.; Mura, U.; Del Corso, A.; Moschini, R. Pathways of 4-Hydroxy-2-Nonenal Detoxification in a Human Astrocytoma Cell Line. *Antioxidants* **2020**, *9*, 385. [CrossRef]
11. Gaikwad, S.; Srivastava, S.K. Role of Phytochemicals in Perturbation of Redox Homeostasis in Cancer. *Antioxidants* **2021**, *10*, 83. [CrossRef]
12. Bushau-Sprinkle, A.; Barati, M.T.; Gagnon, K.B.; Khundmiri, S.J.; Kitterman, K.; Hill, B.G.; Sherwood, A.; Merchant, M.; Rai, S.N.; Srivastava, S.; et al. NHERF1 Loss Upregulates Enzymes of the Pentose Phosphate Pathway in Kidney Cortex. *Antioxidants* **2020**, *9*, 862. [CrossRef] [PubMed]
13. Ulanczyk, Z.; Grabowicz, A.; Cecerska-Heryc, E.; Sleboda-Taront, D.; Krytkowska, E.; Mozolewska-Piotrowska, K.; Safranow, K.; Kawa, M.P.; Dolegowska, B.; Machalinska, A. Dietary and Lifestyle Factors Modulate the Activity of the Endogenous Antioxidant System in Patients with Age-Related Macular Degeneration: Correlations with Disease Severity. *Antioxidants* **2020**, *9*, 954. [CrossRef] [PubMed]
14. Donato, L.; D'Angelo, R.; Alibrandi, S.; Rinaldi, C.; Sidoti, A.; Scimone, C. Effects of A2E-Induced Oxidative Stress on Retinal Epithelial Cells: New Insights on Differential Gene Response and Retinal Dystrophies. *Antioxidants* **2020**, *9*, 307. [CrossRef]
15. Park, J.S.; Piao, J.; Park, G.; Hong, H.S. Substance-P Restores Cellular Activity of ADSC Impaired by Oxidative Stress. *Antioxidants* **2020**, *9*, 978. [CrossRef] [PubMed]
16. Pangrazzi, L.; Meryk, A.; Naismith, E.; Koziel, R.; Lair, J.; Krismer, M.; Trieb, K.; Grubeck-Loebenstein, B. "Inflamm-aging" influences immune cell survival factors in human bone marrow. *Eur. J. Immunol.* **2017**, *47*, 481–492. [CrossRef]
17. Meryk, A.; Grasse, M.; Balasco, L.; Kapferer, W.; Grubeck-Loebenstein, B.; Pangrazzi, L. Antioxidants N-Acetylcysteine and Vitamin C Improve T Cell Commitment to Memory and Long-Term Maintenance of Immunological Memory in Old Mice. *Antioxidants* **2020**, *9*, 1152. [CrossRef] [PubMed]
18. Park, J.H.; Shin, J.M.; Yang, H.W.; Kim, T.H.; Lee, S.H.; Lee, H.M.; Cho, J.G.; Park, I.H. Cigarette Smoke Extract Stimulates MMP-2 Production in Nasal Fibroblasts via ROS/PI3K, Akt, and NF-kappaB Signaling Pathways. *Antioxidants* **2020**, *9*, 739. [CrossRef]
19. Lazado, C.C.; Voldvik, V.; Breiland, M.W.; Osorio, J.; Hansen, M.H.S.; Krasnov, A. Oxidative Chemical Stressors Alter the Physiological State of the Nasal Olfactory Mucosa of Atlantic Salmon. *Antioxidants* **2020**, *9*, 1144. [CrossRef] [PubMed]
20. Cordaro, M.; Cuzzocrea, S.; Crupi, R. An Update of Palmitoylethanolamide and Luteolin Effects in Preclinical and Clinical Studies of Neuroinflammatory Events. *Antioxidants* **2020**, *9*, 216. [CrossRef]
21. Freitas, F.; Tibirica, E.; Singh, M.; Fraser, P.A.; Mann, G.E. Redox Regulation of Microvascular Permeability: IL-1beta Potentiation of Bradykinin-Induced Permeability is Prevented by Simvastatin. *Antioxidants* **2020**, *9*, 1269. [CrossRef]
22. Rius-Perez, S.; Torres-Cuevas, I.; Monsalve, M.; Miranda, F.J.; Perez, S. Impairment of PGC-1 Alpha Up-Regulation Enhances Nitrosative Stress in the Liver during Acute Pancreatitis in Obese Mice. *Antioxidants* **2020**, *9*, 887. [CrossRef] [PubMed]
23. Ciz, M.; Dvorakova, A.; Skockova, V.; Kubala, L. The Role of Dietary Phenolic Compounds in Epigenetic Modulation Involved in Inflammatory Processes. *Antioxidants* **2020**, *9*, 691. [CrossRef] [PubMed]
24. Cordaro, M.; Siracusa, R.; Fusco, R.; D'Amico, R.; Peritore, A.F.; Gugliandolo, E.; Genovese, T.; Scuto, M.; Crupi, R.; Mandalari, G.; et al. Cashew (*Anacardium occidentale* L.) Nuts Counteract Oxidative Stress and Inflammation in an Acute Experimental Model of Carrageenan-Induced Paw Edema. *Antioxidants* **2020**, *9*, 660. [CrossRef]
25. Rizzo, G. The Antioxidant Role of Soy and Soy Foods in Human Health. *Antioxidants* **2020**, *9*, 635. [CrossRef] [PubMed]

26. Centrone, M.; Gena, P.; Ranieri, M.; Di Mise, A.; D'Agostino, M.; Mastrodonato, M.; Venneri, M.; De Angelis, D.; Pavan, S.; Pasqualone, A.; et al. In Vitro and In Vivo Nutraceutical Characterization of Two Chickpea Accessions: Differential Effects on Hepatic Lipid Over-Accumulation. *Antioxidants* **2020**, *9*, 268. [CrossRef] [PubMed]
27. Giordano, M.E.; Caricato, R.; Lionetto, M.G. Concentration Dependence of the Antioxidant and Prooxidant Activity of Trolox in HeLa Cells: Involvement in the Induction of Apoptotic Volume Decrease. *Antioxidants* **2020**, *9*, 1058. [CrossRef] [PubMed]

Article

High Glucose Concentrations Affect Band 3 Protein in Human Erythrocytes

Rossana Morabito [1,†], Alessia Remigante [1,†], Sara Spinelli [1], Giulia Vitale [1], Vincenzo Trichilo [2], Saverio Loddo [2] and Angela Marino [1,*]

[1] Department of Chemical, Biological, Pharmaceutical and Environmental Sciences, University of Messina, Viale F. Stagno D'Alcontres 31-98166, 98122 Messina, Italy; rmorabito@unime.it (R.M.); aremigante@unime.it (A.R.); sara.spinelli1992@libero.it (S.S.); giulia.vitale19@gmail.com (G.V.)

[2] Department of Clinical and Experimental Medicine, AOU Policlinico Universitario "G. Martino", Via Consolare Valeria-98125, 98124 Messina, Italy; vtrichilo@unime.it (V.T.); sloddo@unime.it (S.L.)

* Correspondence: marinoa@unime.it; Tel.: +39-(0)90-676-5214

† These authors contributed equally to this work.

Received: 3 April 2020; Accepted: 26 April 2020; Published: 27 April 2020

Abstract: Hyperglycemia is considered a threat for cell homeostasis, as it is associated to oxidative stress (OS). As erythrocytes are continuously exposed to OS, this study was conceived to verify the impact of either diabetic conditions attested to by glycated hemoglobin (Hb) levels (>6.5% or higher) or treatment with high glucose (15–35 mM, for 24 h) on erythrocyte homeostasis. To this aim, anion exchange capability through the Band 3 protein (B3p) was monitored by the rate constant for SO_4^{2-} uptake. Thiobarbituric acid reactive species (TBARS), membrane sulfhydryl groups mostly belonging to B3p, glutathione reduced (GSH) levels, and B3p expression levels were also evaluated. The rate constant for SO_4^{2-} uptake (0.063 ± 0.001 min^{-1}, 16 min in healthy volunteers) was accelerated in erythrocytes from diabetic volunteers (0.113 ± 0.001 min^{-1}, 9 min) and after exposure to high glucose (0.129 ± 0.001in^{-1}, 7 min), but only in diabetic volunteers was there an increase in TBARS levels and oxidation of membrane sulfhydryl groups, and a decrease in both GSH and B3p expression levels was observed. A combined effect due to the glycated Hb and OS may explain what was observed in diabetic erythrocytes, while in in vitro hyperglycemia, early OS could explain B3p anion exchange capability alterations as proven by the use of melatonin. Finally, measurement of B3p anion exchange capability is a suitable tool to monitor the impact of hyperglycemia on erythrocytes homeostasis, being the first line of high glucose impact before Hb glycation. Melatonin may be useful to counteract hyperglycemia-induced OS at the B3p level.

Keywords: diabetes; glucose exposure; oxidative stress; Band 3 protein; erythrocytes; SO_4^{2-}

---

1. Introduction

Diabetes mellitus type 2 (T2D) is a metabolic disease characterized by hyperglycemia, consisting of high glucose levels chronically present in the blood. According to reports by the World Health Organization (WHO), over 422 million people worldwide are affected by diabetes which is directly responsible for 1.6 million deaths each year [1]. Currently, about 10.3% of the adult population in Europe is estimated to have diabetes, and, as expected, this percentage is going to increase substantially by 2030 due to the fact of obesity and aging [2]. Oxidative stress (OS) is critically involved in diabetes pathogenesis [3] and hyperglycemia is the principal factor in the early stages of development [4]. In addition, hyperglycemic complications, such as glycosylated proteins formation [5], increase reactive oxygen species (ROS) generation, decrease nitric oxide (NO) production, and activation of the protein kinase C (PKC) and polyol pathways may contribute to OS [6].

During their 120 days life span, human erythrocytes are constantly exposed to glucose and other molecules present in the blood (i.e. oxidant compounds), and they have been widely investigated for their important role in different physiological conditions due to the fact of their metabolism and sensitivity to OS [7]. With regard to the effect on hyperglycemia, several studies have been performed. In this context, methylglyoxal concentration, a glucose metabolite involved in advanced glycation end-products formation (AGEs), is abnormally increased in plasma and in erythrocytes, finally exhibiting phosphatidylserine (PS) exposure on the external surface [8] and consequent eryptosis [9]. Hyperglycemia in T2D causes altered lipid–protein interactions, leading to OS and peroxidation [10] which, in turn, alters membrane lipids and membrane-bound enzyme arrangements [11]. Among the consequences of hyperglycemia, glycation of membrane proteins and the resulting decrease in protein activities have also been considered. In this regard, a typical example is provided by glycated hemoglobin (A1c), considered as a standard test to monitor glycemic status [12] since its discovery in the late 1960s.

Oxidative damage linked to hyperglycemic conditions has also been shown to reduce the life span and rheological properties of erythrocytes, altering their shape and deformability [13] which critically correlate with Band 3 protein (B3p) function [14]. Band 3 protein, distributed in millions of copies on erythrocytes' membranes, is responsible for Cl^-/HCO_3^- exchange through erythrocyte membranes for ion balance and gas exchange, thus reflecting erythrocytes' homeostasis as well as homeostasis of the whole organism [15]. Band 3 protein anion exchange capability, measured by evaluating the rate constant for SO_4^{2-} uptake, which is more easily detectable than Cl^- or HCO_3^- exchange [14], is a good tool to verify erythrocyte function under different experimental conditions (i.e., under oxidants or toxins in vitro [16–20]) or in diseases associated to OS [21–23]. In addition, it is useful to prove the possible effects of antioxidants in preventing oxidative damage at the B3p level [24,25]. The attention paid to dietary antioxidants, influencing antioxidant defense by scavenging free radicals and reactive species [26], is important, as they are capable of counteracting OS and preventing diabetic complications [27,28]. Among antioxidants introduced by food or adopted in clinic therapy to counteract OS, melatonin (Mel) was considered in the present study [29,30]. Melatonin, a neuroendocrine hormone primarily produced by the pineal gland, has been shown to exert its antioxidant power on lipids and proteins [31], though its antioxidant activity depends on concentration and cell target [32,33].

Based on this evidence, the aim of the present study was to establish whether high glucose concentrations in the extracellular medium could affect the erythrocytes homeostasis via B3p alterations, as it is widely established that poorly controlled or uncontrolled diabetes leads to hyperglycemia and, consequently, to OS and glycosylation in various tissues. According to this hypothesis, the general goal was to verify whether the anion exchange capability through B3p is affected by hyperglycemia and whether the possible alteration is related to oxidative events. In this context, the following specific aims were: (1) to monitor the anion exchange capability through B3p in erythrocytes deriving from diabetic volunteers with A1c levels higher than in healthy subjects and to verify the presence of related oxidative events; (2) to monitor the anion exchange capability through B3p in in vitro hyperglycemic conditions obtained by exposing erythrocytes from healthy volunteers to increasing concentrations of glucose (15–35 mM) for 24 hours; (3) to prove the effect of OS on the anion exchange capability of B3p under in vitro-induced hyperglycemia; (4) to assay the potential antioxidant effect of Mel (100 μM) in vitro-induced hyperglycemia according to what has previously been demonstrated [32].

2. Materials and Methods

2.1. Solutions and Chemicals

The chemicals used for the present experimental protocols were purchased from Sigma (Milan, Italy). 4,4′-Diisothiocyanato-stilbene-2,2′-disulfonate (DIDS), prepared in DMSO (dimethyl sulfoxide), was diluted from 10 mM stock solution. Melatonin was prepared in 0.5% *v*/*v* ethanol and diluted from a 100 mM stock solution. The DMSO and ethanol were assayed on erythrocytes at their final concentrations to preventively exclude any damage.

2.2. Erythrocytes Preparation

Blood was obtained upon orally informed consent from both healthy and diabetic volunteers. Blood samples were used after all clinical analysis were completed. Blood, collected in tubes containing anticoagulant, was washed with the following isotonic solution: 145 mM NaCl, 5 mM glucose, 5 mM HEPES (4-(2-hydroxyethyl)-1 piperazineethanesulfonic acid), pH 7.4, osmotic pressure 300 mOsm. The samples were centrifuged thrice (ThermoScientific, 1200× *g*, 5 min) to discard both plasma and buffy coat. Erythrocytes were then suspended in the isotonic solution at different hematocrit values, according to the experimental protocols reported below. Erythrocytes samples were handled according to two groups: (i) erythrocytes from diabetic volunteers (A1c > 6.5% or higher) [12]; (ii) erythrocytes withdrawn from healthy volunteers (A1c < 5.7%) [12], and successively incubated at different glucose concentrations (glucose-treated erythrocytes). With regard to this latter point, erythrocytes (3% hematocrit) were incubated for 24 h at 25 °C in isotonic solution containing different glucose concentrations, alternatively 5 mM, 15 mM, or 35 mM and then addressed to the experimental protocols.

Erythrocytes deriving from diabetic volunteers are henceforth referred to as diabetic erythrocytes, while erythrocytes deriving from healthy volunteers and then exposed for 24 h to different glucose concentrations are henceforth referred to as glucose-treated erythrocytes.

2.3. Osmotic Fragility in Diabetic or Glucose-Treated Erythrocytes

To verify the osmotic fragility, diabetic or glucose-treated erythrocytes, after washing in isotonic solution, were suspended at 0.5% hematocrit (isotonic solution) and then centrifuged (ThermoScientific, 25 °C, 1200× *g*, 5 min) and resuspended in a 0.7% *v*/*v* NaCl solution at 0.05% hematocrit. Hemoglobin absorbance was measured at 405 nm wavelength at different time intervals between 5 and 180 min of incubation [34,35]. Diabetic erythrocytes were directly addressed to an osmotic fragility test, while high glucose-exposed erythrocytes were assayed after 24 h incubation in high glucose solution. Analysis of blank (hemolysis solution absorbance) was also performed.

2.4. SO_4^{2-} Uptake Measurement

2.4.1. Control Condition

The SO_4^{2-} uptake through B3p was measured as described elsewhere [21,36]. After washing, erythrocytes, derived from healthy volunteers, were suspended to 3% hematocrit in 35 mL isotonic solution containing SO_4^{2-}, henceforth called SO_4^{2-} medium (118 mM Na_2SO_4, 10 mM HEPES, 5 mM glucose, pH 7.4, osmotic pressure 300 mOsm), and incubated at 25 °C. At fixed-time intervals (5–10–15–30–45–60–90–120 min), 5 mL samples of red blood cell suspensions were transferred in a tube containing DIDS (10 μM), a specific and irreversible inhibitor B3p [37], and kept on ice. At the end of incubation in SO_4^{2-} medium, red blood cells were at least thrice washed in isotonic solution (ThermoScientific, 4 °C, 1200× *g*, 5 min) in order to eliminate SO_4^{2-} from the external medium and then hemolysed by distilled water (1 mL). Proteins were then precipitated by perchloric acid (4% *v*/*v*). After centrifugation (4 °C, 2500× *g*, 10 min), the supernatant containing SO_4^{2-} underwent turbidimetric analysis. The SO_4^{2-} precipitation was performed by mixing the following components: 500 μL supernatant from each sample, 1 mL glycerol previously diluted in distilled water (1:1), 1 mL 4 M NaCl plus HCl (hydrochloric acid 37%) solution (12:1) and, finally, 500 μL 1.24 M $BaCl_2{\cdot}2H_2O$. Each sample was then spectrophotometrically read at 425 nm wavelength. A calibrated standard curve previously obtained by precipitating known SO_4^{2-} concentrations was used to convert the absorption to $[SO_4^{2-}]$ L cells × 10^{-2}. Moreover, the rate constant (min^{-1}) was calculated by the equation: $C_t = C_\infty (1 - e^{-rt}) + C_0$, where C_t, C_∞, and C_0 represent the intracellular SO_4^{2-} concentrations, respectively, at time t, 0, and ∞; *e* is the Neper number (2.7182818); *r* is the rate constant accounting for the process velocity, and *t* is the time fixed for each sample withdrawal (5–10–15–30–45–60–90–120 min). The rate constant is the time needed to reach 63% of total SO_4^{2-} intracellular concentration [21] and [SO_4^{2-}] L cells × 10^{-2} reported in figure stands for SO_4^{2-} micromolar concentration trapped by 10 mL erythrocytes (3% hematocrit). In a separate protocol, in order to assess that SO_4^{2-} was effectively trapped by B3p, red blood cells were suspended in SO_4^{2-} medium (3% hematocrit) and immediately treated with DIDS (10 μM). Then, 5 mL samples at fixed time intervals (5–10–15–30–45–60–90–120 min) were handled as described for control conditions.

2.4.2. SO_4^{2-} Uptake Measurement in Diabetic or Glucose-Treated Erythrocytes

Erythrocytes from diabetic volunteers after washing were centrifuged (ThermoScientific, 4 °C, 1200× *g*, 5 min) to replace supernatant with SO_4^{2-} medium (3% hematocrit). The SO_4^{2-} uptake was successively determined as described for control conditions. With regard to glucose-treated erythrocytes, SO_4^{2-} uptake measurement was performed after 24 h incubation at different glucose concentrations by centrifuging and re-suspending samples in SO_4^{2-} medium at 3% hematocrit.

2.4.3. Preparation of Resealed Ghosts of Diabetic or Glucose-Treated Erythrocytes and SO_4^{2-} Uptake Measurement

Pink resealed ghosts of human erythrocytes, henceforth referred to as resealed ghosts, were prepared, as reported elsewhere [38], with slight modifications. In detail, erythrocytes from either diabetic volunteers or derived from healthy volunteers and incubated for 24 h at different glucose concentrations were washed and re-suspended in 35 mL cold hypoosmotic buffer solution (2.5 mM NaH2PO4, 5 mM HEPES, pH 7.4, 3% hematocrit). After 10 min of gently shaking at 0 °C, intracellular content and hemoglobin were discarded by several centrifugations (Beckman J2-21, 4 °C, 17,000× *g*, 20 min). Then, the supernatant was replaced by 35 mL isotonic resealing medium (145 mM NaCl, 5 mM HEPES, 5 mM Glucose, pH 7.4), previously maintained at 37 °C, and the membranes were incubated for 45 min at 37 °C to allow resealing. Finally, resealed ghosts, containing approximately 10% of the original hemoglobin content, were addressed to SO_4^{2-} uptake measurement, according to the protocol described above for intact erythrocytes in control conditions [16].

2.5. *TBARS Levels in Diabetic or Glucose-Treated Erythrocytes*

In order to prove a possible effect of lipid peroxidation on erythrocytes membrane, TBARS (thiobarbituric acid reactive species) levels, derived from the reaction between TBA (thiobarbituric acid) and MDA (malondialdehyde)—the end product of lipid peroxidation—were determined, as described by other authors [39], with slight modifications. Trichloroacetic acid (TCA, 10% *w/v* final concentration) was added to 1.5 mL of erythrocytes (either from diabetic volunteers or after incubation at different glucose concentrations) suspended at 20% hematocrit. The samples underwent centrifugation (ThermoScientific, 10 min, 3000× *g*) and 1 mL TBA (1% in 0.05 M NaOH) was added to the supernatant. Then, the mixture was heated for 30 min to 95 °C. Finally, TBARS levels were obtained by subtracting 20% of the absorbance at 453 nm from the absorbance at 532 nm (1.56 × 105 M^{-1} cm^{-1} molar extinction coefficient). The results are reported as μM TBARS levels.

2.6. *Membrane Sulfhydryl Groups Determination in Diabetic or Glucose-Treated Erythrocytes*

Concentration of membrane sulfhydryl groups was estimated by the method of Aksenov and Markesbery [40] with some modifications. Briefly, 100 μL of washed erythrocytes (either from diabetic volunteers or after incubation at different glucose concentrations) at 35% hematocrit was added to 1 mL of distilled water. Then, a 50 μL aliquot was diluted to 1 mL phosphate-buffered saline (PBS), pH 7.4, containing 1 mM ethylenediamine tetra acetic acid (EDTA). The addition of 30 μL of 10 mM 5,5′-dithiobis-(2-nitrobenzoic acid) (DTNB) started the reaction, and the samples were incubated in a dark room at 25 °C. Control samples, without protein or DTNB, were simultaneously handled.

After 30 min incubation at 25 °C, the samples were spectrophotometrically read at 412 nm, and the levels of formed TNB were determined by comparison to blanks (DTNB absorbance). The results are reported as μM 3-thio-2-nitro-benzoic acid (TNB)/mg protein and data were normalized to protein content.

2.7. GSH Content Measurement in Diabetic or Glucose-Treated Erythrocytes

The GSH (reduced glutathione) levels were determined on erythrocytes from diabetic volunteers or after incubation at different glucose concentrations by the method of Giustarini and collaborators [41] with some modifications. The assay is based on the oxidation of GSH by the Ellman's reagent DTNB (5,5′-dithiobis (2-nitrobenzoic acid), producing GSSG (oxidized glutathione) and TNB absorbing at 412 nm. The levels of GSH were measured by Cayman's GSH assay kit based on an enzymatic recycling method with glutathione reductase [17]. The amount of GSSG was calculated by the following formula: $1/2\ GSSG = GSH_{total} - GSH_{reduced}$, and the results are expressed as a GSH/GSSG ratio [22].

2.8. Measurement of Glycated Hemoglobin (A1c) in Diabetic or Glucose-Treated Erythrocytes

Glycated hemoglobin levels were determined with A1c liquidirect reagent as previously described by Sompong and collaborators [42] with slight modifications. Erythrocytes samples were lysed with hemolysis buffer (2.5 mM NaH2PO4, 5 mM HEPES, pH 7.4) and incubated with latex reagent for 5 min at 37 °C. Diabetic erythrocytes were immediately addressed to glycated hemoglobin measurement, while high glucose-exposed erythrocytes were assayed after 24 h incubation in high glucose solution. The absorbance was measured at 610 nm and A1c levels calculated from a standard curve using A1c and expressed as % A1c.

2.9. Erythrocytes Membranes Preparation and SDS-PAGE in Diabetic or Glucose-Treated Erythrocytes

Erythrocyte membranes were prepared as described by other authors [43] with slight modifications. Briefly, packed erythrocytes, derived from either diabetic volunteers or after incubation at different glucose concentrations, were diluted into 1.5 mL of 2.5 mM NaH2PO4 (cold hemolysis solution) containing a cocktail of protease and phosphatase inhibitors (1 mM PMSF, 1 mM NaF, 1 mM Na3VO4). Samples were centrifuged several times (Eppendorf, 4 °C, 13,000× *g*, 10 min) to remove hemoglobin. The obtained membranes were solubilized by SDS (sodium dodecyl sulfate, 1% *v*/*v*) and kept for 20 min on ice. Samples were then centrifugated (Eppendorf, 4 °C, 13,000× *g*, 30 min), and the supernatant containing solubilized membrane proteins was used for determination of protein content [44] and stored at 80 °C until use. After thawing, the membranes derived from each experimental condition were solubilized in Laemmli Buffer (1:1 volume ratio) [45], heated at 95 °C for 10 min, and finally the 20 μL of proteins were loaded. The samples were separated on 7.5% polyacrylamide gel before they were transfered to a PVDF (polyvinylidene fluoride) membrane.

2.10. Western Blot Analysis

The membranes were incubated overnight at 4° C with monoclonal anti-Band 3 protein antibody (B9277, 1:5000, Sigma–Aldrich, Milan, Italy), produced in mouse and diluted in Tris-Buffer (TBS) with 5% bovine serum albumin (BSA) and 0.1% Tween 20. Successively, membranes were incubated for 1 h with peroxidase-conjugated goat anti-mouse IgG secondary antibodies (A9044, 1:10,000, Sigma–Aldrich, Milan, Italy) at room temperature. To assess the presence of equal amounts of protein, monoclonal anti-Actin antibody (A1978, 1:1000, Sigma–Aldrich, Milan, Italy) produced in mouse were incubated on the same membranes by using a stripping method, thus allowing to detect different protein targets within a single membrane [46]. A chemiluminescence detection system (Super Signal West Pico Chemiluminescent Substrate, Pierce Thermo Scientific, Rockford, IL, USA) was used to detect signals, and images of them were imported to analysis software (Image Quant TL, v2003). The relative expression of protein bands was determined by densitometry (Bio-Rad ChemiDocTM XRS+).

2.11. Experimental Data and Statistics

Data are expressed as arithmetic means ± SEM for statistical analysis performed by GraphPad Prism software (version 5.00 for Windows; San Diego, CA). Significant differences among means were tested by one-way analysis of variance (ANOVA), followed by Bonferroni's multiple comparison post-hoc test or Student's *t*-test (paired values). Statistically significant differences were assumed at $p < 0.05$; n represents the number of independent experiments.

3. Results

3.1. Diabetic Erythrocytes

3.1.1. Osmotic Fragility Measurement

Figure 1 shows the osmotic fragility in erythrocytes from both healthy (A) and diabetic volunteers (B), reported as absorbance of hemoglobin released at different times of incubation (0–5–15–45–90–180 min) in a 0.7% *v*/*v* NaCl solution. As depicted, the osmotic fragility of erythrocytes from both groups was not significantly different with respect to hemoglobin levels at the following time intervals: 0–5–15–45–90 min. With regard to erythrocytes from diabetic volunteers, osmotic fragility at 180 min was significantly higher than all values at the other time intervals (0–5–45–90 min).

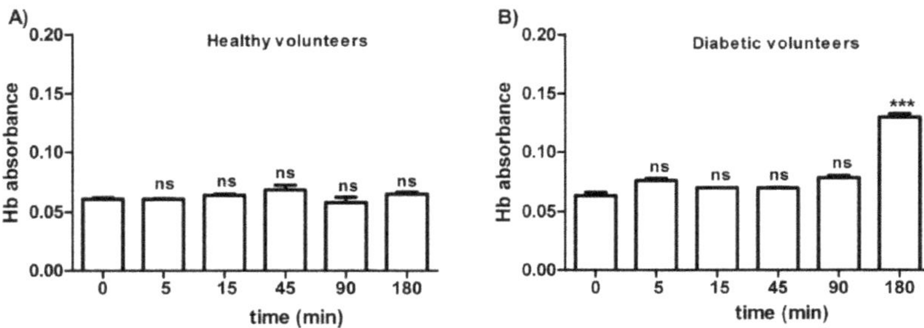

Figure 1. Osmotic fragility measured as hemoglobin (Hb) absorbance (optical density) in erythrocytes from healthy volunteers (**A**) and diabetic volunteers (**B**). *ns* (not significant) versus different time intervals considered (5–15–45–90 min); *** $p < 0.05$ versus different time intervals (0–5–15–45–90 min) as determined by one-way ANOVA followed by Bonferroni's post-hoc test (n = 10).

3.1.2. SO_4^{2-} Uptake Measurement

The curves depicted in Figure 2A describe SO_4^{2-} uptake as a function of time in erythrocytes from healthy (control) and diabetic volunteers. In control conditions, the rate of SO_4^{2-} uptake was augmented until equilibrium within 45 min with a rate constant value of 0.063 ± 0.001 min^{-1}. The rate constant in erythrocytes from diabetic volunteers (0.113 ± 0.001 min^{-1}, * $p < 0.05$, Table 1) was significantly higher than the control. The exposure to 10 µM DIDS since the beginning of incubation in SO_4^{2-} medium inhibited SO_4^{2-} uptake (0.018 ± 0.001 min^{-1}, *** $p < 0.001$, Table 1). The SO_4^{2-} quantity internalized by red blood cells from diabetic volunteers at 45 min of incubation in SO_4^{2-} medium (269.84 ± 16.8) was not significantly different with respect to that one measured in control cells (287.24 ± 20.3, Table 1). In DIDS-treated cells, intracellular SO_4^{2-} quantity at 45 min of incubation (4.75 ± 8.50) was significantly lower than those determined in both the control and in erythrocytes from diabetic volunteers (*** $p < 0.001$, Table 1).

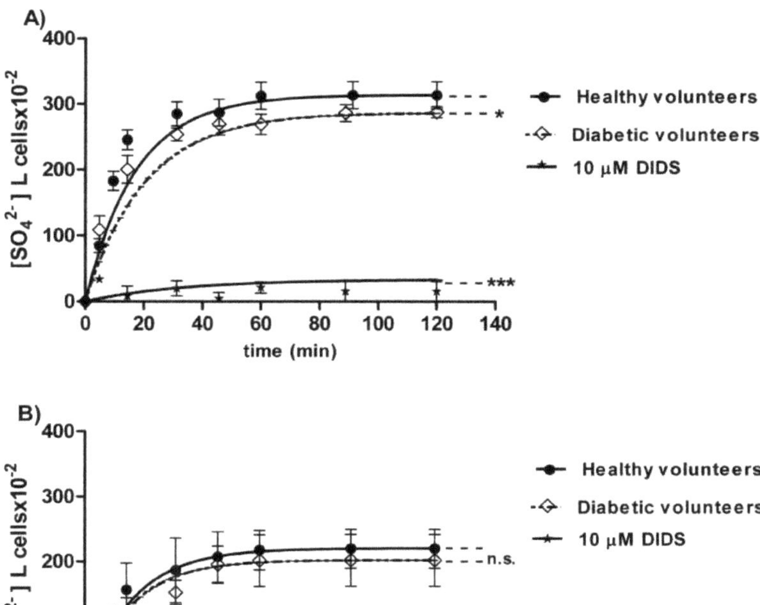

Figure 2. (**A**) Time course of SO_4^{2-} uptake in healthy volunteers (control) and in erythrocytes from diabetic volunteers. * $p < 0.05$, and *** $p < 0.001$ significant versus control as determined by one-way ANOVA followed by Bonferroni's post-hoc test (n = 16). (**B**) Time course of SO_4^{2-} uptake in resealed ghosts from healthy volunteers (control) and from diabetic volunteers. *ns* (not significant) versus control; *** $p < 0.001$ significant versus control as determined by one-way ANOVA followed by Bonferroni's post-hoc test (n = 16).

Table 1. Rate constant for SO_4^{2-} uptake and SO_4^{2-} quantity trapped in both diabetic and glucose-treated erythrocytes. Data are presented as means ± SEM from separate n experiments, where *ns* versus 5 mM glucose (t_{24}) (control) and healthy volunteers; * $p < 0.05$ versus healthy volunteers; ** $p < 0.001$ versus control; *** $p < 0.001$ versus control; §§§ $p < 0.001$ significative different versus 35 mM glucose (t_{24}); $^{\circ\circ\circ}$ $p < 0.001$ significant versus 15 mM glucose (t_{24}) and $^{\$\$\$}$ $p < 0.001$ significant versus 35 mM glucose (t_{24}) as determined by one-way ANOVA followed by Bonferroni's post-hoc test.

	Rate Constant min^{-1}	Time min	*N*	SO_4^{2-} Quantity Trapped at 45 min of Incubation in SO_4^{2-} Medium [SO_4^{2-}] L cells × 10^{-2}
Healthy volunteers	0.063 ± 0.001	16	16	287.24 ± 20.3
Diabetic volunteers	0.113 ± 0.001 *	9	16	269.84 ± 16.8 ^{ns}
5 mM glucose (t_{24})	0.057 ± 0.001	17	16	287.24 ± 18.20
15 mM glucose (t_{24})	0.109 ± 0.001 $^{***,\ §§§}$	9	16	98.42 ± 13.75 ***
35 mM glucose (t_{24})	0.129 ± 0.001 ***	7	16	143.15 ± 24 **
100 μM Mel + 15 mM glucose	0.065 ± 0.001 ***	15	16	310.47 ± 17.7 $^{\circ\circ\circ}$
100 μM Mel + 35 mM glucose	0.061 ± 0.001 $^{ns,\ \$\$\$}$	16	16	269.76 ± 19.7 $^{\$\$\$}$
10 μM DIDS	0.018 ± 0.001 ***	55	10	4.75 ± 8.50 ***

3.1.3. Resealed Ghosts

The curves depicted in Figure 2B describe SO_4^{2-} uptake in resealed ghosts from healthy (control) and diabetic volunteers. In the control resealed ghosts, SO_4^{2-} uptake increased until equilibrium within 45 min, exhibiting a rate constant of 0.063 ± 0.001 min^{-1}. The rate constant in resealed ghosts from diabetic volunteers (0.067 ± 0.002 min^{-1}, Table 2) was not significantly different from the control. The exposure to 10 μM DIDS applied at the beginning of incubation in SO_4^{2-} medium obscured SO_4^{2-} uptake (0.017 ± 0.001 min^{-1}, *** $p < 0.001$, Table 2). The SO_4^{2-} amount internalized by resealed ghosts from diabetic volunteers at 45 min of incubation in SO_4^{2-} medium (197.18 ± 28) was not significantly different with respect to the one measured in control ghosts (287.24 ± 20.3, Table 2). In DIDS-treated ghosts, intracellular SO_4^{2-} at 45 min of incubation (21.45 ± 8.50) was significantly lower than those determined in both the control resealed ghosts and the resealed ghosts from diabetic volunteers (*** $p < 0.001$, Table 2).

Table 2. Rate constant for SO_4^{2-} uptake and quantity of SO_4^{2-} trapped in both resealed ghosts of glucose-treated erythrocytes and in resealed ghosts from diabetic volunteers. Data are presented as means ± SEM from separate *n* experiments, where *ns* versus 5 mM glucose (t_{24}) and healthy volunteers (control); *** $p < 0.001$ versus control as determined by one-way ANOVA followed by Bonferroni's post-hoc test.

	Rate Constant min^{-1}	Time min	*N*	SO_4^{2-} Quantity Trapped at 45 min of Incubation in SO_4^{2-} Medium [SO_4^{2-}] L cells × 10^{-2}
Healthy volunteers	0.063 ± 0.001	16	16	207.01 ± 38
Diabetic volunteers	0.067 ± 0.002 ^{ns}	15	16	197.18 ± 16.8 ^{ns}
5 mM glucose (t_{24})	0.064 ± 0.001	16	16	183.69 ± 24
15 mM glucose (t_{24})	0.065 ± 0.001 ^{ns}	15	16	160.3 ± 20.06 ^{ns}
35 mM glucose (t_{24})	0.066 ± 0.001 ^{ns}	15	16	183.15 ± 14.6 ^{ns}
10 μM DIDS	0.017 ± 0.001 ***	58	10	21.45 ± 8.50 ***

3.1.4. Oxidative Conditions Assessment

The levels of TBARS estimated in erythrocytes from diabetic volunteers were significantly higher than those detected in erythrocytes from healthy volunteers (control) (Figure 3 A, * $p < 0.05$). As depicted in Figure 3B, the membrane sulfhydryl content measured in erythrocytes from diabetic volunteers was significantly lower than the one measured in erythrocytes from healthy volunteers (control) (** $p < 0.01$) as well as the GSH/GSSG ratio (Figure 3 C) (*** $p < 0.001$).

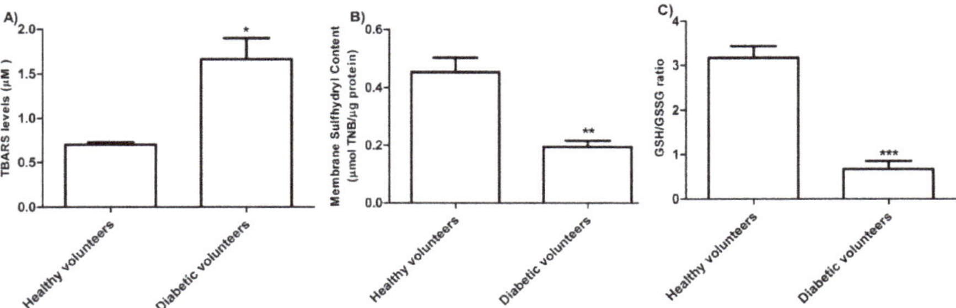

Figure 3. (**A**) TBARS levels (μM) in healthy volunteers (control) and in diabetic volunteers. * $p < 0.05$ significant versus control as determined by the Student's *t*-test ($n = 8$). (**B**) Membrane sulfhydryl content measured as μmol TNB/μg protein in healthy volunteers (control) and diabetic volunteers. ** $p < 0.01$ significant versus control as determined by the Student's *t*-test ($n = 8$). (**C**) Estimation of the GSH/GSSG ratio in erythrocytes from healthy volunteers (control) and diabetic volunteers. *** $p < 0.001$ significant versus control as determined by the Student's *t*-test ($n = 8$).

3.1.5. Band 3 Protein Expression Levels Determination

Figure 4 shows that B3p levels in erythrocytes from diabetic volunteers were significantly lower (*** $p < 0.001$) than those determined in erythrocytes from healthy volunteers (control).

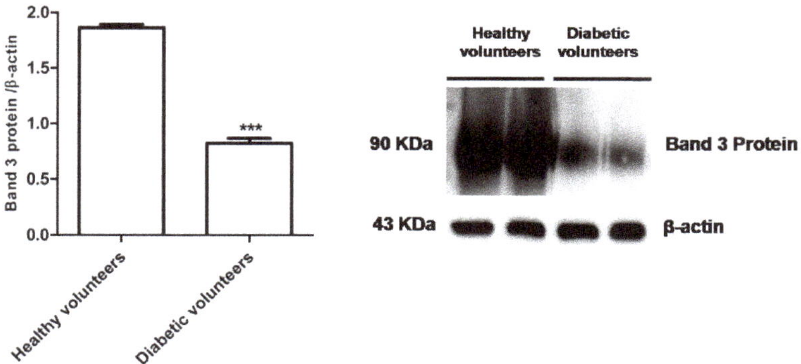

Figure 4. Expression levels of Band 3 protein measured in erythrocytes from healthy volunteers (control) or in diabetic volunteers detected by Western Blot analysis. *** $p < 0.001$ versus control as determined by the Student's *t*-test ($n = 5$).

3.2. Glucose-Treated Erythrocytes

3.2.1. Osmotic Fragility Measurement

Figure 5A–C shows the osmotic fragility reported as hemoglobin absorbance in erythrocytes treated for 24 h with different glucose concentrations (5–15–35 mM). At different times (0–5–15–45–90 min) of incubation in 0.7% *v*/*v* NaCl solution, the osmotic fragility was not significantly different with respect to what was observed at the beginning (0 min) of the test at each glucose concentration considered. After 180 min of incubation with either 15 or 35 mM glucose, respectively, (Figure 5B,C), the osmotic fragility was significantly higher (** $p < 0.05$; *** $p < 0.001$) than what was observed at the other time intervals (0–5–45–90 min).

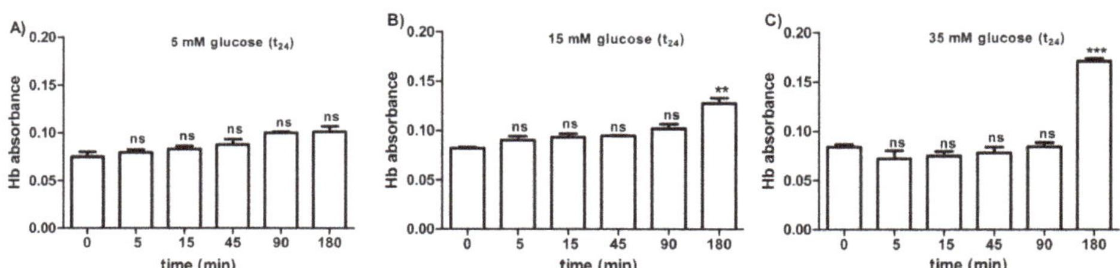

Figure 5. Osmotic fragility measured as hemoglobin (Hb) absorbance (optical density) in erythrocytes after treatment for 24 h with 5 mM glucose (t_{24}) (**A**), 15 mM glucose (t_{24}) (**B**), and 35 mM glucose (t_{24}) (**C**). *ns* (not significant) versus different time intervals (5–15–45–90–180 min); ** $p < 0.01$ and *** $p < 0.001$ versus different time intervals at 15 mM glucose (t_{24}) and 35 mM glucose (t_{24}) as determined by one-way ANOVA followed by Bonferroni's post-hoc test ($n = 10$).

3.2.2. SO_4^{2-} Uptake Measurement

The curves depicted in Figure 6A describe SO_4^{2-} uptake as a function of time. In particular, the rate constant for SO_4^{2-} uptake increased until equilibrium (45 min) with a value of 0.057 ± 0.001 min^{-1} in 5 mM glucose (t_{24}) erythrocytes (control) (Table 1). Erythrocytes treated with 15 mM glucose(t_{24}) and 35 mM glucose(t_{24}) exhibited a rate constant of, respectively, 0.109 ± 0.001 min^{-1} and 0.129 ± 0.001 min^{-1}, both significantly higher than the one measured in the control conditions (*** $p < 0.001$, Table 1). Pre-exposure to 100 μM Mel significantly impaired such an increase, as a rate constant of 0.065 ± 0.001 min^{-1} and 0.061 ± 0.001 min^{-1} was, respectively, observed (°°° $p < 0.001$ and §§§ $p < 0.001$, Table 1) in 15 mM and 35 mM glucose-treated erythrocytes. The exposure to 10 μM DIDS applied at the beginning of incubation in $SO_4{}^{2-}$ medium impaired $SO_4{}^{2-}$ uptake (0.018 ± 0.001 min^{-1}, *** $p < 0.001$, Table 1).

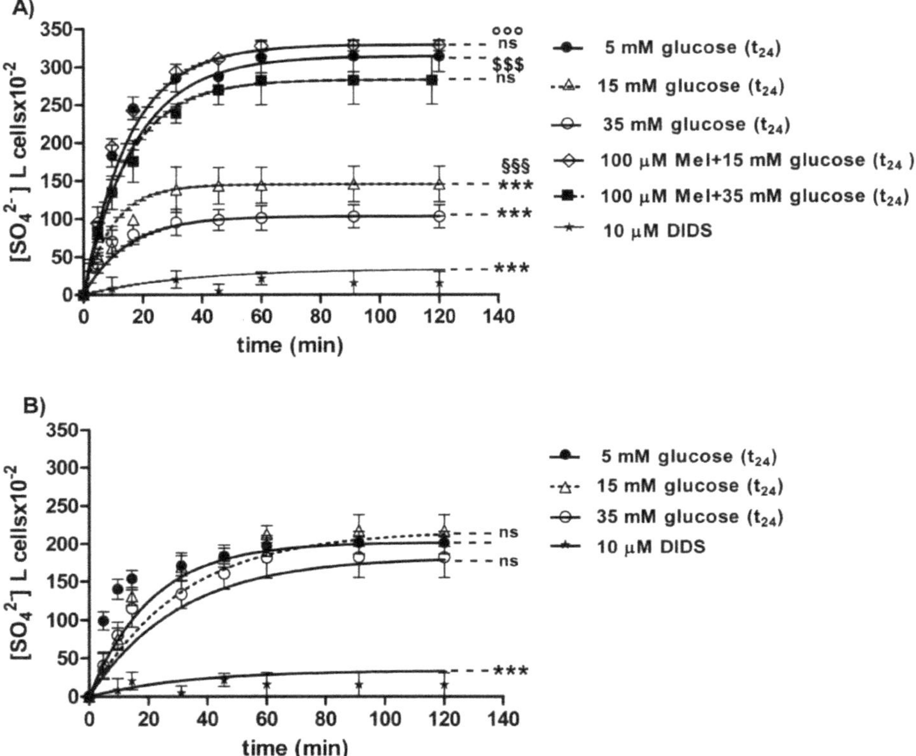

Figure 6. (**A**) Time course of $SO_4{}^{2-}$ uptake in glucose-treated erythrocytes at different glucose concentrations (5–15–35 mM) incubated for 24 h. *ns* versus 5 mM glucose (control); *** $p < 0.001$ versus control; §§§ $p < 0.001$ significant versus 35 mM glucose; °°° $p < 0.001$ significant versus 15 mM glucose and $$$ $p < 0.001$ significant versus 35 mM glucose as determined by one-way ANOVA followed by Bonferroni's post-hoc test ($n = 16$). (**B**) Time course of $SO_4{}^{2-}$ uptake in resealed ghosts from glucose-treated erythrocytes incubated for 24 h at different glucose concentrations (5–15–35 mM). *** $p < 0.001$ versus 5 mM glucose (t_{24}) (control) and *ns* versus control as determined by one-way ANOVA followed by Bonferroni's post-hoc test ($n = 16$).

The $SO_4{}^{2-}$ quantity internalized by red blood cells treated with 15 mM glucose (t_{24}) and 35 mM glucose(t_{24}) at 45 min of incubation in $SO_4{}^{2-}$ medium (98.42 ± 13.75; 143.15 ± 24) was significantly lower than the one measured in the control cells (287.24 ± 18.20) (*** $p < 0.001$ and ** $p < 0.01$, Table 1). Moreover, $SO_4{}^{2-}$ quantity internalized by erythrocytes treated with 100 μM Mel plus either 15 mM or 35 mM glucose was significantly higher (310.47 ± 17.7 and 269.76 ± 19.7) than that one measured at 15 mM glucose (t_{24}) and 35 mM glucose (t_{24}) alone (°°° $p < 0.001$ and §§§ $p < 0.001$, Table 1).

In DIDS-treated cells, the intracellular SO_4^{2-} quantity after 45 min incubation (4.75 ± 8.50) was significantly lower than that one determined in control erythrocytes (*** $p < 0.001$, Table 1).

3.2.3. Resealed Ghosts

The curves depicted in Figure 6B describe SO_4^{2-} uptake as a function of time in resealed ghosts from glucose-treated erythrocytes. In control resealed ghosts, SO_4^{2-} uptake increased until equilibrium (45 min), exhibiting a rate constant of 0.063±0.001 min^{-1}. The rate constant values of 0.065 ± 0.001 min^{-1} and 0.066 ± 0.001 min^{-1} observed in resealed ghosts at, respectively, 15 mM glucose (t_{24}) and 35 mM glucose (t_{24}) were not significantly different with respect to control (Table 2). The exposure to 10 µM DIDS applied at the beginning of incubation in SO_4^{2-} medium inhibited SO_4^{2-} uptake (0.017 ± 0.001 min $^{-1}$, *** $p < 0.001$, Table 2).

The SO_4^{2-} quantity internalized after 45 min incubation in SO_4^{2-} medium by resealed ghosts of erythrocytes previously treated with 15 mM (t_{24}) and 35 mM glucose (t_{24}) (160.3 ± 20.6 and 183.5 ± 14.6) was not significantly different with respect to the one measured in control ghosts (183.69 ± 24, Table 2). In DIDS-treated ghosts, intracellular SO_4^{2-} quantity after 45 min incubation (21.45 ± 8.50) was significantly lower than those determined in both control resealed ghosts and diabetic volunteer resealed ghost erythrocytes (*** $p < 0.001$, Table 2).

3.2.4. Oxidative Conditions Assessment

The levels of TBARS determined in erythrocytes treated with different concentrations of glucose (15–35 mM) were not significantly different with respect to those detected in 5 mM glucose (t_{24}) treated erythrocytes (control) (Figure 7A). As depicted in Figure 7B, membrane sulfhydryl content measured in glucose-treated erythrocytes was not significantly different with respect to that measured in 5 mM (t_{24}) glucose-treated erythrocytes (control). Moreover, the GSH/GSSG ratio (Figure 7C) measured in 15 mM and 35 mM (t_{24}) glucose-treated erythrocytes was significantly lower (*** $p < 0.001$; ** $p < 0.01$) than what measured at 5 mM glucose (t_{24}) (control). Pre-exposure to 100 µM Mel significantly impaired this reduction (§§ $p < 0.01$, ### $p > 0.001$) in both 15 mM and 35 mM glucose-treated erythrocytes.

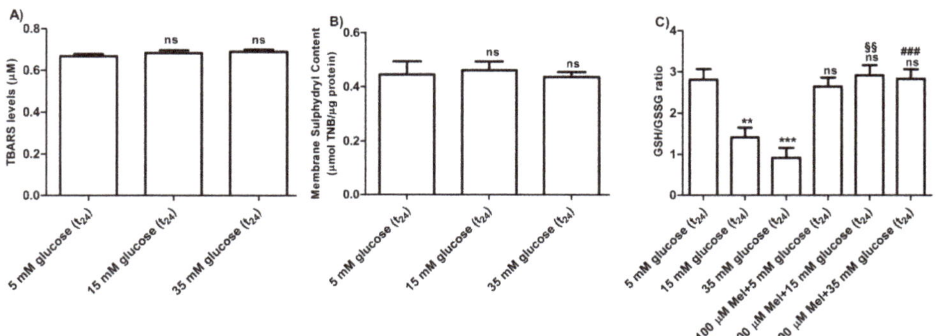

Figure 7. (**A**) TBARS levels (µM) measured in glucose-treated erythrocytes incubated for 24 h at different glucose concentrations (5–15–35 mM). *ns* (not significant) versus 5 mM glucose (t_{24}) (control) as determined by Bonferroni's post-hoc test ($n = 8$). (**B**) Membrane sulfhydryl content measured as µmol TNB/µg protein in glucose-treated erythrocytes (5–15–35 mM glucose) incubated for 24 h. *ns* (not significant) versus control by Bonferroni's post-hoc test (n = 8). (**C**) Estimation of the GSH/GSSG ratio measured in glucose-treated erythrocytes at different concentrations (5–15–35 mM glucose) incubated for 24 h. *ns* (not significant) versus control; *** $p < 0.001$ significant versus control; §§ $p < 0.01$ versus 15 mM glucose and ### $p < 0.001$ significant versus 35 mM glucose as determined by one-way ANOVA followed by Bonferroni's post-hoc test ($n = 8$).

3.2.5. Glycated Hemoglobin (A1c) Measurement

Glycated hemoglobin levels in red blood cells exposed for 24 h to different glucose concentrations (15–35 mM) were not significantly different with respect to those detected in 5 mM (t_{24}) glucose-treated erythrocytes (control) (Figure 8).

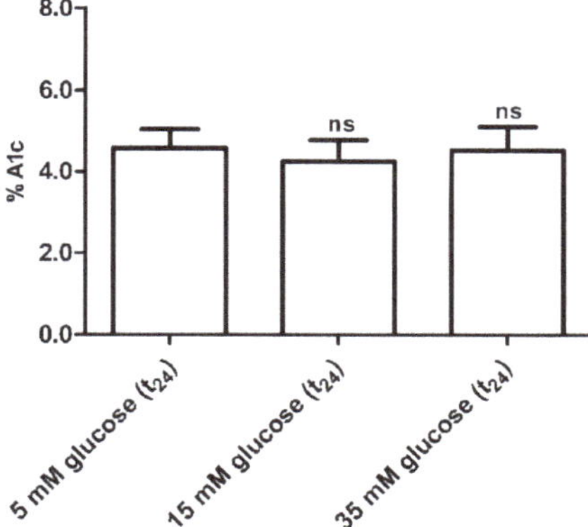

Figure 8. Glycated hemoglobin measurement in glucose-treated erythrocytes incubated for 24 h at different glucose concentrations (5–15–35 mM). *ns* (not significant) versus 5 mM glucose (t24) (control) as determined by one-way ANOVA followed by Bonferroni's post-hoc test ($n = 10$).

3.2.6. Band 3 Protein Expression Levels Determination

Figure 9 shows that B3p levels in erythrocytes incubated with either 15 mM glucose (t_{24}) or 35 mM glucose (t_{24}) were not significantly different with respect to those determined in erythrocytes treated with 5 mM glucose (t_{24}) (control).

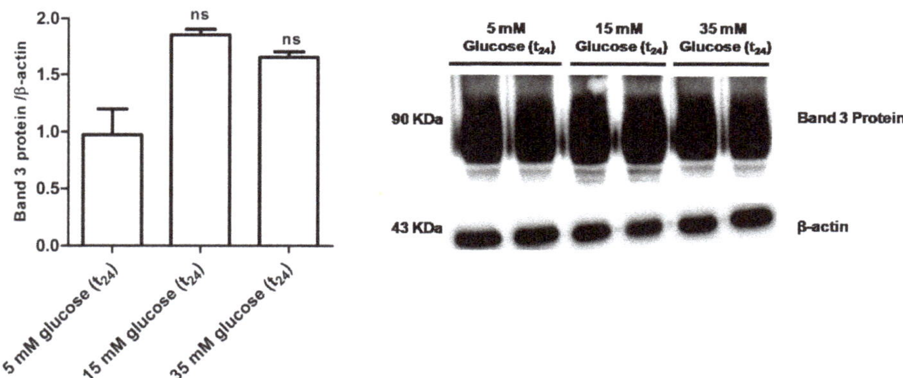

Figure 9. Band 3 protein expression levels measured in glucose- treated erythrocytes incubated for 24 h at different glucose concentrations (5–15–35 mM), detected by Western blot analysis. *ns* versus 5 mM glucose (t_{24}) (control) as determined by one-way ANOVA followed by Bonferroni's post-hoc test ($n = 5$).

4. Discussion

Chronic hyperglycemia is a major factor in diabetic complications development. Clinical evidence and experimental data support the hypothesis that reactive oxygen species (ROS) generation is augmented in T2D and that diabetes development is strictly related to OS with the mechanisms still unclear. Due to the fact of their sensitivity to oxidation, red blood cells represent a unique model to study OS and to possibly verify the efficacy of antioxidant therapies [20] and, in this context, the role of membrane transport systems in erythrocytes from diabetic patients has already been investigated. Some authors reported about Ca_2^+-ATPase activity of erythrocyte membranes decreased in both diabetes [47,48] and in erythrocytes treated ex vivo with glucose [49], probably due to the glycation of Ca_2^+-ATPase [50,51].

Less is known about B3p's function, essential to erythrocytes homeostasis, the assessment of which has already been proven as a good tool to reveal damage in the case of OS-related diseases, such as systemic scleroderma and canine leishmaniasis [22,23], or in other in vitro oxidative conditions [16,32].

Damage due to the fact of OS in hyperglycemic conditions has already been shown to reduce rheological properties, erythrocytes survival, and to affect erythrocytes' shape [13] which is known to critically correlate with B3p function [16,32,52].

On this premise, the present paper aimed to verify whether anion exchange capability through B3p is compromised in erythrocytes from diabetic volunteers or, ex vivo, after 24 h exposure to high glucose concentrations (15–35 mM) in accordance with what was reported by other authors [42].

As an increase of A1c levels is commonly associated to T2D, erythrocytes from diabetic volunteers was considered. Glycated hemoglobin, a minor fraction of Hb, is in vivo produced by non-enzymatic binding of glucose to N-terminal amino acids of Hb A beta chains [53,54].

The first step of the present study was to assess erythrocytes' integrity before conducting measurement of anion exchange capability through B3p. With this purpose, erythrocytes from diabetic volunteers underwent an osmotic fragility test. As no significant release of Hb was seen after 120 min of incubation in a 0.7% *v/v* NaCl solution, SO_4^{2-} uptake measurement was performed. According to what was reported by other groups [42,55], osmotic fragility tests showed that diabetic erythrocytes were more susceptible to hemolysis than those of Healthy volunteers (Figure 1A,B) after 180 min of incubation in 0.7% *v/v* NaCl solution, probably due to the presence of glycated Hb.

Coming back to the assessment of anion exchange capability through B3p, the rate constant for SO_4^{2-} uptake was increased in red blood cells from diabetic volunteers (Figure 2A). To better clarify the mechanisms underlying such acceleration, the technique of resealed ghosts was employed [16]. The evidence that the rate constant for SO_4^{2-} uptake of hemoglobin-free resealed ghosts from diabetic erythrocytes is similar to that one of the intact non-diabetic erythrocytes could suggest that glycation of Hb may affect its cross-link with membrane proteins, namely, B3p, with consequences on anion exchange (Figure 2B).

As expected, the altered anion exchange capability through B3p in erythrocytes from diabetic volunteers was also associated to an increase of intracellular OS, as proven by the loss of membrane asymmetry via formation of thiobarbituric acid reactive substances, oxidation of sulfhydryl groups of membrane proteins, and a decrease in the GSH/GSSG ratio (Figure 3A–C) [56]. Furthermore, a decrease in B3p expression levels was also observed (Figure 4). An increase in lipid peroxidation, a decrease in membrane sulfhydryl groups, and reduced B3p expression levels have previously been associated to a reduced rate of anion exchange capability after exposure to H_2O_2-induced oxidative stress [16]. In this regard, as the present findings show an accelerated rate constant for SO_4^{2-}, the alteration should be ascribed to glycation rather than oxidative damage, though both conditions are associated to diabetes [5,6].

To complete the frame, in an attempt to further focus on the impact of hyperglycemia on anion exchange capability through B3p and to possibly explain why the rate constant for SO_4^{2-} uptake is accelerated in red blood cells exposed to high glucose, hyperglycemia was modeled in vitro and experiments were performed by exposing erythrocytes of healthy volunteers to high glucose levels which mimics early hyperglycemic conditions [55]. Basically, transport via GLUT1 (insulin-independent glucose transporter) makes glucose concentration in red blood cells cytosol similar to plasmatic levels which are normally approximately 5 mM. Nevertheless, such values may increase under hyperglycemia conditions [57].

With regard to such in vitro tests, high glucose concentrations (15 and 35 mM) applied for 24 h make erythrocytes more susceptible to osmotic fragility, similar to what was observed in erythrocytes from diabetic volunteers, i.e., after at least 180 min of incubation in a 0.7% *v*/*v* NaCl solution (Figure 5B,C). Such alteration is also associated to a significant change in anion exchange capability through B3p observed after 24 h incubation at either 15 or 35 mM glucose (Figure 6A). To explain why the rate constant for SO_4^{2-} was accelerated in such treated erythrocytes, resealed ghosts were used, and, in addition, oxidative damage and B3p expression levels were monitored, similarly to what was performed for diabetic erythrocytes.

The first finding was that, similar to what was assessed in diabetic erythrocytes, when erythrocytes were deprived of Hb, the rate constant for SO_4^{2-} totally recovered, suggesting also in this case that 24 h incubation at high glucose concentrations may induce Hb alterations, reflecting on B3p function (Figure 6 B). Similar to the first experimental frame of this work, the impact of oxidative conditions possibly associated to in vitro hyperglycemia was monitored. Our results show that only the GSH/GSSG ratio changed as detected in diabetic erythrocytes, while there were no oxidative membrane sulfhydryl group level or lipid peroxidation events (Figure 7A–C), neither was there a decrease in B3p expression levels associated to the observed accelerated rate constant for SO_4^{2-} uptake (Figure 9). In an attempt to verify whether Hb glycation was involved in this response—based on what was obtained with hemoglobin-free resealed ghosts in studies by other authors [42,58]—the achievement of Hb glycation after 24 h incubation of erythrocytes with 45 mM glucose (lower than glucose concentration here used) was considered. As our results demonstrate, after 24 h exposure to 15 or 35 mM glucose erythrocytes did not show Hb glycation (Figure 8); the alteration in anion exchange capability through B3p could be ascribed to early oxidative damage (shown only by changed GSH levels) rather than to autooxidation leading to glycation of proteins, as shown by the abovementioned studies [42,58]. Such discrepancy may be due to the different glucose concentration employed.

To better support this finding, an antioxidant compound, possibly counteracting the effect of in vitro hyperglycemia, was used. In this regard, Mel has been used on the basis of what was recently reported by our group [32].

The exposure of red blood cells to Mel prevented the increase in the rate constant for SO_4^{2-} uptake observed under high glucose treatment (15 and 35 mM), demonstrating a correlation between Mel, anion exchange capability through B3p, and hyperglycemia in an in vitro model (Figure 6A). This adds new elements to a previous investigation recently performed by our group [32] about the antioxidant effects of Mel. In addition, such a result is in accordance with other authors, reporting that Mel restores the activity of antioxidant enzymes in red blood cells from multiple sclerosis patients which confirms that under OS some antioxidant enzymes of erythrocytes may be selectively activated [59]. In line with these authors, the present data describe Mel's beneficial effect revealed by the GSH/GSSG ratio restoration observed after pre-exposure to this antioxidant. Hence, it could be suggested that oxidative damage by high glucose concentrations could be mainly due to the alterations at the endogen antioxidant system level, namely, intracellular GSH, rather than to an effect directly exerted at the membrane level. Such findings do not seem in line with what was previously demonstrated in an H_2O_2-induced oxidative stress model on human erythrocytes [32] as different mechanisms of action of oxidative damage could occur.

5. Conclusions

In conclusion, the present results suggest that: (i) the B3p function assessment is a sensitive tool to assess the impact of hyperglycemia on red blood cells' homeostasis, namely, in subjects with A1c levels not associated to diabetes but possibly exhibiting a risk of poorly controlled glycemic levels; (ii) in non-diabetic subjects, changes in anion exchange capability through B3p may be one of the first consequences of hyperglycemia; (iii) hyperglycemia complications may be related to antioxidant systems reflecting on B3p function, in addition to glycation of proteins; (iv) in diabetes, the damage at the B3p level is due to the Hb glycation and is exacerbated by OS. Finally, the use of Mel in the in vitro model contributed to: (1) supporting the hypothesis of oxidative damage at the B3p level in early hyperglycemia conditions; (2) adding more knowledge about its antioxidant power, especially in the case of OS associated to early hyperglycemic conditions; and (3) confirming that homeostasis of red blood cells can be protected by Mel against OS.

Author Contributions: R.M. and A.M. conceived and designed the research; R.M., A.R., S.S., G.V., V.T. and S.L. performed the experiments and analyzed the data; R.M., A.R. and A.M. interpreted the results of the experiments; R.M., A.R. prepared the figures; R.M., A.R., and A.M. drafted the manuscript; R.M., A.R. and A.M. edited and revised the manuscript; R.M., A.R., S.S., G.V., V.T., S.L. and A.M. approved the final version of the manuscript. All authors have read and agreed to the published version of the manuscript.

Funding: This research was funded by MIUR, grant number PRIN 2015 (Prot. 20152HKF3Z).

Conflicts of Interest: The authors declare no conflict of interest.

References

1. World Health Organization. Diabetes. Available online: https://www.who.int/health-topics/diabetes#tab=tab_1 (accessed on 25 March 2020).
2. WHO Regional Office for Europe. Diabetes, Data and Statistics. Available online: http://www.euro.who.int/en/health-topics/noncommunicable-diseases/diabetes/data-and-statistics (accessed on 25 March 2020).
3. Fortuno, A.; San Jose, G.; Moreno, M.U.; Beloqui, O.; Diez, J.; Zalba, G. Phagocytic NADPH oxidase overactivity underlies oxidative stress in metabolic syndrome. *Diabetes* **2006**, *55*, 209–215. [CrossRef] [PubMed]
4. Martinelli, I.; Tomassoni, D.; Moruzzi, M.; Roy, P.; Cifani, C.; Amenta, F.; Tayebati, S.K. Cardiovascular Changes Related to Metabolic Syndrome: Evidence in Obese Zucker Rats. *Int. J. Mol. Sci.* **2020**, *21*, 2035. [CrossRef] [PubMed]
5. Zheng, H.; Wu, J.; Jin, Z.; Yan, L.J. Protein Modifications as Manifestations of Hyperglycemic Glucotoxicity in Diabetes and Its Complications. *Biochem. Insights* **2016**, *9*, 1–9. [CrossRef] [PubMed]
6. Aghadavod, E.; Khodadadi, S.; Baradaran, A.; Nasri, P.; Bahmani, M.; Rafieian-Kopaei, M. Role of Oxidative Stress and Inflammatory Factors in Diabetic Kidney Disease. *Iran. J. Kidney Dis.* **2016**, *10*, 337–343. [PubMed]
7. Buehler, P.W.; Alayash, A.I. Redox biology of blood revisited: The role of red blood cells in maintaining circulatory reductive capacity. *Antioxid. Redox Signal.* **2005**, *7*, 1755–1760. [CrossRef]
8. Nicolay, J.P.; Schneider, J.; Niemoeller, O.M.; Artunc, F.; Portero-Otin, M.; Haik, G., Jr.; Thornalley, P.J.; Schleicher, E.; Wieder, T.; Lang, F. Stimulation of suicidal erythrocyte death by methylglyoxal. *Cell. Physiol. Biochem.* **2006**, *18*, 223–232. [CrossRef]
9. Lang, K.S.; Duranton, C.; Poehlmann, H.; Myssina, S.; Bauer, C.; Lang, F.; Wieder, T.; Huber, S.M. Cation channels trigger apoptotic death of erythrocytes. *Cell Death Differ.* **2003**, *10*, 249–256. [CrossRef]
10. Kumar, R. Biochemical changes in erythrocyte membrane in type 2 diabetes mellitus. *Indian J. Med. Sci.* **2012**, *66*, 131–135. [CrossRef]
11. Rodrigo, R.; Bachler, J.P.; Araya, J.; Prat, H.; Passalacqua, W. Relationship between (Na+K)-ATPase activity, lipid peroxidation and fatty acid profile in erythrocytes of hypertensive and normotensive subjects. *Mol. Cell. Biochem.* **2007**, *303*, 73–81. [CrossRef]
12. American Diabetes Association (ADA). Diabetes Overview, Diagnosis. Available online: https://www.diabetes.org/a1c/diagnosis (accessed on 25 March 2020).
13. Verma, N.; Liu, M.; Ly, H.; Loria, A.; Campbell, K.S.; Bush, H.; Kern, P.A.; Jose, P.A.; Taegtmeyer, H.; Bers, D.M.; et al. Diabetic microcirculatory disturbances and pathologic erythropoiesis are provoked by deposition of amyloid-forming amylin in red blood cells and capillaries. *Kidney Int.* **2020**, *97*, 143–155. [CrossRef]
14. Reithmeier, R.A.; Casey, J.R.; Kalli, A.C.; Sansom, M.S.; Alguel, Y.; Iwata, S. Band 3, the human red cell chloride/bicarbonate anion exchanger (AE1, SLC4A1), in a structural context. *Biochim. Biophys. Acta* **2016**, *1858*, 1507–1532. [CrossRef] [PubMed]

15. Steck, T.L. The band 3 protein of the human red cell membrane: A review. *J. Supramol. Struct.* **1978**, *8*, 311–324. [CrossRef] [PubMed]
16. Morabito, R.; Romano, O.; La Spada, G.; Marino, A. H2O2-Induced Oxidative Stress Affects SO4= Transport in Human Erythrocytes. *PLoS ONE* **2016**, *11*, e0146485. [CrossRef] [PubMed]
17. Teti, D.; Crupi, M.; Busa, M.; Valenti, A.; Loddo, S.; Mondello, M.; Romano, L. Chemical and pathological oxidative influences on band 3 protein anion-exchanger. *Cell. Physiol. Biochem.* **2005**, *16*, 77–86. [CrossRef]
18. Morabito, R.R.A.; Arcuri, B.; Marino, A.; Giammanco, M.; La Spada, G.M.A. Effect of cadmium on anion exchange capability through Band 3 protein in human erythrocytes. *J. Biol. Res.* **2018**, *91*, 1–7. [CrossRef]
19. Morabito, R.; Marino, A.; Romano, P.; Rigano, C.; La Spada, G. Sulphate and chloride-dependent potassium transport in human erythrocytes are affected by crude venom from nematocysts of the jellyfish Pelagia noctiluca. *Cell. Physiol. Biochem.* **2013**, *32*, 86–95. [CrossRef]
20. Remigante, A.; Morabito, R.; Marino, A. Natural Antioxidants Beneficial Effects on Anion Exchange through Band 3 Protein in Human Erythrocytes. *Antioxidants* **2019**, *9*, 25. [CrossRef]
21. Romano, L.; Peritore, D.; Simone, E.; Sidoti, A.; Trischitta, F.; Romano, P. Chloride-sulphate exchange chemically measured in human erythrocyte ghosts. *Cell. Mol. Biol. (Noisy Grand)* **1998**, *44*, 351–355.
22. Morabito, R.; Remigante, A.; Cavallaro, M.; Taormina, A.; La Spada, G.; Marino, A. Anion exchange through band 3 protein in canine leishmaniasis at different stages of disease. *Pflug. Arch.* **2017**, *469*, 713–724. [CrossRef]
23. Morabito, R.R.A.; Bagnato, G.; Neal, R.W.; Sciortino, D.; D'Angelo, T.; Iannelli, F.; Iannelli, F.; Cordova, F.; Cirillo, M.; La Spada, G.; et al. Band 3 Protein Function and Oxidative Stress in Erythrocytes from Systemic Sclerosis Patients with Interstitial Lung Disease. *Eur. J. Clin. Biomed. Sci.* **2017**, *3*, 80–84. [CrossRef]
24. Morabito, R.; Remigante, A.; Marino, A. Protective Role of Magnesium against Oxidative Stress on SO4(=) Uptake through Band 3 Protein in Human Erythrocytes. *Cell. Physiol. Biochem.* **2019**, *52*, 1292–1308. [CrossRef] [PubMed]
25. Morabito, R.; Falliti, G.; Geraci, A.; Spada, G.L.; Marino, A. Curcumin Protects -SH Groups and Sulphate Transport after Oxidative Damage in Human Erythrocytes. *Cell. Physiol. Biochem.* **2015**, *36*, 345–357. [CrossRef] [PubMed]
26. Ratliff, B.B.; Abdulmahdi, W.; Pawar, R.; Wolin, M.S. Oxidant Mechanisms in Renal Injury and Disease. *Antioxid. Redox Signal.* **2016**, *25*, 119–146. [CrossRef] [PubMed]
27. Gerardo Yanowsky-Escatell, F.; Andrade-Sierra, J.; Pazarin-Villasenor, L.; Santana-Arciniega, C.; De Jesus Torres-Vazquez, E.; Samuel Chavez-Iniguez, J.; Angel Zambrano-Velarde, M.; Martin Preciado-Figueroa, F. The Role of Dietary Antioxidants on Oxidative Stress in Diabetic Nephropathy. *Iran. J. Kidney Dis.* **2020**, *14*, 81–94. [PubMed]
28. Chambial, S.; Dwivedi, S.; Shukla, K.K.; John, P.J.; Sharma, P. Vitamin C in disease prevention and cure: An overview. *Indian J. Clin. Biochem.* **2013**, *28*, 314–328. [CrossRef] [PubMed]
29. Yalcinkaya, A.S.; Sekeroglu, M.R.; Huyut, Z.; Cokluk, E.; Ozbek, H.; Ozturk, G.; Balahoroglu, R. The levels of nitrite, nitrate and lipid peroxidation in diabetic mouse brain: The effect of melatonin and pentoxifylline. *Arch. Physiol. Biochem.* **2020**, 1–7. [CrossRef]
30. Imenshahidi, M.; Karimi, G.; Hosseinzadeh, H. Effects of melatonin on cardiovascular risk factors and metabolic syndrome: A comprehensive review. *Naunyn Schmiedeberg's Arch. Pharmacol.* **2020**, *393*, 521–536. [CrossRef]
31. da Silva, D.G.H.; Chaves, N.A.; Miyamoto, S.; de Almeida, E.A. Prolonged erythrocyte auto-incubation as an alternative model for oxidant generation system. *Toxicol. In Vitro* **2019**, *56*, 62–74. [CrossRef]
32. Morabito, R.; Remigante, A.; Marino, A. Melatonin Protects Band 3 Protein in Human Erythrocytes against H2O2-Induced Oxidative Stress. *Molecules* **2019**, *24*, 2741. [CrossRef]
33. Chakravarty, S.; Rizvi, S.I. Day and Night GSH and MDA Levels in Healthy Adults and Effects of Different Doses of Melatonin on These Parameters. *Int. J. Cell Biol.* **2011**, *2011*. [CrossRef]
34. Panghal, A.; Sathua, K.B.; Flora, S.J.S. Gallic acid and MiADMSA reversed arsenic induced oxidative/nitrosative damage in rat red blood cells. *Heliyon* **2020**, *6*, e03431. [CrossRef] [PubMed]
35. Veena, C.K.; Josephine, A.; Preetha, S.P.; Varalakshmi, P. Effect of sulphated polysaccharides on erythrocyte changes due to oxidative and nitrosative stress in experimental hyperoxaluria. *Hum. Exp. Toxicol.* **2007**, *26*, 923–932. [CrossRef] [PubMed]

36. Romano, L.; Passow, H. Characterization of anion transport system in trout red blood cell. *Am. J. Physiol.* **1984**, *246*, C330–C338. [CrossRef] [PubMed]
37. Jessen, F.; Sjoholm, C.; Hoffmann, E.K. Identification of the anion exchange protein of Ehrlich cells: A kinetic analysis of the inhibitory effects of 4,4′-diisothiocyano-2,2′-stilbene-disulfonic acid (DIDS) and labeling of membrane proteins with 3H-DIDS. *J. Membr. Biol.* **1986**, *92*, 195–205. [CrossRef]
38. De Luca, G.; Gugliotta, T.; Scuteri, A.; Romano, P.; Rinaldi, C.; Sidoti, A.; Amato, A.; Romano, L. The interaction of haemoglobin, magnesium, organic phosphates and band 3 protein in nucleated and anucleated erythrocytes. *Cell Biochem. Funct.* **2004**, *22*, 179–186. [CrossRef]
39. Mendanha, S.A.; Anjos, J.L.; Silva, A.H.; Alonso, A. Electron paramagnetic resonance study of lipid and protein membrane components of erythrocytes oxidized with hydrogen peroxide. *Braz. J. Med. Biol. Res.* **2012**, *45*, 473–481. [CrossRef]
40. Aksenov, M.Y.; Markesbery, W.R. Changes in thiol content and expression of glutathione redox system genes in the hippocampus and cerebellum in Alzheimer's disease. *Neurosci. Lett.* **2001**, *302*, 141–145. [CrossRef]
41. Giustarini, D.; Dalle-Donne, I.; Milzani, A.; Fanti, P.; Rossi, R. Analysis of GSH and GSSG after derivatization with N-ethylmaleimide. *Nat. Protoc.* **2013**, *8*, 1660–1669. [CrossRef]
42. Sompong, W.; Cheng, H.; Adisakwattana, S. Protective Effects of Ferulic Acid on High Glucose-Induced Protein Glycation, Lipid Peroxidation, and Membrane Ion Pump Activity in Human Erythrocytes. *PLoS ONE* **2015**, *10*, e0129495. [CrossRef]
43. Pantaleo, A.; Ferru, E.; Pau, M.C.; Khadjavi, A.; Mandili, G.; Matte, A.; Spano, A.; De Franceschi, L.; Pippia, P.; Turrini, F. Band 3 Erythrocyte Membrane Protein Acts as Redox Stress Sensor Leading to Its Phosphorylation by p (72) Syk. *Oxid. Med. Cell. Longev.* **2016**, *2016*. [CrossRef]
44. Bradford, M.M. A rapid and sensitive method for the quantitation of microgram quantities of protein utilizing the principle of protein-dye binding. *Anal. Biochem.* **1976**, *72*, 248–254. [CrossRef]
45. Laemmli, U.K. Cleavage of structural proteins during the assembly of the head of bacteriophage T4. *Nature* **1970**, *227*, 680–685. [CrossRef] [PubMed]
46. Yeung, Y.G.; Stanley, E.R. A solution for stripping antibodies from polyvinylidene fluoride immunoblots for multiple reprobing. *Anal. Biochem.* **2009**, *389*, 89–91. [CrossRef]
47. Schaefer, W.; Priessen, J.; Mannhold, R.; Gries, A.F. Ca2+-Mg2+-ATPase activity of human red blood cells in healthy and diabetic volunteers. *Klin. Wochenschr.* **1987**, *65*, 17–21. [CrossRef] [PubMed]
48. Schaefer, W.; Beeker, J.; Gries, F.A. Influence of hyperglycemia on Ca2+-Mg2+-ATPase of red blood cells from diabetic patients. *Klin. Wochenschr.* **1988**, *66*, 443–446. [CrossRef] [PubMed]
49. Gonzalez Flecha, F.L.; Bermudez, M.C.; Cedola, N.V.; Gagliardino, J.J.; Rossi, J.P. Decreased Ca2(+)-ATPase activity after glycosylation of erythrocyte membranes in vivo and in vitro. *Diabetes* **1990**, *39*, 707–711. [CrossRef]
50. Gonzalez Flecha, F.L.; Castello, P.R.; Caride, A.J.; Gagliardino, J.J.; Rossi, J.P. The erythrocyte calcium pump is inhibited by non-enzymic glycation: Studies in situ and with the purified enzyme. *Biochem. J.* **1993**, *293 Pt 2*, 369–375. [CrossRef]
51. Gonzalez Flecha, F.L.; Castello, P.R.; Gagliardino, J.J.; Rossi, J.P. Molecular characterization of the glycated plasma membrane calcium pump. *J. Membr. Biol.* **1999**, *171*, 25–34. [CrossRef]
52. Morabito, R.; Remigante, A.; Di Pietro, M.L.; Giannetto, A.; La Spada, G.; Marino, A. SO4(=) uptake and catalase role in preconditioning after H2O2-induced oxidative stress in human erythrocytes. *Pflug. Arch.* **2017**, *469*, 235–250. [CrossRef]
53. Jeffcoate, S.L. Diabetes control and complications: The role of glycated haemoglobin, 25 years on. *Diabet. Med.* **2004**, *21*, 657–665. [CrossRef]
54. Peterson, K.P.; Pavlovich, J.G.; Goldstein, D.; Little, R.; England, J.; Peterson, C.M. What is hemoglobin A1c? An analysis of glycated hemoglobins by electrospray ionization mass spectrometry. *Clin. Chem.* **1998**, *44*, 1951–1958. [CrossRef] [PubMed]
55. Viskupicova, J.; Blaskovic, D.; Galiniak, S.; Soszynski, M.; Bartosz, G.; Horakova, L.; Sadowska-Bartosz, I. Effect of high glucose concentrations on human erythrocytes in vitro. *Redox Biol.* **2015**, *5*, 381–387. [CrossRef] [PubMed]
56. Jain, S.K.; McVie, R.; Duett, J.; Herbst, J.J. Erythrocyte membrane lipid peroxidation and glycosylated hemoglobin in diabetes. *Diabetes* **1989**, *38*, 1539–1543. [CrossRef] [PubMed]

57. Hernandez-Sanchez, F.; Guzman-Beltran, S.; Herrera, M.T.; Gonzalez, Y.; Salgado, M.; Fabian, G.; Torres, M. High glucose induces O-GlcNAc glycosylation of the vitamin D receptor (VDR) in THP1 cells and in human macrophages derived from monocytes. *Cell Biol. Int.* **2017**, *41*, 1065–1074. [CrossRef]
58. Nandhini, T.A.; Anuradha, C.V. Inhibition of lipid peroxidation, protein glycation and elevation of membrane ion pump activity by taurine in RBC exposed to high glucose. *Clin. Chim. Acta* **2003**, *336*, 129–135. [CrossRef]
59. Miller, E.; Walczak, A.; Majsterek, I.; Kedziora, J. Melatonin reduces oxidative stress in the erythrocytes of multiple sclerosis patients with secondary progressive clinical course. *J. Neuroimmunol.* **2013**, *257*, 97–101. [CrossRef]

© 2020 by the authors. Licensee MDPI, Basel, Switzerland. This article is an open access article distributed under the terms and conditions of the Creative Commons Attribution (CC BY) license (http://creativecommons.org/licenses/by/4.0/).

Article

D-Galactose Decreases Anion Exchange Capability through Band 3 Protein in Human Erythrocytes

Alessia Remigante [1,†], Rossana Morabito [1,†], Sara Spinelli [1], Vincenzo Trichilo [2], Saverio Loddo [2], Antonio Sarikas [3], Silvia Dossena [3] and Angela Marino [1,*]

1 Department of Chemical, Biological, Pharmaceutical and Environmental Sciences, University of Messina, Viale F. Stagno D'Alcontres 31, 98166 Messina, Italy; aremigante@unime.it (A.R.); rmorabito@unime.it (R.M.); sara.spinelli1992@libero.it (S.S.)

2 Department of Clinical and Experimental Medicine, AOU Policlinico Universitario "G. Martino", Via Consolare Valeria, 98125 Messina, Italy; vtrichilo@unime.it (V.T.); sloddo@unime.it (S.L.)

3 Institute of Pharmacology and Toxicology, Paracelsus Medizinische Privatuniversität, Strubergasse 21, Haus C, 5020 Salzburg, Austria; antonio.sarikas@pmu.ac.at (A.S.); silvia.dossena@pmu.ac.at (S.D.)

* Correspondence: marinoa@unime.it; Tel.: +39-(0)90-6765214

† These authors contributed equally to this work.

Received: 30 June 2020; Accepted: 31 July 2020; Published: 2 August 2020

Abstract: D-Galactose (D-Gal), when abnormally accumulated in the plasma, results in oxidative stress production, and may alter the homeostasis of erythrocytes, which are particularly exposed to oxidants driven by the blood stream. In the present investigation, the effect of D-Gal (0.1 and 10 mM, for 3 and 24 h incubation), known to induce oxidative stress, has been assayed on human erythrocytes by determining the rate constant of SO_4^{2-} uptake through the anion exchanger Band 3 protein (B3p), essential to erythrocytes homeostasis. Moreover, lipid peroxidation, membrane sulfhydryl groups oxidation, glycated hemoglobin (% A1c), methemoglobin levels (% MetHb), and expression levels of B3p have been verified. Our results show that D-Gal reduces anion exchange capability of B3p, involving neither lipid peroxidation, nor oxidation of sulfhydryl membrane groups, nor MetHb formation, nor altered expression levels of B3p. D-Gal-induced %A1c, known to crosslink with B3p, could be responsible for rate of anion exchange alteration. The present findings confirm that erythrocytes are a suitable model to study the impact of high sugar concentrations on cell homeostasis; show the first in vitro effect of D-Gal on B3p, contributing to the understanding of mechanisms underlying an in vitro model of aging; demonstrate that the first impact of D-Gal on B3p is mediated by early Hb glycation, rather than by oxidative stress, which may be involved on a later stage, possibly adding more knowledge about the consequences of D-Gal accumulation.

Keywords: D-Galactose; oxidative stress; glycation; Band 3 protein; SO_4^{2-} uptake; anion exchange

---

1. Introduction

D-Galactose (D-Gal) is a monosaccharide sugar, whose metabolism is connected to glucose metabolism [1]. In physiological conditions, it is rapidly metabolized to glucose through specific enzymes and driven from peripheral blood to glycolysis pathway to give energy to the cell [2], keeping thus plasma D-Gal levels very low (plasmatic concentration range between 0.00008 and 0.00018 mM) [3,4]. Abnormally accumulated D-Gal [5,6], associated to galactosemia, can be directly oxidized by galactose oxidase to produce hydrogen peroxide (H_2O_2) or, alternatively, can start a non-enzymatic glycation reactions with free amino acid (mainly lysine and arginine) to form advanced glycation end products (AGEs) [7–9].

When oxidative stress persists, the excessive free radical species accumulation can influence many cellular signaling pathways and damage organic macromolecules, which can be oxidized [10,11].

The overproduction of oxidative stress is a common patho-physiological state underlying many chronic diseases, such as cardiovascular diseases [12,13], neurodegenerative diseases [14], cancer [15], metabolic disorders (diabetes mellitus [16]), and diseases related to aging [17], although the precise mechanisms contributing to the oxidative stress-induced damage are still under investigation. Several studies have exploited the in vivo exposure to D-Gal as a typical model for exploring the mechanisms underlying oxidative stress-related diseases in plasma or blood cells [6,18–21]. However, the effects of D-Gal on oxidative stress production in human erythrocytes have been only partially clarified [22]. Moreover, it has been established that high sugar concentrations are correlated with an increase in glycation of human erythrocytes proteins, resulting in modifications of lipids-proteins interaction. Glycation of membrane proteins could induce membrane fluidity alteration, which reflects not only on membrane bilayer, but also on intracellular environment [22–25].

Human erythrocytes are continuously exposed to oxidative molecules transported within the vascular system and possibly interacting with erythrocytes membrane, and are more susceptible to oxidative stress than other cells [26]. Oxidative damage has been demonstrated to decrease survival and rheological properties of human erythrocytes, affecting their homeostasis, which is strictly linked to Band 3 protein (B3p) function [27,28]. B3p mediates the Cl^-/HCO_3^- exchange across erythrocytes membrane and plays a key role for erythrocytes deformability, ion balance and gas exchange, all essential to tissue oxygenation [29,30]. The uptake of SO_4^{2-}, slower and better detectable than that of Cl^- or HCO_3^- [31,32], has been used to monitor the anion exchange capability through this protein by a turbidimetric method. In particular, this method has been effectively used in vitro to investigate the impact of different conditions on human erythrocytes homeostasis, including H_2O_2-induced oxidative conditions, oxidative stress-related diseases, or high glucose levels [33–35], proving that anion exchange capability of B3p is a suitable tool to evaluate cell response to oxidative conditions. In addition, Pantaleo and collaborators [36] demonstrated that erythrocytes possess a mechanism to sense the oxidative stress via B3p oxidation, docking of Syk kinase and phosphorylation of two B3p Tyr residues, crucial for assuring membrane stability.

The effect of D-Gal on B3p function has never been evaluated. Since an excessive exposure to D-Gal (25 mM) could lead to the overproduction of reactive oxygen species (ROS) and AGEs in erythrocytes, thus contributing to oxidative stress events [22], and considering the current scarce availability of studies in this field, in the present investigation we firstly tried to detect a possible early damage of erythrocyte homeostasis by measuring the anion exchange capability through B3p following short-term exposure (3 and 24 h) of human erythrocytes to 0.1 and 10 mM D-Gal [6] and, secondarily, to verify whether at that stage the potential alterations may be linked to redox balance variations. The present study is included in a wider field of investigation about the relationship, still poorly clarified, between high D-Gal concentrations and membrane transport systems.

2. Materials and Methods

2.1. Solutions and Chemicals

The specific catalase inhibitor 3-amino-1,2,4-triazole (3-AT) [37], freshly prepared H_2O_2 and $NaNO_2$ were dissolved in distilled water and diluted from 3 M, 50 mM and 30% *w/w* stock solutions, respectively. 4,4′-diisothiocyanato-stilbene-2,2′-disulfonate (DIDS) was prepared in dimethyl sulfoxide (DMSO) and diluted from a 10 mM stock solution. N-ethylmaleimide (NEM) was prepared in ethanol and diluted from a 310 mM stock solution. All chemicals were purchased from Sigma (Milan, Italy). DMSO and ethanol were tested on erythrocytes at their final concentration (0.001% *v/v*) to preventively exclude any effect related to the solvent.

2.2. Erythrocytes Preparation

Blood samples were obtained after oral informed consent from adults healthy volunteers (13 ♀, 11 ♂, age 28–45 years) with blood D-Gal concentrations ranging between 0.00008 and 0.00018 mmol/L (physiological concentrations) [4]. Patients with diabetes and cardiovascular disease have been excluded from the study. Blood, collected in tubes containing ethylenediaminetetraacetic acid (EDTA) to prevent coagulation, was washed with an isotonic solution with the following composition (in mM): NaCl 145, glucose 5,4-(2-hydroxyethyl)-1 piperazineethanesulfonic acid (HEPES) 5, pH 7.4, osmotic pressure 300 mOsm/KgH_2O). The samples were centrifuged thrice (ThermoScientific, Milan, Italy, 1200× *g*, 5 min) to discard both plasma and buffy coat. Subsequently, erythrocytes were suspended in isotonic solution at different hematocrit values and addressed to different treatments according to the experimental design reported in Figure 1 (see treatments before time 0). After each treatment, erythrocytes were subjected to SO_4^{2-} uptake measurement, oxidative condition assessment, Western blot analysis, or glycated hemoglobin measurement.

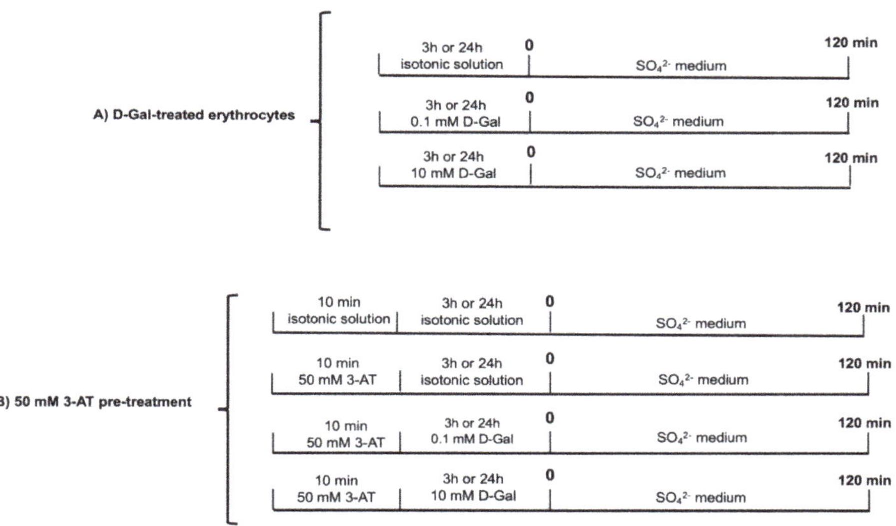

Figure 1. Time course of different experimental conditions before SO_4^{2-} uptake measurement. (**A**) Gal-treated erythrocytes (0.1 or 10 mM, for 3 or 24 h). (**B**) 50 mM 3-AT pretreatment (10 min) plus Gal-treatment.

2.3. SO_4^{2-} Uptake Measurement

2.3.1. Control Condition

SO_4^{2-} uptake via B3p was measured as described elsewhere [31,38]. After washing, erythrocytes samples were suspended to 3% hematocrit and addressed to either control conditions assessment or to incubation with D-Gal. With regard to control conditions, erythrocytes were suspended in 35 mL isotonic solution containing SO_4^{2-}, henceforth called SO_4^{2-} medium (composition in mM: Na_2SO_4 118, HEPES 10, glucose 5, pH 7.4, osmotic pressure 300 mOsm/KgH_2O) and incubated at 25 °C. At fixed time intervals (5, 10, 15, 30, 45, 60, 90 and 120 min), 5 mL erythrocytes suspension were transferred in a tube containing DIDS (10 µM) to block B3p activity [39], and kept on ice. After incubation in SO_4^{2-} medium plus DIDS, samples were washed at least thrice by centrifugation (ThermoScientific, 4 °C, 1200× *g*, 5 min) and resuspension in isotonic solution to eliminate SO_4^{2-} from the external medium, and then hemolyzed in distilled water (1 mL). Proteins were precipitated by perchloric acid (4% *v/v*). After centrifugation (ThermoScientific 4 °C, 2500× *g*, 10 min), the supernatant containing SO_4^{2-} underwent turbidimetric analysis. SO_4^{2-} precipitation was performed by mixing the following components: 500 µL supernatant from each sample, 1 mL glycerol previously diluted in distilled water (1:1), 1 mL 4 M NaCl plus 37% hydrochloric acid (HCl) solution (12:1) and, finally, 500 µL 1.24 M

$BaCl_2{\cdot}2H_2O$. The absorbance of each sample was then read at 425 nm using a spectrophotometer (Eppendorf, BioPhotometer Plus, Milan, Italy).

A calibrated standard curve previously obtained by precipitating known SO_4^{2-} concentrations was used to convert the absorbance to $[SO_4^{2-}]$ L cells × 10^{-2}. Moreover, the rate constant (min^{-1}) was calculated by the equation: $C_t = C_\infty (1 - e^{-rt}) + C_0$, where C_t, C_∞ and C_0 indicate the intracellular SO_4^{2-} concentrations at time t, 0 and ∞ respectively, e is Neper number (2.7182818), r is the rate constant accounting for the process velocity and t is the time of each sample withdrawal. The rate constant is the time needed to reach 63% of total SO_4^{2-} intracellular concentration [31] and $[SO_4^{2-}]$ L cells × 10^{-2} reported in figures stands for SO_4^{2-} micromolar concentration trapped by 5 mL erythrocytes (3% hematocrit).

2.3.2. D-Gal-Treated Erythrocytes

Once washed and suspended to 3% hematocrit and after either 3 or 24 h incubation at different D-Gal concentrations added to the solution (0.1 and 10 mM, at 25 °C), erythrocytes samples were washed and centrifuged (ThermoScientific, 4 °C, 1200× g, 5 min) to replace the supernatant with SO_4^{2-} medium. SO_4^{2-} uptake was successively determined as described for control conditions. In addition, to assay whether D-Gal effect was due to antioxidant system failure, 3-amino-1,2,4-triazole (3-AT), a specific inhibitor of catalase [37], was applied 10 min before both 3 and 24 h treatment with either 0.1 or 10 mM D-Gal. In Figure 1, the time course of all experimental conditions followed by SO_4^{2-} uptake determination is shown.

2.4. *Oxidative Condition Assessment*

2.4.1. Thiobarbituric Acid Reactive Species Determination

To assess a possible lipid peroxidative effect of D-Gal on erythrocytes membranes, thiobarbituric acid reactive species (TBARS) levels, deriving from reaction between thiobarbituric acid (TBA) and malondialdehyde (MDA), i.e., the end products of the lipid peroxidation, were measured as described by Mendanha and collaborators [40], with slight modifications. Trichloroacetic acid (TCA, final concentration 10% *w/v*) was added to 1.5 mL of erythrocytes, which were either untreated or pre-treated with D-Gal with or without 3-AT (Figure 1) and suspended at 20% hematocrit. Samples underwent centrifugation (ThermoScientific, 25° C, 3000× g, 10 min) and 1 mL TBA (1% in 0.05 M NaOH) was added to the supernatant. Then, the mixture was heated for 30 min to 95 °C. Finally, TBARS levels were obtained by subtracting 20% of the absorbance at 453 nm from the absorbance at 532 nm (Eppendorf, BioPhotometer Plus). Erythrocytes samples were also incubated with H_2O_2 (10 mM, for 1 h, at 25 °C), as this compound is known to induce a significant lipid peroxidative effect [41]. Results are indicated as µM TBARS levels (1.56×10^5 M^{-1} cm^{-1} molar extinction coefficient).

2.4.2. Membrane Sulfhydryl Groups Levels Determination

Membrane sulfhydryl groups concentration was estimated by the method of Aksenov and Markesbery [42], with some modifications. Shortly, 100 µL of washed erythrocytes, which were either untreated, or pre-treated with D-Gal with or without 3-AT (Figure 1) and suspended at 35% hematocrit, were added to 1 mL of distilled water. Then, a 50 µL aliquot was diluted with 1 mL phosphate-buffered saline (PBS), pH 7.4, containing 1 mM EDTA. The addition of 30 µL of 10 mM 5,5′-dithiobis-(2-nitrobenzoic acid) (DTNB) began the reaction and samples were incubated for 30 min in a dark room at 25 °C. Control samples, without protein extract or DTNB, were simultaneously handled. Afterwards, samples were spectrophotometrically read at 412 nm (Eppendorf, BioPhotometer Plus), and levels of 3-thio-2-nitro-benzoic acid (TNB) determined by comparison to blank (DTNB absorbance). NEM (2 mM for 1 h incubation, at 25 °C) was used to obtain complete oxidation of membrane sulfhydryl groups [27,43]. Results are reported as µM TNB/mg protein and data were normalized to protein content.

2.4.3. Methemoglobin Levels Determination

Methemoglobin (MetHb) levels have been assayed as previously described [44,45] with some modifications. This assay is based on methemoglobin and (oxy)-hemoglobin determination by spectrophotometry at, respectively, 630 and 540 nm wavelengths. Shortly, 25 μL of erythrocytes at 35% hematocrit, which were untreated or pre-treated D-Gal with or without 3-AT (Figure 1), were lysed in 1975 μL hypotonic buffer (2.5 mM NaH_2PO_4, pH 7.4, 4 °C). Erythrocytes samples were then centrifuged (Eppendorf, 4 °C, 13,000× *g*, 15 min) to discard membranes and the absorbance of the supernatant was read by a spectrophotometer (Eppendorf, BioPhotometer Plus). $NaNO_2$ (4 mM for 1 h incubation at 25 °C) was used to obtain complete Hb oxidation [46]. MetHb percentage (%) was calculated as follows: $\%MetHb = OD_{630}/OD_{540} \times 100$ (OD is optical density).

2.5. Erythrocytes Membranes Preparation

Erythrocyte membranes were prepared as described by other authors [36], with slight modifications. Briefly, packed erythrocytes (untreated or pre-treated with D-Gal with or without 3-AT pre-treatment), were diluted into 1.5 mL of 2.5 mM NaH_2PO_4 (cold hemolysis solution) containing a cocktail of inhibitors (1 mM PMSF, 1 mM NaF, and 1 mM Na_3VO_4). Samples were repeatedly centrifuged (Eppendorf, 4 °C, 18,000× *g*, 10 min) to take out hemoglobin. The obtained membranes were solubilized by sodium dodecyl sulfate (SDS, 1% *v/v*) and kept for 20 min on ice. After centrifugation (Eppendorf, 4 °C, 13,000× *g*, 30 min), the supernatant, containing the solubilized membrane proteins, was stored at −80 °C until use.

2.6. SDS-PAGE Preparation and Western Blot Analysis

After thawing, membranes were solubilized in Laemmli buffer (1:1 volume ratio) [47] and heated for 10 min at 95 °C. The protein samples (20 μL) were separated by 7.5% SDS-polyacrylamide gel electrophoresis and transferred to a polyvinylidene fluoride membrane by applying a constant voltage (75 V) for 2 h at 4 °C. Membranes were blocked for 1 h at room temperature in 5% bovine serum albumin (BSA) diluted in Tris-buffered saline (150 mM NaCl, 15 mM Tris-HCl) containing 0.1% Tween-20 (TBST), and incubated overnight at 4 °C with the primary antibody (monoclonal anti-Band 3 protein antibody, B9277, Sigma-Aldrich, Milan, Italy, produced in mouse and diluted 1:5000 in TBST). Successively, membranes were incubated for 1 h with peroxidase-conjugated goat anti-mouse IgG secondary antibodies (A9044, Sigma-Aldrich, Milan, Italy) diluted 1:10,000 in TBST solution at room temperature. To assess the presence of equal amounts of protein, a monoclonal anti-actin antibody (A1978, Sigma-Aldrich, Milan, Italy), diluted 1:1000 in TBST solution and produced in mouse, was incubated with the same membrane, as suggested by Yeung and collaborators [48]. A chemiluminescence detection system (Super Signal West Pico Chemiluminescent Substrate, Pierce Thermo Scientific, Rockford, IL. USA) was used to detect signals, whose images were imported to analysis software (Image Quant TL, v2003). The intensity of the corresponding protein bands was determined by densitometry (Bio-Rad ChemiDocTM XRS+).

2.7. Glycated Hemoglobin Measurement (%A1c)

The levels of glycated hemoglobin (A1c) were determined with the A1c liquidirect reagent as previously described by Sompong and collaborators [24], with slight modifications. Erythrocytes samples (either untreated or pre-treated with D-Gal with or without 3-AT, Figure 1), were lysed with hemolysis buffer (2.5 mM NaH_2PO_4) and incubated with latex reagent at 37 °C for 5 min. The absorbance was measured at 610 nm (Eppendorf, BioPhotometer Plus). The levels of A1c were calculated from a standard curve obtained by using known concentrations of A1c, and expressed as %A1c.

2.8. Experimental Data and Statistics

Data are expressed as arithmetic means ± S.E.M. Statistical analysis was performed by the GraphPad Prism software (version 6.00 for Windows; San Diego, CA, USA). Significant differences between means were tested by one-way analysis of variance (ANOVA), followed by Bonferroni's multiple comparison *post hoc* test. Statistically significant differences were assumed at $p < 0.05$; n represents the number of independent experiments.

3. Results

3.1. SO_4^{2-} Uptake Measurement

Figure 2A,B describe the SO_4^{2-} uptake as a function of time in untreated erythrocytes (control) and erythrocytes treated with 0.1 or 10 mM D-Gal for 3 (t_3) or 24 (t_{24}) hours respectively, with or without pre-incubation with 50 mM 3-AT. In control conditions, the rate constant of SO_4^{2-} uptake progressively increased and reached equilibrium within 45 min (0.067 ± 0.001 min^{-1}). The rate constant values of SO_4^{2-} uptake in D-Gal (0.1 or 10 mM)-treated erythrocytes at both t_3 and t_{24} (0.066 ± 0.001 min^{-1}, ** $p < 0.01$; 0.065 ± 0.001 min^{-1}, *** $p < 0.001$ and 0.060 ± 0.001 min^{-1}, *** $p < 0.001$; 0.051 ± 0.001 min^{-1}, *** $p < 0.001$; Table 1) were significantly lower than control. The rate constant of SO_4^{2-} uptake in erythrocytes treated with 50 mM 3-AT was not significantly different with respect to control (0.068 ± 0.001 min^{-1}). After pretreatment of erythrocytes with 50 mM 3-AT and subsequent treatment with D-Gal, the rate constant of SO_4^{2-} uptake was significantly lower than both control and D-Gal (0.1 or 10 mM) treatment (0.065 ± 0.001 min^{-1}, ***,§§§, $p < 0.001$; 0.066 ± 0.001 min^{-1}, $^{***,\circ\circ\circ}$, $p < 0.001$ and 0.040 ± 0.001 min^{-1}, ns, ***, $p < 0.001$; 0.038 ± 0.001 min^{-1}, ***, $p < 0.001$, §, $p < 0.05$; Table 1). As expected, SO_4^{2-} uptake was significantly blocked by 10 μM DIDS applied at the beginning of incubation in SO_4^{2-} medium (0.018 ± 0.001 min^{-1}, ***, $p < 0.001$, Table 1). The SO_4^{2-} amount internalized by erythrocytes after 0.1 or 10 mM D-Gal treatment (t_3 and t_{24}) and 45 min of incubation in SO_4^{2-} medium (269.76 ± 39.74 and 203.16 ± 39.74; 191.26 ± 19.90 and 179.97 ± 7.23, ***, $p < 0.001$, Table 1) was significantly lower with respect to that measured in control conditions (313.81 ± 11.09, Table 1). With regard to the treatment with 50 mM 3-AT alone, the amount of internalized SO_4^{2-} was not different with respect to that measured in control conditions (343.43 ± 12.23), while with both 3-AT and 0.1 or 10 mM D-Gal pre-treatment, amounts of internalized SO_4^{2-} were significantly lower than those measured following pre-treatment with 0.1 or 10 mM D-Gal alone (146.85 ± 18.46 and 191.26 ± 15.21, §§§,***, $p < 0.001$; 172.85 ± 18 and 152.85 ± 15.87, §, $p < 0.05$; Table 1). In DIDS-treated cells, the intracellular SO_4^{2-} amount after 45 min of incubation in SO_4^{2-} medium (4.75 ± 8.50) was significantly lower than that determined in control or in D-Gal (0.1 and 10 mM)-treated erythrocytes, or in 50 mM 3-AT pre-treated erythrocytes (***, $p < 0.001$, Table 1).

3.2. Oxidative Conditions Assessment

3.2.1. TBARS Levels

Thiobarbituric-acid-reactive substances (TBARS) levels measured in erythrocytes after a 3 (data not shown) and 24 h incubation with 0.1 or 10 mM D-Gal and in 50 mM 3-AT-incubated erythrocytes were comparable to those detected in control erythrocytes (Figure 3). After 1 h treatment of erythrocytes with 10 mM H_2O_2, known to produce lipid peroxidation, TBARS levels were significantly higher than those measured in control and after 3 (data not shown) or 24 h treatment with D-Gal. TBARS levels in erythrocytes treated with 50 mM 3-AT + 0.1 mM D-Gal (t_{24}) or with 50 mM 3-AT + 10 mM D-Gal (t_{24}) were significantly higher with respect to those measured in control and 0.1 or 10 mM D-Gal-treated erythrocytes. The latter however, were not significantly different with respect to TBARS levels measured in erythrocytes treated with 10 mM H_2O_2 alone.

3.2.2. Membrane Sulfhydryl Groups Content Measurement

Membrane sulfhydryl groups content (μmol TNB/μg protein) measured after a 3 (data not shown) and 24 h (Figure 4) incubation of erythrocytes with 0.1 or 10 mM D-Gal and in 50 mM 3-AT-incubated erythrocytes was not significantly different with respect to those measured in control. As expected, after 1 h of erythrocytes treatment with the alkylating compound NEM (2 mM), sulfhydryl groups content was significantly lower than control. Levels of membrane sulfhydryl groups in erythrocytes treated with 50 mM 3-AT plus 0.1 or 10 mM D-Gal (t_{24}) were significantly reduced with respect to control and 0.1 or 10 mM D-Gal (t_{24}) alone. Erythrocytes treated with 50 mM 3-AT plus 2 mM NEM showed sulfhydryl groups levels lower than control, but not significantly different with respect to 2 mM NEM alone after a 3 (data not shown) and 24 h incubation.

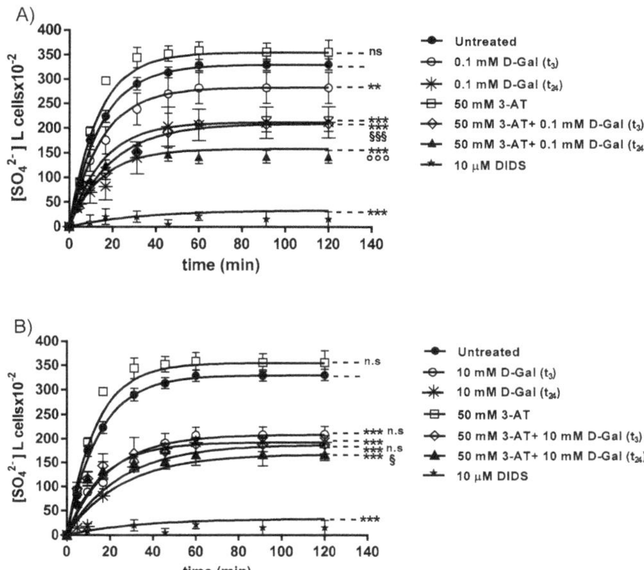

Figure 2. (**A**) Time course of SO_4^{2-} uptake in untreated (control) and in 0.1 mM D-Gal (t_3 and t_{24})-treated erythrocytes. ns, not significant versus control, *** $p < 0.001$, **, $p < 0.01$ versus control, §§§, $p < 0.001$ versus 0.1 mM D-Gal (t_3) and, °°°, $p < 0.001$ versus 0.1 mM D-Gal (t_{24}), as determined by one way ANOVA followed by Bonferroni's post hoc test (n = 6). (**B**) Time course of SO_4^{2-} uptake in untreated (control) and in 10 mM D-Gal (t_3 and t_{24})-treated erythrocytes. ns, not significant versus control, ***, $p < 0.001$ versus control and §, $p < 0.05$ versus 50 mM 3-AT + 10 mM D-Gal (t_3), as determined by one way ANOVA followed by Bonferroni's post hoc test (n = 6).

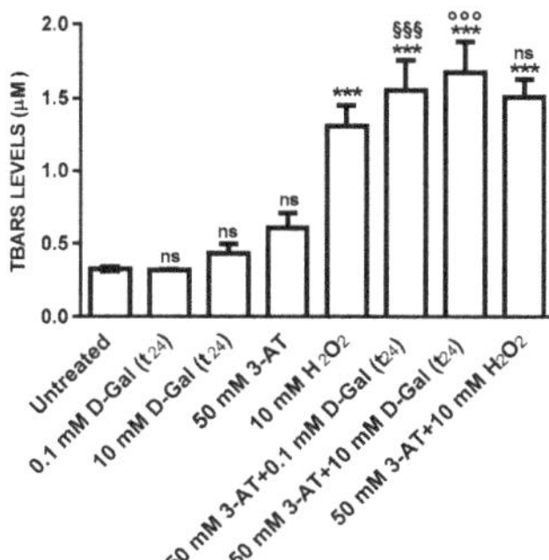

Figure 3. TBARS levels (μM) in untreated (control) and in 0.1 and 10 mM D-Gal (t_{24})-treated erythrocytes, with or without 50 mM 3-AT preincubation. ns, not significant versus control or 10 mM H_2O_2; ***, $p < 0.001$ versus control, §§§ $p < 0.001$ versus 0.1 mM D-Gal (t_{24}) and °°°, $p < 0.001$ versus 10 mM D-Gal (t_{24}) as determined by one way ANOVA followed by Bonferroni's post hoc test ($n = 6$).

Table 1. Rate constant for SO_4^{2-} uptake and amount of SO_4^{2-} trapped in untreated and in D-Gal (t_3 or t_{24})-treated erythrocytes at different concentrations (0.1 or 10 mM), with or without 3-AT pre-treatment.

	Rate Constant (min^{-1})	Time (min)	*n*	SO_4^{2-} Amount Trapped after a 45 min Incubation in SO_4^{2-} Medium [SO_4^{2-}] L cells x10^{-2}
untreated (control)	0.067 ± 0.001	14.92	8	313.81 ± 11.09
0.1 mM D-Gal (t_3)	0.066 ± 0.001 **	15.15	6	269.76 ± 39.74 **
0.1 mM D-Gal (t_{24})	0.065 ± 0.001 ***	15.38	6	203.16 ± 39.74 ***
10 mM D-Gal (t_3)	0.060 ± 0.001 ***	16.66	6	191.26 ± 19.90 ***
10 mM D-Gal (t_{24})	0.051 ± 0.001 ^{***,ns}	19.60	6	179.97 ± 7.23 ***
50 mM 3-AT	0.068 ± 0.001 ^{ns}	14.70	6	343.43 ± 12.23 ^{ns}
50 mM 3-AT + 0.1 mM D-Gal (t_3)	0.065 ± 0.001 ***,§§§	15.38	6	146.85 ± 18.46 §§§
50 mM 3-AT + 0.1 mM D-Gal (t_{24})	0.066 ± 0.001 $^{***,\circ\circ\circ}$	15.15	6	191.26 ± 15.21 $^{\circ\circ\circ}$
50 mM 3-AT + 10 mM D-Gal (t_3)	0.040 ± 0.001 ^{***,ns}	25.00	6	172.85 ± 18 ^{ns}
50 mM 3-AT + 10 mM D-Gal (t_{24})	0.038 ± 0.001 ***,§	26.31	6	152.85 ± 15.87 §
10 μM DIDS	0.018 ± 0.001 ***	55.5	10	4.75 ± 8.50 ***

Time is the reciprocal of the rate constant (see Materials and Methods for definition). Data are presented as means ± SEM from separate *n* experiments, where *ns* is not significant versus untreated,10 mM D-Gal (t_3) and 50 mM 3-AT + 10 mM D-Gal (t_{24}) or 10 mM D-Gal (t_3) and 50 mM 3-AT + 10 mM D-Gal (t_{24}); **, $p < 0.01$ versus untreated (control); ***, $p < 0.001$ versus control, §, $p < 0.05$ significant versus 50 mM 3-AT + 10 mM D-Gal (t_3); §§§, $p < 0.001$ versus 0.1 mM D-Gal (t_3) and, $^{\circ\circ\circ}$, $p < 0.001$ versus 0.1 mM D-Gal (t_{24}), as determined by one way ANOVA followed by Bonferroni's post hoc test ($n = 6$).

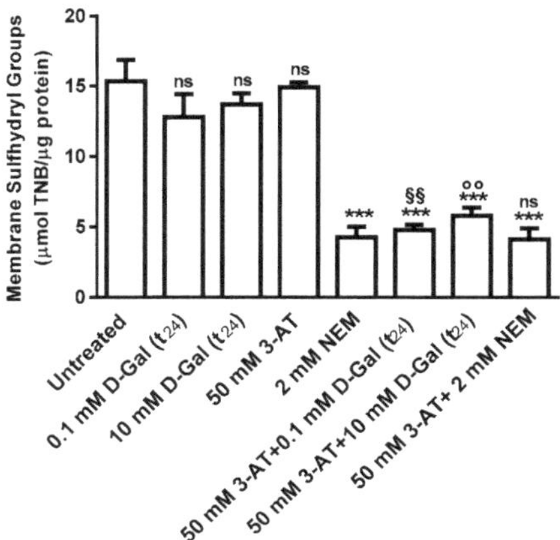

Figure 4. Membrane sulfhydryl groups content (μmol TNB/μg protein) in untreated (control) and in 0.1 or 10 mM D-Gal(t_{24})-treated erythrocytes. ns, not significant versus control or 2 mM NEM, ***, $p < 0.001$ versus control, §§, $p < 0.01$ versus 0.1 mM D-Gal (t_{24}) and °°, $p < 0.01$ versus 10 mM D-Gal (t_{24}) as determined by one way ANOVA followed by Bonferroni's post hoc test ($n = 6$).

3.2.3. Methemoglobin Measurement

Figure 5 shows Methemoglobin levels (% MetHb) in erythrocytes treated with different D-Gal concentrations (0.1 and 10 mM, 24 h incubation) with or without 50 mM 3-AT pre-treatment. MetHb levels measured after 3 h (data not shown) and after 24 h were not significantly different with respect to those detected in untreated erythrocytes (control). After 1 h of erythrocytes treatment with 4 mM $NaNO_2$, a well-known MetHb-forming agent, MetHb levels (%) were significantly higher than control. Erythrocytes treated with 50 mM 3-AT plus 4 mM $NaNO_2$ showed MetHb levels (%) higher than control, but not significantly different with respect to 4 mM $NaNO_2$ alone after both a 3 (data not shown) and 24 h incubation.

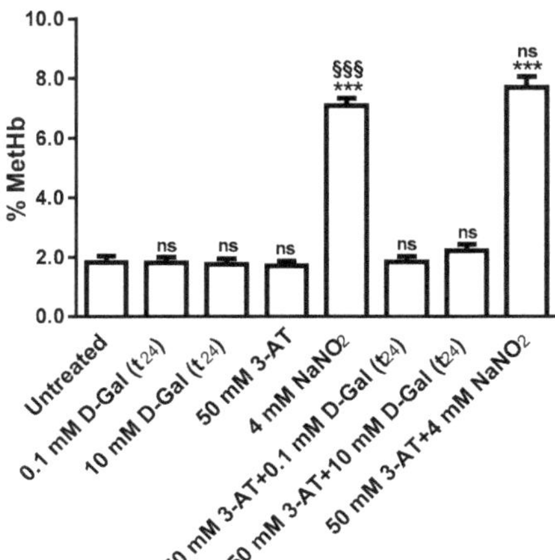

Figure 5. MetHb levels (%) in untreated (control) and in 0.1 or 10 mM D-Gal(t_{24})-treated erythrocytes with or without 50 mM 3-AT. ns, not significant versus untreated erythrocytes (control) and 4 mM $NaNO_2$, ***, $p < 0.001$ versus control, §§§, $p < 0.001$ versus 0.1 or 10 mM D-Gal (t_{24}) with or without 50 mM 3-AT pre-incubation as determined by one way ANOVA followed by Bonferroni's post hoc test ($n = 6$).

3.3. Band 3 Protein Expression Levels Determination

Figure 6A,B show B3p expression levels in erythrocytes incubated for 24 h with 0.1 or 10 mM D-Gal respectively, with or without pre-incubation with 50 mM 3-AT. Proteins expression after 3 (data not shown) and 24 h incubation was not significantly different with respect to that determined in untreated erythrocytes (control).

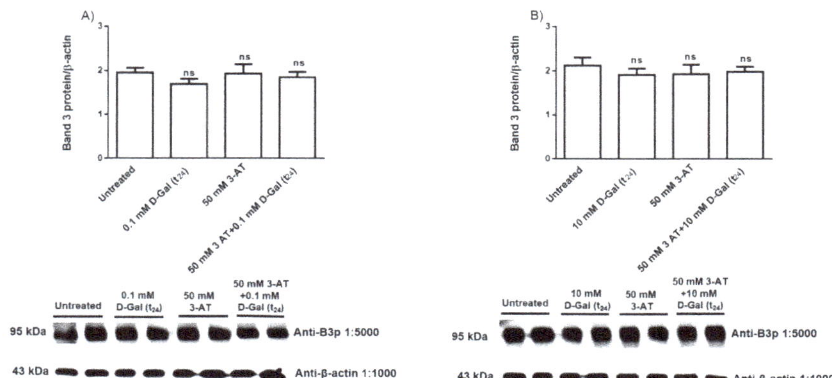

Figure 6. Band 3 protein expression levels measured in untreated (control) and in 0.1 (**A**) or 10 mM (**B**) D-Gal (t_{24})-treated erythrocytes, with or without 50 mM 3-AT pre-incubation, detected by Western blot analysis. ns, not significant versus untreated (control), as determined by one-way ANOVA followed by Bonferroni's post hoc test ($n = 5$).

3.4. Glycated Hemoglobin Measurement

Figure 7 shows glycated hemoglobin levels (% A1c) measured in erythrocytes treated with different D-Gal (0.1 and 10 mM) concentrations for 24 h, with or without 50 mM 3-AT treatment. %A1c levels were unchanged after 3 h exposure to D-Gal (data not shown), while, after 24 h, they were significantly increased with respect to control. Treatment with 50 mM 3-AT alone did not change %A1c levels with respect to those measured in untreated erythrocytes, while %A1c levels in erythrocytes treated with 50 mM 3-AT plus D-Gal (0.1 or 10 mM) were not significantly different with respect to those of erythrocytes treated with D-Gal alone and higher than control.

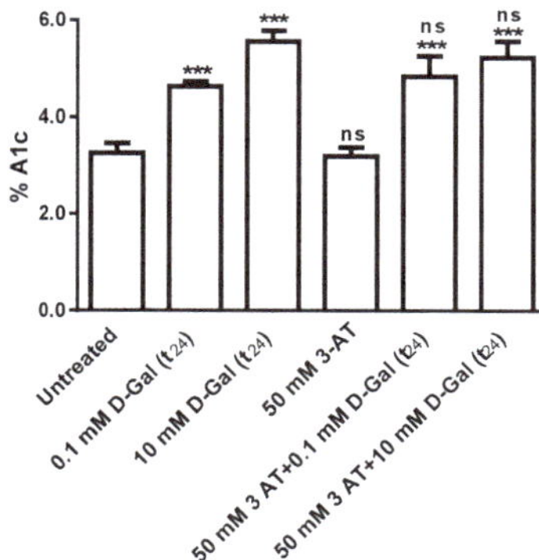

Figure 7. Glycated hemoglobin measurement (% A1c) in D-Gal-treated erythrocytes incubated for 24 h with different D-Gal concentrations (0.1 or 10 mM) with or without 50 mM 3-AT, ***, $p < 0.001$ versus untreated (control); ns, not significant versus both 0.1 mM D-Gal or 10 mM D-Gal, as determined by one way ANOVA followed by Bonferroni's post hoc test ($n = 10$).

4. Discussion

Chronic administration of D-Gal has been widely used as a model to mimic a process very similar to the natural aging, provoking oxidative stress [49] via increased production of ROS and changes in antioxidant enzyme activities in the cell [50]. The decline in cellular homeostatic redox capacity is responsible for macromolecules oxidation, thus compromising their functions [51]. ROS generation may inflict extensive damage to the erythrocyte membranes due to membrane proteins oxidation and lipid peroxidation [52], thus inducing membrane fluidity alteration, which reflects not only on membrane bilayer, but also on intracellular environment. In this regard, specific modified forms of hemoglobin (hemicromes) [53] have been shown to bind to membrane proteins, determining notable alteration of membrane structure, namely on senescent erythrocytes [19].

Though several in vivo experiments have been conducted on plasma or blood cells from animal models [6,18–21], the action mechanism of excessive D-Gal on human erythrocytes remains poorly clarified. D-Galactose-induced superoxide production implies H_2O_2 formation due to superoxide dismutase (SOD). As a consequence, H_2O_2, if abnormally accumulated, can inflict damage to the cell. Under normal conditions, the entire endogenous antioxidant system of erythrocytes, involving catalase, SOD and glutathione peroxidase (GPx) contributes to neutralize H_2O_2, thus minimizing its detrimental oxidative effects. Nevertheless, an unbalance between antioxidant enzymes involved in cell defense against ROS overproduction seems to be a mechanism through which D-Gal induces cell senescence [19]. The re-organization of cell proteins under D-Gal treatment has been also considered as an effect of cell aging. In particular, the oxidant effects of aging on Hb and membrane transport systems have been studied [54,55].

As previously stated, D-Gal oxidant effects have been mainly studied in vivo, while, with regard to in vitro investigations, there is still a lack of knowledge. A few data are available from an in vitro study by Delwing-de Lima and coauthors using D-Gal at final concentrations of 0.1, 0.3, 0.5 and 10 mM on blood withdrawn from 30- or 60-day-old rats. These authors consider D-Gal concentrations comprised in a range including both physiological and pathological values. According to other authors, the normal range in human healthy adults corresponds to 0.000012 mM, which is lower than what measured in newborn plasma [3,4]. On this basis, the aim of the present investigation was to verify the effect of D-Gal at high concentrations (0.1 or 10 mM) in in vitro human erythrocytes model. In particular, the purpose was to evaluate anion exchange capability through B3p and to verify the potential mechanism through which D-Gal may affect this function. As reported elsewhere, B3p is essential to erythrocytes homeostasis, whose assessment is a valid tool to detect damage in case of oxidative stress-related diseases, such as Systemic Sclerodermia [56], Canine Leishmaniasis [45] and hyperglycemic conditions [33], or in other oxidative conditions modeled in vitro [27,34].

The first step of the present research was to evaluate the SO_4^{2-} uptake through B3p [27,31,32] after either 3 or 24 h treatment with D-Gal (0.1 and 10 mM). Under these experimental conditions, the rate constant for SO_4^{2-} uptake. It is reasonably to suggest that, as a consequence of a reduced rate for SO_4^{2-}, the amount of this anion trapped after 45 min of incubation in SO_4^{2-} medium was, in turn, significantly reduced (Figure 2A,B). Similar D-Gal concentrations have been applied by Delwing-de Lima and coauthors, which observed a consequent protective increase in catalase activity in both young and old rat's erythrocytes. In this regard, to verify whether the D-Gal effect was mitigated by the endogenous antioxidant system, human erythrocytes were preventively exposed to 50 mM 3-AT, a specific catalase inhibitor [37]. In these conditions, the D-Gal (0.1 and 10 mM)-induced reduction in the rate constant for SO_4^{2-} uptake was exacerbated, suggesting that antioxidant enzymes, namely catalase, are critically involved in limiting the effect of excessive D-Gal.

The above mentioned results indicate that the modification of anion exchange capability through B3p in erythrocytes is associated to an increase of oxidative stress. Hence, to better explore this effect and according to what reported by other authors [19,57], lipid peroxidation and oxidation of membrane sulfhydryl groups mainly belonging to B3p [58] have been successively evaluated. This choice is also related to our previous studies on the impact of oxidant conditions on B3p efficiency [27,34,45,59,60].

As shown by the present results, D-Gal effect was mediated neither by membrane lipids nor by membrane sulfhydryl groups oxidation, at any concentration and at any time of incubation. Also in this case, the use of the catalase inhibitor before D-Gal incubation produced a significant oxidation of both lipids and membrane proteins (Figures 3 and 5). As a further step, to better clarify the mechanism of action of D-Gal in human erythrocytes, B3p protein abundance has been also determined (Figures 5 and 6), showing that in D-Gal (0.1 or 10 mM)-treated erythrocytes, at both t_3 and t_{24}, the expression levels of B3p were not reduced.

Therefore, following brief exposure of erythrocytes to high concentrations of D-Gal, the anion exchange capability through B3p is lowered, independently on TBARS production, or alteration in oxidative state, or expression levels of the membrane protein responsible for anion exchange. These results are similar to what previously reported by our group on human erythrocytes exposed in vitro to H_2O_2 [27]. Recent scientific evidences have reported that D-Gal accumulation induces oxidative stress, cell senescence and cytotoxicity [61]. In this regard, B3p functions, being crucial in the systemic homeostasis maintenance and affected by oxidative stress, could be correlated to oxidative stress-induced alterations during galactosemia. The hypothesis was that the crosslink between cytoplasmatic domain of B3p and Hb [62] was somehow compromised, thus reflecting on the efficiency of anion exchange, which was lowered. Such change seems not to impair the binding of anti B3p antibody.

This prompted us to take into account Hb in the present investigation, with specific regard to oxidation and glycation processes. In this respect, our results demonstrated that 3 h of incubation with D-Gal (0.1 or 10 mM) are insufficient to induce oxidative damage or glycation processes on Hb (data not shown). On the other hand, a longer incubation (24 h) with D-Gal does not induce oxidative stress events (Figure 5), but allows for the formation of higher glycated Hb levels (%A1c), with respect to untreated erythrocytes (Figure 7). Hence, we could suggest that, unlike what already shown in erythrocytes exposed in vitro to high glucose concentrations [33], D-Gal could generate early Hb glycation [63], which induces direct damage on Hb and, in turn, affects the exchange capability via B3p, cross-linked to Hb. However, we recently reported that in vitro exposure of human erythrocytes to high glucose induces an acceleration of the rate constant of SO_4^{2-} uptake, without A1c formation [33], contrarily to what observed here following D-Gal treatment. This discrepancy could be due to the different action mechanism of the two monosaccharides. Nevertheless, determination of anion exchange capability, in both cases, can be considered a very sensitive tool to detect the effects of exposure to high sugars concentrations [33], which were not revealed by parameters currently used to verify oxidative damage at either lipid or protein level, i.e., TBARS formation, oxidation and expression levels of membrane proteins.

According to other investigations, it seems more likely that in the present experimental conditions glycation of Hb rather than of B3p may occur, putatively due to a faster penetration of D-Gal through erythrocytes membrane at pH 7 than glucose [64] (approximately 3 times), which would let D-Gal induce glycation of Hb more rapidly than glycation of membrane proteins such as B3p. In addition, glycation of B3p should be excluded, since expression levels of this protein are unchanged after D-Gal treatment and, as reported elsewhere [60], the migration of glycated forms of proteins is usually slower. Kenawy and collaborators [63], using D-Gal treatment as a model of aging, stated that 90 days of D-Gal treatment are needed to observe increased levels of glycated Hb in vivo. Therefore, the present study provides novel elements about the first impact of excessive D-Gal on erythrocytes, which, as cited above, seems not associated to oxidative damage but rather to glycation events. In this regard, other authors have investigated proteins glycation induced by different sugars. In particular, the evidence of a D-Gal-dependent glycation has been already provided by Ledesma-Osuna and collaborators [65], reporting that D-Gal is more reactive than glucose or D-lactose, leading to the coupling of 10, 3 and 1 sugar residues, respectively, after 120 min of reaction on bovine serum albumin. Consequently, with regard to the present data and to the present experimental conditions, it is reasonable to suggest that the high reactivity of D-Gal could promote glycation as an early effect.

What recently reported by our group [32] describes that erythrocytes exposed to high glucose or erythrocytes withdrawn from patients with poorly controlled hyperglycemia exhibited an accelerated rate constant of SO_4^{2-} uptake, with the difference that only in the latter condition A1c formation was seen. In light of this report, and taking into account that oxidative conditions along with glycation affect erythrocytes in diabetic patients, it is reasonable to suggest that the extent of A1c formation observed here after D-Gal treatment is not sufficient to provoke an increase in anion exchange capability through B3p comparable to those detected in diabetic patients, but represents a first step of glycation process, which in any case alters the efficiency of the anion exchanger, lowering its rate.

It is interesting to point out that, according to Kenawy [63] and co-authors, 90 days of D-Gal treatment finally led to insulin resistance, which explains oxidative stress and inflammation at brain level. These authors further underscored that insulin resistance reduced glucose utilization, leading to mitochondrial dysfunction and ROS production. As in the present investigation D-Gal exposure was shorter, we could hypothesize that glycation occurred as the first process in our conditions, followed by oxidative stress, both leading to a detrimental effect on B3p. In addition, GSH/GSSG ratio considered as an indicator of oxidative stress is not altered by D-Gal treatment (Figure S1).

At any rate, the in vitro oxidant effect of D-Gal cannot be excluded and higher D-Gal concentrations and longer time of incubation are probably needed to induce oxidative stress events in human erythrocytes, as demonstrated by other studies on senescence models [19,66]. As reported elsewhere, the activities of some anti-oxidants enzymes, such as SOD1 and catalase, decline during aging [67] and, according to other authors, this decrease may depend on an increase in glycation of SOD [68], with consequent increased production of free radicals, responsible for aging. Such observation may further support the hypothesis that, under excessive D-Gal concentrations, oxidative stress follows glycation events. In this latter case, Jafarnejad and collaborators have demonstrated that a reduction in protein glycation can be prevented by acetyl salicylic acid (ASA) [69], known as inhibitor of glycation, thus inducing a decrease in AGEs levels and A1c formation [70]. Therefore, in analogy to what reported by these authors, ASA could be also used to prevent the elevation of glycation levels, specifically for Hb glycation in case of high D-Gal concentrations.

The contribution of the antioxidant system in preventing oxidative damage after production of glycated proteins during D-Gal treatment of erythrocytes is not fully understood [19,71] and future studies will be needed to elucidate the link between glycation and oxidative events.

Finally, it is opportune to mention some limits of this investigation, as well as future perspectives. As a whole, the present study reports for the first time the effect of high D-Gal on erythrocytes by using a validated in vitro model monitoring the anion exchange capability through B3p. At a first stage, high D-Gal seems to reduce the anion exchange capability of erythrocytes as a consequence of Hb glycation, rather than by oxidative action, which can only be detected after inhibition of catalase. This important finding underscores that D-Gal effectively induced oxidative stress, which, however, was mitigated by catalase activity. This would suggest that in vivo and, more specifically, in those pathological conditions characterized by absence or reduction of functioning catalase, the effects of the oxidative action of D-Gal could be more relevant. Future studies should focus on the impact of D-Gal in subjects having a deficiency in erythrocytes catalase such as, for example, patients suffering of acute myocardial infarction. In this context, the use of antioxidants chosen among the most lipophilic radical-scavenging antioxidants, i.e., alpha-tocopherol [72], or low molecular weight antioxidants, such as vitamin C and polyphenols, should also be investigated.

In addition, techniques more sensitive than those employed here (i.e., HPLC to detect MDA levels) to assess the oxidative impact of D-Gal should be also considered, in an attempt to give more details about high D-Gal toxicity and having the present outcomes as a starting point useful to shed light on this topic.

5. Conclusions

Taken together, our findings suggest that: i) B3p function assessment is a sensitive tool to verify the impact of D-Gal on erythrocytes homeostasis; ii) D-Gal (0.1 and 10 mM) induces a reduction in anion exchange capability independent from membrane lipids peroxidation, membrane sulfhydryl groups oxidation, MetHb formation, or changes in B3p expression levels; iii) D-Gal effect, reflecting on B3p efficiency, could develop through a two-phases process, with the first associated to early glycation of Hb, overwhelming the oxidant effect of D-Gal; iv) D-Gal-induced oxidative damage could not be excluded, but is promptly mitigated by the endogenous antioxidant system, specifically catalase.

Finally, our investigation adds first elements to the knowledge about D-Gal toxicity namely at membrane transport systems levels, which is not only associated to in vitro aging models, but also to failure of D-Gal metabolism in humans, due to congenital deficiency of enzymes, leading to D-Gal accumulation. This study could offer a good basis to understand the early effect of excessive D-Gal, shedding more light on possible complications related to undiagnosed galactosemia and highlighting the relationship between AGEs and oxidative events, which need to be better clarified, as possibly involved in the onset of galactosemia symptoms

Supplementary Materials: The following are available online at http://www.mdpi.com/2076-3921/9/8/689/s1, Figure S1: Estimation of GSH/GSSG ratio measured in untreated (control) and D-Gal treated erythrocytes at different concentrations (0.1 and 10 mM) with or without 50 mM 3-AT incubated for 24 h. *ns* not significant *vs* control; *** $p < 0.001$ significant *vs* control; §§§ $p < 0.001$ *vs* 0.1 mM D-Gal; °°° $p < 0.001$ significant *vs* 10 mM D-Gal, as determined by one way ANOVA followed by Bonferroni's *post hoc* test ($n = 3$).

Author Contributions: R.M., S.D. and A.M. conceived and designed the research; A.R., R.M., S.S., V.T. and S.L. performed the experiments and analyzed the data; R.M., A.R., A.S., S.D. and A.M. interpreted the results of the experiments; R.M. and A.R. prepared the figures; R.M., A.R., S.D., V.T., S.L. and A.M. drafted the manuscript; A.R., R.M., S.S., A.S., S.D., V.T., S.L. and A.M. edited and revised the manuscript; All authors have read and agreed to the published version of the manuscript.

Funding: This research was funded by MIUR, grant number PRIN 2015 (Prot. 20152HKF3Z).

Conflicts of Interest: The authors declare no conflict of interest.

References

1. Coelho, A.I.; Berry, G.T.; Rubio-Gozalbo, M.E. Galactose metabolism and health. *Curr. Opin. Clin. Nutr. Metab Care* **2015**, *18*, 422–427. [CrossRef] [PubMed]
2. Bosch, A.M. Classical galactosaemia revisited. *J. Inherit Metab. Dis.* **2006**, *29*, 516–525. [CrossRef] [PubMed]
3. Bossolan, G.; Trindade, C.E.; Barreiros, R.C. Blood galactose and glucose levels in mothers, cord blood, and 48-h-old breast-fed full-term infants. *Neonatology* **2007**, *91*, 121–126. [CrossRef] [PubMed]
4. Schadewaldt, P.; Hammen, H.W.; Loganathan, K.; Bodner-Leidecker, A.; Wendel, U. Analysis of concentration and (13)C enrichment of D-galactose in human plasma. *Clin. Chem.* **2000**, *46*, 612–619. [CrossRef] [PubMed]
5. Gitzelmann, R. Galactose-1-phosphate in the pathophysiology of galactosemia. *Eur. J. Pediatr.* **1995**, *154*, S45–49. [CrossRef] [PubMed]
6. Delwing-de Lima, D.; Hennrich, S.B.; Delwing-Dal Magro, D.; Aurelio, J.G.; Serpa, A.P.; Augusto, T.W.; Pereira, N.R. The effect of d-galactose induced oxidative stress on in vitro redox homeostasis in rat plasma and erythrocytes. *Biomed. Pharmacother.* **2017**, *86*, 686–693. [CrossRef]
7. Bo-Htay, C.; Palee, S.; Apaijai, N.; Chattipakorn, S.C.; Chattipakorn, N. Effects of d-galactose-induced ageing on the heart and its potential interventions. *J. Cell Mol. Med.* **2018**, *22*, 1392–1410. [CrossRef]
8. Luevano-Contreras, C.; Chapman-Novakofski, K. Dietary advanced glycation end products and aging. *Nutrients* **2010**, *2*, 1247–1265. [CrossRef]
9. Ansari, N.A.; Dash, D. Amadori glycated proteins: Role in production of autoantibodies in diabetes mellitus and effect of inhibitors on non-enzymatic glycation. *Aging Dis.* **2013**, *4*, 50–56.
10. Cebe, T.; Yanar, K.; Atukeren, P.; Ozan, T.; Kuruc, A.I.; Kunbaz, A.; Sitar, M.E.; Mengi, M.; Aydin, M.S.; Esrefoglu, M.; et al. A comprehensive study of myocardial redox homeostasis in naturally and mimetically aged rats. *Age (Dordr.)* **2014**, *36*, 9728. [CrossRef]
11. Li, X.; Zhang, Y.; Yuan, Y.; Sun, Y.; Qin, Y.; Deng, Z.; Li, H. Protective Effects of Selenium, Vitamin E, and Purple Carrot Anthocyanins on D-Galactose-Induced Oxidative Damage in Blood, Liver, Heart and Kidney Rats. *Biol. Trace Elem. Res.* **2016**, *173*, 433–442. [CrossRef] [PubMed]
12. Doroszko, A.; Radziwon-Balicka, A.; Skomro, R. Novel Approaches for Diagnosing and Management of Cardiovascular Disorders Mediated by Oxidative Stress. *Oxid. Med. Cell Longev.* **2020**, *2020*, 7096727. [CrossRef] [PubMed]

13. Brunelli, E.; La Russa, D.; Pellegrino, D. Impaired Oxidative Status Is Strongly Associated with Cardiovascular Risk Factors. *Oxid. Med. Cell Longev.* **2017**, *2017*, 6480145. [CrossRef] [PubMed]
14. Ton, A.M.M.; Campagnaro, B.P.; Alves, G.A.; Aires, R.; Coco, L.Z.; Arpini, C.M.; Guerra, E.O.T.; Campos-Toimil, M.; Meyrelles, S.S.; Pereira, T.M.C.; et al. Oxidative Stress and Dementia in Alzheimer's Patients: Effects of Synbiotic Supplementation. *Oxid. Med. Cell Longev.* **2020**, *2020*, 2638703. [CrossRef]
15. Janion, K.; Szczepanska, E.; Nowakowska-Zajdel, E.; Walkiewicz, K.; Strzelczyk, J. Lipid peroxidation and total oxidant/antioxidant status in colorectal cancer patients. *J. Biol. Regul. Homeost. Agents* **2020**, *34*, 239–244. [CrossRef]
16. Zhang, P.; Li, T.; Wu, X.; Nice, E.C.; Huang, C.; Zhang, Y. Oxidative stress and diabetes: Antioxidative strategies. *Front. Med.* **2002**. [CrossRef]
17. Iglesias-Pedraz, J.M.; Comai, L. Measurements of Hydrogen Peroxide and Oxidative DNA Damage in a Cell Model of Premature Aging. *Methods Mol. Biol.* **2020**, *2144*, 245–257. [CrossRef]
18. Hsia, C.H.; Wang, C.H.; Kuo, Y.W.; Ho, Y.J.; Chen, H.L. Fructo-oligosaccharide systemically diminished D-galactose-induced oxidative molecule damages in BALB/cJ. mice. *Br. J. Nutr.* **2012**, *107*, 1787–1792. [CrossRef]
19. Mladenov, M.; Gokik, M.; Hadzi-Petrushev, N.; Gjorgoski, I.; Jankulovski, N. The relationship between antioxidant enzymes and lipid peroxidation in senescent rat erythrocytes. *Physiol. Res.* **2015**, *64*, 891–896. [CrossRef]
20. Song, X.; Bao, M.; Li, D.; Li, Y.M. Advanced glycation in D-galactose induced mouse aging model. *Mech. Ageing Dev.* **1999**, *108*, 239–251. [CrossRef]
21. Haider, S.; Liaquat, L.; Shahzad, S.; Sadir, S.; Madiha, S.; Batool, Z.; Tabassum, S.; Saleem, S.; Naqvi, F.; Perveen, T. A high dose of short term exogenous D-galactose administration in young male rats produces symptoms simulating the natural aging process. *Life Sci.* **2015**, *124*, 110–119. [CrossRef] [PubMed]
22. Inès Birlouez-Aragon, P.S.-M. Marek Morawiec and Majid Shafiezadeh Evidence for a relationship between protein glycation and red blood cell membrane fluidity. *Biochem. Bioph. Res. Commun.* **1990**, *170*, 1107–1113. [CrossRef]
23. Ando, K.; Beppu, M.; Kikugawa, K.; Nagai, R.; Horiuchi, S. Membrane proteins of human erythrocytes are modified by advanced glycation end products during aging in the circulation. *Biochem. Biophys. Res. Commun.* **1999**, *258*, 123–127. [CrossRef] [PubMed]
24. Sompong, W.; Cheng, H.; Adisakwattana, S. Protective Effects of Ferulic Acid on High Glucose-Induced Protein Glycation, Lipid Peroxidation, and Membrane Ion Pump Activity in Human Erythrocytes. *PLoS ONE* **2015**, *10*, e0129495. [CrossRef]
25. Viskupicova, J.; Blaskovic, D.; Galiniak, S.; Soszynski, M.; Bartosz, G.; Horakova, L.; Sadowska-Bartosz, I. Effect of high glucose concentrations on human erythrocytes in vitro. *Redox Biol.* **2015**, *5*, 381–387. [CrossRef]
26. Remigante, A.; Morabito, R.; Marino, A. Natural Antioxidants Beneficial Effects on Anion Exchange through Band 3 Protein in Human Erythrocytes. *Antioxidants (Basel)* **2019**, *9*. [CrossRef]
27. Morabito, R.; Romano, O.; La Spada, G.; Marino, A. H_2O_2-Induced Oxidative Stress Affects SO_4= Transport in Human Erythrocytes. *PLoS ONE* **2016**, *11*, e0146485. [CrossRef]
28. van Zwieten, R.; Verhoeven, A.J.; Roos, D. Inborn defects in the antioxidant systems of human red blood cells. *Free Radic. Biol. Med.* **2014**, *67*, 377–386. [CrossRef]
29. Reithmeier, R.A.; Casey, J.R.; Kalli, A.C.; Sansom, M.S.; Alguel, Y.; Iwata, S. Band 3, the human red cell chloride/bicarbonate anion exchanger (AE1, SLC4A1), in a structural context. *Biochim. Biophys. Acta* **2016**, *1858*, 1507–1532. [CrossRef]
30. Hatae, H.; Inaka, K.; Okamura, R.; Furubayashi, N.; Kamo, M.; Kobayashi, T.; Abe, Y.; Iwata, S.; Hamasaki, N. Crystallization of Human Erythrocyte Band 3, the anion exchanger, at the International Space Station "KIBO". *Anal. Biochem.* **2018**, *559*, 91–93. [CrossRef]
31. Romano, L.; Peritore, D.; Simone, E.; Sidoti, A.; Trischitta, F.; Romano, P. Chloride-sulphate exchange chemically measured in human erythrocyte ghosts. *Cell Mol. Biol. (Noisy-le-grand)* **1998**, *44*, 351–355.
32. Jennings, M.L. Proton fluxes associated with erythrocyte membrane anion exchange. *J. Membr. Biol.* **1976**, *28*, 187–205. [CrossRef] [PubMed]
33. Morabito, R.; Remigante, A.; Spinelli, S.; Vitale, G.; Trichilo, V.; Loddo, S.; Marino, A. High Glucose Concentrations Affect Band 3 Protein in Human Erythrocytes. *Antioxidants (Basel)* **2020**, *9*. [CrossRef] [PubMed]
34. Morabito, R.; Remigante, A.; Di Pietro, M.L.; Giannetto, A.; La Spada, G.; Marino, A. SO_4(=) uptake and catalase role in preconditioning after H2O2-induced oxidative stress in human erythrocytes. *Pflugers Arch.* **2017**, *469*, 235–250. [CrossRef]

35. Morabito, R.; Remigante, A.; Cordaro, M.; Trichilo, V.; Loddo, S.; Dossena, S.; Marino, A. Impact of acute inflammation on Band 3 protein anion exchange capability in human erythrocytes. *Arch. Physiol. Biochem.* **2020**. [CrossRef]
36. Pantaleo, A.; Ferru, E.; Pau, M.C.; Khadjavi, A.; Mandili, G.; Matte, A.; Spano, A.; De Franceschi, L.; Pippia, P.; Turrini, F. Band 3 Erythrocyte Membrane Protein Acts as Redox Stress Sensor Leading to Its Phosphorylation by p (72) Syk. *Oxid. Med. Cell Longev.* **2016**, *2016*, 6051093. [CrossRef]
37. Margoliash, E.; Novogrodsky, A.; Schejter, A. Irreversible reaction of 3-amino-1:2:4-triazole and related inhibitors with the protein of catalase. *Biochem. J.* **1960**, *74*, 339–348. [CrossRef]
38. Romano, L.; Passow, H. Characterization of anion transport system in trout red blood cell. *Am. J. Physiol.* **1984**, *246*, C330–338. [CrossRef]
39. Jessen, F.; Sjoholm, C.; Hoffmann, E.K. Identification of the anion exchange protein of Ehrlich cells: A kinetic analysis of the inhibitory effects of 4,4′-diisothiocyano-2,2′-stilbene-disulfonic acid (DIDS) and labeling of membrane proteins with 3H-DIDS. *J. Membr. Biol.* **1986**, *92*, 195–205. [CrossRef]
40. Mendanha, S.A.; Anjos, J.L.; Silva, A.H.; Alonso, A. Electron paramagnetic resonance study of lipid and protein membrane components of erythrocytes oxidized with hydrogen peroxide. *Braz. J. Med. Biol. Res.* **2012**, *45*, 473–481. [CrossRef]
41. Sokolowska, M.; Oleszek, A.; Wlodek, L. Protective effect of alpha-keto acids on the oxidative hemolysis. *Pol. J. Pharmacol.* **1999**, *51*, 429–434.
42. Aksenov, M.Y.; Markesbery, W.R. Changes in thiol content and expression of glutathione redox system genes in the hippocampus and cerebellum in Alzheimer's disease. *Neurosci. Lett.* **2001**, *302*, 141–145. [CrossRef]
43. Morabito, R.; Falliti, G.; Geraci, A.; Spada, G.L.; Marino, A. Curcumin Protects -SH Groups and Sulphate Transport after Oxidative Damage in Human Erythrocytes. *Cell Physiol. Biochem.* **2015**, *36*, 345–357. [CrossRef] [PubMed]
44. Naoum, P.C.R.; Radispiel, J.; Magaly da Silva, M. Spectrometric measurement of methemoglobin without interference of chemical or enzymatic reagents. *Rev. Bras. Hematol. Hemoter.* **2004**, *26*, 19–22.
45. Morabito, R.; Remigante, A.; Cavallaro, M.; Taormina, A.; La Spada, G.; Marino, A. Anion exchange through band 3 protein in canine leishmaniasis at different stages of disease. *Pflugers Arch.* **2017**, *469*, 713–724. [CrossRef] [PubMed]
46. Zavodnik, I.B.; Lapshina, E.A.; Rekawiecka, K.; Zavodnik, L.B.; Bartosz, G.; Bryszewska, M. Membrane effects of nitrite-induced oxidation of human red blood cells. *Biochim. Biophys. Acta* **1999**, *1421*, 306–316. [CrossRef]
47. Laemmli, U.K. Cleavage of structural proteins during the assembly of the head of bacteriophage T4. *Nature* **1970**, *227*, 680–685. [CrossRef]
48. Yeung, Y.G.; Stanley, E.R. A solution for stripping antibodies from polyvinylidene fluoride immunoblots for multiple reprobing. *Anal. Biochem.* **2009**, *389*, 89–91. [CrossRef]
49. Cebe, T.; Atukeren, P.; Yanar, K.; Kuruc, A.I.; Ozan, T.; Kunbaz, A.; Sitar, M.E.; Mirmaroufizibandeh, R.; Aydin, S.; Cakatay, U. Oxidation scrutiny in persuaded aging and chronological aging at systemic redox homeostasis level. *Exp. Gerontol.* **2014**, *57*, 132–140. [CrossRef]
50. Kayali, R.; Cakatay, U.; Tekeli, F. Male rats exhibit higher oxidative protein damage than females of the same chronological age. *Mech. Ageing Dev.* **2007**, *128*, 365–369. [CrossRef]
51. Cakatay, U.; Kayali, R.; Uzun, H. Relation of plasma protein oxidation parameters and paraoxonase activity in the ageing population. *Clin. Exp. Med.* **2008**, *8*, 51–57. [CrossRef] [PubMed]
52. Yelinova, V.; Glazachev, Y.; Khramtsov, V.; Kudryashova, L.; Rykova, V.; Salganik, R. Studies of human and rat blood under oxidative stress: Changes in plasma thiol level, antioxidant enzyme activity, protein carbonyl content, and fluidity of erythrocyte membrane. *Biochem. Biophys. Res. Commun.* **1996**, *221*, 300–303. [CrossRef] [PubMed]
53. Welbourn, E.M.; Wilson, M.T.; Yusof, A.; Metodiev, M.V.; Cooper, C.E. The mechanism of formation, structure and physiological relevance of covalent hemoglobin attachment to the erythrocyte membrane. *Free Radic. Biol. Med.* **2017**, *103*, 95–106. [CrossRef] [PubMed]
54. Luthra, M.G.; Kim, H.D. Effects of calcium and soluble cytoplasmic activator protein (calmodulin) on various states of (Ca_2+ + Mg_2+)-ATPase activity in isolated membranes of human red cells. *Biochim. Biophys. Acta* **1980**, *600*, 467–479. [CrossRef]
55. Rifkind, J.M.; Nagababu, E. Hemoglobin redox reactions and red blood cell aging. *Antioxid. Redox Signal* **2013**, *18*, 2274–2283. [CrossRef]

56. Morabito, R.; Remigante, A.; Bagnato, G.; Neal, R.W.; Sciortino, D.; D'Angelo, T.; Iannelli, F.; Iannelli, F.; Cordova, F.; Cirillo, M.; et al. Band 3 Protein Function and Oxidative Stress in Erythrocytes from Systemic Sclerosis Patients with Interstitial Lung Disease. *Eur. J. Clin. Biomed. Sci.* **2017**, *3*, 80–84. [CrossRef]
57. Ho, S.C.; Liu, J.H.; Wu, R.Y. Establishment of the mimetic aging effect in mice caused by D-galactose. *Biogerontology* **2003**, *4*, 15–18. [CrossRef]
58. Roy, S.S.; Sen, G.; Biswas, T. Role of sulfhydryl groups in band 3 in the inhibition of phosphate transport across erythrocyte membrane in visceral leishmaniasis. *Arch. Biochem. Biophys.* **2005**, *436*, 121–127. [CrossRef]
59. Morabito, R.; Remigante, A.; Marino, A. Protective Role of Magnesium against Oxidative Stress on $SO_4(=)$ Uptake through Band 3 Protein in Human Erythrocytes. *Cell Physiol. Biochem.* **2019**, *52*, 1292–1308. [CrossRef]
60. Morabito, R.; Remigante, A.; Marino, A. Melatonin Protects Band 3 Protein in Human Erythrocytes against H2O2-Induced Oxidative Stress. *Molecules* **2019**, *24*. [CrossRef]
61. Umbayev, B.; Askarova, S.; Almabayeva, A.; Saliev, T.; Masoud, A.R.; Bulanin, D. Galactose-Induced Skin Aging: The Role of Oxidative Stress. *Oxid. Med. Cell Longev.* **2020**, *2020*, 7145656. [CrossRef] [PubMed]
62. Walder, J.A.; Chatterjee, R.; Steck, T.L.; Low, P.S.; Musso, G.F.; Kaiser, E.T.; Rogers, P.H.; Arnone, A. The interaction of hemoglobin with the cytoplasmic domain of band 3 of the human erythrocyte membrane. *J. Biol. Chem.* **1984**, *259*, 10238–10246. [PubMed]
63. Kenawy, S.; Hegazy, R.; Hassan, A.; El-Shenawy, S.; Gomaa, N.; Zaki, H.; Attia, A. Involvement of insulin resistance in D-galactose-induced age-related dementia in rats: Protective role of metformin and saxagliptin. *PLoS ONE* **2017**, *12*, e0183565. [CrossRef] [PubMed]
64. Faust, R.G. Monosaccharide penetration into human red blood cells by an altered diffusion mechanism. *J. Cell Comp. Physiol.* **1960**, *56*, 103–121. [CrossRef] [PubMed]
65. Ledesma-Osuna, A.I.; Ramos-Clamont, G.; Vazquez-Moreno, L. Characterization of bovine serum albumin glycated with glucose, galactose and lactose. *Acta Biochim. Pol.* **2008**, *55*, 491–497. [CrossRef] [PubMed]
66. Sathiyapriya, V.; Bobby, Z.; Vinod Kumar, S.; Selvaraj, N.; Parthibane, V.; Gupta, S. Evidence for the role of lipid peroxides on glycation of hemoglobin and plasma proteins in non-diabetic asthma patients. *Clin. Chim. Acta* **2006**, *366*, 299–303. [CrossRef]
67. Bartosz, G.; Tannert, C.; Fried, R.; Leyko, W. Superoxide dismutase activity decreases during erythrocyte aging. *Experientia* **1978**, *34*, 1464. [CrossRef]
68. Arai, K.; Maguchi, S.; Fujii, S.; Ishibashi, H.; Oikawa, K.; Taniguchi, N. Glycation and inactivation of human Cu-Zn-superoxide dismutase. Identification of the in vitro glycated sites. *J. Biol. Chem.* **1987**, *262*, 16969–16972.
69. Jafarnejad, A.; Bathaie, S.Z.; Nakhjavani, M.; Hassan, M.Z. Investigation of the mechanisms involved in the high-dose and long-term acetyl salicylic acid therapy of type I diabetic rats. *J. Pharmacol. Exp. Ther.* **2008**, *324*, 850–857. [CrossRef]
70. Rao, G.N.; Lardis, M.P.; Cotlier, E. Acetylation of lens crystallins: A possible mechanism by which aspirin could prevent cataract formation. *Biochem. Biophys. Res. Commun.* **1985**, *128*, 1125–1132. [CrossRef]
71. Jaeschke, H. Glutathione disulfide formation and oxidant stress during acetaminophen-induced hepatotoxicity in mice in vivo: The protective effect of allopurinol. *J. Pharmacol. Exp. Ther.* **1990**, *255*, 935–941. [PubMed]
72. Niki, E.; Traber, M.G. A history of vitamin E. *Ann. Nutr. Metab.* **2012**, *61*, 207–212. [CrossRef]

© 2020 by the authors. Licensee MDPI, Basel, Switzerland. This article is an open access article distributed under the terms and conditions of the Creative Commons Attribution (CC BY) license (http://creativecommons.org/licenses/by/4.0/).

Article

Microvesicle Formation Induced by Oxidative Stress in Human Erythrocytes

Julia Sudnitsyna [1,†], Elisaveta Skverchinskaya [2,†], Irina Dobrylko [2], Elena Nikitina [2], Stepan Gambaryan [2,3] and Igor Mindukshev [2,*]

1 Center for Theoretical Problems of Physico-Chemical Pharmacology, Russian Academy of Sciences, Kosygina st., 4, 119991 Moscow, Russia; sudnitsyna@iephb.ru or julia.sudnitsyna@gmail.com
2 Sechenov Institute of Evolutionary Physiology and Biochemistry, Russian Academy of Sciences, Thorez pr., 44, 194223 Saint-Petersburg, Russia; lisarafail@mail.ru (E.S.); dobrilko@mail.ru (I.D.); elena.nikitina@bk.ru (E.N.); s.gambaryan@klin-biochem.uni-wuerzburg.de (S.G.)
3 Center for Thrombosis and Hemostasis (CTH), University Medical Center of the Johannes Gutenberg-University Mainz, 55131 Mainz, Germany
* Correspondence: iv_mindukshev@mail.ru; Tel.: +7-921-979-4793
† These authors contributed equally to this work.

Received: 30 August 2020; Accepted: 25 September 2020; Published: 28 September 2020

Abstract: Extracellular vesicles (EVs) released by different cell types play an important role in many physiological and pathophysiological processes. In physiological conditions, red blood cell (RBC)-derived EVs compose 4–8% of all circulating EVs, and oxidative stress (OS) as a consequence of different pathophysiological conditions significantly increases the amount of circulated RBC-derived EVs. However, the mechanisms of EV formation are not yet fully defined. To analyze OS-induced EV formation and RBC transformations, we used flow cytometry to evaluate cell esterase activity, caspase-3 activity, and band 3 clustering. Band 3 clustering was additionally analyzed by confocal microscopy. Two original laser diffraction-based approaches were used for the analysis of cell deformability and band 3 activity. Hemoglobin species were characterized spectrophotometrically. We showed that cell viability in *tert*-Butyl hydroperoxide-induced OS directly correlated with oxidant concentration to cell count ratio, and that RBC-derived EVs contained hemoglobin oxidized to hemichrome (HbChr). OS induced caspase-3 activation and band 3 clustering in cells and EVs. Importantly, we showed that OS-induced EV formation is independent of calcium. The presented data indicated that during OS, RBCs eliminated HbChr by vesiculation in order to sacrifice the cell itself, thereby prolonging lifespan and delaying the untimely clearance of in all other respects healthy RBCs.

Keywords: erythrocytes; microparticles; oxidative stress; vesiculation; band 3; *tert*-Bytyl hydroperoxide t-BOOH; nitric oxide donor; calcium ionophore A23187

1. Introduction

Extracellular vesicles (EVs), which consist of microvesicles (MVs), microparticles (MPs), and exosomes, are continuously produced in human blood from different cell types including circulating and endothelial cells. EVs contain various molecules of parent cells such as different proteins, bioactive lipids, and RNAs that can be taken up by recipient cells [1]. EVs are directly involved in different physiological processes such as vasoregulation, thrombosis, hemostasis, and inflammation, acting similar to signaling molecules or by direct transport of their constituents [2,3]. In normal conditions, red blood cell (RBC)-derived EVs compose 4–8% of all circulating EVs [4]. EV formation is triggered by structural alterations of the cell membrane, which are driven by factors that disrupt erythrocyte skeleton/membrane attachment, including aging-associated oxidative damage [5,6], reactive oxygen species (ROS) [7], increased intracellular Ca^{2+} concentration [8,9], adenosine triphosphate depletion [10], and RBC storage in blood banks [11]. RBC-derived EVs are not homogeneous by their size and content. The increase in intracellular calcium concentration induces membrane shedding, producing EVs containing cytoskeleton proteins, glycophorin, band 3 protein (band 3, anion transporter 1, AE1), and acetylcholine esterase, but not hemoglobin (Hb) [12]. During RBC concentrates storage, EVs carry aggregates of Hb, band 3, cytoskeleton proteins, caspases 3 and 8, Cluster of Differentiation (CD) 47, and immunoglobulins G [13]. Throughout the 120 days of RBCs' lifespan, up to 20% of Hb is lost due to EV formation [14]. Formation of such EVs is proposed to rescue RBCs by getting rid of damaged Hb molecules and clustered band 3, which are well-defined markers of senescent RBCs, therefore prolonging the life span of these cells [15]. Two major mechanisms are described for EV formation including an increase in intracellular calcium concentration and oxidative stress (OS)-induced Hb oxidation [16]. Ca^{2+}-activated protease calpain is involved in vesiculation by binding and degrading band 3, band 4.1, and ankyrin, which lead to alteration of cell deformability [9,16–19]. OS triggers membrane and Hb oxidation, clustering, and disruption of band 3-cytoskeleton anchorage, and all these alterations lead to EV formation [7,17,20–23]. All these data indicate that there are various mechanisms of RBC-derived EV formation, however, several questions are still unsolved. In this study, we elucidated the effects of different stimuli such as oxidant concentration and Hb oxidation state, as well as the role of calcium, caspase-3 activation, and band 3 clustering that trigger RBC-derived EV formation.

Here, we showed that cell viability in *tert*-Butyl hydroperoxide-induced OS directly correlated with the ratio of oxidant concentration to cell count, and that RBC-derived EVs contained Hb oxidized to hemichrome (HbChr). OS induced caspase-3 activation and clustering of band 3 protein in both, cells and EVs. Importantly, we showed that OS-induced EV formation was independent of calcium. Presented data indicated that during OS, RBCs eliminate HbChr by vesiculation in order to sacrifice the cell itself, thereby prolonging lifespan and delaying the untimely clearance of in all other respects healthy RBCs.

2. Materials and Methods

2.1. Reagents and Chemicals

Calcein-AM and eosin-5-maleimide (EMA) were from Molecular Probes (Eugene, OR, USA). Annexin-V conjugated with fluorescein isothiocyante (Annexin-V-FITC) was obtained from Biolegend (Amsterdam, Netherlands). Anti-active caspase-3 FITC polyclonal antibodies were from BD Pharmingen (San Diego, CA, USA). *tert*-Butyl hydroperoxide (t-BOOH), S-nitroso-L-cysteine (SNC)—sodium nitrite and L-cysteine hydrochloride monohydrate, calcium ionophore A23187, and basic buffer constituents were from Sigma-Aldrich (Munich, Germany). The buffers were isotonic with osmolality 300 mOsm/kg H_2O controlled by cryoscopic osmometer Osmomat 3000 (Gonotec, Germany), pH 7.4, and had the following composition (in mM): HEPES buffer-NaCl, 140; KCl, 5; 4-(2-hydroxyethyl)-1-piperazineethanesulfonic acid, HEPES, 10; $MgCl_2$, 2; D-glucose, 5; NH_4^+ buffer -NH_4Cl, 140; KCl, 5; HEPES, 10; $MgCl_2$, 2; D-glucose, 5. Ca^{2+} (2 mM) or ethylene glycol-bis(β-aminoethyl ether)-N,N,N′,N′-tetraacetic acid EGTA (2 mM) were added to HEPES buffer, as indicated.

2.2. Methods

2.2.1. RBC Preparation

Human blood was collected from healthy volunteers, who did not take any medication at least 10 days before the experiments, in S-monovette tubes (9NC, Sarstedt, Nümbrecht, Germany) with the addition of 2mM EGTA. The donors provided their informed consents prior the blood draw, which was performed according to our institutional guidelines and the Declaration of Helsinki. Studies using human RBCs were approved by the Sechenov Institute of Evolutionary Physiology and Biochemistry of the Russian Academy of Sciences (IEPHB RAS) (study no. 3-03; 02.03.2019). RBCs were prepared by centrifugation of whole blood at 400× *g* (Centrifuge ELMI-50CM, Elmi, Latvia) in HEPES buffer with EGTA for 3 min at room temperature. Washed RBCs were resuspended in HEPES buffer with EGTA or Ca^{2+}, as indicated, and adjusted to 0.5×10^9 cells/mL (corresponding to Hematocrit 4.0–4.5%). The main blood parameters (red blood cell count and mean cell volume (MCV)) were controlled by the hematological counter Medonic-M20 (Boule Medical A.B., Stockholm, Sweden).

2.2.2. Stress Models

Oxidative stress was induced by *tert*-Butyl hydroperoxide, t-BOOH (0.25, 0.5, 1, 1.5, 2 mM). Calcium stress, as increased intracellular Ca^{2+} concentration, was inducted by calcium ionophore A23187 (1 μM) in HEPES buffer with 2 mM calcium. Nitrosative stress was inducted by S-nitroso-L-cysteine, SNC (500 μM). To evaluate whether t-BOOH effects on RBCs are reversible or not, we incubated RBCs with 1 or 2 mM of t-BOOH for 0.5 h; the cells were divided into two groups, one of which was washed 2 times by HEPES buffer with EGTA ("washed"), and the other left without washing out t-BOOH ("no washing").

2.2.3. Microparticle Isolation and Analysis

In this study, we investigated EVs and divided them by their size, as characterized by flow cytometry, and content, as characterized by spectrophotometry, to MPs, which were smaller and did not contain Hb, and MVs, which were bigger and contained Hb. MP/MV isolation was performed according to the protocol described in detail in [24]. Briefly, for analysis of MPs/MVs, we incubated RBCs with the indicated compounds for 24 h, then the samples were gently centrifuged (50× *g*, 7 min) and the supernatant was collected for the future analysis. RBC supernatant was centrifuged at 20,000× *g* for 30 min (centrifuge 5810R, Eppendorf, Hamburg, Germany) for separation of free Hb and pellet-containing MVs and MPs. The presence of MVs and MPs in pellets was confirmed by flow cytometry, and then MVs and MPs were scanned spectrophotometrically for Hb species analysis.

2.2.4. Spectral Analysis of Hemoglobin Species

Absorption spectra of Hb species were registered by spectrophotometer SPECS SSP-715-M (Spectroscopic Systems LTD, Moscow, Russia) in the wavelength range of 300–700 nm with a step size of 1 nm at 25 °C. To study the effects of different stresses on free Hb, we hypoosmotically lyzed intact cells and added the indicated compounds, and then the spectra were collected at the indicated time. To study the Hb transformation in cells, we incubated the RBCs with the indicated compounds for the indicated time, and then the cells were hypoosmotically lyzed and the free Hb spectra were scanned. To study the Hb species encapsulated in MPs/MVs, we isolated MPs/MVs, as described in Section 2.2.3, and scanned them.

Hemoglobin Species Calculation

The percentage of oxidized Hb in RBC suspensions was determined by spectrophotometry using the millimolar extinction coefficients of the different Hb species (oxyhemoglobin, oxyHb; methemoglobin, metHb; hemichrome, HbChr) according to [25]. Briefly, RBC lysates were scanned from 500 to 700 nm while recording the absorbance values at 560, 577, 630, and 700 nm. These data were used for the calculation of Hb species percentage using the equations presented in [25]. The data are presented as percentage from the sum of all Hb species in the sample taken as 100%.

Induction of Hypoxia

RBC suspension was degassed with argon for 15 min. The oxygen sensor mini-Oksik 3 ("Analitika service" Ltd, Moscow, Russia) was used to control the oxygen content in the hypoxia chamber (Billups-Rothenberg, San Diego, CA, USA), with absorption registered in the range of 300–700 nm. The cuvette was sealed up with wrapping film during the registration of absorption for maintenance of hypoxic conditions.

2.2.5. Characterization of RBC Deformability by Laser Diffraction Method

To estimate the osmotic and ammonium fragility of RBCs, we used the novel laser diffraction method (laser microparticle analyzer LaSca-T, BioMedSystems Ltd., Saint-Petersburg, Russia), adapted for cell physiology, according to Mindukshev et al. [26–28]. The intensity of scattered light was continuously detected by forward scattering at various angles (Figures S1 and S2). The MCV data from hematological counter Medonic-M20 (Boule Medical A.B., Stockholm, Sweden) were used as initial volume values MCV_{300} for the calculation of the MCV changes by the original software of the laser particle analyzer LaSca-TM.

Osmotic Fragility Test (OFT)

RBCs (0.5×10^9 cells/mL) were incubated at indicated concentrations of A23187, SNC, and t-BOOH at indicated times. Then, aliquots (10 µL) of each sample were resuspended in 1 mL of HEPES buffer for osmotic fragility test. Hemolysis curves were registered for a range of osmolality from 210 to 70 mOsm/kg H_2O. For each osmolality step, we added corresponding volume of water and RBCs to the sample to keep RBC concentration constant. The cell volume investigation algorithm was used for the estimation of cell volume changes dynamics and percentage of hemolysis [26–29]. The following parameters were calculated from hemolysis curves: H50, an osmotic fragility variable that represents the saline concentration that induces 50% lysis; W, the distribution width; and MCV_{120}, maximal mean cell volume during the OFT, which is calculated from the hemolysis curve based on MCV_{300}. MCV_{300}, mean cell volume, was controlled by a hematological analyzer. The basic principles of the OFT are described in Figure S1.

Ammonium Stress Test (AST)

RBCs were prepared as for OFT. Control RBCs (10^6 cells/mL) were suspended in 1 mL of HEPES buffer with EGTA and then for AST in 1 mL of NH_4^+ buffer. The following parameters were calculated from the hemolysis curves: V_{hem}, maximal hemolysis rate; %Hem, percentage of lyzed cells; and MCV_{hem}, maximal mean cell volume during the AST, which is calculated from the hemolysis curve based on MCV_{300}. The basic principles of the AST are presented in Figure S2.

2.2.6. Flow Cytometry Analysis

All flow cytometry experiments were performed on flow cytometer Navios (BeckmanCoulter, Brea, CA, USA) with an analysis of no less than 20,000 events.

Size and Structure Analysis

For size and structure analysis, we used forward scatter (FSC)/side scatter (SSC) mode, which provides information about cell size and structure. The intensity of light scattered in a forward direction (FSC) correlates with cell size. The intensity of scattered light measured at a right angle or side scatter (SSC) correlates with internal complexity, granularity, and refractiveness [30–32].

Esterase Activity Analysis

Calcein-AM was used for the evaluation of cell esterase activity. RBCs (0.5×10^7 cells/mL) were incubated with calcein-AM (5 µM, 40 min, 37 °C) then diluted in 300 µL HEPES buffer with the following registration of calcein fluorescence at green detector (fluorescence light sensor 1, FL1) by flow cytometer Navios (BeckmanCoulter, Brea, CA, USA).

Phosphatidylserine (PS) Externalization at the RBC Surface

Annexin-V is a Ca^{2+}-dependent dye, and thus 2 mM Ca^{2+} was added to HEPES buffer for the annexin test. RBCs (0.5×10^9 cells/mL) were incubated with annexin-V (0.1 μg/mL, 15 min, 25 °C), and then fluorescence of annexin was registered at FL1 by flow cytometry.

Analysis of Band 3 Clustering

The contribution of the cytoskeleton in RBC transformations under OS was estimated using the eosin-5-maleimide (EMA)-binding test. RBCs (0.5×10^9 cells/mL) were incubated with EMA in the HEPES buffer (0.07 mM, 40 min) with the following registration of EMA fluorescence intensity at FL1 by flow cytometry.

Microparticle Detection

MPs and MVs were isolated as described in Section 2.2.3. The combined range of the cytometer optical system was used for data collection W2 (Wide-High 9°–19° + Narrow 2°–9° mode). For W2 FSC detection, we used Ultra Rainbow Fluorescent Particles at nominal size 3.0–3.4 μm (BD Biosciences, San Jose, CA, USA) and Latex Beads of 0.5 μm (BeckmanCoulter, Brea, USA) to estimate MP/MV sizes.

Caspase-3 Activation

RBCs at 0.5×10^9 cells/mL were incubated with the indicated concentrations of A23187, SNC, and t-BOOH at indicated time, fixed in 1% formalin (10 min), and were then centrifuged (400× *g*, 3 min). Pellets then were resuspended in phosphate buffered saline containing 1% bovine serum albumin (PBS BSA 1%) and permeabilized by 1% Tween-20 for 10min. Anti-active caspase-3 FITC polyclonal antibodies were added to the samples and incubated for 20 min in the dark. Caspase-3 activity was registered as the mean fluorescence intensity increase at FL1.

2.2.7. Confocal Microscopy

A Leica TCS SP5 MP scanning confocal microscope (Leica Microsystems Inc., Bannockburn, IL, USA) was used for evaluation of EMA-binding/band 3 clustering and MP and MV formation. Erythrocytes were captured with 20 (HCX APO CS 20/0.70; Leica Microsystems, Inc., Buffalo Grove, IL, USA) or 63 (HCX APO CS 63/1.4; Leica Microsystems, Inc., Buffalo Grove, IL, USA) immersion objectives. To resolve fine details (clustering and MVs), we used an additional electronic zoom with a factor of 1.5 to 3.5. For imaging, the emitted fluorescence was acquired at 500 to 600 nm (green region of the spectrum for EMA). Single focal plane images were merged and analyzed with standard Leica LAS AF Software (Leica Microsystems, Inc., Buffalo Grove, IL, USA).

2.2.8. Data Analysis

Laser diffraction data were analyzed by the original software LaSca_32 v.1498 (BioMedSystems Ltd., Saint Petersburg, Russia) of the laser particle analyzer LaSca-TM. Flow cytometry data were analyzed by original software Cytometry List Mode Data Acquisition and Analysis Software v.1.2 for Navios cytometer (BeckmanCoulter, Brea, CA, USA) and by FCS Express Flow 7 (De Novo Software, Pasadena, CA, USA). Differences between groups were analyzed by IBM SPSS Statistics v.26 (IBM Corporation, Armonk, NY, USA). The data are presented as the mean ± SD. Our variables conformed to a normal distribution (Shapiro–Wilk's test, $p > 0.05$). We used one-way analysis of variance (ANOVA) for group comparisons. When the samples were homoscedastic (Levene's test, $p > 0.05$), we used Tukey honestly significant difference (HSD) post hoc analysis. When the equal variances were not assumed, we used Tamhane's T2 post hoc analysis. For paired groups analysis, we used a paired *t*-test. $p < 0.05$ was considered statistically significant.

3. Results

3.1. RBC Viability Strongly Depended on the Ratio of Oxidant to Cell Count

Calcein-AM could be used for the evaluation of RBC viability and OS-induced cytotoxicity [33]. Therefore, in our experiments, we used this test to assess the effects of OS on RBC viability through esterase activity changes. In constant RBC concentration (0.5×10^9 cells/mL), calcein fluorescence intensity significantly decreased with t-BOOH concentration increase (Figure 1A,B), whereas in constant t-BOOH concentration, calcein fluorescence intensity directly correlated with RBC count (Figure 1C). Analysis of these two plots (Figure 1B,C) revealed exponential dependency ($R^2 = 0.98$) between calcein fluorescence intensity, as a marker of OS, and the ratio of oxidant concentration to cell count ([t-BOOH]/RBC) (Figure 1D). Here, we showed that it is very important to consider the ratio ([t-BOOH]/RBC) for characterization of OS effects on RBCs. On the basis of the presented data for all our experiments, we kept RBC concentration (0.5×10^9 cells/mL) constant.

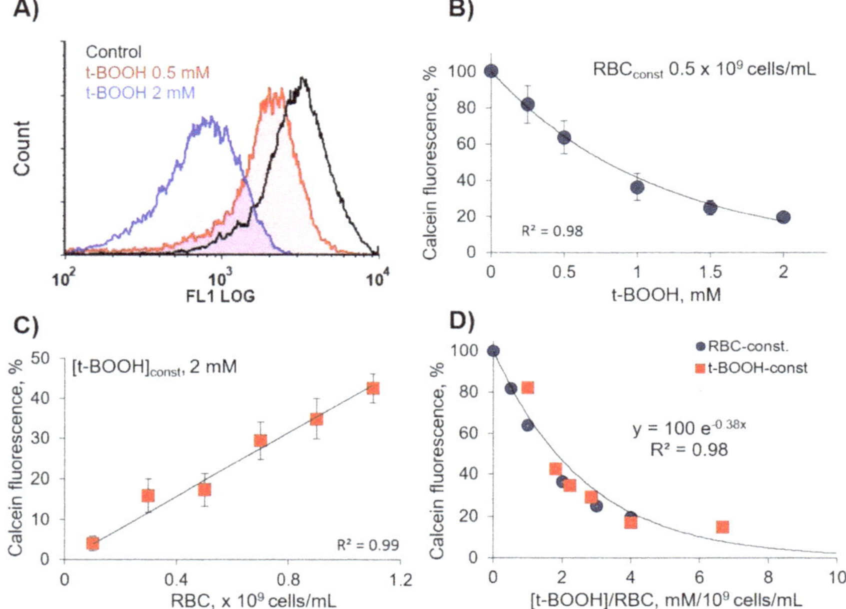

Figure 1. Calcein fluorescence intensity strongly depended on oxidant to cell count ratio. Red blood cell (RBC) suspension was incubated with *tert*-Butyl hydroperoxide (t-BOOH) at indicated concentrations for 1 h, then RBCs were stained with calcein-AM (5 µM, 40 min) and analyzed for calcein fluorescence intensity by flow cytometry at FL1 (logarithmic scale). (**A**) Representative histograms from five independent experiments. (**B**) Dependence of calcein fluorescence intensity in constant RBC count from t-BOOH concentrations. (**C**) Dependence of calcein fluorescence intensity in constant t-BOOH concentration from RBC count. (**D**) Exponential dependency between calcein fluorescence intensity and the ratio of oxidant concentration/cell count ([t-BOOH]/RBC). Data in (**B**,**C**) are presented as means ± SD, $n = 5$.

3.2. t-BOOH Induced RBC Vesiculation

For analysis of RBC transformations, we used flow cytometry protocol according to our previous template [28], and cell volume changes were analyzed by the hematological analyzer (Figures 2 and 3). For the evaluation of different types of RBC transformations, we divided dot plots for four regions.

Gate 1 represents the native cells, gate 2—MPs, gate 3—transformed cells with increased SSC values, and gate 4—MPs with increased SSC values (referred to as microvesicles, MVs) (Figure 2).

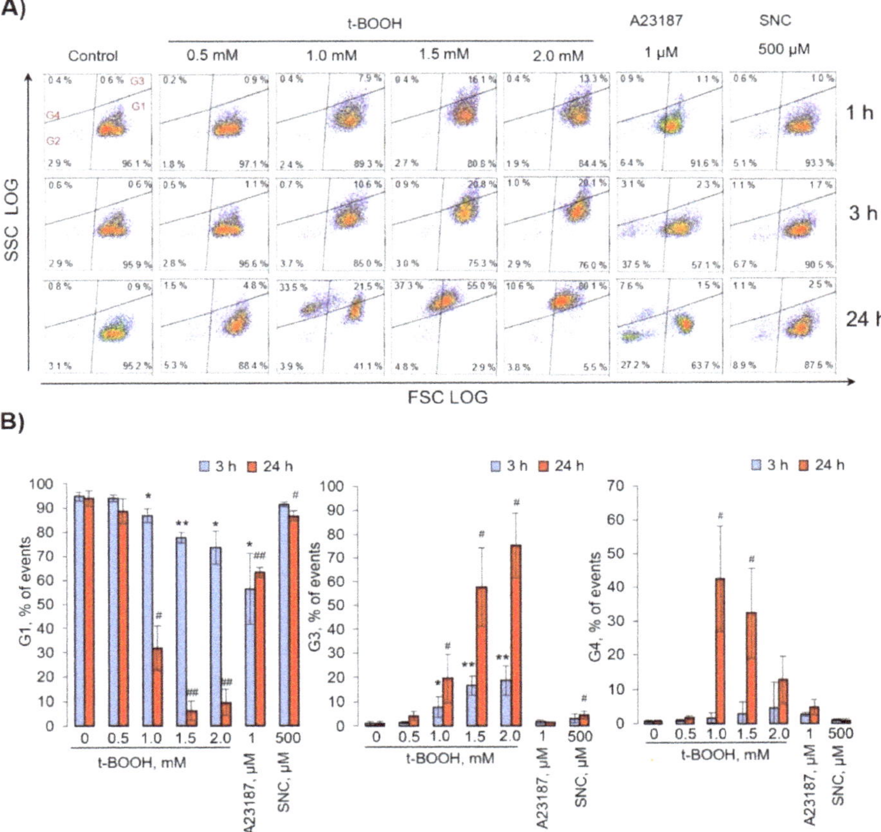

Figure 2. Oxidative stress induced by t-BOOH-triggered RBC transformations and microvesicle formation. RBCs (0.5×10^9 cells/mL) were incubated at indicated concentrations of t-BOOH, A23187, and S-nitroso-L-cysteine (SNC) at the indicated times. Gate 1 represents control RBCs, gate 2—microparticles (MPs), gate 3—transformed RBCs, and gate 4—microvesicles (MVs). (**A**) Representative dot plots out of six independent experiments. (**B**) Quantification of presented data expressed as mean ± SD, $n = 6$. One-way ANOVA, Tamhane T2 (G1, G2 24 h, G4), or Tukey HSD post hoc (G2 3 h) were used where appropriate. * $p < 0.05$, ** $p < 0.001$ compared to control (t-BOOH 0mM, 3 h); # $p < 0.05$, ## $p < 0.001$ compared to control (t-BOOH 0mM, 24 h).

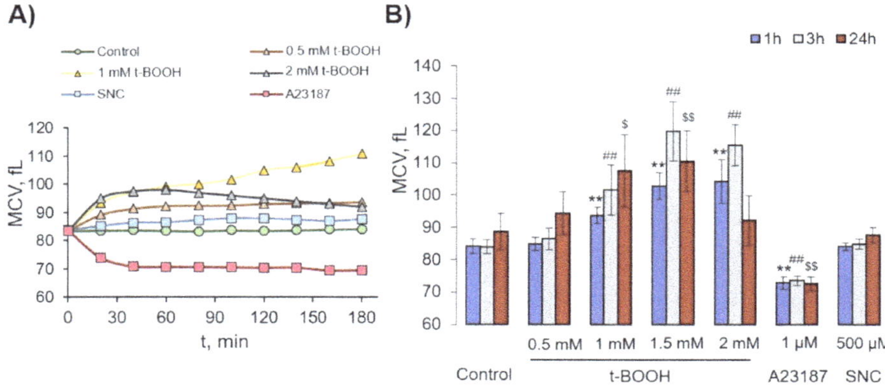

Figure 3. Effects of t-BOOH, A23187, and SNC on RBC volume changes. RBCs (0.5×10^9 cells/mL) were incubated with indicated concentration of t-BOOH, A23187, and SNC for indicated times and were analyzed by hematological analyzer. (**A**) Representative histogram of mean cell volume (MCV) changes for one donor for 3 h; symbols indicate time of analysis. (**B**) Quantitative data from 10 independent experiments (10 donors). Data are presented as mean ± SD ($n = 10$), one-way ANOVA, Levene's test < 0.05, and Tamhane T2 post hoc. ** $p < 0.001$ compared to 1h control; ##, $p < 0.001$ compared to 3 h control; $, $p < 0.05$, $$, $p < 0.001$ compared to 24 h control.

In the control, more than 96% of RBCs were in gate 1 and remained constant for 24 h. t-BOOH for 1- and 3-h slightly increased the FSC (cell volume) concentration-dependently and increased SSC values (structural heterogeneity) without an increase in MP formation. After 24 h of 0.5 mM t-BOOH application, cells slightly increased SSC and FSC values, whereas 1 and 1.5 mM led to the formation of MVs and cells with increased SSC (gates 3 and 4). We also found that 2 mM of t-BOOH transformed RBCs to high SSC (gate 3). The increase of intracellular calcium concentration (facilitated by A23187) time-dependently triggered significant MP formation (gate 2), up to 32 ± 3% ($p < 0.05$, $n = 6$) after 24 h of incubation, whereas no significant MP formation triggered by t-BOOH or SNC application was detected. The effects of SNC on RBCs for 1–24 h (all gates) were similar to t-BOOH at a concentration of 0.5 mM. It is important to underline that t-BOOH induced only MV (gate 4), but not MP (gate 2) formation. Analysis of RBC MCV changes is presented in Figure 3. Low doses of t-BOOH (0.5–1.5 mM) increased cell volume, whereas 2 mM of t-BOOH did not significantly change cell volume. In contrast, A23187 significantly decreased MCV, and SNC had no significant effect on MCV.

3.3. Oxidative Stress Induced Hemoglobin Oxidation to Ferryl (Hemichrome) Forms

To evaluate how Hb oxidation state affects formation of RBC-derived MVs and MPs, we first tested the effects of used compounds on different Hb species formation. To maintain maximum oxygen-carrying capacity, Hb must be kept under reducing conditions in the ferrous (Fe^{2+}) form by an efficient enzymatic machinery [34]. $HbFe^{2+}$ could spontaneously and during OS be oxidized to form $HbFe^{3+}$ (ferric Hb, methemoglobin, metHb), $HbFe^{4+}$ (ferryl Hb), and $\cdot HbFe^{4+}$ (ferryl radical). These oxidatively unstable Hb intermediates (ferryl/ferryl protein radicals refer to as hemichromes, HbChr) oxidize residues within the Hb globin chains (and other proteins within proximity), ultimately leading to Hb degradation and heme loss [35]. We used several compounds that can oxidize Hb and we elucidated their effects on Hb oxidation state (Figure 4).

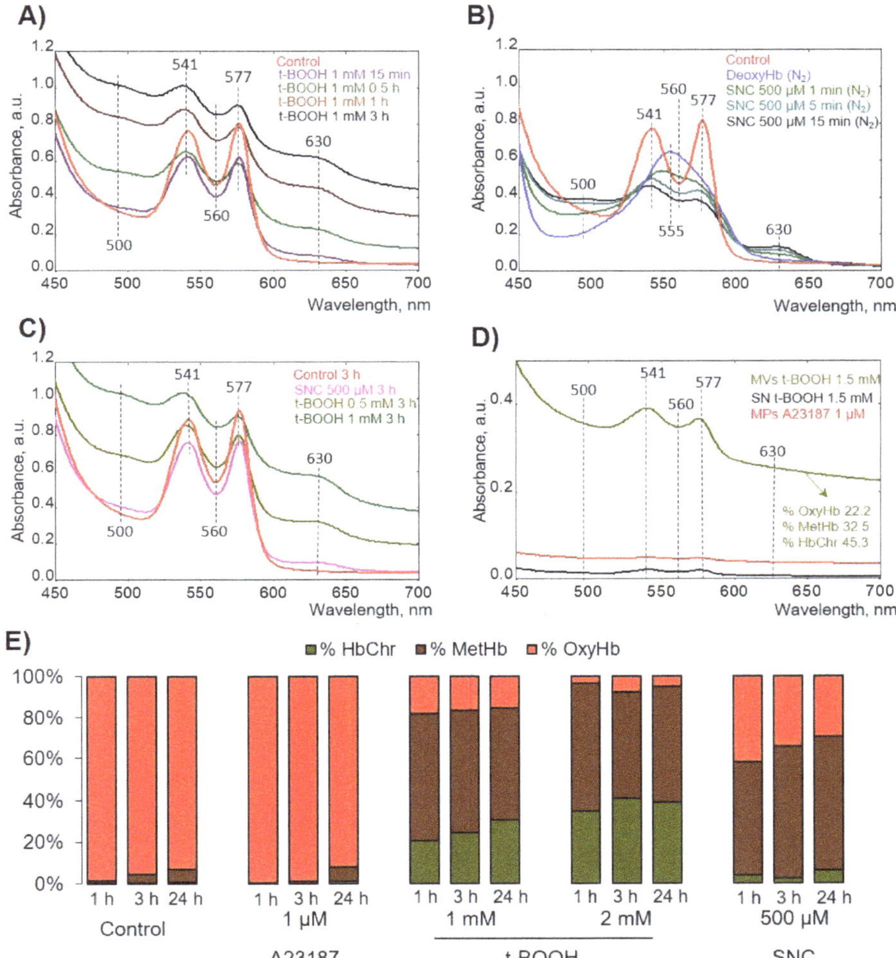

Figure 4. t-BOOH induced hemichrome (HbChr) formation. Spectral scans from 450 to 700 nm captured the different oxidation states of hemoglobin (Hb) identified by characteristic peaks in the visible region. (**A**) Representative spectra of free Hb oxidation by 1mM t-BOOH (ferric, 500 and 630 nm; ferryl/HbChr, 545 nm, 576 nm, and a flattened region between 600 and 700 nm) in comparison with intact Hb spectra (ferrous, 541 and 576 nm) in HEPES-buffer at 25 °C in kinetics. (**B**) Representative spectra of free Hb oxidation by 500 µM SNC in deoxygenated by N_2 HEPES buffer at 25 °C in kinetics. Free oxyHb was deoxygenated by N_2 and then SNC was added for the indicated time. (**C**) Spectra of Hb from hypoosmotically lysed RBCs after 3 h treatment with indicated compounds at indicated concentrations. (**D**) After 24 h of RBC incubation with indicated compounds, we collected the MVs and MPs, as described in the Materials and Methods section, and then MVs/MPs and supernatant (SN) from the last washing step were analyzed. (**E**) Representative bar chart of Hb species calculated from one donor.

First, we tested the effects of used compounds on free Hb (Figure 4A,B), and then intact cells were oxidized and lysed in hypoosmotic conditions, and corresponding spectra were scanned (Figure 4C). t-BOOH firstly oxidized Hb to MetHb, followed by oxidation to HbChr (Figure 4A). To evaluate the effects of NO donor (SNC), we used hypoxic conditions by application of N_2. SNC first formed HbNO and then was oxidized to MetHb without significant formation of HbChr (Figure 4B,C,E and Table 1). MPs and MVs were collected from RBCs after 24 h of the application of indicated compounds and then supernatants and MV- and MP-containing pellets were scanned spectrophotometrically (Figure 4D). t-BOOH-induced MVs contained ferryl forms of Hb, whereas SNC- and A23187-produced MPs did not contain any significant amount of Hb. The percentage of Hb species was calculated according to [25] and is presented in Figure 4E and Table 1. These data indicated that only t-BOOH-induced oxidative stress resulted in HbChr formation, SNC led only to metHb formation, and increase of intracellular calcium concentration had no effects on Hb oxidation.

3.4. t-BOOH Dose-Dependently Decreased RBC Deformability

Next, to characterize OS-induced RBC transformations and possible deformability changes in conditions favorable for MVs and MPs release, we used the osmotic fragility test and ammonium stress test. The osmotic fragility of RBCs is a composite index of their shape; hydration; and, within certain limitations, proneness to in vivo destruction [36,37]. In standard OFT, the percent of hemolysis in increasingly hypotonic solution (0.75%, 0.65%, and 0.60%) is recorded spectrally by optical density. We developed the original automated, easy method for analysis of cells osmotic fragility. Previously, for analysis of band 3 function, we established the original ammonium stress test that is based on band 3 - Rhesus Associated Glycoprotein (RhAG) -facilitated ability of RBCs to swell and lyse in isoosmotic ammonium buffer (NH_4^+ buffer) [27,29]. The basic principles of both tests are described in the Materials and Methods section (Section 2.2.5) and Figures S1 and S2.

The 1 h t-BOOH (0.5–1 mM) treatment led to an increase in H_{50} (the osmolality that triggers hemolysis of 50% cells), indicating elevated fragility, and high doses (2 mM) decreased H_{50}, indicating increased rigidity (Figure 5A) in comparison with the control cells. OS dose-dependently increased cell distribution width (W). Both parameters strongly indicated the significant decrease in RBCs' deformability and increase in RBCs' heterogeneity.

After 1 h of treatment, A23187 led to a significant decrease of MCV_{300} and subsequent MCV_{120}, as well as a significant decrease of H_{50}, indicating decreased deformability. SNC effects were similar to 0.5 mM of t-BOOH, MCV_{300} and MCV_{120}, W, and H_{50} were slightly increased in comparison to control cells. The osmotic fragility test parameters are summarized in Table 2.

Table 1. The percentage of Hb species formation in response to indicated compounds calculated according to Kanias et al. [25].

Hb Species	Control		t-BOOH						A23187		SNC	
			0.5 mM		1 mM		2 mM		1 µM		500 µM	
	1 h	3 h	1 h	3 h	1 h	3 h	1 h	3 h	1 h	3 h	1 h	3 h
% OxyHb	98.0 ± 2.0	95.6 ± 3.0	66.3 ± 4.2 *	45.3 ± 3.9 *	30.7 ± 9.0 **	27.0 ± 13.6 **	20.9 ± 16.4 **	20.7 ± 15.1 **	97.4 ± 3.1	96.6 ± 3.0	47.2 ± 5.9 **	33.1 ± 3.4 **
% MetHb	1.8 ± 0.1	4.0 ± 2.3	30.6 ± 5.1 *	49.1 ± 6.6 *	50.5 ± 8.1 **	47.0 ± 11.8 **	58.8 ± 9.4 **	46.1 ± 6.6 **	2.3 ± 0.5	3.4 ± 0.6	50.8 ± 4.1 **	63.4 ± 4.6 **
% HbChr	0.2 ± 0.6	0.4 ± 0.2	4.9 ± 3.6	5.4 ± 2.1	18.8 ± 5.4 *	25.9 ± 8.1 **	20.3 ± 12.8 *	33.2 ± 9.1 **	0.3 ± 0.2	0.2 ± 0.1	2.5 ± 1.4	3.6 ± 2.5

Data are presented as means ± SD ($n = 7$), one-way ANOVA, Levene's test < 0.05, and Tamhane T2 post hoc. * $p < 0.05$, ** $p < 0.001$, compared to corresponding control (1 h or 3 h).

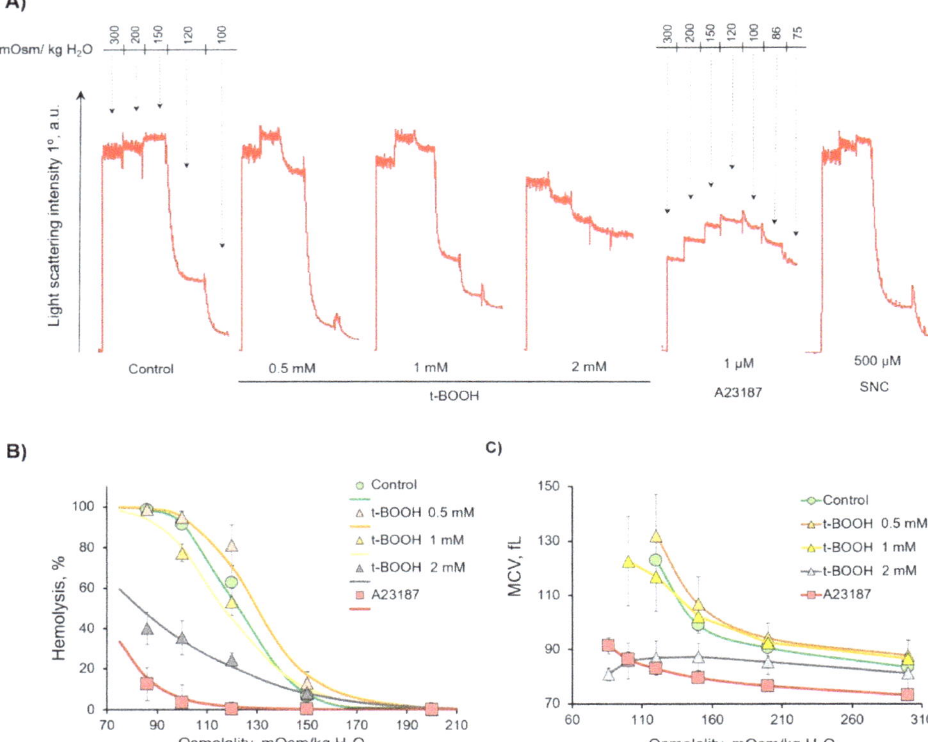

Figure 5. Oxidative stress (OS)-induced decrease in osmotic fragility of RBCs. RBCs (0.5×10^9 cells/mL) were incubated with the indicated substances for 1h, and then aliquots (10 μL, 10^6 cells/mL final concentration) of samples were resuspended in HEPES buffer with EGTA to register light scattering intensity corresponding to control. Then, the osmolality was gradually reduced by H_2O supplementation, from 300 to 70 mOsm/kg H_2O, to maintain the RBC concentration, and the corresponding number of cells was added at each step of H_2O supplementation. (**A**) Representative osmotic hemolysis curves from the osmotic fragility test (OFT). (**B**) Quantification of percentage of hemolysis from osmotic fragility test calculated from six independent experiments. (**C**) Quantification of MCV during osmotic fragility test calculated from six independent experiments.

Table 2. Quantification of osmotic fragility test (OFT) data of six independent experiments (donors).

OFT Characteristics	Control	t-BOOH			A23187	SNC
		0.5 mM	**1 mM**	**2 mM**	**1 μM**	**500 μM**
H_{50}, mOsm/kg H_2O	122 ± 9	130 ± 7	117 ± 8	83 ± 6 **	68 ± 7 **	128 ± 6
W, mOsm/kg H_2O	44 ± 7	52 ± 8	60 ± 9 *	107 ± 10 **	41 ± 6	47 ± 7
MCV_{300}, fL	84 ± 2	88 ± 6	87 ± 7	81 ± 6	73 ± 2 **	86 ± 5
MCV_{120}, fL	124 ± 9	131 ± 8	115 ± 8	88 ± 6 *	83 ± 2 **	127 ± 4

MCV_{300}—data from hematology cell counter. Data are presented as means ± SD (n = 6), one-way ANOVA; if Levene's test < 0.05, Tamhane T2 post hoc (MCV_{300},) was used; if Levene's test > 0.05, Tukey HCD post hoc (H_{50}, W, MCV_{120}) was used. * $p < 0.05$, ** $p < 0.001$, compared to corresponding control.

Ammonium stress test results after 1 h were similar to the osmotic fragility test (Figure 6); quantification of ammonium stress test data is presented in Table 3. %Hem (percentage of hemolyzed cells), V_{hem} (hemolysis maximal rate), and MCV_{hem} (maximal mean cell volume during the ammonium stress test) as the markers of rigidity dose-dependently significantly decreased with the rise in t-BOOH concentration, indicating the increased RBC rigidity and inhibited band 3 protein activity.

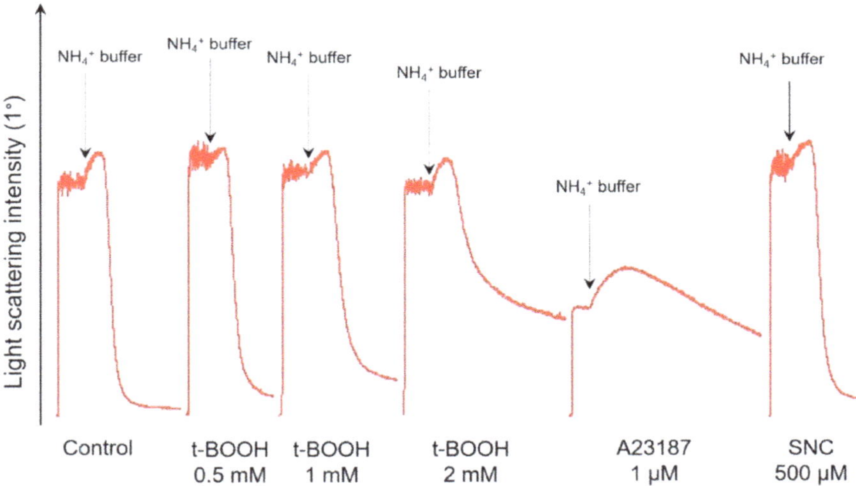

Figure 6. OS dose-dependently decreased RBCs' deformability and inhibited band 3 function. Representative hemolysis curves of ammonium stress test one of eight experiments. RBCs (0.5 × 10^9 cells/mL) were incubated with the indicated compounds for 1 h, and then aliquots (10 μL, 10^6 cells/mL final concentration) of samples were resuspended in HEPES buffer to register light scattering intensity corresponding to control. Then aliquots (10 μL, 10^6 cells/mL final concentration) of samples were resuspended in NH_4^+ buffer for ammonium stress test. Arrows indicate the start of the ammonium stress test. Quantitation of these data is presented in Table 3.

Table 3. Quantification of ammonium stress-test data of five independent experiments (donors).

AST Characteristics	Control	t-BOOH				A23187	SNC
		0.5 mM	1 mM	1.5 mM	2 mM	1 μM	500 μM
V_{hem}	1.00 ± 0.02	1.01 ± 0.02	0.71 ± 0.21	0.57 ± 0.08 *	0.32 ± 0.09 *	0.06 ± 0.02 **	1.04 ± 0.02
%Hem	96.5 ± 0.2	92.1 ± 2.6	80.8 ± 11.4	69.3 ± 9.6 *	47.8 ± 13.2 **	40.0 ± 11.3 **	95.6 ± 1.0
MCV_{300}	84.9 ± 0.9	86.7 ± 2.3	87.5 ± 5.9	87.8 ± 4.3	88.6 ± 10.4	70.5 ± 3.7 *	86.87 ± 1.9
MCV_{hem}	142.3 ± 3.2	143.8 ± 4.8	124.5 ± 9.8	116.0 ± 1.7 *	105.1 ± 1.9 **	96.3 ± 6.5 **	141.0 ± 2.8

MCV_{300}—data from hematology cell counter. Data are presented as means ± SD (n = 5), one-way ANOVA; if Levene's test < 0.05, Tamhane T2 post hoc (V_{hem}, MCV_{300}, MCV_{120}) was used, if Levene's test > 0.05, Tukey HSD post hoc (%Hem) was used. * $p < 0.05$, ** $p < 0.001$, compared to corresponding control.

As described [38,39], an increase in intracellular calcium concentration significantly altered RBCs' deformability. In our experiments, RBC treatment with A23187 led to a significant decrease in V_{hem}, %Hem, and MCV_{120}, indicating the increased RBC rigidity. SNC did not cause any significant deformability alterations compared to control; its effects were similar to those triggered by 0.5 mM of t-BOOH.

Results of both important tests characterizing RBC features revealed that t-BOOH dose-dependently decreased RBCs deformability, accrual of intracellular calcium concentration increased RBC rigidity, and SNC had no significant effect on these parameters.

3.5. OS Induced RBC Transformation and Microvesicle Formation Was Independent on Extracellular Calcium Concentration

Next, we tested whether extracellular calcium plays any role in OS-induced MP or MV formation. It is well known that calcium plays a significant role in RBCs function [40–42] and that it triggers eryptosis [43]; however, the role of calcium in MV formation is not yet clear. All our experiments were performed in both HEPES buffer with EGTA or with 2mM calcium. In normal conditions (intact RBCs, Ca^{2+} - enriched buffer) intracellular calcium concentration maintains less than 60nM [40,44], whereas addition of A23187 equals intra- and extracellular concentrations (2mM) with the following change in features of RBCs [45,46]. Therefore, we compared MV formation, PS surface exposure, and caspase-3 activation in RBCs during OS induced by different t-BOOH concentrations, A23187, and SNC, in HEPES buffer with calcium or with EGTA (Figures 7–9).

For MV formation analysis, we used the same template as in Section 3.2 and calculated events in gates G1, G3, and G4 in HEPES-buffer containing calcium (2 mM) or EGTA (2 mM). Surprisingly, calcium had no significant effect on OS-induced RBC transformations and MV formation (Figure 7A–D). Next, we tested whether PS surface exposure (annexin-V binding) as a marker of eryptosis is dependent on extracellular calcium. Similar to MV formation, calcium had no significant effect on OS-induced annexin V binding (Figure 8A,B). We compared percentage of annexin V-positive cells after A23187, t-BOOH, and SNC treatment (Figure 8C). A23187 and t-BOOH after 3 h significantly increased PS exposure, whereas SNC had no significant effect.

In different conditions, PS surface exposure might be dependent, or independent, upon caspase-3 activation [47,48]. t-BOOH induced strong caspase-3 activation (Figure 9A,B), which, like MV formation and PS surface exposure, was independent of calcium. In all tested conditions, A23187 and SNC did not significantly activate caspase-3 (Figure 9B).

3.6. t-BOOH-Induced OS Triggered Band 3 Clustering

Band 3 plays a significant role in RBC functions, and clustering of this protein was demonstrated in several RBCs disorders [7,21,23]. Additionally, band 3 clustering is one of important signals to eliminate senescent and pathological RBCs [49]; therefore, we next used an EMA test to elucidate the effects of the used compounds on band 3 clustering. Oxidative stress induces band 3 oxidation and dissociation from spectrin skeleton, resulting in enhanced mobility and subsequent band 3 cluster formation [21,23,49]. In our experiments, we tested whether each of the used compounds triggers band 3 clustering by applying the EMA-binding test. As expected, in control cells, EMA showed slight homogeneous fluorescence (Figure 10A and Figure S4). During t-BOOH-induced OS, in both RBCs and RBC-derived MV, we detected clustering of band 3 (Figure 10A and Figure S4). In contrast, A23187 and SNC induced neither MV formation nor band 3 clustering (Figure 10C and Figure S4), and, as expected [50,51], A23187 triggered the formation of narrow spikes on the outer half of the bilayer, the so-called echinocytosis, in RBC (Figure 10A and Figure S4).

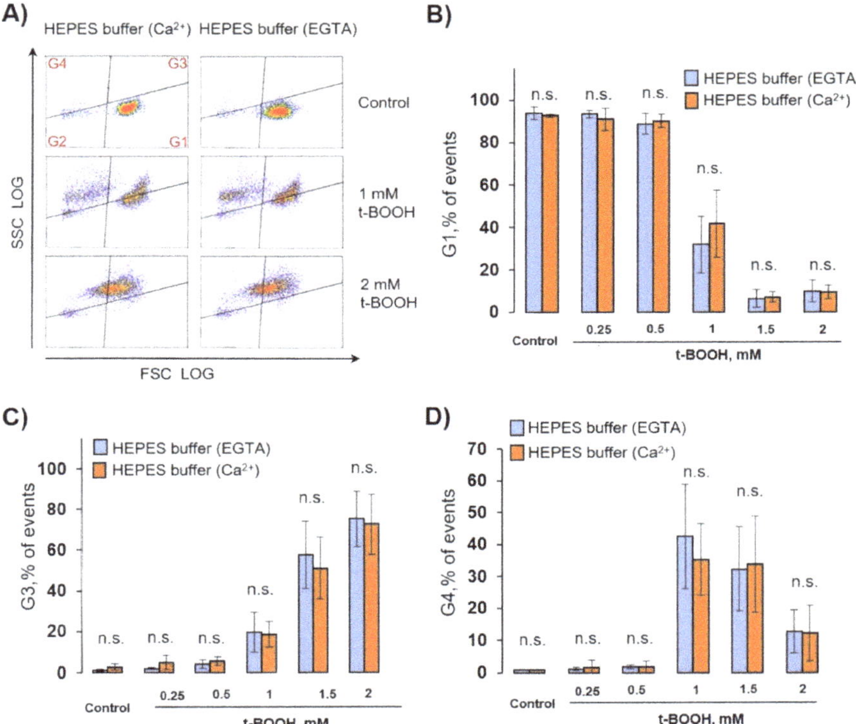

Figure 7. OS-induced RBC transformation and MV formation were calcium-independent. RBCs (0.5×10^9 cells/mL) were incubated with the indicated concentration of t-BOOH in HEPES buffer containing 2 mM calcium, or 2 mM EGTA, for indicated times and were analyzed by flow cytometry. (**A**) Representative SSC/FSC dot plots of one out of six independent experiments for 24 h. Template and gating correspond to Section 3.2. (**B–D**) Calculation of events distributed in the corresponding gates. Data are presented as mean ± SD (n = 7), paired t-test; n.s., not significant.

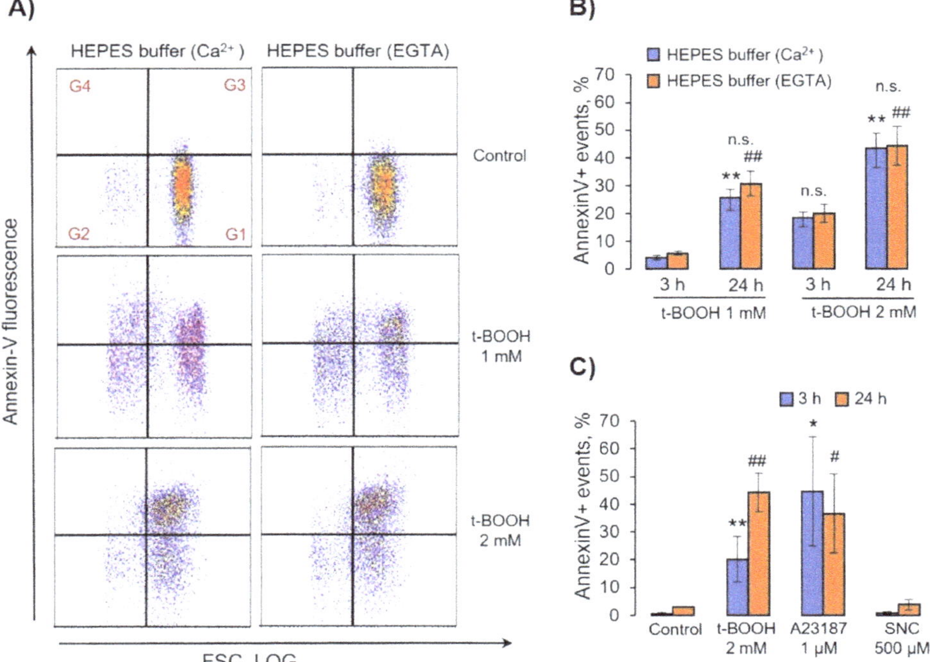

Figure 8. OS-induced annexin-V binding was calcium-independent. RBCs (0.5×10^9 cells/mL) were incubated with indicated concentrations of t-BOOH, A23187, and SNC in HEPES buffer containing 2 mM Ca^{2+} or 2 mM EGTA for indicated times. Annexin-V (0.1 µg/mL, 15 min, 25 °C) was added to treated cells and analyzed by flow cytometry. (**A**) Representative annexin V/FSC dot plots of one out of six independent experiments for 24 h. Gate G1 corresponds to control cells; G2 to annexin-V-negative EVs; G3 to annexin-V-positive cells; G4, annexin-V-positive EVs. (**B,C**) Calculation of annexin-V-positive events in G3 and G4. Data in (**B–D**) are presented as the mean ± SD, in (**B**)—paired *t*-test; ** $p < 0.001$, compared to 3h in HEPES buffer with Ca^{2+}; ## $p < 0.001$, compared to 3 h in HEPES buffer with EGTA; n.s.—not significant. In (**C**)—one-way ANOVA, Levene's test < 0.05, Tamhane's T2 post hoc; * $p < 0.05$, ** $p < 0.001$, compared to 3 h control; # $p < 0.05$, ## $p < 0.001$, compared to 24 h control.

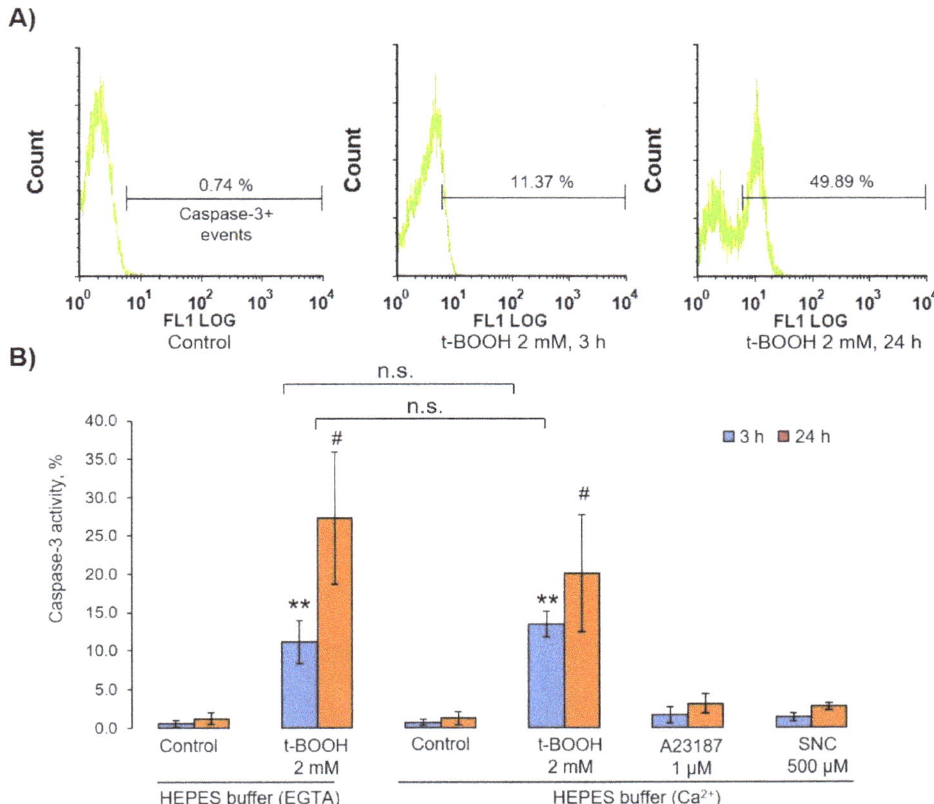

Figure 9. t-BOOH-induced OS activated caspase 3 in RBCs. RBCs (0.5×10^9 cells/mL) were incubated with indicated concentrations of t-BOOH, A23187, and SNC in HEPES buffer containing 2 mM calcium or 2 mM EGTA for indicated times. After we incubated them with indicated compounds, cells were fixed by 1% (final concentration) of methanol-free formaldehyde, permeabilized by 0.5% Tween for 20 min, and then anti-active caspase-3 antibodies were added for 30 min, with caspase-3 activation being measured by flow cytometry according to the manufacturer's instructions. (**A**) Original histograms from one of seven independent experiments. (**B**) Data quantification based on seven independent experiments. Data are presented as the mean ± SD ($n = 7$), one-way ANOVA (HEPES buffer with EGTA both 3 h and 24 h), Tukey HSD post hoc (HEPES buffer with Ca^{2+} 3 h), Tamhane T2 post hoc (HEPES buffer with Ca^{2+} 24 h); paired t-test HEPES buffer (EGTA) and HEPES buffer (Ca^{2+}), ** $p < 0.001$ compared to corresponding control, # $p < 0.05$ compared to corresponding control, n.s.— not significant.

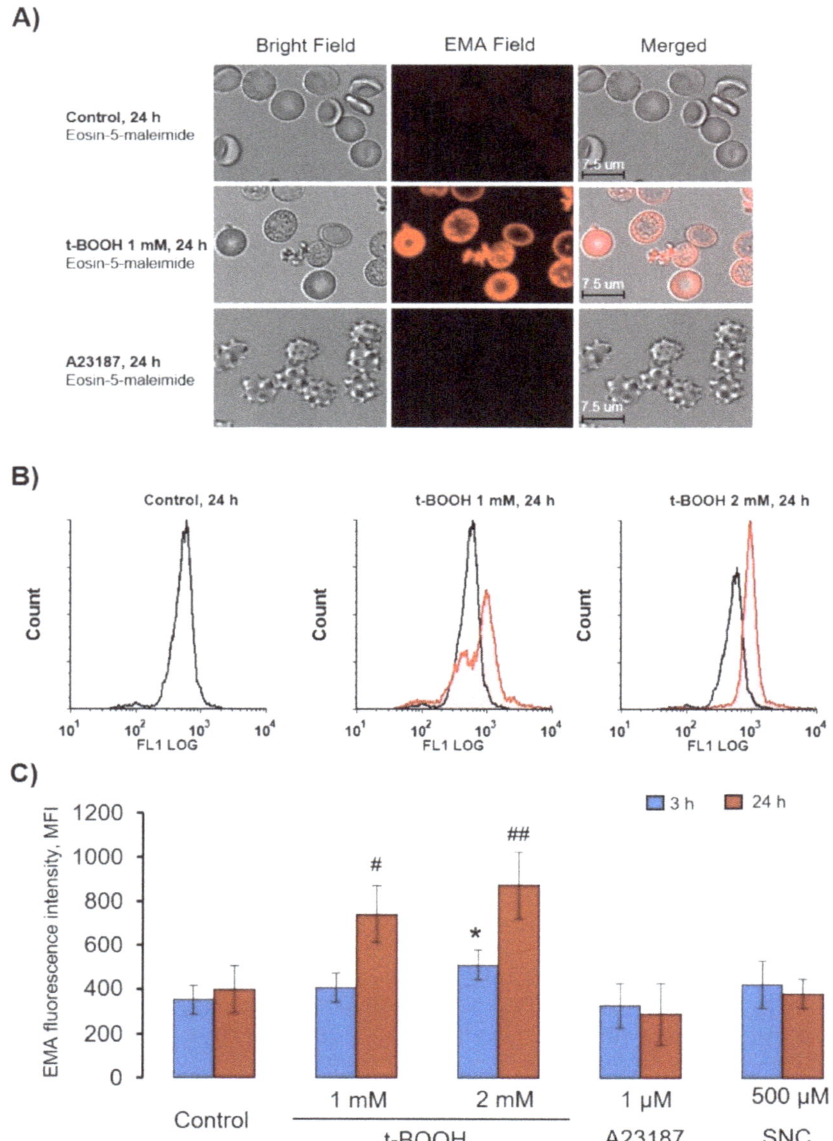

Figure 10. t-BOOH-induced oxidative stress led to band 3 clustering and MV formation. RBCs (0.5×10^9 cells/mL) were incubated as indicated with t-BOOH, A23187, and SNC at the indicated times, followed by EMA staining (0.07 mM, 40 min). (**A**) Representative confocal images of t-BOOH transformed RBCs. RBCs were processed for confocal microscopic analysis as described in the Materials and Methods section (Section 2.2.7). (**B**) Original histograms of EMA fluorescence intensity after 24 h t-BOOH treatment. (**C**) Quantification of flow cytometry data. Data are presented as mean ± SD ($n = 7$), one-way ANOVA, Levene's test > 0.05, Tukey HSD post hoc (3 h) was used, * $p < 0.05$, compared to 3 h control; Levene's test < 0.05, Tamhane T2 post hoc (24 h) was used, # $p < 0.05$, ## $p < 0.001$ compared to 24 h control.

4. Discussion

Extracellular vesicles (EVs) released by different cell types play a significant role in many physiological and pathophysiological processes. The cargo and types of EVs and their functional role strongly depend on the cell of origin and mechanisms of EV formation [52]. It was reported that RBC-derived EVs carry residual Hb, lipids, and proteins that are in charge of known physiological effects of EVs [53,54]. RBC-derived EVs were shown to express phosphatidylserine and cell-specific band 3 epitopes on the surface, as well as to contain enzymes involved in redox homeostasis (glutathione S-transferase, ubiquitin, thioredoxin, peroxiredoxin-1, peroxiredoxin-2) and complement-inhibiting proteins CD55 and CD59. The EV effects in health, disease, and blood transfusion are a matter of continued investigation [3,15,54–58]. However, it has already been defined that MVs could be captured by other circulated blood cells and by endothelial cells; thereby, EVs are involved in coagulation promotion, inflammation, immune modulation, endothelial dysfunction, and vasodilatation impairment, as well as having vasoactive properties potentially altering oxygen delivery homeostasis [17]. Moreover, EVs from RBCs could increase production of tumor necrosis factor-alpha in monocytes and CD4+, as well as CD8+ T-cell proliferation [57].

Two main mechanisms of RBC-derived EV formation connected with an increase of intracellular calcium concentration and Hb oxidation were proposed [58]. The increase in intracellular calcium concentration leads to programmed cell death, so-called eryptosis for RBCs [59], which could be triggered by numerous xenobiotics and exogenous substances [60] as well as several pathological states including diabetes, hepatic failure, sepsis, and chronic kidney disease [61–65]. As expected [38,39,41], intracellular calcium increase mediated by A23187 led to RBC shrinkage (Figures 3 and 5) and subsequent rapid loss of flexibility (Figures 5 and 6), which could be explained by the opening of the Ca^{2+}-sensitive Gardos channels resulting in hyperpolarisation and a loss of K^-, Cl^-, and water [66]. MPs formed during eryptosis by membrane shedding are around 200 nm in diameter and are characterized by surface phosphatidylserine (PS) exposure [9]. However, the question of whether EV formation during eryptosis is connected with classical apoptotic pathways (caspase-3 activation) is still under debate [7,13]. In our experiments, increase in intracellular calcium mediated by application of A23187 led to strong PS surface exposure in both RBCs and EVs but did not activate caspase-3 (Figure 9B). In contrast, OS-triggered PS exposure and EV formation were independent of calcium and caspase-3 activation, which indicated that two different mechanisms, calcium-dependent and -independent mechanisms, are responsible for surface PS exposure and EV formation. The classification of RBC-derived EVs is also not yet clearly defined. Some authors refer to these as microvesicles [13,17], while others use the term microparticles for the same kind of EVs [2,3], without detailed characterization of differences between them. In our study, we divided RBC-derived EVs into two populations: (a) EV, derived from A23187-treated RBCs, which were smaller and did not contain Hb, referred to as MPs, and (b) OS-triggered EVs, which were bigger and contained oxidized Hb, referred to as MVs [22]. Various compounds and pathological states might unquestionably trigger both pathways of RBC-derived EV formation, however, in our study, we focused mainly on important conditions that facilitate OS-induced RBC transformation, MV formation, and characterization of these MVs.

The increase in intracellular calcium concentration is regarded as the main driving force of RBC transformation and EV formation [17]. OS-induced increase of intracellular calcium mobilization was shown in several publications [40,67], however, the question as to whether it is the main trigger of RBC transformation remained open; therefore, in our experiments, we first tested the influence of extracellular calcium on RBC transformation and EV formation, comparing HEPES buffer with 2 mM calcium with HEPES buffer containing 2 mM EGTA. Surprisingly, we found no significant differences in OS-induced RBC transformation and EV formation in these two conditions. Most of our experiments were performed in HEPES buffer with 2 mM EGTA, and similar experiments were performed in HEPES buffer with 2 mM calcium (Figure 7), with there being no significant differences in RBC transformation, annexin-V binding (Figure 8A,B), MV formation, or caspase-3 activation (Figure 9B). On the basis of these data, we concluded that Hb oxidation to HbChr but not increase of intracellular calcium concentration is most likely the main driving force of OS-induced RBC transformation and EV formation.

The next important question raised in our study was connected with oxidant and RBC concentration, as well as time of incubation with oxidant (Figure 1). In most of the literature, t-BOOH is used to induce OS in RBCs, with highly controversial results, from an increase of cell volume to cell shrinkage, content of produced microparticles, and cell deformability. Here, we showed that RBC transformation, deformability, band 3 clustering, and MV formation in t-BOOH-induced OS strongly depended on oxidant concentration/time and the ratio of oxidant to cell count (Figure 1). t-BOOH concentration less than 0.5 mM did not induce significant changes in RBCs; MV formation started at concentrations 1–1.5 mM, whereas at 2 mM, cells rigidity increased with reduced MV formation (Figure 5). We also checked whether the effects of t-BOOH during long-term treatment are reversible or not, finding that even 30 min t-BOOH application is sufficient to make the effects of oxidant irreversible (Figure S3).

Another aspect of OS-induced RBC transformation that was also an important part of our work and should be considered in future studies was connected with the calcein-AM test, a marker of cellular esterase activity. Clearly, esterase activity did not directly correlate with RBC transformation and MV formation (Figures 1 and 2). t-BOOH dose-dependently reduced calcein fluorescence; however, cell transformation and MV formation continued even in the conditions of almost completely inactivated esterase activity after 3 h of incubation with t-BOOH (Figure 2).

Processes such as senescence, hyperthermia, transfusion, increase of intracellular calcium concentration, RBC storage in blood banks, and oxidative stress accelerate RBC vesiculation [11]. Through MV generation, erythrocytes were shown to remove membrane patches containing removal molecules, damaged cells and membrane constituents [15,17,58]. In our experiments, only t-BOOH induced the formation of high SSC MVs, which contained highly oxidized Hb (HbChr), whereas A23187 and SNC did not, indicating that RBC-derived EVs were heterogeneous in content, with or without Hb. These data indicated that during OS, RBCs eliminate HbChr by vesiculation in order to sacrifice the cell itself, thereby prolonging lifespan and delaying the untimely clearance of in all other respects healthy RBCs [6,9,15].

5. Conclusions

OS-triggered RBC transformations and MV formation is mediated by complex processes including Hb oxidation, band 3 clustering, cytoskeleton reorganization, an increase in intracellular calcium concentrations, and other alterations in RBC cellular organization. OS could be developed as consequences of several pathological states such as diabetes, sepsis, chronic kidney disease, and hepatic failure, and the degree of OS and RBC transformations might be significantly variable, being dependent on the conditions. t-BOOH is often used for the analysis of RBCs in OS conditions, however, published data focusing on RBCs' responses to OS are highly variable and depend on concentration/time of t-BOOH administration. Therefore, we first established standards for evaluation of t-BOOH-induced OS and showed that it is especially important to consider not only oxidant concentration but also the ratio of oxidant to cell count. Next, we developed two new original methods, ammonium stress-test and automated osmotic fragility test, based on laser diffraction analysis of RBC transformations, which allow us to characterize RBCs' osmotic and ammonium fragility. Thus, the presented data, methodology, and new methods for the analysis of OS-induced RBC transformations will allow us to characterize these cells more broadly during OS.

Supplementary Materials: The following are available online at http://www.mdpi.com/2076-3921/9/10/929/s1, Figure S1: Evaluation of RBCs' osmotic fragility by laser diffraction method. Figure S2: Basic principles of ammonium stress-test. Figure S3: Effects of 30 min t-BOOH-induced oxidative stress on RBCs were irreversible for 4 and 24 h. Figure S4: t-BOOH-induced oxidative stress led to band 3 clustering and MV formation.

Author Contributions: Conceptualization, I.M., J.S., and E.S.; methodology, I.M. and E.S.; software, J.S. and E.S.; validation, E.S., I.D., and E.N.; formal analysis, J.S.; investigation, E.S. and J.S.; resources, I.M.; data curation, I.M.;

writing—original draft preparation, E.S., J.S., and I.M.; writing—review and editing, S.G. and J.S.; visualization, I.M., J.S., and E.S.; supervision, S.G.; project administration, J.S. and I.M.; funding acquisition, J.S. All authors have read and agreed to the published version of the manuscript.

Funding: This research was funded by the Russian Fund for Basic Researches (grant no. 19-315-60015 to J.S.) and by the State Assignment of Ministry of Science and Higher Education of the Russian Federation (project no. AAAA-A18-118012290371-3 to E.S., I.D., E.N., S.G., and I.M.).

Acknowledgments: The authors would like to thank Sechenov Institute of Evolutionary Physiology and Biochemistry Core Facilities Center for providing the opportunity to use the Navios cytometer.

Conflicts of Interest: The authors declare no conflict of interest. The funders had no role in the design of the study; in the collection, analyses, or interpretation of data; in the writing of the manuscript; or in the decision to publish the results.

References

1. Rosa-Fernandes, L.; Rocha, V.B.; Carregari, V.C.; Urbani, A.; Palmisano, G.A. Perspective on Extracellular Vesicles Proteomics. *Front. Chem.* **2017**, *5*, 102. [CrossRef] [PubMed]
2. Alexandru, N.; Costa, A.; Constantin, A.; Cochior, D.; Georgescu, A. Microparticles: From Biogenesis to Biomarkers and Diagnostic Tools in Cardiovascular Disease. *Curr. Stem Cell Res.* **2017**, *12*, 89–102. [CrossRef] [PubMed]
3. Said, A.S.; Rogers, S.C.; Doctor, A. Physiologic Impact of Circulating RBC Microparticles upon Blood-Vascular Interactions. *Front. Physiol.* **2018**, *8*, 1120. [CrossRef] [PubMed]
4. Shah, M.D.; Bergeron, A.L.; Dong, J.F.; López, J.A. Flow cytometric measurement of microparticles: Pitfalls and protocol modifications. *Platelets* **2008**, *19*, 365–372. [CrossRef]
5. Hattangadi, S.M.; Lodish, H.F. Regulation of erythrocyte lifespan: Do reactive oxygen species set the clock? *J. Clin. Investig.* **2007**, *117*, 2075–2077. [CrossRef]
6. Bosman, G.J.; Lasonder, E.; Luten, M.; Roerdinkholder-Stoelwinder, B.; Novotný, V.M.; Bos, H.; De Grip, W.J. The proteome of red cell membranes and vesicles during storage in blood bank conditions. *Transfusion* **2008**, *48*, 827–835. [CrossRef]
7. Mohanty, J.G.; Nagababu, E.; Rifkind, J.M. Red blood cell oxidative stress impairs oxygen delivery and induces red blood cell aging. *Front. Physiol.* **2014**, *5*, 84. [CrossRef]
8. Bratosin, D.; Estaquier, J.; Petit, F.; Arnoult, D.; Quatannens, B.; Tissier, J.P.; Slomianny, C.; Sartiaux, C.; Alonso, C.; Huart, J.J.; et al. Programmed cell death in mature erythrocytes: A model for investigating death effector pathways operating in the absence of mitochondria. *Cell Death Differ.* **2001**, *8*, 1143–1156. [CrossRef]
9. Nguyen, D.B.; Ly, T.B.; Wesseling, M.C.; Hittinger, M.; Torge, A.; Devitt, A.; Perrie, I.; Bernhardt, I. Characterization of Microvesicles Released from Human Red Blood Cells. *Cell Physiol. Biochem.* **2016**, *38*, 1085–1099. [CrossRef]
10. Lutz, H.U.; Liu, S.C.; Palek, J. Release of spectrin-free vesicles from human erythrocytes during ATP depletion. I. Characterization of spectrin-free vesicles. *J. Cell Biol.* **1977**, *73*, 548–560. [CrossRef]
11. D'Alessandro, A.; Kriebardis, A.G.; Rinalducci, S.; Antonelou, M.H.; Hansen, K.C.; Papassideri, I.S.; Zolla, L. An update on red blood cell storage lesions, as gleaned through biochemistry and omics technologies. *Transfusion* **2015**, *55*, 205–219. [CrossRef] [PubMed]
12. Greenwalt, T.J. The how and why of exocytic vesicles. *Transfusion* **2006**, *46*, 143–152. [CrossRef] [PubMed]
13. Kriebardis, A.G.; Antonelou, M.H.; Stamoulis, K.E.; Economou-Petersen, E.; Margaritis, L.H.; Papassideri, I.S. RBC-derived vesicles during storage: Ultrastructure, protein composition, oxidation, and signaling components. *Transfusion* **2008**, *48*, 1943–1953. [CrossRef] [PubMed]
14. Willekens, F.L.A.; Roerdinkholder-Stoelwinder, B.; Groenen-Döpp, Y.A.M.; Bos, H.J.; Bosman, G.J.C.G.M.; van den Bos, A.G.; Verkleij, A.J.; Were, J.M. Hemoglobin loss from erythrocytes in vivo results from spleen-facilitated vesiculation. *Blood* **2003**, *101*, 747–751. [CrossRef] [PubMed]
15. Willekens, F.L.; Were, J.M.; Groenen-Döpp, Y.A.; Roerdinkholder-Stoelwinder, B.; de Pauw, B.; Bosman, G.J. Erythrocyte vesiculation: A self-protective mechanism? *Br. J. Haematol.* **2008**, *141*, 549–556. [CrossRef]
16. Pollet, H.; Conrard, L.; Cloos, A.S.; Tyteca, D. Plasma Membrane Lipid Domains as Platforms for Vesicle Biogenesis and Shedding? *Biomolecules* **2018**, *8*, 94. [CrossRef]
17. Leal, J.K.F.; Adjobo-Hermans, M.J.W.; Bosman, G.J.C.G.M. Red Blood Cell Homeostasis: Mechanisms and Effects of Microvesicle Generation in Health and Disease. *Front. Physiol.* **2018**, *9*, 703. [CrossRef]

18. Kostova, E.B.; Beuger, B.M.; Klei, T.R.; Halonen, P.; Lieftink, C.; Beijersbergen, R.L.; van den Berg, T.K.; van Bruggen, R. Identification of signalling cascades involved in red blood cell shrinkage and vesiculation. *Biosci. Rep.* **2015**, *35*, e00187. [CrossRef]
19. Cloos, A.-S.; Ghodsi, M.; Stommen, A.; Vanderroost, J.; Dauguet, N.; Pollet, H.; D'Auria, L.; Mignolet, E.; Larondelle, Y.; Terrasi, R.; et al. Interplay Between Plasma Membrane Lipid Alteration, Oxidative Stress and Calcium-Based Mechanism for Extracellular Vesicle Biogenesis from Erythrocytes During Blood Storage. *Front. Physiol.* **2020**, *11*, 712. [CrossRef]
20. Rifkind, J.M.; Nagababu, E. Hemoglobin redox reactions and red blood cell aging. *Antioxid. Redox. Signal* **2013**, *18*, 2274–2283. [CrossRef]
21. Arashiki, N.; Kimata, N.; Manno, S.; Mohandas, N.; Takakuwa, Y. Membrane Peroxidation and Methemoglobin Formation Are Both Necessary for Band 3 Clustering: Mechanistic Insights into Human Erythrocyte Senescence. *Biochemistry* **2013**, *52*, 5760–5769. [CrossRef] [PubMed]
22. Ferru, E.; Pantaleo, A.; Carta, F.; Mannu, F.; Khadjavi, A.; Gallo, V.; Ronzoni, L.; Graziadei, G.; Cappellini, M.D.; Turrini, F. Thalassemic erythrocytes release microparticles loaded with hemichromes by redox activation of p72Syk kinase. *Haematologica* **2014**, *99*, 570–578. [CrossRef] [PubMed]
23. Jana, S.; Strader, M.B.; Meng, F.; Hicks, W.; Kassa, T.; Tarandovskiy, I.; de Paoli, S.; Simak, J.; Heaven, M.R.; Belcher, J.D.; et al. Hemoglobin oxidation-dependent reactions promote interactions with band 3 and oxidative changes in sickle cell-derived microparticles. *JCI Insight* **2018**, *3*, e120451. [CrossRef] [PubMed]
24. Dinkla, S.; Brock, R.; Joosten, I.; Bosman, G.J. Gateway to understanding microparticles: Standardized isolation and identification of plasma membrane-derived vesicles. *Nanomedicine (Lond).* **2013**, *8*, 1657–1668. [CrossRef]
25. Kanias, T.; Acker, J.P. Mechanism of hemoglobin-induced cellular injury in desiccated red blood cells. *Free Radic. Biol. Med.* **2010**, *49*, 539–547. [CrossRef]
26. Mindukshev, I.; Gambaryan, S.; Kehrer, L.; Schuetz, C.; Kobsar, A.; Rukoyatkina, N.; Nikolaev, V.O.; Krivchenko, A.; Watson, S.P.; Walter, U.; et al. Low angle light scattering analysis: A novel quantitative method for functional characterization of human and murine platelet receptors. *Clin. Chem. Lab. Med.* **2012**, *50*, 1253–1262. [CrossRef]
27. Mindukshev, I.; Kudryavtsev, I.; Serebriakova, M.; Trulioff, A.; Gambaryan, S.; Sudnitsyna, J.; Khmelevskoy, D.; Voitenko, N.; Avdonin, P.; Jenkins, R.; et al. Flow cytometry and light scattering technique in evaluation of nutraceuticals. In *Nutraceuticals: Efficacy, Safety and Toxicity*; Gupta, R.C., Ed.; Elsevier: Oxford, UK, 2016; pp. 319–332. [CrossRef]
28. Mindukshev, I.V.; Sudnitsyna, J.S.; Skverchinskaya, E.A.; Andreyeva, A.Y.; Dobrylko, I.A.; Senchenkova, E.Y.; Krivchenko, A.I.; Gambaryan, S.P. Erythrocytes' Reactions to Osmotic, Ammonium, and Oxidative Stress Are Inhibited under Hypoxia. *Biochem (Moscow). Suppl. Ser. A Membr. Cell Biol.* **2019**, *13*, 352–364. [CrossRef]
29. Sudnitsyna, J.S.; Skverchinskaya, E.A.; Dobrylko, I.A.; Nikitina, E.R.; Krivchenko, A.I.; Gambaryan, S.P.; Mindukshev, I.V. Human erythrocyte ammonium transport is mediated by functional interaction of ammonium (RhAG) and anion (AE1) transporters. *Biochem (Moscow). Suppl. Ser. A Membr. Cell Biol.* **2016**, *10*, 301–310. [CrossRef]
30. Darzynkiewicz, Z.; Juan, G.; Li, X.; Gorczyca, W.; Murakami, T.; Traganos, F. Cytometry in cell necrobiology: Analysis of apoptosis and accidental cell death (necrosis). *Cytometry* **1997**, *27*, 1–20. [CrossRef]
31. Luchetti, C.G.; Solano, M.E.; Sander, V.; Arcos, M.L.; Gonzalez, C.D.; Di Girolamo, G.; Chiocchio, S.; Cremaschi, G.A.; Motta, A.B. Effects of dehydroepiandrosterone on ovarian cystogenesis and immune function. *J. Reprod. Immunol.* **2004**, *64*, 59–74. [CrossRef]
32. Tzur, A.; Moore, J.K.; Jorgensen, P.; Shapiro, H.M.; Kirschner, M.W. Optimizing optical flow cytometry for cell volume-based sorting and analysis. *PLoS ONE* **2011**, *6*, e16053. [CrossRef] [PubMed]
33. Bratosin, D.; Mitrofan, L.; Palii, C.; Estaquier, J.; Montreuil, J. Novel fluorescence assay using calcein-AM for the determination of human erythrocyte viability and aging. *Cytometry A* **2005**, *66*, 78–84. [CrossRef] [PubMed]
34. Kassa, T.; Jana, S.; Meng, F.; Alayash, A.I. Differential heme release from various hemoglobin redox states and the upregulation of cellular heme oxygenase-1. *FEBS Open Bio* **2016**, *6*, 876–884. [CrossRef]
35. Chintagari, N.R.; Jana, S.; Alayas, A.I. Oxidized Ferric and Ferryl Forms of Hemoglobin Trigger Mitochondrial Dysfunction and Injury in Alveolar Type I Cells. *Am. J. Respir. Cell Mol. Biol.* **2016**, *55*, 288–298. [CrossRef]

36. King, M.J.; Zanella, A. Hereditary red cell membrane disorders and laboratory diagnostic testing. *Int. J. Lab. Hematol.* **2013**, *35*, 237–243. [CrossRef] [PubMed]
37. Walski, T.; Chludzińska, L.; Komorowska, M.; Witkiewicz, W. Individual osmotic fragility distribution: A new parameter for determination of the osmotic properties of human red blood cells. *Biomed. Res. Int.* **2014**, *2014*, 162102. [CrossRef]
38. Huisjes, R.; Bogdanova, A.; van Solinge, W.W.; Schiffelers, R.M.; Kaestner, L.; van Wijk, R. Squeezing for Life—Properties of Red Blood Cell Deformability. *Front. Physiol.* **2018**, *9*, 656. [CrossRef]
39. Cueff, A.; Seear, R.; Dyrda, A.; Bouyer, G.; Egée, S.; Esposito, A.; Skepper, J.; Tiffert, T. Effects of elevated intracellular calcium on the osmotic fragility of human red blood cells. *Cell Calcium.* **2010**, *47*, 29–36. [CrossRef]
40. Bogdanova, A.; Makhro, A.; Wang, J.; Lipp, P.; Kaestner, L. Calcium in red blood cells—A perilous balance. *Int. J. Mol. Sci.* **2013**, *14*, 9848–9872. [CrossRef]
41. Kuck, L.; Peart, J.N.; Simmonds, M.J. Calcium dynamically alters erythrocyte mechanical response to shear [published online ahead of print, 2020]. *Biochim. Biophys. Acta Mol. Cell Res.* **2020**, *1867*, 118802. [CrossRef]
42. Kuck, L.; Peart, J.N.; Simmonds, M.J. Active modulation of human erythrocyte mechanics. *Am. J. Physiol. Cell Physiol.* **2020**, *319*, C250–C257. [CrossRef] [PubMed]
43. Lang, P.A.; Kaiser, S.; Myssina, S.; Wieder, T.; Lang, F.; Huber, S.M. Role of Ca2+-activated K+ channels in human erythrocyte apoptosis. *Am. J. Physiol Cell Physiol.* **2003**, *285*, C1553–C1560. [CrossRef] [PubMed]
44. Hertz, L.; Huisjes, R.; Llaudet-Planas, E.; Petkova-Kirova, P.; Makhro, A.; Danielczok, J.G.; Egee, S.; Del Mar Mañú-Pereira, M.; van Wijk, R.; Vives Corrons, J.L.; et al. Is Increased Intracellular Calcium in Red Blood Cells a Common Component in the Molecular Mechanism Causing Anemia? *Front. Physiol.* **2017**, *8*, 673. [CrossRef] [PubMed]
45. Dodson, R.A.; Hinds, T.R.; Vincenzi, F.F. Effects of calcium and A23187 on deformability and volume of human red blood cells. *Blood Cells* **1987**, *12*, 555–564. [PubMed]
46. Romero, P.J.; Hernández-Chinea, C. The Action of Red Cell Calcium Ions on Human Erythrophagocytosis in Vitro. *Front. Physiol.* **2017**, *8*, 1008. [CrossRef]
47. Hierso, R.; Waltz, X.; Mora, P.; Romana, M.; Lemonne, N.; Connes, P.; Hardy-Dessources, M.-D. Effects of oxidative stress on red blood cell rheology in sickle cell patients. *Br. J. Haematol.* **2014**, *166*, 601–606. [CrossRef]
48. Raducka-Jaszul, O.; Bogusławska, D.M.; Jędruchniewicz, N.; Sikorski, A.F. Role of Extrinsic Apoptotic Signaling Pathway during Definitive Erythropoiesis in Normal Patients and in Patients with β-Thalassemia. *Int. J. Mol. Sci.* **2020**, *21*, 3325. [CrossRef]
49. Shimo, H.; Arjunan, S.N.; Machiyama, H.; Nishino, T.; Suematsu, M.; Fujita, H.; Tomita, M.; Takahashi, K. Particle Simulation of Oxidation Induced Band 3 Clustering in Human Erythrocytes. *PLoS Comput. Biol.* **2015**, *11*, e1004210. [CrossRef]
50. Gallagher, P.G. Red blood cell membrane disorders. In *Hematology: Basic Principles and Practice*; Hoffman, R., Benz, E.J., Silberstein, L.E., Heslop, H.E., Weitz, J.I., Anastasi, J., Salama, M.E., Abutalib, S., Eds.; Elsevier: Oxford, UK, 2018; pp. 626–647. [CrossRef]
51. Bevers, E.M.; Wiedmer, T.; Comfurius, P.; Shattil, S.J.; Weiss, H.J.; Zwaal, R.F.; Sims, P.J. Defective Ca(2+)-induced microvesiculation and deficient expression of procoagulant activity in erythrocytes from a patient with a bleeding disorder: A study of the red blood cells of Scott syndrome. *Blood* **1992**, *79*, 380–388. [CrossRef]
52. Hafiane, A.; Daskalopoulou, S.S. Extracellular vesicles characteristics and emerging roles in atherosclerotic cardiovascular disease. *Metabolism* **2018**, *85*, 213–222. [CrossRef]
53. Westerman, M.; Porter, J.B. Red blood cell-derived microparticles: An overview. *Blood Cells Mol. Dis.* **2016**, *59*, 134–139. [CrossRef] [PubMed]
54. Bosman, G.J.; Lasonder, E.; Groenen-Döpp, Y.A.; Willekens, F.L.; Were, J.M. The proteome of erythrocyte-derived microparticles from plasma: New clues for erythrocyte aging and vesiculation. *J. Proteom.* **2012**, *76*, 203–210. [CrossRef] [PubMed]
55. Li, K.Y.; Zheng, L.; Wang, Q.; Hu, Y.W. Characteristics of erythrocyte-derived microvesicles and its relation with atherosclerosis. *Atherosclerosis* **2016**, *255*, 140–144. [CrossRef] [PubMed]

56. Noubouossie, D.F.; Henderson, M.W.; Mooberry, M.; Ilich, A.; Ellsworth, P.; Piegore, M.; Skinner, S.C.; Pawlinski, R.; Welsby, I.; Renné, T.; et al. Red blood cell microvesicles activate the contact system, leading to factor IX activation via 2 independent pathways. *Blood* **2020**, *135*, 755–765. [CrossRef]
57. Danesh, A.; Inglis, H.C.; Jackman, R.P.; Wu, S.; Deng, X.; Muench, M.O.; Heitman, J.W.; Norris, P.J. Exosomes from red blood cell units bind to monocytes and induce proinflammatory cytokines, boosting T-cell responses in vitro. *Blood* **2014**, *123*, 687–696. [CrossRef]
58. Tissot, J.D.; Rubin, O.; Canellini, G. Analysis and clinical relevance of microparticles from red blood cells. *Curr. Opin. Hematol.* **2010**, *17*, 571–577. [CrossRef]
59. Lang, E.; Qadri, S.M.; Lang, F. Killing me softly-suicidal erythrocyte death. *Int. J. Biochem. Cell Biol.* **2012**, *44*, 1236–1243. [CrossRef]
60. Föller, M.; Lang, F. Ion Transport in Eryptosis, the Suicidal Death of Erythrocytes. *Front. Cell Dev. Biol.* **2020**, *8*, 597. [CrossRef]
61. Carelli-Alinovi, C.; Misiti, F. Erythrocytes as Potential Link between Diabetes and Alzheimer's Disease. *Front. Aging Neurosci.* **2017**, *9*, 276. [CrossRef]
62. Turpin, C.; Catan, A.; Guerin-Dubourg, A.; Debussche, X.; Bravo, S.B.; Álvarez, E.; van den Elsen, J.; Meilhac, O.; Rondeau, P.; Bourdon, E. Enhanced oxidative stress and damage in glycated erythrocytes. *PLoS ONE* **2020**, *15*, e0235335. [CrossRef]
63. Föller, M.; Huber, S.M.; Lang, F. Erythrocyte programmed cell death. *IUBMB Life* **2008**, *60*, 661–668. [CrossRef] [PubMed]
64. Bateman, R.M.; Sharpe, M.D.; Singer, M.; Ellis, C.G. The Effect of Sepsis on the Erythrocyte. *Int. J. Mol. Sci.* **2017**, *18*, 1932. [CrossRef] [PubMed]
65. Lang, F.; Bissinger, R.; Abed, M.; Artunc, F. Eryptosis—The neglected cause of anemia in end stage renal disease. *Kidney Blood Press. Res.* **2017**, *42*, 749–760. [CrossRef] [PubMed]
66. Bogdanova, A.; Kaestner, L.; Simionato, G.; Wickrema, A.; Makhro, A. Heterogeneity of Red Blood Cells: Causes and Consequences. *Front. Physiol.* **2020**, *11*, 392. [CrossRef]
67. Petrou, T.; Olsen, H.L.; Thrasivoulou, C.; Masters, J.R.; Ashmore, J.F.; Ahmed, A. Intracellular Calcium Mobilization in Response to Ion Channel Regulators via a Calcium-Induced Calcium Release Mechanism. *J. Pharmacol. Exp. Ther.* **2017**, *360*, 378–387. [CrossRef]

© 2020 by the authors. Licensee MDPI, Basel, Switzerland. This article is an open access article distributed under the terms and conditions of the Creative Commons Attribution (CC BY) license (http://creativecommons.org/licenses/by/4.0/).

Article

Syk Kinase Inhibitors Synergize with Artemisinins by Enhancing Oxidative Stress in *Plasmodium falciparum*-Parasitized Erythrocytes

Ioannis Tsamesidis [1,2], Karine Reybier [2], Giuseppe Marchetti [1], Maria Carmina Pau [1], Patrizia Virdis [3], Claudio Fozza [3], Francoise Nepveu [2], Philip S. Low [4], Francesco Michelangelo Turrini [5,†] and Antonella Pantaleo [1,*,†]

1 Department of Biomedical Sciences, University of Sassari, 07100 Sassari, Italy; johntsame@gmail.com (I.T.); gmarchetti@uniss.it (G.M.); paumc81@tiscali.it (M.C.P.)

2 UMR 152 Pharma-Dev, Université de Toulouse, IRD, UPS, 31000 Toulouse, France; karine.reybier-vuattoux@univ-tlse3.fr (K.R.); francoise.nepveu@univ-tlse3.fr (F.N.)

3 Department of Clinical, Surgical and Experimental Sciences, University of Sassari, 07100 Sassari, Italy; patriziavirdis70@gmail.com (P.V.); cfozza@uniss.it (C.F.)

4 Purdue Institute for Drug Discovery and Department of Chemistry, Purdue University, West Lafayette, IN 47907, USA; plow@purdue.edu

5 Department of Oncology, University of Turin, 10126 Turin, Italy; francesco.turrini@unito.it

* Correspondence: apantaleo@uniss.it

† These authors have equal contribution.

Received: 15 July 2020; Accepted: 11 August 2020; Published: 14 August 2020

Abstract: Although artemisinin-based combination therapies (ACTs) treat *Plasmodium falciparum* malaria effectively throughout most of the world, the recent expansion of ACT-resistant strains in some countries of the Greater Mekong Subregion (GMS) further increased the interest in improving the effectiveness of treatment and counteracting resistance. Recognizing that (1) partially denatured hemoglobin containing reactive iron (hemichromes) is generated in parasitized red blood cells (pRBC) by oxidative stress, (2) redox-active hemichromes have the potential to enhance oxidative stress triggered by the parasite and the activation of artemisinin to its pharmaceutically active form, and (3) Syk kinase inhibitors block the release of membrane microparticles containing hemichromes, we hypothesized that increasing hemichrome content in parasitized erythrocytes through the inhibition of Syk kinase might trigger a virtuous cycle involving the activation of artemisinin, the enhancement of oxidative stress elicited by activated artemisinin, and a further increase in hemichrome production. We demonstrate here that artemisinin indeed augments oxidative stress within parasitized RBCs and that Syk kinase inhibitors further increase iron-dependent oxidative stress, synergizing with artemisinin in killing the parasite. We then demonstrate that Syk kinase inhibitors achieve this oxidative enhancement by preventing parasite-induced release of erythrocyte-derived microparticles containing redox-active hemichromes. We also observe that Syk kinase inhibitors do not promote oxidative toxicity to healthy RBCs as they do not produce appreciable amounts of hemichromes. Since some Syk kinase inhibitors can be taken daily with minimal side effects, we propose that Syk kinase inhibitors could evidently contribute to the potentiation of ACTs.

Keywords: *Plasmodium falciparum*; syk kinase inhibitors; artemisinin derivatives; hemichromes; oxidative stress; reactive oxygen species

---

1. Introduction

Following entry into the bloodstream of its human host, *Plasmodium* sporozoites injected by an infected mosquito migrate to the liver and initiate the hepatic stage of the parasite life cycle by invading hepatocytes, within which they multiply and differentiate into schizonts containing thousands of hepatic merozoites. These merozoites are subsequently released into the blood where they initiate the erythrocytic stage by invading and replicating within red blood cells (RBCs), and they multiply ~18-fold every 48 h. During the course of their intra-erythrocyte life, the parasite ingests and digests about 70% of the host cell hemoglobin (Hb) with the obligate release of heme that must be converted into an insoluble crystalline form called hemozoin to prevent oxidative damage to the maturing parasite [1,2] and to maintain the colloid osmotic balance in the parasitized erythrocyte resisting premature lysis [3,4]. Suppression of hemozoin formation is thought to constitute a major mechanism of action of several anti-malaria drugs, including quinine, chloroquine, and mefloquine [5,6]. Artemisinin or one of its synthetic derivatives constitutes the other major component of most anti-malaria therapies [7,8]. Although the mechanisms of action of ACTs (artemisinin combination therapies) are still debated [9,10], most hypotheses invoke involvement of redox reactions in their cytotoxicity, since (i) the essential moiety in all therapeutically active forms of artemisinin is its endoperoxide ring [11], (ii) other components of common ACTs promote the accumulation of oxidative heme [12], (iii) erythrocyte mutations that increase oxidative stress within the parasitized RBC confer partial protection against malaria [13–19] (e.g., sickle cell, β-thalassemia, G6PDH (Glucose-6-phosphate dehydrogenase) deficiency, etc., and (iv) many other oxidizing drugs were found to exhibit antimalarial activity [20,21]. Taken together, these observations suggest that oxidative stress can be detrimental to the parasite, probably by creating an environment that is harmful to parasite maturation, proliferation, or egress. In addition to hemoglobin digestion, from the early stages of their development, parasites also lead to hemoglobin oxidation and partial denaturation products in the host cell [13,22]. In particular, *P. falciparum* intra-erythrocyte growth causes hemoglobin oxidation (methemoglobin) and its partial denaturation (hemichromes), characterized by iron binding to distal histidine in the heme pocket [23–27]. Interestingly, hemichromes (HMCs) are classified as reversible and irreversible. While low-spin state HMCs are poorly reactive, reversible HMCs can still be reduced to their high-spin functional, redox-active form [28]. HMCs bind with high affinity to the N-terminal of the cytoplasmic domain of band 3 [29–34], the major transmembrane erythrocyte protein which forms junctional complexes with the cytoskeleton, assuring mechanical stability of the membrane [35]. Following their binding to band 3, HMCs trigger the oxidation of the two cysteine residues (C201 and C317) lying in the band 3 cytoplasmic domain with the formation of two disulfide intermolecular cross-links between two band 3 molecules [22,36]. Notably, disulfide cross-linked band 3 dimers acquire the capability to dock Syk protein kinase, leading to the phosphorylation of band 3 Tyr 8 and Tyr 21 [37]. This event leads to the uncoupling of the oxidized phosphorylated band 3 from ankyrin, the major link between the membrane and cytoskeleton, facilitating the clustering of the band 3–hemichrome complex to form large aggregates known as Heinz bodies [32,38]. As expected from the essential role of the band 3–ankyrin complex in assuring membrane stability, those aggregates tend to protrude from the erythrocyte surface and are progressively released through membrane vesiculation [39]. As a matter of fact, membrane-bound HMC aggregates and circulating microparticles (MPs) were described in hemolytic diseases such as sickle cell anemia, thalassemias, G6PD deficiency, and malaria [40–42]. Notably, in all cited situations, those changes were attributed to the oxidative stress generated by denatured hemoglobin. As expected, the release of MPs from the membrane of parasitized RBCs determines the depletion of band 3 molecules, leading to pronounced membrane destabilization at the final stages of parasite development. We also demonstrated that counteracting band 3 depletion with Syk inhibitors impedes the final egress of merozoites [43–45]. Due to the recent emergence of malaria strains in Southeast Asia that exhibit resistance to ACTs [46–49], the quest for new anti-malarials that might synergize with artemisinins in preventing the spread of drug resistance is intensifying [50,51]. Recognizing that artemisinin needs to be converted in highly reactive carbon-centered radicals via endoperoxide cleavage and that HMCs containing redox active iron are generated in parasitized erythrocytes, we hypothesized that inducing the retention of HMCs through the inhibition of Syk kinase might both increase oxidative stress and accelerate the activation of artemisinin. In this possible mechanism, we also hypothesized that accumulation of reactive HMCs should also aggravate the oxidative stress caused by artemisinin catalyzing additional peroxidation reactions and inducing further HMC generation with a possible synergistic effect between Syk inhibitors and artemisinin. In this paper, we test multiple aspects of this hypothesis. We firstly show that Syk inhibitors promote increased accumulation of hemichromes within the infected RBC by preventing the release of hemichrome-enriched microparticles. We also observe that this accumulation is strongly augmented by co-administration of artemisinin, enhancing the oxidative stress within the infected erythrocyte. Upon demonstrating this phenomenon, as expected from our hypothesis, the combined effect of a Syk kinase inhibitor plus artemisinin is highly synergistic, creating a combination therapy with superior antimalarial activity to the two antimalarials administered alone (see graphical abstract of the proposed mechanism of action).

2. Material and Methods

Unless otherwise stated, all materials were obtained from Sigma-Aldrich, St. Louis, MO, USA.

2.1. Cultivation of *Plasmodium falciparum*-Infected RBCs (pRBCs)

Freshly drawn blood (Rh+) from healthy adults or from G6PD-deficient patients of both sexes was used. RBCs were separated from plasma and leukocytes by washing three times with RPMI (Roswell Park Memorial Institute) 1640 medium. *Plasmodium falciparum* laboratory strains Palo Alto, FCB1, and It-G (all mycoplasma-free) were used. Parasites were maintained in continuous culture using the method of Träger and Jensen [52]. Specifically, the parasitemia was routinely maintained around 5%, and parasites were synchronized by 5% sorbitol treatment as described by Lambros and Vanderberg [53]. For all experiments, mature parasites (shizonts and segmenters) after percoll separation [54] were added to washed RBCs and, 12 h after the infection (occurring within 6 h), the cultures were ready for the experimental procedures.

2.2. Ethics Statement

Healthy blood donors, all adults, provided written, informed consent before entering the study. The study was conducted in accordance with Good Clinical Practice guidelines and the Declaration of Helsinki. No ethical approval was requested as human blood samples were used only to sustain the parasites in in vitro cultures.

2.3. Drug Susceptibility Assays of Cultured Parasites

All Syk inhibitors were solubilized initially at 10 mM concentration in anhydrous DMSO (Dimethyl sulfoxide) and then serially diluted into anhydrous DMSO prior to the addition to malaria cultures. Untreated cultures were run in parallel with the same final concentration 0.01% (*v*/*v*) of DMSO as the drug-treated cultures. Cultures at the ring stage of *P. falciparum* were treated at 12 h post infection, from 3 to 24 h, with the indicated concentrations of dihydroartemisinin (DHA), artesunate (AS), or artemether (ATH) combined with the desired concentration of one of the following Syk inhibitors: P505-15 (Selleckchem); R406 (Calbiochem, Darmstadt, Germany), entospletinib (Selleckchem), Syk inhibitor II (henceforth abbreviated as SYK II), piceatannol, or imatinib (Santa Cruz Biotechnology).

2.4. Preparation of Iron Chelator Deferasirox

Deferasirox (DFX) (Med Chemexpress) was dissolved in DMSO as a 1 mM stock solution and diluted into culture medium to achieve a final DMSO concentration of 0.001% (*v*/*v*). Stock solutions were then filtered through a Swinnex Millipore filter (pore size 0.2 μm) for sterilization prior to addition to malaria cultures.

2.5. Isobologram Preparation and Combination Index (CI) Measurement

Microsoft Excel was used to plot parasite counts following treatment with different combinations of the above drugs. To characterize the nature of the interaction (i.e., synergy, additivity, antagonism) between any two drugs, the experiments were performed at the IC_{50} (half maximal inhibitory concentration) concentrations obtained in in vitro experiments in pRBCs as previously described. The desired inhibitors were plotted on the x- and y-axes, and the line of additivity was constructed by connecting the two points defining the IC_{50} values of the two monotherapies. As defined in the quantitative combination index theorem of Chow–Talalay [55,56], experimental data points located below, on, or above the additivity line are interpreted to indicate synergy, additivity, or antagonism, respectively.

2.6. Assessment of Parasitemia by Light Microscopy

To evaluate the parasitemia and the parasite stage, thin smears of infected parasite cultures were prepared at the indicated times and stained with Diff-Quick stain (Medion Diagnostics, CH). A minimum of 5000 cells were examined microscopically by three observers. The experiments were done at least in triplicate.

2.7. IC_{50} Measurement

To calculate the half maximal inhibitory concentrations of the different Syk inhibitors, either used as single drugs or in combination with others antimalarial drugs, we used ICEstimator software version 1.2. The program estimates IC_{50} values using a nonlinear regression function of the R software.

2.8. Preparation of Cells for Confocal Microscopy

RBCs and parasitized RBCs (pRBCs) treated for 12 h with P505-15 (0.5 μM) at ring stage were pelleted and washed twice in PBS containing 5 mM glucose and then fixed for 5 min in 0.5% acrolein in PBS (Phosphate-Buffered Saline). Cells were rinsed three times, then permeabilized in PBS containing 0.1 M glycine (rinsing buffer) plus 0.1% Triton X-100 for 5 min, and rinsed again 3× in rinsing buffer. To ensure complete neutralization of unreacted aldehydes, the cells were then incubated in rinsing buffer at room temperature for 30 min. After incubation, all nonspecific binding was blocked by incubation for 60 min in blocking buffer (PBS containing 0.05 mM glycine, 0.2% fish skin gelatin, and 0.05% sodium azide). Resuspended RBCs and pRBCs were allowed to adhere to polylysine-coated cover slips, after which the cover slips were mounted with Aqua-Mount (Lerner Laboratories, New Haven, CT). The autofluorescence of hemichromes was visualized by exciting at 488 nm and observing the emission in the 630–750 nm range (pRBCs stained for nucleic acid by DAPI (4′,6-Diamidino-2-Phenylindole). Samples were imaged using a Bio-Rad MRC1024 (Bio-Rad) confocal microscope equipped with a 60 × 1.4 numerical aperture oil immersion lens.

2.9. Hemichrome Analysis

RBCs and pRBCs were incubated for 3, 6, 12, and 24 h with Syk inhibitors (Imatinib, R406, and P505-15) (0.5 μM) with/without DHA (0.5 nM). Cells were washed with cold PBS, and hypotonic membranes were prepared at 4 °C as previously described [57]. To solubilize the HMCs and to dissociate the cytoskeletal proteins, membranes were treated with 130 mM NaCl, 10 mM HEPES, 1mM EDTA (Ethylenediaminetetraacetic acid), and 1.5% C12E8 [57] and incubated under stirring (1400 rpm) at 37 °C for 15 min (Eppendorf ThermoMixer® C.). To eliminate insoluble aggregates and debris originated from the parasite, detergent-treated membranes were centrifuged for 5 min at 20 °C, 15,000× *g*. To isolate the high-molecular-weight protein aggregate containing HMCs, the supernatant was loaded on a Sepharose CL6B column [57], chromatographic fractions were screened by visible spectroscopy, the fractions were characterized by the absorption spectrum of HMCs, and lack of the absorption peaks of Hb at 520 and 280 nm was collected for the quantitative measurement of HMCs and characterization of its components.

HMCs were quantified in the high-molecular-weight fraction by visible spectrometry using the following equation:

$$-133 \times \text{Abs577} - 144 \times \text{Abs630} + 233 \times \text{Abs560} \quad (1)$$

expressed as nmol/mL of solubilized membranes [58,59].

2.10. Hemoglobin Release Quantification

Following centrifugation at 1000× *g*, hemoglobin concentration was measured in the culture supernatant as previously described [60].

2.11. Electron Paramagnetic Resonance (EPR) Measurements

The detection of free radicals was carried out using *N-tert*-butyl-α-phenylnitrone (PBN) as a spin trap. PBN (1 M stock solution in DMSO) was added to normal RBCs, RBCs treated with 400 μM phenylhydrazine (PHZ), or parasitized (parasitemia 5%) pelleted red blood cells (from 2% hematocrit culture), and the volume was adjusted with PBS after addition of dihydroartemisinin (200 μM in DMSO) or Syk inhibitor (P505-15) (0.5 μM) and different concentrations (100–400 μM) of Deferasirox when needed. The solution was then transferred into a flat quartz cell (FZKI160-5 × 0.3 mm, Magnettech, Berlin, Germany) for EPR analysis. EPR spectra were obtained at room temperature (RT) using the X-band on a Bruker EMX-8/2.7 (9.86 GHz) equipped with a gaussmeter (Bruker, Wissembourg, France) and a high-sensitivity cavity (4119/HS 0205). WINEPR and SIMFONIA software (Bruker, Wissembourg, France) were used for EPR data processing and spectrum simulation. Typical scanning parameters were as follows: scan number, 5; scan rate, 1.2 G/s; modulation frequency, 100 kHz; modulation amplitude, 1 G; microwave power, 20 mW; sweep width, 100 G; sweep time, 83.88 s; time constant, 40.96 ms; and magnetic field, 3460–3560 G. The intensity of the EPR signal was calculated by double integration of the EPR Signal.

2.12. ROS Analysis

Glutathione (GSH) and *N*-acetyl-L-cysteine (NAC) were purchased from Sigma (France).

For the detection of intracellular reactive oxygen species (ROS) levels, we employed the cell-permeable ROS-sensitive probe 2′,7′-dichlorodihydrofluorescein diacetate (CM-H_2DCFDA) which fluorescens at 520 nm (λex = 480 nm) upon oxidation. Oxidation of CM-H_2DCFDA (prepared as a 0.5 mM stock solution in DMSO) in RBCs was monitored by measurement of the fluorescence of the desired RBC suspensions (0.2% Hematocrit) in 96-well black-walled microplates (Corning®, Sigma Aldrich) using an SAFAS Xenius (Monaco). The relative fluorescence is expressed as "% maximal emission" as determined with the software "Xenius", where maximal emission was defined as the fluorescence emission obtained following addition of 3 mM H_2O_2.

2.13. Isolation and Characterization of Microparticles (MPs)

RBCs and pRBCs at ring stage were treated with P505-15 for 12 h. Supernatants containing the RBC-derived MPs were then collected and centrifuged at 25,000× *g* for 10 min at 4 °C to eliminate spontaneously formed red cell ghosts and cell debris. To isolate MPs, the supernatant was then centrifuged at 100,000× *g* for 3 h at 4 °C; the morphology and the size of isolated MPs were documented by confocal microscopy through the characteristic autofluorescence of HMCs (excitation: 488 nm; emission: 630–700 nm) as previously described [39,61].

The presence of HMCs in MPs was further confirmed by visible spectrometry analysis (characteristic absorption spectrum of HMCs [57] and total absence of the absorption peaks of Hb at 520 and 580 nm). Proteins composition of MPs was documented by mass spectrometry analysis as previously described; briefly, MPs obtained from 2 mL of pRBCs were pelleted and solubilized with

2% SDS, solubilized proteins were separated by SDS-PAGE, and, following staining with colloidal Coomasie blue, electrophoretic bands were excised and identified by high-resolution MALDI-TOF peptide mass fingerprinting.

MPs were characterized and quantified by FACS (flow cytometry) analysis as described [44] utilizing a BD FACSCanto™ A flow cytometer (Becton Dickinson Biosciences, San Jose, CA, USA). The RBCs and pRBCs were excited with blue (488 nm, air-cooled, 20 mW solid state) and red (633 nm, 17 mW HeNe) laser light. Data from at least 50,000 events were acquired and analyzed with FACSDiva TM software (Becton Dickinson Biosciences, San Jose, CA, USA). MPs from RBCs and pRBCs were identified by their fluorescence (Glycophorin-A positive cells) and quantified with BD Trucount™ beads (Becton Dickinson Biosciences, San Jose, CA, USA). The MPs localized within the platelet region (R1) were distinguished from platelets (CD61-FITC positive) by their glycophorin-A-positive/CD61-negative response. The number of BD Trucount™ beads was counted, and the absolute numbers of MPs were then calculated with the following formula: (No. of glycophorin A positive events/No. of beads collected) × BD Trucount™ concentration × dilution factor. MPs and platelets were incubated with 20 μL of PBS-G (PBS containing 2 mM glucose) containing 3 μL of PE-conjugated PE Mouse Anti-Human glycophorin A Clone GA-R2 (HIR2) (Becton Dickinson Biosciences, San Jose, CA, USA) monoclonal antibody, alone or with 5 μL of FITC-conjugated Mouse Anti-Human CD61 Clone VI-PL2 (Becton Dickinson Biosciences, San Jose, CA, USA) monoclonal antibody for 30 min at RT (room temperature) in the dark. The MPs were diluted in 400 μL of PBS-G–paraformaldehyde 1% and transferred to flow cytometer tubes preloaded with a known density of fluorescent BD Trucount™ beads (Becton Dickinson Biosciences, San Jose, CA, USA, Catalog No 340334) and quantified.

MPs were also quantified measuring their band 3 content by Western blot analysis as previously described [61].

2.14. Statistical Analysis

We performed comparisons with the use of Student's *t*-test (two-tailed) using the SPSS version 22.0 Statistical package. Descriptive statistics are presented as mean ± standard deviation. Additionally, an independent sample *t*-test was used to compare between means. In all statistical analyses, the level of significance (*p*-value) was set as * $p < 0.05$, ** $p < 0.001$.

3. Results

3.1. Effect of Syk Inhibitors on Accumulation of Hemichromes in Parasitized RBCs

To test the hypothesis that Syk tyrosine kinase inhibitors might increase the hemichrome (HMC) content of malaria-infected erythrocytes (pRBCs) by blocking release of hemichrome-enriched microparticles, we firstly measured the HMC concentrations in pRBCs in the presence and absence of a variety of Syk inhibitors. As shown in Figure 1A, although untreated pRBCs accumulated only low levels of HMCs, incubation with different Syk inhibitors characterized by different IC_{50s} on isolated Syk (P505-15: 1 nM, R406: 41 nM, imatinib: 5 μM) increased accumulation of HMCs, and this accumulation correlated roughly with the potency of the specific Syk inhibitor. Because these inhibitors have no detectable effect on uninfected RBCs, we conclude that Syk inhibitors can enhance accumulation of hemichromes in infected RBCs without promoting hemichrome formation or accumulation in healthy erythrocytes. Next, to determine whether addition of dihydroartemisinin (DHA) might further enhance Syk inhibitor-promoted accumulation of HMCs, the same experiment was repeated using a very low concentration of DHA (0.5 nM) to minimize the cytotoxic effects on *P. falciparum*. This low concentration of DHA did not cause an increase of HMCs in RBCs or in pRBCs. On the contrary, DHA in combination with Syk inhibitors caused a consistent increase of HMC content in pRBCs. No effect on HMC accumulation was measurable in control RBCs (Figure S1, Supplementary Materials). Our motivation for employing such a low concentration of DHA was to obtain an initial indication of possible synergy between DHA and a Syk inhibitor, since the concentration of DHA employed in this study was found to promote no hemichrome formation by itself (Figure 1) and no measurable anti-plasmodial effect. In this regard, it should be noted that, within 12 h of treatment, no morphological changes of parasites were observed with any treatment. On the contrary, the combination of DHA and Syk inhibitor caused a moderate delay of maturation after 24 h. To minimize the bias that may derive from the loss of viability of parasites, data presented in Figure 1 were obtained within 12 h of treatment.

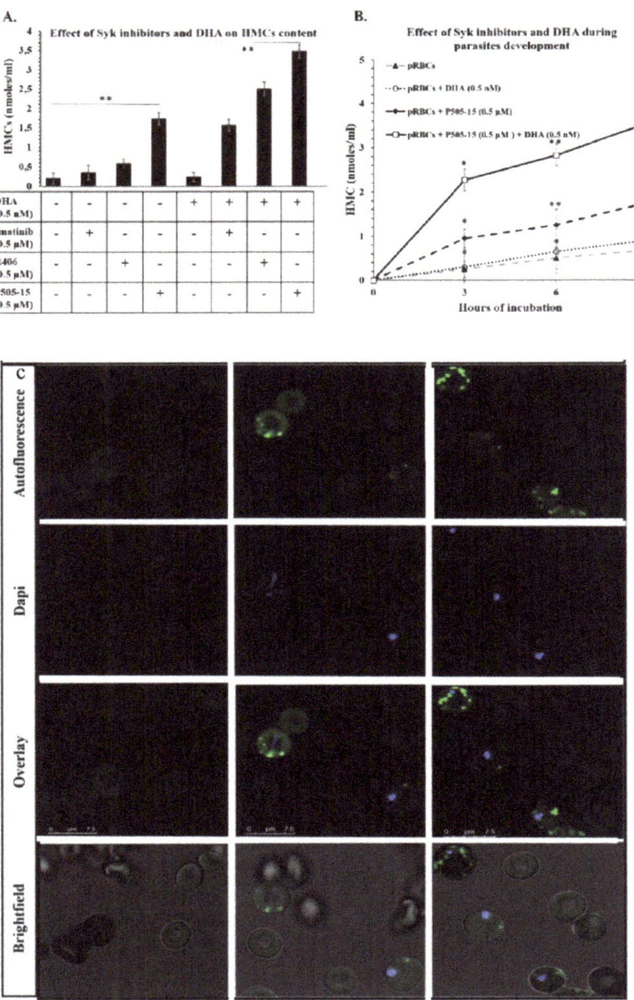

Figure 1. (**A**) Effect of parasitized red blood cell (pRBC) treatments with Syk inhibitors (imatinib, R406, and P505-15) (0.5 µM) and dihydroartemisinin (DHA) (0.5 nM) on the accumulation of HMC after 12 h of incubation. (**B**) Time course of hemichrome (HMC) accumulation contained in pRBCs treated with the representative Syk inhibitor, P505-15 (0.5 µM) and/or DHA (0.5 nM) for 3, 6, and 12 h. Data are the average of five independent experiments ± SD. Significant differences to untreated pRBCs at * $p < 0.05$; ** $p < 0.001$. (**C**) Confocal Images of HMCs contained in pRBCs untreated and treated for 12 h with P505-15 (0.5 µM). HMCs were visualized by their autofluorescence at 488 nm (excitation)/630–750 nm (emission). Images were acquired using the same magnification with a Leica TCS SP5 X (Leica Microsystems, Germany) confocal microscope equipped with a 60 × 1.4 numerical aperture oil immersion lens. The scale bar in the figure is 7.5 µm.

As predicted from our hypothesis, co-administration of the low dose of DHA (0.5 nM) further augmented the hemichrome content of Syk inhibitor-treated RBCs (Figure 1A). Moreover, as shown in Figure 1B, this aggravated accumulation of hemichromes likely begins almost immediately following drug addition, since the pRBC content of hemichromes is prominently increased by 3 h post administration. As Syk inhibitors were added to cultures 12 h post infection, we could notice that hemichrome accumulation was already enhanced at the ring stage. This observation is in accordance with previous reports showing that pro-oxidant hemoglobinopathies such as G6PD deficiency, hemoglobins C and S, and thalassemias can promote hemichrome production in the first half of parasite development [13,23,24,26,34,55]. Taken together, these data suggest that Syk inhibitors and DHA may synergize in enhancing accumulation of reactive hemichromes in pRBCs. We then studied the effect of Syk inhibitors in the presence and absence of DHA during the earlier stages of parasite development. Figure 1B shows the time-dependent accumulation of HMCs in untreated pRBCs and in pRBCs treated with Syk inhibitor (P505-15) in the presence and absence of DHA. Although treatment with DHA had little effect on HMC concentrations at any maturation stage, P505-15 exerted considerable impact on HMC accumulation during all early stages of parasite maturation (from early ring stage cultures to trophozoites), and this effect was further enhanced by co-administration of a low dose (0.5 nM) of DHA. Treatment with the other Syk inhibitors (R406 and imatinib) caused similar degrees of HMC accumulation, both in the presence and in the absence of DHA. Indeed, by imaging the autofluorescence of HMCs upon excitation at 488 nm (λ_{em} = 630–700 nm), the absence of HMCs in healthy RBCs, their low abundance in pRBCs, and their strong accumulation in P505-15-treated pRBCs could be visually confirmed (Figure 1C). HMCs were also measured following their semi-selective extraction from erythrocyte membranes by mild detergent treatment followed by size-exclusion purification to isolate the high-molecular-weight aggregate (>2 × 10^6 kDa) containing HMCs. HMCs were quantified according to their spectral characteristics [28,55]; notably, in the high-molecular-weight fraction, no free oxyHb was detectable (lack of absorption peaks at 488 and 520 nm). The purity of HMCs contained in the high-molecular-weight fraction, measured as a fraction of total protein concentration, was higher than 80%.

3.2. Effect of Syk Inhibitors on MP Production

Next, to test the hypothesis that hemichromes are discharged from pRBCs by blebbing off hemichrome-enriched microparticles (MPs), following a first centrifugation step to eliminate erythrocyte ghosts and debris, we pelleted the MPs at 100,000× *g* from the supernatants of both healthy and parasitized RBC suspensions. Isolated MPs were then resuspended and examined by confocal microscopy exploiting the fluorescence of HMCs upon excitation at 488 nm (λ_{em} = 630–700 nm). As shown in Figure 2A, the supernatant collected from healthy RBCs was essentially void of MPs. In contrast, the supernatant from untreated pRBCs was found to contain considerable hemichrome-loaded MPs, i.e., as evidenced by the autofluorescence of the MPs. More importantly, the supernatant from Syk inhibitor-treated (P505-15) pRBCs displayed essentially the same level of autofluorescent MPs as seen in suspensions of healthy RBCs. The presence of high-molecular-weight protein aggregates in MPs was confirmed utilizing the same method used for pRBC membranes (see previous paragraph). Spectral analysis also revealed the absence of oxyHb, confirming a strong analogy between the high-molecular-weight HMC aggregates isolated from pRBC membranes and MPs. As previously observed [20], MPs isolated from pRBCs supernatants were relatively homogeneous with a diameter lower than 1 μm; moreover, the morphology and the dimensions of the fluorescent aggregates observed in the MPs are compatible with HMC aggregates observed in the pRBC membranes. MPs were characterized and quantified by FACS utilizing anti-glycophorin antibody to label RBC membranes. Western blot analysis using anti-band 3 antibody further confirmed the variations of RBC membrane-derived MPs (Figure 2B–D).

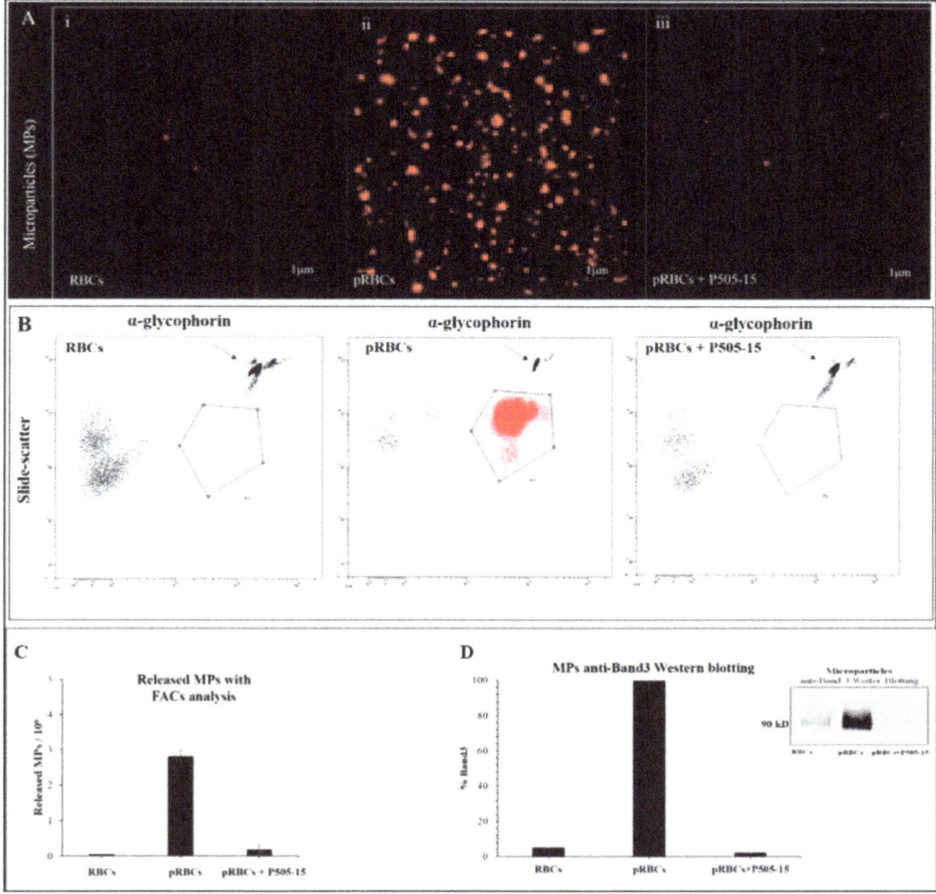

Figure 2. (**A**) Confocal Images of isolated microparticles (MPs) contained in pRBCs untreated and treated for 12 h with P505-15 (0.5 μM). MPs were visualized by their autofluorescence at 488 nm (excitation)/630–750 nm (emission). Images were acquired using the same magnification with a Leica TCS SP5 X (Leica Microsystems, Wetzlar, Germany) confocal microscope equipped with a 60 × 1.4 numerical aperture oil immersion lens. The scale bar in the figure is 1 μm. (**B**) Representative flow cytometric density plot of MPs in supernatant collected from RBCs, pRBCs, and P505-15-treated pRBCs. MPs were identified as glycophorin A-positive events (R1). The number of MPs was quantified using the known density of fluorescent trucount^{TM} beads (arrows). (**C**) Graphical representation of flow cytometric density plot of MPs. (**D**) MP anti-band 3 Western blotting.

To substantiate the role of oxidative stress in MP generation, we treated pRBCs with ROS scavengers reduced glutathione and *N*-acetyl-L-cysteine. As shown in the fifth figure (C,D), those compounds inhibited ROS production and the release of MPs. The relatively high concentrations needed to exert the observed inhibition are plausibly due to the limited membrane permeability of those molecules.

3.3. Hemichromes Accumulation and ROS Production

The next element of our hypothesis was that enhanced accumulation of hemichromes in Syk inhibitor- and DHA-treated pRBCs should increase the oxidative stress within the pRBCs. To test this step in the hypothesis, we quantitated the level of free radicals generated within the RBCs by EPR using a spin trap, α-phenyl-*N-tert*-butylnitrone (PBN), to stabilize any generated free radicals. The six-lined spectra, characteristic of the PBN adduct formed after radical trapping in different samples, are presented in Figure 3A. Simulation of the spectra yielded the following parameters: $g = 2.0059$, $a_N = 14.9$ G, and $a_H = 3.3$ G, indicating that a hydroxyl radical was generated and transformed into a CH_3 radical after reaction with DMSO. Importantly, little or no signal was recordable in either control RBCs or healthy RBCs treated with DHA, P505-15, or both together, i.e., indicating that the combination of DHA + Syk inhibitor does not induce significant oxidative stress in healthy RBCs. In contrast, stimulation of HMC formation in healthy cells by treatment with phenylhydrazine (PHZ) was found to enable the DHA + Syk inhibitor combination to potently induce free radical formation, i.e., suggesting that HMCs must be present to catalyze the antimalarial-induced formation of free radicals. Consistent with these results, although only a low-intensity EPR signal was observed in untreated pRBCs (ring stage), this signal was significantly enhanced upon brief incubation (10 min) with 200 μM DHA (this high concentration was chosen due to the short time of incubation) and then further strengthened by addition of P505-15 (p-value = 0.028) (Figure 3A). Since very similar results were obtained when the ROS-sensitive fluorescent probe, CM-H_2DCFDA (p-value = 0.022), was employed to quantitate oxidative stress (Figure 3B), we conclude that, in the presence of HMCs, induced by phenylhydrazine or parasite growth, intra-erythrocyte oxidative stress can be greatly increased by the addition of Syk inhibitors plus DHA.

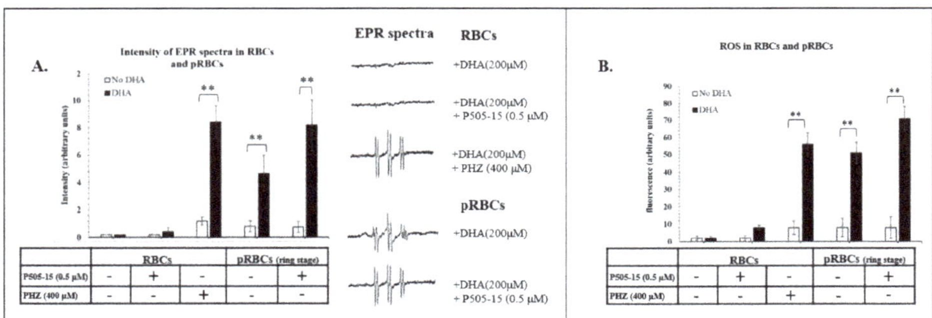

Figure 3. (**A**) Intensity of electron paramagnetic resonance (EPR) spectra (arbitrary units) in the presence of *N-tert*-butyl-α-phenylnitrone (PBN) spin trapping agent in RBCs treated with phenylhydrazine (PHZ) (400 μM) with/without DHA (200 μM). RBCs and pRBCs treated with/out P505-15 (0.5 μM) for 24 h with/without 10-min incubation of DHA (200 μM). (**B**) The same experiments mentioned above (**A**) were performed using the fluorescent probe 2′,7′-dichlorodihydrofluorescein diacetate (CM-H_2DCFDA), a cell-permeable indicator of reactive oxygen species (ROS). Data are the average ± SD of five independent experiments. Significant differences between no-DHA RBCs and pRBCs at ** $p < 0.001$.

Figure 4A, chelation of the heme iron significantly decreased the intensity of the free radical signal (p-value = 0.019), indicating the requirement of accessible iron in the production of activated DHA and oxidative free radicals in the parasitized RBC.

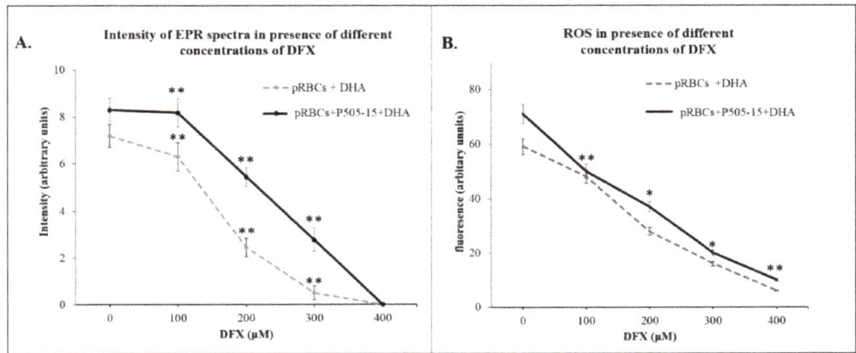

Figure 4. (**A**) Intensity of EPR spectra (arbitrary units) in the presence of PBN spin trapping agent in pRBCs treated with/without P505-15 (0.5 µM) for 24 h with/without 10-min incubation with DHA (200 µM) and with different concentrations (100–400 µM) of Deferasirox (DFX). (**B**) The same experiments mentioned above (**A**) were performed using the probe CM-H_2DCFDA. Data are the average ± SD of five independent experiments. Significant differences between untreated DHA-RBCs and pRBCs at * $p < 0.05$, ** $p < 0.001$.

These data suggest that the availability of reactive iron in hemichrome-rich pRBCs but essentially absent from healthy RBCs is essential for the catalysis of activated DHA and ROS production. As evidenced in the previous paragraph, ROS scavengers decreased the production of ROS in pRBCs (Figure 5A,B) and MPs released from pRBCs (Figure 5C,D).

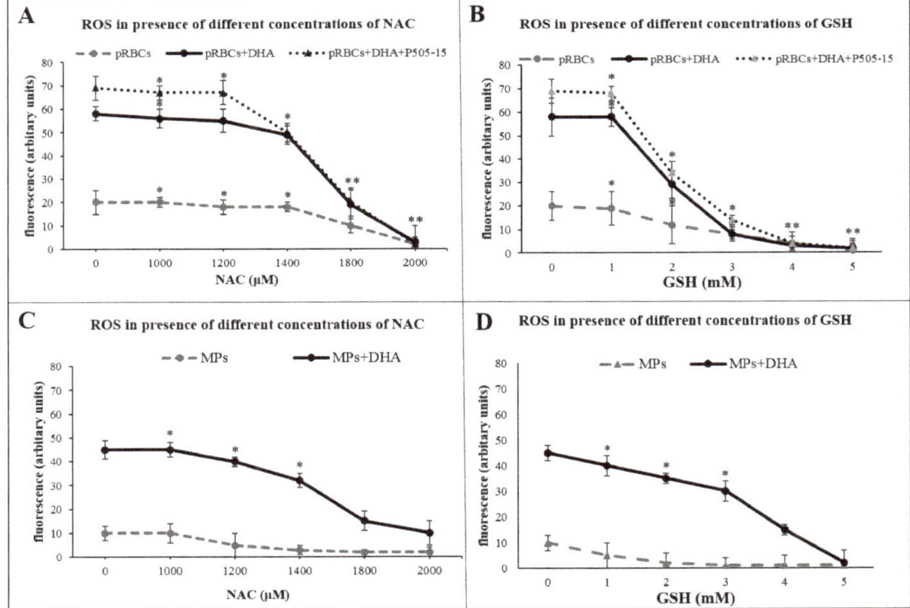

Figure 5. ROS production was performed using the probe CM-H2DCFDA in pRBCs pretreated with/without P505-15 (0.5 µM) and DHA (200 µM) and their MPs for 24 h at different concentrations of *N*-acetyl-L-cysteine (NAC) (**A**,**C**) and GSH (Glutathione) (**B**,**D**). Significant differences between untreated DHA-RBCs/MPs and pRBCs/MPs at * $p < 0.05$, ** $p < 0.001$.

3.4. Evaluation of Synergy between Syk Inhibitors and Artemisinins in Suppression of Parasitemia

The previous paragraphs suggest that DHA and Syk inhibitors act synergistically to induce the formation of ROS in pRBCs. To determine whether this synergistic production of ROS might translate into synergistic suppression of parasitemia, we employed the combination index theorem of Chow–Talalay [55,56] to quantitate the synergy/antagonism between DHA and Syk inhibitors in eliminating *P. falciparum* from cultures of fresh human blood. As noted in Section 3, experimental data points in the required isobolograms that lie below the diagonal line indicate synergy, while those that lie on the diagonal demonstrate additivity, and those that reside above the line are interpreted to indicate antagonism. To construct these isobolograms, ring-stage pRBCs (12 h post infection) were treated with increasing concentrations of the desired drugs for 24 h, and the residual parasitemia was quantitated to determine the IC_{50} value of each Syk inhibitor at the indicated concentration of artemisinins. As seen in Figure 6, regardless of the Syk inhibitor employed, all inhibitors synergized with DHA in suppressing parasitemia.

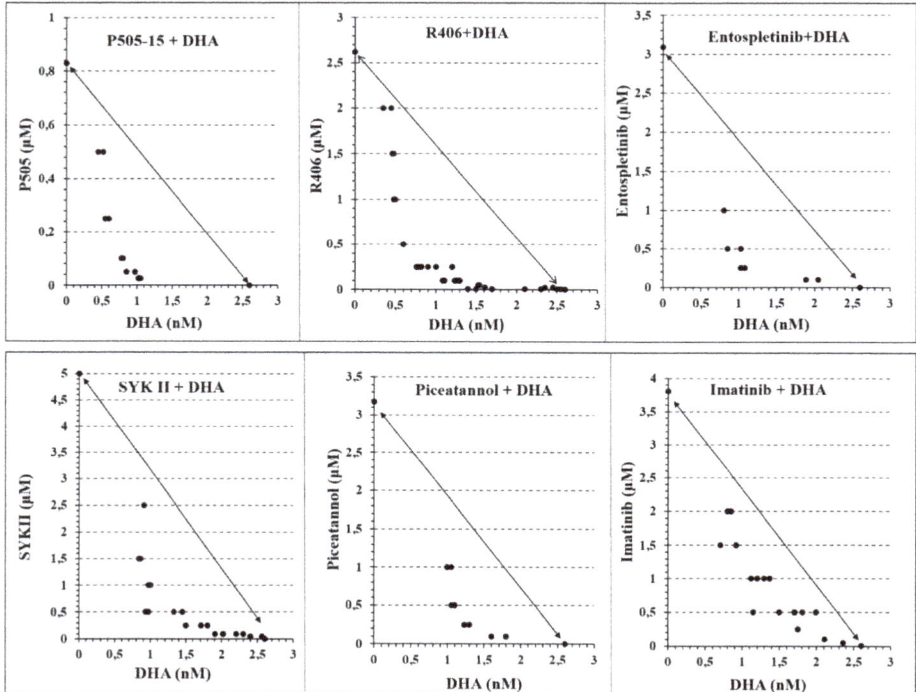

Figure 6. Isobolograms showing the interactions between Syk Inhibitors (P505-15, R406, entospletinib, SYK II, piceatannol, and imatinib) and dihydroartemisinin (DHA), after 24 h of incubation in the *P. falciparum* Palo Alto strain. Synchronized *P. falciparum* cultures were treated for 24 h with different concentrations (from 0.05 to 2.5 µM) of different Syk inhibitors (P505-15, R406, entospletinib, SYK II, piceatannol, and imatinib) in combination with different concentrations of DHA (from 0.6 to 10 nM) at the ring stage. IC_{50} concentrations of all the drug combinations were plotted in isobolograms to determine synergy, additivity, or antagonism.

Moreover, as noted from their quantitative combination index values (Table 1), the strength of the observed synergies corresponded approximately with the potencies of the Syk inhibitors (i.e., combination index (CI) values were 0.42 for P505-15 and R406, followed by entospletinib and SYK II at

0.51, and then piceatannol and imatinib at 0.56 and 0.73, respectively), and similar strong synergy was also observed when other forms of artemisinins were evaluated (Table S1, Supplementary Materials).

Table 1. Combination index value (CI) 24 and 48 h after treatment.

Syk Inhibitors (nM)	Dihydroartemisinin		Artesunate		Artemether	
	24 h	48 h	24 h	48 h	24 h	48 h
			P505-15 [1] (IC_{50} = 1 nM)			
500	0.54 ± 0.03	0.75 ± 0.06	0.97 ± 0.03	0.55 ± 0.07	**0.87 ± 0.03**	0.83 ± 0.01
250	0.51 ± 0.01	0.71 ± 0.03	0.76 ± 0.08	0.51 ± 0.01	0.89 ± 0.01	0.65 ± 0.03
100	**0.42 ± 0.04**	0.55 ± 0.04	0.63 ± 0.03	0.61 ± 0.03	0.95 ± 0.01	0.63 ± 0.04
50	0.43 ± 0.05	**0.49 ± 0.01**	**0.58 ± 0.07**	**0.49 ± 0.05**	0.98 ± 0.04	**0.58 ± 0.01**
			Entospletinib [1] (IC_{50} = 7.7 nM)			
500	0.52 ± 0.01	0.85 ± 0.06	0.87 ± 0.03	0.74 ± 0.04	1.01 ± 0.03	0.83 ± 0.04
250	**0.51 ± 0.07**	0.84 ± 0.07	0.73 ± 0.02	0.62 ± 0.01	**0.95 ± 0.05**	0.68 ± 0.04
100	0.77 ± 0.04	0.78 ± 0.03	0.66 ± 0.01	0.60 ± 0.02	0.98 ± 0.03	**0.63 ± 0.03**
50	0.82 ± 0.03	**0.69 ± 0.05**	**0.62 ± 0.02**	**0.55 ± 0.02**	1.04 ± 0.07	0.72 ± 0.04
			R406 [1] (IC_{50} = 41 nM)			
500	**0.42 ± 0.04**	0.95 ± 0.06	0.82 ± 0.04	0.75 ± 0.02	0.85 ± 0.03	0.93 ± 0.04
250	0.46 ± 0.02	0.81 ± 0.03	**0.63 ± 0.01**	0.72 ± 0.01	0.82 ± 0.01	0.88 ± 0.04
100	0.50 ± 0.03	0.60 ± 0.04	0.76 ± 0.03	0.50 ± 0.02	**0.80 ± 0.04**	0.73 ± 0.03
50	0.60 ± 0.05	**0.58 ± 0.01**	0.92 ± 0.02	**0.44 ± 0.02**	0.82 ± 0.04	**0.72 ± 0.04**
			SYK II [1] (IC_{50} = 41 nM)			
500	**0.51 ± 0.01**	0.92 ± 0.02	0.79 ± 0.04	0.89 ± 0.04	**0.97 ± 0.01**	0.73 ± 0.02
250	0.66 ± 0.01	0.93 ± 0.02	**0.55 ± 0.01**	0.99 ± 0.03	0.99 ± 0.03	0.58 ± 0.04
100	0.77 ± 0.06	0.52 ± 0.03	0.67 ± 0.04	0.96 ± 0.01	1.06 ± 0.05	**0.53 ± 0.05**
50	0.99 ± 0.04	**0.46 ± 0.05**	0.89 ± 0.02	**0.90 ± 0.04**	1.12 ± 0.01	0.62 ± 0.07
			Imatinib [1] (IC_{50} = 5 μM)			
500	0.79 ± 0.02	0.78 ± 0.01	0.67 ± 0.03	0.75 ± 0.02	1.01 ± 0.02	0.83 ± 0.01
250	**0.73 ± 0.06**	0.57 ± 0.07	**0.61 ± 0.04**	0.96 ± 0.04	0.89 ± 0.03	0.78 ± 0.06
100	0.83 ± 0.01	0.58 ± 0.02	0.70 ± 0.01	0.64 ± 0.01	0.91 ± 0.01	**0.73 ± 0.04**
50	0.92 ± 0.05	**0.55 ± 0.05**	0.90 ± 0.02	**0.60 ± 0.02**	**0.85 ± 0.02**	0.82 ± 0.04
			Piceatannol [1] (IC_{50} = 10 μM)			
500	0.57 ± 0.04	0.67 ± 0.02	0.77 ± 0.01	**0.65 ± 0.03**	0.95 ± 0.01	**0.73 ± 0.02**
250	**0.56 ± 0.01**	0.66 ± 0.04	**0.71 ± 0.03**	0.66 ± 0.07	0.78 ± 0.01	0.80 ± 0.01
100	0.68 ± 0.03	0.58 ± 0.05	0.80 ± 0.02	0.84 ± 0.02	0.85 ± 0.02	0.83 ± 0.04
50	0.79 ± 0.02	**0.49 ± 0.03**	085 ± 0.01	0.90 ± 0.04	**0.78 ± 0.03**	0.75 ± 0.01

Combination index (CI) (Chow–Talalay) of different Syk inhibitors (P505-15, R406, entospletinib, SYK II, piceatannol, and imatinib) at different concentrations (50–500 nM) in combination with different artemisinin derivatives (dihydroartemisinin, artesunate, and artemether) after 24 and 48 h of incubation: additive effect (C = 1), synergism (CI < 1), and antagonism (CI > 1). [1] IC_{50} on Syk catalytic subunit.

To add evidence to the role of oxidative stress mediated by HMCs in the mechanism leading to the synergistic interaction between Syk inhibitors and artemisinin, we conducted experiments growing *P. falciparum* in G6PD-deficient RBCs that, due to their reduced capability to produce NADPH (Nicotinamide adenine dinucleotide phosphate), are known to maximize the effects of oxidative stress, promoting the formation of HMCs both in vitro and in G6PD-deficient patients following ingestion of redox-active drugs or fava beans [30,62,63]. As previously described [34], parasites display the same growth rate in control and G6PD-deficient RBCs. In G6PD-deficient RBCs, we observed an increment in HMC content following DHA and Syk inhibitor treatment (Figure S2 Supplementary Materials). In addition, in G6PD-deficient RBCs, we measured a significant ($p < 0.01$) decrease in IC_{50} for DHA (from 2.60 nM to 1.75 nM) and a variable increase in synergistic interaction between DHA and Syk inhibitors. The CI shifted from 0.42 to 0.36 for P505-15 and from 0.73 to 0.33 for imatinib. The change in CI was significant ($p < 0.01$) only for the less specific Syk inhibitor imatinib, possibly indicating that potent Syk inhibitors are capable of determining the maximal synergistic effect on DHA (Figure S3, Supplementary Materials).

It was also important to notice that treatments with DHA + Syk inhibitors did not cause any increase in HMCs in non-parasitized G6PD-deficient RBCs, confirming that this combination does not cause oxidative injury to RBCs. To further investigate this relevant issue, we measured hemolysis as the percentage of total Hb contained in RBCs released in the supernatant following 24 h of treatment with DHA (10 nM) + P505-15 (250 nM). In untreated control and G6PD-deficient RBCs, we could not observe a significant difference in hemolysis (1.21% ± 0.76% and 1.66% ± 0.53%, respectively); following 24 h of treatment, we observed a non-significant increase in hemolysis both in control and in G6PD-deficient RBCs (2.43% ± 1.76% and 3.06% ± 1.53%, respectively).

Because chelation of iron by DFX was found to block the production of ROS by DHA + Syk inhibitor (Figure 6), we then predicted that DFX might similarly prevent the synergistic elimination of parasitemia by DHA + Syk inhibitor. As shown in Figure 7 and Table 2, this prediction was indeed realized.

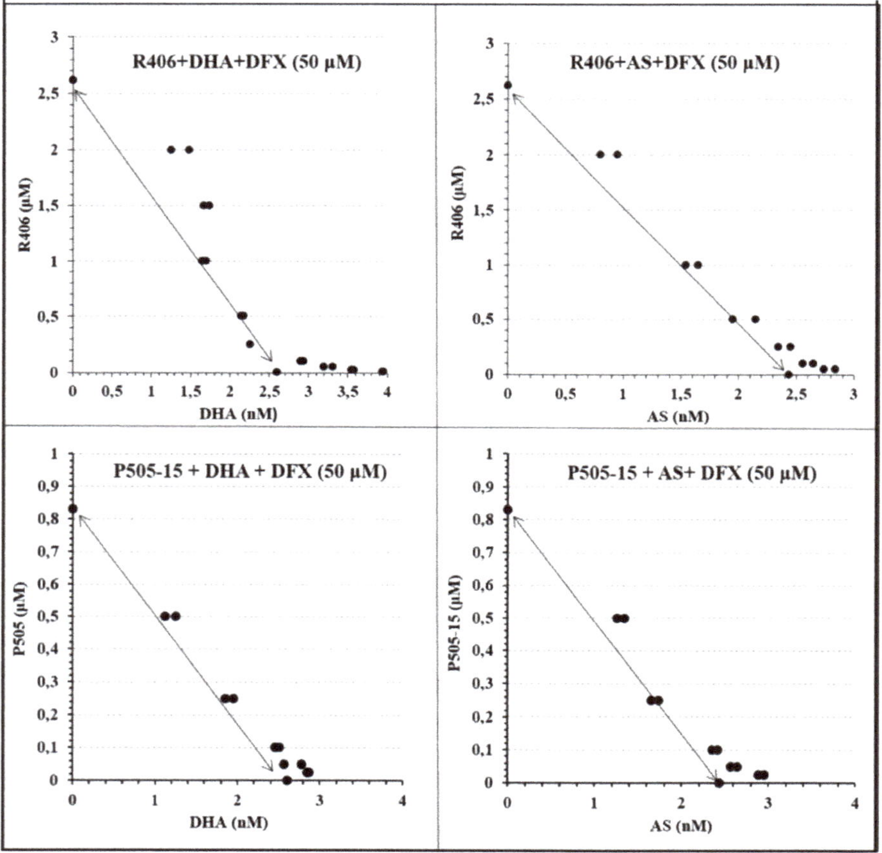

Figure 7. Isobolograms showing the interactions between Syk Inhibitors (P505-15 and R406) and dihydroartemisinin (DHA) and artesunate (AS) in combination with the iron chelator Deferasirox (DFX), after 24 h of incubation in the *P. falciparum* Palo Alto strain. Synchronized *P. falciparum* cultures were treated for 24 h with different concentrations (from 0.05 to 2.5 µM) of representative Syk inhibitors (P505-15 and R406) in combination with different concentrations of DHA and AS (from 0.6 to 10 nM) using a fixed concentration (50 µM) of the iron chelator Deferasirox (DFX) at the ring stage.

Table 2. Combination index value (CI) 24 h after treatment. Combination index (CI) (Chow–Talalay) of different SYK inhibitors (P505-15 and R406) at different concentrations (50–500 nM) in combination with different artemisinin derivatives (dihydroartemisinin and artesunate) using a fixed concentration of the iron chelator Deferasirox (DFX) after 24 and 48 h of incubation.

Syk Inhibitors (μM)	Deferasirox (50 μM)	
P505-15	**Dihydroartemisin**	**Artesunate**
500	1.08 ± 0.05	1.15 ± 0.07
250	1.05 ± 0.04	1.01 ± 0.09
100	1.08 ± 0.10	1.11 ± 0.03
50	1.29 ± 0.08	1.42 ± 0.05
R406		
500	1.03 ± 0.06	1.03 ± 0.02
250	0.96 ± 0.07	1.08 ± 0.04
100	1.17 ± 0.05	1.10 ± 0.06
50	1.29 ± 0.03	1.16 ± 0.08

Thus, regardless of whether dihydroartemisinin or artesunate was combined with P505-15 or R406 (Syk inhibitors), addition of DFX moved all data points back to the diagonal line, indicating the elimination of synergy between artemisinin and Syk inhibitor. These data argue that reactive iron and ROS production are critical to creation of the potent synergy between these two classes of antimalarial drugs.

Finally, to determine whether the combined action of Syk inhibitor plus artemisinin on *P. falciparum* survival might depend on the stage of parasite maturation, we compared the effects of the combination therapy with each monotherapy at different stages of parasite development. Figure 7 shows that, upon adding DHA + P505-15 (2 nM + 250 nM) at different stages of parasite development, the maximal activity was observed between 12 and 36 h post infection corresponding to ring and trophozoite stages. The data in Figure 8B reveal that none of the drugs have a measurable impact on pRBC morphology at 6 h post infection, after which the monotherapies (i.e., DHA or Syk inhibitor alone) display detectable but comparatively mild effects (few pycnotic cells (less than 2%)) on pRBC morphology.

In contrast, by 12 h post infection, the combination therapy is seen to exert a major influence on growth progression and pRBC morphology, essentially blocking the normal pathway of parasite maturation. While other explanations of the synergistic potency of the combination therapy can be envisioned, the data are very consistent with the ability of DHA to enhance the oxidative stress arising from hemichromes accumulated in pRBCs and with the capacity of Syk inhibitors to greatly augment this oxidative stress by preventing the discharge of these oxidative hemichromes in HMC-enriched microparticles.

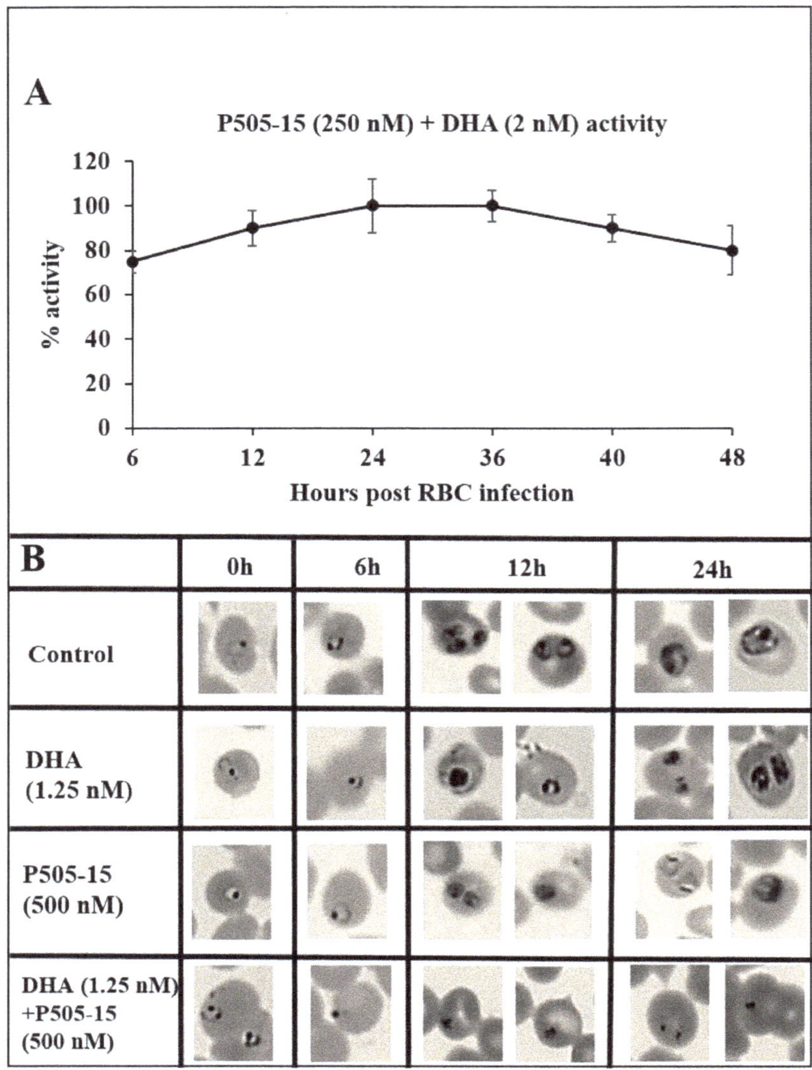

Figure 8. (**A**) Stage dependency of the efficacy (6, 12, 24, 36, 40, and 48 h) expressed as relative activity at one fixed dosage of DHA + P505-15 (2 nM + 250 nM). Relative activity of DHA + P505-15 (2 nM + 250 nM) added at 6, 12, 24, 36, 40, and 48 h of incubation. Values are expressed as percentage of DHA + P505-15 activity (treatment time at 24 h post infection) measured as % activity. Data are the average of five experiments ± SD. (**B**) Morphological changes in *P. falciparum* induced by Syk inhibitor (P505-15) in combination with a fixed concentration (1.25 nM) of dihydroartemisinin after 6, 12, and 24 h of treatment. Representative images of selected damaged parasites at 6, 12, and 24 h of treatment in control and drug-treated cultures selected from Diff-Quik® fix-stained thin blood films. The micrographs were obtained using a Leica DM IRB microscope equipped with a 100× oil planar apochromatic objective with 1.32 numeric aperture and a DFC420C camera and DFC software version 3.3.1 (Leica Microsystems, Wetzlar, Germany). The scale bar in the figure is 7.5 μm.

4. Discussion

Based on all the above data, we wish to propose a mechanism for the effects of artemisinins and Syk inhibitors on *P. falciparum* parasitemia. Following invasion of an erythrocyte, the parasite must digest hemoglobin in order to multiply ~18-fold during its 48-h intra-erythrocyte life cycle [19]. Parasite growth is accompanied by the production of ROS leading to hemoglobin oxidation and the formation of the hemichromes [13,64–66], which are known to be redox-active compounds [28,67,68]. In order to reduce this oxidative toxicity, the parasite promotes erythrocyte membrane weakening via activation of the phosphorylation of band 3 by Syk tyrosine kinase, which in turn enables the release of hemichrome-enriched MPs from the RBC membrane. In the absence of added drugs to enhance this oxidative stress, the parasite apparently copes with these oxidants from the pRBC. However, when a drug is added, it increases the pRBC's oxidative stress, and the parasite's ability to detoxify the added stress is exceeded as shown in Figures 3 and 4, leading eventually to parasite death. Syk kinase inhibitors prevent the discharge of HMCs in membrane-encapsulated microparticles [44] and force the accumulation/retention of redox-active HMC forms, enhancing the oxidative stress within the pRBC and, therefore, activating artemisinins. On the other hand, we observed that artemisinin compounds strongly enhance the generation of HMCs in pRBCs. This apparently cooperative behavior between Syk kinase inhibitors and artemisinins may be at the basis of their synergistic anti-plasmodial interaction.

A mechanism to explain the observed synergy between artemisinins and Syk inhibitors is also implied in the data of Figure 3A,B, where artemisinins are not observed to induce oxidant production in healthy RBCs. We explain this difference between healthy RBCs and pRBCs by noting that redox-active hemichromes and/or traces of free heme released from hemichromes or from the parasite food vacuole are required to catalyze the production of pharmaceutically active artemisinin [69]. In addition, this mechanism could be relevant to determine the high selectivity of artemisinin for parasitized RBCs and, consequently, their effectiveness. The lower effectiveness of artemisinins observed in patients with hemoglobinopathies could, therefore, determine the loss of selectivity due to artemisinin activation in non-parasitized RBCs containing traces of hemichromes and/or free heme.

Based on this pathway, the activation and/or the potency of artemisinins must depend on the abundance of HMCs in parasitized RBCs, which in turn will depend on the ability of the parasite to eliminate these HMCs in discharged microparticles. The release of MPs containing hemichromes was already described in oxidized erythrocytes [70,71], in parasitized RBCs [20,44], and in different hematological diseases characterized by hemichrome formation such as thalassemias and G6PD-deficient RBCs following oxidant treatment [39,66]. Supporting the proposed mechanism of action, iron chelators and ROS scavengers suppressed the anti-plasmodial activity of the Syk inhibitor + artemisinin treatment. On the contrary, G6PD-deficient RBCs enhanced the synergistic combination between Syk inhibitor + artemisinin combination and HMC accumulation. Importantly, no hemichrome formation or hemolysis was observed in non-parasitized G6PD-deficient parasites treated by Syk inhibitors and artemisinins, evidencing the need of the additional ROS generated by malaria parasites to trigger a virtuous cycle between the accumulation of redox-active HMCs and the activation of artemisinin. Those data are also suggestive of a lack of hemolytic effect exerted by Syk inhibitor + artemisinin combination in normal and G6PD-deficient RBCs, although more specific tests will be conducted to exclude the pro-hemolytic activity of this combination.

5. Conclusions

In conclusion, the presented data highlight the complex interactions occurring among ROS production due to parasite metabolism, redox-active HMCs, and artemisinin activation, documenting the role of Syk inhibition as a key element to synergistically improve the activity of artemisinins. Since some Syk kinase inhibitors such as R406 can be administered for long periods with minimal side effects [72,73], we propose that Syk kinase inhibitors could contribute measurably to the potencies of ACTs.

Moreover, classical tyrosine kinase was not identified in the *Plasmodium falciparum* genome [74]. The use of a drug targeting human Syk, an enzyme expressed in erythrocytes, displays the potential advantage of preventing the selection of mutant resistant parasites.

Supplementary Materials: The following are available online at http://www.mdpi.com/2076-3921/9/8/753/s1, Figure S1. Effect of RBC treatment with P505-15 (0.5 μM) and DHA (0.5 nM) on the accumulation of HMC after 12 h of incubation. Data are the average of five experiments ± SD; Figure S2 Effect of pRBC treatments with a representative Syk inhibitor (P505-15) (0.5 μM) and DHA (0.5 nM) on the accumulation of HMC after 3, 6, 12, and 24 h. Data are the average of five experiments ± SD. Significant differences to untreated pRBCs at * $p < 0.05$; ** $p < 0.001$; Figure S3. Isobolograms showing the interactions between Syk Inhibitors (P505-15 and imatinib) and dihydroartemisinin (DHA), after 24 h of incubation in the P. falciparum Palo Alto strain treated with G6PD-d red cells; Table S1. Combination index (CI) (Chow-Talalay) of different SYK inhibitors.

Author Contributions: Conceptualization, K.R., A.P. and F.M.T.; Data curation, I.T., K.R., G.M., M.C.P., C.F., F.N., A.P. and F.M.T.; Formal analysis, I.T., A.P. and F.M.T.; Funding acquisition, K.R., A.P. and F.M.T.; Investigation, I.T., A.P. and F.M.T.; Methodology, I.T., K.R. and F.M.T.; Project administration, F.M.T.; Resources, A.P. and F.M.T.; Software, I.T. and P.V.; Supervision, A.P. and F.M.T.; Validation, G.M.; Visualization, C.F. and F.N.; Writing—original draft, F.M.T. and P.S.L.; Writing—review and editing, I.T., K.R., C.F., F.N., A.P. and F.M.T. All authors have read and agreed to the published version of the manuscript.

Funding: This study was funded by the Fondazione Banco di Sardegna: Prot. U1056.2014/AI.938MGB Prat. 2014.0040.

Acknowledgments: The authors would like to thank Pietro Fresu and Gioacchino Greco and MSc students Alessia Manca, Salvatore Duras, and Cristina D'Avino for their support. This study was supported by grants from the Fondazione di Sardegna Prot. U1056.2014/AI.938MGB Prat. 2014.0040. The funders had no role in study design, data collection and analysis, decision to publish, or preparation of the manuscript.

Conflicts of Interest: The authors declare no conflict of interest.

References

1. Rosenthal, P.J.; Meshnick, S.R. Hemoglobin catabolism and iron utilization by malaria parasites. *Mol. Biochem. Parasitol.* **1996**, *83*, 131–139. [CrossRef]
2. Pandey, A.V.; Tekwani, B.L. Depolymerization of malarial hemozoin: A novel reaction initiated by blood schizontocidal antimalarials. *FEBS Lett.* **1997**. [CrossRef]
3. Lew, V.L.; Tiffert, T.; Ginsburg, H. Excess hemoglobin digestion and the osmotic stability of Plasmodium falciparum—Infected red blood cells. *Blood* **2003**. [CrossRef] [PubMed]
4. Mauritz, J.M.A.; Esposito, A.; Ginsburg, H.; Kaminski, C.F.; Tiffert, T.; Lew, V.L. The Homeostasis of Plasmodium falciparum-Infected Red Blood Cells. *PLoS Comput. Biol.* **2009**. [CrossRef] [PubMed]
5. Coronado, L.M.; Nadovich, C.T.; Spadafora, C. Malarial hemozoin: From target to tool. *Biochim. Biophys. Acta Gen. Subj.* **2014**, *1840*, 2032–2041. [CrossRef]
6. Olafson, K.N.; Nguyen, T.Q.; Rimer, J.D.; Vekilov, P.G. Antimalarials inhibit hematin crystallization by unique drug–surface site interactions. *Proc. Natl. Acad. Sci. USA* **2017**. [CrossRef]
7. WHO. *WHO Malaria Report. 2018*; WHO: Geneva, Switzerland, 2018.
8. Hommel, M. The future of artemisinins: Natural, synthetic or recombinant? *J. Biol.* **2008**, *7*, 38. [CrossRef]
9. O'Neill, P.M.; Barton, V.E.; Ward, S.A. The molecular mechanism of action of artemisinin—The debate continues. *Molecules* **2010**, *15*, 1705–1721. [CrossRef]
10. Cui, L.; Su, X.Z. Discovery, mechanisms of action and combination therapy of artemisinin. *Expert Rev. Anti. Infect. Ther.* **2009**, *7*, 999–1013. [CrossRef]
11. Rudrapal, M.; Chetia, D. Endoperoxide antimalarials: Development, structural diversity and pharmacodynamic aspects with reference to 1,2,4-trioxane-based structural scaffold. *Drug Des. Dev. Ther.* **2016**, *10*, 3575. [CrossRef]
12. Klonis, N.; Crespo-Ortiz, M.P.; Bottova, I.; Abu-Bakar, N.; Kenny, S.; Rosenthal, P.J.; Tilley, L. Artemisinin activity against Plasmodium falciparum requires hemoglobin uptake and digestion. *Proc. Natl. Acad. Sci. USA* **2011**. [CrossRef] [PubMed]
13. Ayi, K.; Turrini, F.; Piga, A.; Arese, P. Enhanced phagocytosis of ring-parasitized mutant erythrocytes: A common mechanism that may explain protection against falciparum malaria in sickle trait and beta-thalassemia trait. *Blood* **2004**. [CrossRef] [PubMed]

14. Pantaleo, A.; Ferru, E.; Carta, F.; Valente, E.; Pippia, P.; Turrini, F. Effect of heterozygous beta thalassemia on the phosphorylative response to Plasmodium falciparum infection. *J. Proteom.* **2012**. [CrossRef] [PubMed]
15. Eridani, S. Sickle cell protection from malaria: A review. *Hematol. Rep. Former. Hematol. Rev.* **2011**. [CrossRef]
16. Mbanefo, E.C.; Ahmed, A.M.; Titouna, A.; Elmaraezy, A.; Trang, N.T.H.; Phuoc Long, N.; Hoang Anh, N.; Diem Nghi, T.; The Hung, B.; Van Hieu, M.; et al. Association of glucose-6-phosphate dehydrogenase deficiency and malaria: A systematic review and meta-analysis. *Sci. Rep.* **2017**. [CrossRef]
17. Roberts, D.J.; Williams, T.N. Haemoglobinopathies and resistance to malaria. *Redox Rep.* **2003**. [CrossRef]
18. López, C.; Saravia, C.; Gomez, A.; Hoebeke, J.; Patarroyo, M.A. Mechanisms of genetically-based resistance to malaria. *Gene* **2010**, *467*, 1–12. [CrossRef]
19. Mohandas, N.; An, X. Malaria and human red blood cells. *Med. Microbiol. Immunol.* **2012**, *201*, 593–598. [CrossRef]
20. Pantaleo, A.; Ferru, E.; Vono, R.; Giribaldi, G.; Lobina, O.; Nepveu, F.; Ibrahim, H.; Nallet, J.P.; Carta, F.; Mannu, F.; et al. New antimalarial indolone-N-oxides, generating radical species, destabilize the host cell membrane at early stages of Plasmodium falciparum growth: Role of band 3 tyrosine phosphorylation. *Free Radic. Biol. Med.* **2012**. [CrossRef]
21. Ehrhardt, K.; Deregnaucourt, C.; Goetz, A.A.; Tzanova, T.; Gallo, V.; Arese, P.; Pradines, B.; Adjalley, S.H.; Bagrel, D.; Blandin, S.; et al. The redox cycler plasmodione is a fast-acting antimalarial lead compound with pronounced activity against sexual and early asexual blood-stage parasites. *Antimicrob. Agents Chemother.* **2016**. [CrossRef]
22. Giribaldi, G.; Ulliers, D.; Mannu, F.; Arese, P.; Turrini, F. Growth of Plasmodium falciparum induces stage-dependent haemichrome formation, oxidative aggregation of band 3, membrane deposition of complement and antibodies, and phagocytosis of parasitized erythrocytes. *Br. J. Haematol.* **2001**. [CrossRef] [PubMed]
23. Fröhlich, B.; Jäger, J.; Lansche, C.; Sanchez, C.P.; Cyrklaff, M.; Buchholz, B.; Soubeiga, S.T.; Simpore, J.; Ito, H.; Schwarz, U.S.; et al. Hemoglobin S and C affect biomechanical membrane properties of P. falciparum-infected erythrocytes. *Commun. Biol.* **2019**. [CrossRef]
24. Méndez, D.; Linares, M.; Diez, A.; Puyet, A.; Bautista, J.M. Stress response and cytoskeletal proteins involved in erythrocyte membrane remodeling upon Plasmodium falciparum invasion are differentially carbonylated in G6PD A- deficiency. *Free Radic. Biol. Med.* **2011**. [CrossRef] [PubMed]
25. Gallo, V.; Schwarzer, E.; Rahlfs, S.; Schirmer, R.H.; van Zwieten, R.; Roos, D.; Arese, P.; Becker, K. Inherited glutathione reductase deficiency and Plasmodium falciparum malaria—A case study. *PLoS ONE* **2009**. [CrossRef]
26. Tokumasu, F.; Nardone, G.A.; Ostera, G.R.; Fairhurst, R.M.; Beaudry, S.D.; Hayakawa, E.; Dvorak, J.A. Altered membrane structure and surface potential in homozygous hemoglobin C erythrocytes. *PLoS ONE* **2009**. [CrossRef]
27. Sherman, I.W.; Eda, S.; Winograd, E. Erythrocyte aging and malaria. *Cell. Mol. Biol. (Noisy-le-grand)* **2004**, *50*, 159–169.
28. Welbourn, E.M.; Wilson, M.T.; Yusof, A.; Metodiev, M.V.; Cooper, C.E. The mechanism of formation, structure and physiological relevance of covalent hemoglobin attachment to the erythrocyte membrane. *Free Radic. Biol. Med.* **2017**. [CrossRef]
29. Turrini, F.; Arese, P.; Yuan, J.; Low, P.S. Clustering of integral membrane proteins of the human erythrocyte membrane stimulates autologous IgG binding, complement deposition, and phagocytosis. *J. Biol. Chem.* **1991**, *266*, 23611–23617.
30. Tokumasu, F.; Fairhurst, R.M.; Ostera, G.R.; Brittain, N.J.; Hwang, J.; Wellems, T.E.; Dvorak, J.A. Band 3 modifications in Plasmodium falciparum-infected AA and CC erythrocytes assayed by autocorrelation analysis using quantum dots. *J. Cell Sci.* **2005**. [CrossRef]
31. Low, P.S.; Waugh, S.M.; Zinke, K.; Drenckhahn, D. The Role of Hemoglobin Denaturation and Band 3 Clustering in Red Blood Cell Aging. *Science* **1985**. [CrossRef]
32. Pantaleo, A.; Ferru, E.; Giribaldi, G.; Mannu, F.; Carta, F.; Matte, A.; de Franceschi, L.; Turrini, F. Oxidized and poorly glycosylated band 3 is selectively phosphorylated by Syk kinase to form large membrane clusters in normal and G6PD-deficient red blood cells. *Biochem. J.* **2009**. [CrossRef] [PubMed]

33. Ferru, E.; Giger, K.; Pantaleo, A.; Campanella, E.; Grey, J.; Ritchie, K.; Vono, R.; Turrini, F.; Low, P.S. Regulation of membrane-cytoskeletal interactions by tyrosine phosphorylation of erythrocyte band 3. *Blood* **2011**. [CrossRef] [PubMed]
34. Cappadoro, M.; Giribaldi, G.; O'Brien, E.; Turrini, F.; Mannu, F.; Ulliers, D.; Simula, G.; Luzzatto, L.; Arese, P. Early phagocytosis of glucose-6-phosphate dehydrogenase (G6PD)-deficient erythrocytes parasitized by Plasmodium falciparum may explain malaria protection in G6PD deficiency. *Blood* **1998**. [CrossRef]
35. Mohandas, N.; Gallagher, P.G. Red cell membrane: Past, present, and future. *Blood* **2008**. [CrossRef] [PubMed]
36. Thevenin, B.J.M.; Willardson, B.M.; Low, P.S. The redox state of cysteines 201 and 317 of the erythrocyte anion exchanger is critical for ankyrin binding. *J. Biol. Chem.* **1989**, *264*, 15886–15892.
37. Pantaleo, A.; Ferru, E.; Pau, M.C.; Khadjavi, A.; Mandili, G.; Mattè, A.; Spano, A.; De Franceschi, L.; Pippia, P.; Turrini, F. Band 3 erythrocyte membrane protein acts as redox stress sensor leading to its phosphorylation by p72 Syk. *Oxid. Med. Cell. Longev.* **2016**. [CrossRef]
38. Turrini, F.; Giribaldi, G.; Carta, F.; Mannu, F.; Arese, P. Mechanisms of band 3 oxidation and clustering in the phagocytosis of *Plasmodium falciparum* -infected erythrocytes. *Redox Rep.* **2003**. [CrossRef]
39. Ferru, E.; Pantaleo, A.; Carta, F.; Mannu, F.; Khadjavi, A.; Gallo, V.; Ronzoni, L.; Graziadei, G.; Cappellini, M.D.; Turrini, F. Thalassemic erythrocytes release microparticles loaded with hemichromes by redox activation of p72Syk kinase. *Haematologica* **2014**. [CrossRef]
40. Leal, J.K.F.; Adjobo-Hermans, M.J.W.; Bosman, G.J.C.G.M. Red blood cell homeostasis: Mechanisms and effects of microvesicle generation in health and disease. *Front. Physiol.* **2018**, *9*, 703. [CrossRef]
41. Aharon, A.; Rebibo-Sabbah, A.; Tzoran, I.; Levin, C. Extracellular Vesicles in Hematological Disorders. *Rambam Maimonides Med. J.* **2014**. [CrossRef]
42. Westerman, M.; Pizzey, A.; Hirschman, J.; Cerino, M.; Weil-Weiner, Y.; Ramotar, P.; Eze, A.; Lawrie, A.; Purdy, G.; Mackie, I.; et al. Microvesicles in haemoglobinopathies offer insights into mechanisms of hypercoagulability, haemolysis and the effects of therapy. *Br. J. Haematol.* **2008**. [CrossRef] [PubMed]
43. Kesely, K.R.; Pantaleo, A.; Turrini, F.M.; Olupot-Olupot, P.; Low, P.S. Inhibition of an erythrocyte tyrosine kinase with imatinib prevents Plasmodium falciparum egress and terminates parasitemia. *PLoS ONE* **2016**. [CrossRef] [PubMed]
44. Pantaleo, A.; Kesely, K.R.; Pau, M.C.; Tsamesidis, I.; Schwarzer, E.; Skorokhod, O.A.; Chien, H.D.; Ponzi, M.; Bertuccini, L.; Low, P.S.; et al. Syk inhibitors interfere with erythrocyte membrane modification during P falciparum growth and suppress parasite egress. *Blood* **2017**, *130*. [CrossRef] [PubMed]
45. Pantaleo, A.; Ferru, E.; Carta, F.; Mannu, F.; Giribaldi, G.; Vono, R.; Lepedda, A.J.; Pippia, P.; Turrini, F. Analysis of changes in tyrosine and serine phosphorylation of red cell membrane proteins induced by P. falciparum growth. *Proteomics* **2010**. [CrossRef]
46. Amato, R.; Pearson, R.D.; Almagro-Garcia, J.; Amaratunga, C.; Lim, P.; Suon, S.; Sreng, S.; Drury, E.; Stalker, J.; Miotto, O.; et al. Origins of the current outbreak of multidrug-resistant malaria in southeast Asia: A retrospective genetic study. *Lancet Infect. Dis.* **2018**. [CrossRef]
47. Dondorp, A.M.; Nosten, F.; Yi, P.; Das, D.; Phyo, A.P.; Tarning, J.; Lwin, K.M.; Ariey, F.; Hanpithakpong, W.; Lee, S.J.; et al. Artemisinin resistance in Plasmodium falciparum malaria. *N. Engl. J. Med.* **2009**. [CrossRef]
48. Hanboonkunupakarn, B.; White, N.J. *Travel Medicine and Infectious Disease*; Elsevier: Amsterdam, The Netherlands, 2016.
49. Pantaleo, A.; Pau, M.C.; Chien, H.D.; Turrini, F. Artemisinin resistance, some facts and opinions. *J. Infect. Dev. Ctries.* **2015**. [CrossRef]
50. Blasco, B.; Leroy, D.; Fidock, D.A. Antimalarial drug resistance: Linking Plasmodium falciparum parasite biology to the clinic. *Nat. Med.* **2017**, *23*, 917. [CrossRef]
51. Cui, L.; Mharakurwa, S.; Ndiaye, D.; Rathod, P.K.; Rosenthal, P.J. Antimalarial drug resistance: Literature review and activities and findings of the ICEMR network. *Am. J. Trop. Med. Hyg.* **2015**, *93*, 57–68. [CrossRef]
52. Trager, W.; Jensen, J.B. Human malaria parasites in continuous culture. *Science* **1976**. [CrossRef]
53. Lambros, C.; Vanderberg, J.P. Synchronization of Plasmodium falciparum Erythrocytic Stages in Culture. *J. Parasitol.* **1979**. [CrossRef]
54. Rivadeneira, E.M.; Wasserman, M.; Espinal, C.T. Separation and Concentration of Schizonts of Plasmodium falciparum by Percoll Gradients. *J. Protozool.* **1983**. [CrossRef] [PubMed]

55. Chou, T.C. Drug combination studies and their synergy quantification using the chou-talalay method. *Cancer Res.* **2010**, *70*, 440–446. [CrossRef] [PubMed]
56. Chou, T.-C. Theoretical Basis, Experimental Design, and Computerized Simulation of Synergism and Antagonism in Drug Combination Studies. *Pharmacol. Rev.* **2006**. [CrossRef] [PubMed]
57. Mannu, F.; Arese, P.; Cappellini, M.D.; Fiorelli, G.; Cappadoro, M.; Giribaldi, G.; Turrini, F. Role of hemichrome binding to erythrocyte membrane in the generation of band-3 alterations in beta-thalassemia intermedia erythrocytes. *Blood* **1995**, *86*, 2014–2020. [CrossRef]
58. Creek, D.J.; Ryan, E.; Charman, W.N.; Chiu, F.C.K.; Prankerd, R.J.; Vennerstrom, J.L.; Charman, S.A. Stability of peroxide antimalarials in the presence of human hemoglobin. *Antimicrob. Agents Chemother.* **2009**. [CrossRef]
59. Winterbourn, C.C. Free-radical production and oxidative reactions of hemoglobin. *Environ. Health Perspect.* **1985**. [CrossRef]
60. Antonini, E.; Brunori, M. *"Hemoglobin and Methemoglobin" in the Red Blood*; Surgenor, D.M.E., Ed.; NY Acad. Press: New York, NY, USA, 1975; Volume 2, pp. 753–797.
61. Pantaleo, A.; Ferru, E.; Carta, F.; Mannu, F.; Simula, L.F.; Khadjavi, A.; Pippia, P.; Turrini, F. Irreversible AE1 tyrosine phosphorylation leads to membrane vesiculation in G6PD deficient red cells. *PLoS ONE* **2011**. [CrossRef]
62. Arese, P.; Gallo, V.; Pantaleo, A.; Turrini, F. Life and death of glucose-6-phosphate dehydrogenase (G6PD) deficient erythrocytes-role of redox stress and band 3 modifications. *Transfus. Med. Hemotherapy* **2012**, *39*, 328–334. [CrossRef]
63. Luzzatto, L.; Arese, P. Favism and glucose-6-phosphate dehydrogenase deficiency. *N. Engl. J. Med.* **2018**, *378*, 60–71. [CrossRef]
64. Becker, K.; Tilley, L.; Vennerstrom, J.L.; Roberts, D.; Rogerson, S.; Ginsburg, H. Oxidative stress in malaria parasite-infected erythrocytes: Host-parasite interactions. *Int. J. Parasitol.* **2004**. [CrossRef] [PubMed]
65. Kavishe, R.A.; Koenderink, J.B.; Alifrangis, M. Oxidative stress in malaria and artemisinin combination therapy: Pros and Cons. *FEBS J.* **2017**, *284*, 2579–2591. [CrossRef] [PubMed]
66. Scott, M.D.; Eaton, J.W. Thalassaemic erythrocytes: Cellular suicide arising from iron and glutathione-dependent oxidation reactions? *Br. J. Haematol.* **1995**. [CrossRef] [PubMed]
67. Bayer, S.B.; Low, F.M.; Hampton, M.B.; Winterbourn, C.C. Interactions between peroxiredoxin 2, hemichrome and the erythrocyte membrane. *Free Radic. Res.* **2016**. [CrossRef] [PubMed]
68. Voskou, S.; Aslan, M.; Fanis, P.; Phylactides, M.; Kleanthous, M. Oxidative stress in β-thalassaemia and sickle cell disease. *Redox Biol.* **2015**, *6*, 226–239. [CrossRef]
69. Zhang, S.; Gerhard, G.S. Heme activates artemisinin more efficiently than hemin, inorganic iron, or hemoglobin. *Bioorganic Med. Chem.* **2008**. [CrossRef]
70. De Franceschi, L.; Bertoldi, M.; Matte, A.; Santos Franco, S.; Pantaleo, A.; Ferru, E.; Turrini, F. Oxidative stress and β -thalassemic erythroid cells behind the molecular defect. *Oxid. Med. Cell. Longev.* **2013**. [CrossRef]
71. Tsamesidis, I.; Perio, P.; Pantaleo, A.; Reybier, K. Oxidation of erythrocytes enhance the production of Reactive species in the presence of Artemisinins. *Int. J. Mol. Sci.* **2020**, *21*, 4799. [CrossRef]
72. Baluom, M.; Grossbard, E.B.; Mant, T.; Lau, D.T.W. Pharmacokinetics of fostamatinib, a spleen tyrosine kinase (SYK) inhibitor, in healthy human subjects following single and multiple oral dosing in three phase I studies. *Br. J. Clin. Pharmacol.* **2013**. [CrossRef]
73. Pritchard, E.M.; Stewart, E.; Zhu, F.; Bradley, C.; Griffiths, L.; Yang, L.; Suryadevara, P.K.; Zhang, J.; Freeman, B.B.; Guy, R.K.; et al. Pharmacokinetics and efficacy of the spleen tyrosine kinase inhibitor R406 after ocular delivery for retinoblastoma. *Pharm. Res.* **2014**. [CrossRef]
74. Doerig, C.; Abdi, A.; Bland, N.; Eschenlauer, S.; Dorin-Semblat, D.; Fennell, C.; Halbert, J.; Holland, Z.; Nivez, M.P.; Semblat, J.P.; et al. Malaria: Targeting parasite and host cell kinomes. *Biochim. Biophys. Acta Proteins Proteom.* **2010**, *1804*, 604–612. [CrossRef] [PubMed]

© 2020 by the authors. Licensee MDPI, Basel, Switzerland. This article is an open access article distributed under the terms and conditions of the Creative Commons Attribution (CC BY) license (http://creativecommons.org/licenses/by/4.0/).

Review

Non-Coding RNAs as Sensors of Oxidative Stress in Neurodegenerative Diseases

Ana Gámez-Valero [1], Anna Guisado-Corcoll [1,†], Marina Herrero-Lorenzo [1,†], Maria Solaguren-Beascoa [1,†] and Eulàlia Martí [1,2,*]

1 Department de Biomedicina, Facultat de Medicina i Ciències de la Salut, Institut de Neurociències, Universitat de Barcelona, C/Casanova 143, 08036 Barcelona, Spain; a.gamez@ub.edu (A.G.-V.); annaguisado@ub.edu (A.G.-C.); marina.herrero@ub.edu (M.H.-L.); m.solaguren-beascoa@ub.edu (M.S.-B.)

2 Centro de Investigación Biomédica en Red de Epidemiología y Salud Pública (CIBERESP), Ministerio de Ciencia Innovación y Universidades, 28046 Madrid, Spain

* Correspondence: Eulalia.marti@ub.edu

† These authors contributed equally in this work.

Received: 29 September 2020; Accepted: 6 November 2020; Published: 8 November 2020

Abstract: Oxidative stress (OS) results from an imbalance between the production of reactive oxygen species and the cellular antioxidant capacity. OS plays a central role in neurodegenerative diseases, where the progressive accumulation of reactive oxygen species induces mitochondrial dysfunction, protein aggregation and inflammation. Regulatory non-protein-coding RNAs (ncRNAs) are essential transcriptional and post-transcriptional gene expression controllers, showing a highly regulated expression in space (cell types), time (developmental and ageing processes) and response to specific stimuli. These dynamic changes shape signaling pathways that are critical for the developmental processes of the nervous system and brain cell homeostasis. Diverse classes of ncRNAs have been involved in the cell response to OS and have been targeted in therapeutic designs. The perturbed expression of ncRNAs has been shown in human neurodegenerative diseases, with these changes contributing to pathogenic mechanisms, including OS and associated toxicity. In the present review, we summarize existing literature linking OS, neurodegeneration and ncRNA function. We provide evidences for the central role of OS in age-related neurodegenerative conditions, recapitulating the main types of regulatory ncRNAs with roles in the normal function of the nervous system and summarizing up-to-date information on ncRNA deregulation with a direct impact on OS associated with major neurodegenerative conditions.

Keywords: oxidative stress; neurodegeneration; ncRNA; miRNA; tRNA fragments; lncRNA; circRNA

1. Introduction

Neurodegenerative disorders are a heterogeneous class of diseases characterized by a slow, progressive neuronal dysfunction and loss. The etiology of this class of diseases has not been fully elucidated, with highly complex and multifactorial causes participating in disease onset and progression. Nevertheless, several lines of evidence suggest that oxidative stress (OS) has a central role in this type of disorders. OS is the result of an imbalance between the increase in reactive oxygen species (ROS) and the production of antioxidants to eliminate them. A variety of mechanisms involving oxidative damage have been described in neurodegenerative conditions, including mitochondrial dysfunction, the oxidation of nucleic acids, proteins and lipids, activation of glial cells, apoptosis, cytokines production and inflammatory responses. The identification of molecular modifiers of the endogenous antioxidant activity is key to boost novel therapeutic designs [1].

The advent of high-throughput sequencing technologies revealed that the human genome is pervasively transcribed; with a large proportion corresponding to non-coding RNA (ncRNA) transcripts, the most abundant class of RNAs in cells [2,3]. ncRNAs have been classified according to their length and activity [4] with many classes revealed as essential regulators of gene expression through a variety of mechanisms [5].

ncRNAs are especially abundant and diverse in the central nervous system (CNS), showing a highly specific and dynamic temporal and spatial expression pattern that guides neurodevelopmental processes, including neural stem cell proliferation and differentiation [6,7]. In addition to their role in developmental processes, ncRNAs are key for the maintenance and normal functioning of adult, post-mitotic neurons. Most studies have been focused on microRNAs (miRNAs), the best-known class of ncRNAs. Although the roles for a progressively increasing number of ncRNAs are continuously being uncovered, the biological functions of the vast majority remain unknown.

While the precise expression levels of ncRNAs are key for the normal function of the CNS, the deregulation of ncRNA pathways have been linked to human disease, including neurodegenerative conditions [8–11]. In the present manuscript, we will provide an overview on the functional association between ncRNAs and the response to OS, a major hallmark in neurodegenerative disorders, and we will review the OS–ncRNA axis in major neurodegenerative conditions.

2. Oxidative Stress in Neurodegenerative Diseases

2.1. Reactive Oxygen Species Production and Activities

Reactive oxygen species (ROS), including hydrogen peroxide (H_2O_2) and free radicals such as superoxide anion (O_2^-) and hydroxyl radical (OH^-), are the consequence of aerobic metabolism and endogenously produced as by-products of diverse enzymatic reactions, including those of the mitochondrial respiratory chain. During mitochondrial electron transport, O_2 reduction results in superoxide ($O_2^{\bullet-}$) that is detoxified to H_2O_2 by the mitochondrial manganese superoxide dismutase, and H_2O_2 can be converted to the hydroxyl radical (OH^-). Another important source of ROS is the β-oxidation occurring in peroxisomes, whose main product is H_2O_2 [12,13]. Endogenously, ROS are generated following immune cell activation and ischemia or in ageing.

ROS can be also produced by non-enzymatic reactions, as those initiated by ionizing radiation, or as a result of the interaction between redox-active metals and oxygen species. Free iron (Fe^{2+}) reacts with H_2O_2 (Fenton reaction) generating very reactive and damaging hydroxyl radicals. Other chemical reactions involving superoxide lead to Fe^{2+} production, which in turn affects redox recycling [14]. Calcium, which is an important signaling molecule, is indirectly involved in the generation of ROS by enhancing the electric flow into the respiratory chain and by stimulating the nitric oxide synthetase that produces nitric oxide. In addition, exogenous sources of ROS exist, including air pollution, cigarette smoke or heavy metals. These exogenous products are incorporated by cells and metabolized into free radicals [15,16].

At low concentrations, ROS can serve as relevant second messengers in cell signaling and are vital to human health. Examples of activities regulated by ROS at low to moderate concentrations include the support of cell proliferation and survival pathways in host defense cells [17] and the regulation of ATP production through the activation of uncoupling proteins [18]. Nevertheless, the increased and sustained production of ROS induces a pathological chain reaction that results in damage to DNA, lipids and proteins, overall producing progressive cell dysfunction [19,20]. To maintain physiological redox balance, cells have endogenous antioxidant defenses regulated at the transcriptional level by Nrf2/ARE [21]. The imbalance between the formation of ROS and the antioxidant capacity of cells to detoxify ROS or repair the resulting damage may eventually lead to many chronic diseases, such as atherosclerosis, cancer, diabetics, and neurodegenerative diseases.

2.2. *Oxidative Stress in Neurodegeneration*

Neurodegenerative diseases are characterized by progressive and chronic cell dysfunction and death that frequently starts affecting specific regions of the central nervous system (CNS), worsening over time and impacting more regions in a predictable fashion. Neuronal loss is often associated with protein misfolding and aggregation both in the extracellular space and intracellularly, in specific cells (Figure 1). While many neurons cope with sustained ROS increases, there are particular cell populations that show varying degrees of susceptibility to OS, a phenomenon that has been called differential vulnerability [22].

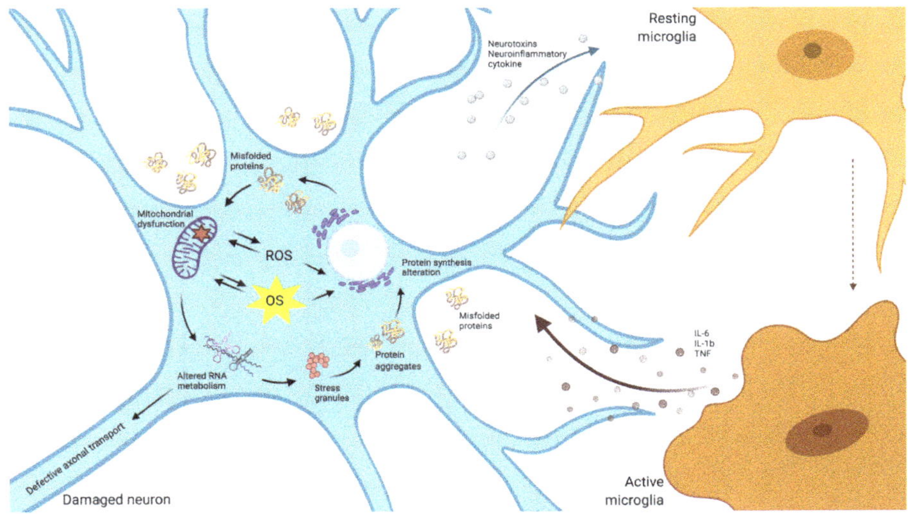

Figure 1. Oxidative stress in neurodegenerative disorders. Dysfunction in neural protein synthesis and processing entails mitochondrial defects, the alteration of RNA metabolism and the deposition of protein aggregates, which ultimately, encompass the generation of ROS and OS. At the same time, ROS production implies the alteration of protein folding and the production of neurotoxins which activates neuroimmune cells and enhances the production of neuroinflammatory cytokines.

Although differential vulnerability is influenced by the etiology of each neurodegenerative condition; shared unique features generally render neuronal cells particularly susceptible. Compared to other cells, neurons are more dependent on mitochondrial oxidative phosphorylation, to fulfill their high-energy demands. Neurons are exposed to high O_2 concentrations and consume 20% of the body oxygen [23]. About 1–2% of the consumed O_2 is converted into ROS, with this percentage dramatically increasing with age [24]. In addition, neurons are enriched in metal ions, which are accumulated along aging and catalyze ROS formation; and are also rich in fatty acids that are prone to oxidation. This is accompanied by a decreased concentration of antioxidant enzymes in neurons compared with other cells and tissues, resulting in impaired antioxidant capacities in the brain [25].

OS is considered an important age-dependent factor. During the process of ageing, the capacity of neurons to counteract ROS accumulation diminishes, rendering them more susceptible to neurodegenerative conditions such as Alzheimer's disease (AD) and Parkinson's disease (PD) [26]. The excess of ROS produces the oxidation and misfunction of biomolecules [21,27,28]. The ROS mediated oxidation of amino-acid residues in proteins results in the introduction of carbonyl groups, which is a good estimate of OS extent [29–31]. ROS also produce lipid peroxidation and peroxy-radicals initiate a chain reaction leading to the formation of breakdown products such as 4-hydroxy-2, 3-nonenal (HNE) [32]. HNE can modify proteins and produce detrimental effects, including the dysregulation of intracellular calcium signaling which ultimately induces an apoptotic cascade in neurons [33–35]. OS damage in DNA involves hydroxylation, carbonylation and nitration [36,37] and is characterized by increased levels of 8-hydroxy-2-deoxyguanosine and 8-hydroxyguanosine [38,39]. OS can produce DNA strand breaks, which agrees with increased carbonyls in the nuclei, in neurodegenerative diseases. A non-enzymatic reaction of sugars with protein deposits produces advanced glycation end-products that are neurotoxic and pro-inflammatory molecules [40,41]. The accumulation of these products in neurodegenerative diseases is caused by an accelerated oxidation of glycated proteins [42].

The aggregation of misfolded proteins forming high-ordered insoluble fibrils is common to diverse neurodegenerative diseases and participates in neurotoxicity [43], although the underlying mechanism is still uncertain [44]. Abnormal folding is thought to be influenced both by genetic and environmental factors, with compelling evidence supporting a tight effect of OS [45]. Proteins modified by accumulated ROS form aggregates, which inhibit the proteasome, the major machinery for the removal of misfolded and aberrant proteins [46]. The normal function of the proteasome is essential for the timely removal of oxidized proteins and the maintenance of normal cell homeostasis, but the disruption of this process leads to protein accumulation and cell death [47,48].

Finally, the generation of ROS is regulated by redox-sensitive metals and therefore the disruption of metal homeostasis in the brain may also lead to the uncontrolled formation of ROS. Specifically, increased exposure to metals like iron, copper and zinc has been associated to neurodegenerative conditions through the disruption of the mechanisms that regulate their activity.

3. Regulatory Non-Coding RNAs in the Normal Function of the Nervous System

The human genome is almost transcribed in its entirety, from both strands-sense and reverse-, in a very tissue-, cell-, organ-environmental-specific way. Nevertheless, as descried in the Human Genome Project, only the 1–5% of it is transcribed into protein-coding genes [49]. The phase III (2012–2017) of the ENCODE (Encyclopedia of DNA Elements) project identified around 20,000 protein-coding genes and almost 40,000 are non-coding genes transcribed into ncRNA [50]. The increasing number of functional ncRNA in complex organisms has been proposed as the answer to the inconsistent correlation between the complexity of an organism and the number of protein-coding genes [51].

NcRNAs are classified as *housekeeping* RNAs and *regulatory* ncRNAs. Housekeeping ncRNA are highly expressed constitutively and ubiquitously and play crucial roles in routine cell maintenance; they include transfer RNAs (tRNAs), ribosomal RNAs (rRNAs), small nuclear RNAs (snRNAs) and small nucleolar RNAs (snoRNAs). Regulatory ncRNAs, which are the focus of the present review, are characterized by a highly regulated and specific expression pattern, by their activity as transcriptional and post-transcriptional gene expression modulators and their high sensitivity to cell needs [52]. Regulatory ncRNAs can be classified based on their length as long non-coding RNAs (lncRNAs; >200 nucleotides) and short/small non-coding RNAs (sncRNAs; <200 nucleotides, usually between 18 and 40 nt) [11]. Discoveries over the last 40 years have elucidated diverse functions for ncRNAs involving the regulation of mRNA expression stability and translation, including splicing regulation, chromosome maintenance and segregation, and even the transmission of regulatory signals.

In particular, RNA biology is of paramount significance in the CNS. The development of CNS and brain morphogenesis is a very complex spatiotemporal-organized process and requires an exquisite regulation of stem cell proliferation and differentiation. The brain is the organ showing the highest diversity of regulatory ncRNAs [7,53,54], and their dynamic expression and activity underlies CNS complex functions including neurodevelopment, synapse function, cell polarity and maturation, and the responses to intra- and extracellular stimuli [6]. In this section, we outline the role of the main classes of regulatory ncRNAs in the development and correct function of the nervous system. We have focused on main regulatory ncRNAs with (1) reported roles in OS management and (2) perturbed activity in neurodegenerative conditions.

3.1. microRNAs

microRNAs (miRNAs, ~21 nt), the most studied class of regulatory ncRNAs, were discovered in 1993 in *C. elegans* [55]. They are one of the main negative post-transcriptional regulators of gene expression. This regulation is achieved by the direct, incomplete, base pair targeting of mRNAs, which results in the recruitment of diverse RNA-decay factors, or in translation repression.

miRNAs can be generated from both intronic and exonic regions of the genome. Transcribed by RNA polymerase II as long double-stranded primary miRNAs, they are processed by Drosha RNase III enzyme and DGCR8 protein in the nucleus. Exported into the cytosol as pre-miRNA, RNase III Dicer-1 processes it into a mature ~21 nt double stranded RNAs. After strands' separation, the so-called guide strand will associate to the Argonaute (AGO) proteins forming the RNA induced silencing complex (RISC), the main player in gene translation repression by miRNAs [56,57].

Although miRNA expression and correct biogenesis is important in the whole organism, they appear as indispensable molecules for CNS development and adult brain function [58,59]. In fact, the depletion of miRNA biogenesis-related proteins, Dicer-1 or DGCR8, strongly affects embryonic and neural development [58,60–62]. Chmielarz and colleagues recently reported that the specific removal of Dicer in dopaminergic neurons leads to the progressive degeneration of these cells in mice striatum and consequent locomotor affectation [63]. In the same work, the authors stimulated miRNA biogenesis using enoxacin-Dicer stimulator—in Dicer-KO cultured cells obtaining a neuroprotective effect and rescuing neuron viability [63]. miRNAs also orchestrate neural signaling processes including neurotransmitter release and dendrites and spines generation. During synaptogenesis, local mRNA translation and protein synthesis are mediated through miRNA regulation (reviewed by Hu and Li 2018 [64]). Specifically, diverse evidence points to miR-137 as an important regulator of synaptic function and pre-synaptic protein expression [65–67]. Together with miR-146a-5p, miR125b and miR-223, miR-137 also controls post-synaptic responses as they target different glutamate receptors [68,69]. Deep sequencing analysis revealed that miR-135 and miR-191 are involved in spine remodeling and plasticity [70]. Among others, miR-16-5p, miR-134 and miR-132 are pivotal for dendritic growth [71–74].

miRNAs have likewise been studied in glial cells. Several works on murine cells and animal models have reported the role of specific miRNAs in gliogenesis [75]. The specific deletion of Dicer1 in mouse oligodendrocyte precursors revealed the role of miR-219 in their differentiation [76]. Additionally, other miRNAs such as miR-125b and miR-124 are implicated in cerebral immune response through the regulation of glial cell proliferation and microglial polarization [77,78], as reviewed by Guo et al. [79].

3.2. tRNA-Derived Fragments

tRNAs-derived fragments (tRFs) are being revealed as a new class of regulatory ncRNAs. tRFs (<40 nt) are bioactive molecules classified according to their size and the part of the mature tRNA from where they derive, as previously reviewed in Kumar et al. [80]. tRFs can be constitutively produced and/or induced as a result of cellular stress response pathways. Diverse nucleases are responsible for the biogenesis of tRFs in vertebrates including RNase Z, Dicer and angiogenin (ANG). Under stress conditions, ANG produces a subtype of tRNA halves (tRNA-derived stress-induced RNAs or tiRNAs) [81], and Dicer is involved in the biogenesis of tRFS derived from the D loop or the 3′-end of mature tRNAs [82]. However, the role of Dicer in tRF biogenesis may be limited, as its depletion has little impact on tRF levels in multiple model organisms [83].

tRFs are highly abundant in the brain, where they accumulate during ageing [84,85], pointing these species as risk factors in neurodegenerative conditions. Furthermore, tRF expression dynamics have been associated to stem cell differentiation [86], denoting that the biogenesis of specific tRFs is an indicator of cell status and participates in biological processes associated to particular physiological or pathological conditions.

Although tRFs' heterogenous functions are not entirely understood, they regulate mRNA processing and translation, controlling processes such as cell growth and differentiation [87]. The discovery of stable tRFs–AGO complexes led to the hypothesis that tRFs could work as miRNA inhibiting gene expression. Recently, a tRF derived from a tRNA–Gly with miRNA structural and functional features, was shown to bind to AGO and repress the expression of the Replication Protein A1 (RPA1) gene [88]. However, a recent large-scale meta-analysis of CLASH sequencing data (cross-linking ligation and sequencing of hybrids) [89] suggests a more complex scenario, since tRF-RNA pairing in AGO1 occurs in the 3′-UTR, 5′-UTR and intronic regions of genes. tRFs can also regulate gene expression by competitively binding to AGO proteins, thus affecting the silencing activity on target genes [90].

Furthermore, tRFs repress translation through binding to ribosomal factors, emphasizing their implication in multiple layers of gene expression regulation [91]. Specific tiRNAs, derived from tRNA-Ala and tRNA-Cys, have been shown to inhibit protein translation by assembling into a G-quadruplex-like structure (G4-motif). G4–tRFs displace eIF4G/A from caped and uncapped mRNAs, inhibit translation initiation, and induce the assembly of stress granules, which are dense cytosolic aggregates where RNA-binding proteins control the utilization of mRNA during stress [92]. However, G4-tRFs do not impair translation from internal ribosomic entry sites (IRES), underlying survival and apoptosis pathways [93,94]. tiRNAs cooperate with the translator inhibitor YB-1 to promote stress granules' formation [95], promoting survival under acute stress conditions in U2OS or MCF7 cells. These studies highlight a mechanism by which the OS-induced tRNA cleavage inhibits protein synthesis and activates a cytoprotective stress response program. Nevertheless, while these studies have been performed in cell lines undergoing acute stress, the role of tiRNAs and stress granules in neuronal pathology involving chronic stress needs to be clarified.

3.3. Long Non-Coding RNAs

The ENCODE project established the existence of 9000–50,000 lncRNAs in the human genome. Transcribed from intergenic regions, from gene regulatory regions and from specific chromosomal regions by RNA polmymerase II, they are usually subjected to capping and alternative splicing [7,96]. lncRNAs' main functions encompass the control of gene expression at different levels—both transcriptional and translational—the control of chromatin folding and the recruitment of chromatin modifiers [97,98]. Due to this wide repertoire of functions, lncRNAs can be located in the nucleus maintaining the nuclear architecture, in the cytoplasm regulating gene expression and in the mitochondria where they participate in signaling and regulate cell-energy and apoptosis [99,100]. The function of lncRNAs is associated with their unique subcellular localization [101], but how cells sort different lncRNAs to specific subcellular compartments remains unclear. Furthermore, although several lncRNAs have been well characterized to date, further research is still required to understand their myriad of functions.

Forty percent of described lncRNAs have been reported in the CNS [102], where they participate in gene expression, imprinting and pluripotency regulation, as reported by numerous studies [103–106]. One of the first described lncRNAs was H19, a paternally imprinted gene, whose importance in embryonic development and neural viability has been established [107]. Furthermore, H19 is involved in glia activation during epilepsy [108]. Other lncRNAs, such as Evf2 and MALAT1, are highly abundant in neurons showing a dynamic expression pattern in cell differentiation. Evf2 regulates neural differentiation [109] and MALAT1 plays an essential role in synapse formation and dendritic development in hippocampal neurons [104]. NeuroLNC, a nuclear lncRNA, has been recently involved in synapsis establishment. Its interaction with the well-known RNA-binding protein TDP-43 appears to be essential for the correct release of neurotransmitters [106]. In addition, RMST1 functions as a guide-RNA for SOX2 transcription factor towards neurogenic genes [110] promoting neural differentiation although it is also expressed in astrocytes [111]. Similarly, Pnky lncRNA, a highly conserved lncRNA, is very abundant in neural stem cells, regulating alternative splicing and cell fate decisions. Ramos and colleagues described it as essential for neural cell differentiation and renewal [112]. Another highly conserved lncRNA, Cyrano, is as well enriched in the nervous system, being a very important molecule for embryonic stem cells maintenance. Kleaveland et al. determined a whole functional circuit for Cyrano in which it represses miR-7 expression and avoids CDR1-AS destruction, a circular RNA known to regulate neuronal activity [113]. On the other hand, NEAT1, a lncRNA mostly expressed in oligodendrocytes and astrocytes, regulates alternative splicing and has been related to the neural response to stress [99].

3.4. Circular RNAs (circRNAs)

Circular RNAs (circRNAs), are the only type of ncRNA with covalently linked ends. Usually derived from protein-coding genes and single-stranded, they are produced by a non-canonical form of splicing, the so-called back-splicing [114,115]. This extraordinarily stable class of ncRNAs have been found in diverse species, from prokaryotes to eukaryotes, being expressed in a tissue-specific manner. circRNAs are generally found in the cytoplasm where they mostly act as miRNA sponges [116]. Other attributed circRNA functions comprise the regulation of riboproteins and RNA interaction, the transport of miRNAs inside the cells or the regulation of cell differentiation and proliferation [116].

Similar to other ncRNAs, circRNAs are over-represented in the nervous system, showing a differential expression pattern along neurodevelopmental stages and in the distinct brain regions. As recently reviewed somewhere else [117], the high expression of circRNAs during neurogenesis and nervous system development has been already reported [117,118] and their role in synaptic transmission has been widely explored. In particular, CDR1-AS, a miR-7 sponge, was analyzed in a mouse model where synaptic malfunction and an adverse sensorimotor effect was observed after *Cdr1as* KO [119]. Moreover, hippocampal neurons present an increased expression of circHomer1_a among other circRNAs, in their post-synaptic terminals, contributing to synapses homeostasis and function [118]. Another example of the important regulatory role of circRNAs is CircDYM, shown to regulate cerebral immune response by decreasing microglial activation [120].

4. NcRNAs and Oxidative Stress Management in Neurodegenerative Diseases

Diverse classes of ncRNAs are involved in OS signaling pathways. The general oxidation of RNAs, including the most abundant classes of ncRNAs (rRNAs and tRNAs) is an early event in major neurodegenerative diseases (reviewed in [121]). However, the understanding of the impact of oxidized rRNA on cell physiology is still very limited [122]. Other classes of housekeeping ncRNAs (snRNAs and snoRNAs) are also altered in response to ROS accumulation [123], with specific species modulating the cell response to OS in cancer [124]. Interestingly, defects in snoRNAs are associated with neurodevelopmental disorders such as Prader–Willi syndrome [125]. Moreover, several works have identified small RNA molecules generated through the processing of snoRNAs (sno-derived RNAs), specifically involved in gene silencing [126]. Nevertheless, although snRNAs and snoRNAs are altered in neurodegenerative conditions, their direct involvement in OS management and neuronal dysfunction needs further research efforts.

Compelling evidence exists involving the regulatory ncRNAs in OS-neurodegeneration pathways. Regulatory ncRNA expression is sensitive to ROS accumulation, with these perturbations contributing to neuronal dysfunction and microglial activation. miRNAs are the most intensively studied class, and different types are up- and downregulated in response to OS; however, analogous studies regarding other classes of ncRNAs are scarce. The dynamic changes in the concentration of specific ncRNA that target important genes for the control of ROS homeostasis and metabolism, together with the intrinsic differential susceptibility of cells to ROS-damage, underlie the cell capability to cope with sustained ROS production. This is particularly relevant in neuronal cells with increased susceptibility to OS as explained in the previous section.

Specific ncRNA pathways regulating the cell response to OS are similarly altered in diverse neurodegenerative conditions, which may underlie commonly perturbed pathways, including mitochondrial dysfunction and the spread of insoluble protein inclusions. However, the altered activity of ncRNAs specifically targeting disease-related genes provides a scenario highlighting the differences in the development and progression of the different neurodegenerative diseases (Figure 2, Table S1).

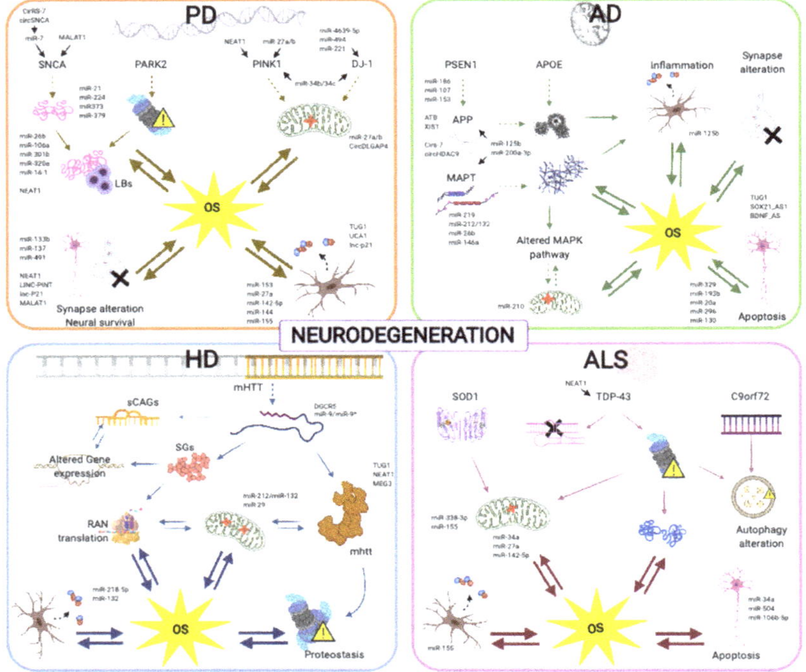

Figure 2. Regulatory ncRNAs are involved in oxidative stress (OS) management in diverse neurodegenerative diseases. Experimentally evidenced ncRNAs related to proteosomal and mitochondrial dysfunction, protein metabolism, reactive oxygen species (ROS) and OS generation and neural viability for Alzheimer's disease (AD), Parkinson's disease (PD), Huntington's disease (HD) and amyotrophic lateral sclerosis (ALS) are shown.

In this section, we will focus on regulatory ncRNA deregulation with direct implications in OS in the context of major neurodegenerative conditions: Alzheimer's disease (AD), Parkinson's disease (PD), Huntington's disease (HD) and amyotrophic lateral sclerosis (ALS). Most studies report miRNA alterations and we will provide the latest evidence for the involvement of lncRNAs, and other less known classes.

4.1. Alzheimer's Disease

AD is a progressive neurodegenerative disease associated with memory deficits and cognitive decline. AD is the most common form of dementia, accounting for 60 to 80% of cases, with less than half expected to be pure AD and the majority expected to be mixed dementias [127]. The main hallmarks of AD are the extracellular accumulation of amyloid-β (Aβ) peptides and the hyperphosphorylation of the microtubule-associated Tau protein, leading to the accumulation of senile plagues and neurofibrillary tangles, respectively. These changes are accompanied by progressive synaptic loss, and severe neuronal demise. OS plays a major role in AD, believed to be stronger than in other neurodegenerative diseases [128]. Despite the vast amount of knowledge about the pathology, the causes of AD development are still not clear. Mutations in PSEN1, PSEN2 and APP genes are associated to rare autosomal, dominant, early-onset AD; [129] and variants of the APOE gene are a strong genetic risk factor for both early-onset and late-onset AD [130].

Diverse classes of ncRNAs are involved in AD pathogenesis [131]. Numerous studies report the participation of miRNAs in AD, by regulating the target genes involved in the accumulation of Aβ and Tau phosphorylation [132–135]. In addition to these mechanisms, miRNAs regulate the cell response to OS, which is key in AD progression, causing early, chronic inflammation and underlying mitochondrial dysfunction, abnormal changes in lipids and proteins, the oxidative damage of nucleic acids and changes in gene expression [128]. Loops may be generated in which ncRNAs change their expression under OS and numerous proteins involved in OS management are targeted by miRNAs.

As previously mentioned, aging is a major risk factor in neurodegenerative conditions, where ROS are progressively accumulated [26]. In line with this, ageing-associated decreased levels of miR-186 may contribute to AD pathogenesis since miR-186 targets 3′UTR of BACE1 encoding the β-secretase enzyme that processes APP to Aβ peptides and protects from OS-pathogenesis in AD [136].

Through studying ROS-modulated miRNAs in oxidative stressed primary hippocampal neurons, a group of miRNAs (miR-329, miR-193b, miR-20a, miR-296, and miR-130b) was upregulated [137]. Analogous perturbations were found in the hippocampus of senescence-accelerated mouse strains, and enrichment pathway analysis of genes targeted by co-regulated miRNAs highlighted pathways analogously affected in AD, including apoptosis and MAPK signaling [138]. These data suggest an involvement of these miRNAs in the frame ncRNAs–OS–AD [137]. Specific miRNAs can also promote OS. For instance, miR-125b has been tagged as an important factor in AD promoting APP, BACE1 and Tau overexpression and hyperphosphorylation [139]. In a neuroblastoma cell model, the overexpression of miR-125b induced OS and stimulated apoptosis. Furthermore, supporting a link between miRNA deregulation, OS and neuronal dysfunction, miR-125b stimulated the production of pro-inflammatory cytokines [140].

In AD, ROS-generating soluble Aβ [141] induces the expression of miR-134, miR-145 and miR-210 and reduces the expression of miR-107 [142]. Decreased expression of miR-107, which targets BACE1 mRNA, is associated with early stages of AD [142]. Furthermore, carriers of APOE4 allele, show decreased levels of miR-107 [143,144] and a parallel increased production of Aβ peptides, which in turn produce Tp53 gene expression deregulation involved in cell death in neurodegenerative conditions, including AD [143]. Overall, these data suggest a positive feed-back loop in which the early decrease in miR-107 contributes to sustained ROS biogenesis in AD. In addition, miR-153 that targets APP and APLP2, an APP homologue, is decreased in a human neuronal cell line upon exposure to Aβ peptides and H_2O_2 [145], further linking ncRNAs with OS-pathology in AD.

ROS equally influences Tau phosphorylation and the associated neurofibrillary tangles accumulation in AD. Numerous studies correlate miRNA activity with Tau phosphorylation. For instance, miR-200a-3p reduces Aβ accumulation and Tau hyperphosphorylation by targeting BACE1 and PRKACB, respectively [146]. miR-219 is overexpressed in AD brains and in a neuroblastoma cell line, where decreased phosphorylated Tau is observed due to the miR-219 targeting of TTBK1 and GSK-3β [147]. Other miRNAs show an AD deregulation pattern that may contribute to Tau-phosphorylation pathology. miR-132-212 is amongst the top downregulated miRNAs in AD, and has been revealed as a master regulator of neuronal health, through a direct regulation of the Tau modifiers acetyltransferase EP300, kinase GSK3β, RNA-binding protein Rbfox1, and proteases Calpain 2 and Caspases 3/7 [148]. In addition, miR-132-212 directly targets Tau mRNA [133] overall suggesting that the deficiency of this miRNA contributes to neuronal dysfunction in AD and other tauopathies. Another miRNA involved in Tau metabolism is miR-26b which is upregulated in AD brains at early stages of the disease [149]. Induced miR-26b overexpression in rat brains resulted in increased Tau-phosphorylation and apoptotic neuronal death. Moreover, miR-26 blockage in cell culture was neuroprotective in front of OS [149]. Furthermore, miR-146a upregulation in AD may have detrimental consequences by targeting the ROCK1 kinase. ROCK1 decreased activity is associated with the inhibition of phosphorylation of the phosphatase and tensin homolog (PTEN), a pathway involved in neuronal Tau hyperphosphorylation in AD [150].

tRFs are another type of regulatory small ncRNAs included in the present review, with emerging roles in the physiology and pathology of the nervous system. There is little information directly correlating the activity of specific tRFs in AD pathology. However, 13 tRFs are deregulated in the senescence-accelerated mouse prone 8 (SAMP8) as an AD model and other aging related neurodegenerative diseases [85]. The pathway enrichment analysis of the mRNAs putatively targeted by deregulated small tRFs highlighted brain functions, such as vesicle synapse formation and vesicle cycle. In addition, tRFs accumulate during neuronal OS [151], a major contributor to AD pathology. These results, the fact that tRFs increase in aging in the nervous system [84,85] and the activity of specific species in the cell response to stress strongly suggests a role in AD pathology.

In addition to sncRNAs, the most recent studies show that the manipulation of diverse lncRNAs regulates the cell response to OS in cell and mouse models of AD, suggesting they should be considered in therapeutic developments. Mechanistically, lncRNAs modulate the expression of specific genes involved in cell survival and/or in the management of OS. Silencing SOX21-AS1 lncRNA has been proposed to alleviate neuronal OS and suppress neuronal apoptosis in AD mice through the upregulation of FZD3/5 and the subsequent activation of the Wnt signaling pathway [152]. Other studies have shown that silencing BDNF-AS [153] and ATB [154] lncRNAs in PC12 neuronal cell model attenuates apoptosis and OS induced by Aβ.

A common mechanism of action of lncRNAs is the specific sponging of miRNAs that results in their reduced activity. It has been shown that Aβ treatment increased the expression of XIST in hippocampal neurons. XIST knockdown ameliorated the toxicity, oxidative stress, and apoptosis induced by Aβ in hippocampal neurons through a mechanism involving miR-132 targeting [155]. In an Aβ-induced mouse model of AD, the increased expression of SNHG1 was detected, and its knockdown could reverse Aβ toxicity. Besides, SNHG1 sponged miR-361-3p, which targets ZNF217, suggesting that SNHG1 detrimental effects are the result of the deregulation of the miR-361-3p/ZNF217 axis [156]. Another recent study using an analogous AD mouse model showed that the lncRNA TUG1 that binds miR-15a was upregulated in response to Aβ intraventricular injection. TUG1 silencing and miR-15a upregulation improved the spatial learning ability and memory ability, ameliorated pathological injury, depressed neuronal apoptosis, and strengthened the antioxidant ability of hippocampal neurons in AD mice [157]. Although these studies point to a role of lncRNAs in OS-induced neurotoxicity, data on the deregulation pattern of the aforementioned lncRNAs in AD in humans are lacking and therefore the involvement of these species in AD evolution needs further investigation.

Through the regulation of miRNA availability, circRNAs provide a novel, little explored source of bioactive ncRNAs, whose dysregulation may impact neurodegenerative processes, including AD [158]. circRNA deregulation is detected in the cerebrospinal fluid of AD patients [158] and in AD cortices, showing associations with AD diagnosis and neuropathological severity [158]. Furthermore, changes were reported at pre-symptomatic stages in autosomal, dominant AD, suggesting a possible involvement in disease progression [159]. Specifically, ciRS-7 is downregulated in the grey matter of AD patients' cortex [160,161]. ciRS-7 sponges miR-7 and its downregulation results in the increased activity of miR-7. Because miR-7 targets Ubiquitin Ligase 2A (UBE2A) involved in the autophagic clearance of Aβ [162], it is tempting to speculate about the possible involvement of ciRS-7-miR-7-UBE2A pathway in Aβ metabolism and OS. ciRS-7 has been recently shown to regulate the levels of APP and BACE1 protein by promoting their degradation via proteasome and lysosome [163]. This would point to ciRS-7 being a neuroprotective. In addition, APP reduced the level of ciRS-7, revealing a mutual regulation of ciRS-7 and APP, which would explain the reduced ciRS-7 expression in AD [163]. Another recent study shows that the expression of circHDAC9, which is a miR-138 sponge is inversely correlated with that of miR-138 in Aβ-treated N2a cells and APP/PS1 mice. Decreased miR-138 results in excessive Aβ production induced by miR-138, in vitro. The authors show decreased levels of circHDAC9 in the serum of AD patients; suggesting that the pathway circHDAC9/miR-138 could participate in APP processing in AD [164].

4.2. Parkinson's Disease

PD is the most common movement disorder, affecting approximately 1% of the population over 60 years old and reaching 4–5% of the population over 85 years old. Clinically, PD is mainly described by motor symptoms, which include a tremor at rest, bradykinesia, stooped posture and a characteristic festinating walking. Non-motor symptoms such as depression, sleep disorders, and dementia precede the appearance of motor symptoms and worsen with progression of PD [165]. Pathologically, PD is defined by the loss of dopaminergic (DA) neurons in the substantia nigra (SN) pars compacta, corpus striatum and brain cortex [166]. This loss is accompanied by the presence of cytoplasmic protein inclusions, named Lewy bodies (LB), and enlarged aberrant Lewy neurites (LN). The causes leading to PD are not well understood. While genetic mutations in specific genes account for 5–10% of cases, in the majority of cases the etiology is unknown [167,168]. Familial and idiopathic forms of the disease share common pathogenic pathways, including mitochondrial dysfunction, energy production imbalance disruption of the ubiquitin–proteasome system and OS [166,169,170].

Major sources of OS in the highly vulnerable DA neurons include dopamine metabolism, the impairment of the antioxidant system, mitochondrial dysfunction and α-synuclein aggregation [170]. Excess of oxidized dopamine in the cytosol produces ROS, which was proposed to underlie selective vulnerability of DA neurons. miRNA-133b regulates cytosolic concentration of dopamine, by indirectly targeting the vesicular monoamine transporter 2 (VMAT2). miR-133b downregulates the transcription factor Pitx3, resulting in decreased VMAT2 expression [171]. The dopamine transporter DAT is another target of miR-133b that can regulate cytosolic dopamine concentration and ROS production in DA neurons [172]. Its downregulation in the PD midbrain suggests its possible participation in PD neuropathology and OS [173].

Diverse genes involved in the correct function of mitochondria are mutated in familial forms of PD, including DJ-1, Parkin and PINK1 [169]. However, altered levels or the activity of normal versions of these genes underlie mitochondrial dysfunction and OS also in sporadic, non-familial cases of PD. In this scenario, perturbations of ncRNAs that target these genes may contribute to PD neuropathology. Precisely, PINK1 protects neurons from OS [174] and its loss of function increases ROS biogenesis in dopaminergic neurons [175]. miR-27a and miR-27b target PINK1 and regulate the autophagic clearance of damaged mitochondria [176]. Furthermore, miR-27a regulates the expression of subunits of the mitochondrial complex [177], overall suggesting that their manipulation could be considered in therapeutic designs.

DJ-1 protein is an important player of mitochondrial activity and cell response to OS [178–180]. miR-4639 inhibits DJ-1 expression, and subsequent OS and neuronal death. Early miR-4639-5p upregulation has been recently shown in the plasma of PD patients [181], thus identifying this miRNA as a putative biomarker and a target for therapeutic development. Moreover, a recent study has shown that DJ-1 downregulation results in miR-221 decreased expression in mouse and cell models [182]. miR-221, with a relevant role in neurite outgrowth and neuronal differentiation, regulates the expression of diverse pro-apoptotic proteins and prevents OS-induced BIM expression. In addition, miR-221 protects neuronal DA cells knocked down for DJ-1 from MPP+-induced OS and cell death. Together, these data point to an important role of miR-221 perturbed expression in the neuronal response to OS. Additionally, the upregulation of miR-494 contributes to OS-induced neuronal death by inhibiting the expression of DJ-1 [183].

The downregulated expression of miR-34b and miR-34c has been likewise reported in PD brains, at pre-motor, early stages of PD [184]. The reduced activity of miR-34b and miR-34c using a complementary antisense oligonucleotide in a DA neuronal cell line resulted in ROS induction, mitochondrial dysfunction and cell death, and a parallel decreased expression of Parkin and DJ-1. The deregulation at pre-motor stages of PD brains can be extended to diverse classes of ncRNAs [185], suggesting that ncRNA perturbations occur early in the course of the disease, and participate in disease evolution.

In PD, deregulated miRNAs may also contribute to an impaired antioxidant system. OS induces the translocation of the transcription factor Nrf2 to the nucleus, where it activates the expression of genes involved in the response to OS. Keap-1 is an inhibitor of Nrf2 and is regulated by diverse miRNAs, including miR-153, miR-27a, miR-142-5p, and miR-144 and miR-7 [186–189]. Among them, miR-7 downregulation is of particular interest since it has been shown to be decreased in the substantia nigra of PD patients [187]. In PD, decreased miR-7 may contribute to the impaired expression of antioxidant genes, through the Keap-1, Nrf2 pathway.

A recent study showed that, in addition to miRNAs, diverse types of sncRNAs are deregulated in PD human brains [185]. Among the different classes, tRFs were revealed as the species optimally classifying pre-motor cases of PD versus control patients; suggesting that tRFs are early perturbed in PD. The mechanisms underlying the altered expression and activities of deregulated tRFs have not been elucidated. ANG is directly involved in the stress-induced biogenesis of tiRNAs [81]. Specific ANG mutations have been found in PD patients [190], and ANG is decreased in a α-syn mouse model of PD [191], suggesting that ANG dysfunction could contribute to tRFs perturbations. Mechanistically, ANG-generated tiRNAs have been proposed as protective [90]; however, the virally mediated overexpression of ANG in the substantia nigra failed to protect from DA neuronal loss in a neurotoxin-based mouse model of PD [192]. Other studies confirm a link between ANG-induced tiRNAs and the neuronal response to stress [193]. The lack of NSun2 that methylates two cytosine residues in tRNAs results in an increase in stress-induced ANG-mediated 5′-tiRNAs accumulation. The increase in these species contributes to the death of hippocampal and striatal neurons, through the inhibition of protein translation and cell stress [193]. These studies suggest a context-dependent effect of ANG–tiRNAs as neuroprotective or neurotoxic molecules. Moreover, a high-throughput functional analysis, that included antisense oligonucleotides targeting specific tRFs deregulated in PD brains, identified specific types involved in the response of dopaminergic SH-SY5Y cells to MPP+ OS [194]. In brains from motor cases of PD, a tRNA-Tyr derived fragment is upregulated [185]. This tRF sensitizes neuronal cells to OS-induced p53-dependent cell death [195], suggesting a possible contribution of TyR-tRF biogenesis to the neuronal dysfunction in PD.

On the other hand, numerous studies have recently highlighted the participation of lncRNAs in OS pathways in the context of PD. MPTP (1-methyl-4-phenyl-1,2,3,6-tetrahydropyridine) and paraquat, that damage the substantia nigra and are widely used to model PD [196], influence the activity of the transcription factor Nrf2 and alter the expression of diverse lncRNAs. In a rat model of PD induced by 6-OHDA, the lncRNA UCA1 was increased in the rat midbrain [197]. The siRNA-mediated decrease in UCA1 upregulated the expression of tyrosine hydroxylase (TH) positive cells, reduced apoptosis and OS in neurons of the substantia nigra and improved the neuroinflammatory response [197]. Several works report the role of MALAT1 lncRNA in apoptosis regulation in PD pathology. The downregulation of MALAT1 in MPP+-induced animal and cell models inhibited apoptosis through the sponging of miR-124 [198]. TUG1 that sponges miR-152-3p is increased in mouse and cell models of PD [199]. In these models, TUG1 silencing protected dopaminergic SH-SY5Y cells and neurons from OS and neuroinflammation. In cells submitted to MPP+ stress, TUG1 inhibition was accompanied by a suppression of PTEN and cleaved caspase-3 expression and an increase in TH. The protective effects of TUG1-decrease were reversed in the presence of a miR-152-3p inhibitor, suggesting that the miR-152-3p/PTEN pathway regulated TUG1 effects.

In addition, the linc-p21 is upregulated in PD [200], and highly expressed in SH-SY5Y cells undergoing MPP+ stress [201]. The knockdown of linc-p21 attenuated cytotoxicity, OS, cell apoptosis and neuroinflammation induced by MPP+ in this model. Moreover, the overexpression of linc-p21 produced opposite effects, overall suggesting that linc-p21 is an OS responsive and dosage sensitive lncRNA likely involved in PD. The linc-p21-regulated response to MPP+ depends on its activity as a sponge of miR-625, which in turn regulates TRPM2 expression. In the substantia nigra of PD patients and in diverse PD model systems, the neuron-specific lncRNA NEAT1 is also upregulated [202]. However, NEAT1 expression has been shown to be neuroprotective, since the interference-mediated depletion of NEAT1 aggravated death in the cells exposed to paraquat in a LRRK2-mediated manner. A study focused on identifying deregulated lncRNAs in the substantia nigra of PD patients described a pronounced increase in the Long Intergenic Non-Protein Coding RNA, p53-Induced Transcript, LINC-PINT, that was also observed in other models of PD and other neurodegenerative conditions [203]. Experimentally, the depletion of LINC-PINT exacerbated the death of cultured neuronal cells exposed to OS, suggesting a neuroprotective role of this transcript in diverse neurodegenerative disorders.

Studies about the possible involvement of circRNAs in PD are limited. In the SN, the age-dependent accumulation of circRNAs observed in healthy individuals was lost in PD individuals [204]. However, in the same brain area, increased levels of CircSLC8A1 were found, with this circRNA showing binding sites for miR-128. Hence, miR-128 targeted genes were increased in the PD SN, suggesting a regulation of miR-128 activity by CircSLC8A1. In addition, CircSLC8A1 expression was sensitive to OS status in culture cells, overall suggesting an OS-mediated deregulation of the CircSLC8A1/miR-128 pathway in PD. circRNAs have also been shown to be deregulated in multiple brain regions in the MPTP mouse model of PD [205]. The enrichment pathway analysis of deregulated circRNAs highlighted diverse biological functions, including synapse function, neuronal differentiation, axon guidance and PD metabolism. Specifically, the authors suggest that the mmu_circRNA_0003292-miRNA-132-Nr4a2 pathway can be involved in the regulation of pathogenic mechanisms [205]. Using the same mouse model of PD, a recent report has shown a decreased expression of circDLGAP4 upon MPTP treatment and an analogous deregulation pattern in MPP+ treated cell models [206]. In SH-SY5Y and MN9D cells, circDLGAP4 was neuroprotective, exerting its function via the regulation of miR-134-5p activity, which in turn regulates the cAMP response element-binding protein (CREB). The circDLGAP4/miR-134-5p axis modulates the CREB-dependent signaling which affects the expression of genes involved in neuronal survival, providing the mechanistic scenario for the protective activity of circDLGAP4 and the detrimental consequences of its decrease in PD. A direct role of ciRS-7 in PD has not been established. However, the ciRS-7 targeting of miR-7 activity results in increased levels of miR-7-regulated genes, including α-syn [161]. Another recently reported sponge of miR-7 is circSNCA [207]. In the MPP+ PD cell model, the dopamine receptor agonist pramipexole, used in PD treatment, downregulated circSNCA, resulting in an miR-7 increase, an α-syn decrease and reduced cell apoptosis. These results point to a better understanding of PD pathology and to possible therapeutic designs contemplating the manipulation of circRNA activity.

4.3. Huntington's Disease

HD is a dominant rare hereditary disorder (1:10,000–1:20,000), caused by an expanded CAG trinucleotide repeat in the exon 1 of the HTT gene (reviewed in McColgan et al. [208]). CAG repeat length correlates with age onset symptoms that include motor, cognitive psychiatric alterations, combined with sleep and circadian disturbances. A major neuropathological feature is the loss of striatal GABAergic medium spiny neurons and the cortical neurons projecting to them, accompanied by astrogliosis and microglia activation. The progressive atrophy of the striatum and cerebral cortex leads to patient death at 15–20 years from the disease onset [208,209].

The expanded CAG repeat produces an abnormal protein with an expanded track of glutamines that has been proposed to underlie a number of toxic effects [209,210]. In addition, the RNA has been recently highlighted as a direct pathogenic contributor, involving gene silencing, the sequestrations of proteins with affinity for the CAG repeat RNAs and repeat associated non-ATG (RAN) translation resulting in homopolymeric proteins prone to aggregation [211,212].

Strong miRNA deregulation is detected in brain samples of HD patients and HD mouse models. In humans, miRNAs and miRNA-variants (IsomiRs) are dysregulated both in the frontal cortex and putamen, the most affected areas in HD [213]. Although the effect of the isomiRs is uncertain, the perturbations of those involving the seed region of the miRNA may likely vary the gene targeting profile, thus increasing the complexity in the miRNA depending effects. The strong perturbations in miRNA expression described in HD could be explained, at least in part, by the accumulation of AGO2 in aggregates [214], through a mechanism involving impaired autophagy [215]. AGO2 is a major executor of the miRNA-mediated gene silencing, and its accumulation in HTT-aggregates results in abnormal miRNA activity and expression levels. Furthermore, reduced levels of Drosha and Dicer, involved in the biogenesis and activity of miRNAs, are downregulated in mouse models of HD [216], providing further evidences for a general dysregulation of miRNA pathways. Specifically, Packer et al. reported that miR-9/miR-9* is downregulated in HD, with this resulting in the mislocalization of the transcriptional repressor REST that controls the expression of neural genes and is a key mediator of the transcriptional changes occurring in HD [217]. Specific miRNAs are altered at pre-motor stages of the disease in a knock-in mouse model, indicating that miRNA perturbations are an early event [218]. This study showed CAG length-dependent miRNA expression changes in the brain, especially in the striatum, revealing a molecular signature that correlates with the onset of disease. In comparing mRNA and miRNA transcriptomic data from the same animals, the authors highlighted a number of miRNAs with potential relevance in the regulation of HD mRNA perturbations, including miR-212/miR-132, miR-218 and miR-128 [218].

HD-perturbed miRNAs have been shown to regulate the cell response to OS. For instance, the sequence-specific inhibition of miR212/miR132 induces apoptosis in cultured primary neurons, whereas their overexpression is neuroprotective against OS in neurodegenerative scenarios [219]. In addition, miR-218 downregulation protects from oxidative-glucose deprivation/re-oxygenation in PC12 cells, through reducing inflammatory cytokines secretion, OS status, and apoptosis rate [220]. Furthermore, a number of the HD-deregulated miRNAs have been identified as general regulators of the neuronal response to OS, in a high-throughput functional screening [194]. These data strongly suggest a contribution of miRNA perturbations in the OS management, linked to HD.

lncRNA involvement in OS in the context of HD has been little explored; nevertheless, specific lncRNAs are perturbed in HD and likely contribute to pathogenesis. An analysis of microarray data reported diverse deregulated lncRNAs, including DGCR5, a target of the transcriptional repressor REST [221]. In addition, the upregulation of TUG1 and NEAT1 and the downregulation of MEG3 are detected in the HD caudate [221]. The upregulation of the TUG1 and NEAT1 mouse orthologs has been validated in the brain of diverse HD mouse models [222]. The inhibition of NEAT1 and MEG3 results in a significant decrease in HTT aggregates and Tp53 expression inhibition in a cell model. Although an OS readout is not provided, alterations in Tp3 and protein aggregation have direct links with cell OS.

4.4. Amyotrophic Lateral Sclerosis

ALS is a progressive disorder characterized by the progressive degeneration of motor neurons in the brain and spinal cord leading to the loss of the voluntary control of movements. It begins with focal weakness but spreads inexorably to involve most muscles. Death due to respiratory paralysis typically occurs in 3 to 5 years. The total number of cases is approximately 3 to 5 per 100,000 (reviewed in [223]). The degeneration of motor neurons is accompanied by neuroinflammatory processes, with the activation and proliferation of astroglia, microglia, and oligodendroglial cells [224,225]. SOD1 mutations are present in about 15 to 20% of families with ALS [226]. In addition, a hexanucleotide repeat expansion in the first intron/promoter region of the lncRNA C9ORF72 is the most common genetic cause of ALS, present in 40% of familiar cases [227]. Mechanistically, the RNA toxicity of C9ORF72-related ALS is based on the direct sequestration of important RNA binding proteins by the hexanucleotide repeat and RAN translation into dipeptide repeats.

Familial and sporadic ALS forms commonly show the aggregation of specific proteins in the cytoplasm, especially in motor neurons. These include the nuclear TAR DNA-binding protein 43 (TDP-43), showing cleaved forms, hyperphosphorylation, and mislocalization to the cytoplasm [228]. Ubiquilin 2 aggregates are also common [229], similar to intracytoplasmic deposits of wild-type superoxide dismutase 1 (SOD1) in sporadic ALS [230]. The SOD1 gene encodes the cytosolic copper- and zinc-dependent superoxide dismutase (Cu, ZnSOD) that converts O^{2-} into H_2O_2; however, several observations make it unlikely that motor neuron death occurs because of a loss of dismutase function [231] and it is generally accepted that mutant Cu, ZnSOD presents a novel, cytotoxic activity of unknown nature. However, diverse evidence supports the involvement of OS in ALS pathology (reviewed in [231]).

Studies of miRNAs in ALS have been centered in biomarker discovery [232–237] and miRNA activity manipulation for therapeutic designs [234,236,238,239]. For example, miR-27a, miR-34a, miR-155, miR-142-5p, and miR-338-3p have been proposed as biomarkers and potential ALS therapeutic targets, directly or indirectly involved in OS [234,236,238,239]. A study identified 15 deregulated miRNAs in ALS and induced pluripotent stem cells differentiated to motor neurons [235]. In this study, gene ontology and molecular pathway enrichment analysis indicated that the predicted target genes of the deregulated miRNAs are involved in neurodegeneration-related pathways. miR-34a and miR-504 are especially interesting as they are involved in the p53 pathway. Specifically, miR-34a targets SIRT1, previously reported as involved in neural survival in an ALS mouse model [240]. In addition, a microarray analysis identified the upregulation of miR-29a, miR-29b and miR-338-3p in ALS brains [241]. miR-338-3p deregulation is particularly interesting since it targets nuclear-encoded mitochondrial mRNAs encoding subunits of the oxidative phosphorylation machinery [242]. A general downregulation of miRNA levels is found in multiple forms of human ALS. ALS-causing mutations are sufficient to inhibit miRNA biogenesis. The underlying mechanism involves stress granule formation and perturbations in Dicer and AGO2 interactions with their partners. Enhancing Dicer activity with enoxacin appeared to be beneficial for neuromuscular function in ALS mouse models, suggesting that targeting miRNA activity in future therapeutic designs would be interesting [243].

ER stress leading to apoptosis has been strongly associated with motor neuron degeneration in ALS. Nrf2, involved in redox reactions, participates in ER stress-induced apoptosis and together with ATF4 repress miR-106b-25 cluster. It has been shown that the Nrf2/ATF4/miR-106b-25 cluster may be relevant for ALS pathogenesis [244]. In addition, miR-142-5p, that is downregulated in the CSF of ALS patients, has been shown to reduce OS via the upregulation of the Nrf2 signaling pathway [236,245].

The involvement of tRFs in ALS is supported by ANG mutations occurring in familial and sporadic forms of ALS [246]. ANG mutations are associated to a decreased ANG ribonuclease activity and impaired formation of protective 5′-tiRNAs that inhibit the translation initiation and trigger the assembly of stress granules, as explained in Section 3.2 [247,248]. In a YB-1-dependent manner, cells treated with 5′-tiDNA-Ala modestly, but significantly, enhanced motor neuron survival in response to OS [93]. These data suggest that ANG dysfunction and the impaired formation of specific tiRNAs are involved in motor neuron dysfunction. However, the effect of other tRFs formed in the ALS context needs to be evaluated in more detail. A recent study has defined a serum-based ncRNA diagnostic biomarker signature containing two tRFs [249]. Whether these changes are reflected in motor neurons and the functional consequences of its perturbed biogenesis needs further research.

The mutation of the lncRNA C9ORF72 in familial cases of ALS underlie OS and senescence in induced pluripotent stem cells differentiated to astrocytes [250]. Mutant astrocytes showed the downregulated secretion of several antioxidant proteins and, conditioned media from C9ORF72-astrocytes increased OS in wild-type motor neurons. Other lncRNAs including NEAT1 and NEAT1_2 have also been related to ALS. The binding of NEAT1_2 to TDP-43 and FUS/TL, which are ALS-associated RNA-binding proteins, is an early phenomenon in ALS pathogenesis [251]. Electron microscopy showed a specific assembly of NEAT1_2 lncRNA around the interchromatin granule-associated zone in the nucleus of ALS spinal motor neurons. These results suggested that the lncRNA NEAT1_2 may act as a scaffold of RNA binding proteins in the nuclei of ALS motor neurons, thereby modulating their functions during the early phase of ALS. NEAT1, has been recently shown to regulate TDP-43 function in the response to arsenite-induced stress [252]. The formation of TDP-43-nuclear bodies in response to stress alleviates TDP-43-mediated cytotoxicity. TDP-43 nuclear bodies partially colocalized with paraspeckles, in which NEAT1, which is upregulated in stressed neurons, acts as a scaffold. NEAT1 promotes TDP-43 separation in vitro. The authors provide evidence that ALS-associated TDP-43 mutation impairs the NEAT1-mediated TDP-43 separation, with this resulting in the translocation of TDP-43 to the cytoplasm to form stress granules. Overall, these data suggest a stress-protective role for TDP-43 nuclear bodies, regulated by NEAT1.

5. Conclusions

The above reviewed results emphasize the relevance of regulatory ncRNAs as bioactive molecules regulating the management of OS in neurodegenerative conditions. Nevertheless, we are aware that additional ncRNA classes present in the CNS can be affected by and associated to OS and neurodegeneration, including highly abundant rRNAs.

The overall conclusions from the works presented here are: (1) OS plays a major and early role in neurodegenerative diseases; (2) regulatory ncRNAs expression and activity dynamics regulate, directly and/or indirectly, gene expression networks involved in ROS homeostasis and buffering; (3) miRNAs are the best known class of ncRNAs, but other types of regulatory ncRNAs, including lncRNAs, tRFs, and circRNAs, appear as novel classes with a direct impact in OS modulation and neurodegeneration.

These evidences have boosted pre-clinical research, encouraging the modification of the activity of ncRNA in therapeutic designs to control OS [253,254]. This involves the use of anti-sense oligonucleotides targeting ncRNAs for their inhibition, or ncRNA mimic molecules for their activation. However, there is a number of experimental and technical considerations that need to be fully controlled for translational development, including the use of appropriate control sequences and the effective concentration of an oligonucleotide to be used for in vivo experiments. In addition, results obtained with mimic molecules should be interpreted with caution, since they are used at diverse orders of magnitude higher than normally observed in vivo, and can produce unspecific effects [255]. Experimental research to understand the functions of specific ncRNAs in OS, particularly miRNAs, has been done in neuronal or glial cell lines and non-human animal models of neurodegenerative diseases. However, cell-type-specific ncRNA profiles in immortalized, primary cultured cells or animal models differ [256], and analogous differences may be found for the ncRNA-targeted genes.

In addition, caution should be taken in translating the results from cell and animal models to the human context. Many neuroprotective treatments successfully working in rodent and primate models, have failed in human patients [257]. Furthermore, due to the heterogeneity and complexity of neurodegenerative disorders, animal models reproduce particular features of the disease process but often fail to fully recapitulate neuropathology, including the characteristic spatial pattern of neuronal loss [258]. While the manipulation of specific ncRNAs in these models produces a particular OS readout, analogous results in the human cells of the nervous system should be validated. Evolutionary conserved ncRNAs may analogously regulate OS pathways in human cells. However, the human nervous system contains the largest diversity of ncRNAs, with many human-specific species showing a highly precise expression pattern in particular neuronal and glial cell types [259–261]. Moreover, the functional interaction between different classes of ncRNAs (for instance lncRNAs or circRNAs buffering specific miRNAs) provides an additional layer of complexity, in which multiple ncRNA-signaling pathways orchestrate the response to OS.

Patient-derived induced pluripotent stem cells (iPSCs) differentiated to specific neuronal populations have been implemented as a human cell model to study biochemical and molecular aspects of neurodegeneration [262,263]. However, differentiation protocols are challenging [262] and cultured neurons in artificial media may present differences in ncRNA expression and the regulation of OS. Three-dimensional human brain organoids offer a hopeful strategy to study OS and neurodegeneration since they recapitulate the interaction between different cell types, including glial cells known to modulate the effects of OS [264]. The development of experimental strategies to reliably monitor levels of oxidative damage in human iPSCs-derived cell models or brain organoids [265] should help in understanding the temporal dynamics of ncRNA activity modulation, OS, mitochondrial dysfunction, protein aggregates and cell survival.

Studies aimed at understanding the involvement of ncRNAs in neurodegeneration have focused on (1) identifying disease-deregulated ncRNAs and (2) reproducing the deregulation pattern in cell and/or animal models to evaluate possible detrimental consequences. The identification of ncRNA-deregulated species at pre-symptomatic stages is particularly interesting, as their perturbed activity can be an early pathogenic driver that may be targeted in therapeutic designs. Nevertheless, the majority of these studies are focused on a particular ncRNA, and do not take into account that multiple species are functionally altered in neurodegenerative diseases. This raises the challenge of targeting the activity of multiple ncRNAs to better protect from OS damage.

Another layer of activity of ncRNAs is produced by their release to the extracellular space both within vesicles, free or bound to RNA binding proteins, rendering these molecules as paracrine signaling mediators [266]. Although the majority of studies on the activity of extracellular ncRNAs have been centered in miRNAs, the most recent reports point to additional species as bioactive extracellular molecules, including lncRNAs and tRFs. In neurodegenerative diseases, the profiling of extracellular ncRNAs in biofluids has been pointed as a new source of biomarkers [267,268]. However, a new field of study is required to understand the modulation of ncRNA release and their regulatory activities in different aspects of neuropathology, including OS management.

Supplementary Materials: The following are available online at http://www.mdpi.com/2076-3921/9/11/1095/s1, Table S1. List of ncRNAs involved in OS pathways in the context of neurodegenerative conditions. Experimental works proving the involvement of ncRNAs in oxidative stress along AD, PD, HD and ALS are exposed.

Author Contributions: A.G.-V. wrote the manuscript, designed both figures and revised the table; A.G.-C., M.H.-L. and M.S.-B. designed and completed the table; and E.M. conceived and wrote the manuscript, and revised both the table and figure design. All authors revised the manuscript prior to submission and approved the final version. All authors have read and agreed to the published version of the manuscript

Funding: This work was supported by the Spanish Ministry of Economy and Competitiveness and FEDER funds (SAF2017-88452-R).

Acknowledgments: We acknowledge the support of the Spanish Ministry of Science Innovation and Universities, Maria Maeztu Unit of Excellence Programme. Figures were created with BioRender.com.

Conflicts of Interest: The authors declare that they have no competing interests.

References

1. Yaribeygi, H.; Panahi, Y.; Javadi, B.; Sahebkar, A. The underlying role of oxidative stress in neurodegeneration: A mechanistic review. *CNS Neurol. Disord.-Drug Targets* **2018**, *17*, 207–215. [CrossRef] [PubMed]
2. Feingold, E.A.; Good, P.J.; Guyer, M.S.; Kamholz, S.; Liefer, L.; Wetterstrand, K.; Collins, F.S.; Gingeras, T.R.; Kampa, D.; Sekinger, E.A.; et al. The ENCODE (ENCyclopedia of DNA Elements) project. *Science* **2004**, *306*, 636–640.
3. ENCODE Project Consortium. An integrated encyclopedia of DNA elements in the human genome. *Nature* **2012**, *489*, 57–74. [CrossRef] [PubMed]
4. Nie, L.; Wu, H.-J.; Hsu, J.-M.; Chang, S.-S.; Labaff, A.M.; Li, C.-W.; Wang, Y.; Hsu, J.L.; Hung, M.-C. Long non-coding RNAs: Versatile master regulators of gene expression and crucial players in cancer. *Am. J. Transl. Res.* **2012**, *4*, 127–150. [PubMed]
5. Hombach, S.; Kretz, M. Non-coding RNAs: Classification, biology and functioning. *Adv. Exp. Med. Biol.* **2016**, *937*, 3–17. [CrossRef] [PubMed]
6. Salta, E.; De Strooper, B. Non-coding RNAs with essential roles in neurodegenerative disorders. *Lancet Neurol.* **2012**, *11*, 189–200. [CrossRef]
7. Qureshi, I.A.; Mehler, M.F. Emerging roles of non-coding RNAs in brain evolution, development, plasticity and disease. *Nat. Rev. Neurosci.* **2012**, *13*, 528–541. [CrossRef] [PubMed]
8. Wu, Y.-Y.; Kuo, H.-C. Functional roles and networks of non-coding RNAs in the pathogenesis of neurodegenerative diseases. *J. Biomed. Sci.* **2020**, *27*, 1–23. [CrossRef]

9. Ma, F.; Zhang, X.; Yin, K.-J. MicroRNAs in central nervous system diseases: A prospective role in regulating blood-brain barrier integrity. *Exp. Neurol.* **2020**, *323*, 113094. [CrossRef]
10. Duran, R.C.-D.; Wei, H.; Kim, D.H.; Wu, J.Q. Invited review: Long non-coding RNAs: Important regulators in the development, function and disorders of the central nervous system. *Neuropathol. Appl. Neurobiol.* **2019**, *45*, 538–556. [CrossRef]
11. Esteller, M. Non-coding RNAs in human disease. *Nat. Rev. Genet.* **2011**, *12*, 861–874. [CrossRef] [PubMed]
12. Dröge, W. Free radicals in the physiological control of cell function. *Physiol. Rev.* **2002**, *82*, 47–95. [CrossRef]
13. Beckman, K.B.; Ames, B.N. The free radical theory of aging matures. *Physiol. Rev.* **1998**, *78*, 547–581. [CrossRef]
14. Bush, A.I. Metals and neuroscience. *Curr. Opin. Chem. Biol.* **2000**, *4*, 184–191. [CrossRef]
15. Jan, A.T.; Azam, M.; Siddiqui, K.; Ali, A.; Choi, I.; Haq, Q.M.R. Heavy metals and human health: Mechanistic insight into toxicity and counter defense system of antioxidants. *Int. J. Mol. Sci.* **2015**, *16*, 29592–29630. [CrossRef]
16. Calderón-Garcidueñas, L.; Leray, E.; Heydarpour, P.; Torres-Jardón, R.; Reis, J. Air pollution, a rising environmental risk factor for cognition, neuroinflammation and neurodegeneration: The clinical impact on children and beyond. *Rev. Neurol.* **2016**, *172*. [CrossRef]
17. Harwell, B. Biochemistry of oxidative stress. *Biochem Soc Trans.* **2007**, *35 (Pt 5)*, 1147–1150.
18. Echtay, K.S.; Roussel, D.; St-Pierre, J.; Jekabsons, M.B.; Cadenas, S.; Stuart, J.A.; Harper, J.A.; Roebuck, S.J.; Morrison, A.; Pickering, S.; et al. Superoxide activates mitochondrial uncoupling proteins. *Nat. Cell Biol.* **2002**, *415*, 96–99. [CrossRef] [PubMed]
19. Halliwell, B. Free radicals, antioxidants, and human disease: Curiosity, cause, or consequence? *Lancet* **1994**, *344*, 721–724. [CrossRef]
20. Fulda, S.; Gorman, A.M.; Hori, O.; Samali, A. Cellular stress responses: Cell survival and cell death. *Int. J. Cell Biol.* **2010**, *2010*, 1–23. [CrossRef]
21. Adibhatla, R.M.; Hatcher, J.F. Lipid oxidation and peroxidation in cns health and disease: From molecular mechanisms to therapeutic opportunities. *Antioxid. Redox Signal.* **2010**, *12*, 125–169. [CrossRef] [PubMed]
22. Fu, H.; Hardy, J.; Duff, K.E. Selective vulnerability in neurodegenerative diseases. *Nat. Neurosci.* **2018**, *21*, 1350–1358. [CrossRef]
23. Smith, D.G.; Cappai, R.; Barnham, K.J. The redox chemistry of the Alzheimer's disease amyloid β peptide. *Biochim. Biophys. Acta (BBA)-Biomembr.* **2007**, *1768*, 1976–1990. [CrossRef]
24. Dinkova-Kostova, A.T.; Talalay, P.; Sharkey, J.; Zhang, Y.; Holtzclaw, W.D.; Wang, X.J.; David, E.; Schiavoni, K.H.; Finlayson, S.; Mierke, D.F.; et al. An exceptionally potent inducer of cytoprotective enzymes: Elucidation of the structural features that determine inducer potency and reactivity with Keap1. *J. Biol. Chem.* **2010**, *285*, 33747–33755. [CrossRef]
25. Ghafouri-Fard, S.; Shoorei, H.; Taheri, M. Non-coding RNAs are involved in the response to oxidative stress. *Biomed. Pharmacother.* **2020**, *127*, 110228. [CrossRef] [PubMed]
26. Finkel, T.; Holbrook, N.J. Oxidants, oxidative stress and the biology of ageing. *Nat. Cell Biol.* **2000**, *408*, 239–247. [CrossRef]
27. Martinez, A.; Pamplona, R.; Ferrer, I.; Portero-Otin, M. Protein targets of oxidative damage in human neurodegenerative diseases with abnormal protein aggregates. *Brain Pathol.* **2010**, *20*, 281–297. [CrossRef]
28. Sedelnikova, O.A.; Redon, C.E.; Dickey, J.S.; Nakamura, A.J.; Georgakilas, A.G.; Bonner, W.M. Role of oxidatively induced DNA lesions in human pathogenesis. *Mutat. Res./Rev. Mutat. Res.* **2010**, *704*, 152–159. [CrossRef]
29. Korolainen, M.A.; Nyman, T.A.; Nyyssonen, P.; Hartikainen, E.S.; Pirttila, T. Multiplexed proteomic analysis of oxidation and concentrations of cerebrospinal fluid proteins in alzheimer disease. *Clin. Chem.* **2007**, *53*, 657–665. [CrossRef]
30. Smith, M.A.; Sayre, L.M.; Anderson, V.E.; Harris, P.L.; Beal, M.F.; Kowall, N.; Perry, G. Cytochemical demonstration of oxidative damage in alzheimer disease by immunochemical enhancement of the carbonyl reaction with 2,4-dinitrophenylhydrazine. *J. Histochem. Cytochem.* **1998**, *46*, 731–735. [CrossRef]
31. Davies, M.J. The oxidative environment and protein damage. *Biochim. Biophys. Acta (BBA)-Proteins Proteom.* **2005**, *1703*, 93–109. [CrossRef]
32. Peña-Bautista, C.; Vento, M.; Baquero, M.; Cháfer-Pericás, C. Lipid peroxidation in neurodegeneration. *Clin. Chim. Acta* **2019**, *497*, 178–188. [CrossRef] [PubMed]

33. Xiao, M.; Zhong, H.; Xia, L.; Tao, Y.; Yin, H. Pathophysiology of mitochondrial lipid oxidation: Role of 4-hydroxynonenal (4-HNE) and other bioactive lipids in mitochondria. *Free Radic. Biol. Med.* **2017**, *111*, 316–327. [CrossRef]
34. Bruce-Keller, A.J.; Li, Y.J.; Lovell, M.A.; Kraemer, P.J.; Gary, D.S.; Brown, R.R.; Markesbery, W.R.; Mattson, M.P. 4-Hydroxynonenal, a product of lipid peroxidation, damages cholinergic neurons and impairs visuospatial memory in rats. *J. Neuropathol. Exp. Neurol.* **1998**, *57*, 257–267. [CrossRef]
35. Keller, J.N.; Pang, Z.; Geddes, J.W.; Begley, J.G.; Germeyer, A.; Waeg, G.; Mattson, M.P. Impairment of glucose and glutamate transport and induction of mitochondrial oxidative stress and dysfunction in synaptosomes by amyloid β-peptide: Role of the lipid peroxidation product 4-hydroxynonenal. *J. Neurochem.* **2002**, *69*, 273–284. [CrossRef]
36. Markesbery, W.R.; Lovell, M.A. DNA oxidation in Alzheimer's disease. *Antioxid. Redox Signal.* **2006**, *8*, 2039–2045. [CrossRef]
37. Lovell, M.A.; Markesbery, W.R. Oxidative DNA damage in mild cognitive impairment and late-stage Alzheimer's disease. *Nucleic Acids Res.* **2007**, *35*, 7497–7504. [CrossRef]
38. Lovell, M.A.; Gabbita, S.P.; Markesbery, W.R. Increased DNA oxidation and decreased levels of repair products in Alzheimer's disease ventricular CSF. *J. Neurochem.* **1999**, *72*, 771–776. [CrossRef]
39. Nunomura, A.; Perry, G.; Pappolla, M.A.; Wade, R.; Hirai, K.; Chiba, S.; Smith, M.A. RNA oxidation is a prominent feature of vulnerable neurons in Alzheimer's disease. *J. Neurosci.* **1999**, *19*, 1959–1964. [CrossRef]
40. Byun, K.; Yoo, Y.; Son, M.; Lee, J.; Jeong, G.-B.; Park, Y.M.; Salekdeh, G.H.; Lee, B. Advanced glycation end-products produced systemically and by macrophages: A common contributor to inflammation and degenerative diseases. *Pharmacol. Ther.* **2017**, *177*, 44–55. [CrossRef]
41. Li, J.; Liu, D.; Sun, L.; Lu, Y.; Zhang, Z. Advanced glycation end products and neurodegenerative diseases: Mechanisms and perspective. *J. Neurol. Sci.* **2012**, *317*, 1–5. [CrossRef]
42. Münch, G.; Cunningham, A.M.; Riederer, P.; Braak, H. Advanced glycation endproducts are associated with Hirano bodies in Alzheimer's disease. *Brain Res.* **1998**, *796*, 307–310. [CrossRef]
43. Gandhi, J.; Antonelli, A.C.; Afridi, A.; Vatsia, S.; Joshi, G.; Romanov, V.; Murray, I.V.; Khan, S.A. Protein misfolding and aggregation in neurodegenerative diseases: A review of pathogeneses, novel detection strategies, and potential therapeutics. *Rev. Neurosci.* **2019**, *30*, 339–358. [CrossRef]
44. Kumar, V.; Sami, N.; Kashav, T.; Islam, A.; Ahmad, F.; Hassan, M.I. Protein aggregation and neurodegenerative diseases: From theory to therapy. *Eur. J. Med. Chem.* **2016**, *124*, 1105–1120. [CrossRef]
45. Lévy, E.; El Banna, N.; Baïlle, D.; Heneman-Masurel, A.; Truchet, S.; Rezaei, H.; Huang, M.-E.; Béringue, V.; Martin, D.; Vernis, L. Causative links between protein aggregation and oxidative stress: A review. *Int. J. Mol. Sci.* **2019**, *20*, 3896. [CrossRef] [PubMed]
46. Rousseau, A.; Bertolotti, A. Regulation of proteasome assembly and activity in health and disease. *Nat. Rev. Mol. Cell Biol.* **2018**, *19*, 697–712. [CrossRef] [PubMed]
47. Seifert, U.; Bialy, L.P.; Ebstein, F.; Bech-Otschir, D.; Voigt, A.; Schröter, F.; Prozorovski, T.; Lange, N.; Steffen, J.; Rieger, M.; et al. Immunoproteasomes preserve protein homeostasis upon interferon-induced oxidative stress. *Cell* **2010**, *142*, 613–624. [CrossRef]
48. Ciechanover, A.; Brundin, P. The ubiquitin proteasome system in neurodegenerative diseases: Sometimes the chicken, sometimes the egg. *Neuron* **2003**, *40*, 427–446. [CrossRef]
49. Green, E.D.; Watson, J.D.; Collins, F. Human genome project: Twenty-five years of big biology. *Nat. Cell Biol.* **2015**, *526*, 29–31. [CrossRef] [PubMed]
50. The ENCODE Project Consortium; Snyder, M.P.; Gingeras, T.R.; Moore, J.E.; Weng, Z.; Gerstein, M.B.; Ren, B.; Hardison, R.C.; Stamatoyannopoulos, J.A.; Graveley, B.R.; et al. Perspectives on ENCODE. *Nature* **2020**, *583*, 693–698.
51. Mattick, J.S. Deconstructing the dogma: A new view of the evolution and genetic programming of complex organisms. *Ann. N. Y. Acad. Sci.* **2009**. [CrossRef]
52. Yang, J.X.; Rastetter, R.H.; Wilhelm, D. Non-coding RNAs: An introduction. *Adv. Exp. Med. Biol.* **2016**, *886*, 13–32.
53. Quan, Z.; Zheng, D.; Qing, H. Regulatory roles of long non-coding rnas in the central nervous system and associated neurodegenerative diseases. *Front. Cell. Neurosci.* **2017**, *11*, 175. [CrossRef]
54. Mercer, T.R.; Dinger, M.E.; Sunkin, S.M.; Mehler, M.F.; Mattick, J.S. Specific expression of long noncoding RNAs in the mouse brain. *Proc. Natl. Acad. Sci. USA* **2008**, *105*, 716–721. [CrossRef]

55. Lee, R.C.; Feinbaum, R.L.; Ambros, V. The *C. elegans* heterochronic gene lin-4 encodes small RNAs with antisense complementarity to lin-14. *Cell* **1993**, *75*, 843–854. [CrossRef]
56. Bhaskaran, M.; Mohan, M. MicroRNAs: History, biogenesis, and their evolving role in animal development and disease. *Vet. Pathol.* **2014**, *51*, 759–774. [CrossRef]
57. Zlotorynski, E. Insights into the kinetics of microRNA biogenesis and turnover. *Nat. Rev. Mol. Cell Biol.* **2019**, *20*, 511. [CrossRef]
58. Cuellar, T.L.; Davis, T.H.; Nelson, P.T.; Loeb, G.B.; Harfe, B.D.; Ullian, E.; McManus, M.T. Dicer loss in striatal neurons produces behavioral and neuroanatomical phenotypes in the absence of neurodegeneration. *Proc. Natl. Acad. Sci. USA* **2008**, *105*, 5614–5619. [CrossRef] [PubMed]
59. Rajman, M.; Schratt, G. MicroRNAs in neural development: From master regulators to fine-tuners. *Development* **2017**, *144*, 2310–2322. [CrossRef]
60. Fineberg, S.K.; Kosik, K.S.; Davidson, B.L. MicroRNAs potentiate neural development. *Neuron* **2009**, *64*, 303–309. [CrossRef]
61. Bernstein, E.; Kim, S.Y.; Carmell, A.M.; Murchison, E.P.; Alcorn, H.L.; Li, M.Z.; Mills, A.A.; Elledge, S.J.; Anderson, K.V.; Hannon, G.J. Dicer is essential for mouse development. *Nat. Genet.* **2003**, *35*, 215–217. [CrossRef]
62. Davis, T.H.; Cuellar, T.L.; Koch, S.M.; Barker, A.J.; Harfe, B.D.; McManus, M.T.; Ullian, E.M. Conditional loss of dicer disrupts cellular and tissue morphogenesis in the cortex and hippocampus. *J. Neurosci.* **2008**, *28*, 4322–4330. [CrossRef]
63. Chmielarz, P.; Konovalova, J.; Najam, S.S.; Alter, H.; Piepponen, T.P.; Erfle, H.; Sonntag, K.C.; Schütz, G.; Vinnikov, I.A.; Domanskyi, A. Dicer and microRNAs protect adult dopamine neurons. *Cell Death Dis.* **2017**, *8*, e2813. [CrossRef]
64. Hu, Z.; Li, Z. MiRNAs in synapse development and synaptic plasticity. *Curr. Opin. Neurobiol.* **2017**, *45*, 24–31. [CrossRef] [PubMed]
65. Strazisar, M.; Cammaerts, S.; Van Der Ven, K.; Forero, A.D.; Lenaerts, A.-S.; Nordin, A.; Almeida-Souza, L.; Genovese, G.; Timmerman, V.; Liekens, A.; et al. MIR137 variants identified in psychiatric patients affect synaptogenesis and neuronal transmission gene sets. *Mol. Psychiatry* **2015**, *20*, 472–481. [CrossRef]
66. The Schizophrenia Psychiatric Genome-Wide Association Study (GWAS) Consortium. Genome-wide association study identifies five new schizophrenia loci. *Nat. Genet.* **2011**, *43*, 969–976. [CrossRef] [PubMed]
67. Loohuis, N.F.O.; Ba, W.; Stoerchel, P.H.; Kos, A.; Jager, A.; Schratt, G.; Martens, G.J.; Van Bokhoven, H.; Kasri, N.N.; Aschrafi, A.; et al. MicroRNA-137 controls ampa-receptor-mediated transmission and mGluR-Dependent LTD. *Cell Rep.* **2015**, *11*, 1876–1884. [CrossRef]
68. Harraz, M.M.; Eacker, S.M.; Wang, X.; Dawson, T.M.; Dawson, V.L. MicroRNA-223 is neuroprotective by targeting glutamate receptors. *Proc. Natl. Acad. Sci. USA* **2012**, *109*, 18962–18967. [CrossRef]
69. Edbauer, D.; Neilson, J.R.; Foster, K.A.; Wang, C.-F.; Seeburg, D.P.; Batterton, M.N.; Tada, T.; Dolan, B.M.; Sharp, P.A.; Sheng, M. Regulation of Synaptic Structure and Function by FMRP-Associated MicroRNAs miR-125b and miR-132. *Neuron* **2010**, *65*, 373–384. [CrossRef]
70. Hu, Z.; Yu, D.; Gu, Q.-H.; Yang, Y.; Tu, K.; Zhu, J.; Li, Z. miR-191 and miR-135 are required for long-lasting spine remodelling associated with synaptic long-term depression. *Nat. Commun.* **2014**, *5*, 1–17. [CrossRef]
71. Schratt, G.M.; Tuebing, F.; Nigh, E.A.; Kane, C.G.; Sabatini, M.E.; Kiebler, M.; Greenberg, M.E. A brain-specific microRNA regulates dendritic spine development. *Nat. Cell Biol.* **2006**, *439*, 283–289. [CrossRef]
72. Antoniou, A.; Khudayberdiev, S.; Idziak, A.; Bicker, S.; Jacob, R.; Schratt, G. The dynamic recruitment of TRBP to neuronal membranes mediates dendritogenesis during development. *EMBO Rep.* **2017**, *19*, e44853. [CrossRef]
73. Fiore, R.; Khudayberdiev, S.; Christensen, M.; Siegel, G.; Flavell, S.W.; Kim, T.-K.; Greenberg, M.E.; Schratt, G.M. Mef2-mediated transcription of the miR379–410 cluster regulates activity-dependent dendritogenesis by fine-tuning Pumilio2 protein levels. *EMBO J.* **2009**, *28*, 697–710. [CrossRef] [PubMed]
74. Marler, K.J.; Suetterlin, P.; Dopplapudi, A.; Rubikaite, A.; Adnan, J.; Maiorano, N.A.; Lowe, A.S.; Thompson, I.D.; Pathania, M.; Bordey, A.; et al. BDNF promotes axon branching of retinal ganglion cells via miRNA-132 and p250GAP. *J. Neurosci.* **2014**, *34*, 969–979. [CrossRef]
75. Zhao, X.; He, X.; Han, X.; Yu, Y.; Ye, F.; Chen, Y.; Hoang, T.; Xu, X.; Mi, Q.-S.; Xin, M.; et al. MicroRNA-Mediated control of oligodendrocyte differentiation. *Neuron* **2010**, *65*, 612–626. [CrossRef]

76. Dugas, J.C.; Cuellar, T.L.; Scholze, A.; Ason, B.; Ibrahim, A.; Emery, B.; Zamanian, J.L.; Foo, L.C.; McManus, M.T.; Barres, B.A. Dicer1 and miR-219 are required for normal oligodendrocyte differentiation and myelination. *Neuron* **2010**, *65*, 597–611. [CrossRef]
77. Pogue, A.; Cui, J.; Li, Y.; Zhao, Y.; Culicchia, F.; Lukiw, W. Micro RNA-125b (miRNA-125b) function in astrogliosis and glial cell proliferation. *Neurosci. Lett.* **2010**, *476*, 18–22. [CrossRef]
78. Taj, S.H.; Kho, W.; Riou, A.; Wiedermann, D.; Hoehn, M. MiRNA-124 induces neuroprotection and functional improvement after focal cerebral ischemia. *Biomaterials* **2016**, *91*, 151–165. [CrossRef]
79. Guo, Y.; Hong, W.; Wang, X.; Zhang, P.; Körner, H.; Tu, J.; Wei, W. MicroRNAs in microglia: How do MicroRNAs affect activation, inflammation, polarization of microglia and mediate the interaction between microglia and glioma? *Front. Mol. Neurosci.* **2019**, *12*, 125. [CrossRef]
80. Kumar, P.; Kuscu, C.; Dutta, A. biogenesis and function of transfer RNA-related fragments (tRFs). *Trends Biochem. Sci.* **2016**, *41*, 679–689. [CrossRef] [PubMed]
81. Yamasaki, S.; Ivanov, P.; Hu, G.-F.; Anderson, P. Angiogenin cleaves tRNA and promotes stress-induced translational repression. *J. Cell Biol.* **2009**, *185*, 35–42. [CrossRef]
82. Kumar, P.; Mudunuri, S.B.; Anaya, J.; Dutta, A. TRFdb: A database for transfer RNA fragments. *Nucleic Acids Res.* **2015**, *43*, D141–D145. [CrossRef]
83. Guzzi, N.; Bellodi, C. Novel insights into the emerging roles of tRNA-derived fragments in mammalian development. *RNA Biol.* **2020**, *17*, 1214–1222. [CrossRef]
84. Karaiskos, S.; Grigoriev, A. Dynamics of tRNA fragments and their targets in aging mammalian brain. *F1000Research* **2016**, *5*, 2758. [CrossRef]
85. Zhang, S.; Li, H.; Zheng, L.; Li, H.; Feng, C.; Zhang, W. Identification of functional tRNA-derived fragments in senescence-accelerated mouse prone 8 brain. *Aging* **2019**, *11*, 10485–10498. [CrossRef]
86. Krishna, S.; Yim, D.G.; Lakshmanan, V.; Tirumalai, V.; Koh, J.L.; Park, J.E.; Cheong, J.K.; Low, J.L.; Lim, M.J.; Sze, S.K.; et al. Dynamic expression of tRNA-derived small RNAs define cellular states. *EMBO Rep.* **2019**, *20*. [CrossRef]
87. Yu, M.; Lu, B.; Zhang, J.; Ding, J.; Liu, P.; Lu, Y. tRNA-derived RNA fragments in cancer: Current status and future perspectives. *J. Hematol. Oncol.* **2020**, *13*, 1–14. [CrossRef]
88. Maute, R.L.; Schneider, C.; Sumazin, P.; Holmes, A.; Califano, A.; Basso, K.; Dalla-Favera, R. tRNA-derived microRNA modulates proliferation and the DNA damage response and is down-regulated in B cell lymphoma. *Proc. Natl. Acad. Sci. USA* **2013**, *110*, 1404–1409. [CrossRef]
89. Guan, L.; Karaiskos, S.; Grigoriev, A. Inferring targeting modes of Argonaute-loaded tRNA fragments. *RNA Biol.* **2019**, *17*, 1070–1080. [CrossRef]
90. Kumar, P.; Anaya, J.; Mudunuri, S.B.; Dutta, A. Meta-analysis of tRNA derived RNA fragments reveals that they are evolutionarily conserved and associate with AGO proteins to recognize specific RNA targets. *BMC Biol.* **2014**, *12*, 78. [CrossRef] [PubMed]
91. Gonskikh, Y.; Gerstl, M.; Kos, M.; Borth, N.; Schosserer, M.; Grillari, J.; Polacek, N. Modulation of mammalian translation by a ribosome-associated tRNA half. *RNA Biol.* **2020**, *17*, 1125–1136. [CrossRef] [PubMed]
92. Wolozin, B.; Ivanov, P. Stress granules and neurodegeneration. *Nat. Rev. Neurosci.* **2019**, *20*, 649–666. [CrossRef]
93. Ivanov, P.; O'Day, E.; Emara, M.M.; Wagner, G.; Lieberman, J.; Anderson, P.J. G-quadruplex structures contribute to the neuroprotective effects of angiogenin-induced tRNA fragments. *Proc. Natl. Acad. Sci. USA* **2014**, *111*, 18201–18206. [CrossRef]
94. Ivanov, P.; Emara, M.M.; Villen, J.; Gygi, S.P.; Anderson, P. Angiogenin-induced tRNA fragments inhibit translation initiation. *Mol. Cell* **2011**, *43*, 613–623. [CrossRef]
95. Lyons, S.M.; Achorn, C.; Kedersha, N.L.; Anderson, P.J.; Ivanov, P. YB-1 regulates tiRNA-induced Stress Granule formation but not translational repression. *Nucleic Acids Res.* **2016**, *44*, 6949–6960. [CrossRef]
96. Salvatori, B.; Biscarini, S.; Morlando, M. Non-coding RNAs in nervous system development and disease. *Front. Cell Dev. Biol.* **2020**, *8*, 8. [CrossRef]
97. Ulitsky, I.; Bartel, D.P. LincRNAs: Genomics, evolution, and mechanisms. *Cell* **2013**, *154*, 26–46. [CrossRef]
98. Wang, K.C.; Chang, H.Y. Molecular mechanisms of long noncoding RNAs. *Mol. Cell* **2011**, *43*, 904–914. [CrossRef]
99. An, H.; Williams, N.G.; Shelkovnikova, T.A. NEAT1 and paraspeckles in neurodegenerative diseases: A missing lnc found? *Non-Coding RNA Res.* **2018**, *3*, 243–252. [CrossRef]

100. Zhao, Y.; Sun, L.; Wang, R.R.; Hu, J.-F.; Cui, J. The effects of mitochondria-associated long noncoding RNAs in cancer mitochondria: New players in an old arena. *Crit. Rev. Oncol.* **2018**, *131*, 76–82. [CrossRef]
101. Chen, L.-L. Linking long noncoding RNA localization and function. *Trends Biochem. Sci.* **2016**, *41*, 761–772. [CrossRef]
102. Briggs, J.A.; Wolvetang, E.J.; Mattick, J.S.; Rinn, J.L.; Barry, G. Mechanisms of long non-coding RNAs in mammalian nervous system development, plasticity, disease, and evolution. *Neuron* **2015**, *88*, 861–877. [CrossRef]
103. Pollard, K.S.; Salama, S.R.; Lambert, N.; Lambot, M.-A.; Coppens, S.; Pedersen, J.S.; Katzman, S.; King, B.; Onodera, C.; Siepel, A.; et al. An RNA gene expressed during cortical development evolved rapidly in humans. *Nat. Cell Biol.* **2006**, *443*, 167–172. [CrossRef]
104. Bernard, D.; Prasanth, K.V.; Tripathi, V.; Colasse, S.; Nakamura, T.; Xuan, Z.; Zhang, M.Q.; Sedel, F.; Jourdren, L.; Coulpier, F.; et al. A long nuclear-retained non-coding RNA regulates synaptogenesis by modulating gene expression. *EMBO J.* **2010**, *29*, 3082–3093. [CrossRef]
105. Hart, R.P.; Goff, L.A. Long noncoding RNAs: Central to nervous system development. *Int. J. Dev. Neurosci.* **2016**, *55*, 109–116. [CrossRef]
106. Keihani, S.; Kluever, V.; Mandad, S.; Bansal, V.; Rahman, R.; Fritsch, E.; Gomes, L.C.; Gärtner, A.; Kügler, S.; Urlaub, H.; et al. The long noncoding RNA neuroLNC regulates presynaptic activity by interacting with the neurodegeneration-associated protein TDP-43. *Sci. Adv.* **2019**, *5*, eaay2670. [CrossRef]
107. Yang, C.; Tang, R.; Ma, X.; Wang, Y.; Luo, D.; Xu, Z.; Zhu, Y.; Yang, L. Tag SNPs in long non-coding RNA H19 contribute to susceptibility to gastric cancer in the Chinese Han population. *Oncotarget* **2015**, *6*, 15311–15320. [CrossRef]
108. Han, C.-L.; Ge, M.; Liu, Y.-P.; Zhao, X.-M.; Wang, K.-L.; Chen, N.; Meng, W.-J.; Hu, W.; Zhang, J.-G.; Li, L.; et al. LncRNA H19 contributes to hippocampal glial cell activation via JAK/STAT signaling in a rat model of temporal lobe epilepsy. *J. Neuroinflamm.* **2018**, *15*, 1–9. [CrossRef]
109. Bond, A.M.; VanGompel, M.J.W.; Sametsky, E.A.; Clark, M.F.; Savage, J.C.; Disterhoft, J.F.; Kohtz, J.D. Balanced gene regulation by an embryonic brain ncRNA is critical for adult hippocampal GABA circuitry. *Nat. Neurosci.* **2009**, *12*, 1020–1027. [CrossRef]
110. Ng, S.-Y.; Bogu, G.K.; Soh, B.S.; Stanton, L.W. The long noncoding RNA RMST interacts with SOX2 to regulate neurogenesis. *Mol. Cell* **2013**, *51*, 349–359. [CrossRef]
111. Reddy, A.S.; O'Brien, D.; Pisat, N.; Weichselbaum, C.T.; Sakers, K.; Lisci, M.; Dalal, J.S.; Dougherty, J.D. A comprehensive analysis of cell type–specific nuclear RNA from neurons and glia of the brain. *Biol. Psychiatry* **2017**, *81*, 252–264. [CrossRef]
112. Ramos, A.D.; Andersen, R.E.; Liu, S.J.; Nowakowski, T.J.; Hong, S.J.; Gertz, C.C.; Salinas, R.D.; Zarabi, H.; Kriegstein, A.R.; Lim, D.A. The long noncoding RNA pnky regulates neuronal differentiation of embryonic and postnatal neural stem cells. *Cell Stem Cell* **2015**, *16*, 439–447. [CrossRef]
113. Kleaveland, B.; Shi, C.Y.; Stefano, J.; Bartel, D.P. A network of noncoding regulatory RNAs Acts in the mammalian brain. *Cell* **2018**, *174*, 350–362.e17. [CrossRef]
114. Jeck, W.R.; Sorrentino, J.A.; Wang, K.; Slevin, M.K.; Burd, C.E.; Liu, J.; Marzluff, W.F.; Sharpless, N.E. Circular RNAs are abundant, conserved, and associated with ALU repeats. *RNA* **2013**, *19*, 141–157. [CrossRef]
115. Kristensen, L.S.; Andersen, M.S.; Stagsted, L.V.W.; Ebbesen, K.K.; Hansen, T.B.; Kjems, J. The biogenesis, biology and characterization of circular RNAs. *Nat. Rev. Genet.* **2019**, *20*, 675–691. [CrossRef]
116. Yu, C.-Y.; Kuo, H.-C. The emerging roles and functions of circular RNAs and their generation. *J. Biomed. Sci.* **2019**, *26*, 1–12. [CrossRef] [PubMed]
117. Mehta, S.L.; Dempsey, R.J.; Vemuganti, R. Role of circular RNAs in brain development and CNS diseases. *Prog. Neurobiol.* **2020**, *186*, 101746. [CrossRef]
118. You, X.; Vlatkovic, I.; Babic, A.; Will, T.J.; Epstein, I.; Tushev, G.; Akbalik, G.; Wang, M.; Glock, C.; Quedenau, C.; et al. Neural circular RNAs are derived from synaptic genes and regulated by development and plasticity. *Nat. Neurosci.* **2015**, *18*, 603–610. [CrossRef]
119. Piwecka, M.; Glažar, P.; Hernandez-Miranda, L.R.; Memczak, S.; Wolf, S.A.; Rybak-Wolf, A.; Filipchyk, A.; Klironomos, F.; Jara, C.A.C.; Fenske, P.; et al. Loss of a mammalian circular RNA locus causes miRNA deregulation and affects brain function. *Science* **2017**, *357*. [CrossRef] [PubMed]

120. Zhang, Y.; Du, L.; Bai, Y.; Han, B.; He, C.; Gong, L.; Huang, R.; Shen, L.; Chao, J.; Liu, P.; et al. CircDYM ameliorates depressive-like behavior by targeting miR-9 to regulate microglial activation via HSP90 ubiquitination. *Mol. Psychiatry* **2018**, *25*, 1175–1190. [CrossRef]
121. Liu, Z.; Chen, X.; Li, Z.; Ye, W.; Ding, H.; Li, P.; Aung, L.H.H. Role of RNA oxidation in neurodegenerative diseases. *Int. J. Mol. Sci.* **2020**, *21*, 5022. [CrossRef]
122. Shcherbik, N.; Pestov, D.G. The impact of oxidative stress on ribosomes: From injury to regulation. *Cells* **2019**, *8*, 1379. [CrossRef]
123. Leisegang, M.S.; Schröder, K.; Brandes, R.P. Redox regulation and noncoding RNAs. *Antioxidants Redox Signal.* **2018**, *29*, 793–812. [CrossRef] [PubMed]
124. Chu, L.; Su, M.Y.; Maggi, L.B., Jr.; Lu, L.; Mullins, C.; Crosby, S.; Huang, G.; Chng, W.J.; Vij, R.; Tomasson, M.H. Multiple myeloma–associated chromosomal translocation activates orphan snoRNA ACA11 to suppress oxidative stress. *J. Clin. Investig.* **2012**, *122*, 2793–2806. [CrossRef]
125. Sahoo, T.; Del Gaudio, D.; German, J.R.; Shinawi, M.; Peters, S.U.; Person, R.E.; Garnica, A.D.; Cheung, S.W.; Beaudet, A.L. Prader-Willi phenotype caused by paternal deficiency for the HBII-85 C/D box small nucleolar RNA cluster. *Nat. Genet.* **2008**, *40*, 719–721. [CrossRef]
126. Taft, R.J.; Glazov, E.A.; Lassmann, T.; Hayashizaki, Y.; Carninci, P.; Mattick, J.S. Small RNAs derived from snoRNAs. *RNA* **2009**, *15*, 1233–1240. [CrossRef]
127. Barker, W.W.; Luis, C.A.; Kashuba, A.; Luis, M.; Harwood, D.G.; Loewenstein, D.; Waters, C.; Jimison, P.; Shepherd, E.; Sevush, S.; et al. Relative frequencies of alzheimer disease, lewy body, vascular and frontotemporal dementia, and hippocampal sclerosis in the state of florida brain bank. *Alzheimer Dis. Assoc. Disord.* **2002**, *16*, 203–212. [CrossRef] [PubMed]
128. Tönnies, E.; Trushina, E. Oxidative stress, synaptic dysfunction, and Alzheimer's disease. *J. Alzheimer's Dis.* **2017**, *57*, 1105–1121. [CrossRef]
129. Lanoiselée, H.-M.; Nicolas, G.; Wallon, D.; Rovelet-Lecrux, A.; Lacour, M.; Rousseau, S.; Richard, A.-C.; Pasquier, F.; Rollin-Sillaire, A.; Martinaud, O.; et al. APP, PSEN1, and PSEN2 mutations in early-onset Alzheimer disease: A genetic screening study of familial and sporadic cases. *PLoS Med.* **2017**, *14*, e1002270. [CrossRef]
130. Van Cauwenberghe, C.; Van Broeckhoven, C.; Sleegers, K. The genetic landscape of Alzheimer disease: Clinical implications and perspectives. *Genet. Med.* **2016**, *18*, 421–430. [CrossRef] [PubMed]
131. Millan, M.J. Linking deregulation of non-coding RNA to the core pathophysiology of Alzheimer's disease: An integrative review. *Prog. Neurobiol.* **2017**, *156*, 1–68. [CrossRef]
132. Zhao, J.; Yue, D.; Zhou, Y.; Jia, L.; Wang, H.; Guo, M.; Xu, H.; Chen, C.; Zhang, J.; Xu, L. The role of MicroRNAs in Aβ deposition and tau phosphorylation in Alzheimer's disease. *Front. Neurol.* **2017**, *8*. [CrossRef]
133. Smith, P.Y.; Hernandez-Rapp, J.; Jolivette, F.; Lecours, C.; Bisht, K.; Goupil, C.; Dorval, V.; Parsi, S.; Morin, F.; Planel, E.; et al. miR-132/212 deficiency impairs tau metabolism and promotes pathological aggregation in vivo. *Hum. Mol. Genet.* **2015**, *24*, 6721–6735. [CrossRef]
134. Sierksma, A.; Lu, A.; Salta, E.; vanden Eynden, E.; Callaerts-Vegh, Z.; D'Hooge, R.; Blum, D.; Buée, L.; Fiers, M.; De Strooper, B. Deregulation of neuronal miRNAs induced by amyloid-β or TAU pathology. *Mol. Neurodegener.* **2018**, *13*, 1–15. [CrossRef]
135. Amakiri, N.; Kubosumi, A.; Tran, J.; Reddy, P.H. Amyloid beta and MicroRNAs in Alzheimer's disease. *Front. Neurosci.* **2019**, *13*, 430. [CrossRef] [PubMed]
136. Kim, J.; Yoon, H.; Chung, D.-E.; Brown, J.L.; Belmonte, K.C.; Kim, J. MiR-186 is decreased in aged brain and suppresses BACE1 expression. *J. Neurochem.* **2016**, *137*, 436–445. [CrossRef]
137. Zhang, R.; Zhang, Q.; Niu, J.; Lu, K.; Xie, B.; Cui, D.; Xu, S. Screening of microRNAs associated with Alzheimer's disease using oxidative stress cell model and different strains of senescence accelerated mice. *J. Neurol. Sci.* **2014**, *338*, 57–64. [CrossRef]
138. Zhou, Y.; Wang, Z.-F.; Li, W.; Hong, H.; Chen, J.; Tian, Y.; Liu, Z.-Y. Protective effects of microRNA-330 on amyloid β-protein production, oxidative stress, and mitochondrial dysfunction in Alzheimer's disease by targeting VAV1 via the MAPK signaling pathway. *J. Cell. Biochem.* **2018**, *119*, 5437–5448. [CrossRef]
139. Zhang, L.; Dong, H.; Si, Y.; Wu, N.; Cao, H.; Mei, B.; Meng, B. MiR-125b promotes tau phosphorylation by targeting the neural cell adhesion molecule in neuropathological progression. *Neurobiol. Aging* **2019**, *73*, 41–49. [CrossRef]

140. Jin, Y.; Tu, Q.; Liu, M. MicroRNA-125b regulates Alzheimer's disease through SphK1 regulation. *Mol. Med. Rep.* **2018**, *18*, 2373–2380. [CrossRef]
141. Li, J.J.; Dolios, G.; Wang, R.; Liao, F.-F. Soluble beta-amyloid peptides, but not insoluble fibrils, have specific effect on neuronal MicroRNA expression. *PLoS ONE* **2014**, *9*, e90770. [CrossRef] [PubMed]
142. Wang, W.-X.; Rajeev, B.W.; Stromberg, A.J.; Ren, N.; Tang, G.; Huang, Q.; Rigoutsos, I.; Nelson, P.T. The expression of MicroRNA miR-107 decreases early in Alzheimer's disease and may accelerate disease progression through regulation of β-site amyloid precursor protein-cleaving enzyme 1. *J. Neurosci.* **2008**, *28*, 1213–1223. [CrossRef]
143. De La Monte, S.M.; Wands, J.R. Molecular indices of oxidative stress and mitochondrial dysfunction occur early and often progress with severity of Alzheimer's disease. *J. Alzheimer's Dis.* **2006**, *9*, 167–181. [CrossRef]
144. Prendecki, M.; Florczak-Wyspianska, J.; Kowalska, M.; Ilkowski, J.; Grzelak, T.; Bialas, K.; Kozubski, W.; Dorszewska, J. APOE genetic variants and apoE, miR-107 and miR-650 levels in Alzheimer's disease. *Folia Neuropathol.* **2019**, *57*, 106–116. [CrossRef] [PubMed]
145. Liang, C.; Zhu, H.; Xu, Y.; Huang, L.; Ma, C.; Deng, W.; Liu, Y.; Qin, C. MicroRNA-153 negatively regulates the expression of amyloid precursor protein and amyloid precursor-like protein 2. *Brain Res.* **2012**, *1455*, 103–113. [CrossRef]
146. Wang, L.; Liu, J.; Wang, Q.; Jiang, H.; Zeng, L.; Li, Z.; Liu, R. MicroRNA-200a-3p mediates neuroprotection in Alzheimer-Related deficits and attenuates amyloid-beta overproduction and tau hyperphosphorylation via coregulating BACE1 and PRKACB. *Front. Pharmacol.* **2019**, *10*, 806. [CrossRef]
147. Li, J.; Chen, W.; Yi, Y.; Tong, Q. miR-219-5p inhibits tau phosphorylation by targeting TTBK1 and GSK-3β in Alzheimer's disease. *J. Cell. Biochem.* **2018**, *120*, 9936–9946. [CrossRef]
148. El Fatimy, R.; Li, S.; Chen, Z.; Mushannen, T.; Gongala, S.; Wei, Z.; Balu, D.T.; Rabinovsky, R.; Cantlon, A.; Elkhal, A.; et al. MicroRNA-132 provides neuroprotection for tauopathies via multiple signaling pathways. *Acta Neuropathol.* **2018**, *136*, 537–555. [CrossRef]
149. Absalon, S.; Kochanek, D.M.; Raghavan, V.; Krichevsky, A.M. MiR-26b, upregulated in Alzheimer's disease, activates cell cycle entry, tau-phosphorylation, and apoptosis in postmitotic neurons. *J. Neurosci.* **2013**, *33*, 14645–14659. [CrossRef]
150. Wang, G.; Huang, Y.; Wang, L.-L.; Zhang, Y.-F.; Xu, J.; Zhou, Y.; Lourenco, G.F.; Zhang, B.; Wang, Y.; Ren, R.-J.; et al. MicroRNA-146a suppresses ROCK1 allowing hyperphosphorylation of tau in Alzheimer's disease. *Sci. Rep.* **2016**, *6*, 26697. [CrossRef]
151. Thompson, D.M.; Lu, C.; Green, P.J.; Parker, R. TRNA cleavage is a conserved response to oxidative stress in eukaryotes. *RNA* **2008**, *14*, 2095–2103. [CrossRef]
152. Zhang, L.; Fang, Y.; Cheng, X.; Lian, Y.-J.; Xu, H.-L. Silencing of long noncoding RNA SOX21-AS1 relieves neuronal oxidative stress injury in mice with Alzheimer's disease by upregulating FZD3/5 via the wnt signaling pathway. *Mol. Neurobiol.* **2018**, *56*, 3522–3537. [CrossRef]
153. Guo, C.-C.; Jiao, C.-H.; Gao, Z.-M. Silencing of LncRNA BDNF-AS attenuates Aβ25-35-induced neurotoxicity in PC12 cells by suppressing cell apoptosis and oxidative stress. *Neurol. Res.* **2018**, *40*, 795–804. [CrossRef] [PubMed]
154. Wang, J.; Zhou, T.; Wang, T.; Wang, B. Suppression of lncRNA-ATB prevents amyloid-β-induced neurotoxicity in PC12 cells via regulating miR-200/ZNF217 axis. *Biomed. Pharmacother.* **2018**, *108*, 707–715. [CrossRef]
155. Wang, X.; Wang, C.; Geng, C.; Zhao, K. LncRNA XIST knockdown attenuates Aβ25-35-induced toxicity, oxidative stress, and apoptosis in primary cultured rat hippocampal neurons by targeting miR-132. *Int. J. Clin. Exp. Pathol.* **2018**, *11*, 3915.
156. Gao, Y.; Zhang, N.; Lv, C.; Li, N.; Li, X.; Li, W. lncRNA SNHG1 knockdown alleviates amyloid-β-induced neuronal injury by regulating ZNF217 via sponging miR-361-3p in Alzheimer's disease. *J. Alzheimer's Dis.* **2020**. [CrossRef]
157. Li, X.; Wang, S.-W.; Li, X.-L.; Yu, F.-Y.; Cong, H.-M. Knockdown of long non-coding RNA TUG1 depresses apoptosis of hippocampal neurons in Alzheimer's disease by elevating microRNA-15a and repressing ROCK1 expression. *Inflamm. Res.* **2020**, *69*, 897–910. [CrossRef]
158. Akhter, R. Circular RNA and Alzheimer's disease. *Adv. Exp. Med. Biol.* **2018**, *1087*, 239–243.
159. Li, Y.; Fan, H.; Sun, J.; Ni, M.; Zhang, L.; Chen, C.; Hong, X.; Fang, F.; Zhang, W.; Ma, P. Circular RNA expression profile of Alzheimer's disease and its clinical significance as biomarkers for the disease risk and progression. *Int. J. Biochem. Cell Biol.* **2020**, *123*, 105747. [CrossRef]

160. Dube, U.; Del-Aguila, J.L.; Li, Z.; Budde, J.P.; Jiang, S.; Hsu, S.; Ibanez, L.; Fernandez, M.V.; Farias, F.; Norton, J.; et al. An atlas of cortical circular RNA expression in Alzheimer disease brains demonstrates clinical and pathological associations. *Nat. Neurosci.* **2019**, *22*, 1903–1912. [CrossRef]
161. Hansen, T.B.; Jensen, T.I.; Clausen, B.H.; Bramsen, J.B.; Finsen, B.; Damgaard, C.K.; Kjems, J. Natural RNA circles function as efficient microRNA sponges. *Nature* **2013**, *495*, 384–388. [CrossRef] [PubMed]
162. Lukiw, W.J. Circular RNA (circRNA) in Alzheimer's disease (AD). *Front. Genet.* **2013**, *4*, 307. [CrossRef]
163. Shi, Z.; Chen, T.; Yao, Q.; Zheng, L.; Zhang, Z.; Wang, J.; Hu, Z.; Cui, H.; Han, Y.; Han, X.; et al. The circular RNA ci RS-7 promotes APP and BACE 1 degradation in an NF-κB-dependent manner. *FEBS J.* **2017**, *284*, 1096–1109. [CrossRef]
164. Lu, Y.; Tan, L.; Wang, X. Circular HDAC9/microRNA-138/Sirtuin-1 pathway mediates synaptic and amyloid precursor protein processing deficits in Alzheimer's disease. *Neurosci. Bull.* **2019**, *35*, 877–888. [CrossRef] [PubMed]
165. Jellinger, K.A. Neuropathobiology of non-motor symptoms in Parkinson disease. *J. Neural Transm.* **2015**, *122*, 1429–1440. [CrossRef]
166. Przedborski, S. The two-century journey of Parkinson disease research. *Nature Rev. Neurosci.* **2017**, *18*, 251. [CrossRef]
167. Tysnes, O.-B.; Storstein, A. epidemiology of Parkinson's disease. *J. Neural Transm.* **2017**, *124*, 901–905. [CrossRef]
168. Inzelberg, R.; Schecthman, E.; Paleacu, D.; Zach, L.; Bonwitt, R.; Carasso, R.L.; Nisipeanu, P. Onset and progression of disease in familial and sporadic Parkinson's disease. *Am. J. Med Genet.* **2003**, 255–258. [CrossRef]
169. Eblesa, J.; Etrigo-Damas, I.; Equiroga-Varela, A.; Jackson-Lewis, V.R. Oxidative stress and Parkinson's disease. *Front. Neuroanat.* **2015**, *9*, 91. [CrossRef]
170. Wei, Z.; Li, X.; Li, X.; Liu, Q.; Cheng, Y. Oxidative stress in Parkinson's disease: A systematic review and meta-analysis. *Front. Mol. Neurosci.* **2018**, *11*, 236. [CrossRef]
171. Caudle, W.M.; Richardson, J.R.; Wang, M.Z.; Taylor, T.N.; Guillot, T.S.; McCormack, A.L.; Colebrooke, R.E.; Di Monte, D.A.; Emson, P.C.; Miller, G.W. Reduced vesicular storage of dopamine causes progressive nigrostriatal neurodegeneration. *J. Neurosci.* **2007**, *27*, 8138–8148. [CrossRef]
172. Hwang, D.-Y.; Hong, S.; Jeong, J.-W.; Choi, S.; Kim, H.; Kim, J.; Kim, K.-S. Vesicular monoamine transporter 2 and dopamine transporter are molecular targets of Pitx3 in the ventral midbrain dopamine neurons. *J. Neurochem.* **2009**, *111*, 1202–1212. [CrossRef]
173. Kim, J.; Inoue, K.; Ishii, J.; Vanti, W.B.; Voronov, S.V.; Murchison, E.; Hannon, G.; Abeliovich, A. A microRNA feedback circuit in midbrain dopamine neurons. *Science* **2007**, *317*, 1220–1224. [CrossRef] [PubMed]
174. Wang, H.-L.; Chou, A.-H.; Wu, A.-S.; Chen, S.-Y.; Weng, Y.-H.; Kao, Y.-C.; Yeh, T.-H.; Chu, P.-J.; Lu, C.-S. PARK6 PINK1 mutants are defective in maintaining mitochondrial membrane potential and inhibiting ROS formation of substantia nigra dopaminergic neurons. *Biochim. Biophys. Acta (BBA)-Mol. Basis Dis.* **2011**, *1812*, 674–684. [CrossRef]
175. Wood-Kaczmar, A.; Gandhi, S.; Yao, Z.; Abramov, A.S.Y.; Miljan, E.A.; Keen, G.; Stanyer, L.; Hargreaves, I.; Klupsch, K.; Deas, E.; et al. PINK1 Is necessary for long term survival and mitochondrial function in human dopaminergic neurons. *PLoS ONE* **2008**, *3*, e2455. [CrossRef]
176. Kim, J.; Fiesel, F.C.; Belmonte, K.C.; Hudec, R.; Wang, W.-X.; Kim, C.; Nelson, P.T.; Springer, W.; Kim, J. MiR-27a and miR-27b regulate autophagic clearance of damaged mitochondria by targeting PTEN-induced putative kinase 1 (PINK1). *Mol. Neurodegener.* **2016**, *11*, 1–16. [CrossRef]
177. Prajapati, P.; Sripada, L.; Singh, R.; Bhatelia, K.; Singh, R.; Singh, R. TNF-α regulates miRNA targeting mitochondrial complex-I and induces cell death in dopaminergic cells. *Biochim. Biophys. Acta (BBA)-Mol. Basis Dis.* **2015**, *1852*, 451–461. [CrossRef]
178. Hayashi, T.; Ishimori, C.; Takahashi-Niki, K.; Taira, T.; Kim, Y.-C.; Maita, H.; Maita, C.; Ariga, H.; Iguchi-Ariga, S.M. DJ-1 binds to mitochondrial complex I and maintains its activity. *Biochem. Biophys. Res. Commun.* **2009**, *390*, 667–672. [CrossRef]
179. Ariga, H.; Takahashi-Niki, K.; Kato, I.; Maita, H.; Niki, T.; Iguchi-Ariga, S.M.M. Neuroprotective function of DJ-1 in Parkinson's disease. *Oxidative Med. Cell. Longev.* **2013**, *2013*, 1–9. [CrossRef] [PubMed]

180. Choi, J.; Sullards, M.C.; Olzmann, J.A.; Rees, H.D.; Weintraub, S.T.; Bostwick, D.E.; Gearing, M.; Levey, A.I.; Chin, L.-S.; Li, L. Oxidative damage of DJ-1 is linked to sporadic parkinson and Alzheimer diseases. *J. Biol. Chem.* **2006**, *281*, 10816–10824. [CrossRef]
181. Chen, Y.; Gao, C.; Sun, Q.; Pan, H.; Huang, P.; Ding, J.; Chen, S. MicroRNA-4639 is a regulator of DJ-1 expression and a potential early diagnostic marker for Parkinson's disease. *Front. Aging Neurosci.* **2017**, *9*, 232. [CrossRef]
182. Oh, S.E.; Park, H.-J.; He, L.; Skibiel, C.; Junn, E.; Mouradian, M.M. The Parkinson's disease gene product DJ-1 modulates miR-221 to promote neuronal survival against oxidative stress. *Redox Biol.* **2018**, *19*, 62–73. [CrossRef]
183. Xiong, R.; Wang, Z.; Zhao, Z.; Li, H.; Chen, W.; Zhang, B.; Wang, L.; Wu, L.; Li, W.; Ding, J.; et al. MicroRNA-494 reduces DJ-1 expression and exacerbates neurodegeneration. *Neurobiol. Aging* **2014**, *35*, 705–714. [CrossRef] [PubMed]
184. Miñones-Moyano, E.; Porta, S.; Escaramís, G.; Rabionet, R.; Iraola, S.; Kagerbauer, B.; Espinosa-Parrilla, Y.; Ferrer, I.; Estivill, X.; Martí, E. MicroRNA profiling of Parkinson's disease brains identifies early downregulation of miR-34b/c which modulate mitochondrial function. *Hum. Mol. Genet.* **2011**, *20*, 3067–3078. [CrossRef]
185. Pantano, L.; Friedländer, M.R.; Escaramís, G.; Lizano, E.; Pallarès-Albanell, J.; Ferrer, I.; Estivill, X.; Martí, E. Specific small-RNA signatures in the amygdala at premotor and motor stages of Parkinson's disease revealed by deep sequencing analysis. *Bioinformatice* **2015**, *32*, 673–681. [CrossRef] [PubMed]
186. Junn, E.; Lee, K.W.; Byeong, S.J.; Chan, T.W.; Im, J.Y.; Mouradian, M.M. Repression of α-synuclein expression and toxicity by microRNA-7. *Proc. Natl. Acad. Sci. USA* **2009**. [CrossRef] [PubMed]
187. McMillan, K.J.; Murray, T.K.; Bengoa-Vergniory, N.; Cordero-Llana, O.; Cooper, J.; Buckley, A.; Wade-Martins, R.; Uney, J.B.; O'Neill, M.J.; Wong, L.F.; et al. Loss of MicroRNA-7 regulation leads to α-synuclein accumulation and dopaminergic neuronal loss in vivo. *Mol. Ther.* **2017**, *25*, 2404–2414. [CrossRef]
188. Kabaria, S.; Choi, D.C.; Chaudhuri, A.D.; Jain, M.R.; Li, H.; Junn, E. MicroRNA-7 activates Nrf2 pathway by targeting Keap1 expression. *Free. Radic. Biol. Med.* **2015**, *89*, 548–556. [CrossRef]
189. Narasimhan, M.; Patel, D.; Vedpathak, D.; Rathinam, M.; Henderson, G.; Mahimainathan, L. Identification of novel microRNAs in post-transcriptional control of Nrf2 expression and redox homeostasis in neuronal, SH-SY5Y cells. *PLoS ONE* **2012**, *7*, e51111. [CrossRef]
190. Van Es, M.A.; Schelhaas, H.J.; Van Vught, P.W.J.; Ticozzi, N.; Andersen, P.M.; Groen, E.J.N.; Schulte, C.; Blauw, H.M.; Koppers, M.; Diekstra, F.P.; et al. Angiogenin variants in Parkinson disease and amyotrophic lateral sclerosis. *Ann. Neurol.* **2011**, *70*, 964–973. [CrossRef]
191. Steidinger, T.U.; Standaert, D.G.; Yacoubian, T.A. A neuroprotective role for angiogenin in models of Parkinson's disease. *J. Neurochem.* **2010**, *116*, 334–341. [CrossRef]
192. Steidinger, T.U.; Slone, S.R.; Ding, H.; Standaert, D.G.; Yacoubian, T.A. Angiogenin in Parkinson disease models: Role of akt phosphorylation and evaluation of AAV-mediated angiogenin expression in MPTP treated mice. *PLoS ONE* **2013**, *8*, e56092. [CrossRef]
193. Blanco, S.; Dietmann, S.; Flores, J.V.; Hussain, S.; Kutter, C.; Humphreys, P.; Lukk, M.; Lombard, P.; Treps, L.; Popis, M.; et al. Aberrant methylation of t RNA s links cellular stress to neuro-developmental disorders. *EMBO J.* **2014**, *33*, 2020–2039. [CrossRef]
194. Pallarès-Albanell, J.; Zomeño-Abellán, M.T.; Escaramís, G.; Pantano, L.; Soriano, A.; Segura, M.F.; Martí, E. A high-throughput screening identifies microRNA inhibitors that influence neuronal maintenance and/or response to oxidative stress. *Mol. Ther.-Nucleic Acids* **2019**, *17*, 374–387. [CrossRef]
195. Hanada, T.; Weitzer, S.; Mair, B.; Bernreuther, C.; Wainger, B.J.; Ichida, J.K.; Hanada, R.; Orthofer, M.; Cronin, S.J.; Komnenovic, V.; et al. CLP1 links tRNA metabolism to progressive motor-neuron loss. *Nature* **2013**, *495*, 474–480. [CrossRef]
196. Wang, L.; Yang, H.; Wang, Q.; Zhang, Q.; Wang, Z.; Zhang, Q.; Wu, S.; Li, H. Paraquat and MPTP induce alteration in the expression profile of long noncoding RNAs in the substantia nigra of mice: Role of the transcription factor Nrf2. *Toxicol. Lett.* **2018**, *291*, 11–28. [CrossRef]

197. Cai, L.; Tu, L.; Li, T.; Yang, X.; Ren, Y.; Gu, R.; Zhang, Q.; Yao, H.; Qu, X.; Wang, Q.; et al. Downregulation of lncRNA UCA1 ameliorates the damage of dopaminergic neurons, reduces oxidative stress and inflammation in Parkinson's disease through the inhibition of the PI3K/Akt signaling pathway. *Int. Immunopharmacol.* **2019**, *75*, 105734. [CrossRef]
198. Liu, W.; Zhang, Q.; Zhang, J.; Pan, W.; Zhao, J.; Xu, Y. Long non-coding RNA MALAT1 contributes to cell apoptosis by sponging miR-124 in Parkinson disease. *Cell Biosci.* **2017**, *7*, 1–9. [CrossRef]
199. Zhai, K.; Liu, B.; Gao, L. Long Non-coding RNA TUG1 Promotes Parkinson's disease via modulating MiR-152-3p/PTEN pathway. *Hum. Gene Ther.* **2020**. [CrossRef]
200. Kraus, T.F.J.; Haider, M.; Spanner, J.; Steinmaurer, M.; Dietinger, V.; Kretzschmar, H.A. Altered long noncoding RNA expression precedes the course of Parkinson's disease—A preliminary report. *Mol. Neurobiol.* **2016**, *54*, 2869–2877. [CrossRef]
201. Ding, X.-M.; Zhao, L.-J.; Qiao, H.-Y.; Wu, S.-L.; Wang, X.-H. Long non-coding RNA-p21 regulates MPP+-induced neuronal injury by targeting miR-625 and derepressing TRPM2 in SH-SY5Y cells. *Chem. Interact.* **2019**, *307*, 73–81. [CrossRef] [PubMed]
202. Simchovitz, A.; Hanan, M.; Niederhoffer, N.; Madrer, N.; Yayon, N.; Bennett, E.R.; Greenberg, D.S.; Kadener, S.; Soreq, H. NEAT1 is overexpressed in Parkinson's disease substantia nigra and confers drug-inducible neuroprotection from oxidative stress. *FASEB J.* **2019**, *33*, 11223–11234. [CrossRef]
203. Simchovitz, A.; Hanan, M.; Yayon, N.; Lee, S.; Bennett, E.R.; Greenberg, D.S.; Kadener, S.; Soreq, H. A lncRNA survey finds increases in neuroprotective LINC-PINT in Parkinson's disease substantia nigra. *Aging Cell* **2020**, *19*, e13115. [CrossRef]
204. Hanan, M.; Simchovitz, A.; Yayon, N.; Vaknine, S.; Cohen-Fultheim, R.; Karmon, M.; Madrer, N.; Rohrlich, T.M.; Maman, M.; Bennett, E.R.; et al. A Parkinson's disease Circ RNA s resource reveals a link between circ SLC 8A1 and oxidative stress. *EMBO Mol. Med.* **2020**, *12*. [CrossRef]
205. Jia, E.; Zhou, Y.; Liu, Z.; Wang, L.; Ouyang, T.; Pan, M.; Liu, Z.; Ge, Q. transcriptomic profiling of circular RNA in different brain regions of Parkinson's disease in a mouse model. *Int. J. Mol. Sci.* **2020**, *21*, 3006. [CrossRef]
206. Feng, Z.; Zhang, L.; Wang, S.; Hong, Q. Circular RNA circDLGAP4 exerts neuroprotective effects via modulating miR-134-5p/CREB pathway in Parkinson's disease. *Biochem. Biophys. Res. Commun.* **2020**, *522*, 388–394. [CrossRef]
207. Sang, Q.; Liu, X.; Wang, L.; Qi, L.; Sun, W.; Wang, W.; Sun, Y.; Zhang, H. CircSNCA downregulation by pramipexole treatment mediates cell apoptosis and autophagy in Parkinson's disease by targeting miR-7. *Aging* **2018**, *10*, 1281–1293. [CrossRef] [PubMed]
208. McColgan, P.; Tabrizi, S.J. Huntington's disease: A clinical review. *Eur. J. Neurol.* **2018**, *25*, 24–34. [CrossRef] [PubMed]
209. Ross, C.A.; Tabrizi, S.J. Huntington's disease: From molecular pathogenesis to clinical treatment. *Lancet Neurol.* **2011**, *10*, 83–98. [CrossRef]
210. Bañez-Coronel, M.; Ayhan, F.; Tarabochia, A.D.; Zu, T.; Perez, B.A.; Tusi, S.K.; Pletnikova, O.; Borchelt, D.R.; Ross, C.A.; Margolis, R.L.; et al. RAN translation in Huntington disease. *Neuron* **2015**, *88*, 667–677. [CrossRef]
211. Báñez-Coronel, M.; Porta, S.; Kagerbauer, B.; Mateu-Huertas, E.; Pantano, L.; Ferrer, I.; Guzmán, M.; Estivill, X.; Martí, E. A Pathogenic mechanism in Huntington's disease involves small CAG-Repeated RNAs with neurotoxic activity. *PLoS Genet.* **2012**, *8*, e1002481. [CrossRef]
212. Martí, E. RNA toxicity induced by expanded CAG repeats in Huntington's disease. *Brain Pathol.* **2016**, *26*, 779–786. [CrossRef]
213. Martí, E.; Pantano, L.; Bañez-Coronel, M.; Llorens, F.; Miñones-Moyano, E.; Porta, S.; Sumoy, L.; Ferrer, I.; Estivill, X. A myriad of miRNA variants in control and Huntington's disease brain regions detected by massively parallel sequencing. *Nucleic Acids Res.* **2010**, *38*, 7219–7235. [CrossRef]
214. Savas, J.N.; Makusky, A.; Ottosen, S.; Baillat, D.; Then, F.; Krainc, D.; Shiekhattar, R.; Markey, S.P.; Tanese, N. Huntington's disease protein contributes to RNA-mediated gene silencing through association with Argonaute and P bodies. *Proc. Nat.Acad. Sci. USA* **2008**. [CrossRef]
215. Pircs, K.; Petri, R.; Madsen, S.; Brattås, P.L.; Vuono, R.; Ottosson, D.R.; St-Amour, I.; Hersbach, B.A.; Matusiak-Brückner, M.; Lundh, S.H.; et al. Huntingtin aggregation impairs autophagy, leading to argonaute-2 accumulation and global microRNA dysregulation. *Cell Rep.* **2018**, *24*, 1397–1406. [CrossRef]

216. Lee, S.-T.; Chu, K.; Im, W.-S.; Yoon, H.-J.; Im, J.-Y.; Park, J.-E.; Park, K.-H.; Jung, K.-H.; Lee, S.K.; Kim, M.; et al. Altered microRNA regulation in Huntington's disease models. *Exp. Neurol.* **2011**, *227*, 172–179. [CrossRef]
217. Packer, A.N.; Xing, Y.; Harper, S.Q.; Jones, L.; Davidson, B.L. The bifunctional microRNA miR-9/miR-9* regulates REST and coREST and is downregulated in Huntington's disease. *J. Neurosci.* **2008**, *28*, 14341–14346. [CrossRef] [PubMed]
218. Langfelder, P.; Gao, F.; Wang, N.; Howland, D.; Kwak, S.; Vogt, T.F.; Aaronson, J.S.; Rosinski, J.; Coppola, G.; Horvath, S.; et al. MicroRNA signatures of endogenous Huntingtin CAG repeat expansion in mice. *PLoS ONE* **2018**, *13*, e0190550. [CrossRef] [PubMed]
219. Wong, A.H.-K.; Veremeyko, T.; Patel, N.; Lemere, C.A.; Walsh, M.M.; Esau, C.; Vanderburg, C.R.; Krichevsky, A.M. De-repression of FOXO3a death axis by microRNA-132 and -212 causes neuronal apoptosis in Alzheimer's disease. *Hum. Mol. Genet.* **2013**, *22*, 3077–3092. [CrossRef]
220. Zhu, H.; Wang, X.; Chen, S. Downregulation of MiR-218-5p protects against oxygen-glucose deprivation/reperfusion-induced injuries of PC12 cells via upregulating n-myc downstream regulated gene 4 (NDRG4). *Med Sci. Monit. Int. Med. J. Exp. Clin. Res.* **2020**, *26*, e920101-1. [CrossRef]
221. Johnson, R. Long non-coding RNAs in Huntington's disease neurodegeneration. *Neurobiol. Dis.* **2012**, *46*, 245–254. [CrossRef]
222. Chanda, K.; Das, S.; Chakraborty, J.; Bucha, S.; Maitra, A.; Chatterjee, R.; Mukhopadhyay, D.; Bhattacharyya, N.P. Altered levels of long NcRNAs Meg3 and Neat1 in cell and animal models of Huntington's disease. *RNA Biol.* **2018**, *15*, 1348–1363. [CrossRef] [PubMed]
223. Brown, R.H.; Al-Chalabi, A. Amyotrophic lateral sclerosis. *N. Engl. J. Med.* **2001**, *344*, 1688–1700.
224. Kang, S.H.; Li, Y.; Fukaya, M.; Lorenzini, I.; Cleveland, D.W.; Ostrow, L.W.; Rothstein, J.D.; Bergles, D.E. Degeneration and impaired regeneration of gray matter oligodendrocytes in amyotrophic lateral sclerosis. *Nat. Neurosci.* **2013**, *16*, 571–579. [CrossRef]
225. Philips, T.; Rothstein, J.D. Glial cells in amyotrophic lateral sclerosis. *Exp. Neurol.* **2014**, *262*, 111–120. [CrossRef]
226. Rosen, D.R.; Siddique, T.; Patterson, D.; Figlewicz, D.A.; Sapp, P.C.; Hentati, A.; Donaldson, D.H.; Goto, J.; O'Regan, J.P.; Deng, H.-X.; et al. Mutations in Cu/Zn superoxide dismutase gene are associated with familial amyotrophic lateral sclerosis. *Nature* **1993**, *362*, 59–62. [CrossRef]
227. Douglas, A.G.L. Non-coding RNA in C9orf72-related amyotrophic lateral sclerosis and frontotemporal dementia: A perfect storm of dysfunction. *Non-Coding RNA Res.* **2018**, *3*, 178–187. [CrossRef]
228. Neumann, M.; Sampathu, D.M.; Kwong, L.K.; Truax, A.C.; Micsenyi, M.C.; Chou, T.T.; Bruce, J.; Schuck, T.; Grossman, M.; Clark, C.M.; et al. Ubiquitinated TDP-43 in frontotemporal lobar degeneration and amyotrophic lateral sclerosis. *Science* **2006**, *314*, 130–133. [CrossRef]
229. Deng, H.-X.; Chen, W.; Hong, S.-T.; Boycott, K.M.; Gorrie, G.H.; Siddique, N.; Yang, Y.; Fecto, F.; Shi, Y.; Zhai, H.; et al. Mutations in UBQLN2 cause dominant X-linked juvenile and adult-onset ALS and ALS/dementia. *Nature* **2011**, *477*, 211–215. [CrossRef]
230. Bosco, D.A.; Morfini, G.; Karabacak, N.M.; Song, Y.; Gros-Louis, F.; Pasinelli, P.; Goolsby, H.; Fontaine, B.A.; Lemay, N.; McKenna-Yasek, D.; et al. Wild-type and mutant SOD1 share an aberrant conformation and a common pathogenic pathway in ALS. *Nat. Neurosci.* **2010**, *13*, 1396–1403. [CrossRef]
231. Robberecht, W. Oxidative stress in amyotrophic lateral sclerosis. *J. Neurol.* **2000**, *247*. [CrossRef] [PubMed]
232. Joilin, G.; Leigh, P.N.; Newbury, S.F.; Hafezparast, M. An overview of microRNAs as biomarkers of ALS. *Front. Neurol.* **2019**, *10*, 186. [CrossRef]
233. Kovanda, A.; Leonardis, L.; Zidar, J.; Koritnik, B.; Dolenc-Groselj, L.; Kovacic, S.R.; Curk, T.; Rogelj, B. Differential expression of microRNAs and other small RNAs in muscle tissue of patients with ALS and healthy age-matched controls. *Sci. Rep.* **2018**, *8*, 5609. [CrossRef]
234. Ricci, C.; Marzocchi, C.; Battistini, S. MicroRNAs as biomarkers in amyotrophic lateral sclerosis. *Cells* **2018**, *7*, 219. [CrossRef]
235. Rizzuti, M.; Filosa, G.; Melzi, V.; Calandriello, L.; Dioni, L.; Bollati, V.; Bresolin, N.; Comi, G.P.; Barabino, S.; Nizzardo, M.; et al. MicroRNA expression analysis identifies a subset of downregulated miRNAs in ALS motor neuron progenitors. *Sci. Rep.* **2018**, *8*, 1–12. [CrossRef] [PubMed]
236. Waller, R.; Wyles, M.; Heath, P.R.; Kazoka, M.; Wollff, H.; Shaw, P.J.; Kirby, J. Small RNA sequencing of sporadic amyotrophic lateral sclerosis cerebrospinal fluid reveals differentially expressed miRNAs Related to neural and glial activity. *Front. Neurosci.* **2018**, *11*, 731. [CrossRef]

237. Ravnik-Glavač, M.; Glavač, D. Circulating RNAs as potential biomarkers in amyotrophic lateral sclerosis. *Int. J. Mol. Sci.* **2020**, *21*, 1714. [CrossRef]
238. Koval, E.D.; Shaner, C.; Zhang, P.; Du Maine, X.; Fischer, K.; Tay, J.; Chau, B.N.; Wu, G.F.; Miller, T.M. Method for widespread microRNA-155 inhibition prolongs survival in ALS-model mice. *Hum. Mol. Genet.* **2013**, *22*, 4127–4135. [CrossRef] [PubMed]
239. Li, C.; Wei, Q.; Gu, X.; Chen, Y.; Chen, X.; Cao, B.; Ou, R.; Shang, H.-F. Decreased glycogenolysis by miR-338-3p promotes regional glycogen accumulation within the spinal cord of amyotrophic lateral sclerosis mice. *Front. Mol. Neurosci.* **2019**, *12*, 114. [CrossRef]
240. Kim, D.; Nguyen, M.D.; Dobbin, M.M.; Fischer, A.; Sananbenesi, F.; Rodgers, J.T.; Delalle, I.; Baur, J.A.; Sui, G.; Armour, S.M.; et al. SIRT1 deacetylase protects against neurodegeneration in models for Alzheimer's disease and amyotrophic lateral sclerosis. *EMBO J.* **2007**, *26*, 3169–3179. [CrossRef]
241. Shioya, M.; Obayashi, S.; Tabunoki, H.; Arima, K.; Saito, Y.; Ishida, T.; Satoh, J. Aberrant microRNA expression in the brains of neurodegenerative diseases: MiR-29a decreased in Alzheimer disease brains targets neurone navigator 3. *Neuropathol. Appl. Neurobiol.* **2010**, *36*, 320–330. [CrossRef]
242. Aschrafi, A.; Kar, A.N.; Natera-Naranjo, O.; MacGibeny, M.A.; Gioio, A.E.; Kaplan, B.B. MicroRNA-338 regulates the axonal expression of multiple nuclear-encoded mitochondrial mRNAs encoding subunits of the oxidative phosphorylation machinery. *Cell. Mol. Life Sci.* **2012**, *69*, 4017–4027. [CrossRef]
243. Emde, A.; Eitan, C.; Liou, L.; Libby, R.T.; Rivkin, N.; Magen, I.; Reichenstein, I.; Oppenheim, H.; Eilam, R.; Silvestroni, A.; et al. Dysregulated mi RNA biogenesis downstream of cellular stress and ALS-causing mutations: A new mechanism for ALS. *EMBO J.* **2015**, *34*, 2633–2651. [CrossRef] [PubMed]
244. Paladino, S.; Conte, A.; Caggiano, R.; Pierantoni, G.M.; Faraonio, R. Nrf2 pathway in age-related neurological disorders: Insights into microRNAs. *Cell. Physiol. Biochem.* **2018**, *47*, 1951–1976. [CrossRef]
245. Wang, N.; Zhang, L.; Lu, Y.; Zhang, M.; Zhang, Z.; Wang, K.; Lv, J. Down-regulation of microRNA-142-5p attenuates oxygen-glucose deprivation and reoxygenation-induced neuron injury through up-regulating Nrf2/ARE signaling pathway. *Biomed. Pharmacother.* **2017**, *89*, 1187–1195. [CrossRef] [PubMed]
246. Greenway, M.J.; Andersen, P.M.; Russ, C.; Ennis, S.; Cashman, S.; Donaghy, C.; Patterson, V.; Swingler, R.; Kieran, D.; Prehn, J.; et al. ANG mutations segregate with familial and 'sporadic' amyotrophic lateral sclerosis. *Nat. Genet.* **2006**, *38*, 411–413. [CrossRef]
247. Wolozin, B. Physiological protein aggregation run amuck: Stress granules and the genesis of neurodegenerative disease. *Discov. Med.* **2014**, *17*, 47–52.
248. Wolozin, B. Regulated protein aggregation: Stress granules and neurodegeneration. *Mol. Neurodegener.* **2012**, *7*, 56. [CrossRef]
249. Joilin, G.; Gray, E.; Thompson, A.G.; Bobeva, Y.; Talbot, K.; Weishaupt, J.; Ludolph, A.; Malaspina, A.; Leigh, P.N.; Newbury, S.F.; et al. Identification of a potential non-coding RNA biomarker signature for amyotrophic lateral sclerosis. *Brain Commun.* **2020**, *2*, fcaa053. [CrossRef]
250. Birger, A.; Ben-Dor, I.; Ottolenghi, M.; Turetsky, T.; Gil, Y.; Sweetat, S.; Perez, L.; Belzer, V.; Casden, N.; Steiner, D.; et al. Human iPSC-derived astrocytes from ALS patients with mutated C9ORF72 show increased oxidative stress and neurotoxicity. *EBioMedicine* **2019**, *50*, 274–289. [CrossRef] [PubMed]
251. Nishimoto, Y.; Nakagawa, S.; Hirose, T.; Okano, H.J.; Takao, M.; Shibata, S.; Suyama, S.; Kuwako, K.-I.; Imai, T.; Murayama, S.; et al. The long non-coding RNA nuclear-enriched abundant transcript 1_2 induces paraspeckle formation in the motor neuron during the early phase of amyotrophic lateral sclerosis. *Mol. Brain* **2013**, *6*, 31. [CrossRef]
252. Wang, C.; Duan, Y.; Duan, G.; Wang, Q.; Zhang, K.; Deng, X.; Qian, B.; Gu, J.; Ma, Z.; Zhang, S.; et al. Stress induces dynamic, cytotoxicity-antagonizing TDP-43 nuclear bodies via paraspeckle lncRNA NEAT1-mediated liquid-liquid phase separation. *Mol. Cell* **2020**, *79*, 443–458.e7. [CrossRef]
253. Gupta, P.; Bhattacharjee, S.; Sharma, A.R.; Sharma, G.; Lee, S.-S.; Chakraborty, C. miRNAs in Alzheimer disease—A therapeutic perspective. *Curr. Alzheimer Res.* **2017**, *14*, 1198–1206. [CrossRef]
254. Adams, B.D.; Parsons, C.; Walker, L.; Zhang, W.C.; Slack, F.J. Targeting noncoding RNAs in disease. *J. Clin. Investig.* **2017**, *127*, 761–771. [CrossRef]
255. Jin, H.Y.; Gonzalez-Martin, A.; Miletic, A.V.; Lai, M.; Knight, S.; Sabouri-Ghomi, M.; Head, S.R.; Macauley, M.S.; Rickert, R.C.; Xiao, C. Transfection of microRNA mimics should be used with caution. *Front. Genet.* **2015**, *6*, 340. [CrossRef]

256. He, M.; Liu, Y.; Wang, X.; Zhang, M.Q.; Hannon, G.J.; Huang, Z.J. Cell-type-based analysis of microRNA profiles in the mouse brain. *Neuron* **2012**, *73*, 35–48. [CrossRef]
257. Espay, A.J.; Brundin, P.; Lang, A.E. Precision medicine for disease modification in Parkinson disease. *Nat. Rev. Neurol.* **2017**, *13*, 119–126. [CrossRef] [PubMed]
258. Dawson, T.M.; Golde, T.E.; Lagier-Tourenne, C. Animal models of neurodegenerative diseases. *Nat. Neurosci.* **2018**, *21*, 1370–1379. [CrossRef]
259. Derrien, T.; Johnson, R.; Bussotti, G.; Tanzer, A.; Djebali, S.; Tilgner, H.; Guernec, G.; Martin, D.; Merkel, A.; Knowles, D.G.; et al. The GENCODE v7 catalog of human long noncoding RNAs: Analysis of their gene structure, evolution, and expression. *Genome Res.* **2012**, *22*, 1775–1789. [CrossRef] [PubMed]
260. Cao, X.; Yeo, G.; Muotri, A.R.; Kuwabara, T.; Gage, F.H. Noncoding rnas in the mammalian central nervous system. *Annu. Rev. Neurosci.* **2006**, *29*, 77–103. [CrossRef]
261. Friedländer, M.R.; Lizano, E.; Houben, A.J.S.; Bezdan, D.; Báñez-Coronel, M.; Kudla, G.; Mateu-Huertas, E.; Kagerbauer, B.; González, J.; Chen, K.C.; et al. Evidence for the biogenesis of more than 1000 novel human microRNAs. *Genome Biol.* **2014**, *15*, R57. [CrossRef] [PubMed]
262. Engle, S.J.; Blaha, L.; Kleiman, R.J. Best practices for translational disease modeling using human iPSC-derived neurons. *Neuron* **2018**, *100*, 783–797. [CrossRef]
263. Tolosa, E.; Botta-Orfila, T.; Morató, X.; Calatayud, C.; Ferrer-Lorente, R.; Martí, M.-J.; Fernández, M.; Gaig, C.; Raya, Á.; Consiglio, A.; et al. MicroRNA alterations in iPSC-derived dopaminergic neurons from Parkinson disease patients. *Neurobiol. Aging* **2018**, *69*, 283–291. [CrossRef]
264. Grenier, K.; Kao, J.; Diamandis, P. Three-dimensional modeling of human neurodegeneration: Brain organoids coming of age. *Mol. Psychiatry* **2019**, *25*, 254–274. [CrossRef]
265. Katerji, M.; Filippova, M.; Duerksen-Hughes, P. Approaches and methods to measure oxidative stress in clinical samples: Research applications in the cancer field. *Oxidative Med. Cell. Longev.* **2019**, *2019*, 1–29. [CrossRef]
266. O'Brien, K.; Breyne, K.; Ughetto, S.; Laurent, L.C.; Breakefield, X.O. RNA delivery by extracellular vesicles in mammalian cells and its applications. *Nat. Rev. Mol. Cell Biol.* **2020**, *21*, 585–606. [CrossRef]
267. van den Berg, M.; Krauskopf, J.; Ramaekers, J.G.; Kleinjans, J.C.S.; Prickaerts, J.; Briedé, J.J. Circulating microRNAs as potential biomarkers for psychiatric and neurodegenerative disorders. *Prog. Neurobiol.* **2020**, *185*, 101732. [CrossRef]
268. Brennan, S.; Keon, M.; Liu, B.; Su, Z.; Saksena, N.K. Panoramic visualization of circulating microRNAs across Neurodegenerative diseases in humans. *Mol. Neurobiol.* **2019**, *56*, 7380–7407. [CrossRef]

Publisher's Note: MDPI stays neutral with regard to jurisdictional claims in published maps and institutional affiliations.

© 2020 by the authors. Licensee MDPI, Basel, Switzerland. This article is an open access article distributed under the terms and conditions of the Creative Commons Attribution (CC BY) license (http://creativecommons.org/licenses/by/4.0/).

Review

Seizure-Induced Oxidative Stress in Status Epilepticus: Is Antioxidant Beneficial?

Tsu-Kung Lin [1,2,3,†], Shang-Der Chen [1,4,†], Kai-Jung Lin [2] and Yao-Chung Chuang [1,3,4,5,6,*]

1 Department of Neurology, Kaohsiung Chang Gung Memorial Hospital, Kaohsiung 83301, Taiwan; tklin@adm.cgmh.org.tw (T.-K.L.); chensd@adm.cgmh.org.tw (S.-D.C.)
2 Center for Mitochondrial Research and Medicine, Kaohsiung Chang Gung Memorial Hospital, Kaohsiung 833, Taiwan; b101101092@tmu.edu.tw
3 College of Medicine, Chang Gung University, Taoyuan 33302, Taiwan
4 Institute for Translation Research in Biomedicine, Kaohsiung Chang Gung Memorial Hospital, Kaohsiung 83301, Taiwan
5 Department of Neurology, School of Medicine, College of Medicine, Kaohsiung Medical University, Kaohsiung 80708, Taiwan
6 Department of Biological Science, National Sun Yat-sen University, Kaohsiung 80424, Taiwan
* Correspondence: ycchuang@cgmh.org.tw; Tel.: +886-7-7317123
† These authors contributed equally to this work.

Received: 9 September 2020; Accepted: 19 October 2020; Published: 22 October 2020

Abstract: Epilepsy is a common neurological disorder which affects patients physically and mentally and causes a real burden for the patient, family and society both medically and economically. Currently, more than one-third of epilepsy patients are still under unsatisfied control, even with new anticonvulsants. Other measures may be added to those with drug-resistant epilepsy. Excessive neuronal synchronization is the hallmark of epileptic activity and prolonged epileptic discharges such as in status epilepticus can lead to various cellular events and result in neuronal damage or death. Unbalanced oxidative status is one of the early cellular events and a critical factor to determine the fate of neurons in epilepsy. To counteract excessive oxidative damage through exogenous antioxidant supplements or induction of endogenous antioxidative capability may be a reasonable approach for current anticonvulsant therapy. In this article, we will introduce the critical roles of oxidative stress and further discuss the potential use of antioxidants in this devastating disease.

Keywords: epilepsy; status epilepticus; oxidative stress; antioxidant

---

1. Introduction

Epilepsy is a common neurological disorder and estimated with a prevalence of 0.5–1.5% in the general population varying between developed countries and developing countries [1–3]. The term "epilepsy" originated from the Greek verb "epilambanein" (επιλαμϖανειν), which means "to be attacked or to be seized". Epilepsy as a medical condition is known throughout human history and was described among famous people such as Fyodor Dostoyevsky (Russian writer), Gaius Julius Caesar (Roman general and the great strategist), and Napoleon Bonaparte (French general and Emperor) [4–6]. According to the International League Against Epilepsy (ILAE), an epileptic seizure is caused by abnormal, excessive, and synchronous neuronal activity in the different brain areas and presenting various signs and/or symptoms accordingly [7]. Epilepsy is typified by repeated seizures and the aberrant firing of neurons located mainly in the cerebral cortex and often under unprovoked conditions. The burst firing neurons accompanied with epileptic discharges could result in a variety of changes at the cellular level, e.g., activation of glutamate receptors, alteration of γ-aminobutyric acid (GABA) receptor, activation of cytokine expression, increased oxidative stress, modification of neurogenesis, adjustment of plasticity or stimulation of some late cell death pathways [8–10].

Status epilepticus, defined with prolonged or continuous epileptic seizures, is a neurological emergency associated with significant morbidity and mortality [11–13]. Clinical studies revealed that patients with epilepsy are often associated with behavioral and cognitive decline from multifactorial contributions [14–16]. About 4–16% of people with epilepsy have experienced at least one attack of status epilepticus [17]. A recent study showed status epilepticus with prominent motor phenomena was 24 per 100,000 adults per year [18]. Status epilepticus has a 10–20% mortality in previous studies [13,17], although more rapid treatment regimens at an early stage reduce disease-related mortality in recent studies [12,19]. Human and animal studies also revealed that status epilepticus causes substantial cerebral damage, raises the risk to develop succeeding epileptic episodes, and accompanies with a distinctive pattern of neuronal death in the hippocampus, such as in dentate gyrus hilus, CA3, and CA1 subfield [20,21].

It was gradually recognized that excessive oxidative stress plays a crucial role in the pathophysiology of status epilepticus [22]. While in prolonged seizure or status epilepticus, reactive oxygen species (ROS) are generated mainly by nicotinamide adenine dinucleotide phosphate (NADPH) oxidase through NMDA receptor activation [23,24]. The inflammatory response following status epilepticus causes activation of inducible nitric oxide synthase (iNOS) and iNOS-derived nitric oxide (NO) reacts with O_2^- to form the peroxynitrite and contributes the severity of oxidative stress in kainic-acid induced experimental status epilepticus in our previous studies [25–28]. Several studies with a various animal model of status epilepticus such as pilocarpine-induced, pentylenetetrazole-induced, diisopropylfluorophosphate, lithium-pilocarpine model, also demonstrated the critical role of NO-related pathway and neuro-inflammation contributing to neuronal damage [29–36].

Excessive production of ROS/reactive nitrogen species (RNS) can damage macromolecules in cells such as with lipid peroxidation, DNA damage, oxidative damage with enzyme inhibition and result in neuronal death [22]. As excessive ROS/RNS may affect neuronal fate through diverse mechanisms, modification of those detrimental pathways may have beneficial clinical implications. In this review, we will focus on the critical role of ROS/RNS in status epilepticus which can further impair mitochondrial function, enhance the severity of oxidative stress, and cause neuronal damage or death. The potential use of antioxidants in this devastating disease will be further discussed.

2. The Critical Cole of ROS/RNS in Status Epilepticus

It is known that the burst-firing neurons with epileptic discharges in prolonged seizure can lead to substantial changes in neuronal cells. These include toxic activation of glutamate receptors, decreasing the reuptake of glutamate, alteration of GABA receptor, an upsurge of cytokine expression, escalation of oxidative stress, modification in neuroplasticity, and later to initiate cell death pathways [37].

Diverse sources of ROS exist in living cells that include 5-lipoxygenase, NADPH oxidase, cytochrome P450 enzymes, and mitochondria [38–40]. Balanced ROS are indispensable in the preservation of redox homeostasis and making numerous cellular signaling pathways functional [41]. Living organisms can generate unwarranted ROS under various stressful conditions such as with cytokine stimulation, serum deprivation, and hypoxia [40,41]. To keep a low level of ROS/RNS is vital for continuing the process of regular neuronal function [42]. Overproduction of ROS/RNS may proceed with a detrimental effect and generate wide-ranging damage to lipid membranes, protein components, and DNA structures that can evolve to cell death. This is implicated in the pathogenesis of various neurodegenerative disorders such as Alzheimer's disease (AD), Parkinson's disease (PD), and Huntington's disease (HD) [43–45]. This is no exception for seizure disorder, one of the common neurological disorders, which also need to deal with excessive oxidative stress.

Conventionally, it was assumed that mitochondria are the main source of ROS during epileptiform activity [46,47]. It was gradually appreciated with recent studies that in prolonged seizure or status epilepticus, ROS are generated mainly by NADPH oxidase through NMDA receptor activation [23,24].

Using live-cell imaging techniques in neuron-glia cultures, it was demonstrated that prolonged seizure-like activity raises ROS generation in an NMDA receptor-dependent manner rather than calcium- and mitochondria-dependent ROS pathway [23]. It was reported that ROS derived from NADPH oxidase in the pilocarpine-induced temporal lobe epilepsy and apocynin, an NADPH oxidase inhibitor, attenuated ROS production and neurodegeneration from status epilepticus [48] and prevent seizure-induced neuronal death [49]. Using the perforant path model of epilepsy, inhibition ROS production by 4-(2-Aminomethyl)benzenesulfonyl fluoride hydrochloride (AEBSF), another NADPH oxidase inhibitor, significantly lessened seizure-induced cell death [50]. In a recent study, it was demonstrated the critical role of NMDA receptor-mediated NADPH oxidase-induced oxidative stress and targeting NADPH oxidase may offer an innovative treatment for epilepsies [51]. In a human study with surgically resected epileptic tissue from drug-resistant patients, the hippocampus showed cytoplasmic positivity for p47(phox) and p67(phox) in neurons and glial cells which indicates the activation of NOX2 [52]. Inhibition of NADPH oxidase activation by apocynin involves hippocampal neurogenesis which revealed increased neuronal survival and neuroblast production [53]. It was reported that pilocarpine rats had increased NADPH oxidase 2 expressions and decreased superoxide dismutase (SOD) expression [54]. With in vivo and in vitro pilocarpine model of epilepsy, hyperactivation of NMDA receptors and subsequent activation of NADPH oxidase are crucial contributors for the development of epileptogenesis [55]. The oxidative stress induced by NADPH oxidase activation also plays a pivotal role in the pentylenetetrazol-induced kindling process as well as in pentylenetetrazol kindling-induced hippocampal CA1 autophagy [56]. A study with a combination of antioxidant therapy including NADPH oxidase inhibition and the endogenous antioxidant system activation such as nuclear factor erythroid 2-related factor 2 (Nrf2) can prevent epileptogenesis and modifies chronic epilepsy [57]. These studies revealed the critical role of NADPH oxidase, particularly, NADPH oxidase 2, in oxidative stress-induced neuronal damage and involved neurogenesis, autophagy and epileptogenesis as well.

The production of RNS has closely related ROS. RNS is derived from nitric oxide (NO) and with superoxide (O_2), a reactive oxygen species, to form the peroxynitrite. Peroxynitrite, a highly reactive species, can directly or indirectly react with other cellular components such as DNA, lipids or proteins and induced neuronal death [58]. We have shown before the findings of upregulation of iNOS in the hippocampal CA3 subfield following kainic acid-induced experimental status epilepticus in the rat [27]. Experimental findings revealed the increased level of O_2^- and peroxynitrite and sequential changes in time were well correlated accordingly which lead to mitochondrial apoptosis in the hippocampal CA3 subfield [26]. We further provide evidence that activation of NF-kappaB (NF-κB), a transcriptional factor critical for inflammatory genes expression, heightens iNOS gene expression in hippocampal CA3 neurons following status epilepticus in an animal model [25].

It was generally appreciated that glutamate-mediated excitotoxicity causing necrotic changes in the cerebral cortex are the major morphological findings among dying neuronal cells after status epilepticus [59]. There has been evidence that revealed apoptotic cell death also plays a crucial role in seizure-induced brain damage in animal models of experimental status epilepticus [27,60–62]. It has long been aware that oxidative stress, apoptosis, and neuroinflammation have a key role in various neurological diseases such as ischemic stroke, neurodegenerative diseases or epilepsy [63–67]. Apparently, neuroinflammation and oxidative stress are mutually affected with each other, contribute to immediate and longstanding sequelae of status epilepticus, and indicate a potential therapeutic target.

Several inflammatory mediators were shown to have critical roles in status epilepticus such as Toll-like receptor 4 (TLR4), high mobility group box 1(HMGB1), NF-kB, NLRP3, and iNOS which are induced during status epilepticus and predict the development of neuronal cell loss, cognitive deficits, mortality, the progress of chronic epilepsy [68]. Mounting evidence revealed the signaling pathways involving neuroinflammation affecting status epilepticus. TLR4 and receptor for advanced glycation end product (RAGE) are activated by HMGB1 and critical for seizure generation [69,70]. IL-1 receptor type 1 (IL-1R1)/TLR4 pathway is a potential therapeutic target for disease-modifications in patients with epilepsy [71]. Inhibitors of the IL-1R/TLR signaling exert the anticonvulsant effects in various seizures models which indicate the potential to reduce seizure frequency in currently pharmaco-resistant epilepsies [72]. Seizure-induced neuroinflammation impairs the integrity of the blood-brain barrier (BBB) [73]. Inhibition of the prostaglandin E2 receptor 2 can reduce inflammation, restore BBB function, and decrease mortality following status epilepticus [74]. It was realized that excessive oxidative stress and neuroinflammatory response occur together during status epilepticus and persist afterward. These results may further damage the mitochondrial function and aggravate the severity of oxidative stress. We will further discuss mitochondrial dysfunction after status epilepticus under these signaling cascades toward the neuronal death in the next section.

3. Mitochondrial Dysfunction Further Aggravates the Extent of Oxidative Stress Following Status Epilepticus

Mitochondria are membrane-bound cellular organelles widely distributed in the cytoplasm of cells in most eukaryotic organisms. Emerging studies revealed the multifaceted roles of mitochondria which involve various pathophysiological functions such as energy production, ROS generation, apoptotic process, mitochondrial biogenesis, dynamics, mitophagy, and inflammation [75]. Mitochondrial oxidative phosphorylation comprises five enzyme complexes placed in the mitochondrial inner membrane [76–78]. These are NADH-ubiquinone oxidoreductase (Complex I), succinate-ubiquinone oxidoreductase (Complex II), ubiquinone-cytochrome c oxidoreductase (Complex III), and cytochrome c oxidase (Complex IV) which work together to produce a proton motive force and drive the generation of ATP by complex IV (F1F0-ATP synthase).

As status epilepticus can cause excessive oxidative stress accompanying the neuroinflammatory response and augment the formation of peroxynitrite from NOS II derived NO and O_2^- which may impair the mitochondrial function. Previous studies revealed that prolonged epileptic seizures will alter the redox status, reduce the production of ATP, and lead to energy failure in the brain [79,80]. There has been evidence to show mitochondrial dysfunction in epilepsy in animal and human studies [28,81–84]. We have shown before that in kainic acid-induced status epilepticus, enzyme assay for the mitochondrial respiratory chain showed a notable depression of the activity of Complex I + III in the dentate gyrus and subfield of CA1 and CA3 in the hippocampus [28]. In contrast, the activities of Complex II + III and Complex IV remained unchanged. These findings were accompanied by swelling of mitochondrial spaces including cristae and a variable degree of mitochondrial membrane disruption. These findings stressed the pivotal role of Complex I dysfunction in mitochondria and damage of ultrastructure in the mitochondria membrane in the hippocampus of kainic acid-induced status epilepticus in the rat [28]. In a pilocarpine-induced status epilepticus, selective decline of Complex I and IV function in the respiratory chain was observed in hippocampal CA1 and CA3 subfields and suggested that seizure activity can downregulate the mitochondrial-encoded enzyme's activity of oxidative phosphorylation [85]. Inhibition of mitochondrial Complex I activity was also reported in other studies using different animal models [81,83]. In a human study, it was observed that mitochondrial Complex I dysfunction in the CA3 subfield in patients with refractory temporal lobe epilepsy [84]. Although the underlying mechanism is not well elucidated, it appeared that Complex I of the mitochondria is more vulnerable to ROS and RNS than in other respiratory chain complexes [86].

It is known that status epilepticus can cause mitochondrial respiratory chain dysfunction and result in energy failure. These consequences can further aggravate the severity of oxidative stress and cause neuronal damage or death in the hippocampus. Although glutamate-mediated excitotoxicity often causes necrotic changes in the cerebral cortex after status epilepticus [59], several studies including ours showed that apoptotic cell death also plays an important role in seizure-induced brain damage in experimental status epilepticus [27,60–62]. The existence of apoptosis may lend some time to intervene and lessen neuronal damage, such as with antioxidant treatment. The apoptosis process includes the successive activation of a stream of cysteine proteases (caspases) [87]. A long list of apoptosis-related proteins from mitochondria was entailed such as apoptosis-inducing factor (AIF), cytochrome c, endonuclease G, Smac/DIABLO, and HtrA2/OMI [37,88]. Cells also comprise both pro-apoptotic and anti-apoptotic "Bcl-2 family" proteins and are engaged to the apoptotic pathway under various clinical conditions including brain disease. Detailed apoptosis mechanisms are beyond the scope of this review; some of the excellent reviews available can be helpful for readers to gain a further understanding [88–92].

Except for the well-known pathways involving mitochondria-related oxidative stress, apoptosis, and neuronal damage, the emerging role of the sirtuin family, or the special diet formulation such as ketogenic diet (KD), are also revealed as crucial players in status epilepticus and worth mentioning here. In mammals, the sirtuin family include seven sirtuins: sirtuin1–7, and belong to a class of nicotinamide adenine dinucleotide (NAD)-consuming enzymes that are involved in various biological pathways [93]. The beneficial effect of sirtuins on longevity is related to the capability to modulate on the metabolic pathways which present a potential target to treat human diseases [93,94]. Limited studies concerning the sirtuin family and status epilepticus were reported. It was shown that sirtuin1 activation increases the peroxisome proliferator-activated receptor gamma coactivator 1-alpha (PGC-1α) expression and enhances mitochondrial antioxidant system in status epilepticus [95]. We have demonstrated that the downregulation of sirtuin1 can reduce PGC-1α expression, impair mitochondrial biogenesis, augment Complex I dysfunction, heighten the extent of oxidized proteins, increase caspase-3 expression, and promote neuronal cell damage in the hippocampus in kainic acid-induced status epilepticus [96]. It was also revealed that microRNA-199a-5p regulates the sirtuin1-p53 pathway in pilocarpine-induced status epilepticus and targeting of microRNA-199a-5p exerts a beneficial effect in seizure disorder [97]. In a recent study, it was shown that sirtuin3, a major mitochondria NAD^+-dependent deacetylase, through regulating manganese SOD to lessen excessive oxidative stress from mitochondria can reduce neuronal damage in hippocampal cells after status epilepticus [98]. With these studies, it is noteworthy to further explore the vital role of the sirtuin family in seizure disorders, and this may open an avenue to develop innovative therapy in epilepsy in the future.

The KD is a special diet formulation with high fat, low carbohydrate, controlled protein regime that has been used since one hundred years ago for the treatment of epilepsy [99]. The mechanisms of KD may act through the effects of glucose restriction with ketones formation and interactions with receptors, channels, and metabolic enzymes [100,101]. It can reduce mitochondrial ROS/RNS due to the alteration of the source of energy by using more fat-derived ketone bodies and with less consumption from carbohydrates [102–104]. There are several different forms of the ketogenic diet such as the classical diet, medium-chain triglyceride diet, modified Atkins diet, and modified ketogenic diet, and low glycaemic index treatment [100–105]. These diet formulas may help to reduce the number or severity of seizures in patients with poorly controlled seizures after two more anti-epileptic drugs used. KD is a well-known therapeutic option for children with poor control epilepsy [105]. Mounting evidence also reveals the effectiveness of KD for adults with status epilepticus which denotes the usefulness through dietary adjustment [106–108]. The diet modification offers another treatment option in patients with refractory epilepsy and warrants to further testify the crucial role of critical condition as in status epilepticus.

It is crucial to deal with excessive ROS production from NADPH oxidase caused by NMDA receptor activation in prolonged seizure or status epilepticus [23,24]. The oxidative stress may further impair mitochondrial ultrastructure of the membrane and mitochondrial respiratory chain and aggravate the severity of oxidant damage to neuronal cells [37,109]. In Figure 1, we briefly sketch the signaling pathway about status epilepticus causing the excessive generation of ROS and further aggravating mitochondrial function, increasing the severity of oxidative stress, and causing neuronal apoptotic death. Except for the anticonvulsants for status epilepticus, adding antioxidants in this scenario may have the potential to improve the therapeutic effect. We will further discuss the significance of antioxidants used in status epilepticus in the following section.

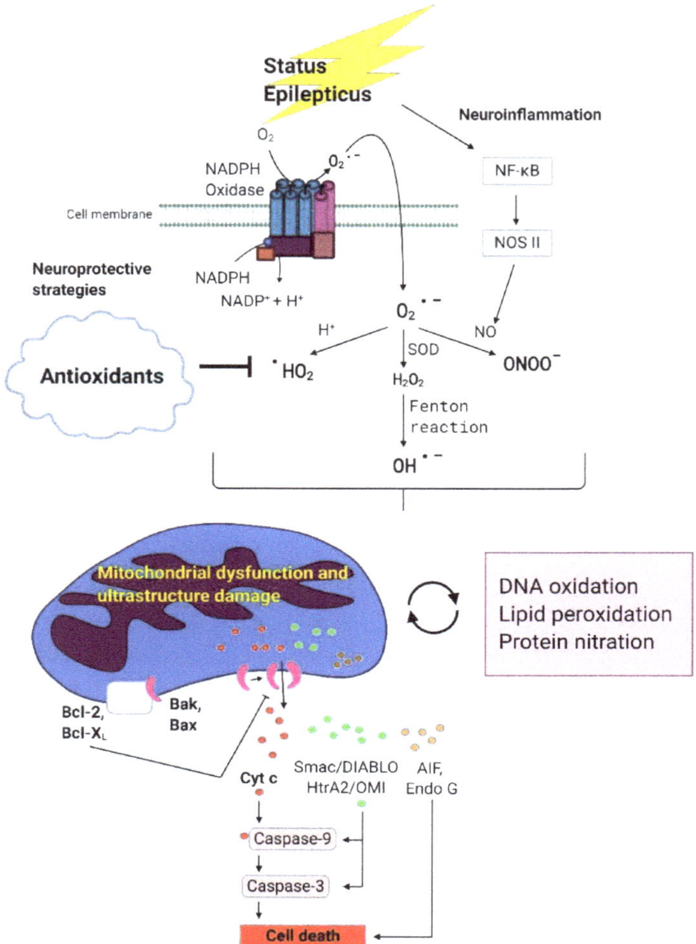

Figure 1. The potential beneficial effects of antioxidants on status epilepticus induced oxidative stress, mitochondrial damage and neurodegeneration. During status epilepticus, superoxide are generated mainly by nicotinamide adenine dinucleotide phosphate (NADPH) oxidase through NMDA receptor activation. Once produced, superoxide can interact with various molecules and further generate secondary radicals. Superoxide can react with nitric oxide and produce peroxynitrite. Superoxide can also be converted into hydrogen peroxide by superoxide dismutase, and then to hydroxyl radicals under the famous Fenton reaction. Interaction of superoxide with protons can produce hydroperoxyl radicals. All these reactive oxygen species/reactive nitrogen species (ROS/RNS) are highly reactive and can cause oxidative stress via alterations of macromolecules. Overproduction of these ROS/RNS due to prolonged epileptic seizures will lead to alteration of mitochondrial redox status, Complex I dysfunction and ultrastructure damage in mitochondria membrane, reduce ATP production and eventually the induction of mitochondrial dependent apoptotic pathway involving the release of a serious of apoptosis-related proteins from mitochondria intermembrane space including apoptosis-inducing factor (AIF), cytochrome c, endonuclease G, Smac/DIABLO and HtrA2/OMI. Protection of mitochondria from bioenergetic failure and oxidative and nitrosative stress with antioxidants may be considered as a potential neuroprotective strategy in neuronal degeneration associated with status epilepticus.

4. The Potentials of Antioxidants in Status Epilepticus

As status epilepticus or prolonged seizure can cause excessive oxidative stress with the various sequential molecular events to cause neuronal damage. Apart from using antiepileptic drugs to reduce or stop the seizure attack, adding antioxidants may have some potential effects to lessen neuronal injury from excessive oxidative stress. Although there are many new antiepileptic drugs with varying mechanisms developed and available in recent decades, more than one-third of patients remained poorly controlled in seizure frequency [110]. It is obvious that there are still unmet needs for drug-resistant epilepsy which highlights the demand for an innovative and maybe an unconventional approach for those patients [111,112]. Studies with drugs or chemical compounds possessing antioxidant characters used in animal models of status epilepticus were selected to discuss and may provide some clues in this regard. However, we do not intend to make a comprehensive review for all kind of antioxidants in epilepsy studies but just to offer a perspective concerning the potential and possibility among drugs or chemical compounds which may be used concurrently with the antiepileptic drugs to further decrease seizure occurrence and improve the outcome of epilepsy. To consider the significance of translation medicine in preclinical studies, we also searched from ClinicalTrials.gov for some of these antioxidants to offer information on the potential clinical use in brain diseases including epilepsy. Selected drugs or chemical compounds with antioxidant characters used in a various animal model of status epilepticus will be presented in Table 1. Relevant information for clinical trials in these compounds and drugs is included and the identifier numbers of ClinicalTrials.gov for epilepsy are provided for readers to more easily locate.

Table 1. Selected references of drugs or chemical compounds with antioxidant characters used in various animal model of status epilepticus. Information concerning clinical trials and various brain diseases including epilepsy and these drugs/chemical compounds from ClinicalTrials.gov are provided. (Abbreviations: Traumatic brain injury, TBI; Alzheimer disease, AD; Parkinson disease, PD; Huntington's disease HD; Peroxisome proliferator-activated receptor gamma, PPARγ; Ataxia-telangiectasia, AT; Malondialdehyde, MDA; Superoxide dismutase, SOD; Glutathione peroxidase, GSH-Px; progressive supranuclear palsy, PSP; Thiobarbituric acid reacting substances, TBARS; Attention deficit hyperactivity disorder, ADHD; Inducible nitric oxide synthase, iNOS).

Name	Relevant Antioxidative Effects or Mechanisms	Pre-Clinical Models	Reference	Relevant Information about Brain Diseases from ClinicalTrials.Gov
N-acetylcysteine	mimetic of glutathione	pentylenetetrazol -induced model	[113]	Trials with neurodegenerative diseases, TBI, brain tumor, and vascular diseases. With one clinical trial concerning epilepsy (ClinicalTrials.gov Identifier: NCT02054949)
Ubiquinone (CoQ10)	potent antioxidant potentiates the antiepileptic effects of phenytoin treatment restores mitochondrial enzyme complex activities, reduce inflammation	pilocarpine-induced model pentylenetetrazol-induced model	[114,115]	Trials with migraine, stroke, AD, PD, HD, epilepsy, ataxia, mitochondrial disease. With one clinical trial concerning epilepsy (Identifier: NCT04488172)
Naringenin	recovery of glutathione content and antioxidant enzymes activity	pilocarpine-induced model	[116]	No clinical trial was found.
EUK-134	synthetic of superoxide dismutase/catalase mimetic, prevent excitotoxic neuronal injury	kainic acid-induced model	[117]	No clinical trial was found.
pioglitazone	PPAR γ agonist reduction of inflammatory responses increased UCP2 expression, reduced oxidant overproduction and improved mitochondrial function	genetically epileptic EL mice kainic acid-induced model	[118,119]	Trials with neurodegenerative diseases, stroke, brain tumor, atherosclerosis, adrenomyeloneuropathy, Friedreich's Ataxia, and AT. No clinical trial involving epilepsy was found.

Table 1. *Cont.*

Name	Relevant Antioxidative Effects or Mechanisms	Pre-Clinical Models	Reference	Relevant Information about Brain Diseases from ClinicalTrials.Gov
Aryl semicarbazides	reduced formation of MDA and increased formation of SOD and GSH-Px	pentylenetetrazol -induced model	[120]	No clinical trial was found.
α-lipoic acid	reduced lipid peroxidation level, nitrite content, augmented the SOD, catalase and GSH-Px activities	pilocarpine-induced model	[121,122]	Trials with AD, PD, mucopolysaccharidosis disorders, ischemic stroke, brain injury, adrenomyeloneuropathy, and PSP. No clinical trial concerning epilepsy was found.
Melatonin	scavenger of hydroxyl radical decreased nitrite content	kainic acid-induced model pentylenetetrazole -induced model	[123,124]	Trials with neurodegenerative diseases, TBI, brain tumor, stroke, birth-related brain injury, migraine, sleep disorders. Four trials concerning "epilepsy and melatonin" were found (Identifier: NCT02195661, NCT00965575, NCT01161108, and NCT01370486)
α-tocopherol (vitamin E)	oxidative stress with chaperone-mediated autophagy decreased TBARS production and total protein carbonylation	pilocarpine-induced model pentylenetetrazol- and methylmalonic acid-induced model	[125,126]	Trials with premature birth-related intraventricular hemorrhage, migraine and sleep disorders. No clinical trial concerning epilepsy was found.
Ascorbic acid (Vitamin C)	decreased lipid peroxidation and increased catalase activity increase GSH level and the decrease in lipid peroxidation level	pilocarpine-induced model penicillin -induced model	[127,128]	Trials with AD, PD, stroke, brain tumor, carotid atherosclerosis, ischemic or, hepatic encephalopathy, cerebral palsy, and ADHD. One study concerning epilepsy (Identifier: NCT02369822)
Selenium and sildenafil	lipid peroxides and nitrotyrosine levels, concomitantly with iNOS inhibition, normalization of TrxR activity and HO-1 expression, and evident neo-angiogenesis	pentylenetetrazol -induced model	[129]	For selenium, trials with phenylketonurias, AD, HD, stroke, cerebral palsy. Two trials concerning epilepsy (Identifier: NCT01764516, NCT01795170)
Selenium and Topiramate	inhibiting free radical supporting antioxidant redox system	pentylenetetrazol -induced model	[130]	As above information
Resveratrol	reduced markers both of oxidative stress and mitochondrial dysfunction suppressing oxidative stress and inflammation	lithium-pilocarpine-induced model kainic acid-induced model	[131,132]	Trials with AD, PD, HD, Friedreich ataxia and hypoxic brain. No clinical trial concerning epilepsy was found.
RTA 408	KEAP1 inhibition, Nrf2 activation	kainic acid-induced model	[133]	One clinical trial concerning "brain disease and RTA 408" for Friedreich's Ataxia". No clinical trial involving epilepsy was found.
AEBSF	NADPH oxidase inhibitor	perforant path stimulation model	[50]	No clinical trial was found.
RTA 408+ AEBSF	KEAP1 inhibition, Nrf2 activation + NADPH oxidase inhibitor	kainic acid induced model	[57]	No clinical trial was found
1400W	selective inhibitor of inducible nitric oxide synthase	kainic acid-induced model	[134]	No clinical trial was found.
Apocynin	an NADPH oxidase assembly inhibitor	pilocarpine-induced model	[49]	No clinical trial was found.
AEOL10150	metalloporphyrin catalytic antioxidant, scavenges peroxynitrite, inhibits lipid peroxidation with SOD and catalase-like activities	pilocarpine-induced model	[135,136]	No clinical trial was found.

N-acetylcysteine (NAC) or acetylcysteine, a medication often used to loosen thick mucus in individuals with chronic lung disease, can replenish glutathione stores, help the formation of glutathione in the body and exert the antioxidative effect [137,138]. NAC was found to be safe with favorable evidence in many psychiatric and neurological diseases in a systemic review [139]. In pentylenetetrazole-induced seizures, adding NAC exhibited an anticonvulsant effect [113]. More than 30 clinical trials concerning "brain disease and N-acetylcysteine" were searched from ClinicalTrials.gov. Brain diseases include various neurodegenerative diseases, traumatic brain injury (TBI), epilepsy, brain tumor, or vascular diseases. However, no data available from epilepsy and N-acetylcysteine trial due to no eligible subjects located.

Ubiquinone, also known as coenzyme Q, is a coenzyme family and widely distributed in living organisms. In the human being, the most common form is coenzyme Q10. As one component in the electron transport chain, ubiquinone plays a crucial role in aerobic cellular respiration and produces ATP for energy need in the body. It also is an essential antioxidant and a modulator of the permeability transition pore and critical in various pathophysiological effects [140–142]. In pilocarpine-induced seizures, coenzyme Q10 lessened the severity of oxidative stress. Furthermore, it enhanced the antiepileptic effects of phenytoin treatment [114]. With coenzyme Q10 combined with minocycline in pentylenetetrazol-induced epilepsy, the ability to reverse oxidative damage and restore mitochondrial enzyme complex activities is better than control and individual effect of coenzyme Q10 or minocycline alone [115]. There were 35 clinical trials that were found concerning "brain disease and coenzyme Q10" that were found on ClinicalTrials.gov. Various brain diseases patients recruited for the study include migraine, stroke, AD, PD, HD, ataxia, mitochondrial disease, and epilepsy. One study including coenzyme Q10 as one of multi-vitamin supplementation in epilepsy subjects is still recruiting and no result available yet.

Naringenin is a flavanone and was found in various fruits and herbs [143]. It has significant antioxidant properties, affected plasma lipid levels and antioxidant activity [144], protected mice against radiation-induced DNA damage [145], or defended ferric iron-induced oxidative damage in HepG2 cells [146]. In pilocarpine-induced status epilepticus, reduction of antioxidant enzymes such as catalase, superoxide dismutase (SOD), and glutathione reductase was found in the hippocampus, and naringenin can reduce ROS formation, restore antioxidant enzymes and lessen seizure severity [116]. No clinical trial was found regarding "brain disease and Naringenin" from ClinicalTrials.gov.

EUK-134 is a salen-manganese complex with potent catalase/SOD activity and exerted beneficial effect to counteract excessive oxidative stress in ischemic brain injury [147]. It was able to extend the life-span in Caenorhabditis elegans by augmenting the natural antioxidant defenses system [148]. In kainic acid-induced seizures, EUK-134 showed a significant reduction of protein nitration, NF-κB-binding activity, and neuronal damage from seizure activity over the limbic system [117]. This may imply the potential of EUK-134 to prevent excitotoxic neuronal injury from various neurological diseases including epilepsy. No clinical trial was found about "brain disease and EUK-134" from ClinicalTrials.gov.

Pioglitazone, a peroxisome proliferator-activated receptor-gamma (PPARγ) agonist used clinically as an anti-diabetic medication, is known to exert various biological functions such as reducing inflammation and oxidative injury and may use for stroke or other neurodegenerative diseases [149]. We have shown before that pioglitazone can exert anti-oxidative and anti-apoptotic effects against ischemic neuronal injury [150]. We also demonstrated that pioglitazone decreased reactive oxygen species, increased mitochondrial uncoupling protein 2 (UCP2) expression, lessened mitochondrial apoptotic process, stabilized the mitochondrial transmembrane potential, and decreased neuronal death in the hippocampus following kainic acid-induced seizures [119]. Pioglitazone also exerts a beneficial effect over genetically epileptic EL mice with a reduction of inflammatory responses in the brain [118]. There are 15 clinical trials concerning "brain disease and pioglitazone" that were searched from ClinicalTrials.gov. The brain diseases include various neurodegenerative diseases, ischemic and hemorrhagic stroke, brain tumor, atherosclerosis, adrenomyeloneuropathy, Friedreich's Ataxia, and Ataxia-Telangiectasia (AT). No clinical trial involving epilepsy and pioglitazone was found in ClinicalTrials.gov.

Semicarbazide is a water-soluble white solid derivative from urea with the formula OC(NH2)(N2H3). Aryl semicarbazones, the product of semicarbazide, was found to have excellent activity as an anticonvulsant with low neurotoxicity in experimental animals [151]. It can prevent seizure-induced lipid peroxidation and augment antioxidant enzyme activities such as SOD and glutathione peroxidase (GSH-Px) in the brain homogenate of mice [120]. No clinical trials concerning "brain disease and aryl semicarbazones" was found in ClinicalTrials.gov.

α-lipoic acid is a naturally occurring compound and is critical for energy metabolism, redox regulation, and cell signaling [152]. In a systemic review concerning the use of α-lipoic acid for the treatment of psychiatric and neurological conditions, it revealed that α-lipoic acid improves schizophrenia symptoms with fewer side effects from antipsychotic medication. It also showed the effectiveness in the prevention of Alzheimer's disease progression and the ability to improve clinical parameters and oxidative imbalance in stroke patients [153]. In a clinical trial, α-lipoic acid was found to stimulate cAMP production in healthy control and patients of multiple sclerosis [154]. α-lipoic acid takes part in mitochondrial oxidative metabolism and acting as biological antioxidants with potential therapeutic use in various chronic diseases [155,156]. It was shown in pilocarpine-induced status epilepticus that α-lipoic acid decreased seizure episodes by reducing oxidative stress in rat hippocampus and striatum [121,122]. There are 11 clinical trials concerning "brain disease and α-lipoic acid" that were found from ClinicalTrials.gov. The brain diseases include AD, PD, mucopolysaccharidosis disorders, ischemic stroke, brain injury, adrenomyeloneuropathy, and progressive supranuclear palsy (PSP). No clinical trial concerning epilepsy and α-lipoic acid was found in ClinicalTrials.gov.

Melatonin is a widely existing molecule in living organisms implicated in various biological and physiological effects among different tissues or organs. It possesses the capability to cross the BBB, exerts notable antioxidant effects, performs as a free radical scavenger, increases antioxidant enzymes, lessens mitochondrial electron leakage, and interferes with pro-inflammatory signaling pathways [157]. These effects of melatonin highlight the potential clinical use in various neurodegenerative diseases such as AD, PD, amyotrophic lateral sclerosis (ALS), and multiple sclerosis (MS) [158,159]. In pentylenetetrazole-induced seizures, treatment with melatonin notably altered neurotransmitters and nitrite levels, decreased seizure occurrence, and mortality. The results indicate that the anticonvulsant property of melatonin involves both brain amino acids and NO production [124]. In another study, it was revealed that co-treatment with melatonin in kainic acid-induced seizures, it can lessen the hydroxyl radicals, reduce the damage of mtDNA, decrease lipid peroxidation in the brain, hence lessen the severity of seizures from kainic acid [123]. More than 50 clinical trials concerning "brain disease and melatonin" were found from ClinicalTrials.gov. Brain diseases include various neurodegenerative diseases, traumatic brain injury, epilepsy, brain tumor, cerebrovascular diseases, premature birth-related brain injury, migraine, sleep disorders. Four trials concerning "epilepsy and melatonin" were found for various epilepsy patients including Lennox-Gastaut Syndrome.

α-tocopherol, a type of vitamin E, is engaged in various biochemical processes closely related to lipid and lipoprotein homeostasis. It functions as a hydroperoxyl radical scavenger and is critical in a variety of oxidative stress-related disease conditions in humans [160]. In pilocarpine-induced status epilepticus in rats, it was revealed that pretreatment with α-tocopherol reduced oxidative stress and partially reduced chaperone-mediated autophagy in the hippocampus at 1 day after status epilepticus versus vehicle [125]. With two different chemical compounds, methylmalonic acid and pentylenetetrazol-induced seizure, both revealed the increased content of thiobarbituric acid-reactive substances (TBARS) and total protein carbonylation in striatum and α-tocopherol showed the ability to attenuate the severity in a dose-dependent manner [126]. There are 3 clinical trials concerning "brain disease and α-tocopherol" were found from ClinicalTrials.gov. The brain diseases include premature birth-related intraventricular hemorrhage, common chronic health conditions including migraine and sleep disorders, and dietary habits and consumption including check blood α-tocopherol level in dementia patients (ClinicalTrials.gov Identifier: NCT01193270, NCT02758990, NCT03794141). No clinical trial concerning epilepsy and α-tocopherol was found in ClinicalTrials.gov.

Ascorbic acid, also known as vitamin C, is found in various foods and used as a dietary supplement in daily life. Ascorbic acid is involved in various physiological functions of the body including the nervous system, such as supporting the structure of the neurons, engaging the processes of differentiation, maturation, and survival in the neuron, and aiding and modulating the synthesis of neurotransmission [161]. Ascorbic acid is crucial to maintain the balance of redox state and possesses the potential effect to treat various neurodegenerative diseases, autoimmune diseases, and cancer [161–163]. In pilocarpine-induced status epilepticus, it was revealed that ascorbic acid delays the onset of seizure and reduces the mortality rate, and decreases lipid peroxidation levels [127]. In another study, ascorbic acid can restore the decrease of glutathione levels and counteract the increased lipid peroxidation level caused by penicillin-induced epileptiform activity [128]. There are 17 clinical trials concerning "brain disease and ascorbic acid" that were found from ClinicalTrials.gov. The brain diseases include AD, PD, ischemic and hemorrhagic stroke, brain tumor, carotid atherosclerosis, ischemic encephalopathy, hepatic encephalopathy, cerebral palsy, attention deficit hyperactivity disorder (ADHD), and epilepsy. There is one study concerning epilepsy and ascorbic acid entitled "Trial of Vitamin C as Add on Therapy for Children With Idiopathic Epilepsy", but no result available until now.

Selenium is a trace element and an essential micronutrient for animals. It is integrated into antioxidant enzymes such as glutathione peroxidases (GSH-Px) and thioredoxin reductases (TrxR) and plays an important role in normal brain function [164,165]. Deficiency of selenium may involve with mentality decline and selenoproteins are beneficial to neurodegenerative diseases such as AD, PD, or HD through redox signaling regulation [164,165]. Selenium deficiency also involves the risk of seizures attack and supplementation of selenium may reduce seizures [166]. Increased nitrosative/oxidative stress in hippocampal tissue such as enhanced iNOS expression along with TrxR activity induction have been revealed in a pentylenetetrazol-induced seizure. Selenium treatment helped to reduce the expression of 4-hydroneonenal, nitrotyrosine, iNOS, and concomitantly normalized TrxR and HO-1 activity. Adding sildenafil, a selective phosphodiesterase 5 inhibitor, can further enhance these effects [129]. It was also reported that selenium with topiramate, an antiepileptic drug, has protective effects on the pentylenetetrazol-induced seizure by inhibiting free radicals and supporting antioxidant redox systems [130]. There are 12 clinical trials concerning "brain disease and selenium" that were found from ClinicalTrials.gov. The brain diseases include phenylketonurias, AD, HD, ischemic and hemorrhagic stroke, carotid atherosclerosis, cerebral palsy, and epilepsy. Two trials concerning "epilepsy and selenium" were found but without final results available till now.

Resveratrol, a nonflavonoid polyphenol, is mainly found in red grapes/wine [167]. It can offer protective effects against various cardiovascular diseases, neurodegenerative diseases, stroke, and epilepsy through diverse mechanisms [167–171]. In a lithium-pilocarpine-induced status epilepticus, it was shown that the application of resveratrol can reduce oxidative stress and mitochondrial dysfunction. This denotes the potential of resveratrol, combined with antiepileptic drugs, to target oxidative stress and may provide a further beneficial effect to control epilepsy [131]. It is also shown in kainic acid-induced status epilepticus that in addition to lessening neurodegeneration and aberrant neurogenesis, resveratrol also suppresses the severity of oxidative stress and inflammation in the hippocampus [132]. There are 16 clinical trials concerning "brain disease and resveratrol" that were found from ClinicalTrials.gov. The brain diseases include mainly neurodegenerative diseases such as AD, PD, HD, Friedreich ataxia, and hypoxic brain. No clinical trial concerning epilepsy and resveratrol was found in ClinicalTrials.gov.

RTA 408 (omaveloxolone) holds the characters of both antioxidative and anti-inflammatory activities [172,173], and possesses the capability to improve mitochondrial bioenergetics [174]. The mechanism of RTA 408 has been suggested to involve a combination of activation of the antioxidative transcription factor nuclear factor erythroid 2-related factor 2 (Nrf2) and inhibition of the pro-inflammatory transcription factor NF-κB [172]. In a kainic acid-induced status epilepticus in rats,

RTA 408 inhibits kelch like ECH associated protein 1 (KEAP1) expression and activates Nrf2 to increase endogenous antioxidant and eventually reduce cellular ROS production and neuronal death. RTA 408 also significantly reduced the frequency of late spontaneous seizures following status epilepticus [133]. To date, one clinical trial concerning "brain disease and RTA 408" was found from ClinicalTrials.gov (ClinicalTrials.gov Identifier: NCT02255435). The title is "A Phase 2 Study of the Safety, Efficacy, and Pharmacodynamics of RTA 408 in the Treatment of Friedreich's Ataxia". No clinical trial involving epilepsy and RTA 408 was found in ClinicalTrials.gov.

AEBSF is a water-soluble, broad-spectrum, and irreversible inhibitor of serine proteases and can also avoid the activation of NADPH oxidase, the ROS generator [175,176]. In the perforant path stimulation model of status epilepticus, it was demonstrated that inhibition of excessive generation of ROS by AEBSF, significantly lessened seizure-induced cell death. These results indicate that NADPH oxidase inhibition can utilize as a new strategy of treatment to prevent brain damage in status epilepticus and related seizure disorders [50]. In a recent study, it was shown that combining AEBSF and RTA 408 in kainic acid-induced status epilepticus can reduce the occurrence of spontaneous seizures in animals following status epilepticus and modified the severity of epilepsy. It was also revealed that AEBSF and RTA 408 can reduce the generation of ROS through NADPH oxidase inhibition and increase the endogenous antioxidant system through Nrf2 activation respectively and preclude unwarranted ROS accumulation, mitochondrial membrane depolarization, and decrease neuronal death in *in vitro* seizure models [57]. No clinical trial was found regarding "brain disease and AEBSF" from ClinicalTrials.gov.

N-[3-(aminomethyl)benzyl] acetamidine or 1400 W, is an aralkylamine and acts as a highly selective inhibitor of iNOS [177]. With the kainic acid-induced status epilepticus, 1400 W, by decreasing the epileptiform spike rate in the first three days of after status epilepticus and can potentially modify the development of epileptogenesis and reduce the severity of chronic epilepsy [134]. Thus, 1400W is also revealed to modulate proteins involving neuroinflammatory responses such as heat shock protein beta-1, glial fibrillary acidic protein, and CD44 antigen in kainic acid-induced status epilepticus [178]. No clinical trial was found regarding "brain disease and 1400 W" from ClinicalTrials.gov.

Apocynin, a cell-permeable phenol, is recognized as a powerful inhibitor of NADPH oxidase [49]. It can block superoxide and peroxynitrite formation in murine macrophages [179] and increase glutathione synthesis via the activation of activation protein 1 (AP-1) [180]. In a pilocarpine-induced status epilepticus, it was demonstrated that apocynin reduced the production of ROS, lessened the extent of lipid peroxidation, diminished seizure-induced BBB disintegration, declined microglial activation and neutrophil infiltration, and further reduced neuronal death in the hippocampus [49]. This may suggest the therapeutic potential for apocynin in seizure-induced neuronal damage. No clinical trial was found regarding "brain disease and apocynin" from ClinicalTrials.gov.

AEOL10150, a metalloporphyrin with broad-spectrum catalytic antioxidant, possesses the ability to scavenge free radicals and inhibit lipid peroxidation [181–183]. It can cross the BBB to attain therapeutic concentrations in the brain, which is useful for this antioxidant to provide antioxidative capability sufficient to attenuate oxidative stress in pilocarpine-induced neurological damage and reduce mortality of experimental animals [135,136]. The beneficial effects depend on the ability of metalloporphyrin to scavenge peroxynitrite, lessen lipid peroxidation, with SOD and catalase-like activities [135]. Until now, no clinical trial was found about "brain disease and AEOL10150" from ClinicalTrials.gov.

Although this article focuses on the potential drugs and chemical compounds with the antioxidant character for status epilepticus, we need to stress that the treatment of choice for epilepsy is still the anti-epileptic drugs. However, taking into accounts the capability of antioxidant character in some anti-epileptic drugs, the choice of medications for epilepsy would add additional consideration and need to balance the gain and loss [151]. For example, different classes of antiepileptic drugs can influence the intima thickness of the common carotid artery in epileptic patients [184,185]. Several studies have shown that certain antiepileptic drugs exhibit antioxidant effects through modifying the activity of various enzymes system under epilepsy. A previous study showed that levetiracetam, a widely used anti-epileptic drug, exerted beneficial effects to reduce oxidative stress and lessen the severity of mitochondrial dysfunction after status epilepticus [83]. Phenytoin was found to have a neuroprotective effect over survival-promoting and antioxidant genes such as Akt1 and glutathione reductase in a DNA microarrays study [186]. Topiramate can inhibit free radical production, attenuate lipid peroxidation, and supporting the antioxidant redox system to protect pentylenetetrazol-induced brain injury [187]. In a recent study, it was shown that children with valproic acid monotherapy has significant antioxidant effects and decreases seizure activity [188]. Valproic acid may increase levels of glutathione and exert its neuroprotective effect [189]. Zonisamide is useful in animal and human studies of epilepsy. Zonisamide exerts antioxidant character by scavenging hydroxyl and nitric oxide radicals [190]. These examples just remind us to think about the drug choice in a more broaden way if possible with extra benefits in patients with epilepsy.

5. Conclusions and Future Perspective

Epilepsy is a common neurological disease with a prevalence of 0.5–1.5% in the general population. Although antiepileptic drug or anticonvulsants is a standard treatment for patients with epilepsy, more than one-third of patients remain in poor control of seizure frequency, the so-called drug-resistant epilepsy even with the several new antiepileptic drugs used in recent decades. Status epilepticus, in patients with prolonged or continuous epileptic seizures, is often associated with significant morbidity and mortality. Emerging evidence reveals the significant role of ROS/RNS in status epilepticus. It is reasonable to speculate that besides antiepileptic drugs, the addition of antioxidants as a combination therapy may exert an additional beneficial effect, lessen the neuronal damage, and improve clinical outcomes. Here, we provide some evidence for the potential application of antioxidants in epileptic patients. However, more studies are still mandatory both pre-clinically and clinically in order to achieve this goal. Although mounting evidence revealed the success of antioxidants treatment in status epilepticus in pre-clinical studies as summarized in Table 1, the difficulty of translation into clinical practice is obvious from the information of ClinicITrials.gov. There are some molecules, such as ascorbic acid, α-tocopherol, coenzyme Q10, lipoic acid, and melatonin that may represent an endogenous defensive mechanism to counteract oxidative stress, and other chemical compounds or drugs need to deal with the capability to enter the BBB if used in diseases of the central nervous system including epilepsy. Other possible reasons for lack of efficacy in clinical trials may stem from the underlying pathophysiological mechanisms. Status epilepticus or refractory epilepsy may trigger multiple pathophysiological and biochemical changes, as shown in Figure 1. These cascades reveal that excessive oxidative stress and inflammatory response from status epilepticus can cause highly reactive ROS/RNS, impair mitochondrial function, reduce the ATP supply, result in the apoptotic process, and to neuronal death. The combination therapy with one more drug to aim at various targets such as together with antioxidant, anti-inflammatory drugs or medications altering apoptotic pathways may yield additional or even synergistic effects. However, the current drug trial mainly uses one drug only to compare with placebo-control. This may also explain the lack of efficiency in various clinical trials due to the failure to cope with the complexity of pathophysiological changes in various diseases including epilepsy. This problem is warranted to further explored in clinical trials in the future.

Author Contributions: T.-K.L.: contributed to concept generation, data interpretation, drafting of the manuscript; S.-D.C.: contributed to concept generation, data interpretation, drafting of the manuscript; K.-J.L.: contributed to concept generation, graphic drawing and approval of the article; Y.-C.C.: contributed to concept generation, data interpretation, approval of the article. All authors have read and agreed to the published version of the manuscript.

Funding: This work was supported by the Ministry of Science and Technology in Taiwan (MOST 108-2314-B-182A-149 – to Shang-Der Chen), Chang Gung Medical Foundation (CMRPG8I0051 and CMRPG8I0051 to Shang-Der Chen; CMRPG8J0591 and CMRPG8J0592 to Yao-Chung Chuang).

Acknowledgments: We would like to acknowledge that Figure 1 was created using ©BioRender–biorender.com platform.

Conflicts of Interest: The authors declare no conflict of interest.

References

1. Fiest, K.M.; Sauro, K.M.; Wiebe, S.; Patten, S.B.; Kwon, C.S.; Dykeman, J.; Pringsheim, T.; Lorenzetti, D.L.; Jette, N. Prevalence and incidence of epilepsy: A systematic review and meta-analysis of international studies. *Neurology* **2017**, *88*, 296–303. [CrossRef] [PubMed]
2. Ngugi, A.K.; Bottomley, C.; Kleinschmidt, I.; Sander, J.W.; Newton, C.R. Estimation of the burden of active and life-time epilepsy: A meta-analytic approach. *Epilepsia* **2010**, *51*, 883–890. [CrossRef] [PubMed]
3. Hauser, W.A.; Annegers, J.F.; Kurland, L.T. Prevalence of epilepsy in Rochester, Minnesota: 1940–1980. *Epilepsia* **1991**, *32*, 429–445. [CrossRef]
4. Chirchiglia, D.; Chirchiglia, P.; Marotta, R. When History Meets Neurology: Neurological Diseases of Famous People. *Neuroscientist* **2019**, *25*, 388–393. [CrossRef] [PubMed]
5. Iniesta, I. Epilepsy in Dostoevsky. *Prog. Brain Res.* **2013**, *205*, 277–293. [CrossRef]
6. Lascaratos, J.; Zis, P.V. The epilepsy of Emperor Michael IV, Paphlagon (1034–1041 A.D.): Accounts of Byzantine historians and physicians. *Epilepsia* **2000**, *41*, 913–917. [CrossRef]
7. Fisher, R.S.; Cross, J.H.; French, J.A.; Higurashi, N.; Hirsch, E.; Jansen, F.E.; Lagae, L.; Moshe, S.L.; Peltola, J.; Roulet Perez, E.; et al. Operational classification of seizure types by the International League Against Epilepsy: Position Paper of the ILAE Commission for Classification and Terminology. *Epilepsia* **2017**, *58*, 522–530. [CrossRef]
8. Becker, A.J. Review: Animal models of acquired epilepsy: Insights into mechanisms of human epileptogenesis. *Neuropathol. Appl. Neurobiol.* **2018**, *44*, 112–129. [CrossRef]
9. Henshall, D.C.; Simon, R.P. Epilepsy and apoptosis pathways. *J. Cereb. Blood Flow Metab.* **2005**, *25*, 1557–1572. [CrossRef]
10. Engelborghs, S.; D'Hooge, R.; De Deyn, P.P. Pathophysiology of epilepsy. *Acta Neurol. Belg.* **2000**, *100*, 201–213.
11. Giovannini, G.; Monti, G.; Tondelli, M.; Marudi, A.; Valzania, F.; Leitinger, M.; Trinka, E.; Meletti, S. Mortality, morbidity and refractoriness prediction in status epilepticus: Comparison of STESS and EMSE scores. *Seizure* **2017**, *46*, 31–37. [CrossRef] [PubMed]
12. Neligan, A.; Shorvon, S.D. Prognostic factors, morbidity and mortality in tonic-clonic status epilepticus: A review. *Epilepsy Res.* **2011**, *93*, 1–10. [CrossRef] [PubMed]
13. Lowenstein, D.H.; Alldredge, B.K. Status epilepticus. *N. Engl. J. Med.* **1998**, *338*, 970–976. [CrossRef] [PubMed]
14. Breuer, L.E.; Boon, P.; Bergmans, J.W.; Mess, W.H.; Besseling, R.M.; de Louw, A.; Tijhuis, A.G.; Zinger, S.; Bernas, A.; Klooster, D.C.; et al. Cognitive deterioration in adult epilepsy: Does accelerated cognitive ageing exist? *Neurosci. Biobehav. Rev.* **2016**, *64*, 1–11. [CrossRef]
15. Badawy, R.A.; Johnson, K.A.; Cook, M.J.; Harvey, A.S. A mechanistic appraisal of cognitive dysfunction in epilepsy. *Neurosci. Biobehav. Rev.* **2012**, *36*, 1885–1896. [CrossRef] [PubMed]
16. Sutula, T.P. Mechanisms of epilepsy progression: Current theories and perspectives from neuroplasticity in adulthood and development. *Epilepsy Res.* **2004**, *60*, 161–171. [CrossRef]
17. Hauser, W.A. Status epilepticus: Epidemiologic considerations. *Neurology* **1990**, *40*, 9–13.
18. Leitinger, M.; Trinka, E.; Giovannini, G.; Zimmermann, G.; Florea, C.; Rohracher, A.; Kalss, G.; Neuray, C.; Kreidenhuber, R.; Hofler, J.; et al. Epidemiology of status epilepticus in adults: A population-based study on incidence, causes, and outcomes. *Epilepsia* **2019**, *60*, 53–62. [CrossRef]
19. Neligan, A.; Walker, M.C. Falling status epilepticus mortality rates in England and Wales: 2001–2013? *Epilepsia* **2016**, *57*, e121–e124. [CrossRef]
20. Santos, V.R.; Melo, I.S.; Pacheco, A.L.D.; Castro, O.W. Life and death in the hippocampus: What's bad? *Epilepsy Behav.* **2019**, 106595. [CrossRef]
21. Riban, V.; Bouilleret, V.; Pham-Le, B.T.; Fritschy, J.M.; Marescaux, C.; Depaulis, A. Evolution of hippocampal epileptic activity during the development of hippocampal sclerosis in a mouse model of temporal lobe epilepsy. *Neuroscience* **2002**, *112*, 101–111. [CrossRef]
22. Shekh-Ahmad, T.; Kovac, S.; Abramov, A.Y.; Walker, M.C. Reactive oxygen species in status epilepticus. *Epilepsy Behav.* **2019**, *101*, 106410. [CrossRef] [PubMed]

23. Kovac, S.; Domijan, A.M.; Walker, M.C.; Abramov, A.Y. Seizure activity results in calcium- and mitochondria-independent ROS production via NADPH and xanthine oxidase activation. *Cell Death Dis.* **2014**, *5*, e1442. [CrossRef] [PubMed]
24. Patel, M.; Li, Q.Y.; Chang, L.Y.; Crapo, J.; Liang, L.P. Activation of NADPH oxidase and extracellular superoxide production in seizure-induced hippocampal damage. *J. Neurochem.* **2005**, *92*, 123–131. [CrossRef]
25. Chuang, Y.C.; Chen, S.D.; Lin, T.K.; Chang, W.N.; Lu, C.H.; Liou, C.W.; Chan, S.H.; Chang, A.Y. Transcriptional upregulation of nitric oxide synthase II by nuclear factor-kappaB promotes apoptotic neuronal cell death in the hippocampus following experimental status epilepticus. *J. Neurosci. Res.* **2010**, *88*, 1898–1907. [CrossRef]
26. Chuang, Y.C.; Chen, S.D.; Liou, C.W.; Lin, T.K.; Chang, W.N.; Chan, S.H.; Chang, A.Y. Contribution of nitric oxide, superoxide anion, and peroxynitrite to activation of mitochondrial apoptotic signaling in hippocampal CA3 subfield following experimental temporal lobe status epilepticus. *Epilepsia* **2009**, *50*, 731–746. [CrossRef]
27. Chuang, Y.C.; Chen, S.D.; Lin, T.K.; Liou, C.W.; Chang, W.N.; Chan, S.H.; Chang, A.Y. Upregulation of nitric oxide synthase II contributes to apoptotic cell death in the hippocampal CA3 subfield via a cytochrome c/caspase-3 signaling cascade following induction of experimental temporal lobe status epilepticus in the rat. *Neuropharmacology* **2007**, *52*, 1263–1273. [CrossRef]
28. Chuang, Y.C.; Chang, A.Y.; Lin, J.W.; Hsu, S.P.; Chan, S.H. Mitochondrial dysfunction and ultrastructural damage in the hippocampus during kainic acid-induced status epilepticus in the rat. *Epilepsia* **2004**, *45*, 1202–1209. [CrossRef]
29. Peng, J.; Wang, K.; Xiang, W.; Li, Y.; Hao, Y.; Guan, Y. Rosiglitazone polarizes microglia and protects against pilocarpine-induced status epilepticus. *CNS Neurosci. Ther.* **2019**, *25*, 1363–1372. [CrossRef]
30. Meskinimood, S.; Rahimi, N.; Faghir-Ghanesefat, H.; Gholami, M.; Sharifzadeh, M.; Dehpour, A.R. Modulatory effect of opioid ligands on status epilepticus and the role of nitric oxide pathway. *Epilepsy Behav.* **2019**, *101*, 106563. [CrossRef]
31. Putra, M.; Sharma, S.; Gage, M.; Gasser, G.; Hinojo-Perez, A.; Olson, A.; Gregory-Flores, A.; Puttachary, S.; Wang, C.; Anantharam, V.; et al. Inducible nitric oxide synthase inhibitor, 1400W, mitigates DFP-induced long-term neurotoxicity in the rat model. *Neurobiol. Dis.* **2020**, *133*, 104443. [CrossRef] [PubMed]
32. Wang, A.; Si, Z.; Xue, P.; Li, X.; Liu, J. Tacrolimus protects hippocampal neurons of rats with status epilepticus through suppressing oxidative stress and inhibiting mitochondrial pathway of apoptosis. *Brain Res.* **2019**, *1715*, 176–181. [CrossRef]
33. Sharma, S.; Carlson, S.; Puttachary, S.; Sarkar, S.; Showman, L.; Putra, M.; Kanthasamy, A.G.; Thippeswamy, T. Role of the Fyn-PKCdelta signaling in SE-induced neuroinflammation and epileptogenesis in experimental models of temporal lobe epilepsy. *Neurobiol. Dis.* **2018**, *110*, 102–121. [CrossRef] [PubMed]
34. Eslami, S.M.; Ghasemi, M.; Bahremand, T.; Momeny, M.; Gholami, M.; Sharifzadeh, M.; Dehpour, A.R. Involvement of nitrergic system in anticonvulsant effect of zolpidem in lithium-pilocarpine induced status epilepticus: Evaluation of iNOS and COX-2 genes expression. *Eur. J. Pharmacol.* **2017**, *815*, 454–461. [CrossRef] [PubMed]
35. Eslami, S.M.; Moradi, M.M.; Ghasemi, M.; Dehpour, A.R. Anticonvulsive Effects of Licofelone on Status Epilepticus Induced by Lithium-pilocarpine in Wistar Rats: A Role for Inducible Nitric Oxide Synthase. *J. Epilepsy Res.* **2016**, *6*, 51–58. [CrossRef] [PubMed]
36. Liu, J.; Wang, A.; Li, L.; Huang, Y.; Xue, P.; Hao, A. Oxidative stress mediates hippocampal neuron death in rats after lithium-pilocarpine-induced status epilepticus. *Seizure* **2010**, *19*, 165–172. [CrossRef]
37. Chen, S.D.; Chang, A.Y.; Chuang, Y.C. The potential role of mitochondrial dysfunction in seizure-associated cell death in the hippocampus and epileptogenesis. *J. Bioenerg. Biomembr.* **2010**, *42*, 461–465. [CrossRef]
38. Veith, A.; Moorthy, B. Role of Cytochrome P450s in the Generation and Metabolism of Reactive Oxygen Species. *Curr. Opin. Toxicol.* **2018**, *7*, 44–51. [CrossRef]
39. Novo, E.; Parola, M. The role of redox mechanisms in hepatic chronic wound healing and fibrogenesis. *Fibrogenesis Tissue Repair* **2012**, *5*, S4. [CrossRef]
40. Bae, Y.S.; Oh, H.; Rhee, S.G.; Yoo, Y.D. Regulation of reactive oxygen species generation in cell signaling. *Mol. Cells* **2011**, *32*, 491–509. [CrossRef]
41. Dan Dunn, J.; Alvarez, L.A.; Zhang, X.; Soldati, T. Reactive oxygen species and mitochondria: A nexus of cellular homeostasis. *Redox Biol.* **2015**, *6*, 472–485. [CrossRef]

42. Numakawa, T.; Matsumoto, T.; Numakawa, Y.; Richards, M.; Yamawaki, S.; Kunugi, H. Protective Action of Neurotrophic Factors and Estrogen against Oxidative Stress-Mediated Neurodegeneration. *J. Toxicol.* **2011**, *2011*, 405194. [CrossRef]
43. Liu, Z.; Zhou, T.; Ziegler, A.C.; Dimitrion, P.; Zuo, L. Oxidative Stress in Neurodegenerative Diseases: From Molecular Mechanisms to Clinical Applications. *Oxid. Med. Cell. Longev.* **2017**, *2017*, 2525967. [CrossRef]
44. Jiang, T.; Sun, Q.; Chen, S. Oxidative stress: A major pathogenesis and potential therapeutic target of antioxidative agents in Parkinson's disease and Alzheimer's disease. *Prog. Neurobiol.* **2016**, *147*, 1–19. [CrossRef]
45. Salim, S. Oxidative Stress and the Central Nervous System. *J. Pharmacol. Exp. Ther.* **2017**, *360*, 201–205. [CrossRef]
46. Waldbaum, S.; Patel, M. Mitochondrial dysfunction and oxidative stress: A contributing link to acquired epilepsy? *J. Bioenerg. Biomembr.* **2010**, *42*, 449–455. [CrossRef]
47. Kovacs, R.; Schuchmann, S.; Gabriel, S.; Kann, O.; Kardos, J.; Heinemann, U. Free radical-mediated cell damage after experimental status epilepticus in hippocampal slice cultures. *J. Neurophysiol.* **2002**, *88*, 2909–2918. [CrossRef]
48. Pestana, R.R.; Kinjo, E.R.; Hernandes, M.S.; Britto, L.R. Reactive oxygen species generated by NADPH oxidase are involved in neurodegeneration in the pilocarpine model of temporal lobe epilepsy. *Neurosci. Lett.* **2010**, *484*, 187–191. [CrossRef]
49. Kim, J.H.; Jang, B.G.; Choi, B.Y.; Kim, H.S.; Sohn, M.; Chung, T.N.; Choi, H.C.; Song, H.K.; Suh, S.W. Post-treatment of an NADPH oxidase inhibitor prevents seizure-induced neuronal death. *Brain Res.* **2013**, *1499*, 163–172. [CrossRef]
50. Williams, S.; Hamil, N.; Abramov, A.Y.; Walker, M.C.; Kovac, S. Status epilepticus results in persistent overproduction of reactive oxygen species, inhibition of which is neuroprotective. *Neuroscience* **2015**, *303*, 160–165. [CrossRef]
51. Malkov, A.; Ivanov, A.I.; Latyshkova, A.; Bregestovski, P.; Zilberter, M.; Zilberter, Y. Activation of nicotinamide adenine dinucleotide phosphate oxidase is the primary trigger of epileptic seizures in rodent models. *Ann. Neurol.* **2019**, *85*, 907–920. [CrossRef] [PubMed]
52. Pecorelli, A.; Natrella, F.; Belmonte, G.; Miracco, C.; Cervellati, F.; Ciccoli, L.; Mariottini, A.; Rocchi, R.; Vatti, G.; Bua, A.; et al. NADPH oxidase activation and 4-hydroxy-2-nonenal/aquaporin-4 adducts as possible new players in oxidative neuronal damage presents in drug-resistant epilepsy. *Biochim. Biophys. Acta* **2015**, *1852*, 507–519. [CrossRef] [PubMed]
53. Lee, S.H.; Choi, B.Y.; Kho, A.R.; Jeong, J.H.; Hong, D.K.; Kang, D.H.; Kang, B.S.; Song, H.K.; Choi, H.C.; Suh, S.W. Inhibition of NADPH Oxidase Activation by Apocynin Rescues Seizure-Induced Reduction of Adult Hippocampal Neurogenesis. *Int. J. Mol. Sci.* **2018**, *19*, 3087. [CrossRef]
54. Hirotsu, C.; Matos, G.; Tufik, S.; Andersen, M.L. Changes in gene expression in the frontal cortex of rats with pilocarpine-induced status epilepticus after sleep deprivation. *Epilepsy Behav.* **2013**, *27*, 378–384. [CrossRef]
55. Di Maio, R.; Mastroberardino, P.G.; Hu, X.; Montero, L.; Greenamyre, J.T. Pilocapine alters NMDA receptor expression and function in hippocampal neurons: NADPH oxidase and ERK1/2 mechanisms. *Neurobiol. Dis.* **2011**, *42*, 482–495. [CrossRef] [PubMed]
56. Zhu, X.; Shen, K.; Bai, Y.; Zhang, A.; Xia, Z.; Chao, J.; Yao, H. NADPH oxidase activation is required for pentylenetetrazole kindling-induced hippocampal autophagy. *Free Radic Biol. Med.* **2016**, *94*, 230–242. [CrossRef]
57. Shekh-Ahmad, T.; Lieb, A.; Kovac, S.; Gola, L.; Christian Wigley, W.; Abramov, A.Y.; Walker, M.C. Combination antioxidant therapy prevents epileptogenesis and modifies chronic epilepsy. *Redox Biol.* **2019**, *26*, 101278. [CrossRef]
58. Ramdial, K.; Franco, M.C.; Estevez, A.G. Cellular mechanisms of peroxynitrite-induced neuronal death. *Brain Res. Bull.* **2017**, *133*, 4–11. [CrossRef]
59. Fujikawa, D.G. Prolonged seizures and cellular injury: Understanding the connection. *Epilepsy Behav.* **2005**, *7* (Suppl. 3), S3–S11. [CrossRef]
60. Baille, V.; Clarke, P.G.; Brochier, G.; Dorandeu, F.; Verna, J.M.; Four, E.; Lallement, G.; Carpentier, P. Soman-induced convulsions: The neuropathology revisited. *Toxicology* **2005**, *215*, 1–24. [CrossRef]

61. Shinoda, S.; Araki, T.; Lan, J.Q.; Schindler, C.K.; Simon, R.P.; Taki, W.; Henshall, D.C. Development of a model of seizure-induced hippocampal injury with features of programmed cell death in the BALB/c mouse. *J. Neurosci. Res.* **2004**, *76*, 121–128. [CrossRef] [PubMed]
62. Liu, H.; Cao, Y.; Basbaum, A.I.; Mazarati, A.M.; Sankar, R.; Wasterlain, C.G. Resistance to excitotoxin-induced seizures and neuronal death in mice lacking the preprotachykinin A gene. *Proc. Natl. Acad. Sci. USA* **1999**, *96*, 12096–12101. [CrossRef] [PubMed]
63. Ulamek-Koziol, M.; Czuczwar, S.J.; Januszewski, S.; Pluta, R. Proteomic and Genomic Changes in Tau Protein, Which Are Associated with Alzheimer's Disease after Ischemia-Reperfusion Brain Injury. *Int. J. Mol. Sci.* **2020**, *21*, 892. [CrossRef]
64. Leech, T.; Chattipakorn, N.; Chattipakorn, S.C. The beneficial roles of metformin on the brain with cerebral ischaemia/reperfusion injury. *Pharmacol. Res.* **2019**, *146*, 104261. [CrossRef]
65. Yaribeygi, H.; Panahi, Y.; Javadi, B.; Sahebkar, A. The Underlying Role of Oxidative Stress in Neurodegeneration: A Mechanistic Review. *CNS Neurol. Disord. Drug Targets* **2018**, *17*, 207–215. [CrossRef] [PubMed]
66. Uttara, B.; Singh, A.V.; Zamboni, P.; Mahajan, R.T. Oxidative stress and neurodegenerative diseases: A review of upstream and downstream antioxidant therapeutic options. *Curr. Neuropharmacol.* **2009**, *7*, 65–74. [CrossRef]
67. Chong, Z.Z.; Kang, J.Q.; Maiese, K. Essential cellular regulatory elements of oxidative stress in early and late phases of apoptosis in the central nervous system. *Antioxid. Redox Signal.* **2004**, *6*, 277–287. [CrossRef]
68. Terrone, G.; Frigerio, F.; Balosso, S.; Ravizza, T.; Vezzani, A. Inflammation and reactive oxygen species in status epilepticus: Biomarkers and implications for therapy. *Epilepsy Behav.* **2019**, *101*, 106275. [CrossRef]
69. Iori, V.; Maroso, M.; Rizzi, M.; Iyer, A.M.; Vertemara, R.; Carli, M.; Agresti, A.; Antonelli, A.; Bianchi, M.E.; Aronica, E.; et al. Receptor for Advanced Glycation Endproducts is upregulated in temporal lobe epilepsy and contributes to experimental seizures. *Neurobiol. Dis.* **2013**, *58*, 102–114. [CrossRef]
70. Maroso, M.; Balosso, S.; Ravizza, T.; Liu, J.; Aronica, E.; Iyer, A.M.; Rossetti, C.; Molteni, M.; Casalgrandi, M.; Manfredi, A.A.; et al. Toll-like receptor 4 and high-mobility group box-1 are involved in ictogenesis and can be targeted to reduce seizures. *Nat. Med.* **2010**, *16*, 413–419. [CrossRef]
71. Iori, V.; Iyer, A.M.; Ravizza, T.; Beltrame, L.; Paracchini, L.; Marchini, S.; Cerovic, M.; Hill, C.; Ferrari, M.; Zucchetti, M.; et al. Blockade of the IL-1R1/TLR4 pathway mediates disease-modification therapeutic effects in a model of acquired epilepsy. *Neurobiol. Dis.* **2017**, *99*, 12–23. [CrossRef] [PubMed]
72. Vezzani, A.; Maroso, M.; Balosso, S.; Sanchez, M.A.; Bartfai, T. IL-1 receptor/Toll-like receptor signaling in infection, inflammation, stress and neurodegeneration couples hyperexcitability and seizures. *Brain Behav. Immun.* **2011**, *25*, 1281–1289. [CrossRef] [PubMed]
73. Librizzi, L.; Noe, F.; Vezzani, A.; de Curtis, M.; Ravizza, T. Seizure-induced brain-borne inflammation sustains seizure recurrence and blood-brain barrier damage. *Ann. Neurol.* **2012**, *72*, 82–90. [CrossRef]
74. Jiang, J.; Quan, Y.; Ganesh, T.; Pouliot, W.A.; Dudek, F.E.; Dingledine, R. Inhibition of the prostaglandin receptor EP2 following status epilepticus reduces delayed mortality and brain inflammation. *Proc. Natl. Acad. Sci. USA* **2013**, *110*, 3591–3596. [CrossRef]
75. Yang, J.L.; Mukda, S.; Chen, S.D. Diverse roles of mitochondria in ischemic stroke. *Redox Biol.* **2018**, *16*, 263–275. [CrossRef] [PubMed]
76. Signes, A.; Fernandez-Vizarra, E. Assembly of mammalian oxidative phosphorylation complexes I-V and supercomplexes. *Essays Biochem.* **2018**, *62*, 255–270. [CrossRef] [PubMed]
77. van der Bliek, A.M.; Sedensky, M.M.; Morgan, P.G. Cell Biology of the Mitochondrion. *Genetics* **2017**, *207*, 843–871. [CrossRef] [PubMed]
78. Hatefi, Y. The mitochondrial electron transport and oxidative phosphorylation system. *Annu. Rev. Biochem.* **1985**, *54*, 1015–1069. [CrossRef]
79. Walker, M.C. Pathophysiology of status epilepticus. *Neurosci. Lett.* **2018**, *667*, 84–91. [CrossRef]
80. Wasterlain, C.G.; Fujikawa, D.G.; Penix, L.; Sankar, R. Pathophysiological mechanisms of brain damage from status epilepticus. *Epilepsia* **1993**, *34* (Suppl. 1), S37–S53. [CrossRef]
81. Folbergrova, J.; Jesina, P.; Haugvicova, R.; Lisy, V.; Houstek, J. Sustained deficiency of mitochondrial complex I activity during long periods of survival after seizures induced in immature rats by homocysteic acid. *Neurochem. Int.* **2010**, *56*, 394–403. [CrossRef] [PubMed]

82. Chuang, Y.C.; Lin, J.W.; Chen, S.D.; Lin, T.K.; Liou, C.W.; Lu, C.H.; Chang, W.N. Preservation of mitochondrial integrity and energy metabolism during experimental status epilepticus leads to neuronal apoptotic cell death in the hippocampus of the rat. *Seizure* **2009**, *18*, 420–428. [CrossRef] [PubMed]
83. Gibbs, J.E.; Walker, M.C.; Cock, H.R. Levetiracetam: Antiepileptic properties and protective effects on mitochondrial dysfunction in experimental status epilepticus. *Epilepsia* **2006**, *47*, 469–478. [CrossRef] [PubMed]
84. Kunz, W.S.; Kudin, A.P.; Vielhaber, S.; Blumcke, I.; Zuschratter, W.; Schramm, J.; Beck, H.; Elger, C.E. Mitochondrial complex I deficiency in the epileptic focus of patients with temporal lobe epilepsy. *Ann. Neurol.* **2000**, *48*, 766–773. [CrossRef]
85. Kudin, A.P.; Kudina, T.A.; Seyfried, J.; Vielhaber, S.; Beck, H.; Elger, C.E.; Kunz, W.S. Seizure-dependent modulation of mitochondrial oxidative phosphorylation in rat hippocampus. *Eur. J. Neurosci.* **2002**, *15*, 1105–1114. [CrossRef]
86. Cadenas, E.; Davies, K.J. Mitochondrial free radical generation, oxidative stress, and aging. *Free Radic Biol. Med.* **2000**, *29*, 222–230. [CrossRef]
87. Shalini, S.; Dorstyn, L.; Dawar, S.; Kumar, S. Old, new and emerging functions of caspases. *Cell Death Differ.* **2015**, *22*, 526–539. [CrossRef]
88. Tower, J. Programmed cell death in aging. *Ageing Res. Rev.* **2015**, *23*, 90–100. [CrossRef]
89. Fan, J.; Dawson, T.M.; Dawson, V.L. Cell Death Mechanisms of Neurodegeneration. *Adv. Neurobiol.* **2017**, *15*, 403–425. [CrossRef]
90. Siddiqui, W.A.; Ahad, A.; Ahsan, H. The mystery of BCL2 family: Bcl-2 proteins and apoptosis: An update. *Arch. Toxicol.* **2015**, *89*, 289–317. [CrossRef] [PubMed]
91. Radi, E.; Formichi, P.; Battisti, C.; Federico, A. Apoptosis and oxidative stress in neurodegenerative diseases. *J. Alzheimers Dis.* **2014**, *42* (Suppl. 3), S125–S152. [CrossRef]
92. Ghavami, S.; Shojaei, S.; Yeganeh, B.; Ande, S.R.; Jangamreddy, J.R.; Mehrpour, M.; Christoffersson, J.; Chaabane, W.; Moghadam, A.R.; Kashani, H.H.; et al. Autophagy and apoptosis dysfunction in neurodegenerative disorders. *Prog. Neurobiol.* **2014**, *112*, 24–49. [CrossRef]
93. Yamamoto, H.; Schoonjans, K.; Auwerx, J. Sirtuin functions in health and disease. *Mol. Endocrinol.* **2007**, *21*, 1745–1755. [CrossRef] [PubMed]
94. Yu, J.; Auwerx, J. The role of sirtuins in the control of metabolic homeostasis. *Ann. N. Y. Acad. Sci.* **2009**, *1173* (Suppl. 1), E10–E19. [CrossRef] [PubMed]
95. Wang, S.J.; Zhao, X.H.; Chen, W.; Bo, N.; Wang, X.J.; Chi, Z.F.; Wu, W. Sirtuin 1 activation enhances the PGC-1alpha/mitochondrial antioxidant system pathway in status epilepticus. *Mol. Med. Rep.* **2015**, *11*, 521–526. [CrossRef] [PubMed]
96. Chuang, Y.C.; Chen, S.D.; Jou, S.B.; Lin, T.K.; Chen, S.F.; Chen, N.C.; Hsu, C.Y. Sirtuin 1 Regulates Mitochondrial Biogenesis and Provides an Endogenous Neuroprotective Mechanism Against Seizure-Induced Neuronal Cell Death in the Hippocampus Following Status Epilepticus. *Int. J. Mol. Sci.* **2019**, *20*, 3588. [CrossRef] [PubMed]
97. Wang, D.; Li, Z.; Zhang, Y.; Wang, G.; Wei, M.; Hu, Y.; Ma, S.; Jiang, Y.; Che, N.; Wang, X.; et al. Targeting of microRNA-199a-5p protects against pilocarpine-induced status epilepticus and seizure damage via SIRT1-p53 cascade. *Epilepsia* **2016**, *57*, 706–716. [CrossRef] [PubMed]
98. Cho, I.; Jeong, K.H.; Zhu, J.; Choi, Y.H.; Cho, K.H.; Heo, K.; Kim, W.J. Sirtuin3 Protected Against Neuronal Damage and Cycled into Nucleus in Status Epilepticus Model. *Mol. Neurobiol.* **2019**, *56*, 4894–4903. [CrossRef]
99. Wheless, J.W. History of the ketogenic diet. *Epilepsia* **2008**, *49* (Suppl. 8), 3–5. [CrossRef]
100. Zhang, Y.; Xu, J.; Zhang, K.; Yang, W.; Li, B. The Anticonvulsant Effects of Ketogenic Diet on Epileptic Seizures and Potential Mechanisms. *Curr. Neuropharmacol.* **2018**, *16*, 66–70. [CrossRef]
101. Boison, D. New insights into the mechanisms of the ketogenic diet. *Curr. Opin. Neurol.* **2017**, *30*, 187–192. [CrossRef] [PubMed]
102. Hartman, A.L.; Zheng, X.; Bergbower, E.; Kennedy, M.; Hardwick, J.M. Seizure tests distinguish intermittent fasting from the ketogenic diet. *Epilepsia* **2010**, *51*, 1395–1402. [CrossRef] [PubMed]
103. Kim, D.Y.; Rho, J.M. The ketogenic diet and epilepsy. *Curr. Opin. Clin. Nutr. Metab. Care* **2008**, *11*, 113–120. [CrossRef] [PubMed]
104. Hartman, A.L.; Gasior, M.; Vining, E.P.; Rogawski, M.A. The neuropharmacology of the ketogenic diet. *Pediatr. Neurol.* **2007**, *36*, 281–292. [CrossRef] [PubMed]

105. van der Louw, E.; van den Hurk, D.; Neal, E.; Leiendecker, B.; Fitzsimmon, G.; Dority, L.; Thompson, L.; Marchio, M.; Dudzinska, M.; Dressler, A.; et al. Ketogenic diet guidelines for infants with refractory epilepsy. *Eur. J. Paediatr. Neurol.* **2016**, *20*, 798–809. [CrossRef]
106. Mahmoud, S.H.; Ho-Huang, E.; Buhler, J. Systematic review of ketogenic diet use in adult patients with status epilepticus. *Epilepsia Open* **2020**, *5*, 10–21. [CrossRef]
107. Nelson, S.E.; Varelas, P.N. Status Epilepticus, Refractory Status Epilepticus, and Super-refractory Status Epilepticus. *Continuum* **2018**, *24*, 1683–1707. [CrossRef]
108. Thakur, K.T.; Probasco, J.C.; Hocker, S.E.; Roehl, K.; Henry, B.; Kossoff, E.H.; Kaplan, P.W.; Geocadin, R.G.; Hartman, A.L.; Venkatesan, A.; et al. Ketogenic diet for adults in super-refractory status epilepticus. *Neurology* **2014**, *82*, 665–670. [CrossRef]
109. Geronzi, U.; Lotti, F.; Grosso, S. Oxidative stress in epilepsy. *Expert Rev. Neurother.* **2018**, *18*, 427–434. [CrossRef]
110. Chen, Z.; Brodie, M.J.; Liew, D.; Kwan, P. Treatment Outcomes in Patients With Newly Diagnosed Epilepsy Treated With Established and New Antiepileptic Drugs: A 30-Year Longitudinal Cohort Study. *JAMA Neurol.* **2018**, *75*, 279–286. [CrossRef]
111. Edward, H.R. Drug Resistant Epilepsy and New AEDs: Two Perspectives. *Epilepsy Curr.* **2018**, *18*, 304–306. [CrossRef]
112. Kwan, P.; Schachter, S.C.; Brodie, M.J. Drug-resistant epilepsy. *N. Engl. J. Med.* **2011**, *365*, 919–926. [CrossRef]
113. Uma Devi, P.; Pillai, K.K.; Vohora, D. Modulation of pentylenetetrazole-induced seizures and oxidative stress parameters by sodium valproate in the absence and presence of N-acetylcysteine. *Fundam. Clin. Pharmacol.* **2006**, *20*, 247–253. [CrossRef]
114. Tawfik, M.K. Coenzyme Q10 enhances the anticonvulsant effect of phenytoin in pilocarpine-induced seizures in rats and ameliorates phenytoin-induced cognitive impairment and oxidative stress. *Epilepsy Behav.* **2011**, *22*, 671–677. [CrossRef] [PubMed]
115. Bhardwaj, M.; Kumar, A. Neuroprotective mechanism of Coenzyme Q10 (CoQ10) against PTZ induced kindling and associated cognitive dysfunction: Possible role of microglia inhibition. *Pharmacol. Rep.* **2016**, *68*, 1301–1311. [CrossRef] [PubMed]
116. Shakeel, S.; Rehman, M.U.; Tabassum, N.; Amin, U.; Mir, M.U.R. Effect of Naringenin (A naturally occurring flavanone) Against Pilocarpine-induced Status Epilepticus and Oxidative Stress in Mice. *Pharmacogn. Mag.* **2017**, *13*, S154–S160. [CrossRef]
117. Rong, Y.; Doctrow, S.R.; Tocco, G.; Baudry, M. EUK-134, a synthetic superoxide dismutase and catalase mimetic, prevents oxidative stress and attenuates kainate-induced neuropathology. *Proc. Natl. Acad. Sci. USA* **1999**, *96*, 9897–9902. [CrossRef] [PubMed]
118. Okada, K.; Yamashita, U.; Tsuji, S. Ameliorative effect of pioglitazone on seizure responses in genetically epilepsy-susceptible EL mice. *Brain Res.* **2006**, *1102*, 175–178. [CrossRef]
119. Chuang, Y.C.; Lin, T.K.; Huang, H.Y.; Chang, W.N.; Liou, C.W.; Chen, S.D.; Chang, A.Y.; Chan, S.H. Peroxisome proliferator-activated receptors gamma/mitochondrial uncoupling protein 2 signaling protects against seizure-induced neuronal cell death in the hippocampus following experimental status epilepticus. *J. Neuroinflamm.* **2012**, *9*, 184. [CrossRef]
120. Azam, F.; Alkskas, I.A.; Khokra, S.L.; Prakash, O. Synthesis of some novel N4-(naphtha[1,2-*d*]thiazol-2-yl)semicarbazides as potential anticonvulsants. *Eur. J. Med. Chem.* **2009**, *44*, 203–211. [CrossRef]
121. de Freitas, R.M. Lipoic acid alters delta-aminolevulinic dehydratase, glutathione peroxidase and Na^+,K^+-ATPase activities and glutathione-reduced levels in rat hippocampus after pilocarpine-induced seizures. *Cell Mol. Neurobiol.* **2010**, *30*, 381–387. [CrossRef] [PubMed]
122. Militao, G.C.; Ferreira, P.M.; de Freitas, R.M. Effects of lipoic acid on oxidative stress in rat striatum after pilocarpine-induced seizures. *Neurochem. Int.* **2010**, *56*, 16–20. [CrossRef] [PubMed]
123. Mohanan, P.V.; Yamamoto, H.A. Preventive effect of melatonin against brain mitochondria DNA damage, lipid peroxidation and seizures induced by kainic acid. *Toxicol. Lett.* **2002**, *129*, 99–105. [CrossRef]
124. Bikjdaouene, L.; Escames, G.; Leon, J.; Ferrer, J.M.; Khaldy, H.; Vives, F.; Acuna-Castroviejo, D. Changes in brain amino acids and nitric oxide after melatonin administration in rats with pentylenetetrazole-induced seizures. *J. Pineal Res.* **2003**, *35*, 54–60. [CrossRef]
125. Cao, L.; Chen, R.; Xu, J.; Lin, Y.; Wang, R.; Chi, Z. Vitamin E inhibits activated chaperone-mediated autophagy in rats with status epilepticus. *Neuroscience* **2009**, *161*, 73–77. [CrossRef]

126. Ribeiro, M.C.; de Avila, D.S.; Schneider, C.Y.; Hermes, F.S.; Furian, A.F.; Oliveira, M.S.; Rubin, M.A.; Lehmann, M.; Krieglstein, J.; Mello, C.F. alpha-Tocopherol protects against pentylenetetrazol- and methylmalonate-induced convulsions. *Epilepsy Res.* **2005**, *66*, 185–194. [CrossRef]
127. Xavier, S.M.; Barbosa, C.O.; Barros, D.O.; Silva, R.F.; Oliveira, A.A.; Freitas, R.M. Vitamin C antioxidant effects in hippocampus of adult Wistar rats after seizures and status epilepticus induced by pilocarpine. *Neurosci. Lett.* **2007**, *420*, 76–79. [CrossRef]
128. Ayyildiz, M.; Coskun, S.; Yildirim, M.; Agar, E. The effects of ascorbic acid on penicillin-induced epileptiform activity in rats. *Epilepsia* **2007**, *48*, 1388–1395. [CrossRef] [PubMed]
129. Tawfik, K.M.; Moustafa, Y.M.; El-Azab, M.F. Neuroprotective mechanisms of sildenafil and selenium in PTZ-kindling model: Implications in epilepsy. *Eur. J. Pharmacol.* **2018**, *833*, 131–144. [CrossRef]
130. Kutluhan, S.; Naziroglu, M.; Celik, O.; Yilmaz, M. Effects of selenium and topiramate on lipid peroxidation and antioxidant vitamin levels in blood of pentylentetrazol-induced epileptic rats. *Biol. Trace Elem. Res.* **2009**, *129*, 181–189. [CrossRef]
131. Folbergrova, J.; Jesina, P.; Kubova, H.; Otahal, J. Effect of Resveratrol on Oxidative Stress and Mitochondrial Dysfunction in Immature Brain during Epileptogenesis. *Mol. Neurobiol.* **2018**, *55*, 7512–7522. [CrossRef] [PubMed]
132. Mishra, V.; Shuai, B.; Kodali, M.; Shetty, G.A.; Hattiangady, B.; Rao, X.; Shetty, A.K. Resveratrol Treatment after Status Epilepticus Restrains Neurodegeneration and Abnormal Neurogenesis with Suppression of Oxidative Stress and Inflammation. *Sci. Rep.* **2015**, *5*, 17807. [CrossRef] [PubMed]
133. Shekh-Ahmad, T.; Eckel, R.; Dayalan Naidu, S.; Higgins, M.; Yamamoto, M.; Dinkova-Kostova, A.T.; Kovac, S.; Abramov, A.Y.; Walker, M.C. KEAP1 inhibition is neuroprotective and suppresses the development of epilepsy. *Brain* **2018**, *141*, 1390–1403. [CrossRef] [PubMed]
134. Puttachary, S.; Sharma, S.; Verma, S.; Yang, Y.; Putra, M.; Thippeswamy, A.; Luo, D.; Thippeswamy, T. 1400W, a highly selective inducible nitric oxide synthase inhibitor is a potential disease modifier in the rat kainate model of temporal lobe epilepsy. *Neurobiol. Dis.* **2016**, *93*, 184–200. [CrossRef]
135. Pearson-Smith, J.N.; Liang, L.P.; Rowley, S.D.; Day, B.J.; Patel, M. Oxidative Stress Contributes to Status Epilepticus Associated Mortality. *Neurochem. Res.* **2017**, *42*, 2024–2032. [CrossRef]
136. Pearson, J.N.; Rowley, S.; Liang, L.P.; White, A.M.; Day, B.J.; Patel, M. Reactive oxygen species mediate cognitive deficits in experimental temporal lobe epilepsy. *Neurobiol. Dis.* **2015**, *82*, 289–297. [CrossRef]
137. Aldini, G.; Altomare, A.; Baron, G.; Vistoli, G.; Carini, M.; Borsani, L.; Sergio, F. N-Acetylcysteine as an antioxidant and disulphide breaking agent: The reasons why. *Free Radic Res.* **2018**, *52*, 751–762. [CrossRef]
138. Rushworth, G.F.; Megson, I.L. Existing and potential therapeutic uses for N-acetylcysteine: The need for conversion to intracellular glutathione for antioxidant benefits. *Pharmacol. Ther.* **2014**, *141*, 150–159. [CrossRef]
139. Deepmala; Slattery, J.; Kumar, N.; Delhey, L.; Berk, M.; Dean, O.; Spielholz, C.; Frye, R. Clinical trials of N-acetylcysteine in psychiatry and neurology: A systematic review. *Neurosci. Biobehav. Rev.* **2015**, *55*, 294–321. [CrossRef]
140. Martelli, A.; Testai, L.; Colletti, A.; Cicero, A.F.G. Coenzyme Q10: Clinical Applications in Cardiovascular Diseases. *Antioxidants* **2020**, *9*, 341. [CrossRef]
141. Acosta, M.J.; Vazquez Fonseca, L.; Desbats, M.A.; Cerqua, C.; Zordan, R.; Trevisson, E.; Salviati, L. Coenzyme Q biosynthesis in health and disease. *Biochim. Biophys. Acta* **2016**, *1857*, 1079–1085. [CrossRef] [PubMed]
142. Matthews, R.T.; Yang, L.; Browne, S.; Baik, M.; Beal, M.F. Coenzyme Q10 administration increases brain mitochondrial concentrations and exerts neuroprotective effects. *Proc. Natl. Acad. Sci. USA* **1998**, *95*, 8892–8897. [CrossRef] [PubMed]
143. Yanez, J.A.; Andrews, P.K.; Davies, N.M. Methods of analysis and separation of chiral flavonoids. *J. Chromatogr. B Analyt Technol. Biomed. Life Sci.* **2007**, *848*, 159–181. [CrossRef]
144. Gorinstein, S.; Leontowicz, H.; Leontowicz, M.; Krzeminski, R.; Gralak, M.; Delgado-Licon, E.; Martinez Ayala, A.L.; Katrich, E.; Trakhtenberg, S. Changes in plasma lipid and antioxidant activity in rats as a result of naringin and red grapefruit supplementation. *J. Agric. Food Chem.* **2005**, *53*, 3223–3228. [CrossRef]
145. Kumar, S.; Tiku, A.B. Biochemical and Molecular Mechanisms of Radioprotective Effects of Naringenin, a Phytochemical from Citrus Fruits. *J. Agric. Food Chem.* **2016**, *64*, 1676–1685. [CrossRef]
146. Jagetia, G.C.; Reddy, T.K.; Venkatesha, V.A.; Kedlaya, R. Influence of naringin on ferric iron induced oxidative damage in vitro. *Clin. Chim. Acta* **2004**, *347*, 189–197. [CrossRef]

147. Baker, K.; Marcus, C.B.; Huffman, K.; Kruk, H.; Malfroy, B.; Doctrow, S.R. Synthetic combined superoxide dismutase/catalase mimetics are protective as a delayed treatment in a rat stroke model: A key role for reactive oxygen species in ischemic brain injury. *J. Pharmacol. Exp. Ther.* **1998**, *284*, 215–221. [PubMed]
148. Melov, S.; Ravenscroft, J.; Malik, S.; Gill, M.S.; Walker, D.W.; Clayton, P.E.; Wallace, D.C.; Malfroy, B.; Doctrow, S.R.; Lithgow, G.J. Extension of life-span with superoxide dismutase/catalase mimetics. *Science* **2000**, *289*, 1567–1569. [CrossRef]
149. Bordet, R.; Ouk, T.; Petrault, O.; Gele, P.; Gautier, S.; Laprais, M.; Deplanque, D.; Duriez, P.; Staels, B.; Fruchart, J.C.; et al. PPAR: A new pharmacological target for neuroprotection in stroke and neurodegenerative diseases. *Biochem. Soc. Trans.* **2006**, *34*, 1341–1346. [CrossRef]
150. Chuang, Y.C.; Lin, T.K.; Yang, D.I.; Yang, J.L.; Liou, C.W.; Chen, S.D. Peroxisome proliferator-activated receptor-gamma dependent pathway reduces the phosphorylation of dynamin-related protein 1 and ameliorates hippocampal injury induced by global ischemia in rats. *J. Biomed. Sci.* **2016**, *23*, 44. [CrossRef]
151. Azam, F.; Prasad, M.V.; Thangavel, N. Targeting oxidative stress component in the therapeutics of epilepsy. *Curr. Top. Med. Chem.* **2012**, *12*, 994–1007. [CrossRef] [PubMed]
152. Packer, L.; Cadenas, E. Lipoic acid: Energy metabolism and redox regulation of transcription and cell signaling. *J. Clin. Biochem. Nutr.* **2011**, *48*, 26–32. [CrossRef]
153. de Sousa, C.N.S.; da Silva Leite, C.M.G.; da Silva Medeiros, I.; Vasconcelos, L.C.; Cabral, L.M.; Patrocinio, C.F.V.; Patrocinio, M.L.V.; Mouaffak, F.; Kebir, O.; Macedo, D.; et al. Alpha-lipoic acid in the treatment of psychiatric and neurological disorders: A systematic review. *Metab. Brain Dis.* **2019**, *34*, 39–52. [CrossRef] [PubMed]
154. Fiedler, S.E.; Yadav, V.; Kerns, A.R.; Tsang, C.; Markwardt, S.; Kim, E.; Spain, R.; Bourdette, D.; Salinthone, S. Lipoic Acid Stimulates cAMP Production in Healthy Control and Secondary Progressive MS Subjects. *Mol. Neurobiol.* **2018**, *55*, 6037–6049. [CrossRef]
155. Gomes, M.B.; Negrato, C.A. Alpha-lipoic acid as a pleiotropic compound with potential therapeutic use in diabetes and other chronic diseases. *Diabetol. Metab. Syndr.* **2014**, *6*, 80. [CrossRef] [PubMed]
156. Rochette, L.; Ghibu, S.; Richard, C.; Zeller, M.; Cottin, Y.; Vergely, C. Direct and indirect antioxidant properties of alpha-lipoic acid and therapeutic potential. *Mol. Nutr. Food Res.* **2013**, *57*, 114–125. [CrossRef]
157. Salehi, B.; Sharopov, F.; Fokou, P.V.T.; Kobylinska, A.; Jonge, L.; Tadio, K.; Sharifi-Rad, J.; Posmyk, M.M.; Martorell, M.; Martins, N.; et al. Melatonin in Medicinal and Food Plants: Occurrence, Bioavailability, and Health Potential for Humans. *Cells* **2019**, *8*, 681. [CrossRef]
158. Wongprayoon, P.; Govitrapong, P. Melatonin as a mitochondrial protector in neurodegenerative diseases. *Cell. Mol. Life Sci.* **2017**, *74*, 3999–4014. [CrossRef]
159. Miller, E.; Morel, A.; Saso, L.; Saluk, J. Melatonin redox activity. Its potential clinical applications in neurodegenerative disorders. *Curr. Top. Med. Chem.* **2015**, *15*, 163–169. [CrossRef]
160. Rigotti, A. Absorption, transport, and tissue delivery of vitamin E. *Mol. Aspects Med.* **2007**, *28*, 423–436. [CrossRef]
161. Figueroa-Mendez, R.; Rivas-Arancibia, S. Vitamin C in Health and Disease: Its Role in the Metabolism of Cells and Redox State in the Brain. *Front. Physiol.* **2015**, *6*, 397. [CrossRef] [PubMed]
162. Moretti, M.; Fraga, D.B.; Rodrigues, A.L.S. Ascorbic Acid to Manage Psychiatric Disorders. *CNS Drugs* **2017**, *31*, 571–583. [CrossRef] [PubMed]
163. Harrison, F.E.; Bowman, G.L.; Polidori, M.C. Ascorbic acid and the brain: Rationale for the use against cognitive decline. *Nutrients* **2014**, *6*, 1752–1781. [CrossRef] [PubMed]
164. Dominiak, A.; Wilkaniec, A.; Wroczynski, P.; Adamczyk, A. Selenium in the Therapy of Neurological Diseases. Where is it Going? *Curr. Neuropharmacol.* **2016**, *14*, 282–299. [CrossRef]
165. Solovyev, N.D. Importance of selenium and selenoprotein for brain function: From antioxidant protection to neuronal signalling. *J. Inorg. Biochem.* **2015**, *153*, 1–12. [CrossRef]
166. Pillai, R.; Uyehara-Lock, J.H.; Bellinger, F.P. Selenium and selenoprotein function in brain disorders. *IUBMB Life* **2014**, *66*, 229–239. [CrossRef]
167. Del Rio, D.; Rodriguez-Mateos, A.; Spencer, J.P.; Tognolini, M.; Borges, G.; Crozier, A. Dietary (poly)phenolics in human health: Structures, bioavailability, and evidence of protective effects against chronic diseases. *Antioxid. Redox Signal.* **2013**, *18*, 1818–1892. [CrossRef]

168. Chuang, Y.C.; Chen, S.D.; Hsu, C.Y.; Chen, S.F.; Chen, N.C.; Jou, S.B. Resveratrol Promotes Mitochondrial Biogenesis and Protects against Seizure-Induced Neuronal Cell Damage in the Hippocampus Following Status Epilepticus by Activation of the PGC-1alpha Signaling Pathway. *Int. J. Mol. Sci.* **2019**, *20*, 998. [CrossRef]
169. Magalingam, K.B.; Radhakrishnan, A.K.; Haleagrahara, N. Protective Mechanisms of Flavonoids in Parkinson's Disease. *Oxid. Med. Cell. Longev.* **2015**, *2015*, 314560. [CrossRef]
170. Wang, X.; Ouyang, Y.Y.; Liu, J.; Zhao, G. Flavonoid intake and risk of CVD: A systematic review and meta-analysis of prospective cohort studies. *Br. J. Nutr.* **2014**, *111*, 1–11. [CrossRef]
171. Sun, A.Y.; Wang, Q.; Simonyi, A.; Sun, G.Y. Resveratrol as a therapeutic agent for neurodegenerative diseases. *Mol. Neurobiol.* **2010**, *41*, 375–383. [CrossRef] [PubMed]
172. Reisman, S.A.; Lee, C.Y.; Meyer, C.J.; Proksch, J.W.; Sonis, S.T.; Ward, K.W. Topical application of the synthetic triterpenoid RTA 408 protects mice from radiation-induced dermatitis. *Radiat. Res.* **2014**, *181*, 512–520. [CrossRef] [PubMed]
173. Reisman, S.A.; Lee, C.Y.; Meyer, C.J.; Proksch, J.W.; Ward, K.W. Topical application of the synthetic triterpenoid RTA 408 activates Nrf2 and induces cytoprotective genes in rat skin. *Arch. Dermatol. Res.* **2014**, *306*, 447–454. [CrossRef]
174. Neymotin, A.; Calingasan, N.Y.; Wille, E.; Naseri, N.; Petri, S.; Damiano, M.; Liby, K.T.; Risingsong, R.; Sporn, M.; Beal, M.F.; et al. Neuroprotective effect of Nrf2/ARE activators, CDDO ethylamide and CDDO trifluoroethylamide, in a mouse model of amyotrophic lateral sclerosis. *Free Radic Biol. Med.* **2011**, *51*, 88–96. [CrossRef]
175. Powers, J.C.; Asgian, J.L.; Ekici, O.D.; James, K.E. Irreversible inhibitors of serine, cysteine, and threonine proteases. *Chem. Rev.* **2002**, *102*, 4639–4750. [CrossRef] [PubMed]
176. Diatchuk, V.; Lotan, O.; Koshkin, V.; Wikstroem, P.; Pick, E. Inhibition of NADPH oxidase activation by 4-(2-aminoethyl)-benzenesulfonyl fluoride and related compounds. *J. Biol. Chem.* **1997**, *272*, 13292–13301. [CrossRef]
177. Garvey, E.P.; Oplinger, J.A.; Furfine, E.S.; Kiff, R.J.; Laszlo, F.; Whittle, B.J.; Knowles, R.G. 1400W is a slow, tight binding, and highly selective inhibitor of inducible nitric-oxide synthase in vitro and in vivo. *J. Biol. Chem.* **1997**, *272*, 4959–4963. [CrossRef]
178. Tse, K.; Hammond, D.; Simpson, D.; Beynon, R.J.; Beamer, E.; Tymianski, M.; Salter, M.W.; Sills, G.J.; Thippeswamy, T. The impact of postsynaptic density 95 blocking peptide (Tat-NR2B9c) and an iNOS inhibitor (1400W) on proteomic profile of the hippocampus in C57BL/6J mouse model of kainate-induced epileptogenesis. *J. Neurosci. Res.* **2019**, *97*, 1378–1392. [CrossRef]
179. Muijsers, R.B.; van Den Worm, E.; Folkerts, G.; Beukelman, C.J.; Koster, A.S.; Postma, D.S.; Nijkamp, F.P. Apocynin inhibits peroxynitrite formation by murine macrophages. *Br. J. Pharmacol.* **2000**, *130*, 932–936. [CrossRef]
180. Lapperre, T.S.; Jimenez, L.A.; Antonicelli, F.; Drost, E.M.; Hiemstra, P.S.; Stolk, J.; MacNee, W.; Rahman, I. Apocynin increases glutathione synthesis and activates AP-1 in alveolar epithelial cells. *FEBS Lett.* **1999**, *443*, 235–239. [CrossRef]
181. Day, B.J. Catalytic antioxidants: A radical approach to new therapeutics. *Drug Discov. Today* **2004**, *9*, 557–566. [CrossRef]
182. Day, B.J.; Batinic-Haberle, I.; Crapo, J.D. Metalloporphyrins are potent inhibitors of lipid peroxidation. *Free Radic Biol. Med.* **1999**, *26*, 730–736. [CrossRef]
183. Day, B.J.; Crapo, J.D. A metalloporphyrin superoxide dismutase mimetic protects against paraquat-induced lung injury in vivo. *Toxicol. Appl. Pharmacol.* **1996**, *140*, 94–100. [CrossRef] [PubMed]
184. Chuang, Y.C.; Chuang, H.Y.; Lin, T.K.; Chang, C.C.; Lu, C.H.; Chang, W.N.; Chen, S.D.; Tan, T.Y.; Huang, C.R.; Chan, S.H. Effects of long-term antiepileptic drug monotherapy on vascular risk factors and atherosclerosis. *Epilepsia* **2012**, *53*, 120–128. [CrossRef] [PubMed]
185. Tan, T.Y.; Lu, C.H.; Chuang, H.Y.; Lin, T.K.; Liou, C.W.; Chang, W.N.; Chuang, Y.C. Long-term antiepileptic drug therapy contributes to the acceleration of atherosclerosis. *Epilepsia* **2009**, *50*, 1579–1586. [CrossRef] [PubMed]
186. Mariotti, V.; Melissari, E.; Amar, S.; Conte, A.; Belmaker, R.H.; Agam, G.; Pellegrini, S. Effect of prolonged phenytoin administration on rat brain gene expression assessed by DNA microarrays. *Exp. Biol. Med.* **2010**, *235*, 300–310. [CrossRef]

187. Naziroglu, M.; Kutluhan, S.; Yilmaz, M. Selenium and topiramate modulates brain microsomal oxidative stress values, Ca^{2+}-ATPase activity, and EEG records in pentylentetrazol-induced seizures in rats. *J. Membr. Biol.* **2008**, *225*, 39–49. [CrossRef]
188. Beltran-Sarmiento, E.; Arregoitia-Sarabia, C.K.; Floriano-Sanchez, E.; Sandoval-Pacheco, R.; Galvan-Hernandez, D.E.; Coballase-Urrutia, E.; Carmona-Aparicio, L.; Ramos-Reyna, E.; Rodriguez-Silverio, J.; Cardenas-Rodriguez, N. Effects of Valproate Monotherapy on the Oxidant-Antioxidant Status in Mexican Epileptic Children: A Longitudinal Study. *Oxid. Med. Cell. Longev.* **2018**, *2018*, 7954371. [CrossRef]
189. Cui, J.; Shao, L.; Young, L.T.; Wang, J.F. Role of glutathione in neuroprotective effects of mood stabilizing drugs lithium and valproate. *Neuroscience* **2007**, *144*, 1447–1453. [CrossRef]
190. Mori, A.; Noda, Y.; Packer, L. The anticonvulsant zonisamide scavenges free radicals. *Epilepsy Res* **1998**, *30*, 153–158. [CrossRef]

Publisher's Note: MDPI stays neutral with regard to jurisdictional claims in published maps and institutional affiliations.

© 2020 by the authors. Licensee MDPI, Basel, Switzerland. This article is an open access article distributed under the terms and conditions of the Creative Commons Attribution (CC BY) license (http://creativecommons.org/licenses/by/4.0/).

Review

Management of Traumatic Brain Injury: From Present to Future

Rosalia Crupi [1,†], Marika Cordaro [2,†], Salvatore Cuzzocrea [3,4,*] and Daniela Impellizzeri [3]

1 Department of Veterinary Science, University of Messina, 98168 Messina, Italy; rcrupi@unime.it
2 Department of Biomedical and Dental Sciences and Morphofunctional Imaging, University of Messina, Via Consolare Valeria 1, 98100 Messina, Italy; cordarom@unime.it
3 Department of Chemical, Biological, Pharmacological and Environmental Sciences, Messina University, Viale F. Stagno D'Alcontres 31, 98166 Messina, Italy; dimpellizzeri@unime.it
4 Department of Pharmacological and Physiological Science, Saint Louis University, Saint Louis, MO 63104, USA
* Correspondence: salvator@unime.it; Tel.: +390-906-765-208
† The authors equally contributed to this work.

Received: 3 March 2020; Accepted: 31 March 2020; Published: 2 April 2020

Abstract: TBI (traumatic brain injury) is a major cause of death among youth in industrialized societies. Brain damage following traumatic injury is a result of direct and indirect mechanisms; indirect or secondary injury involves the initiation of an acute inflammatory response, including the breakdown of the blood–brain barrier (BBB), brain edema, infiltration of peripheral blood cells, and activation of resident immunocompetent cells, as well as the release of numerous immune mediators such as interleukins and chemotactic factors. TBI can cause changes in molecular signaling and cellular functions and structures, in addition to tissue damage, such as hemorrhage, diffuse axonal damages, and contusions. TBI typically disturbs brain functions such as executive actions, cognitive grade, attention, memory data processing, and language abilities. Animal models have been developed to reproduce the different features of human TBI, better understand its pathophysiology, and discover potential new treatments. For many years, the first approach to manage TBI has been treatment of the injured tissue with interventions designed to reduce the complex secondary-injury cascade. Several studies in the literature have stressed the importance of more closely examining injuries, including endothelial, microglia, astroglia, oligodendroglia, and precursor cells. Significant effort has been invested in developing neuroprotective agents. The aim of this work is to review TBI pathophysiology and existing and potential new therapeutic strategies in the management of inflammatory events and behavioral deficits associated with TBI.

Keywords: neuroinflammation; traumatic brain injury; palmitoylethanolamide (PEA); therapeutic strategies; oxidative stress.

---

1. Introduction

Traumatic brain injury (TBI) is defined as damage to the brain sustained after the application of external physical force that causes temporary or permanent functional or structural damage to the brain. TBI stands as a major cause of death among youth in industrialized societies [1]. Brain injury can be mild, moderate, and severe. It is not a distinct unit but a heterogeneous group of pathologies that are initiated by diverse mechanisms and have different survival consequences. Head injuries can be typically classified as closed or penetrating. A closed head injury is normally used to describe automobile accidents, assaults, and falls, while a penetrating injury usually results from gunshot or stab wounds. The use of explosive devices in military conflict has generated a category known as blast injury, which is rare in injury pattern and consideration [2]. The early injury resulting from an external force creates brain tissue destruction with parenchymal impairment, intracerebral hemorrhage, and axonal cutting. Likewise, the primary insult provokes secondary neurometabolic and neurochemical events, including inflammation, cerebral edema, disruption of the blood–brain barrier (BBB), oxidative stress, excitotoxicity, and mitochondrial and metabolic dysfunctions, that can extremely modify the outcome and the recovery patterns, persisting for months to years post-injury [3]. While animal models do not replicate all the physiological, anatomical, and neurobehavioral qualities of human TBI, they do provide important insight into pathophysiological mechanisms and provide the opportunity for translational research and development of efficacious neurotherapeutic interventions [3]. Animal TBI models can be catalogued as penetrating or non-penetrating with the principal difference being the occurrence of a direct deformation of the brain in penetrating injuries, thus provoking a focal or diffused damage at the injury site. Several experimental TBI models that have been designed are listed in Table 1 [4].

Table 1. Animal models of traumatic brain injury (TBI).

Model	Injury
FPI	Focal/diffuse
Lateral	Mixed
Middle	Mixed
CCI	Primarily focal
PBBI	Primarily focal
Blast	Primarily diffuse
Weight Drop	Focal/diffuse
Repeated Mild	Primarily diffuse

FPI: fluid percussion injury; CCI: controlled cortical impact; PBBI: penetrating ballistic-like brain injury.

The golden age of TBI research has been encouraged, thanks to the prominence of repetitive concussions or mild TBIs (mTBIs). Because of the failure of translational therapies focused on moderate to severe TBI, novel therapies have developed, defining two typical approaches. The traditional neuroprotection-based approach is based on the identification of key actions implicated in the advancement of secondary injury whether in mild or severe TBI. In this method, treatment is started as soon as possible after injury. Another methodology, more studied in clinical trials of mTBI patients, is one of targeting symptoms such as vestibular/oculomotor disturbances, headache, sleep illnesses, post-traumatic stress disorder (PTSD), cognitive dysfunction, or others [3,5]. Based on these findings, in this review, we describe the current therapeutic strategies and new therapeutic approaches for the treatment of neuroinflammatory phenomena and TBI symptom management.

2. The Pathophysiology of TBI

The pathological manifestations of TBI are characterized by BBB alteration arising from cerebral ischemia, inflammation, and redox imbalances [6]. The early phase of trauma is characterized by disruption of the BBB, reduced or altered blood flow, and neuronal and glial damage [6]. Secondary injury starts from this primary injury and emerges hours, days, or months later, involves various events such as oxidative stress, modified calcium homeostasis, inflammation, and axonal damage, terminating in cellular degeneration, disturbed neural circuits, and impaired synaptic transmission and synaptic plasticity [6]. Behaviorally, these alterations manifest as post traumatic headache, depression, individuality changes, anxiety, aggression, and deficits in attentiveness, cognition, sensory processing, and communication [7–9].

3. TBI and Neurotoxicity

The neuroinflammation process that characterizes TBI progression can mobilize astrocytes, cytokines, chemokines, and immune cells toward the inflamed area, generating a pro-inflammatory reaction against insult in the brain. Nevertheless, in the chronic step, excessive activation of inflammatory mediators contributes to an injury in the brain circuit, which mainly co-occurs with secondary cell death in TBI. Different secondary cell death mechanisms drive TBI. Among these, excitotoxicity is a process characterized by increased levels of neurotransmitters and glutamate in the synaptic space that stimulate the surrounding nerve cells' *N*-methyl-D-aspartate (NMDA) and α-amino-3-hydroxy-5-methyl-4-isoxazolepropionic acid (AMPA) receptors [10]. These receptors remain activated, favoring the influx of both sodium and calcium ions into cells [10]. In the cytosol, a high concentration of calcium ions determines the activation of protein phosphatases, phospholipases, and proteases. This activation damages DNA, structures, and membranes. In addition to apoptosis and necrosis, other forms of cell death may be active such as necroptosis, autophagy, etc. Overexcitement of glutamate receptors stimulates the production of nitrogen oxide (NO), free radicals, and pro-death transcription factors [11].

4. TBI and Oxidative Stress

Another cell death episode that happens shortly after a TBI is oxidative stress, accompanied by accumulation of both reactive nitrogen species and reactive oxygen species (RNS and ROS) [12]. High ROS levels cause lipoperoxidation of the cellular membrane, leading to dysfunction of mitochondria and oxidizing proteins, which may cause the alteration in the structure of membrane pores [13]. After the primary injury, endogenous inflammatory responses are activated with the invasion of monocytes, neutrophils, and lymphocytes through the BBB [14]. These produce prostaglandins, proinflammatory cytokines, free radicals, and several inflammatory elements, which up-regulate the levels of cell adhesion molecules and chemokines [14]. TBI activates microglia cells, which release proinflammatory cytokines and astrocytes that can up-regulate brain-derived neurotrophic factors. These, in turn, support and guide axons in recovery, increase cell production, and stimulate the long-term persistence of neurons by stopping cell death [15]. Moreover, extracellular glutamate levels are regulated by astrocytes, which also reduce glutamate excitotoxicity to neurons and other cells [16]. The pathophysiological heterogeneity detected in TBI patients may result from the nature, severity, and location of the primary injury, as well as conditions such as age, gender, genetics, and medications [17].

5. Biomarkers in TBI

The development of biomarkers that reveal the pathogenicity of TBI could be clinically useful to establish both diagnosis and prognosis. In particular, blood levels of the neuronal marker ubiquitin C-terminal hydrolaseL1 (UCHL1) and the astroglial marker glial fibrillary acidic protein (GFAP) represent important TBI biomarkers to support drug development and efficacy. Neurofilaments (NFs) are a major element of the axonal cytoskeleton, and play a fundamental role in structural support and regulating axon diameter [18]. Several works suggested that a phosphorylated axonal form of the heavy neurofilament (pNF-H) subunit is released from damaged neurons and might be a sensitive marker of axonal injury following TBI. In that regard, serum pNF-H was reported as a diagnostic tool to predict injury severity in patients who have suffered mild TBI, and it was helpful in understanding which patients required further CT imaging. In a recent report, simvastatin monotherapy supported neurological and functional recovery after experimental TBI, maybe via decreasing axonal injury as verified by a significant increase in the density of NF-H-positive profiles [18]. Recently, another type of pharmacodynamic/response biomarker was identified, specifically, cerebrospinal fluid (CSF) pharmaco/metabolomics are used to evaluate the influence of the combination of antioxidant N-acetylcysteine (NAC) and probenecid on the glutathione pathway after severe TBI in children [5]. Although NAC crosses the BBB, its CSF levels were only a small portion of those in blood. This is partly because NAC is speedily transported back into blood by the organic acid transporters 1 and 3. Probenecid is able to inhibit those transporters, improving brain NAC levels. Thus, the combination of probenecid and NAC produced a significant change in the CSF metabolomic markers of TBI [5]. However, the most important mTBI biomarkers are summarized in Table 2 [19].

Table 2. Biomarkers in TBI.

Biomarkers	Injury Field	Models	References
CSF/serum albumin ratio	BBB dysfunction	patients with severe TBI	[20]
Tight junction proteins		mTBI in rats ischemic stroke in rats	[21,22]
S100B		patients with minor head injury patients with extracranial pathology	[23,24]
Plasma-soluble prion protein PrPc		rat model of concussion concussed athletes	[25,26]
Tau proteins	Axonal injury	patients with acute TBI	[27]
UCHL1		patients with mild or moderate TBI patients with severe TBI concussed athletes	[28–30]
Neurofilaments (NFs)		rat models of TBI patients with mTBI patients with amyotrophic lateral sclerosis	[18,31,32]
NSE		Patients with severe TBI	[33]
GFAP		Rat TBI models Patients with severe TBI	[18,34]
MBP		Children with TBI	[35]
αII and βII-Spectrin breakdown products		Patients with severe TBI	[36]
NGAL		Rat model of TBI Patients with severe TBI	[37,38]
IL-6, IL-8, IL-10	Neuroinflammation	Animal and clinical models of TBI	[39,40]
MMP		mTBI patients	[41]
MBG			[42]
APOE	Genetic variations	mTBI patients	[43]
BDNF			[44]

CSF, cerebrospinal fluid; UCHL1, deubiquitinase ubiquitin carboxyl-terminal hydrolase isoenzyme L1; NSE, glycolytic enzyme neuron-specific enolase; MBP, myelin basic protein; NGAL, neutrophil gelatinase-associated lipocalin; MBG, marinobufagenin; APOE, apolipoprotein E; BDNF, brain-derived neurotrophic factor, MMP, metalloproteinase, mTBI, mild traumatic brain injury.

6. Review of Existing Drug Interventions

The main contributor to secondary injury is the neuroinflammatory process principally characterized by chronic microglial stimulation, astrocytes activation, pro-inflammatory cytokines release, and oxidative stress. It was reported that it is fundamental to start the therapeutic interventions immediately following TBI, in particular within 4 h post-injury, to realize the best promising neuroprotective outcome [45]. Different therapeutic drugs with anti-inflammatory action in some experimental TBI studies are summarized in Table 3.

Table 3. Therapeutic drugs with anti-inflammatory action for TBI.

Drug	Route of Administration	Preclinical Model	Inflammatory Events	References
Dexamethasone	I.P.	WD	⇓ Microglia (CD68, MHC II) ⇓ Microglia (Endothelial-Monocyte Activating Polypeptide II, P2X4 Receptor, Iba-1)	[46,47]
Meloxicam	I.P.	M-WD	⇓ Lipid Peroxidation GSSH Nakatpase	[48]
Etazolate	I.P.	WD	⇓ IL-1β ⇓ Microglia (CD11b)	[49]
Carpofen	S.C.	WD	⇓ Microglia (Iba-1) ⇓ IL-1β, ⇓ IL-6 ⇔ IL-4, ⇔ IL-10	[50]
Indomethacin	I.P.	M-WD	⇓ IL-1β, ⇓ 6-Keto PGF1α	[51,52]
	I.P.	WD		
Nimesulide	I.P.	WD	⇓ 6-Keto PGF1a	[53]
Celecoxib	I.P.	WD	⇓ Il-1β, ⇔ IL-10	[52]
Meloxicam	I.P	WD	⇓ 6-Keto PGF1a	[52]
Etanercept	I.P.	FPI	⇓ TNF-α	[54]
Dexamethasone Melatonin	I.P.	CCI	⇓ MMP-2, ⇓ MMP-9, ⇓ Inos	[55]
Lipoxin A4	I.C.V.	WD	⇓ IL-1β, ⇓ IL-6, ⇓ TNFα, ⇓ GFAP ⇓ Microglia (CD11b)	[56]
Ibuprofen	I.P.	FPI	⇔ IL-4, ⇔ IL-10 ⇔ TNFα ⇔ IL-1α⇔ IL-6	[57]
Minocycline	I.P.	WD	⇓ microglia (CD11b)	[58]
	I.P.	WD	⇓ microglia, ⇓ IL-1β	[59]
Fenofibrate	P.O.	LFP	⇓ cerebral oedema ⇓ ICAM-1, ⇓ brain lesion	[60]
Pioglitazone and Rosiglitazone	I.P.	CCI	⇓ activated microglia (OX-42)	[61]
N-acetylcysteine	I.P.	WD	⇓ NF-kB, ⇓ IL-1β ⇓ IL-6, ⇓ TNF-α	[62]
Flavopiridol	I.P.	LFP	⇓ GFAP, ⇓ microglia	[63]
Roscovitine	I.C.V.	CCI	⇓ microglia (Iba-1)	[63]
Erythropoietin	I.P.	CCI	⇓ NF-kB, ⇓ ICAM-1, ⇓ IL-1β ⇓ TNF-α, ⇔ IL-6	[64]
	I.P.	WD	⇓ CCL-2 ⇓ microglia (CD-68)	[65]
Simvastatin	P.O.	CCI	⇓ TLR4, ⇓ NF-κB ⇓ IL-1β, ⇓ TNFα ⇓ Il-6, ⇓ ICAM-1	[66]
			⇓ Il-1β, ⇓ GFAP ⇓ IL-6, ⇓ TNF-α microglia (CD68)	[67]
Progesterone	I.P.	WD	COX-2, ⇓ PGE2, ⇓ NF-κB	[68]
	I.P./S.C.	CCI	⇓ IL-6, ⇓ COX-2, ⇓ NF-κB	[69]

⇑, increase; ⇓, decrease; ⇔, no change, I.P., intraperitoneal; S.C., subcutaneous; I.C.V., intracerebroventricular; P.O., oral; FPI, fluid percussion injury; CCI, controlled cortical impact; WD, weight drop; M-WD, Marmarou weight drop; MHC, major histocompatibility complex; CD68, cluster of differentiation protein 68; IL, interleukin; TNF, tumor necrosis factor; LFP, lateral fluid percussion; ICAM-1, intercellular adhesion molecule, MMP, metalloproteinase, COX-2, cyclooxygenase-2; NF-kB, nuclear factor-kB; GSSH, oxidized glutathione; CCL2, C–C motif chemokine ligand 2.

In particular, minocycline, a tetracycline derivative, is pharmaceutically efficient in many models of central nervous system (CNS) illnesses and reduces inflammatory and apoptotic processes [70].

A single dose of minocycline decreases lesion volume and ameliorates neurological outcomes linked to diminished microgliosis and interleukin-1β expression in a murine model of closed head injury [59]. Administration of minocycline reduces cerebral edema and improves long-term neurological retrieval [58]. Synthetic peroxisome proliferator-activated receptor (PPAR) agonists also serve as a potent anti-inflammatory, therapeutic agents for TBI [71,72]. Fenofibrate, a PPARα receptor agonist, diminishes cerebral edema, oxidative stress, and inflammation by reducing behavioral deficits after TBI induction [60]. Pioglitazone and rosiglitazone, also PPARγ receptor agonists, diminish microglial activation, enhance neuroprotective antioxidant proteins, and change histological and behavioral outcomes after TBI [61]. Another TBI treatment approach is to block glial proliferation by cell cycle inhibition. Throughout cyclic-dependent kinase (CDK) inhibition, flavopiridol is effective at reducing lesion volume and promoting sensorimotor cognition and recovery after TBI [73]. Roscovitine, another cell cycle inhibitor, also modulates CDKs and has been shown to moderate post-injury neuroinflammation and neurodegeneration [74]. In addition, among anti-oxidants, NAC could also act as an anti-inflammatory drug. Interestingly, NAC repressed NF-κB, IL-1β, TNFα, IL-6, edema and breakdown of the BBB three days after TBI [62].

6.1. Clinical Trials of Drugs with Anti-Inflammatory Effect

Of the therapeutic strategies reported above in Table 3 for TBI management, some have already progressed into clinical trials. Erythropoietin (EPO) demonstrated potential neuroprotective proprieties in most animal models of TBI [75]. However, in a clinical trial with 200 patients presenting severe TBI, EPO administration failed to improve outcomes at 6 months [76]. Thus, although EPO has proven neuroprotective effects in preclinical animal models of TBI, its helpfulness as a medical approach is questionable [75]. In addition, a phase I/II clinical trial also showed the safety and usefulness of minocycline administration for human TBI (NCT01058395) [77,78]. Furthermore, statins, which inhibit cholesterol biosynthesis, also can promote functional recovery following TBI in rodents [79]. Simvastatin inhibits caspase-3 activation and apoptotic cell death, thereby increasing neuronal rescue after TBI [80]. Simvastatin enhances the expression of several growth factors and stimulates neurogenesis, controlling restoration of mental function [81] and supporting functional improvement after TBI (3 months) [82] in rats. However, the United States Food and Drug Administration reported cognitive side effects associated with statins treatment [83]. Given these conflicting findings, more clinical trials are needed to confirm the neuroprotective benefits of statin treatment after TBI. The effects of rosuvastatin on TBI-stimulated cytokine alteration were evaluated in a phase I/II trial (NCT00990028) [77].

A previous study also reported that the TNF-α antagonist, etanercept, has been given perispinally for back pain and sciatica treatment [84]. Twelve patients with chronic neurological dysfunction after TBI who were treated with etanercept showed improvements in many parameters of motor, cognitive, sensory, and psychological performance at several time points [85]. A case report also showed that a single dose of perispinal etanercept produced an important improvement in a patient with neurological dysfunction present for more than 3 years after acute brain injury [86]. Importantly, NAC also has shown potential in preventing sequelae from blast-induced mild TBI, apparently via its antioxidant capacity in the brain [87]. The safety and potential therapeutic efficacy of NAC was effectively evaluated in 41 military personnel who had a mild blast-induced TBI [87]. A phase I randomized clinical trial reported the effects of NAC in combination with an adjuvant probenecid for treatment of severe TBI in children [88].

Progesterone has also demonstrated helpful actions in animal models of TBI and clinical improvement in two phase II randomized, controlled trials [89]. Despite positive effects from preclinical studies and from two positive phase II clinical trials, two big phase III clinical trials of progesterone treatment of acute TBI ended with negative data, respectively, SYNAPSE (NCT01143064) and ProTECT III (NCT00822900) [89]; therefore, the results continue to fail in the field of TBI clinical trials.

6.2. Therapeutic Strategies to manage Neuronal Recovery and Neurobehavioral Sequelae after Injury

TBI progression affects the quality of life of a lot of people causing anxiety, agitation, memory deficiencies, and behavioral changes. Pharmacological compounds that increase cyclic 3′,5′-adenosine monophosphate (cAMP) signaling such as phosphodiesterase (PDE) inhibitors (rolipram, dipyridamole, BC11-38) [90,91], selective serotonin reuptake inhibitors (e.g., fluoxetine) [92], and serotonin-dopamine reuptake inhibitors (e.g., UWA-121), could help in brain repair, recovery of neuronal function [93], and alleviation of disabilities after injury including cognitive deficits, changes in personality, and increased rates of psychiatric illness. Table 4 gives an overview of the most frequently used treatments in the management of neuropsychiatric, neurocognitive, and neurobehavioral sequelae of injury to the brain [94].

Table 4. Current drugs for neurobehavioral disorders after TBI.

Class	Drug	Mechanism	Effect
CNS stimulants	Methylphenidate	acts as an NDRI	amplified the speed of information processing in many neuropsychological tests
CNS stimulants	Modafinil	unknown	raised alertness by the modulation of both noradrenergic and dopaminergic systems
Atypical antidepressant	Agomelatine	a melatonin receptor agonist and serotonin 5-HT2C and 5-HT2B	led to better sleep efficacy
Antiviral	Amantadine	reflect a growth in synthesis and discharge of dopamine	decreased the incidence and gravity of irritability
Antidepressant of the selective SNRI class	Venlafaxine	acts as an SNDRI	increased obsessive behaviors, irritability, and sadness symptoms
Anticonvulsant	Valproate	blockade of voltage-gated sodium channels and increased brain levels of GABA	had benign neuropsychological effects, and is a safe drug to control recognized seizures or stabilize mood
Acetylcholinesterase inhibitor	Rivastigmine	inhibits butyrylcholinesterase and acetylcholinesterase	encouraging in the subgroup of patients with moderate/severe memory weakening
Cholinesterase inhibitor	Galantamine	allosteric potentiating ligand of human nicotinic acetylcholine receptors (nAChRs) $\alpha4\beta2$, $\alpha3\beta4$, and $\alpha6\beta4$ and chicken/mouse nAChRs $\alpha7$/5-HT3 in some brain areas	refined fatigue, memory, attention, and initiative
Cholinesterase inhibitor	Donepezil	develops cholinergic function	indorsed clinical improvement and metabolism

CNS, central nervous system; NDRI, norepinephrine–dopamine reuptake inhibitor; SNDRI, serotonin-norepinephrine-dopamine reuptake inhibitor; GABA, gamma-aminobutyric acid; SNRI., serotonin-norepinephrine reuptake inhibitor.

7. New Therapeutic Strategies

Studying strategies to treat TBI-induced neuroinflammation requires understanding the usual mechanisms, including the tendency to protect against inflammation. Chronic inflammatory events can initiative a program of resolution that involves the release of lipid mediators capable of extinguishing inflammation [95]. Among these are fatty acid amides *N*-acylethanolamines (NAEs), including *N*-arachidonoylethanolamine (endocannabinoid), and the congeners *N*-stearoylethanolamine,

N-oleoylethanolamine, and plus *N*-palmitoylethanolamine (PEA or palmitoylethanolamide) [96]. Several studies documented the positive effects of exogenously dispensed PEA in experimental models of TBI, spinal cord injuries, pain, cerebral ischemia, and Parkinson's and Alzheimer diseases [97]. PEA has no unfavorable effects at pharmacological doses [97]. In addition, several experimental works showed the beneficial effects of new PEA formulations (micronized or ultramicronized) in carrageenan-induced inflammation [98] on cognitive decline associated to neuropathic pain [99] in an Alzheimer disease model [100], tibia fracture mouse model [101], and diabetic neuropathy [102]. Recent observational clinical studies reported the beneficial use of ultramicronized PEA as an add-on therapy in patients suffering from low back pain [103] as well as in patients suffering from fibromyalgia syndrome (FMS) [104]. In addition, interestingly, a co-ultramicronized PEA/luteolin (PEALUT) composite (10:1 mass ratio) displayed important neuroprotective effects in preclinical studies for various conditions (e.g., TBI, arthritis, depression, neurogenesis, Parkinson's and Alzheimer's diseases, and spinal cord injury) and, more recently, in experimental models of autism and vascular dementia [105–113]. In addition, Caltagirone et al. [114] showed that co-ultra PEALUT reduced brain injury in a rat model of Middle Cerebral Artery Occlusion (MCAO) and, more interestingly, in a clinical study. A group of 250 patients with stroke was administered a pharmaceutical formulation of co-ultraPEALut (Glialia®) for 60 days. At baseline and after 30 days of treatment, the patients showed improved neurological status, cognitive functions, spasticity, pain, and ability to perform activities of daily living. Despite its observational nature, the authors of [114] first described administration of co-ultraPEALut to human stroke patients, resulting in important clinical improvements. Inhibition of PEA degradation by affecting its primary catabolic enzyme, NAE-hydrolyzing acid amidase (NAAA), can also present an unconventional method to manage neuroinflammation. A recent study reported that pharmacological modulation and not blocking specific amidases for nacylamides, such as NAAA, can serve as a new approach to preserve the PEA function of maintaining cellular homeostasis through its quick, on-demand synthesis and correspondingly fast degradation. The most recent investigations reported that pharmacological changes in NAAA can be found with the oxazoline of PEA (PEA-OXA) [115]. In rat paws and the carrageenan (CAR) model, PEA-OXA accomplishes significantly better anti-inflammatory action than PEA [116]. In addition, Impellizzeri et al. [117] demonstrated the neuroprotective effects of PEA-OXA in spinal and brain injuries. PEA and new formulations of PEA, therefore, can present new therapeutic strategies in the management of neuroinflammation related to TBI and other CNS disorders.

8. Biologics

In addition to pharmacologic interventions for TBI, promising, innovative developments based on preclinical findings draw on the practice of biologics (e.g., gene therapy, eRNA, DNA, microRNA, antagomirs, peptide therapy, stem cells, exogenous growth factors, and peptides) [118]. Neural and mesenchymal stem cell therapy displays neuroregenerative and neurorestorative potential [119]. A recent work discussed novel associations in drug therapies that have been examined together with stem cells to overcome the restrictions allied with stem cell transplantation and to progress functional recuperation and brain repair post-TBI. To date, progesterone (clinical trials phase I and II), statins, and erythropoietin demonstrated the most encouraging results for the endogenous stem-cells-mediated repair [3].

Growth factors, moreover, draw significant attention for their neuroprotective and neuroregenerative efficiency. In particular, vascular endothelial growth factor (VEGF), human fibroblast growth factor 2 (FGF2), and brain-derived neurotrophic factor have been shown to improve neuronal survival when accompanying transplanted stem cells in unhealthy and injured models [120]. VEGF and FGF2 also improve functional outcomes in the preclinical model of TBI [121], while nerve growth factor decreases brain edema and reduces beta-amyloid production after TBI [122,123]. In addition, gene therapy and viral and non-viral-mediated delivery systems mark progress in managing neuronal injury. Adeno-associated viral vectors present attractive instruments for re-expressing and over-expressing genes in neurodegenerative disorders [124]. Micelles, polyethyleneimine (PEI)-coated micelles, and further micelle-like nanoparticles also might contain genetic material (DNA or RNA) and be an appealing approach for gene therapy due to their low or no immunogenicity. They can also be inserted into the brain via intranasal delivery, eliminating the need for direct intracerebral drug delivery. Nanoparticles, such as micelles, have been studied in a preclinical model of TBI to distribute DNA intranasally [125,126].

9. Neuropsychological Rehabilitation (NR) and Neurotherapy

TBI typically disturbs brain functions such as executive actions, cognitive grade, attention, memory, data processing, and language skills. Neuropsychological rehabilitation (NR) is aimed at ameliorating cognitive, emotional, psychosocial, and behavioral deficits caused by an insult to the brain. The NR of TBI patients represents a global problem, one with which modern medicine is grappling [127]. One of the central motives is the deficiency of strictly delineated theoretic supports for therapy and the means for the incessant repressing of their effects. Every brain damage causes conflicts with the so-called electric and chemical brain language, altering the space of prevailing networks and the action of neurotransmitters, which provoke a decline of the brain systems. Some studies confirmed that neurotherapy, also called neurofeedback therapy (NFT), can promote neuroplasticity [128]. NFT was shown to excite meaningful variants in structural and functional connectivity among young patients moderate TBI [127]. Transcranial magnetic stimulation (TMS) as a way of non-invasive direct modulation of neuronal activity is verysuitable for the treatment of TBI [127]. Recently, new tools for the evaluation of brain neuromarkers in TBI were developed. These include quantitative electroencephalography (EEG) to detect cortical self-regulation of the brain and event-related potentials for the flow of information in the brain [127]. Nevertheless, despite neurotherapy being very important for TBI management, more research projects are needed (Figure 1).

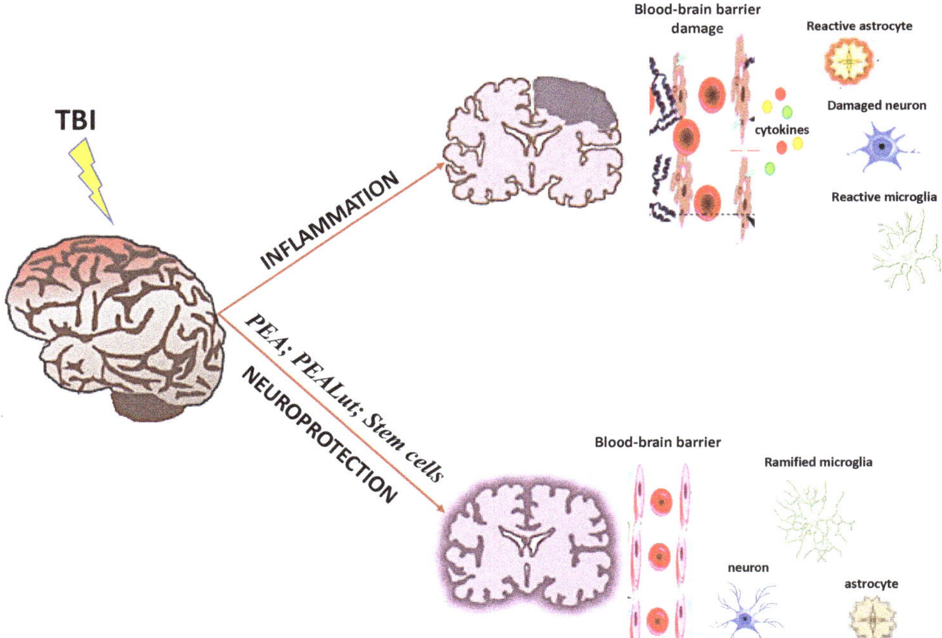

Figure 1. Pathophysiological heterogeneity detected in TBI.

10. Conclusions

Neuroprotective approaches are the focus for TBI management, particularly methods to classify and target specific mechanisms involved in the complex secondary-injury cascade. The literature shows that neuroprotective approaches historically have been dominated by a neurocentric view, making alteration of neuronal-based injury mechanisms the primary or exclusive focus of neuroprotective strategies. The data in the literature, therefore, stress the relevance of more broadly viewing injury as comprising endothelial, microglia, astroglia, oligodendroglia, and precursor cells. Recent neuroprotection methods describe this multifaceted structure and interplay, highlighting therapeutic approaches that stimulate the recovery and optimal functioning of non-neuronal cells and inhibit the underlying mechanism of neuronal cell death. Several encouraging, recently developed treatments include neuroprotective, neurorestorative, and anti-inflammatory agents (for example PEA formulations or biologics). In addition, researchers have reported the need for developing new neurothechnologies and the neuromarkers of brain injuries to enable a correct diagnosis and, as a result, appropriate selection of methods for neuropsychological rehabilitation including neurotherapy. However, due to the difficulty and heterogeneity of brain injuries, post-TBI neural therapies are still facing several challenges.

Author Contributions: All authors have read and agreed to the published version of the manuscript. R.C. and D.I. were involved in the design and intellectual concept of the study; M.C. performed the literature search. S.C. designed the study and critically revised the manuscript.

Funding: This research received no external funding.

Acknowledgments: The writers would like to thank Maria Antonietta Medici for brilliant methodical assistance during this research, Francesco Soraci for office support and assistance, and also Valentina Malvagni for editorial assistance with the text.

Conflicts of Interest: Salvatore Cuzzocrea is a co-inventor on patent WO2013121449 A8 (Epitech Group Srl), which deals with methods and compositions for the modulation of amidases capable of hydrolyzing *N*-acylethanolamines that are employable in the treatment of inflammatory diseases. This invention is wholly unrelated to the present study. Moreover, Salvatore Cuzzocrea is also, with the Epitech Group, a co-inventor on the following patents: EP 2 821 083; MI2014 A001495; 102015000067344, that are unrelated to the study. The remaining authors report no conflict of interest.

References

1. Kline, A.E.; Leary, J.B.; Radabaugh, H.L.; Cheng, J.P.; Bondi, C.O. Combination therapies for neurobehavioral and cognitive recovery after experimental traumatic brain injury: Is more better? *Prog. Neurobiol.* **2016**, *142*, 45–67. [CrossRef] [PubMed]
2. Logsdon, A.F.; Lucke-Wold, B.P.; Turner, R.C.; Huber, J.D.; Rosen, C.L.; Simpkins, J.W. Role of Microvascular Disruption in Brain Damage from Traumatic Brain Injury. *Compr. Physiol.* **2015**, *5*, 1147–1160. [CrossRef] [PubMed]
3. Zibara, K.; Ballout, N.; Mondello, S.; Karnib, N.; Ramadan, N.; Omais, S.; Nabbouh, A.; Caliz, D.; Clavijo, A.; Hu, Z.; et al. Combination of drug and stem cells neurotherapy: Potential interventions in neurotrauma and traumatic brain injury. *Neuropharmacology* **2019**, *145*, 177–198. [CrossRef] [PubMed]
4. Xiong, Y.; Mahmood, A.; Chopp, M. Animal models of traumatic brain injury. *Nat. Rev. Neurosci.* **2013**, *14*, 128–142. [CrossRef]
5. Kochanek, P.M.; Jackson, T.C.; Jha, R.; Clark, R.S.B.; Okonkwo, D.O.; Bayir, H.; Poloyac, S.M.; Wagner, A.M.D.; Empey, P.E.; Conley, Y.P.; et al. Paths to successful translation of new therapies for severe TBI in the golden age of traumatic brain injury research: A Pittsburgh vision. *J. Neurotrauma* **2018**. [CrossRef]
6. Greve, M.W.; Zink, B.J. Pathophysiology of traumatic brain injury. *Mt. Sinai J. Med.* **2009**, *76*, 97–104. [CrossRef]
7. Chan, T.L.H.; Woldeamanuel, Y.W. Exploring naturally occurring clinical subgroups of post-traumatic headache. *J. Headache Pain* **2020**, *21*, 12. [CrossRef]
8. Larsen, E.L.; Ashina, H.; Iljazi, A.; Al-Khazali, H.M.; Seem, K.; Ashina, M.; Ashina, S.; Schytz, H.W. Acute and preventive pharmacological treatment of post-traumatic headache: A systematic review. *J. Headache Pain* **2019**, *20*, 98. [CrossRef]

9. Bedaso, A.; Geja, E.; Ayalew, M.; Oltaye, Z.; Duko, B. Post-concussion syndrome among patients experiencing head injury attending emergency department of Hawassa University Comprehensive specialized hospital, Hawassa, southern Ethiopia. *J. Headache Pain* **2018**, *19*, 112. [CrossRef]
10. Niyonkuru, C.; Wagner, A.K.; Ozawa, H.; Amin, K.; Goyal, A.; Fabio, A. Group-based trajectory analysis applications for prognostic biomarker model development in severe TBI: A practical example. *J. Neurotrauma* **2013**, *30*, 938–945. [CrossRef]
11. Wang, Y.; Qin, Z.H. Molecular and cellular mechanisms of excitotoxic neuronal death. *Apoptosis Int. J. Program. Cell Death* **2010**, *15*, 1382–1402. [CrossRef] [PubMed]
12. Dasuri, K.; Zhang, L.; Keller, J.N. Oxidative stress, neurodegeneration, and the balance of protein degradation and protein synthesis. *Free Radic. Biol. Med.* **2013**, *62*, 170–185. [CrossRef] [PubMed]
13. Mutinati, M.; Pantaleo, M.; Roncetti, M.; Piccinno, M.; Rizzo, A.; Sciorsci, R.L. Oxidative stress in neonatology: A review. *Reprod. Domest. Anim.* **2014**, *49*, 7–16. [CrossRef] [PubMed]
14. Fluiter, K.; Opperhuizen, A.L.; Morgan, B.P.; Baas, F.; Ramaglia, V. Inhibition of the membrane attack complex of the complement system reduces secondary neuroaxonal loss and promotes neurologic recovery after traumatic brain injury in mice. *J. Immunol.* **2014**, *192*, 2339–2348. [CrossRef]
15. Wu, L.Y.; Bao, X.Q.; Sun, H.; Zhang, D. Scavenger receptor on astrocytes and its relationship with neuroinflammation. *Acta Acad. Med. Sin.* **2014**, *36*, 330–335. [CrossRef]
16. Kumar, A.; Loane, D.J. Neuroinflammation after traumatic brain injury: Opportunities for therapeutic intervention. *Brain Behav. Immun.* **2012**, *26*, 1191–1201. [CrossRef]
17. Margulies, S.; Hicks, R. Combination therapies for traumatic brain injury: Prospective considerations. *J. Neurotrauma* **2009**, *26*, 925–939. [CrossRef]
18. Yang, Z.; Zhu, T.; Mondello, S.; Akel, M.; Wong, A.T.; Kothari, I.M.; Lin, F.; Shear, D.A.; Gilsdorf, J.S.; Leung, L.Y.; et al. Serum-Based Phospho-Neurofilament-Heavy Protein as Theranostic Biomarker in Three Models of Traumatic Brain Injury: An Operation Brain Trauma Therapy Study. *J. Neurotrauma* **2019**, *36*, 348–359. [CrossRef]
19. Kim, H.J.; Tsao, J.W.; Stanfill, A.G. The current state of biomarkers of mild traumatic brain injury. *JCI Insight* **2018**, *3*. [CrossRef]
20. Hagos, F.T.; Empey, P.E.; Wang, P.; Ma, X.; Poloyac, S.M.; Bayir, H.; Kochanek, P.M.; Bell, M.J.; Clark, R.S.B. Exploratory Application of Neuropharmacometabolomics in Severe Childhood Traumatic Brain Injury. *Crit. Care Med.* **2018**, *46*, 1471–1479. [CrossRef]
21. Higashida, T.; Kreipke, C.W.; Rafols, J.A.; Peng, C.; Schafer, S.; Schafer, P.; Ding, J.Y.; Dornbos, D., 3rd; Li, X.; Guthikonda, M.; et al. The role of hypoxia-inducible factor-1alpha, aquaporin-4, and matrix metalloproteinase-9 in blood-brain barrier disruption and brain edema after traumatic brain injury. *J. Neurosurg.* **2011**, *114*, 92–101. [CrossRef] [PubMed]
22. Pan, R.; Yu, K.; Weatherwax, T.; Zheng, H.; Liu, W.; Liu, K.J. Blood Occludin Level as a Potential Biomarker for Early Blood Brain Barrier Damage Following Ischemic Stroke. *Sci. Rep.* **2017**, *7*, 40331. [CrossRef] [PubMed]
23. Zongo, D.; Ribereau-Gayon, R.; Masson, F.; Laborey, M.; Contrand, B.; Salmi, L.R.; Montaudon, D.; Beaudeux, J.L.; Meurin, A.; Dousset, V.; et al. S100-B protein as a screening tool for the early assessment of minor head injury. *Ann. Emerg. Med.* **2012**, *59*, 209–218. [CrossRef] [PubMed]
24. Neher, M.D.; Keene, C.N.; Rich, M.C.; Moore, H.B.; Stahel, P.F. Serum biomarkers for traumatic brain injury. *South. Med. J.* **2014**, *107*, 248–255. [CrossRef]
25. Kochanek, P.M.; Dixon, C.E.; Shellington, D.K.; Shin, S.S.; Bayir, H.; Jackson, E.K.; Kagan, V.E.; Yan, H.Q.; Swauger, P.V.; Parks, S.A.; et al. Screening of biochemical and molecular mechanisms of secondary injury and repair in the brain after experimental blast-induced traumatic brain injury in rats. *J. Neurotrauma* **2013**, *30*, 920–937. [CrossRef]
26. Pham, N.; Akonasu, H.; Shishkin, R.; Taghibiglou, C. Plasma soluble prion protein, a potential biomarker for sport-related concussions: A pilot study. *PLoS ONE* **2015**, *10*, e0117286. [CrossRef]
27. Rubenstein, R.; Chang, B.; Yue, J.K.; Chiu, A.; Winkler, E.A.; Puccio, A.M.; Diaz-Arrastia, R.; Yuh, E.L.; Mukherjee, P.; Valadka, A.B.; et al. Comparing Plasma Phospho Tau, Total Tau, and Phospho Tau-Total Tau Ratio as Acute and Chronic Traumatic Brain Injury Biomarkers. *JAMA Neurol.* **2017**, *74*, 1063–1072. [CrossRef]

28. Papa, L.; Lewis, L.M.; Silvestri, S.; Falk, J.L.; Giordano, P.; Brophy, G.M.; Demery, J.A.; Liu, M.C.; Mo, J.; Akinyi, L.; et al. Serum levels of ubiquitin C-terminal hydrolase distinguish mild traumatic brain injury from trauma controls and are elevated in mild and moderate traumatic brain injury patients with intracranial lesions and neurosurgical intervention. *J. Trauma Acute Care Surg.* **2012**, *72*, 1335–1344. [CrossRef]
29. Mondello, S.; Linnet, A.; Buki, A.; Robicsek, S.; Gabrielli, A.; Tepas, J.; Papa, L.; Brophy, G.M.; Tortella, F.; Hayes, R.L.; et al. Clinical utility of serum levels of ubiquitin C-terminal hydrolase as a biomarker for severe traumatic brain injury. *Neurosurgery* **2012**, *70*, 666–675. [CrossRef]
30. Puvenna, V.; Brennan, C.; Shaw, G.; Yang, C.; Marchi, N.; Bazarian, J.J.; Merchant-Borna, K.; Janigro, D. Significance of ubiquitin carboxy-terminal hydrolase L1 elevations in athletes after sub-concussive head hits. *PLoS ONE* **2014**, *9*, e96296. [CrossRef]
31. Gatson, J.W.; Barillas, J.; Hynan, L.S.; Diaz-Arrastia, R.; Wolf, S.E.; Minei, J.P. Detection of neurofilament-H in serum as a diagnostic tool to predict injury severity in patients who have suffered mild traumatic brain injury. *J. Neurosurg.* **2014**, *121*, 1232–1238. [CrossRef] [PubMed]
32. Rossi, D.; Volanti, P.; Brambilla, L.; Colletti, T.; Spataro, R.; La Bella, V. CSF neurofilament proteins as diagnostic and prognostic biomarkers for amyotrophic lateral sclerosis. *J. Neurol.* **2018**, *265*, 510–521. [CrossRef] [PubMed]
33. Bohmer, A.E.; Oses, J.P.; Schmidt, A.P.; Peron, C.S.; Krebs, C.L.; Oppitz, P.P.; D'Avila, T.T.; Souza, D.O.; Portela, L.V.; Stefani, M.A. Neuron-specific enolase, S100B, and glial fibrillary acidic protein levels as outcome predictors in patients with severe traumatic brain injury. *Neurosurgery* **2011**, *68*, 1624–1630. [CrossRef] [PubMed]
34. Lei, J.; Gao, G.; Feng, J.; Jin, Y.; Wang, C.; Mao, Q.; Jiang, J. Glial fibrillary acidic protein as a biomarker in severe traumatic brain injury patients: A prospective cohort study. *Crit. Care* **2015**, *19*, 362. [CrossRef]
35. Berger, R.P.; Adelson, P.D.; Pierce, M.C.; Dulani, T.; Cassidy, L.D.; Kochanek, P.M. Serum neuron-specific enolase, S100B, and myelin basic protein concentrations after inflicted and noninflicted traumatic brain injury in children. *J. Neurosurg.* **2005**, *103*, 61–68. [CrossRef]
36. Mondello, S.; Robicsek, S.A.; Gabrielli, A.; Brophy, G.M.; Papa, L.; Tepas, J.; Robertson, C.; Buki, A.; Scharf, D.; Jixiang, M.; et al. alphaII-spectrin breakdown products (SBDPs): Diagnosis and outcome in severe traumatic brain injury patients. *J. Neurotrauma* **2010**, *27*, 1203–1213. [CrossRef]
37. Zhao, J.; Xi, G.; Wu, G.; Keep, R.F.; Hua, Y. Deferoxamine Attenuated the Upregulation of Lipocalin-2 Induced by Traumatic Brain Injury in Rats. *Acta Neurochir. Suppl.* **2016**, *121*, 291–294. [CrossRef]
38. Zhao, J.; Chen, H.; Zhang, M.; Zhang, Y.; Qian, C.; Liu, Y.; He, S.; Zou, Y.; Liu, H. Early expression of serum neutrophil gelatinase-associated lipocalin (NGAL) is associated with neurological severity immediately after traumatic brain injury. *J. Neurol. Sci.* **2016**, *368*, 392–398. [CrossRef]
39. Semple, B.D.; Bye, N.; Rancan, M.; Ziebell, J.M.; Morganti-Kossmann, M.C. Role of CCL2 (MCP-1) in traumatic brain injury (TBI): Evidence from severe TBI patients and CCL2-/- mice. *J. Cereb. Blood Flow Metab. Off. J. Int. Soc. Cereb. Blood Flow Metab.* **2010**, *30*, 769–782. [CrossRef]
40. Buttram, S.D.; Wisniewski, S.R.; Jackson, E.K.; Adelson, P.D.; Feldman, K.; Bayir, H.; Berger, R.P.; Clark, R.S.; Kochanek, P.M. Multiplex assessment of cytokine and chemokine levels in cerebrospinal fluid following severe pediatric traumatic brain injury: Effects of moderate hypothermia. *J. Neurotrauma* **2007**, *24*, 1707–1717. [CrossRef]
41. Berger, R.P.; Ta'asan, S.; Rand, A.; Lokshin, A.; Kochanek, P. Multiplex assessment of serum biomarker concentrations in well-appearing children with inflicted traumatic brain injury. *Pediatric Res.* **2009**, *65*, 97–102. [CrossRef] [PubMed]
42. Oliver, J.; Abbas, K.; Lightfoot, J.T.; Baskin, K.; Collins, B.; Wier, D.; Bramhall, J.P.; Huang, J.; Puschett, J.B. Comparison of Neurocognitive Testing and the Measurement of Marinobufagenin in Mild Traumatic Brain Injury: A Preliminary Report. *J. Exp. Neurosci.* **2015**, *9*, 67–72. [CrossRef] [PubMed]
43. Lee, H.H.; Yeh, C.T.; Ou, J.C.; Ma, H.P.; Chen, K.Y.; Chang, C.F.; Lai, J.H.; Liao, K.H.; Lin, C.M.; Lin, S.Y.; et al. The Association of Apolipoprotein E Allele 4 Polymorphism with the Recovery of Sleep Disturbance after Mild Traumatic Brain Injury. *Acta Neurol. Taiwanica* **2017**, *26*, 13–19.
44. Hayes, J.P.; Reagan, A.; Logue, M.W.; Hayes, S.M.; Sadeh, N.; Miller, D.R.; Verfaellie, M.; Wolf, E.J.; McGlinchey, R.E.; Milberg, W.P.; et al. BDNF genotype is associated with hippocampal volume in mild traumatic brain injury. *Genes Brain Behav.* **2018**, *17*, 107–117. [CrossRef] [PubMed]

45. Sullivan, P.G.; Sebastian, A.H.; Hall, E.D. Therapeutic window analysis of the neuroprotective effects of cyclosporine A after traumatic brain injury. *J. Neurotrauma* **2011**, *28*, 311–318. [CrossRef] [PubMed]
46. Holmin, S.; Mathiesen, T. Dexamethasone and colchicine reduce inflammation and delayed oedema following experimental brain contusion. *Acta Neurochir.* **1996**, *138*, 418–424. [CrossRef]
47. Zhang, Z.; Zhang, Z.; Artelt, M.; Burnet, M.; Schluesener, H.J. Dexamethasone attenuates early expression of three molecules associated with microglia/macrophages activation following rat traumatic brain injury. *Acta Neuropathol.* **2007**, *113*, 675–682. [CrossRef]
48. Hakan, T.; Toklu, H.Z.; Biber, N.; Ozevren, H.; Solakoglu, S.; Demirturk, P.; Aker, F.V. Effect of COX-2 inhibitor meloxicam against traumatic brain injury-induced biochemical, histopathological changes and blood-brain barrier permeability. *Neurol. Res.* **2010**, *32*, 629–635. [CrossRef]
49. Siopi, E.; Cho, A.H.; Homsi, S.; Croci, N.; Plotkine, M.; Marchand-Leroux, C.; Jafarian-Tehrani, M. Minocycline restores sAPPalpha levels and reduces the late histopathological consequences of traumatic brain injury in mice. *J. Neurotrauma* **2011**, *28*, 2135–2143. [CrossRef]
50. Thau-Zuchman, O.; Shohami, E.; Alexandrovich, A.G.; Trembovler, V.; Leker, R.R. The anti-inflammatory drug carprofen improves long-term outcome and induces gliogenesis after traumatic brain injury. *J. Neurotrauma* **2012**, *29*, 375–384. [CrossRef]
51. Chao, P.K.; Lu, K.T.; Jhu, J.Y.; Wo, Y.Y.; Huang, T.C.; Ro, L.S.; Yang, Y.L. Indomethacin protects rats from neuronal damage induced by traumatic brain injury and suppresses hippocampal IL-1beta release through the inhibition of Nogo-A expression. *J. Neuroinflamm.* **2012**, *9*, 121. [CrossRef] [PubMed]
52. Girgis, H.; Palmier, B.; Croci, N.; Soustrat, M.; Plotkine, M.; Marchand-Leroux, C. Effects of selective and non-selective cyclooxygenase inhibition against neurological deficit and brain oedema following closed head injury in mice. *Brain Res.* **2013**, *1491*, 78–87. [CrossRef] [PubMed]
53. Clond, M.A.; Lee, B.S.; Yu, J.J.; Singer, M.B.; Amano, T.; Lamb, A.W.; Drazin, D.; Kateb, B.; Ley, E.J.; Yu, J.S. Reactive oxygen species-activated nanoprodrug of Ibuprofen for targeting traumatic brain injury in mice. *PLoS ONE* **2013**, *8*, e61819. [CrossRef] [PubMed]
54. Cheong, C.U.; Chang, C.P.; Chao, C.M.; Cheng, B.C.; Yang, C.Z.; Chio, C.C. Etanercept attenuates traumatic brain injury in rats by reducing brain TNF- alpha contents and by stimulating newly formed neurogenesis. *Mediat. Inflamm.* **2013**, *2013*, 620837. [CrossRef]
55. Campolo, M.; Ahmad, A.; Crupi, R.; Impellizzeri, D.; Morabito, R.; Esposito, E.; Cuzzocrea, S. Combination therapy with melatonin and dexamethasone in a mouse model of traumatic brain injury. *J. Endocrinol.* **2013**, *217*, 291–301. [CrossRef]
56. Luo, C.L.; Li, Q.Q.; Chen, X.P.; Zhang, X.M.; Li, L.L.; Li, B.X.; Zhao, Z.Q.; Tao, L.Y. Lipoxin A4 attenuates brain damage and downregulates the production of pro-inflammatory cytokines and phosphorylated mitogen-activated protein kinases in a mouse model of traumatic brain injury. *Brain Res.* **2013**, *1502*, 1–10. [CrossRef]
57. Harrison, J.L.; Rowe, R.K.; O'Hara, B.F.; Adelson, P.D.; Lifshitz, J. Acute over-the-counter pharmacological intervention does not adversely affect behavioral outcome following diffuse traumatic brain injury in the mouse. *Exp. Brain Res.* **2014**, *232*, 2709–2719. [CrossRef]
58. Homsi, S.; Piaggio, T.; Croci, N.; Noble, F.; Plotkine, M.; Marchand-Leroux, C.; Jafarian-Tehrani, M. Blockade of acute microglial activation by minocycline promotes neuroprotection and reduces locomotor hyperactivity after closed head injury in mice: A twelve-week follow-up study. *J. Neurotrauma* **2010**, *27*, 911–921. [CrossRef]
59. Bye, N.; Habgood, M.D.; Callaway, J.K.; Malakooti, N.; Potter, A.; Kossmann, T.; Morganti-Kossmann, M.C. Transient neuroprotection by minocycline following traumatic brain injury is associated with attenuated microglial activation but no changes in cell apoptosis or neutrophil infiltration. *Exp. Neurol.* **2007**, *204*, 220–233. [CrossRef]
60. Besson, V.C.; Chen, X.R.; Plotkine, M.; Marchand-Verrecchia, C. Fenofibrate, a peroxisome proliferator-activated receptor alpha agonist, exerts neuroprotective effects in traumatic brain injury. *Neurosci. Lett.* **2005**, *388*, 7–12. [CrossRef]
61. Sauerbeck, A.; Gao, J.; Readnower, R.; Liu, M.; Pauly, J.R.; Bing, G.; Sullivan, P.G. Pioglitazone attenuates mitochondrial dysfunction, cognitive impairment, cortical tissue loss, and inflammation following traumatic brain injury. *Exp. Neurol.* **2011**, *227*, 128–135. [CrossRef] [PubMed]

62. Chen, G.; Shi, J.; Hu, Z.; Hang, C. Inhibitory effect on cerebral inflammatory response following traumatic brain injury in rats: A potential neuroprotective mechanism of N-acetylcysteine. *Mediat. Inflamm.* **2008**, *2008*, 716458. [CrossRef] [PubMed]
63. Bergold, P.J. Treatment of traumatic brain injury with anti-inflammatory drugs. *Exp. Neurol.* **2016**, *275 Pt 3*, 367–380. [CrossRef]
64. Chen, G.; Shi, J.X.; Hang, C.H.; Xie, W.; Liu, J.; Liu, X. Inhibitory effect on cerebral inflammatory agents that accompany traumatic brain injury in a rat model: A potential neuroprotective mechanism of recombinant human erythropoietin (rhEPO). *Neurosci. Lett.* **2007**, *425*, 177–182. [CrossRef] [PubMed]
65. Xu, F.; Yu, Z.Y.; Ding, L.; Zheng, S.Y. Experimental studies of erythropoietin protection following traumatic brain injury in rats. *Exp. Med.* **2012**, *4*, 977–982. [CrossRef]
66. Chen, G.; Zhang, S.; Shi, J.; Ai, J.; Qi, M.; Hang, C. Simvastatin reduces secondary brain injury caused by cortical contusion in rats: Possible involvement of TLR4/NF-kappaB pathway. *Exp. Neurol.* **2009**, *216*, 398–406. [CrossRef]
67. Li, B.; Mahmood, A.; Lu, D.; Wu, H.; Xiong, Y.; Qu, C.; Chopp, M. Simvastatin attenuates microglial cells and astrocyte activation and decreases interleukin-1beta level after traumatic brain injury. *Neurosurgery* **2009**, *65*, 179–185. [CrossRef]
68. Si, D.; Li, J.; Liu, J.; Wang, X.; Wei, Z.; Tian, Q.; Wang, H.; Liu, G. Progesterone protects blood-brain barrier function and improves neurological outcome following traumatic brain injury in rats. *Exp. Med.* **2014**, *8*, 1010–1014. [CrossRef]
69. Cutler, S.M.; Cekic, M.; Miller, D.M.; Wali, B.; VanLandingham, J.W.; Stein, D.G. Progesterone improves acute recovery after traumatic brain injury in the aged rat. *J. Neurotrauma* **2007**, *24*, 1475–1486. [CrossRef]
70. Orsucci, D.; Calsolaro, V.; Mancuso, M.; Siciliano, G. Neuroprotective effects of tetracyclines: Molecular targets, animal models and human disease. *CNS Neurol. Disord. Drug Targets* **2009**, *8*, 222–231. [CrossRef]
71. Villapol, S.; Yaszemski, A.K.; Logan, T.T.; Sanchez-Lemus, E.; Saavedra, J.M.; Symes, A.J. Candesartan, an angiotensin II AT(1)-receptor blocker and PPAR-gamma agonist, reduces lesion volume and improves motor and memory function after traumatic brain injury in mice. *Neuropsychopharmacol. Off. Publ. Am. Coll. Neuropsychopharmacol.* **2012**, *37*, 2817–2829. [CrossRef] [PubMed]
72. Mandrekar-Colucci, S.; Sauerbeck, A.; Popovich, P.G.; McTigue, D.M. PPAR agonists as therapeutics for CNS trauma and neurological diseases. *ASN Neuro* **2013**, *5*, e00129. [CrossRef] [PubMed]
73. Di Giovanni, S.; Movsesyan, V.; Ahmed, F.; Cernak, I.; Schinelli, S.; Stoica, B.; Faden, A.I. Cell cycle inhibition provides neuroprotection and reduces glial proliferation and scar formation after traumatic brain injury. *Proc. Natl. Acad. Sci. USA* **2005**, *102*, 8333–8338. [CrossRef] [PubMed]
74. Kabadi, S.V.; Stoica, B.A.; Byrnes, K.R.; Hanscom, M.; Loane, D.J.; Faden, A.I. Selective CDK inhibitor limits neuroinflammation and progressive neurodegeneration after brain trauma. *J. Cereb. Blood Flow Metab. Off. J. Int. Soc. Cereb. Blood Flow Metab.* **2012**, *32*, 137–149. [CrossRef]
75. Peng, W.; Xing, Z.; Yang, J.; Wang, Y.; Wang, W.; Huang, W. The efficacy of erythropoietin in treating experimental traumatic brain injury: A systematic review of controlled trials in animal models. *J. Neurosurg.* **2014**, *121*, 653–664. [CrossRef]
76. Robertson, C.S.; Hannay, H.J.; Yamal, J.M.; Gopinath, S.; Goodman, J.C.; Tilley, B.C.; Epo Severe, T.B.I.T.I.; Baldwin, A.; Rivera Lara, L.; Saucedo-Crespo, H.; et al. Effect of erythropoietin and transfusion threshold on neurological recovery after traumatic brain injury: A randomized clinical trial. *JAMA* **2014**, *312*, 36–47. [CrossRef]
77. Xiong, Y.; Zhang, Y.; Mahmood, A.; Chopp, M. Investigational agents for treatment of traumatic brain injury. *Expert Opin. Investig. Drugs* **2015**, *24*, 743–760. [CrossRef]
78. Meythaler, J.; Fath, J.; Fuerst, D.; Zokary, H.; Freese, K.; Martin, H.B.; Reineke, J.; Peduzzi-Nelson, J.; Roskos, P.T. Safety and feasibility of minocycline in treatment of acute traumatic brain injury. *Brain Inj.* **2019**, *33*, 679–689. [CrossRef]
79. Lu, D.; Goussev, A.; Chen, J.; Pannu, P.; Li, Y.; Mahmood, A.; Chopp, M. Atorvastatin reduces neurological deficit and increases synaptogenesis, angiogenesis, and neuronal survival in rats subjected to traumatic brain injury. *J. Neurotrauma* **2004**, *21*, 21–32. [CrossRef]
80. Wu, H.; Lu, D.; Jiang, H.; Xiong, Y.; Qu, C.; Li, B.; Mahmood, A.; Zhou, D.; Chopp, M. Increase in phosphorylation of Akt and its downstream signaling targets and suppression of apoptosis by simvastatin after traumatic brain injury. *J. Neurosurg.* **2008**, *109*, 691–698. [CrossRef]

81. Wu, H.; Lu, D.; Jiang, H.; Xiong, Y.; Qu, C.; Li, B.; Mahmood, A.; Zhou, D.; Chopp, M. Simvastatin-mediated upregulation of VEGF and BDNF, activation of the PI3K/Akt pathway, and increase of neurogenesis are associated with therapeutic improvement after traumatic brain injury. *J. Neurotrauma* **2008**, *25*, 130–139. [CrossRef] [PubMed]
82. Mahmood, A.; Goussev, A.; Kazmi, H.; Qu, C.; Lu, D.; Chopp, M. Long-term benefits after treatment of traumatic brain injury with simvastatin in rats. *Neurosurgery* **2009**, *65*, 187–191. [CrossRef] [PubMed]
83. Head, B.P.; Patel, H.H.; Insel, P.A. Interaction of membrane/lipid rafts with the cytoskeleton: Impact on signaling and function: Membrane/lipid rafts, mediators of cytoskeletal arrangement and cell signaling. *Biochim. Biophys. Acta* **2014**, *1838*, 532–545. [CrossRef] [PubMed]
84. Tobinick, E. Perispinal etanercept: A new therapeutic paradigm in neurology. *Expert Rev. Neurother.* **2010**, *10*, 985–1002. [CrossRef] [PubMed]
85. Tobinick, E.; Kim, N.M.; Reyzin, G.; Rodriguez-Romanacce, H.; DePuy, V. Selective TNF inhibition for chronic stroke and traumatic brain injury: An observational study involving 629 consecutive patients treated with perispinal etanercept. *CNS Drugs* **2012**, *26*, 1051–1070. [CrossRef]
86. Tobinick, E.; Rodriguez-Romanacce, H.; Levine, A.; Ignatowski, T.A.; Spengler, R.N. Immediate neurological recovery following perispinal etanercept years after brain injury. *Clin. Drug Investig.* **2014**, *34*, 361–366. [CrossRef]
87. Hoffer, M.E.; Balaban, C.; Slade, M.D.; Tsao, J.W.; Hoffer, B. Amelioration of acute sequelae of blast induced mild traumatic brain injury by N-acetyl cysteine: A double-blind, placebo controlled study. *PLoS ONE* **2013**, *8*, e54163. [CrossRef]
88. Clark, R.S.B.; Empey, P.E.; Bayir, H.; Rosario, B.L.; Poloyac, S.M.; Kochanek, P.M.; Nolin, T.D.; Au, A.K.; Horvat, C.M.; Wisniewski, S.R.; et al. Phase I randomized clinical trial of N-acetylcysteine in combination with an adjuvant probenecid for treatment of severe traumatic brain injury in children. *PLoS ONE* **2017**, *12*, e0180280. [CrossRef]
89. Stein, D.G. Embracing failure: What the Phase III progesterone studies can teach about TBI clinical trials. *Brain Inj.* **2015**, *29*, 1259–1272. [CrossRef]
90. Atkins, C.M.; Oliva, A.A., Jr.; Alonso, O.F.; Pearse, D.D.; Bramlett, H.M.; Dietrich, W.D. Modulation of the cAMP signaling pathway after traumatic brain injury. *Exp. Neurol.* **2007**, *208*, 145–158. [CrossRef]
91. Ceyhan, O.; Birsoy, K.; Hoffman, C.S. Identification of biologically active PDE11-selective inhibitors using a yeast-based high-throughput screen. *Chem. Biol.* **2012**, *19*, 155–163. [CrossRef] [PubMed]
92. Kaminska, K.; Golembiowska, K.; Rogoz, Z. Effect of risperidone on the fluoxetine-induced changes in extracellular dopamine, serotonin and noradrenaline in the rat frontal cortex. *Pharmacol. Rep. Pr.* **2013**, *65*, 1144–1151. [CrossRef]
93. Huot, P.; Johnston, T.H.; Lewis, K.D.; Koprich, J.B.; Reyes, M.G.; Fox, S.H.; Piggott, M.J.; Brotchie, J.M. UWA-121, a mixed dopamine and serotonin re-uptake inhibitor, enhances L-DOPA anti-parkinsonian action without worsening dyskinesia or psychosis-like behaviours in the MPTP-lesioned common marmoset. *Neuropharmacology* **2014**, *82*, 76–87. [CrossRef] [PubMed]
94. Chew, E.; Zafonte, R.D. Pharmacological management of neurobehavioral disorders following traumatic brain injury–a state-of-the-art review. *J. Rehabil. Res. Dev.* **2009**, *46*, 851–879. [CrossRef] [PubMed]
95. Buckley, C.D.; Gilroy, D.W.; Serhan, C.N.; Stockinger, B.; Tak, P.P. The resolution of inflammation. *Nat. Rev. Immunol.* **2013**, *13*, 59–66. [CrossRef]
96. Pacher, P.; Batkai, S.; Kunos, G. The endocannabinoid system as an emerging target of pharmacotherapy. *Pharmacol. Rev.* **2006**, *58*, 389–462. [CrossRef]
97. Skaper, S.D.; Facci, L.; Giusti, P. Mast cells, glia and neuroinflammation: Partners in crime? *Immunology* **2014**, *141*, 314–327. [CrossRef]
98. Impellizzeri, D.; Bruschetta, G.; Cordaro, M.; Crupi, R.; Siracusa, R.; Esposito, E.; Cuzzocrea, S. Micronized/ultramicronized palmitoylethanolamide displays superior oral efficacy compared to nonmicronized palmitoylethanolamide in a rat model of inflammatory pain. *J. Neuroinflamm.* **2014**, *11*, 136. [CrossRef]
99. Boccella, S.; Marabese, I.; Iannotta, M.; Belardo, C.; Neugebauer, V.; Mazzitelli, M.; Pieretti, G.; Maione, S.; Palazzo, E. Metabotropic Glutamate Receptor 5 and 8 Modulate the Ameliorative Effect of Ultramicronized Palmitoylethanolamide on Cognitive Decline Associated with Neuropathic Pain. *Int. J. Mol. Sci.* **2019**, 20. [CrossRef]

100. Scuderi, C.; Bronzuoli, M.R.; Facchinetti, R.; Pace, L.; Ferraro, L.; Broad, K.D.; Serviddio, G.; Bellanti, F.; Palombelli, G.; Carpinelli, G.; et al. Ultramicronized palmitoylethanolamide rescues learning and memory impairments in a triple transgenic mouse model of Alzheimer's disease by exerting anti-inflammatory and neuroprotective effects. *Transl. Psychiatry* **2018**, *8*, 32. [CrossRef]
101. Fusco, R.; Gugliandolo, E.; Campolo, M.; Evangelista, M.; Di Paola, R.; Cuzzocrea, S. Effect of a new formulation of micronized and ultramicronized N-palmitoylethanolamine in a tibia fracture mouse model of complex regional pain syndrome. *PLoS ONE* **2017**, *12*, e0178553. [CrossRef] [PubMed]
102. Impellizzeri, D.; Peritore, A.F.; Cordaro, M.; Gugliandolo, E.; Siracusa, R.; Crupi, R.; D'Amico, R.; Fusco, R.; Evangelista, M.; Cuzzocrea, S.; et al. The neuroprotective effects of micronized PEA (PEA-m) formulation on diabetic peripheral neuropathy in mice. *FASEB J.* **2019**, *33*, 11364–11380. [CrossRef] [PubMed]
103. Passavanti, M.B.; Fiore, M.; Sansone, P.; Aurilio, C.; Pota, V.; Barbarisi, M.; Fierro, D.; Pace, M.C. The beneficial use of ultramicronized palmitoylethanolamide as add-on therapy to Tapentadol in the treatment of low back pain: A pilot study comparing prospective and retrospective observational arms. *BMC Anesth.* **2017**, *17*, 171. [CrossRef] [PubMed]
104. Schweiger, V.; Martini, A.; Bellamoli, P.; Donadello, K.; Schievano, C.; Balzo, G.D.; Sarzi-Puttini, P.; Parolini, M.; Polati, E. Ultramicronized Palmitoylethanolamide (um-PEA) as Add-on Treatment in Fibromyalgia Syndrome (FMS): Retrospective Observational Study on 407 Patients. *CNS Neurol. Disord. Drug Targets* **2019**, *18*, 326–333. [CrossRef] [PubMed]
105. Crupi, R.; Paterniti, I.; Ahmad, A.; Campolo, M.; Esposito, E.; Cuzzocrea, S. Effects of palmitoylethanolamide and luteolin in an animal model of anxiety/depression. *CNS Neurol. Disord. Drug Targets* **2013**, *12*, 989–1001. [CrossRef] [PubMed]
106. Bertolino, B.; Crupi, R.; Impellizzeri, D.; Bruschetta, G.; Cordaro, M.; Siracusa, R.; Esposito, E.; Cuzzocrea, S. Beneficial Effects of Co-Ultramicronized Palmitoylethanolamide/Luteolin in a Mouse Model of Autism and in a Case Report of Autism. *CNS Neurosci. Ther.* **2017**, *23*, 87–98. [CrossRef]
107. Crupi, R.; Impellizzeri, D.; Bruschetta, G.; Cordaro, M.; Paterniti, I.; Siracusa, R.; Cuzzocrea, S.; Esposito, E. Co-Ultramicronized Palmitoylethanolamide/Luteolin Promotes Neuronal Regeneration after Spinal Cord Injury. *Front. Pharmacol.* **2016**, *7*, 47. [CrossRef]
108. Impellizzeri, D.; Esposito, E.; Di Paola, R.; Ahmad, A.; Campolo, M.; Peli, A.; Morittu, V.M.; Britti, D.; Cuzzocrea, S. Palmitoylethanolamide and luteolin ameliorate development of arthritis caused by injection of collagen type II in mice. *Arthritis Res. Ther.* **2013**, *15*, R192. [CrossRef]
109. Paterniti, I.; Cordaro, M.; Campolo, M.; Siracusa, R.; Cornelius, C.; Navarra, M.; Cuzzocrea, S.; Esposito, E. Neuroprotection by association of palmitoylethanolamide with luteolin in experimental Alzheimer's disease models: The control of neuroinflammation. *CNS Neurol. Disord. Drug Targets* **2014**, *13*, 1530–1541. [CrossRef]
110. Paterniti, I.; Impellizzeri, D.; Di Paola, R.; Navarra, M.; Cuzzocrea, S.; Esposito, E. A new co-ultramicronized composite including palmitoylethanolamide and luteolin to prevent neuroinflammation in spinal cord injury. *J. Neuroinflamm.* **2013**, *10*, 91. [CrossRef]
111. Siracusa, R.; Paterniti, I.; Impellizzeri, D.; Cordaro, M.; Crupi, R.; Navarra, M.; Cuzzocrea, S.; Esposito, E. The Association of Palmitoylethanolamide with Luteolin Decreases Neuroinflammation and Stimulates Autophagy in Parkinson's Disease Model. *CNS Neurol. Disord. Drug Targets* **2015**, *14*, 1350–1365. [CrossRef] [PubMed]
112. Cordaro, M.; Impellizzeri, D.; Paterniti, I.; Bruschetta, G.; Siracusa, R.; De Stefano, D.; Cuzzocrea, S.; Esposito, E. Neuroprotective Effects of Co-UltraPEALut on Secondary Inflammatory Process and Autophagy Involved in Traumatic Brain Injury. *J. Neurotrauma* **2016**, *33*, 132–146. [CrossRef] [PubMed]
113. Siracusa, R.; Impellizzeri, D.; Cordaro, M.; Crupi, R.; Esposito, E.; Petrosino, S.; Cuzzocrea, S. Anti-Inflammatory and Neuroprotective Effects of Co-UltraPEALut in a Mouse Model of Vascular Dementia. *Front. Neurol.* **2017**, *8*, 233. [CrossRef] [PubMed]
114. Caltagirone, C.; Cisari, C.; Schievano, C.; Di Paola, R.; Cordaro, M.; Bruschetta, G.; Esposito, E.; Cuzzocrea, S.; Stroke Study Group. Co-ultramicronized Palmitoylethanolamide/Luteolin in the Treatment of Cerebral Ischemia: From Rodent to Man. *Transl. Stroke Res.* **2016**, *7*, 54–69. [CrossRef] [PubMed]
115. Impellizzeri, D.; Cordaro, M.; Bruschetta, G.; Crupi, R.; Pascali, J.; Alfonsi, D.; Marcolongo, G.; Cuzzocrea, S. 2-pentadecyl-2-oxazoline: Identification in coffee, synthesis and activity in a rat model of carrageenan-induced hindpaw inflammation. *Pharmacol. Res.* **2016**, *108*, 23–30. [CrossRef] [PubMed]

116. Petrosino, S.; Campolo, M.; Impellizzeri, D.; Paterniti, I.; Allara, M.; Gugliandolo, E.; D'Amico, R.; Siracusa, R.; Cordaro, M.; Esposito, E.; et al. 2-Pentadecyl-2-Oxazoline, the Oxazoline of Pea, Modulates Carrageenan-Induced Acute Inflammation. *Front. Pharmacol.* **2017**, *8*, 308. [CrossRef]
117. Impellizzeri, D.; Cordaro, M.; Bruschetta, G.; Siracusa, R.; Crupi, R.; Esposito, E.; Cuzzocrea, S. N-Palmitoylethanolamine-Oxazoline as a New Therapeutic Strategy to Control Neuroinflammation: Neuroprotective Effects in Experimental Models of Spinal Cord and Brain Injury. *J. Neurotrauma.* **2017**, *34*, 2609–2623. [CrossRef]
118. Mouhieddine, T.H.; Kobeissy, F.H.; Itani, M.; Nokkari, A.; Wang, K.K. Stem cells in neuroinjury and neurodegenerative disorders: Challenges and future neurotherapeutic prospects. *Neural Regen. Res.* **2014**, *9*, 901–906. [CrossRef]
119. Liao, G.P.; Harting, M.T.; Hetz, R.A.; Walker, P.A.; Shah, S.K.; Corkins, C.J.; Hughes, T.G.; Jimenez, F.; Kosmach, S.C.; Day, M.C.; et al. Autologous bone marrow mononuclear cells reduce therapeutic intensity for severe traumatic brain injury in children. *Pediatr. Crit. Care Med. J. Soc. Crit. Care Med. World Fed. Pediatr. Intensive Crit. Care Soc.* **2015**, *16*, 245–255. [CrossRef]
120. Blaya, M.O.; Tsoulfas, P.; Bramlett, H.M.; Dietrich, W.D. Neural progenitor cell transplantation promotes neuroprotection, enhances hippocampal neurogenesis, and improves cognitive outcomes after traumatic brain injury. *Exp. Neurol.* **2015**, *264*, 67–81. [CrossRef]
121. Thau-Zuchman, O.; Shohami, E.; Alexandrovich, A.G.; Leker, R.R. Combination of vascular endothelial and fibroblast growth factor 2 for induction of neurogenesis and angiogenesis after traumatic brain injury. *J. Mol. Neurosci.* **2012**, *47*, 166–172. [CrossRef] [PubMed]
122. Lv, Q.; Fan, X.; Xu, G.; Liu, Q.; Tian, L.; Cai, X.; Sun, W.; Wang, X.; Cai, Q.; Bao, Y.; et al. Intranasal delivery of nerve growth factor attenuates aquaporins-4-induced edema following traumatic brain injury in rats. *Brain Res.* **2013**, *1493*, 80–89. [CrossRef] [PubMed]
123. Tian, L.; Guo, R.; Yue, X.; Lv, Q.; Ye, X.; Wang, Z.; Chen, Z.; Wu, B.; Xu, G.; Liu, X. Intranasal administration of nerve growth factor ameliorate beta-amyloid deposition after traumatic brain injury in rats. *Brain Res.* **2012**, *1440*, 47–55. [CrossRef] [PubMed]
124. Murlidharan, G.; Samulski, R.J.; Asokan, A. Biology of adeno-associated viral vectors in the central nervous system. *Front. Mol. Neurosci.* **2014**, *7*, 76. [CrossRef]
125. Das, M.; Wang, C.; Bedi, R.; Mohapatra, S.S.; Mohapatra, S. Magnetic micelles for DNA delivery to rat brains after mild traumatic brain injury. *Nanomed. Nanotechnol. Biol. Med.* **2014**, *10*, 1539–1548. [CrossRef]
126. Harmon, B.T.; Aly, A.E.; Padegimas, L.; Sesenoglu-Laird, O.; Cooper, M.J.; Waszczak, B.L. Intranasal administration of plasmid DNA nanoparticles yields successful transfection and expression of a reporter protein in rat brain. *Gene Ther.* **2014**, *21*, 514–521. [CrossRef]
127. Chantsoulis, M.; Mirski, A.; Rasmus, A.; Kropotov, J.D.; Pachalska, M. Neuropsychological rehabilitation for traumatic brain injury patients. *Ann. Agric. Environ. Med. AAEM* **2015**, *22*, 368–379. [CrossRef]
128. Rostami, R.; Salamati, P.; Yarandi, K.K.; Khoshnevisan, A.; Saadat, S.; Kamali, Z.S.; Ghiasi, S.; Zaryabi, A.; Ghazi Mir Saeid, S.S.; Arjipour, M.; et al. Effects of neurofeedback on the short-term memory and continuous attention of patients with moderate traumatic brain injury: A preliminary randomized controlled clinical trial. *Chin. J. Traumatol.* **2017**, *20*, 278–282. [CrossRef]

© 2020 by the authors. Licensee MDPI, Basel, Switzerland. This article is an open access article distributed under the terms and conditions of the Creative Commons Attribution (CC BY) license (http://creativecommons.org/licenses/by/4.0/).

Review

Molecular and Cellular Mechanisms Associated with Effects of Molecular Hydrogen in Cardiovascular and Central Nervous Systems

Miroslav Barancik [1], Branislav Kura [1,2], Tyler W. LeBaron [1,3,4], Roberto Bolli [5], Jozef Buday [6] and Jan Slezak [1,*]

1 Centre of Experimental Medicine, Slovak Academy of Sciences, 84104 Bratislava, Slovakia; miroslav.barancik@savba.sk (M.B.); branislav.kura@savba.sk (B.K.); lebaront@molecularhydrogeninstitute.com (T.W.L.)
2 Faculty of Medicine, Institute of Physiology, Comenius University in Bratislava, 84215 Bratislava, Slovakia
3 Molecular Hydrogen Institute, Enoch, UT 84721, USA
4 Department of Kinesiology and Outdoor Recreation, Southern Utah University, Cedar City, UT 84720, USA
5 Department of Medicine, Institute of Molecular Cardiology, University of Louisville, Louisville, KY 40292, USA; roberto.bolli@louisville.edu
6 Department of Psychiatry, First Faculty of Medicine, Charles University in Prague and General University Hospital in Prague, 12108 Prague, Czech Republic; Jozef.Buday@vfn.cz
* Correspondence: jan.slezak@savba.sk; Tel.: +42-19-03-620-181

Received: 14 October 2020; Accepted: 13 December 2020; Published: 15 December 2020

Abstract: The increased production of reactive oxygen species and oxidative stress are important factors contributing to the development of diseases of the cardiovascular and central nervous systems. Molecular hydrogen is recognized as an emerging therapeutic, and its positive effects in the treatment of pathologies have been documented in both experimental and clinical studies. The therapeutic potential of hydrogen is attributed to several major molecular mechanisms. This review focuses on the effects of hydrogen on the cardiovascular and central nervous systems, and summarizes current knowledge about its actions, including the regulation of redox and intracellular signaling, alterations in gene expressions, and modulation of cellular responses (e.g., autophagy, apoptosis, and tissue remodeling). We summarize the functions of hydrogen as a regulator of nuclear factor erythroid 2-related factor 2 (Nrf2)-mediated redox signaling and the association of hydrogen with mitochondria as an important target of its therapeutic action. The antioxidant functions of hydrogen are closely associated with protein kinase signaling pathways, and we discuss possible roles of the phosphoinositide 3-kinase/protein kinase B (PI3K/Akt) and Wnt/β-catenin pathways, which are mediated through glycogen synthase kinase 3β and its involvement in the regulation of cellular apoptosis. Additionally, current knowledge about the role of molecular hydrogen in the modulation of autophagy and matrix metalloproteinases-mediated tissue remodeling, which are other responses to cellular stress, is summarized in this review.

Keywords: molecular hydrogen; oxidative stress; autophagy; matrix metalloproteinases

---

1. Introduction

1.1. Molecular Hydrogen and Its Potential Use in Therapy

Hydrogen is a colorless and odorless, diatomic gas, which, in mammals, is produced by intestinal bacteria. The hydrogen molecule is small (molecular weight 2 Da), electrically neutral, and nonpolar. These properties allow for its easy entrance into cells and rapid diffusion across all biological cell membranes. In this way, molecular hydrogen can reach the subcellular compartments, such as mitochondria and endo/sarcoplasmic reticulum, and nuclei, which are the primary sites of reactive oxygen species (ROS) generation and DNA damage, respectively. Moreover, it can easily penetrate several barriers, such as the blood-brain barrier, the placental barrier, and the testis barrier.

Molecular hydrogen is currently recognized as an emerging therapeutic, as its supplemental application exerts protective effects in cardiovascular diseases [1,2], neurodegenerative diseases [3], inflammatory diseases [4], neuromuscular disorders [5], metabolic syndrome [6] diabetes [7,8], kidney disorders [9,10], and cancer [11]. The protective effects of molecular hydrogen are largely related to its antiapoptotic, anti-inflammatory, and antioxidative actions (Figure 1).

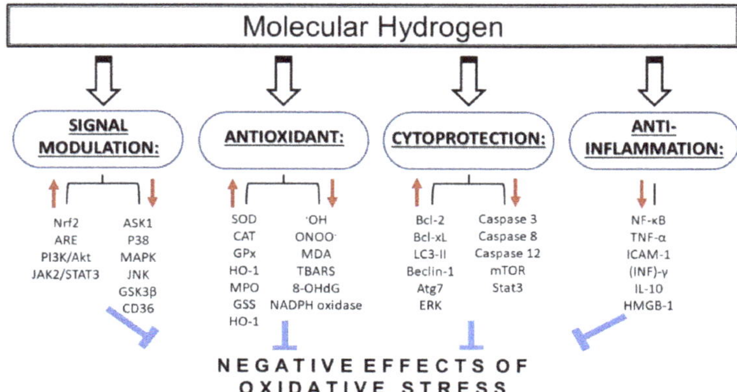

Figure 1. Mechanisms of action of molecular hydrogen in conditions of increased oxidative stress. Molecular hydrogen has been shown to provide protective effects via several mechanisms including antioxidant, anti-inflammatory, and cytoprotective action, as well as via signal modulation. MDA—malondialdehyde; TBARS—thiobarbituric acid reactive substances; 8-OHdG—8-hydroxy-desoxyguanosine; SOD—superoxide dismutase; CAT—catalase; GPx—glutathione peroxidase; HO-1—heme oxygenase 1; MPO—myeloperoxidase; GSS—glutathione synthetase; ASK1—apoptotic signal-regulated kinase 1; MAPK—mitogen-activated protein kinase; JNK—c-Jun N-terminal kinase; CD36—cyclin-dependent kinase 36; Nrf2—nuclear factor-erythroid-2-related factor 2; ARE—antioxidant response elements; NF-κB—nuclear factor kappa B; TNF-α—tumor necrosis factor alpha; ICAM-1—intercellular cell adhesion molecule-1; (INF)γ—interferon gamma; IL-1β—interleukin beta; HMGB-1—high-mobility group box protein 1; mTOR—mammalian target of rapamycin; Stat3—signal transducer and activator of transcription 3; LC3-II—microtubule-associated protein 1A/1B-light chain 3; ERK—extracellular signal-regulated kinase; Atg7—autophagy related 7.

Molecular hydrogen has no known negative side effects on cells. Its use does not disturb cellular metabolic redox reactions, intracellular signaling (e.g., the signaling role of reactive oxygen species—ROS) [12], or physiological metabolic and enzymatic reactions. Hydrogen has very low reactivity with other gases at therapeutic concentrations and lacks reactivity to nitric oxide ($NO^{\bullet}$). This allows its administration with other therapeutic gases, including inhaled anesthetic agents, and enables the concomitant administration of hydrogen with $NO^{\bullet}$. The administration of molecular hydrogen can be performed in several ways. The most common methods are inhalation of molecular hydrogen as a gas [13], application of a hydrogen-rich solution [3], or administration of hydrogen-loaded eye drops [14]. Another method is the use of hydrogen-rich water, which is more convenient for long-term treatment.

The positive effects of molecular hydrogen have been demonstrated not only in animal experiments, but also in clinical trials. In a single-center prospective, open-label, blinded study, Katsumata et al. [15]

investigated the feasibility and effects of hydrogen inhalation on infarct size and adverse left ventricular remodeling after a primary percutaneous coronary intervention (PCI) for ST-elevated myocardial infarction (STEMI). They found that inhalation of 1.3% H_2 during PCI resulted in left ventricular (LV) reverse remodeling at six months after STEMI. The positive effects of H_2 were also demonstrated in metabolic syndrome ($n = 60$) using a double-blinded, placebo-controlled trial [6]. The consumption of H_2-rich water for 24 weeks led to a significant reduction in blood cholesterol, glucose levels, attenuated serum hemoglobin A1c, and improved biomarkers of inflammation and redox homeostasis as compared to the placebo group. Similarly, an earlier study by Kajiyama et al. [16] reported that drinking H_2-rich water for eight weeks (with a 12-week washout period) significantly decreased the levels of modified low-density lipoprotein (LDL) cholesterol, small dense LDL, urinary 8-isoprostanes, serum concentrations of oxidized LDL and free fatty acids, and increased plasma levels of adiponectin and extracellular-superoxide dismutase in patients with diabetes mellitus type 2. After H_2 therapy, normalization of the oral glucose tolerance test occurred in four out of six patients with impaired glucose tolerance. In another randomized, double-blinded, placebo-controlled trial, the efficacy of drinking of hydrogen water for 48 weeks in Japanese patients with levodopa-medicated Parkinson's disease (PD) was investigated [17]. Despite the small number of patients and the short duration of the trial, the results clearly demonstrated the beneficial effects of hydrogen water. Drinking hydrogen water was shown to be safe and well tolerated, and was associated with a significant improvement in Total Unified Parkinson's Disease Rating Scale (UPDRS) scores for patients with PD. Sakai et al. [18] showed that molecular hydrogen can be a useful modulator of blood vessel function. The data observed in their study documented that the vasculature of volunteers drinking daily water containing a high concentration of hydrogen was protective against shear stress-derived detrimental ROS. The protective effects of hydrogen may have been mediated by a reduction of detrimental ROS, and were associated with preserving the bioavailability of nitric oxide (NO) and maintaining the NO-mediated vasomotor response.

Importantly, from the perspective of clinical use, the administration of molecular hydrogen is safe and is not associated with undesirable toxic effects.

1.2. Molecular Hydrogen and The Cardiovascular System

Diseases of the cardiovascular system are among the most serious medical problems, and represent a major cause of health complications and morbidity in modern society [19]. Increased production of ROS and oxidative stress are important factors that contribute to the development of cardiovascular diseases (CVD) such as hypertension [20], cardiac hypertrophy [21,22], and heart failure [23]. A very important and preventable cause of CVD is hypertension, which may, without appropriate treatment, lead to cardiac remodeling followed by left ventricular hypertrophy, and potentially also heart failure [24].

Ischemia/reperfusion (I/R) injury also plays an important role in the induction of cardiac remodeling. Reperfusion is induced by the blood supply returning to the heart after a period of ischemia, and is associated with the induction of oxidative stress injury, calcium overload, inflammation, and apoptosis [25–28]. This can impair cardiac function and lead to myocardial infarction and malignant arrhythmias.

Various studies have employed several potential strategies for the prevention, control, and treatment of cardiovascular diseases including the reduction of increased ROS production and oxidative stress, as well as targeting the signaling pathways modulated by ROS [29–32]. The antioxidative, anti-inflammatory, and antiapoptotic properties of molecular hydrogen may explain why its clinical application may result in the improvement of oxidative stress-related cardiovascular diseases.

The positive effects of molecular hydrogen on diseases related to the cardiovascular system have been reported in several studies. Hydrogen inhalation significantly improved cardiac and brain function in a rat model of cardiac arrest [1], and chronic treatment of spontaneous hypertensive rats with hydrogen-rich saline (HRS) attenuated left ventricular hypertrophy development in these animals [33]. The protective effects of hydrogen on left ventricular function were also observed in other studies demonstrating its ability to attenuate left ventricular remodeling induced by intermittent hypoxia [34] or ischemia/reperfusion (I/R) injury [13]. The positive role of molecular hydrogen treatment in modulating myocardial responses to ischemia/reperfusion injury has been demonstrated in several other studies where different methods of hydrogen application, such as inhalation of hydrogen gas [13] or intraperitoneal application of hydrogen-rich saline, were used [35]. Inhalation of hydrogen gas during reperfusion reduced infarct size in models of cardiac I/R injury in rats [13] as well as in dogs [36]. In the canine model, it was suggested that the cardioprotective effects of hydrogen were realized via the opening of mitochondrial ATP-sensitive potassium channels (mitK-ATP) and the subsequent inhibition of mitochondrial permeability transition pores [36]. mitK(ATP) channels were found to be involved in myocardial responses to ischemia/repefusion [37], and pretreatment with diazoxide, mitK(ATP) channel opener, was found to protect rat heart against ischemia/reperfusion injury [38]. An investigation of the in vivo effects of hydrogen on myocardial I/R injury in rats found that intraperitoneal application of HRS reduced infarct size and also improved cardiac dysfunction. I/R injury caused excessive release of pro-inflammatory molecules (TNF-α, IL-1β, IL-6, and HMGB1), and the hydrogen-mediated cardioprotection was associated with a reduction of these I/R-induced inflammatory responses in myocardial tissue.

Another study demonstrated that molecular hydrogen potentiates the beneficial infarct-sparing effect of hypoxic postconditioning (HPostC) in isolated rat hearts [2]. The infusion of Krebs-Henseleit buffer with molecular hydrogen during HPostC further decreased infarct size, attenuated severe arrhythmias, and significantly restored heart function compared with HPostC alone. Interestingly, one group [39] found that hydrogen gas can attenuate myocardial I/R injury in rats, independent of postconditioning. Compared with ischemic postconditioning, hydrogen had a better protective effect on I/R injury; this was associated with the attenuation of endoplasmic reticulum stress and the downregulation of excessive autophagy [39]. Treatment with HRS was also found to attenuate the myocardial injury and apoptosis in heart tissue induced by a cardiopulmonary bypass (CPB). The available evidence indicated that the protective effects of HRS involved opposite effects on two distinct signaling pathways, i.e., attenuation of the PI3K/Akt pathway [40] and upregulation of JAK2/STAT3 signaling [41].

1.3. Molecular Hydrogen and the Central Nervous System

The nonpolar nature and low molecular weight (2 Da) of molecular hydrogen allow it to easily penetrate biological membranes. This is important in the central nervous system (CNS), because hydrogen can penetrate the membranes that make up the blood-brain barrier (BBB), which plays a pivotal role in the protection of the CNS. Several lines of evidence point to oxidative stress, the activation of matrix metalloproteinases (MMPs), and inflammation as mechanisms linking some pathological conditions, such as cardiovascular diseases and hypertension, to BBB breakdown [42,43].

Vital to the regulation of BBB permeability is the integrity of the endothelial cells. Disruption of this integrity can lead to a dysfunction of the BBB, which can cause neurological disorders such as brain injury and neurodegenerative disorders, and may play a significant role in the pathogenesis of vascular dementia [44,45]. Disruption of BBB function is followed by blood-to-brain extravasation of circulating neuro-inflammatory molecules, which may increase the risk of brain injury. Some cytokines and chemokines, such as IL-6 and TNF-α, are known to be transported across the BBB from the blood into the brain [46]. Moreover, some studies have shown that circulating peripheral immune cells, i.e., macrophages, invade the CNS [47,48]. The crosstalk between the signaling cascades underlying oxidative stress and the inflammatory responses may be a critical factor in neurodegenerative disorders [49,50].

The penetration of molecular hydrogen through the membranes of the BBB and its unrestricted access to the CNS are unique characteristics, shared by only a few therapeutics. It has been found that hydrogen gas inhalation suppressed redox stress and BBB disruption by reducing mast cell activation and degranulation [51]. Moreover, the observed effects of hydrogen were also associated with the suppression of brain edema and neurological deficits [51]. HRS was found to ameliorate brain edema and decrease infarct volume also in the neonatal brain injury in mice. Other studies have demonstrated that supplemental molecular hydrogen improved clinical features in neuromuscular and neurodegenerative diseases [17,52].

The protective effects of molecular hydrogen in the central nervous system are related to the modulation of cellular responses to stress conditions, and are realized through several cellular mechanisms. In 2007, Ohsawa et al. [53] reported that gaseous molecular hydrogen acts as a therapeutic and preventive antioxidant by selectively reducing the levels of strong oxidants, such as the hydroxyl radical (•OH) and peroxynitrite (ONOO–), in cells [53]. The protective effects of molecular hydrogen resulted in the suppression of I/R injury in the brain. Molecular hydrogen selectively reduces the levels of the highly toxic hydroxyl radicals and peroxynitrite, but not superoxide, hydrogen peroxide, or nitric oxide [53].

When contemplating the mechanisms for the anti-inflammatory effects of molecular hydrogen in the brain, both the neuro-immunological interactions and crosstalk with oxidative stress need to be considered. Important hydrogen-mediated protective effects include the buffering of oxidative stress, downregulation of endoplasmic reticulum (ER), stress, inhibition of apoptosis, suppression of inflammatory responses, and regulation of autophagy machinery.

2. Mechanisms and Cellular Systems Involved in the Actions of Molecular Hydrogen

The effects of molecular hydrogen on various diseases may be attributed to several molecular mechanisms. Hydrogen was first reported to be a selective scavenger of •OH and peroxynitrite [53]; these reactive molecules are primary direct targets of hydrogen. However, mounting evidence suggests that hydrogen can also function as a signal modulator [54–56], and several molecules are mediators that are secondarily changed by the administration of hydrogen. The radical scavenging and signal modulating activities of molecular hydrogen are closely related to modulation (regulation) of redox signaling and alterations in gene expressions [54]. Next, we will focus on the role of molecular hydrogen in modulating redox status as well as intracellular protein signaling and the consequences thereof on gene expression, autophagy, and matrix metalloproteinases.

2.1. Molecular Hydrogen as Regulator of Redox Signaling

Molecular hydrogen has been reported to be an antioxidant that protects cells against oxidative stress through selectively decreasing cellular levels of hydroxyl radicals (•OH) and peroxynitrite (ONOO−) [53]. The stoichiometric reaction between H_2 and hydroxyl radicals is:

$$H_2 + 2^{\bullet}OH => 2\,H_2O$$

Although hydrogen cannot eliminate peroxynitrite as efficiently as hydroxyl radicals, in rodents, it was found that hydrogen can efficiently reduce nitro-tyrosine formation, which is induced by nitric oxide ($NO^{\bullet}$) via the production of peroxynitrite [57,58]. $NO^{\bullet}$ is a gaseous molecule that also exerts therapeutic effects including relaxation of blood vessels and inhibition of platelet aggregation [59]. $NO^{\bullet}$, however, can be toxic at higher concentrations because it leads to peroxynitrite-mediated production of nitrotyrosine, which compromises protein functions. Part of the effects of hydrogen may thus be attributed to the reduced production of nitrotyrosine [58].

Molecular hydrogen reduces oxidative stress not only directly, but also indirectly by inducing antioxidation systems, including heme oxygenase-1 (HO-1) [60,61], superoxide dismutase (SOD) [7,9], catalase [62], and myeloperoxidase (Figure 1) [62,63]. In a rat model of traumatic brain injury, it was observed that the beneficial effects of hydrogen inhalation were mediated by the reduction of oxidative stress and the stimulation of enzymatic activities of the endogenous antioxidants SOD and catalase [64]. Beneficial effects of hydrogen on the activities of antioxidant enzymes were also observed by Guan et al. [9], who discovered that molecular hydrogen protected against renal dysfunction induced by chronic intermittent hypoxia. Hydrogen was shown to alleviate oxidative damage by promoting the upregulation of SOD and glutathione peroxidase (GSH-Px) activities and increasing the GSH/GSSG ratio. The effects of hydrogen were also associated with a reduction of malondialdehyde (MDA) content (a product of oxidative stress). Other studies support the antioxidant functions of molecular hydrogen through the Nrf2/ARE [54,65,66] pathway. This pathway plays an essential role in protection against oxidative stress and in the transcriptional regulation of numerous other antioxidant and cytoprotective proteins [49]. Nrf2 (nuclear factor erythroid 2-related factor 2) is a transcription factor playing an important role in the redox-sensitive regulation of the expression of several endogenous antioxidants and detoxification enzymes [67,68]. Under normal conditions, Nrf2 is inhibited by Kelch-like ECH-associated protein 1 (Keap1), which mediates the Cullin3/Rbx1-dependent polyubiquitination of Nrf2 and its subsequent proteosomal degradation [69]. After exposure of cells to stress, the cysteine residues of Keap1 are modified by oxidative/electrophilic molecules, and Nrf2 escapes Keap1-mediated repression. When Nrf2 is not ubiquitinated, it translocates in the nucleus, where it forms heterodimers with small MAF or JUN proteins and binds to the antioxidant response element (ARE), i.e., the upstream promoter region of many antioxidative genes, and initiates their transcription [19,70]. Regulation of the Nrf2/ARE pathway is generally dependent on the duration and intensity of the oxidative stress. The above mentioned effects are realized above all during acute stress. Prolonged stress induces negative modulation or downregulation of Nrf2 activity, and decreases or arrests antioxidant and detoxification responses. Glycogen synthase kinase 3β (GSK-3β) plays an important role in this modulation by phosphorylating threonine residues of Fyn kinases. The Fyn kinase then translocates into the nucleus where it phosphorylates Nrf2, which leads to Nrf2 export out of the nucleus to the cytoplasm, where it is exposed to ubiquitination and proteasome degradation [71].

The role of Nrf2 in mediating the effects of hydrogen is supported by the results of a study showing that hydrogen gas reduced hyperoxic lung injury via the Nrf2 pathway, and through the induction of Nrf2-dependent genes, such as HO-1 [66]. It was also demonstrated that hydrogen can play a significant antioxidative role in the brain after focal cerebral ischemia reperfusion through upregulation of HO-1 levels [72]. Moreover, recent data indicated that the Nrf2/ARE signaling pathway and upregulation of HO-1 are involved in the neuroprotective effects of hydrogen-rich saline in mice with experimental autoimmune encephalomyelitis [73].

2.2. Molecular Hydrogen and Mitochondria

Mitochondria are important organelles involved in several essential cellular functions, such as energy production (ATP), cell differentiation, regulation of calcium homeostasis, and signal transduction [74–76]. They also significantly contribute to cellular stress responses associated with cell death. The mitochondria-mediated regulation of apoptosis [77,78] and autophagy [79] is an important biological process. Mitochondrial dysfunction contributes to various human diseases. Mitochondria are known as a major sources of cellular ROS production. The process of oxidative phosphorylation leads to the conversion of oxygen (O_2) into water (H_2O) by four-electron reduction. However, a small percentage of O_2 is converted into superoxide anion radicals by one-electron reduction. Superoxide is decomposed with mitochondrial superoxide dismutase (SOD) and converted to O_2 and hydrogen peroxide (H_2O_2).

The physical properties of molecular hydrogen allow its effective penetration through subcellular compartments, such as mitochondria [80]. Mitochondria are an important therapeutic target, and so the small hydrogen molecule could be applicable for interventions in diseases related to mitochondria. The potential effects of molecular hydrogen have been investigated in several studies. In one model of isolated mitochondria, it was found that molecular hydrogen could suppress superoxide generation in complex I [81]. The same authors demonstrated that the presence of molecular hydrogen in culture media reduced the membrane potential in living human lung cells (A549) [81]. Based on the results of both in vitro and in vivo studies, the authors assumed that electrons released by molecular hydrogen could be donated to the iron-sulfur N2 cluster in complex I of the respiratory chain. In this way, H_2 may trigger a conformational change in this complex and affect transmembrane proton transfer and/or uncoupling of the membrane potential. Therefore, it was proposed that H_2 may function as a rectifier of mitochondrial electron flow in the disordered or pathological states when the accumulation of electrons leads to ROS production [82].

Positive effects of molecular hydrogen on mitochondria through the activation of mitochondrial unfolded protein response (mtUPR) pathways have also been demonstrated. mtUPR is a defense mechanism that is activated by stress occurring in the mitochondrial matrix, when damaged proteins accumulate in the mitochondrial matrix and exceed the maximal capacity of the protein folding apparatus [83]. It was found that molecular hydrogen induced this mitochondrial defense mechanism by induction of mtUPR-related proteins expression and via histone 3 (H3) demethylase induction and modification of H3 methylation on lysine 27 (H3K27) [66,84]. The positive effects of hydrogen were also documented by Iuchi et al. [85], who found that molecular hydrogen can suppress tert-butyl hydroperoxide-induced cell death by reducing mitochondrial dysfunction and fatty acid peroxidation [85].

These possible mechanisms involved in the effects of molecular hydrogen may explain the results in recent studies showing protective effects of hydrogen-rich saline against experimental diabetic peripheral neuropathy in rats, which was associated with the activation of mitochondrial ATP-sensitive potassium channels [86]. Moreover, the application of 5-hydroxydecanoate, a mitochondrial ATP-sensitive potassium channels inhibitor, eliminated the neuroprotective effects of hydrogen-rich saline treatment. ATP-sensitive potassium channels can be found in the plasma membrane and the inner membrane of mitochondria [87]. It was demonstrated that these mitochondrial channels play an important role in the protection of myocardial cells against injuries [88], and that their activation can inhibit the apoptosis induced by hydrogen peroxide [89]. Nrf2 is an important regulator of redox signaling. A recent study showed that hydrogen-rich saline can alleviate mitochondrial dysfunction via the Nrf2 pathway [90]. The authors found that sepsis-associated encephalopathy (SAE) led to mitochondrial dysfunction. Hydrogen-rich saline improved the function of mitochondria, as demonstrated by an increase of the mitochondrial membrane potential (MMP), respiratory control ratio (RCR), and ATP release. In addition, hydrogen-rich saline alleviated SAE-induced changes and the production of ROS. The relationship between hydrogen and the Nrf2 pathway was confirmed by findings that the protective effects of hydrogen occurred in wild type but not Nrf2-knockout mice.

Gvozdjakova et al. [91] demonstrated that molecular hydrogen stimulates myocardial mitochondrial function in rats. Drinking molecular hydrogen-rich water (HRW) resulted in increased levels of ATP production at Complexes I and II in the rat cardiac mitochondria. Similarly, coenzyme Q9 levels in the rat plasma, myocardial tissue, and mitochondria were increased after HRW administration.

2.3. *Molecular Hydrogen and Modulation of Intracellular Protein Kinases Signal Transduction*

Molecular hydrogen plays a role not only as a potential free radical scavenger during oxidative stress, but it can also act as a modulator of intracellular signaling mediated by protein kinases. This signaling role is facilitated by the physical properties of molecular hydrogen, namely, the fact that it can easily diffuse throughout tissues and cells. In this way, hydrogen represents a gaseous-signaling molecule, similar to $NO^{\bullet}$. Importantly, molecular hydrogen can reduce oxidative stress, but it cannot directly eliminate functionally important signaling ROS [12,53]. Several studies have documented that molecular hydrogen may exert its effects by targeting several protein kinases [9,55,92–94]. The effects of hydrogen on signaling are not unidirectional, and both stimulatory and inhibitory roles of molecular hydrogen in the activation of distinct protein kinase cascades have been demonstrated [54].

The cellular effects of molecular hydrogen seem to be mediated by several intracellular protein kinase pathways. We will focus on the PI3K/Akt and Wnt/β-catenin (Wingless-type mouse mammary tumor virus integration site family member/beta catenin) pathways, which are associated with glycogen synthase kinase 3β (GSK3β), an enzyme that plays an important role in the regulation of cellular endogenous apoptosis.

2.3.1. Molecular Hydrogen and the PI3K/Akt/GSK-3β Signaling Pathway

The PI3K/Akt kinase signaling pathway is a key player in the regulation of the protective induction of Nrf2/ARE during acute oxidative stress. However, in long-term stress, Nrf2 is deactivated to various degrees by kinase cascades associated with the activation of glycogen synthase kinase-3β (GSK3β), as explained above via phosphorylating Fyn kinase, which results in the attenuation or complete cessation of the antioxidant and detoxication response [19,95]. GSK3β is a serine/threonine kinase and the major downstream target of the PI3K/Akt pathway. This enzyme is phosphorylated by Akt kinase and can directly phosphorylate Nrf2 [96]. GSK3β also regulates the cellular endogenous apoptosis pathway, and its downstream target MCL-1 (myeloid lymphoma 1) in the Bcl-2 family [97]. The PI3K/Akt/GSK3β signaling pathway mediates cell survival and plays an important role in I/R injury in the heart [98], brain [99], and kidney [100]. Moreover, this pathway is involved in regulating the function of cerebral blood vessels, and inhibiting its function may aggravate neuronal injury. GSK-3β is considered a therapeutic target for a variety of nervous system diseases. Several studies have indicated that GSK-3β may regulate neuronal apoptosis through the activation of NF-κB or β-catenin signaling pathways [101–103].

Several lines of evidence indicate that at least part of the protective effects of molecular hydrogen are related to the regulation of the PI3K/Akt/GSK3β signaling pathway. HRS was found to protect cerebral microvascular endothelial cells from apoptosis after hypoxia/reoxygenation by inhibiting the PI3K/Akt/GSK3β pathway [104]. HRS was also protective against brain injury induced by cardiopulmonary bypass; this protection was associated with the inhibition of apoptosis through the PI3K/Akt/GSK3β pathway [104]. This inhibition of apoptosis by HRS was realized through downregulation of Akt and GSK3β activity, which are core components of this pathway, and through the inhibition of proapoptotic caspase-3 and Bax expression levels. Inactivation of the Akt kinase pathway also played a role in HRS-induced attenuation of vascular smooth muscle cell proliferation and neointimal hyperplasia. In addition to the Akt kinase pathway, the effects of HRS were also associated with inhibition of ROS production and inactivation of the Ras-ERK1/2-MEK1/2 pathway [105].

The fact that the effects of hydrogen on signaling pathways are not always unidirectional is further confirmed in the study of Wang et al. [61], where hydrogen exerted neuroprotection in a cellular in vitro model of traumatic brain injury through activation of the miR-21/PI3K/Akt/GSK-3β signaling pathway. Activation of the PI3K/Akt kinase pathway has also been found to be related to the neuroprotective effects of molecular hydrogen against neurologic damage and apoptosis in early brain injury induced by subarachnoid hemorrhage [106]. Similarly, protection of mouse hearts against I/R injury by molecular hydrogen involved the activation of the PI3K/Akt pathway [107]. A recent study showed that the Akt kinase pathway plays a critical role in the neuroprotective ability of hydrogen-rich saline (HRS). HRS restored behavioral deficits following hypoxia-ischemia injury in neonatal mice via the activation of Akt kinase pathway [108].

2.3.2. Effects of Molecular Hydrogen on Wnt/β-catenin Signaling

Wnts are secreted glycoproteins. In mammals, the family of Wnt ligands consist of 19 members [109]. Activation of the Wnt signaling system includes canonical and noncanonical pathways which regulate a variety of cellular activities.

Two central components of the canonical Wnt pathway are β-catenin and Axin1. Axin1 plays crucial roles in both the destruction of the β-catenin complex and activation of the LRP6 signaling complex [110]. Wnt ligands in this pathway (Wnt1, Wnt2a, Wnt3a, WNT6, and Wnt8a, WNT9b) bind to the frizzled (FZD) and low-density-lipoprotein-related protein 5/6 (LRP5/6) receptor complex. This complex activates the Dishevelled protein, which, in turn, inhibits the degradation complex that destroys the synthetized β-catenin. The β-catenin destruction complex consists of the central scaffold protein Axin and three other components: (i) tumor-suppressor protein adenomatous polyposis coli (APC), (ii) glycogen synthase kinase-3β (GSK-3β), and (iii) casein kinase-1 (CKI) [111]. Inhibition of the degradation complex leads to the stabilization and translocation of β-catenin into the nucleus. The activation of canonical Wnt signaling depends on the nuclear localization of β-catenin. In the nucleus, β-catenin acts as a transcriptional coactivator which, together with the T-cell factor (TCF) and the lymphoid enhancer factor (LEF) transcription factors, initiates the Wnt transcriptional program [111–113]. Potential Wnt/β-catenin downstream target genes are angiogenic factors such as vascular endothelial growth factor (VEGF) [114] and interleukin-8 (IL-8) [115]. Reports have also demonstrated that canonical Wnt signaling can regulate the expression of neurospecific growth factors such as neurotrophin-3 (NT-3) [116] and brain-derived neurotrophic factor (BDNF) [117]. Another possible target of Wnt/β-catenin signaling is the voltage-gated Na^+ channel NaV1.5 encoded by the SCN5a gene. It was found that the activation of Wnt/β-catenin signaling by hydrogen peroxide inhibited the NaV1.5 expression at the transcriptional level [72].

Canonical Wnt/β-catenin signaling controls cell proliferation and differentiation by regulating the expression of target genes. This cascade plays important roles in the regulation of many cellular functions [112,118,119]. Aberrant activation of this signaling pathway is associated with a number of diseases, including cancers, metabolic, and degenerative diseases [120,121]. Moreover, dysfunction of Wnt signaling has been implicated in age-related diseases [122], and sustained activation of Wnt signaling plays a key role in the pathogenesis of a variety of tissue fibrosis [123]. GSK-3β plays an important role in the destabilization of the Wnt signaling component, β-catenin. Activation of GSK-3β and the disturbed function of Wnt/β-catenin signaling are linked to tissue fibrosis, such as lung fibrosis [124], liver fibrosis [125], and cardiac fibrosis [126].

Lin et al. [55] demonstrated that molecular hydrogen suppressed abnormally activated Wnt/β-catenin signaling by promoting phosphorylation and degradation of β-catenin in cancer L and HeLa cell lines. Hydrogen had no effect on the basal endogenous β-catenin level and inhibited only β-catenin accumulation induced by Wnt3a and GSK3 inhibitors. The protective effects of molecular hydrogen occurred only in situations of aberrant activation of the Wnt/β-catenin pathway, and the data observed using a GSK3 inhibitor pointed to an important role of this kinase in the modulation of β-catenin function. Moreover, the mechanisms of hydrogen-mediated suppression of Wnt/β-catenin signaling did not involve the scavenging of hydroxyl radicals or peroxynitrite. Modulation of the β-catenin pathway by hydrogen was also found in melanocytes [127]. Hydrogen reversed the hydrogen peroxide-induced apoptosis and dysfunction in melanocytes, and the data indicated that the hydrogen-mediated beneficial effects involved Wnt/β-catenin-mediated activation of Nrf2 signaling [127].

Noncanonical Wnt signaling is activated by stimulation of the Frizzled receptor by noncanonical Wnt ligands, such as Wnt4, Wnt5, and Wnt11 [128–130]. The activation is β-catenin-independent and may trigger gene transcription by activating a planar cell polarity pathway (PCP) [131,132] and a calcium-dependent pathway (Wnt/$Ca^{2+)}$ [128]. Downstream effectors of the Wnt/PCP pathway may involve the small GTP-binding protein RhoA, c-Jun N-terminal protein kinases (JNK) [133,134], and downstream proteins of the Wnt/Ca^{2+} may involve several kinases, including protein kinase C (PKC) and calcium/calmodulin-dependent kinase (CaMKII) [128,135].

Several studies have shown the effects of molecular hydrogen on components of noncanonical Wnt signaling. It has been documented that a hydrogen-rich medium decreased expression of downstream RhoA effector, Rho-associated coiled-coil protein kinase (ROCK); this was associated with attenuation of LPS-induced vascular hyperpermeability and vascular endothelial-cadherin disruption [136]. Supression of RhoA activity by hydrogen was also found in the human colon cancer cell line, Caco-2; this was connected with hydrogen-mediated amelioration of LPS-induced barrier dysfunction [137]. Zhang et al. [94] described the positive effects of intraperitoneal hydrogen injection on the prevention of isoproterenol-induced cardiac hypertrophy and improvement of cardiac function in mice. Hydrogen exerted its protective effects through blocking several protein kinase pathways (ERK, p38-MAPK), including JNK signaling.

2.4. Effects of Hydrogen on Gene Expression Regulation

Hydrogen was found to be involved in regulating the expression of various genes; however, it is not clear whether these regulations are the cause or consequence of the effects of hydrogen against oxidative stress [54]. The primary molecular targets of molecular hydrogen remain unknown, and there is no evidence that hydrogen directly reacts with factors involved in transcriptional regulation. Several lines of evidence indicate that molecular hydrogen can realize its effects in various pathological situations indirectly, through the up- or down- regulation of the expression of distinct genes. The antiapoptotic effects of hydrogen seem to be associated with the upregulation of antiapoptotic factors, and/or the downregulation of proapoptotic factors (Figure 1). It has been observed that hydrogen induced expressions of the antiapoptotic factors Bcl-2 and Bcl-xL [138], and suppressed the expressions of various proapoptotic factors, including caspase 3 [139,140], caspase 8 [138], and caspase 12 [139].

An important factor in modulating responses to oxidative stress, and a major mechanism underlying the cellular protective effects of hydrogen, is the upregulation of the expression of genes that encode several antioxidant enzymes (Figure 1) [7,9,62]. Kawamura et al. [66] reported that inhalation of hydrogen gas during exposure to hyperoxia improved blood oxygenation, reduced inflammation, and induced HO-1 expression in the lung. The HO-1 enzyme participates in the cellular defense against oxidative stress, and its transcription is regulated by nuclear factor erythroid 2-related factor 2 (Nrf2). Nrf-2 seems to play a very important role in mediating the effects of molecular hydrogen on gene expression. Furthermore, Chen et al. [141] reported that molecular hydrogen attenuates the inflammatory response during sepsis by activating the Nrf2-mediated HO-1 signaling pathway.

Hydrogen gas enhances gene expression by promoting the translocation of Nrf2 into the nucleus [142]. A consequence of nuclear Nrf2 translocation is the upregulation of its downstream effectors such as HO-1, SOD, and catalase. The relationship between the effects of molecular hydrogen and Nrf2 was also confirmed by data obtained using a septic model in mice. It was demonstrated that hydrogen therapy protected against intestinal injury induced by oxidative stress and inflammation, and increased the survival rate in WT septic mice, but not in Nrf2-KO mice [143].

Other downstream indirect targets of molecular hydrogen that can be up- or down- regulated include several other genes, such as MMP-2, MMP-9 [144], MMP3, MMP13 [57], cyclooxygenase-2 [145], and connexins [146]. In addition, it has been proposed that molecular hydrogen may affect the expression of several protein kinase signaling pathways [93,147]. However, these molecules are likely downstream and indirectly regulated by H_2, as the direct targets of H_2 have yet to be elucidated [54].

2.5. Molecular Hydrogen and Autophagy

Autophagy is an adaptive response of cells during conditions of cellular stress, such as nutrient limitation, increased production of ROS, accumulation of protein aggregates or damaged organelles, or presence of extracellular pathogens [148,149]. Through autophagy, the machinery and cytosolic components that are damaged or dysfunctional, along with extracellular pathogens, are degraded via the autophagosome, which fuses with lysosomes to degrade and recycle the sequestered substrates. There are at least three types of autophagy: macroautophagy, microautophagy, and chaperone-mediated autophagy. Basal autophagy helps cells survive by reducing stress sources, and is beneficial to keeping normal cellular functions during the early stage of disease [150,151]. However, excessive autophagy leads to autophagic cell death and is implicated in disease progression [152]. The importance of autophagy for cellular function is emphasized by the fact that autophagy dysfunction can result in impaired mitochondrial function, ROS accumulation, and oxidative stress [153]. A consequence of excessive ROS production could be glutathione depletion and stress of the ER. Functional autophagy plays an important role in protecting cells against death induced by ER stress [151], and its activation may attenuate the ER stress-mediated development of injury. This is also supported by the results of a study showing the protection of the brain from an ischemic insult associated with increased autophagy and reduction of ER stress [150]. Autophagy and ER stress are associated with several pathological conditions, such as neurodegeneration [154], diabetes [155], and hypoxia [156]. Increasing evidence indicates that autophagy may share common molecular inducers and regulatory mechanisms with apoptosis, and a switch in cellular responses from apoptosis to autophagy can lead to cell survival. For example, beclin-1, a protein with a central role in autophagy, may interact with antiapoptotic Bcl-2 family members [157].

Several studies have documented the relationship between hydrogen and autophagy [158,159]. However, consensus regarding the precise role of molecular hydrogen in autophagy regulation has not been reached. Some studies have shown that hydrogen can downregulate autophagy [39,158,159], while others have demonstrated that it induces autophagy [3]. Treatment with hydrogen significantly attenuated neuronal injury in the hippocampal *cornu ammonis* 1 sector. This was associated with autophagy inhibition, and, as a consequence, there was a reduction of brain edema following 24 h of reperfusion [158]. Similarly, hydrogen-mediated downregulation of autophagy was also demonstrated in cardiac cells [159]. In this case, pretreatment with a hydrogen-rich medium suppressed isoproterenol-induced, excessive autophagy in H9c2 cardiomyocytes [159]. In addition, the authors used a mouse model of cardiac hypertrophy and demonstrated that intraperitoneal administration of hydrogen significantly blocked β adrenoceptor agonist-mediated, excessive autophagy in vivo [159]. In contrast to these studies demonstrating the downregulation of autophagy by hydrogen, other studies have documented the opposite effects. It was found that the application of HRS exerted neuroprotection against hypoxic-ischemic brain damage in neonatal mice, and that these effects were mediated in part by the upregulation of autophagy machinery and the downregulation of ER stress [3]. In these cases, the effects of HRS on autophagy pathways included increased LC3B and Beclin-1 expression, and decreased phosphorylation of mTOR and STAT3. Moreover, these changes were associated with the phosphorylation of extracellular-signal regulated protein kinases. Similarly, it was reported that an important role mediating the neuroprotective effects of hydrogen-rich water in a rat model of vascular dementia was the stimulation of FoxO1-mediated autophagy [160]. Additionally, activation of p53-mediated autophagy was shown to play a positive role in HRS-mediated attenuation of acute kidney injury after liver transplantation [161].

Wang et al. [162] found that in a neuropathic pain model, the potent analgesic effects of HRS were associated with the activation of cell autophagy via inducing hypoxia-inducible factor-1α (HIF-1α). This pathway plays a pivotal role in regulating gene expression to maintain oxygen homeostasis [163], and in initiating the transcription of target genes that are important for cellular adaption to hypoxic stimuli. Hypoxia influences autophagy in part via the activation of the HIF1α -dependent pathways [164] (Figure 2).

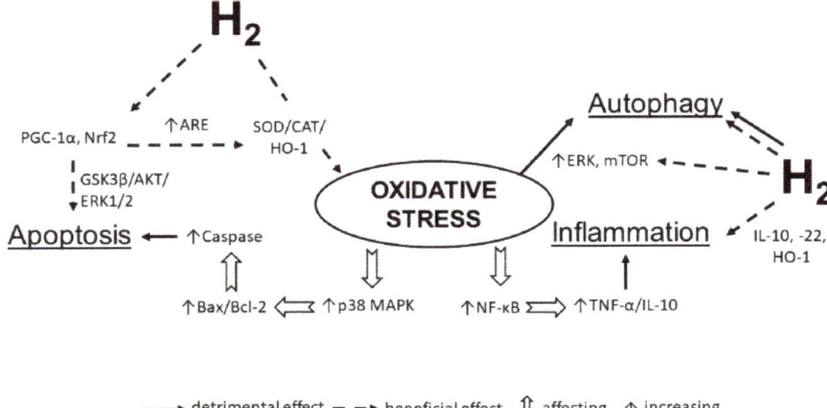

Figure 2. Molecular hydrogen alleviates the negative impact of oxidative stress on the heart. Hydrogen increases the expression of antioxidant enzymes (SOD, catalase, etc.), anti-inflammatory molecules (IL-10, IL-22, HO-1), activates PGC-1α and Nrf2 transcription to decreases apoptosis and regulate autophagy by influencing ERK and mTOR. SOD—superoxide dismutase; IL—interleukin; NF-κB—nuclear factor kappa B; TNF-α—tumor necrosis factor alpha; HO-1—heme oxygenase 1; ERK—extracellular signal-regulated kinase; mTOR—mammalian target of rapamycin; PGC-1α—peroxisome proliferator-activated receptor-gamma coactivator–1 alpha.

2.6. Molecular Hydrogen and Matrix Metalloproteinases

The extracellular matrix (ECM) plays an important role in the development of cardiac fibrosis, which is also one of the typical clinical features of the aged heart [165]. This process is associated with the modulation of ECM, changes in the gene expression of cardiac fibroblasts, and stimulation of proliferation and phenotypic differentiation of myofibroblasts into fibroblasts [166]. A consequence of cardiac fibroblasts activation is the expression of contractile proteins and the secretion of ECM components, which influence the development of pathophysiological processes [167]. Matrix metalloproteinases (MMPs) play an important role in the regulation and modulation of the ECM. Importantly, it has been documented that molecular hydrogen can regulate the expression of several MMPs, including MMP-2, MMP-9, MMP-3, and MMP-13 [57,144]. Other data suggest that hydrogen-rich water inhibits the migration of smooth muscle cells into vein grafts; this is at least partially explained by lowering the expression of MMP-2 and MMP-9 [168]. MMP-9 was found to promote hemorrhagic infarction by disrupting cerebral vessels in a rat model of middle cerebral artery occlusion (MCAO), but inhalation of molecular hydrogen significantly reduced the MMP-9 expression [144,169]. ROS/RNS play an important mechanistic role in regulating MMP activities/expression, especially peroxynitrite (ONOO–). Therefore, the fact that molecular hydrogen reduces peroxynitrite and other toxic ROS will contribute to the favorable regulation of MMPs.

3. Conclusions

The therapeutic potential of molecular hydrogen in the treatment of various diseases may be attributed to several molecular mechanisms. Current information indicates that the cellular protective effects mediated by molecular hydrogen are attributable to the modulation of cellular antioxidant defenses (antioxidant and cytoprotective genes), including intracellular and extracellular redox signaling.

However, the effects of hydrogen on signaling pathways and adaptive cellular responses (e.g., autophagy) are not always unidirectional; both stimulatory and inhibitory effects of molecular hydrogen have been demonstrated.

Further studies are essential to understand the details underlying the regulatory function of molecular hydrogen and the precise mechanisms by which it affects cellular functions in pathological conditions.

Author Contributions: J.S. supervised the writing project of the manuscript; M.B., B.K. and T.W.L. prepared the manuscript and wrote the draft together; B.K. and J.S. prepared the figures; R.B., J.B., J.S. and T.W.L. reviewed and edited the manuscript. All authors have read and agreed to the published version of the manuscript.

Funding: This research was funded by grants from Slovak Research and Development Agency (APVV-0241-11, APVV-15-0376, APVV-19-0317), grant from the Slovak Academy of Sciences (VEGA 2/0063/18), grant from European Union Structural funds (ITMS 26230120009), grant (2018/7838:1-26C0), and grant from Ministry of Health of The Slovak Republic (2019-CEMSAV-1).

Conflicts of Interest: T.W.L. has received travel reimbursement, honoraria, and speaking and consultancy fees from various academic and commercial entities regarding molecular hydrogen. All other authors report no conflict of interest.

References

1. Hayashida, K.; Sano, M.; Kamimura, N.; Yokota, T.; Suzuki, M.; Maekawa, Y.; Kawamura, A.; Abe, T.; Ohta, S.; Fukuda, K.; et al. H_2 Gas Improves Functional Outcome After Cardiac Arrest to an Extent Comparable to Therapeutic Hypothermia in a Rat Model. *J. Am. Heart Assoc.* **2012**, *1*, 1–13. [CrossRef]
2. Zálešák, M.; Kura, B.; Graban, J.; Farkašová, V.; Slezák, J.; Ravingerová, T. Molecular hydrogen potentiates beneficial anti-infarct effect of hypoxic postconditioning in isolated rat hearts: A novel cardioprotective intervention. *Can. J. Physiol. Pharmacol.* **2017**, *95*, 888–893. [CrossRef] [PubMed]
3. Bai, X.; Liu, S.; Yuan, L.; Xie, Y.; Li, T.; Wang, L.; Wang, X.; Zhang, T.; Qin, S.; Song, G.; et al. Hydrogen-rich saline mediates neuroprotection through the regulation of endoplasmic reticulum stress and autophagy under hypoxia-ischemia neonatal brain injury in mice. *Brain Res.* **2016**, *1646*, 410–417. [CrossRef] [PubMed]
4. Yang, L.; Guo, Y.; Fan, X.; Chen, Y.; Yang, B.; Liu, K.X.; Zhou, J. Amelioration of Coagulation Disorders and Inflammation by Hydrogen-Rich Solution Reduces Intestinal Ischemia/Reperfusion Injury in Rats through NF- κ B/NLRP3 Pathway. *Mediat. Inflamm.* **2020**, *2020*, 1–12. [CrossRef] [PubMed]
5. Hasegawa, S.; Ito, M.; Fukami, M.; Hashimoto, M.; Hirayama, M.; Ohno, K. Molecular hydrogen alleviates motor deficits and muscle degeneration in mdx mice. *Redox Rep.* **2017**, *22*, 26–34. [CrossRef]
6. LeBaron, T.; Singh, R.; Fatima, G.; Kartikey, K.; Sharma, J.P.; Ostojic, S.; Gvozdjakova, A.; Kura, B.; Noda, M.; Mojto, V.; et al. The Effects of 24-Week, High-Concentration Hydrogen-Rich Water on Body Composition, Blood Lipid Profiles and Inflammation Biomarkers in Men and Women with Metabolic Syndrome: A Randomized Controlled Trial. *Diabetes Metab. Syndr. Obes. Targets Ther.* **2020**, *13*, 889–896. [CrossRef]
7. Ming, Y.; Ma, Q.; Han, X.; Li, H. Molecular hydrogen improves type 2 diabetes through inhibiting oxidative stress. *Exp. Ther. Med.* **2020**, *20*, 359–366. [CrossRef]
8. Qiu, X.; Ye, Q.; Sun, M.; Wang, L.; Tan, Y.; Wu, G. Saturated hydrogen improves lipid metabolism disorders and dysbacteriosis induced by a high-fat diet. *Exp. Biol. Med.* **2020**, *245*, 512–521. [CrossRef]
9. Guan, P.; Sun, Z.-M.; Luo, L.-F.; Zhou, J.; Yang, S.; Zhao, Y.-S.; Yu, F.-Y.; An, J.-R.; Wang, N.; Ji, E.-S. Hydrogen protects against chronic intermittent hypoxia induced renal dysfunction by promoting autophagy and alleviating apoptosis. *Life Sci.* **2019**, *225*, 46–54. [CrossRef]
10. Li, J.; Hong, Z.; Liu, H.; Zhou, J.; Cui, L.; Yuan, S.; Chu, X.; Yu, P. Hydrogen-rich saline promotes the recovery of renal function after ischemia/reperfusion injury in rats via anti-apoptosis and anti-inflammation. *Front. Pharmacol.* **2016**, *7*, 106. [CrossRef]
11. Meng, J.; Liu, L.; Wang, D.; Yan, Z.; Chen, G. Hydrogen gas represses the progression of lung cancer via down-regulating CD47. *Biosci. Rep.* **2020**, *40*. [CrossRef] [PubMed]
12. Ohta, S. Recent Progress Toward Hydrogen Medicine: Potential of Molecular Hydrogen for Preventive and Therapeutic Applications. *Curr. Pharm. Des.* **2011**, *17*, 2241–2252. [CrossRef] [PubMed]
13. Hayashida, K.; Sano, M.; Ohsawa, I.; Shinmura, K.; Tamaki, K.; Kimura, K.; Endo, J.; Katayama, T.; Kawamura, A.; Kohsaka, S.; et al. Inhalation of hydrogen gas reduces infarct size in the rat model of myocardial ischemia–reperfusion injury. *Biochem. Biophys. Res. Commun.* **2008**, *373*, 30–35. [CrossRef] [PubMed]

14. Oharazawa, H.; Igarashi, T.; Yokota, T.; Fujii, H.; Suzuki, H.; Machide, M.; Takahashi, H.; Ohta, S.; Ohsawa, I. Protection of the retina by rapid diffusion of hydrogen: Administration of hydrogen-loaded eye drops in retinal ischemia-reperfusion injury. *Investig. Ophthalmol. Vis. Sci.* **2010**, *51*, 487–492. [CrossRef] [PubMed]
15. Katsumata, Y.; Sano, F.; Abe, T.; Tamura, T.; Fujisawa, T.; Shiraishi, Y.; Kohsaka, S.; Ueda, I.; Homma, K.; Suzuki, M.; et al. The Effects of Hydrogen Gas Inhalation on Adverse Left Ventricular Remodeling After Percutaneous Coronary Intervention for ST-Elevated Myocardial Infarction—First Pilot Study in Humans—. *Circ. J.* **2017**, *81*, 940–947. [CrossRef]
16. Kajiyama, S.; Hasegawa, G.; Asano, M.; Hosoda, H.; Fukui, M.; Nakamura, N.; Kitawaki, J.; Imai, S.; Nakano, K.; Ohta, M.; et al. Supplementation of hydrogen-rich water improves lipid and glucose metabolism in patients with type 2 diabetes or impaired glucose tolerance. *Nutr. Res.* **2008**, *28*, 137–143. [CrossRef]
17. Yoritaka, A.; Takanashi, M.; Hirayama, M.; Nakahara, T.; Ohta, S.; Hattori, N. Pilot study of H_2 therapy in Parkinson's disease: A randomized double-blind placebo-controlled trial. *Mov. Disord.* **2013**, *28*, 836–839. [CrossRef]
18. Sakai, T.; Sato, B.; Hara, K.; Hara, Y.; Naritomi, Y.; Koyanagi, S.; Hara, H.; Nagao, T.; Ishibashi, T. Consumption of water containing over 3.5 mg of dissolved hydrogen could improve vascular endothelial function. *Vasc. Health Risk Manag.* **2014**, *10*, 591–597.
19. Barancik, M.; Gresova, L.; Bartekova, M.; Dovinova, I. Nrf2 as a Key Player of Redox Regulation in Cardiovascular Diseases. *Physiol. Res.* **2016**, *65*, S1–S10. [CrossRef]
20. Montezano, A.C.; Tsiropoulou, S.; Dulak-Lis, M.; Harvey, A.; Camargo, L.D.L.; Touyz, R.M. Redox signaling, Nox5 and vascular remodeling in hypertension. *Curr. Opin. Nephrol. Hypertens.* **2015**, *24*, 425–433. [CrossRef]
21. Li, J.-M.; Gall, N.P.; Grieve, D.J.; Chen, M.; Shah, A.M. Activation of NADPH Oxidase During Progression of Cardiac Hypertrophy to Failure. *Hypertension* **2002**, *40*, 477–484. [CrossRef] [PubMed]
22. Zhang, G.-X.; Ohmori, K.; Nagai, Y.; Fujisawa, Y.; Nishiyama, A.; Abe, Y.; Kimura, S. Role of AT1 receptor in isoproterenol-induced cardiac hypertrophy and oxidative stress in mice. *J. Mol. Cell. Cardiol.* **2007**, *42*, 804–811. [CrossRef] [PubMed]
23. Giordano, F.J. Oxygen, oxidative stress, hypoxia, and heart failure. *J. Clin. Investig.* **2005**, *115*, 500–508. [CrossRef] [PubMed]
24. van Zwieten, P. The influence of antihypertensive drug treatment on the prevention and regression of left ventricular hypertrophy. *Cardiovasc. Res.* **2000**, *45*, 82–91. [CrossRef]
25. Braunersreuther, V.; Montecucco, F.; Ashri, M.; Pelli, G.; Galan, K.; Frias, M.; Burger, F.; Quinderé, A.L.G.; Montessuit, C.; Krause, K.-H.; et al. Role of NADPH oxidase isoforms NOX1, NOX2 and NOX4 in myocardial ischemia/reperfusion injury. *J. Mol. Cell. Cardiol.* **2013**, *64*, 99–107. [CrossRef]
26. Han, J.; Su, G.; Wang, Y.; Lu, Y.; Zhao, H.; Shuai, X. 18β-Glycyrrhetinic Acid Improves Cardiac Diastolic Function by Attenuating Intracellular Calcium Overload. *Curr. Med. Sci.* **2020**, *40*, 654–661. [CrossRef]
27. Shanmugam, K.; Ravindran, S.; Kurian, G.A.; Rajesh, M. Fisetin Confers Cardioprotection against Myocardial Ischemia Reperfusion Injury by Suppressing Mitochondrial Oxidative Stress and Mitochondrial Dysfunction and Inhibiting Glycogen Synthase Kinase 3 β Activity. *Oxid. Med. Cell. Longev.* **2018**, *2018*, 1–16. [CrossRef]
28. Wallert, M.; Ziegler, M.; Wang, X.; Maluenda, A.; Xu, X.; Yap, M.L.; Witt, R.; Giles, C.; Kluge, S.; Hortmann, M.; et al. α-Tocopherol preserves cardiac function by reducing oxidative stress and inflammation in ischemia/reperfusion injury. *Redox Biol.* **2019**, *26*. [CrossRef]
29. Dolinsky, V.W.; Soltys, C.-L.M.; Rogan, K.J.; Chan, A.Y.M.; Nagendran, J.; Wang, S.; Dyck, J.R.B. Resveratrol prevents pathological but not physiological cardiac hypertrophy. *J. Mol. Med.* **2015**, *93*, 413–425. [CrossRef]
30. de Britto, R.M.; da Silva-Neto, J.A.; Mesquita, T.R.R.; de Vasconcelos, C.M.L.; de Almeida, G.K.M.; de Jesus, I.C.G.; dos Santos, P.H.; Souza, D.S.; Miguel-dos-Santos, R.; de Sá, L.A.; et al. Myrtenol protects against myocardial ischemia-reperfusion injury through antioxidant and anti-apoptotic dependent mechanisms. *Food Chem. Toxicol.* **2018**, *111*, 557–566. [CrossRef]
31. Jin, L.; Sun, S.; Ryu, Y.; Piao, Z.H.; Liu, B.; Choi, S.Y.; Kim, G.R.; Kim, H.-S.; Kee, H.J.; Jeong, M.H. Gallic acid improves cardiac dysfunction and fibrosis in pressure overload-induced heart failure. *Sci. Rep.* **2018**, *8*, 1–11. [CrossRef] [PubMed]
32. Porcu, E.P.; Cossu, M.; Rassu, G.; Giunchedi, P.; Cerri, G.; Pourová, J.; Najmanová, I.; Migkos, T.; Pilařová, V.; Nováková, L.; et al. Aqueous injection of quercetin: An approach for confirmation of its direct in vivo cardiovascular effects. *Int. J. Pharm.* **2018**, *541*, 224–233. [CrossRef] [PubMed]

33. Yu, Y.S.; Zheng, H. Chronic hydrogen-rich saline treatment reduces oxidative stress and attenuates left ventricular hypertrophy in spontaneous hypertensive rats. *Mol. Cell. Biochem.* **2012**, *365*, 233–242. [CrossRef] [PubMed]
34. Kato, R.; Nomura, A.; Sakamoto, A.; Yasuda, Y.; Amatani, K.; Nagai, S.; Sen, Y.; Ijiri, Y.; Okada, Y.; Yamaguchi, T.; et al. Hydrogen gas attenuates embryonic gene expression and prevents left ventricular remodeling induced by intermittent hypoxia in cardiomyopathic hamsters. *Am. J. Physiol. Circ. Physiol.* **2014**, *307*, H1626–H1633. [CrossRef]
35. Yao, L.; Chen, H.; Wu, Q.; Xie, K. Hydrogen-rich saline alleviates inflammation and apoptosis in myocardial I/R injury via PINK-mediated autophagy. *Int. J. Mol. Med.* **2019**, *44*, 1048–1062. [CrossRef]
36. Yoshida, A.; Asanuma, H.; Sasaki, H.; Sanada, S.; Yamazaki, S.; Asano, Y.; Shinozaki, Y.; Mori, H.; Shimouchi, A.; Sano, M.; et al. H_2 Mediates Cardioprotection Via Involvements of KATP Channels and Permeability Transition Pores of Mitochondria in Dogs. *Cardiovasc. Drugs Ther.* **2012**, *26*, 217–226. [CrossRef]
37. Colareda, G.A.; Ragone, M.I.; Bonazzola, P.; Consolini, A.E. The mKATP Channels and protein-kinase C Are Involved in the Cardioprotective Effects of Genistein on Estrogen-Deficient Rat Hearts Exposed to Ischemia/Reperfusion. *J. Cardiovasc. Pharmacol.* **2020**, *75*, 460–474. [CrossRef]
38. Simoncíková, P.; Ravingerová, T.; Andelová, E.; Tribulová, N.; Barancík, M. Changes in rat myocardium associated with modulation of ischemic tolerance by diazoxide. *Gen. Physiol. Biophys.* **2007**, *26*, 75–85.
39. Gao, Y.; Yang, H.; Chi, J.; Xu, Q.; Zhao, L.; Yang, W.; Liu, W.; Yang, W. Hydrogen Gas Attenuates Myocardial Ischemia Reperfusion Injury Independent of Postconditioning in Rats by Attenuating Endoplasmic Reticulum Stress-Induced Autophagy. *Cell. Physiol. Biochem.* **2017**, *43*, 1503–1514. [CrossRef]
40. Song, D.; Liu, X.; Diao, Y.; Sun, Y.; Gao, G.; Zhang, T.; Chen, K.; Pei, L. Hydrogen-rich solution against myocardial injury and aquaporin expression via the PI3K/Akt signaling pathway during cardiopulmonary bypass in rats. *Mol. Med. Rep.* **2018**, *18*, 1925–1938. [CrossRef]
41. Chen, K.; Sun, Y.; Diao, Y.; Zhang, T.; Dong, W. Hydrogen-rich solution attenuates myocardial injury caused by cardiopulmonary bypass in rats via the Janus-activated kinase 2/signal transducer and activator of transcription 3 signaling pathway. *Oncol. Lett.* **2018**, *16*, 167–178. [CrossRef] [PubMed]
42. Haorah, J.; Ramirez, S.H.; Schall, K.; Smith, D.; Pandya, R.; Persidsky, Y. Oxidative stress activates protein tyrosine kinase and matrix metalloproteinases leading to blood?brain barrier dysfunction. *J. Neurochem.* **2007**, *101*, 566–576. [CrossRef] [PubMed]
43. Yang, Y.; Rosenberg, G.A. Blood–Brain Barrier Breakdown in Acute and Chronic Cerebrovascular Disease. *Stroke* **2011**, *42*, 3323–3328. [CrossRef] [PubMed]
44. Ueno, M.; Chiba, Y.; Matsumoto, K.; Murakami, R.; Fujihara, R.; Kawauchi, M.; Miyanaka, H.; Nakagawa, T. Blood-brain barrier damage in vascular dementia. *Neuropathology* **2016**, *36*, 115–124. [CrossRef]
45. Wang, M.; Norman, J.E.; Srinivasan, V.J.; Rutledge, J.C. Metabolic, inflammatory, and microvascular determinants of white matter disease and cognitive decline. *Am. J. Neurodegener. Dis.* **2016**, *5*, 171–177.
46. Rochfort, K.D.; Cummins, P.M. The blood–brain barrier endothelium: A target for pro-inflammatory cytokines. *Biochem. Soc. Trans.* **2015**, *43*, 702–706. [CrossRef]
47. Arima, Y.; Harada, M.; Kamimura, D.; Park, J.-H.; Kawano, F.; Yull, F.E.; Kawamoto, T.; Iwakura, Y.; Betz, U.A.K.; Márquez, G.; et al. Regional Neural Activation Defines a Gateway for Autoreactive T Cells to Cross the Blood-Brain Barrier. *Cell* **2012**, *148*, 447–457. [CrossRef]
48. Reboldi, A.; Coisne, C.; Baumjohann, D.; Benvenuto, F.; Bottinelli, D.; Lira, S.; Uccelli, A.; Lanzavecchia, A.; Engelhardt, B.; Sallusto, F. C-C chemokine receptor 6–regulated entry of TH-17 cells into the CNS through the choroid plexus is required for the initiation of EAE. *Nat. Immunol.* **2009**, *10*, 514–523. [CrossRef]
49. Jazwa, A.; Cuadrado, A. Targeting Heme Oxygenase-1 for Neuroprotection and Neuroinflammation in Neurodegenerative Diseases. *Curr. Drug Targets* **2010**, *11*, 1517–1531. [CrossRef]
50. Papa, S.; Zazzeroni, F.; Pham, C.; Bubici, C.; Franzoso, G. Linking JNK signaling to NF- B: A key to survival. *J. Cell Sci.* **2004**, *117*, 5197–5208. [CrossRef]
51. Manaenko, A.; Lekic, T.; Ma, Q.; Zhang, J.H.; Tang, J. Hydrogen Inhalation Ameliorated Mast Cell–Mediated Brain Injury After Intracerebral Hemorrhage in Mice. *Crit. Care Med.* **2013**, *41*, 1266–1275. [CrossRef] [PubMed]
52. Ito, M.; Ibi, T.; Sahashi, K.; Ichihara, M.; Ito, M.; Ohno, K. Open-label trial and randomized, double-blind, placebo-controlled, crossover trial of hydrogen-enriched water for mitochondrial and inflammatory myopathies. *Med. Gas Res.* **2011**, *1*, 24. [CrossRef] [PubMed]

53. Ohsawa, I.; Ishikawa, M.; Takahashi, K.; Watanabe, M.; Nishimaki, K.; Yamagata, K.; Katsura, K.; Katayama, Y.; Asoh, S.; Ohta, S. Hydrogen acts as a therapeutic antioxidant by selectively reducing cytotoxic oxygen radicals. *Nat. Med.* **2007**, *13*, 688–694. [CrossRef] [PubMed]
54. LeBaron, T.W.; Kura, B.; Kalocayova, B.; Tribulova, N.; Slezak, J. A new approach for the prevention and treatment of cardiovascular disorders. Molecular hydrogen significantly reduces the effects of oxidative stress. *Molecules* **2019**, *24*, 2076. [CrossRef]
55. Lin, Y.; Ohkawara, B.; Ito, M.; Misawa, N.; Miyamoto, K.; Takegami, Y.; Masuda, A.; Toyokuni, S.; Ohno, K. Molecular hydrogen suppresses activated Wnt/β-catenin signaling. *Sci. Rep.* **2016**, *6*. [CrossRef]
56. Sobue, S.; Yamai, K.; Ito, M.; Ohno, K.; Ito, M.; Iwamoto, T.; Qiao, S.; Ohkuwa, T.; Ichihara, M. Simultaneous oral and inhalational intake of molecular hydrogen additively suppresses signaling pathways in rodents. *Mol. Cell. Biochem.* **2015**, *403*, 231–241. [CrossRef]
57. Hanaoka, T.; Kamimura, N.; Yokota, T.; Takai, S.; Ohta, S. Molecular hydrogen protects chondrocytes from oxidative stress and indirectly alters gene expressions through reducing peroxynitrite derived from nitric oxide. *Med. Gas Res.* **2011**, *1*, 1–9. [CrossRef]
58. Kiyoi, T.; Liu, S.; Takemasa, E.; Nakaoka, H.; Hato, N.; Mogi, M. Constitutive hydrogen inhalation prevents vascular remodeling via reduction of oxidative stress. *PLoS ONE* **2020**, *15*, e0227582. [CrossRef]
59. Liu, Y.; Croft, K.D.; Hodgson, J.M.; Mori, T.; Ward, N.C. Mechanisms of the protective effects of nitrate and nitrite in cardiovascular and metabolic diseases. *Nitric Oxide* **2020**, *96*, 35–43. [CrossRef]
60. Huang, C.S.; Kawamura, T.; Toyoda, Y.; Nakao, A. Recent advances in hydrogen research as a therapeutic medical gas. *Free Radic. Res.* **2010**, *44*, 971–982. [CrossRef]
61. Wang, P.; Zhao, M.; Chen, Z.; Wu, G.; Fujino, M.; Zhang, C.; Zhou, W.; Zhao, M.; Hirano, S.; Li, X.-K.; et al. Hydrogen Gas Attenuates Hypoxic-Ischemic Brain Injury via Regulation of the MAPK/HO-1/PGC-1a Pathway in Neonatal Rats. *Oxid. Med. Cell. Longev.* **2020**, *2020*, 1–16. [CrossRef] [PubMed]
62. Ge, Y.-S.; Zhang, Q.-Z.; Li, H.; Bai, G.; Jiao, Z.-H.; Wang, H.-B. Hydrogen-rich saline protects against hepatic injury induced by ischemia-reperfusion and laparoscopic hepatectomy in swine. *Hepatobiliary Pancreat. Dis. Int.* **2019**, *18*, 48–61. [CrossRef] [PubMed]
63. Zou, R.; Wang, M.-H.; Chen, Y.; Fan, X.; Yang, B.; Du, J.; Wang, X.-B.; Liu, K.-X.; Zhou, J. Hydrogen-Rich Saline Attenuates Acute Lung Injury Induced by Limb Ischemia/Reperfusion via Down-Regulating Chemerin and NLRP3 in Rats. *Shock* **2019**, *52*, 134–141. [CrossRef] [PubMed]
64. Ji, X.; Liu, W.; Xie, K.; Liu, W.; Qu, Y.; Chao, X.; Chen, T.; Zhou, J.; Fei, Z. Beneficial effects of hydrogen gas in a rat model of traumatic brain injury via reducing oxidative stress. *Brain Res.* **2010**, *1354*, 196–205. [CrossRef]
65. Kura, B.; Bagchi, A.K.; Singal, P.K.; Barancik, M.; Lebaron, T.W.; Valachova, K.; Šoltés, L.; Slezák, J. Molecular hydrogen: Potential in mitigating oxidative-stress-induced radiation injury. *Can. J. Physiol. Pharmacol.* **2019**, *97*, 287–292. [CrossRef]
66. Kawamura, T.; Wakabayashi, N.; Shigemura, N.; Huang, C.-S.; Masutani, K.; Tanaka, Y.; Noda, K.; Peng, X.; Takahashi, T.; Billiar, T.R.; et al. Hydrogen gas reduces hyperoxic lung injury via the Nrf2 pathway in vivo. *Am. J. Physiol. Lung Cell Mol. Physiol.* **2013**, *304*, 646–656. [CrossRef]
67. Pall, M.L.; Levine, S. Nrf2, a master regulator of detoxification and also antioxidant, anti-inflammatory and other cytoprotective mechanisms, is raised by health promoting factors. *Sheng Li Xue Bao* **2015**, *67*, 1–18.
68. Huang, Y.; Li, W.; Su, Z.; Kong, A.-N.T. The complexity of the Nrf2 pathway: Beyond the antioxidant response. *J. Nutr. Biochem.* **2015**, *26*, 1401–1413. [CrossRef]
69. Espinosa-Diez, C.; Miguel, V.; Mennerich, D.; Kietzmann, T.; Sánchez-Pérez, P.; Cadenas, S.; Lamas, S. Antioxidant responses and cellular adjustments to oxidative stress. *Redox Biol.* **2015**, *6*, 183–197. [CrossRef]
70. Buendia, I.; Michalska, P.; Navarro, E.; Gameiro, I.; Egea, J.; León, R. Nrf2–ARE pathway: An emerging target against oxidative stress and neuroinflammation in neurodegenerative diseases. *Pharmacol. Ther.* **2016**, *157*, 84–104. [CrossRef]
71. Jain, A.K.; Jaiswal, A.K. GSK-3β acts upstream of Fyn kinase in regulation of nuclear export and degradation of NF-E2 related factor 2. *J. Biol. Chem.* **2007**, *282*, 16502–16510. [CrossRef] [PubMed]
72. Wang, N.; Huo, R.; Cai, B.; Lu, Y.; Ye, B.; Li, X.; Li, F.; Xu, H. Activation of Wnt/β-catenin signaling by hydrogen peroxide transcriptionally inhibits NaV1.5 expression. *Free Radic. Biol. Med.* **2016**, *96*, 34–44. [CrossRef] [PubMed]

73. Liu, Y.; Dong, F.; Guo, R.; Zhang, Y.; Qu, X.; Wu, X.; Yao, R. Hydrogen-Rich Saline Ameliorates Experimental Autoimmune Encephalomyelitis in C57BL/6 Mice Via the Nrf2-ARE Signaling Pathway. *Inflammation* **2019**, *42*, 586–597. [CrossRef] [PubMed]
74. Benard, G.; Bellance, N.; Jose, C.; Melser, S.; Nouette-Gaulain, K.; Rossignol, R. Multi-site control and regulation of mitochondrial energy production. *Biochim. Biophys. Acta Bioenerg.* **2010**, *1797*, 698–709. [CrossRef] [PubMed]
75. Smith, R.A.J.; Hartley, R.C.; Cochemé, H.M.; Murphy, M.P. Mitochondrial pharmacology. *Trends Pharmacol. Sci.* **2012**, *33*, 341–352. [CrossRef]
76. Raffaello, A.; Mammucari, C.; Gherardi, G.; Rizzuto, R. Calcium at the Center of Cell Signaling: Interplay between Endoplasmic Reticulum, Mitochondria, and Lysosomes. *Trends Biochem. Sci.* **2016**, *41*, 1035–1049. [CrossRef]
77. Wang, C.; Youle, R.J. The Role of Mitochondria in Apoptosis. *Annu. Rev. Genet.* **2009**, *43*, 95–118. [CrossRef]
78. Morciano, G.; Giorgi, C.; Bonora, M.; Punzetti, S.; Pavasini, R.; Wieckowski, M.R.; Campo, G.; Pinton, P. Molecular identity of the mitochondrial permeability transition pore and its role in ischemia-reperfusion injury. *J. Mol. Cell. Cardiol.* **2015**, *78*, 142–153. [CrossRef]
79. Shen, J.; Xu, S.; Zhou, H.; Liu, H.; Jiang, W.; Hao, J.; Hu, Z. IL-1β induces apoptosis and autophagy via mitochondria pathway in human degenerative nucleus pulposus cells. *Sci. Rep.* **2017**, *7*, 1–12. [CrossRef]
80. Ohta, S. Molecular hydrogen is a novel antioxidant to efficiently reduce oxidative stress with potential for the improvement of mitochondrial diseases. *Biochim. Biophys. Acta Gen. Subj.* **2012**, *1820*, 586–594. [CrossRef]
81. Ishihara, G.; Kawamoto, K.; Komori, N.; Ishibashi, T. Molecular hydrogen suppresses superoxide generation in the mitochondrial complex I and reduced mitochondrial membrane potential. *Biochem. Biophys. Res. Commun.* **2020**, *522*, 965–970. [CrossRef] [PubMed]
82. Ishibashi, T. Therapeutic Efficacy of Molecular Hydrogen: A New Mechanistic Insight. *Curr. Pharm. Des.* **2019**, *25*, 946–955. [CrossRef] [PubMed]
83. Jovaisaite, V.; Mouchiroud, L.; Auwerx, J. The mitochondrial unfolded protein response, a conserved stress response pathway with implications in health and disease. *J. Exp. Biol.* **2014**, *217*, 137–143. [CrossRef] [PubMed]
84. Sobue, S.; Inoue, C.; Hori, F.; Qiao, S.; Murate, T.; Ichihara, M. Molecular hydrogen modulates gene expression via histone modification and induces the mitochondrial unfolded protein response. *Biochem. Biophys. Res. Commun.* **2017**, *493*, 318–324. [CrossRef] [PubMed]
85. Iuchi, K.; Nishimaki, K.; Kamimura, N.; Ohta, S. Molecular hydrogen suppresses free-radical-induced cell death by mitigating fatty acid peroxidation and mitochondrial dysfunction. *Can. J. Physiol. Pharmacol.* **2019**, *97*, 999–1005. [CrossRef] [PubMed]
86. Jiao, Y.; Yu, Y.; Li, B.; Gu, X.; Xie, K.; Wang, G.; Yu, Y. Protective effects of hydrogen-rich saline against experimental diabetic peripheral neuropathy via activation of the mitochondrial ATP-sensitive potassium channel channels in rats. *Mol. Med. Rep.* **2020**, *21*, 282–290. [CrossRef]
87. Szewczyk, A. The ATP-regulated K+ channel in mitochondria: Five years after its discovery. *Acta Biochim. Pol.* **1996**, *43*, 713–720. [CrossRef]
88. Nozawa, Y.; Miura, T.; Miki, T.; Ohnuma, Y.; Yano, T.; Shimamoto, K. Mitochondrial K ATP channel-dependent and -independent phases of ischemic preconditioning against myocardial infarction in the rat. *Basic Res. Cardiol.* **2003**, *98*, 50–58. [CrossRef]
89. Bajgar, R.; Seetharaman, S.; Kowaltowski, A.J.; Garlid, K.D.; Paucek, P. Identification and Properties of a Novel Intracellular (Mitochondrial) ATP-sensitive Potassium Channel in Brain. *J. Biol. Chem.* **2001**, *276*, 33369–33374. [CrossRef]
90. Xie, K.; Zhang, Y.; Wang, Y.; Meng, X.; Wang, Y.; Yu, Y.; Chen, H. Hydrogen attenuates sepsis-associated encephalopathy by NRF2 mediated NLRP3 pathway inactivation. *Inflamm. Res.* **2020**, *69*, 697–710. [CrossRef]
91. Gvozdjáková, A.; Kucharská, J.; Kura, B.; Vančová, O.; Rausová, Z.; Sumbalová, Z.; Uličná, O.; Slezák, J. A new insight into the molecular hydrogen effect on coenzyme Q and mitochondrial function of rats. *Can. J. Physiol. Pharmacol.* **2020**, *98*, 29–34. [CrossRef] [PubMed]
92. Lee, J.; Yang, G.; Kim, Y.-J.; Tran, Q.H.; Choe, W.; Kang, I.; Kim, S.S.; Ha, J. Hydrogen-rich medium protects mouse embryonic fibroblasts from oxidative stress by activating LKB1-AMPK-FoxO1 signal pathway. *Biochem. Biophys. Res. Commun.* **2017**, *491*, 733–739. [CrossRef] [PubMed]

93. Li, L.; Li, X.; Zhang, Z.; Liu, L.; Liu, T.; Li, S.; Liu, S.; Zhou, Y.; Liu, F. Effects of Hydrogen-rich Water on the PI3K/AKT Signaling Pathway in Rats with Myocardial Ischemia-reperfusion Injury. *Curr. Mol. Med.* **2020**, *20*, 396–406. [CrossRef] [PubMed]
94. Zhang, Y.; Xu, J.; Long, Z.; Wang, C.; Wang, L.; Sun, P.; Li, P.; Wang, T. Hydrogen (H_2) inhibits isoproterenol-induced cardiac hypertrophy via antioxidative pathways. *Front. Pharmacol.* **2016**, *7*. [CrossRef]
95. Cuadrado, A.; Kügler, S.; Lastres-Becker, I. Pharmacological targeting of GSK-3 and NRF2 provides neuroprotection in a preclinical model of tauopathy. *Redox Biol.* **2018**, *14*, 522–534. [CrossRef]
96. Manning, B.D.; Toker, A. AKT/PKB Signaling: Navigating the Network. *Cell* **2017**, *169*, 381–405. [CrossRef]
97. Teng, L.; Meng, Q.; Lu, J.; Xie, J.; Wang, Z.; Liu, Y.; Wang, D. Liquiritin modulates ERK- and AKT/GSK-3β-dependent pathways to protect against glutamate-induced cell damage in differentiated PC12 cells. *Mol. Med. Rep.* **2014**, *10*, 818–824. [CrossRef]
98. Yu, H.; Zhang, H.; Zhao, W.; Guo, L.; Li, X.; Li, Y.; Zhang, X.; Sun, Y. Gypenoside Protects against Myocardial Ischemia-Reperfusion Injury by Inhibiting Cardiomyocytes Apoptosis via Inhibition of CHOP Pathway and Activation of PI3K/Akt Pathway in Vivo and in Vitro. *Cell. Physiol. Biochem.* **2016**, *39*, 123–136. [CrossRef]
99. Li, P.; Zhang, Y.; Liu, H. The role of Wnt/β-catenin pathway in the protection process by dexmedetomidine against cerebral ischemia/reperfusion injury in rats. *Life Sci.* **2019**, *236*. [CrossRef]
100. Liu, T.; Fang, Y.; Liu, S.; Yu, X.; Zhang, H.; Liang, M.; Ding, X. Limb ischemic preconditioning protects against contrast-induced acute kidney injury in rats via phosphorylation of GSK-3β. *Free Radic. Biol. Med.* **2015**, *81*, 170–182. [CrossRef]
101. Sathiya Priya, C.; Vidhya, R.; Kalpana, K.; Anuradha, C.V. Indirubin-3′-monoxime prevents aberrant activation of GSK-3β/NF-κB and alleviates high fat-high fructose induced Aβ-aggregation, gliosis and apoptosis in mice brain. *Int. Immunopharmacol.* **2019**, *70*, 396–407. [CrossRef] [PubMed]
102. Jiang, P.; Li, G.; Zhou, X.; Wang, C.; Qiao, Y.; Liao, D.; Shi, D. Chronic fluoride exposure induces neuronal apoptosis and impairs neurogenesis and synaptic plasticity: Role of GSK-3β/β-catenin pathway. *Chemosphere* **2019**, *214*, 430–435. [CrossRef] [PubMed]
103. Wang, L.; Yin, Z.; Wang, F.; Han, Z.; Wang, Y.; Huang, S.; Hu, T.; Guo, M.; Lei, P. Hydrogen exerts neuroprotection by activation of the miR-21/PI3K/AKT/GSK-3β pathway in an in vitro model of traumatic brain injury. *J. Cell. Mol. Med.* **2020**, *24*, 4061–4071. [CrossRef] [PubMed]
104. Chen, K.; Wang, N.; Diao, Y.; Dong, W.; Sun, Y.; Liu, L.; Wu, X. Hydrogen-Rich Saline Attenuates Brain Injury Induced by Cardiopulmonary Bypass and Inhibits Microvascular Endothelial Cell Apoptosis Via the PI3K/Akt/GSK3β Signaling Pathway in Rats. *Cell. Physiol. Biochem.* **2017**, *43*, 1634–1647. [CrossRef]
105. Chen, Y.; Jiang, J.; Miao, H.; Chen, X.; Sun, X.; Li, Y. Hydrogen-rich saline attenuates vascular smooth muscle cell proliferation and neointimal hyperplasia by inhibiting reactive oxygen species production and inactivating the Ras-ERK1/2-MEK1/2 and Akt pathways. *Int. J. Mol. Med.* **2013**, *31*, 597–606. [CrossRef]
106. Hong, Y.; Shao, A.W.; Wang, J.; Chen, S.; Wu, H.J.; McBride, D.W.; Wu, Q.; Sun, X.J.; Zhang, J.M. Neuroprotective effect of hydrogen-rich saline against neurologic damage and apoptosis in early brain injury following subarachnoid hemorrhage: Possible role of the Akt/GSK3β signaling pathway. *PLoS ONE* **2014**, *9*, e96212. [CrossRef]
107. Chen, O.; Cao, Z.; Li, H.; Ye, Z.; Zhang, R.; Zhang, N.; Huang, J.; Zhang, T.; Wang, L.; Han, L.; et al. High-concentration hydrogen protects mouse heart against ischemia/reperfusion injury through activation of thePI3K/Akt1 pathway. *Sci. Rep.* **2017**, *7*, 1–14. [CrossRef]
108. Ke, H.; Liu, D.; Li, T.; Chu, X.; Xin, D.; Han, M.; Wang, S.; Wang, Z. Hydrogen-Rich Saline Regulates Microglial Phagocytosis and Restores Behavioral Deficits Following Hypoxia-Ischemia Injury in Neonatal Mice via the Akt Pathway. *Drug Des. Devel. Ther.* **2020**, *14*, 3827–3839. [CrossRef]
109. Voloshanenko, O.; Schwartz, U.; Kranz, D.; Rauscher, B.; Linnebacher, M.; Augustin, I.; Boutros, M. β-catenin-independent regulation of Wnt target genes by RoR2 and ATF2/ATF4 in colon cancer cells. *Sci. Rep.* **2018**, *8*, 1–14. [CrossRef]
110. Mao, J.; Wang, J.; Liu, B.; Pan, W.; Farr, G.H.; Flynn, C.; Yuan, H.; Takada, S.; Kimelman, D.; Li, L.; et al. Low-Density Lipoprotein Receptor-Related Protein-5 Binds to Axin and Regulates the Canonical Wnt Signaling Pathway. *Mol. Cell* **2001**, *7*, 801–809. [CrossRef]
111. MacDonald, B.T.; Tamai, K.; He, X. Wnt/β-Catenin Signaling: Components, Mechanisms, and Diseases. *Dev. Cell* **2009**, *17*, 9–26. [CrossRef] [PubMed]

112. Nusse, R.; Clevers, H. Wnt/β-Catenin Signaling, Disease, and Emerging Therapeutic Modalities. *Cell* **2017**, *169*, 985–999. [CrossRef] [PubMed]
113. Tao, H.; Yang, J.-J.; Shi, K.-H.; Li, J. Wnt signaling pathway in cardiac fibrosis: New insights and directions. *Metabolism* **2016**, *65*, 30–40. [CrossRef] [PubMed]
114. Easwaran, V.; Lee, S.H.; Inge, L.; Guo, L.; Goldbeck, C.; Garrett, E.; Wiesmann, M.; Garcia, P.D.; Fuller, J.H.; Chan, V.; et al. beta-Catenin regulates vascular endothelial growth factor expression in colon cancer. *Cancer Res.* **2003**, *63*, 3145–3153.
115. Lévy, L.; Neuveut, C.; Renard, C.-A.; Charneau, P.; Branchereau, S.; Gauthier, F.; Van Nhieu, J.T.; Cherqui, D.; Petit-Bertron, A.-F.; Mathieu, D.; et al. Transcriptional Activation of Interleukin-8 by β-Catenin-Tcf4. *J. Biol. Chem.* **2002**, *277*, 42386–42393. [CrossRef]
116. Patapoutian, A.; Backus, C.; Kispert, A.; Reichardt, L.F. Regulation of Neurotrophin-3 Expression by Epithelial-Mesenchymal Interactions: The Role of Wnt Factors. *Science* **1999**, *283*, 1180–1183. [CrossRef]
117. Yi, H.; Hu, J.; Qian, J.; Hackam, A.S. Expression of brain-derived neurotrophic factor is regulated by the Wnt signaling pathway. *Neuroreport* **2012**, *23*, 189–194. [CrossRef]
118. Clevers, H.; Nusse, R. Wnt/β-Catenin Signaling and Disease. *Cell* **2012**, *149*, 1192–1205. [CrossRef]
119. Edeling, M.; Ragi, G.; Huang, S.; Pavenstädt, H.; Susztak, K. Developmental signalling pathways in renal fibrosis: The roles of Notch, Wnt and Hedgehog. *Nat. Rev. Nephrol.* **2016**, *12*, 426–439. [CrossRef] [PubMed]
120. Chang, B.; Chen, W.; Zhang, Y.; Yang, P.; Liu, L. Tripterygium wilfordii mitigates hyperglycemia-induced upregulated Wnt/β-catenin expression and kidney injury in diabetic rats. *Exp. Ther. Med.* **2018**, *15*, 3874–3882.
121. Luo, J.; Chen, J.; Deng, Z.-L.; Luo, X.; Song, W.-X.; Sharff, K.A.; Tang, N.; Haydon, R.C.; Luu, H.H.; He, T.-C. Wnt signaling and human diseases: What are the therapeutic implications? *Lab. Investig.* **2007**, *87*, 97–103. [CrossRef]
122. Miao, J.; Liu, J.; Niu, J.; Zhang, Y.; Shen, W.; Luo, C.; Liu, Y.; Li, C.; Li, H.; Yang, P.; et al. Wnt/β-catenin/RAS signaling mediates age-related renal fibrosis and is associated with mitochondrial dysfunction. *Aging Cell* **2019**, *18*, e13004. [CrossRef] [PubMed]
123. Lehmann, M.; Baarsma, H.A.; Königshoff, M. WNT Signaling in Lung Aging and Disease. *Ann. Am. Thorac. Soc.* **2016**, *13*, S411–S416. [CrossRef] [PubMed]
124. Zhang, X.; Liu, Y.; Shao, R.; Li, W. Cdc42-interacting protein 4 silencing relieves pulmonary fibrosis in STZ-induced diabetic mice via the Wnt/GSK-3β/β-catenin pathway. *Exp. Cell Res.* **2017**, *359*, 284–290. [CrossRef] [PubMed]
125. Zhuang, S.; Hua, X.; He, K.; Zhou, T.; Zhang, J.; Wu, H.; Ma, X.; Xia, Q.; Zhang, J. Inhibition of GSK-3β induces AP-1-mediated osteopontin expression to promote cholestatic liver fibrosis. *FASEB J.* **2018**, *32*, 4494–4503. [CrossRef] [PubMed]
126. Guo, Y.; Gupte, M.; Umbarkar, P.; Singh, A.P.; Sui, J.Y.; Force, T.; Lal, H. Entanglement of GSK-3β, β-catenin and TGF-β1 signaling network to regulate myocardial fibrosis. *J. Mol. Cell. Cardiol.* **2017**, *110*, 109–120. [CrossRef] [PubMed]
127. Fang, W.; Tang, L.; Wang, G.; Lin, J.; Liao, W.; Pan, W.; Xu, J. Molecular Hydrogen Protects Human Melanocytes from Oxidative Stress by Activating Nrf2 Signaling. *J. Investig. Dermatol.* **2020**, *S0022-202X*, 31206–31209. [CrossRef] [PubMed]
128. Kühl, M.; Sheldahl, L.C.; Park, M.; Miller, J.R.; Moon, R.T. The Wnt/Ca2+ pathway. *Trends Genet.* **2000**, *16*, 279–283. [CrossRef]
129. Bergmann, M.W. WNT Signaling in Adult Cardiac Hypertrophy and Remodeling. *Circ. Res.* **2010**, *107*, 1198–1208. [CrossRef]
130. Uehara, S.; Udagawa, N.; Mukai, H.; Ishihara, A.; Maeda, K.; Yamashita, T.; Murakami, K.; Nishita, M.; Nakamura, T.; Kato, S.; et al. Protein kinase N3 promotes bone resorption by osteoclasts in response to Wnt5a-Ror2 signaling. *Sci. Signal.* **2017**, *10*, eaan0023. [CrossRef]
131. Mulligan, K.A.; Cheyette, B.N.R. Wnt Signaling in Vertebrate Neural Development and Function. *J. Neuroimmune Pharmacol.* **2012**, *7*, 774–787. [CrossRef] [PubMed]
132. Gao, C.; Chen, Y.-G. Dishevelled: The hub of Wnt signaling. *Cell. Signal.* **2010**, *22*, 717–727. [CrossRef]
133. Santos, A.; Bakker, A.D.; de Blieck-Hogervorst, J.M.A.; Klein-Nulend, J. WNT5A induces osteogenic differentiation of human adipose stem cells via rho-associated kinase Rock. *Cytotherapy* **2010**, *12*, 924–932. [CrossRef] [PubMed]

134. Hu, Y.; Liu, J.; Lu, J.; Wang, P.; Chen, J.; Guo, Y.; Han, F.; Wang, J.; Li, W.; Liu, P. sFRP1 protects H9c2 cardiac myoblasts from doxorubicin-induced apoptosis by inhibiting the Wnt/PCP-JNK pathway. *Acta Pharmacol. Sin.* **2020**, *41*, 1150–1157. [CrossRef] [PubMed]
135. Sheldahl, L.C.; Slusarski, D.C.; Pandur, P.; Miller, J.R.; Kühl, M.; Moon, R.T. Dishevelled activates Ca2+ flux, PKC, and CamKII in vertebrate embryos. *J. Cell Biol.* **2003**, *161*, 769–777. [CrossRef] [PubMed]
136. Xie, K.; Wang, W.; Chen, H.; Han, H.; Liu, D.; Wang, G.; Yu, Y. Hydrogen-rich medium attenuated lipopolysaccharide-induced monocyte-endothelial cell adhesion and vascular endothelial permeability via rho-associated coiled-coil protein kinase. *Shock* **2015**, *44*, 58–64. [CrossRef]
137. Yang, T.; Wang, L.; Sun, R.; Chen, H.; Zhang, H.; Yu, Y.; Wang, Y.; Wang, G.; Yu, Y.; Xie, K. Hydrogen-rich medium ameliorates lipopolysaccharide-induced barrier dysfunction via rhoa-mdia1 signaling in caco-2 cells. *Shock* **2016**, *45*, 228–237. [CrossRef]
138. Kawamura, T.; Huang, C.S.; Tochigi, N.; Lee, S.; Shigemura, N.; Billiar, T.R.; Okumura, M.; Nakao, A.; Toyoda, Y. Inhaled hydrogen gas therapy for prevention of lung transplant-induced ischemia/reperfusion injury in rats. *Transplantation* **2010**, *90*, 1344–1351. [CrossRef]
139. Cai, J.; Kang, Z.; Liu, W.W.; Luo, X.; Qiang, S.; Zhang, J.H.; Ohta, S.; Sun, X.; Xu, W.; Tao, H.; et al. Hydrogen therapy reduces apoptosis in neonatal hypoxia–ischemia rat model. *Neurosci. Lett.* **2008**, *441*, 167–172. [CrossRef]
140. Sun, Q.; Kang, Z.; Cai, J.; Liu, W.; Liu, Y.; Zhang, J.H.; Denoble, P.J.; Tao, H.; Sun, X. Hydrogen-Rich Saline Protects Myocardium Against Ischemia/Reperfusion Injury in Rats. *Exp. Biol. Med.* **2009**, *234*, 1212–1219. [CrossRef]
141. Chen, H.; Xie, K.; Han, H.; Li, Y.; Liu, L.; Yang, T.; Yu, Y. Molecular hydrogen protects mice against polymicrobial sepsis by ameliorating endothelial dysfunction via an Nrf2/HO-1 signaling pathway. *Int. Immunopharmacol.* **2015**, *28*, 643–654. [CrossRef] [PubMed]
142. Liu, L.; Xie, K.; Chen, H.; Dong, X.; Li, Y.; Yu, Y.; Wang, G.; Yu, Y. Inhalation of hydrogen gas attenuates brain injury in mice with cecal ligation and puncture via inhibiting neuroinflammation, oxidative stress and neuronal apoptosis. *Brain Res.* **2014**, *1589*, 78–92. [CrossRef] [PubMed]
143. Yu, Y.; Yang, Y.; Bian, Y.; Li, Y.; Liu, L.; Zhang, H.; Xie, K.; Wang, G.; Yu, Y. Hydrogen Gas Protects Against Intestinal Injury in Wild Type But Not NRF2 Knockout Mice With Severe Sepsis by Regulating HO-1 and HMGB1 Release. *Shock* **2017**, *48*, 364–370. [CrossRef] [PubMed]
144. Chen, C.H.; Manaenko, A.; Zhan, Y.; Liu, W.W.; Ostrowki, R.P.; Tang, J.; Zhang, J.H. Hydrogen gas reduced acute hyperglycemia-enhanced hemorrhagic transformation in a focal ischemia rat model. *Neuroscience* **2010**, *169*, 402–414. [CrossRef] [PubMed]
145. Varga, V.; Németh, J.; Oláh, O.; Tóth-Szűki, V.; Kovács, V.; Remzső, G.; Domoki, F. Molecular hydrogen alleviates asphyxia-induced neuronal cyclooxygenase-2 expression in newborn pigs. *Acta Pharmacol. Sin.* **2018**, *39*, 1273–1283. [CrossRef] [PubMed]
146. Hugyecz, M.; Mracskó, É.; Hertelendy, P.; Farkas, E.; Domoki, F.; Bari, F. Hydrogen supplemented air inhalation reduces changes of prooxidant enzyme and gap junction protein levels after transient global cerebral ischemia in the rat hippocampus. *Brain Res.* **2011**, *1404*, 31–38. [CrossRef] [PubMed]
147. Yang, J.; Wu, S.; Zhu, L.; Cai, J.; Fu, L. Hydrogen-containing saline alleviates pressure overload-induced interstitial fibrosis and cardiac dysfunction in rats. *Mol. Med. Rep.* **2017**, *16*, 1771–1778. [CrossRef]
148. Levine, B.; Klionsky, D.J. Development by Self-Digestion. *Dev. Cell* **2004**, *6*, 463–477. [CrossRef]
149. Ohsumi, Y. Historical landmarks of autophagy research. *Cell Res.* **2014**, *24*, 9–23. [CrossRef]
150. Carloni, S.; Albertini, M.C.; Galluzzi, L.; Buonocore, G.; Proietti, F.; Balduini, W. Increased autophagy reduces endoplasmic reticulum stress after neonatal hypoxia–ischemia: Role of protein synthesis and autophagic pathways. *Exp. Neurol.* **2014**, *255*, 103–112. [CrossRef]
151. Ogata, M.; Hino, S.; Saito, A.; Morikawa, K.; Kondo, S.; Kanemoto, S.; Murakami, T.; Taniguchi, M.; Tanii, I.; Yoshinaga, K.; et al. Autophagy Is Activated for Cell Survival after Endoplasmic ReticulumStress. *Mol. Cell. Biol.* **2006**, *26*, 9220–9231. [CrossRef] [PubMed]
152. Levine, B.; Kroemer, G. Autophagy in the Pathogenesis of Disease. *Cell* **2008**, *132*, 27–42. [CrossRef] [PubMed]
153. Kawakami, T.; Gomez, I.G.; Ren, S.; Hudkins, K.; Roach, A.; Alpers, C.E.; Shankland, S.J.; D'Agati, V.D.; Duffield, J.S. Deficient Autophagy Results in Mitochondrial Dysfunction and FSGS. *J. Am. Soc. Nephrol.* **2015**, *26*, 1040–1052. [CrossRef] [PubMed]

154. Thangaraj, A.; Sil, S.; Tripathi, A.; Chivero, E.T.; Periyasamy, P.; Buch, S. Targeting endoplasmic reticulum stress and autophagy as therapeutic approaches for neurological diseases. *Int. Rev. Cell Mol. Biol.* **2020**, *350*, 285–325. [PubMed]
155. Shi, W.; Guo, Z.; Yuan, R. Testicular Injury Attenuated by Rapamycin Through Induction of Autophagy and Inhibition of Endoplasmic Reticulum Stress in Streptozotocin- Induced Diabetic Rats. *Endocr. Metab. Immune Disord. Drug Targets* **2019**, *19*, 665–675. [CrossRef]
156. Guan, G.; Yang, L.; Huang, W.; Zhang, J.; Zhang, P.; Yu, H.; Liu, S.; Gu, X. Mechanism of interactions between endoplasmic reticulum stress and autophagy in hypoxia/reoxygenation-induced injury of H9c2 cardiomyocytes. *Mol. Med. Rep.* **2019**, *20*, 350–358. [CrossRef]
157. Xu, H.-D.; Qin, Z.-H. Beclin 1, Bcl-2 and Autophagy. In *Autophagy: Biology and Diseases*; Xu, H.D., Qin, Z.H., Eds.; Springer: Singapore, 2019; pp. 109–126.
158. Nagatani, K.; Wada, K.; Takeuchi, S.; Kobayashi, H.; Uozumi, Y.; Otani, N.; Fujita, M.; Tachibana, S.; Nawashiro, H. Effect of Hydrogen Gas on the Survival Rate of Mice Following Global Cerebral Ischemia. *Shock* **2012**, *37*, 645–652. [CrossRef]
159. Zhang, Y.; Long, Z.; Xu, J.; Tan, S.; Zhang, N.; Li, A.; Wang, L.; Wang, T. Hydrogen inhibits isoproterenol-induced autophagy in cardiomyocytes in vitro and in vivo. *Mol. Med. Rep.* **2017**, *16*, 8253–8258. [CrossRef]
160. Jiang, X.; Niu, X.; Guo, Q.; Dong, Y.; Xu, J.; Yin, N.; Qi, Q.; Jia, Y.; Gao, L.; He, Q.; et al. FoxO1-mediated autophagy plays an important role in the neuroprotective effects of hydrogen in a rat model of vascular dementia. *Behav. Brain Res.* **2019**, *356*, 98–106. [CrossRef] [PubMed]
161. Du, H.; Sheng, M.; Wu, L.; Zhang, Y.; Shi, D.; Weng, Y.; Xu, R.; Yu, W. Hydrogen-Rich Saline Attenuates Acute Kidney Injury After Liver Transplantation via Activating p53-Mediated Autophagy. *Transplantation* **2016**, *100*, 563–570. [CrossRef]
162. Wang, H.; Huo, X.; Chen, H.; Li, B.; Liu, J.; Ma, W.; Wang, X.; Xie, K.; Yu, Y.; Shi, K. Hydrogen-Rich Saline Activated Autophagy via HIF-1 α Pathways in Neuropathic Pain Model. *Biomed Res. Int.* **2018**, *2018*, 1–13. [CrossRef] [PubMed]
163. Jaakkola, P.; Mole, D.R.; Tian, Y.-M.; Wilson, M.I.; Gielbert, J.; Gaskell, S.J.; Kriegsheim, A.V.; Hebestreit, H.F.; Mukherji, M.; Schofield, C.J.; et al. Targeting of HIF-alpha to the von Hippel-Lindau Ubiquitylation Complex by O2-Regulated Prolyl Hydroxylation. *Science* **2001**, *292*, 468–472. [CrossRef] [PubMed]
164. Fang, Y.; Tan, J.; Zhang, Q. Signaling pathways and mechanisms of hypoxia-induced autophagy in the animal cells. *Cell Biol. Int.* **2015**, *39*, 891–898. [CrossRef] [PubMed]
165. Murtha, L.A.; Morten, M.; Schuliga, M.J.; Mabotuwana, N.S.; Hardy, S.A.; Waters, D.W.; Burgess, J.K.; Ngo, D.T.; Sverdlov, A.L.; Knight, D.A.; et al. The Role of Pathological Aging in Cardiac and Pulmonary Fibrosis. *Aging Dis.* **2019**, *10*, 419–428. [CrossRef]
166. Travers, J.G.; Kamal, F.A.; Robbins, J.; Yutzey, K.E.; Blaxall, B.C. Cardiac Fibrosis: The Fibroblast Awakens. *Circ. Res.* **2016**, *118*, 1021–1040. [CrossRef]
167. Zhao, X.; Kwan, J.Y.Y.; Yip, K.; Liu, P.P.; Liu, F.-F. Targeting metabolic dysregulation for fibrosis therapy. *Nat. Rev. Drug Discov.* **2020**, *19*, 57–75. [CrossRef]
168. Sun, Q.; Kawamura, T.; Masutani, K.; Peng, X.; Sun, Q.; Stolz, D.B.; Pribis, J.P.; Billiar, T.R.; Sun, X.; Bermudez, C.A.; et al. Oral intake of hydrogen-rich water inhibits intimal hyperplasia in arterialized vein grafts in rats. *Cardiovasc. Res.* **2012**, *94*, 144–153. [CrossRef]
169. Yang, Y.; Estrada, E.Y.; Thompson, J.F.; Liu, W.; Rosenberg, G.A. Matrix Metalloproteinase-Mediated Disruption of Tight Junction Proteins in Cerebral Vessels is Reversed by Synthetic Matrix Metalloproteinase Inhibitor in Focal Ischemia in Rat. *J. Cereb. Blood Flow Metab.* **2007**, *27*, 697–709. [CrossRef]

Publisher's Note: MDPI stays neutral with regard to jurisdictional claims in published maps and institutional affiliations.

© 2020 by the authors. Licensee MDPI, Basel, Switzerland. This article is an open access article distributed under the terms and conditions of the Creative Commons Attribution (CC BY) license (http://creativecommons.org/licenses/by/4.0/).

Review

Oxidative Stress and Preeclampsia-Associated Prothrombotic State

Cha Han [1,2,†], Pengzhu Huang [1,2,†], Meilu Lyu [1,2] and Jingfei Dong [3,4,*]

1 Department of Obstetrics and Gynecology, Tianjin Medical University General Hospital, Tianjin 300052, China; tjhancha@tmu.edu.cn (C.H.); huangpz2019@163.com (P.H.); lyumeilu@tmu.edu.cn (M.L.)
2 Tianjin Key Laboratory of Female Reproductive Health and Eugenic, Tianjin 300052, China
3 Bloodworks Research Institute, Seattle, WA 98102, USA
4 Division of Hematology, Department of Medicine, School of Medicine, University of Washington, Seattle, WA 98102, USA
* Correspondence: jfdong@bloodworksnw.org; Tel.: +1-206-398-5914
† The authors contributed equally to the manuscript.

Received: 8 October 2020; Accepted: 11 November 2020; Published: 17 November 2020

Abstract: Preeclampsia (PE) is a common obstetric disease characterized by hypertension, proteinuria, and multi-system dysfunction. It endangers both maternal and fetal health. Although hemostasis is critical for preventing bleeding complications during pregnancy, delivery, and post-partum, PE patients often develop a severe prothrombotic state, potentially resulting in life-threatening thrombosis and thromboembolism. The cause of this thrombotic complication is multi-factorial, involving endothelial cells, platelets, adhesive ligands, coagulation, and fibrinolysis. Increasing evidence has shown that hemostatic cells and factors undergo oxidative modifications during the systemic inflammation found in PE patients. However, it is largely unknown how these oxidative modifications of hemostasis contribute to development of the PE-associated prothrombotic state. This knowledge gap has significantly hindered the development of predictive markers, preventive measures, and therapeutic agents to protect women during pregnancy. Here we summarize reports in the literature regarding the effects of oxidative stress and antioxidants on systemic hemostasis, with emphasis on the condition of PE.

Keywords: preeclampsia; hemostasis; platelets; coagulation; oxidative stress

1. Introduction

Preeclampsia (PE) is a gestational disease that severely endangers maternal and fetal health and can develop into more severe complications (e.g., eclampsia) with long-term consequences. PE is defined as newly onset hypertension or proteinuria after 20 weeks of gestation without a prior history of hypertension, or newly onset hypertension with any of the following systemic manifestations: thrombocytopenia, renal insufficiency, impaired liver function, pulmonary edema, and severe headache without alternative diagnoses or visual impairments [1]. As the leading cause of maternal and perinatal morbidity and mortality, PE affects approximately 2% to 8% of pregnancies worldwide [2]. In an age-adjusted cohort study, the rate of PE in the US increased from 3.4% in 1980 to 3.8% in 2010 but the rate of severe PE increased by 322% during the same period [3]. This drastic increase was believed to be caused primarily by the age-cohort effect but it also highlights the need for improving pregnancy care. The current treatment of PE in clinical practice is delivery in time but antihypertension and spasmolysis may be used to control the progression of PE, prevent severe complications like eclampsia, and prolong the gestational week to improve maternal and fetal survival.

The maternal complications of PE include cerebrovascular diseases, acute renal failure, and subcapsular hematoma of the liver; the adverse perinatal outcomes include preterm delivery, fetal growth restriction (FGR), and fetal death [4]. PE patients also have a higher risk for long-term complications such as hypertension, ischemic heart disease, stroke, and venous thromboembolism [5]. Women with normal pregnancies often develop a systemic hypercoagulable state that progresses into a prothrombotic state in PE patients. In this review, we summarize the current knowledge of the role of oxidative stress in the development of this PE-associated prothrombotic state. This review focuses on linking oxidative stress to the prothrombotic state associated with PE. While both conditions develop frequently in patients with PE, they have been studied individually, without taking their causal relationship into consideration. Studying both conditions together will allow us to define this causal relationship, thus developing more accurate predictive markers and new targeted therapeutics for the patients.

The literature search was conducted in PubMed of the National Library of Medicine using the following keywords, either alone or in different combinations: placenta development, preeclampsia, HELLP syndrome, platelets, endothelial cells, coagulation factors, fibrinolysis, oxidative stress, and antioxidants. These keywords identified 1091 articles, of which 169 were considered relevant. The findings from these relevant reports were reviewed and discussed in the manuscript.

2. Placental Development

The placenta serves as a transient organ supplying nutrients and oxygen to and removing wastes from the fetus through the umbilical cord. It consists of the basal decidua, chorion frondosum, and amnion. Before trophoblast invasion, the distal tips of the spiral arteries are obstructed by the endovascular trophoblast plugs, and the maternal arterial connection with the intervillous space is restricted to the network of tortuous intercellular space. As a result, blood in the intervillous space has low oxygen tension, which is necessary to protect embryos from free radical-induced damage [6–8]. After the spiral artery plugs are gradually loosened and removed, maternal arterial blood flows into the intervillous space, increasing the oxygen tension from <20 mmHg at 8 weeks of gestation to >50 mmHg at 12 weeks. This transition results in a burst of oxidative stress in the placenta [9].

During early placental development, cytotrophoblast cells either fuse into syncytiotrophoblasts, which are exposed to maternal blood for maternal–fetal exchange or invade the decidua and myometrium [10]. The balance between trophoblast proliferation and invasion is regulated by oxygen. Cytotrophoblast cells proliferate in 2% oxygen, which mimics the uterine environment before 10 weeks of gestation, but begin to differentiate at 20% oxygen [11]. Extravillous trophoblast cells have a higher invasive activity at 17% oxygen than at 3% oxygen [12]. If these punctual and stage-specific rates of oxygenation become dysregulated during early pregnancy, placental development can be disrupted or impaired, resulting in various pathologies including PE [13].

3. Pathogenesis of PE

Placental villous lesions are found in 45.2% of PE patients compared with 14.6% of women with normal pregnancies, with vascular lesions being most common [14]. However, maternal conditions such as vascular disease, obesity, and autoimmune disease also contribute substantially to poor placentation and remodeling of the spiral arteries, leading to placental tissue ischemia and oxidative stress [15]. Placental tissue hypoxia is a hallmark event of PE, characterized by the overexpression of hypoxia-inducible transcription factor (HIF) [16] and poor angiogenesis in the placenta. It is associated with hypertension, proteinuria, and FGR [17].

The poor placental angiogenesis can result from dysregulation of growth factors such as vascular endothelial cell growth factor (VEGF) and the associated intracellular signaling pathways. For example, VEGF binds the fms-like tyrosine kinase receptor (Flt) to trigger proangiogenic signals. However, PE patients have significantly enhanced expression of soluble Flt 1 (sFlt-1) [18], an alternatively spliced variant of Flt that lacks the transmembrane and intracellular domains [19]. sFlt-1 binds VEGF and placental growth factor (PLGF) with a high affinity to competitively block VEGF binding to membrane-bound Flt and the resulting angiogenesis [19]. Placental biopsies have indeed shown that placentas from the majority of PE patients have poor trophoblastic invasion and vasculopathy of the spiral arteries that includes fibrin deposition, acute atherosis, and thrombosis [20]. More importantly, these local placental lesions can disseminate systemically by releasing soluble factors and extracellular vesicles into the maternal circulation [21,22], resulting in systemic endothelial injury and a prothrombotic state, as seen in PE patients.

There are two forms of PE: early-onset PE develops prior to 34 weeks of gestation and late-onset PE occurs after 34 gestational weeks. While both forms of PE are caused by syncytiotrophoblast stress, early-onset PE is closely associated with insufficient remodeling of spiral arteries and the resulting poor placentation, whereas late-onset PE often results from the restriction of placenta growth [15]. Early-onset PE is associated with more fetal death (adjusted odds ratios (OR) 5.8; 95% CI 4.0–8.3 vs. adjusted OR 1.3; 95% CI 0.8–2.0) and perinatal death/morbidity (adjusted OR 16.4; 95% CI 14.5–18.6 vs. adjusted OR 2.0; 95% CI 1.8–2.3) than late-onset PE [23].

4. The PE-Associated Hypercoagulable State

Pregnant women, especially during their late pregnancies, develop a systemic hypercoagulable state that is physiologically important for preventing excessive bleeding during delivery and postpartum. The cause of this hypercoagulable state is multi-factorial. For example, the syncytiotrophoblast, which is directly exposed to maternal circulation, lacks proliferative capacity and is maintained by fusion of the underlying cytotrophoblast cells, with the assistance of syncytin. Syncytiotrophoblastic cells therefore undergo constant apoptosis [24–26]. These apoptotic cells express anionic phospholipids on their surface. Anionic phospholipids such as phosphatidylserine (PS) and phosphatidylethanolamine are normally present on the inner leaflet, whereas neutral phospholipids such as phosphatidylcholine (PC) and sphingomyelin are on the external leaflet of a cell membrane bilayer. This asymmetrical distribution of membrane phospholipids is maintained by enzymatic transporters [27,28] but is lost during apoptosis [29], leading to the exposure of PS on the outer membrane. The anionic phospholipids are highly procoagulant because they promote and accelerate the formation of tenase to generate thrombin [30]. Apoptotic syncytiotrophoblast cells can also shed from the placenta into the circulation to form large multinucleated syncytial knots, which not only initiate coagulation but also invoke the immune-mediated inflammatory response. These syncytial knots can be removed by macrophages [31]. In addition, the apoptotic syncytiotrophoblastic cells also release PS-exposed extracellular vesicles to the maternal circulation [21,22] to induce hypercoagulation and inflammation, as syncytial knots do.

This protective hypercoagulable state is significantly exaggerated in pregnant women with placenta-mediated complications, such as PE, becoming a pathological prothrombotic state or thrombosis [32]. Feldstein et al. [33] studied 348 PE patients and reported that 16.1% had abnormal coagulation profiles defined by changes in prothrombin time (PT), international normalization ratio (INR), activated partial thromboplastin time (aPTT), and/or plasma fibrinogen. However, the actual rate of coagulation dysregulation is likely to be significantly higher because these tests (1) are calibrated to detect bleeding diathesis and are insensitive for detecting prothrombotic states, and (2) could not accurately measure the procoagulant activity of activated platelets and endothelial cells. Consistent with this finding, Moldenhauer et al. [34] reported that PE patients have significantly higher rates of intervillous thrombi (OR 1.95; 95% CI 1.0–3.7), central tissue infarction (OR 5.9; 95% CI 3.1–11.1), and thrombi in the fetal circulation (OR 2.8; 95% CI 1.2–6.6) than women with normotensive pregnancies. In addition to PE, dysregulated hemostasis is also found in patients with hemolysis, elevated liver enzymes, and low platelet count (HELLP syndrome), which is considered to be a severe complication of PE, but its pathogenesis is not identical to that of PE [35]. Women with early-onset HELLP syndrome have a lower prevalence of proteinuria, hypertriglyceridemia, hyperglycemia, and thrombophilia but higher levels of the inflammatory mediator homocysteine than those with early-onset PE [36]. Patients with HELLP syndrome are therefore more likely to develop severe inflammation [37] and thus are likely to suffer more severe oxidative stress. Patients with PE and those with HELLP syndrome are known to be in a hypercoagulable state [38] but whether the hypercoagulable state differs in level and nature between the two conditions remains to be studied. There is no published report on whether oxidative stress differentially regulates the hypercoagulable state associated with PE and that with HELLP syndrome.

Multiple factors contribute to the transition from the hypercoagulable state found in most normal pregnancies to the prothrombotic state developed primarily in PE. For example, both tissue factor, which initiates the extrinsic coagulation following vascular injury, and plasminogen activator inhibitor (PAI)-1, which blocks fibrinolysis, are elevated in the blood and placental tissues of PE patients, likely due to injuries to trophoblast cells undergoing repeated cycles of hypoxia–reoxygenation [39]. PE patients also have an enhanced rate of thrombin generation [40], which proteolytically promotes fibrin formation and activates platelets and endothelial cells to express procoagulant activity [41–43]. Consistent with these observations, more activated platelets are detected in PE patients [44,45], which increases platelet–leukocyte aggregation in circulation, a common marker for the prothrombotic state in PE [46]. These platelet–leukocyte aggregates upregulate the expression of sFlt-1, propagating endothelial dysfunction and inflammation in PE [47]. We have recently shown that placenta-derived extracellular vesicles (pcEVs) are detectable in the peripheral blood of pregnant women but the levels are significantly higher in PE patients [22]. These pcEVs are highly procoagulant because they express anionic phospholipids such phosphatidylserine (PS) on their surfaces and are associated with the prothrombotic state of PE patients [22]. In mouse models, we further showed that these pcEVs induce hypertension and proteinuria not only in pregnant mice but also in non-pregnant mice [21], demonstrating the importance of pcEVs in the pathogenesis of PE. These pcEVs can cause PE through multiple pathways [48] by (1) promoting coagulation through PS exposed on their surfaces [21,22,49], (2) expressing tissue factor to trigger extrinsic coagulation [49], and (3) activating platelets and endothelial cells to release procoagulant extracellular vesicles [21]. Together, these findings demonstrate that PE is initiated by placental factors and propagated by maternal factors in the systemic proinflammatory and hypercoagulable states, which are closely associated with oxidative stress.

5. PE and Oxidative Stress

The placental hypoperfusion caused by poor transformation of the spiral arteries and the resulting ischemia and reperfusion damage can cause the severe oxidative stress found in placentas from PE patients. This oxidative stress develops because of an imbalance between the production and clearance of oxidants. Nicotinamide adenine dinucleotide phosphate oxidase (NADPH oxidase), which catalyzes the production of superoxide free radicals, is found to be hyperactive in placentas from PE patients, especially on the surface of the syncytiotrophoblast microvilli [50,51], suggesting that PE placentas produce more superoxide. In contrast, the NADPH oxidase isoform NOX4, which protects vascular function [52], is downregulated in the placental villous tissue of PE patients [53]. The activity of xanthine oxidase (XO), which uses O_2 as the electron acceptor to generate reactive oxygen species (ROS) [54], is also higher in the plasma and cytotrophoblast cells of PE patients than women with normal pregnancies [55,56].

Mitochondria, which provide 90% of the cellular metabolic energy through aerobic energy production [57], can become a major source of intracellular and extracellular oxidative stress when they become dysfunctional, as found in the placentas of PE patients [58]. Placental ischemia reduces mitochondrial respiration, increasing ROS production during PE [59–61] and leading to mitochondrial oxidative stress in placental cells [6,60]. However, the mitochondrial oxidative stress is not limited to placental cells; it can also occur in maternal organs such as the kidneys [60]. In addition to intracellular mitochondrial dysfunction, morphologically intact extracellular mitochondria (exMTs) have been increasingly detected in diseased states [62–66]. We have recently shown that exMTs released during acute phase reactions remain metabolically competent and capable of generating ATP and ROS [67].

However, the evidence directly linking oxidative stress to the PE-induced prothrombotic state remains circumstantial, as we discuss in the following sections.

6. Oxidative Stress in the PE-Induced Prothrombotic State

Human hemostasis consists of four distinct but closely related components: endothelial cells, platelets and adhesive ligands, coagulation, and fibrinolysis (Figure 1). Their dysfunction, either alone or in different combinations, can result in bleeding diathesis or thrombosis in the veins and arteries. The PE-associated prothrombotic state involves all four components, but how oxidative stress modifies these hemostatic components remains poorly understood.

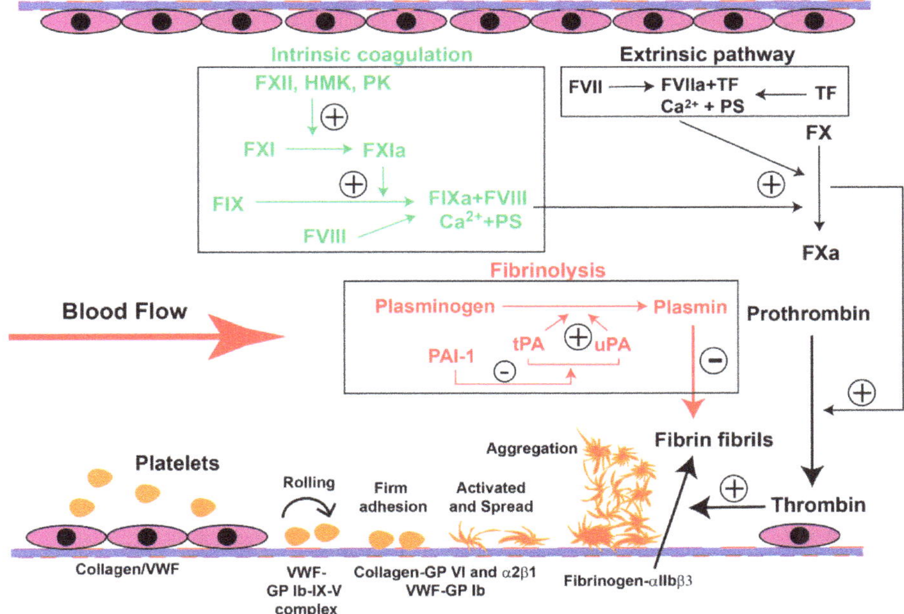

Figure 1. Schematic illustration of human hemostasis. Hemostasis begins at the site of vascular injury. Platelets first adhere to the exposed subendothelium and, through multiple ligand–receptor interactions, are activated and aggregate by fibrinogen crosslinking to form a plug to seal the wound. Anionic phospholipids exposed on the surface of activated platelets also trigger coagulation through a series of enzymatic actions of coagulation factors (e.g., inactive factor X (FX) to activated FX (Fxa)), resulting in the eventual activation of thrombin, which cleaves fibrinogen to generate fibrin fibrils that crosslink and stabilize the platelet plug to arrest bleeding. The fibrin fibrils then activate the fibrinolysis pathway that generates plasmin to cleave these fibrils to re-establish circulation. VWF: von Willebrand factor; TF: tissue factor; PS: phosphatidylserine; HMK: high molecular weight kininogen; PK: prekallibrein; tPA: tissue plasminogen activator; uPA: urokinase.

6.1. Oxidative Stress on the Endothelium

The intact endothelium lining the vascular lumen is an anticoagulant and antithrombotic surface because it is covered by a protective layer of proteoglycans called the glycocalyx [68]. The endothelium exposed to oxidative stress and inflammatory stimuli loses the glycocalyx and becomes highly procoagulant by increasing its permeability, reducing the bioavailability of the vascular relaxing factor nitro-oxide (NO) and the antiplatelet factor prostaglandin [69], tethering platelets and leukocytes to its surface [70], and promoting the deposition of fibrin generated through coagulation. Exogenous superoxide activates the endothelial cells to release proinflammatory mediators (e.g., interleukin (IL)-8, tumor necrosis factor (TNF)-α) and express adhesion molecules (e.g., Intercellular Adhesion Molecule 1 and P-selectin) and tissue factor [71,72].

6.2. Oxidative Stress on Platelets

Platelets adhere to and aggregate on the subendothelial matrix to form a hemostatic plug at the site of vascular injury. Their reactivity to common agonists is regulated by endogenous and exogenous redox signals [73]. Platelets produce and release ROS upon activation by common agonists [74] and can also be activated [75] or sensitized by ROS for activation by other platelet stimuli [76]. For example, the subendothelial matrix protein collagen stimulates platelets to produce ROS by binding to the glycoprotein (GP) VI and subsequent signaling [77,78]. Platelets produce ROS primarily through NAD(P)H-oxidase (NOX) after exposure to agonists; the ROS thus produced regulate the activation of αIIbβ3 integrin but have a minimal impact on the secretion from α- and dense-granules and the shape changes of platelets [74]. Human platelets express both NOX1 and NOX2 [79], which differentially regulate platelet reactivity [80,81]. Platelets produce O_2^- when they are exposed to anoxia–reoxygenation conditions and aggregate [82]. Superoxide can be converted to hydrogen peroxide (H_2O_2), which also induces platelet aggregation [76] and enhances the aggregation induced by collagen and arachidonic acid [78,83]. H_2O_2 sensitizes platelets for collagen-induced activation by oxidizing thiols in the SH2 domain-containing protein tyrosine phosphatase-2 in the GP VI-mediated signaling pathway [78,84,85]. Platelet reactivity is attenuated by the redox factor NO [86] and inhibited by superoxide dismutase (SOD), catalase, and the hydroxyl radical scavengers mannitol and deoxyribose. The redox states of thiols in key platelet receptors also regulate the platelets' reactivity to agonists [87]. For example, the platelet-derived protein disulfide isomerase (PDI), which is detected on the surface of resting platelets [88], regulates the redox states of active thiols on αIIbβ3 integrin for binding fibrinogen [89,90]. PDI also mediates the disulfide exchange required for collagen binding to its platelet receptor α2β1 [91]. It is therefore not surprising that platelets deficient in PDI and its homologs activate and aggregate poorly in response to agonists [92].

In addition to soluble oxidants, metabolically active exMTs released from injured cells can also become a significant and persistent source of oxidative stress [67] because blood is only mildly reducing compared with the highly reducing environment of the cytoplasm [93]. These exMTs bind platelets through the lipid receptor CD36 to activate them in an ROS-dependent manner [67]. By binding to target cells, ROS produced by exMTs can be localized, concentrated, and potentially shielded from the antioxidant activity of the blood.

Oxidative stress has been well established as a key pathology of PE but evidence of how oxidative stress regulates platelets in PE patients remains circumstantial. Although oxidized and free cysteine and homocysteine are significantly increased, the ratio of free to oxidized cysteine is lower in the blood of PE patients than in that of women with normotensive pregnancies [94], suggesting a systemic oxidative stress that could affect the redox state of platelet surface thiols [93]. Platelets from PE patients also have reduced L-arginine transport, increased protein carbonyl content, and reduced catalase activity, compared with women with normotensive pregnancies [95]. Consistent with this notion, we detected significantly more platelet-derived extracellular vesicles in PE patients than in women with normotensive pregnancies [22]. These EVs can be released from platelets exposed to oxidative stress [67,96,97]. This finding is reproducible in mouse models of PE [21]. Platelets activated by oxidative stress can be rapidly removed from circulation, resulting in thrombocytopenia, as reported for mice deficient in endothelial nitric oxide synthase (eNOS) in the presence of excess sFlt-1 [98]. Furthermore, patients with HELLP syndrome, a severe complication of PE [35], develop thrombocytopenia primarily due to thrombotic microangiopathy [99,100], which is similar to that found in patients with thrombotic thrombocytopenic purpura. This finding not only implicates platelets in the pathogenesis of PE and its complications but also suggests that von Willebrand factor (VWF), a key adhesive ligand that mediates platelet hemostasis [101], is dysfunctional in patients with PE and in those with HELLP syndrome.

6.3. Oxidative Stress on Platelet Adhesive Ligands

Oxidative stress regulates not only platelet reactivity but also the adhesive ligands that bind and activate platelets. Among these adhesive ligands, VWF has been studied most extensively for redox regulations. VWF is synthesized in megakaryocytes and endothelial cells as large multimers, which can interact simultaneously with multiple receptors on the same cell (receptor crosslinking) or different cells (cell coupling). Once synthesized, VWF multimers are either constitutively released into the circulation or stored in the Weibel–Palade bodies of endothelial cells and the α-granules of megakaryocytes/platelets, where multimerization continues to generate ultra-large (UL) VWF multimers [101]. These ULVWF multimers are highly prothrombotic because they are intrinsically hyperadhesive and spontaneously bind circulating platelets, endothelial cells, and other cells [102–105]. They are released when endothelial cells and platelets are activated by inflammatory and oxidative stimuli and are anchored to endothelial cells, where they are rapidly cleaved at the Y1605–M1606 peptide bond in the A2 domain of VWF by the metalloprotease ADAMTS-13 (a disintegrin and metalloprotease with a thrombospondin Type 1 motif, Member 13) into smaller multimers, which are then released into circulation [106–108]. These smaller circulating VWF multimers are significantly less adhesive (inert), binding platelets poorly unless they are immobilized on subendothelial collagen at the site of vascular injury or activated by high fluid shear stress [109]. However, oxidative stress can activate these inert VWF multimers or enhance their adhesive activity through several means. First, it oxidizes the methionine residues of VWF, especially at the cleavage site (M1606) to make it resistant to cleavage by ADAMTS-13 [110]. VWF oxidized by peroxynitrite is also resistant to ADAMTS-13 cleavage [111]. Second, multiple cysteine thiols on VWF can be oxidized to form laterally associated hyperadhesive fibrils [112,113]. Third, the oxidation of two vicinal cysteines in the A2 domain induces a conformational change to expose the platelet-binding A1 domain and activate VWF [114]. Fourth, methionine oxidation also makes ADAMTS-13 less active in cleaving VWF [115]. Finally, thiols at the C-terminus of ADAMTS-13 can be oxidized to abolish their disulfide bond-reducing activity [116].

While this has not been specifically reported, several lines of circumstantial evidence suggest that the activities of both VWF and ADAMTS-13 are changed in PE patients. First, systemic and placental oxidative stress has been demonstrated extensively in the condition of PE. Second, plasma VWF antigens and their adhesive activities are both significantly elevated in patients with PE and those with HELLP syndrome [22,117–119]. Third, plasma ADAMTS-13 antigens and activity are reduced in patients with PE or HELLP syndrome [117,120], resulting in a kinetic ADAMTS-13 deficiency and hyperadhesive VWF.

6.4. Oxidative Stress on Coagulation and Fibrinolysis

At the site of vascular injury, platelets adhere and aggregate on the subendothelial matrix to form a plug that seals the wound (Figure 1). This platelet plug is stabilized by crosslinked fibrin, which is generated from fibrinogen cleaved by thrombin. When hemostasis is achieved, the fibrin polymers are lysed by plasmin to reestablish circulation occluded by the hemostatic plug. Both coagulation and fibrinolysis are achieved through serial enzymatic reactions involving multiple coagulation and fibrinolytic factors.

Thrombin was reported to be inactivated irreversibly by increasing levels of oxygen in 1949 [121], providing the first evidence of oxidative regulation of coagulation. Both the intrinsic and extrinsic pathways are susceptible to shifts in the redox balance, with the intrinsic pathway being more sensitive to oxidation, whereas the extrinsic pathway is more sensitive to reduction [122]. Tissue factor contains a surface-exposed allosteric disulfide bond in the membrane-proximal fibronectin Type III domain [123]. This allosteric disulfide bond stabilizes the carboxyl-terminal domain involved in interactions with coagulation factors VII and X on a PS-rich surface to trigger coagulation [124,125]. PDI can modify this allosteric disulfide bond by reduction, S-nitrosylation, and glutathionation [124]. Both prothrombin and fibrinogen also undergo carbonylation under oxidative stress [126]. Fibrinogen crosslinks αIIbβ3 integrin to aggregate platelets and, upon cleavage by thrombin, polymerizes into fibrin fibrils. It can be oxidized at multiple sites, reducing the rate of fibrin polymerization [127] and the stability and strength of the fibrin network [128,129]. Oxidized fibrinogen also suppresses platelet aggregation induced by collagen, VWF, and adenosine diphosphate [127]. The prothrombotic genetic variants factor V Leiden and prothrombin G20210A have been associated with PE [130,131] and HELLP syndrome [132,133] but the association has not been consistently detected [134–136]. There is no report on oxidative modification of both factors in PE or other settings.

In contrast, the impact of oxidation on fibrinolysis is far less known and remains controversial. On the one hand, oxidized fibrin serves as a better substrate for plasmin to cleave during fibrinolysis and enhances the activation of pro-urokinase [137]. On the other hand, hydroxyl radicals generated from ferric ions convert fibrinogen to aberrant fibrin clots that are resistant to fibrinolytic degradation [138]. Leukocyte-derived oxidant chloramine-T, which preferentially oxidizes methionine residues, reduces the activity of tissue plasminogen activator by 40% [139]. Physiological concentrations of chloramines at 1 mM or higher can oxidize fibrinogen and coagulation factors V, VIII, and X in a dose-dependent manner to reduce their activities and inhibit platelet aggregation [140]. The inhibitory effects of chloramines can be blocked by methionine, cysteine, or ascorbic acid but not by mannitol, superoxide dismutase, or catalase.

Oxidative stress modifies not only proteins but also lipids [141]. Lipid peroxidation could regulate hemostasis during pregnancy through at least two pathways. First, membrane lipids such as PS, which initiates and propagates both intrinsic and extrinsic coagulation, are highly sensitive to peroxidation [142]. Enzymatically oxidized phospholipids restore hemostasis in humans and mice with pathological bleeding, suggesting that lipid peroxidation enhances the procoagulant activity of these anionic phospholipids. Oxidized PS has indeed been shown to enhance the formation of extrinsic tenase, intrinsic tenase, and prothrombinase [143], resulting in a hypercoagulable state. Second, patients with PE often develop dyslipidemia, with increased levels of serum triglyceride, decreased levels of high-density lipoprotein (HDL), reduced sizes of low-density lipoprotein (LDL), and a reduced ratio of LDL cholesterol to apolipoprotein B (LDLc–apo B) in the third trimester [144]. These smaller LDL particles are more susceptible to oxidation [145]. In one report, women with high concentrations of oxidized LDL (≥50 U/L) had a 2.9-fold increased risk of PE over those with less oxidized LDL (95% CI 1.4–5.9) [146].

7. Antioxidant Therapies in PE

7.1. Endogenous Antioxidants

The primary defense against oxidative stress are endogenous antioxidants such as SOD, catalase, and glutathione peroxidase that neutralize ROS. The level of SOD in the normal placental villous tissues increases from 8 weeks to 20 weeks of gestation [147], presumably to enhance antioxidant defense. The synthesis and enzymatic activity of all three enzymes are lower in placentas from PE patients than those from women with normotensive pregnancies [148–150]. Both SOD and catalase are also lower in the blood of PE patients than in that of normotensive pregnant women [151]. The rate-limiting enzyme heme oxygenase-1 (HO-1) catalyzes the potent oxidant heme into carbon monoxide, biliverdin, and free iron, thus serving as an antioxidant against cellular stresses like hypoxia and inflammation. The expression of HO-1 is reduced in placentas from PE patients [152,153]. Nuclear factor erythroid 2-related factor 2 (Nrf2) is a transcription activator that is expressed primarily in cytotrophoblast cells and regulates the expression of antioxidant proteins in these cells. Its expression is upregulated in the placentas of PE patients compared with gestation-matched normotensive subjects [154]. The chemical element selenium is an essential component of the antioxidants glutathione peroxidase and thioredoxin reductase. A meta-analysis by Xu et al. [155] reported that patients with PE have lower levels of plasma selenium than healthy pregnant women.

7.2. Therapeutic Antioxidants

Nanoscale selenium (Nano-Se) has been shown to ameliorate hyper-homocysteinemia-induced endothelial injury by inhibiting mitochondrial oxidative damage and apoptosis [156]. Selenium could also protect trophoblast cells from mitochondrial oxidative stress and ROS-mediated apoptosis [157,158], reducing the rate of PE [155]. Women who received 200 mg daily of CoQ10, a proton transfer agent in the mitochondrial electron transport chain, between 20 weeks of gestation and delivery, reduced their risk of PE compared with placebo controls [159]. CoQ10 has also been shown to improve specifically endothelial function [160]. Ergothioneine reduces the risk of PE [161], primarily by mitigating iron-induced oxidative stress. In a rat model of PE, ergothioneine ameliorated hypertension, reduced mitochondrial-specific H_2O_2 in the kidneys, and increased the fetal weight [162].

Both antiplatelet aspirin and anticoagulant heparin have been increasingly recommended for women at high risk of preeclampsia [163]. Apart from its conventional antithrombotic activity, aspirin has a modest ability of scavenging superoxide and it also contributes to the release of NO from the endothelium, which might result from the direct acetylation of eNOS [164]. Aspirin-triggered lipoxins inhibit the NADPH oxidase-mediated generation of ROS and nitrotyrosine in the endothelial cells [165,166]. Heparin has also been reported to reduce the plasma level of peroxides and increase the activity of SOD and catalase in red blood cells [167]. Low molecular weight heparin is shown to protect endothelial cells exposed to H_2O_2 [168]. However, it is not known whether aspirin and heparin reduce oxidative stress directly or by improving the systemic inflammation and tissue ischemia associated with the prothrombotic state of PE patients. Both the antioxidants pyrrolidine dithiocarbamate and N-acetylcysteine reduce the procoagulant activity of TNF-α-stimulated endothelial cells [72]. Antioxidants and NOX inhibitors significantly reduce platelet activation, aggregation, and thrombus formation [79,169].

8. Conclusions

Oxidative stress plays a physiological role in placenta development but it can also cause placental pathologies and systemic conditions that initiate or propagate PE (Figure 2). Oxidative stress modifies the cells and molecules involved in all four components of hemostasis, resulting in hypercoagulable and prothrombotic states. These oxidative modifications are well documented for their biochemistry and cellular impacts but remain poorly understood for their specific roles in the pathogenesis of PE, especially regarding their impact on hemostasis in the condition of PE. Elucidating the role of oxidative stress in the pathogenesis of the PE-induced prothrombotic state could lead to new predictive markers and therapeutic targets for this severe pregnancy complication.

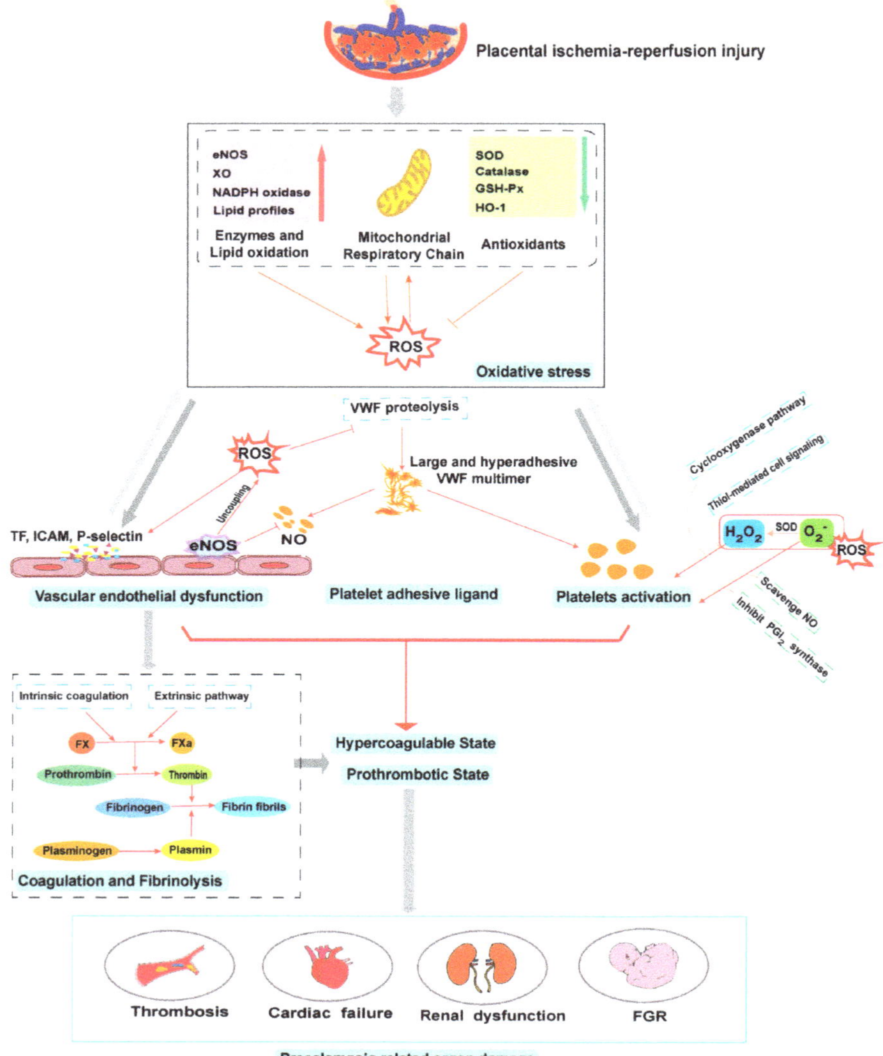

Figure 2. Pathogenesis of the preeclampsia (PE)-associated prothrombotic state. Placental ischemia–reperfusion injury induces mitochondrial dysfunction and oxidative stress and reduces the antioxidative defense, producing excessive reactive oxygen species (ROS). This oxidative stress propagates placental injury and also disseminates to systemic circulation, where it injures endothelial cells and activates platelets to express procoagulant activities and to produce prothrombotic and proinflammatory extracellular vesicles. The procoagulant cells and molecules initiate and propagate the hypercoagulable and prothrombotic states that result in arterial and venous thrombosis, renal dysfunction, and fetal growth restriction (FGR). eNOS: endothelial nitric oxide synthase; XO: xanthine oxidase; SOD: superoxide dismutase; GSH-Px: glutathione peroxidase; HO-1: heme oxygenase-1; TF: tissue factor; ICAM: Intercellular Adhesion Molecule; NO: nitric oxide; PGI_2: Prostaglandin I_2.

Author Contributions: C.H., P.H., M.L., and J.D. conducted the literature search and wrote the manuscript. All authors have read and agreed to the published version of the manuscript.

Funding: This work is supported by Grant HL152200 from the National Institute of Health, USA; Grant 82071674 from the Natural Science Foundation of China; and Grant 20JCYBJC00440 from the Natural Science Foundation of Tianjin Municipal Science and Technology.

Conflicts of Interest: The authors declare no conflict of interest.

References

1. American College of Obstetricians and Gynecologists' Committee on Practice Bulletins—Obstetrics. Gestational hypertension and preeclampsia: Acog practice bulletin, number 222. *Obstet. Gynecol.* **2020**, *135*, e237–e260. [CrossRef]
2. Steegers, E.A.; von Dadelszen, P.; Duvekot, J.J.; Pijnenborg, R. Pre-eclampsia. *Lancet* **2010**, *376*, 631–644. [CrossRef]
3. Ananth, C.V.; Keyes, K.M.; Wapner, R.J. Pre-eclampsia rates in the united states, 1980–2010: Age-period-cohort analysis. *BMJ (Clin. Res. Ed.)* **2013**, *347*, f6564. [CrossRef]
4. Chaiworapongsa, T.; Chaemsaithong, P.; Yeo, L.; Romero, R. Pre-eclampsia part 1: Current understanding of its pathophysiology. *Nat. Rev. Nephrol.* **2014**, *10*, 466–480. [CrossRef]
5. Bellamy, L.; Casas, J.P.; Hingorani, A.D.; Williams, D.J. Pre-eclampsia and risk of cardiovascular disease and cancer in later life: Systematic review and meta-analysis. *BMJ (Clin. Res. Ed.)* **2007**, *335*, 974. [CrossRef]
6. Schoots, M.H.; Gordijn, S.J.; Scherjon, S.A.; van Goor, H.; Hillebrands, J.L. Oxidative stress in placental pathology. *Placenta* **2018**, *69*, 153–161. [CrossRef]
7. Burton, G.J.; Jauniaux, E.; Watson, A.L. Maternal arterial connections to the placental intervillous space during the first trimester of human pregnancy: The boyd collection revisited. *Am. J. Obstet. Gynecol.* **1999**, *181*, 718–724. [CrossRef]
8. Burton, G.J.; Hempstock, J.; Jauniaux, E. Oxygen, early embryonic metabolism and free radical-mediated embryopathies. *Reprod. Biomed. Online* **2003**, *6*, 84–96. [CrossRef]
9. Jauniaux, E.; Watson, A.L.; Hempstock, J.; Bao, Y.P.; Skepper, J.N.; Burton, G.J. Onset of maternal arterial blood flow and placental oxidative stress. A possible factor in human early pregnancy failure. *Am. J. Pathol.* **2000**, *157*, 2111–2122. [CrossRef]
10. Ji, L.; Brkić, J.; Liu, M.; Fu, G.; Peng, C.; Wang, Y.L. Placental trophoblast cell differentiation: Physiological regulation and pathological relevance to preeclampsia. *Mol. Asp. Med.* **2013**, *34*, 981–1023. [CrossRef]
11. Genbacev, O.; Zhou, Y.; Ludlow, J.W.; Fisher, S.J. Regulation of human placental development by oxygen tension. *Science* **1997**, *277*, 1669–1672. [CrossRef] [PubMed]
12. Crocker, I.P.; Wareing, M.; Ferris, G.R.; Jones, C.J.; Cartwright, J.E.; Baker, P.N.; Aplin, J.D. The effect of vascular origin, oxygen, and tumour necrosis factor alpha on trophoblast invasion of maternal arteries in vitro. *J. Pathol.* **2005**, *206*, 476–485. [CrossRef] [PubMed]
13. Caniggia, I.; Winter, J.; Lye, S.J.; Post, M. Oxygen and placental development during the first trimester: Implications for the pathophysiology of pre-eclampsia. *Placenta* **2000**, *21* (Suppl. A), S25–S30. [CrossRef] [PubMed]
14. Falco, M.L.; Sivanathan, J.; Laoreti, A.; Thilaganathan, B.; Khalil, A. Placental histopathology associated with pre-eclampsia: Systematic review and meta-analysis. *Ultrasound Obstet. Gynecol.* **2017**, *50*, 295–301. [CrossRef] [PubMed]
15. Staff, A.C. The two-stage placental model of preeclampsia: An update. *J. Reprod. Immunol.* **2019**, *134–135*, 1–10. [CrossRef]
16. Rajakumar, A.; Brandon, H.M.; Daftary, A.; Ness, R.; Conrad, K.P. Evidence for the functional activity of hypoxia-inducible transcription factors overexpressed in preeclamptic placentae. *Placenta* **2004**, *25*, 763–769. [CrossRef]
17. Tal, R.; Shaish, A.; Barshack, I.; Polak-Charcon, S.; Afek, A.; Volkov, A.; Feldman, B.; Avivi, C.; Harats, D. Effects of hypoxia-inducible factor-1alpha overexpression in pregnant mice: Possible implications for preeclampsia and intrauterine growth restriction. *Am. J. Pathol.* **2010**, *177*, 2950–2962. [CrossRef]
18. Levine, R.J.; Maynard, S.E.; Qian, C.; Lim, K.H.; England, L.J.; Yu, K.F.; Schisterman, E.F.; Thadhani, R.; Sachs, B.P.; Epstein, F.H.; et al. Circulating angiogenic factors and the risk of preeclampsia. *N. Engl. J. Med.* **2004**, *350*, 672–683. [CrossRef]

19. Kendall, R.L.; Thomas, K.A. Inhibition of vascular endothelial cell growth factor activity by an endogenously encoded soluble receptor. *Proc. Natl. Acad. Sci. USA* **1993**, *90*, 10705–10709. [CrossRef]
20. Pijnenborg, R.; Anthony, J.; Davey, D.A.; Rees, A.; Tiltman, A.; Vercruysse, L.; van Assche, A. Placental bed spiral arteries in the hypertensive disorders of pregnancy. *Br. J. Obstet. Gynaecol.* **1991**, *98*, 648–655. [CrossRef]
21. Han, C.; Wang, C.; Chen, Y.; Wang, J.; Xu, X.; Hilton, T.; Cai, W.; Zhao, Z.; Wu, Y.; Li, K.; et al. Placenta-derived extracellular vesicles induce preeclampsia in mouse models. *Haematologica* **2020**, *105*, 1686–1694. [CrossRef] [PubMed]
22. Chen, Y.; Huang, P.; Han, C.; Li, J.; Liu, L.; Zhao, Z.; Gao, Y.; Qin, Y.; Xu, Q.; Yan, Y.; et al. Association of placenta-derived extracellular vesicles with preeclampsia and associated hypercoagulability: A clinical observational study. *BJOG Int. J. Obstet. Gynaecol.* **2020**. [CrossRef] [PubMed]
23. Lisonkova, S.; Joseph, K.S. Incidence of preeclampsia: Risk factors and outcomes associated with early- versus late-onset disease. *Am. J. Obstet. Gynecol.* **2013**, *209*, 544.e1–544.e12. [CrossRef] [PubMed]
24. Nelson, D.M. Apoptotic changes occur in syncytiotrophoblast of human placental villi where fibrin type fibrinoid is deposited at discontinuities in the villous trophoblast. *Placenta* **1996**, *17*, 387–391. [CrossRef]
25. Huppertz, B.; Frank, H.G.; Reister, F.; Kingdom, J.; Korr, H.; Kaufmann, P. Apoptosis cascade progresses during turnover of human trophoblast: Analysis of villous cytotrophoblast and syncytial fragments in vitro. *Lab. Investig.* **1999**, *79*, 1687–1702.
26. Smith, S.C.; Baker, P.N. Placental apoptosis is increased in post-term pregnancies. *Br. J. Obstet. Gynaecol.* **1999**, *106*, 861–862. [CrossRef]
27. Devaux, P.F. Protein involvement in transmembrane lipid asymmetry. *Annu. Rev. Biophys. Biomol. Struct.* **1992**, *21*, 417–439. [CrossRef]
28. Suzuki, J.; Umeda, M.; Sims, P.J.; Nagata, S. Calcium-dependent phospholipid scrambling by tmem16f. *Nature* **2010**, *468*, 834–838. [CrossRef]
29. Zwaal, R.F.; Comfurius, P.; van Deenen, L.L. Membrane asymmetry and blood coagulation. *Nature* **1977**, *268*, 358–360. [CrossRef]
30. Lentz, B.R. Exposure of platelet membrane phosphatidylserine regulates blood coagulation. *Prog. Lipid Res.* **2003**, *42*, 423–438. [CrossRef]
31. Abumaree, M.H.; Stone, P.R.; Chamley, L.W. The effects of apoptotic, deported human placental trophoblast on macrophages: Possible consequences for pregnancy. *J. Reprod. Immunol.* **2006**, *72*, 33–45. [CrossRef] [PubMed]
32. Gris, J.C.; Bouvier, S.; Cochery-Nouvellon, É.; Mercier, É.; Mousty, È.; Pérez-Martin, A. The role of haemostasis in placenta-mediated complications. *Thromb. Res.* **2019**, *181* (Suppl. 1), S10–S14. [CrossRef]
33. Feldstein, O.; Kovo, M. The association between abnormal coagulation testing in preeclampsia, adverse pregnancy outcomes, and placental histopathology. *Hypertens. Pregnancy* **2019**, *38*, 176–183. [CrossRef] [PubMed]
34. Moldenhauer, J.S.; Stanek, J.; Warshak, C.; Khoury, J.; Sibai, B. The frequency and severity of placental findings in women with preeclampsia are gestational age dependent. *Am. J. Obstet. Gynecol.* **2003**, *189*, 1173–1177. [CrossRef]
35. Abildgaard, U.; Heimdal, K. Pathogenesis of the syndrome of hemolysis, elevated liver enzymes, and low platelet count (hellp): A review. *Eur. J. Obstet. Gynecol. Reprod. Biol.* **2013**, *166*, 117–123. [CrossRef]
36. Sep, S.; Verbeek, J.; Koek, G.; Smits, L.; Spaanderman, M.; Peeters, L. Clinical differences between early-onset hellp syndrome and early-onset preeclampsia during pregnancy and at least 6 months postpartum. *Am. J. Obstet. Gynecol.* **2010**, *202*, 271.e1–271.e5. [CrossRef]
37. van Runnard Heimel, P.J.; Kavelaars, A.; Heijnen, C.J.; Peters, W.H.; Huisjes, A.J.; Franx, A.; Bruinse, H.W. Hellp syndrome is associated with an increased inflammatory response, which may be inhibited by administration of prednisolone. *Hypertens. Pregnancy* **2008**, *27*, 253–265. [CrossRef]
38. George, J.N.; Nester, C.M.; McIntosh, J.J. Syndromes of thrombotic microangiopathy associated with pregnancy. *Hematol. Am. Soc. Hematol. Educ. Program* **2015**, *2015*, 644–648. [CrossRef]
39. Teng, Y.C.; Lin, Q.D.; Lin, J.H.; Ding, C.W.; Zuo, Y. Coagulation and fibrinolysis related cytokine imbalance in preeclampsia: The role of placental trophoblasts. *J. Perinat. Med.* **2009**, *37*, 343–348. [CrossRef]
40. Macey, M.G.; Bevan, S.; Alam, S.; Verghese, L.; Agrawal, S.; Beski, S.; Thuraisingham, R.; MacCallum, P.K. Platelet activation and endogenous thrombin potential in pre-eclampsia. *Thromb. Res.* **2010**, *125*, e76–e81. [CrossRef]

41. Posma, J.J.; Posthuma, J.J.; Spronk, H.M. Coagulation and non-coagulation effects of thrombin. *J. Thromb. Haemost. JTH* **2016**, *14*, 1908–1916. [CrossRef] [PubMed]
42. Rabiet, M.J.; Plantier, J.L.; Dejana, E. Thrombin-induced endothelial cell dysfunction. *Br. Med Bull.* **1994**, *50*, 936–945. [CrossRef] [PubMed]
43. Goligorsky, M.S.; Menton, D.N.; Laszlo, A.; Lum, H. Nature of thrombin-induced sustained increase in cytosolic calcium concentration in cultured endothelial cells. *J. Biol. Chem.* **1989**, *264*, 16771–16775. [PubMed]
44. Hutt, R.; Ogunniyi, S.O.; Sullivan, M.H.; Elder, M.G. Increased platelet volume and aggregation precede the onset of preeclampsia. *Obstet. Gynecol.* **1994**, *83*, 146–149. [CrossRef]
45. Janes, S.L.; Kyle, P.M.; Redman, C.; Goodall, A.H. Flow cytometric detection of activated platelets in pregnant women prior to the development of pre-eclampsia. *Thromb. Haemost.* **1995**, *74*, 1059–1063. [CrossRef]
46. Lukanov, T.H.; Bojinova, S.I.; Popova, V.S.; Emin, A.L.; Veleva, G.L.; Gecheva, S.P.; Konova, E.I. Flow cytometric investigation of cd40-cd40 ligand system in preeclampsia and normal pregnancy. *Clin. Appl. Thromb./Hemost.* **2010**, *16*, 306–312. [CrossRef]
47. Major, H.D.; Campbell, R.A.; Silver, R.M.; Branch, D.W.; Weyrich, A.S. Synthesis of sflt-1 by platelet-monocyte aggregates contributes to the pathogenesis of preeclampsia. *Am. J. Obstet. Gynecol.* **2014**, *210*, e541–e547. [CrossRef]
48. Han, C.; Han, L.; Huang, P.; Chen, Y.; Wang, Y.; Xue, F. Syncytiotrophoblast-derived extracellular vesicles in pathophysiology of preeclampsia. *Front. Physiol.* **2019**, *10*, 1236. [CrossRef]
49. Owens, A.P., 3rd; Mackman, N. Microparticles in hemostasis and thrombosis. *Circ. Res.* **2011**, *108*, 1284–1297. [CrossRef]
50. Raijmakers, M.T.; Peters, W.H.; Steegers, E.A.; Poston, L. Nad(p)h oxidase associated superoxide production in human placenta from normotensive and pre-eclamptic women. *Placenta* **2004**, *25* (Suppl. A), S85–S89. [CrossRef]
51. Matsubara, S.; Sato, I. Enzyme histochemically detectable nad(p)h oxidase in human placental trophoblasts: Normal, preeclamptic, and fetal growth restriction-complicated pregnancy. *Histochem. Cell Biol.* **2001**, *116*, 1–7. [CrossRef] [PubMed]
52. Schröder, K.; Zhang, M.; Benkhoff, S.; Mieth, A.; Pliquett, R.; Kosowski, J.; Kruse, C.; Luedike, P.; Michaelis, U.R.; Weissmann, N.; et al. Nox4 is a protective reactive oxygen species generating vascular nadph oxidase. *Circ. Res.* **2012**, *110*, 1217–1225. [CrossRef] [PubMed]
53. Lapaire, O.; Grill, S.; Lalevee, S.; Kolla, V.; Hösli, I.; Hahn, S. Microarray screening for novel preeclampsia biomarker candidates. *Fetal Diagn. Ther.* **2012**, *31*, 147–153. [CrossRef] [PubMed]
54. Granger, D.N.; Kvietys, P.R. Reperfusion injury and reactive oxygen species: The evolution of a concept. *Redox Biol.* **2015**, *6*, 524–551. [CrossRef] [PubMed]
55. Karabulut, A.B.; Kafkasli, A.; Burak, F.; Gozukara, E.M. Maternal and fetal plasma adenosine deaminase, xanthine oxidase and malondialdehyde levels in pre-eclampsia. *Cell Biochem. Funct.* **2005**, *23*, 279–283. [CrossRef]
56. Many, A.; Hubel, C.A.; Fisher, S.J.; Roberts, J.M.; Zhou, Y. Invasive cytotrophoblasts manifest evidence of oxidative stress in preeclampsia. *Am. J. Pathol.* **2000**, *156*, 321–331. [CrossRef]
57. Brealey, D.; Brand, M.; Hargreaves, I.; Heales, S.; Land, J.; Smolenski, R.; Davies, N.A.; Cooper, C.E.; Singer, M. Association between mitochondrial dysfunction and severity and outcome of septic shock. *Lancet* **2002**, *360*, 219–223. [CrossRef]
58. McCarthy, C.M.; Kenny, L.C. Mitochondrial [dys]function; culprit in pre-eclampsia? *Clin. Sci.* **2016**, *130*, 1179–1184. [CrossRef]
59. Myatt, L. Review: Reactive oxygen and nitrogen species and functional adaptation of the placenta. *Placenta* **2010**, *31*, S66–S69. [CrossRef]
60. Vaka, V.R.; McMaster, K.M.; Cunningham, M.W., Jr.; Ibrahim, T.; Hazlewood, R.; Usry, N.; Cornelius, D.C.; Amaral, L.M.; LaMarca, B. Role of mitochondrial dysfunction and reactive oxygen species in mediating hypertension in the reduced uterine perfusion pressure rat model of preeclampsia. *Hypertension* **2018**, *72*, 703–711. [CrossRef]
61. Covarrubias, A.E.; Lecarpentier, E.; Lo, A.; Salahuddin, S.; Gray, K.J.; Karumanchi, S.A.; Zsengeller, Z.K. Ap39, a modulator of mitochondrial bioenergetics, reduces antiangiogenic response and oxidative stress in hypoxia-exposed trophoblasts: Relevance for preeclampsia pathogenesis. *Am. J. Pathol.* **2019**, *189*, 104–114. [CrossRef] [PubMed]

62. Cairns, C.B.; Moore, F.A.; Haenel, J.B.; Gallea, B.L.; Ortner, J.P.; Rose, S.J.; Moore, E.E. Evidence for early supply independent mitochondrial dysfunction in patients developing multiple organ failure after trauma. *J. Trauma* **1997**, *42*, 532–536. [CrossRef] [PubMed]
63. Smeding, L.; Plotz, F.B.; Groeneveld, A.B.; Kneyber, M.C. Structural changes of the heart during severe sepsis or septic shock. *Shock* **2012**, *37*, 449–456. [CrossRef] [PubMed]
64. Xiong, Y.; Peterson, P.L.; Verweij, B.H.; Vinas, F.C.; Muizelaar, J.P.; Lee, C.P. Mitochondrial dysfunction after experimental traumatic brain injury: Combined efficacy of snx-111 and u-101033e. *J. Neurotrauma* **1998**, *15*, 531–544. [CrossRef]
65. Azbill, R.D.; Mu, X.; Bruce-Keller, A.J.; Mattson, M.P.; Springer, J.E. Impaired mitochondrial function, oxidative stress and altered antioxidant enzyme activities following traumatic spinal cord injury. *Brain Res.* **1997**, *765*, 283–290. [CrossRef]
66. Zhao, Z.; Wang, M.; Tian, Y.; Hilton, T.; Salsbery, B.; Zhou, E.Z.; Wu, X.; Thiagarajan, P.; Boilard, E.; Li, M.; et al. Cardiolipin-mediated procoagulant activity of mitochondria contributes to traumatic brain injury-associated coagulopathy in mice. *Blood* **2016**, *127*, 2763–2772. [CrossRef]
67. Zhao, Z.; Zhou, Y.; Hilton, T.; Li, F.; Han, C.; Liu, L.; Yuan, H.; Li, Y.; Xu, X.; Wu, X.; et al. Extracellular mitochondria released from traumatized brains induced platelet procoagulant activity. *Haematologica* **2020**, *105*, 209–217. [CrossRef]
68. Jedlicka, J.; Becker, B.F.; Chappell, D. Endothelial glycocalyx. *Crit. Care Clin.* **2020**, *36*, 217–232. [CrossRef]
69. Kao, C.K.; Morton, J.S.; Quon, A.L.; Reyes, L.M.; Lopez-Jaramillo, P.; Davidge, S.T. Mechanism of vascular dysfunction due to circulating factors in women with pre-eclampsia. *Clin. Sci.* **2016**, *130*, 539–549. [CrossRef]
70. Dong, J.F.; Moake, J.L.; Nolasco, L.; Bernardo, A.; Arceneaux, W.; Shrimpton, C.N.; Schade, A.J.; McIntire, L.V.; Fujikawa, K.; Lopez, J.A. Adamts-13 rapidly cleaves newly secreted ultralarge von willebrand factor multimers on the endothelial surface under flowing conditions. *Blood* **2002**, *100*, 4033–4039. [CrossRef]
71. Jacobi, J.; Kristal, B.; Chezar, J.; Shaul, S.M.; Sela, S. Exogenous superoxide mediates pro-oxidative, proinflammatory, and procoagulatory changes in primary endothelial cell cultures. *Free Radic. Biol. Med.* **2005**, *39*, 1238–1248. [CrossRef] [PubMed]
72. Szotowski, B.; Antoniak, S.; Goldin-Lang, P.; Tran, Q.V.; Pels, K.; Rosenthal, P.; Bogdanov, V.Y.; Borchert, H.H.; Schultheiss, H.P.; Rauch, U. Antioxidative treatment inhibits the release of thrombogenic tissue factor from irradiation- and cytokine-induced endothelial cells. *Cardiovasc. Res.* **2007**, *73*, 806–812. [CrossRef] [PubMed]
73. Arthur, J.F.; Gardiner, E.E.; Kenny, D.; Andrews, R.K.; Berndt, M.C. Platelet receptor redox regulation. *Platelets* **2008**, *19*, 1–8. [CrossRef] [PubMed]
74. Begonja, A.J.; Gambaryan, S.; Geiger, J.; Aktas, B.; Pozgajova, M.; Nieswandt, B.; Walter, U. Platelet nad(p)h-oxidase-generated ros production regulates alphaiibbeta3-integrin activation independent of the no/cgmp pathway. *Blood* **2005**, *106*, 2757–2760. [CrossRef] [PubMed]
75. Krotz, F.; Sohn, H.Y.; Pohl, U. Reactive oxygen species: Players in the platelet game. *Arterioscler. Thromb. Vasc. Biol.* **2004**, *24*, 1988–1996. [CrossRef]
76. Freedman, J.E. Oxidative stress and platelets. *Arterioscler. Thromb. Vasc. Biol.* **2008**, *28*, s11–s16. [CrossRef]
77. Arthur, J.F.; Qiao, J.; Shen, Y.; Davis, A.K.; Dunne, E.; Berndt, M.C.; Gardiner, E.E.; Andrews, R.K. Itam receptor-mediated generation of reactive oxygen species in human platelets occurs via syk-dependent and syk-independent pathways. *J. Thromb. Haemost. JTH* **2012**, *10*, 1133–1141. [CrossRef]
78. Pietraforte, D.; Vona, R.; Marchesi, A.; de Jacobis, I.T.; Villani, A.; Del Principe, D.; Straface, E. Redox control of platelet functions in physiology and pathophysiology. *Antioxid. Redox Signal.* **2014**, *21*, 177–193. [CrossRef]
79. Vara, D.; Campanella, M.; Pula, G. The novel nox inhibitor 2-acetylphenothiazine impairs collagen-dependent thrombus formation in a gpvi-dependent manner. *Br. J. Pharmacol.* **2013**, *168*, 212–224. [CrossRef]
80. Fuentes, E.; Gibbins, J.M.; Holbrook, L.M.; Palomo, I. Nadph oxidase 2 (nox2): A key target of oxidative stress-mediated platelet activation and thrombosis. *Trends Cardiovasc. Med.* **2018**, *28*, 429–434. [CrossRef]
81. Masselli, E.; Pozzi, G.; Vaccarezza, M.; Mirandola, P.; Galli, D.; Vitale, M.; Carubbi, C.; Gobbi, G. Ros in platelet biology: Functional aspects and methodological insights. *Int. J. Mol. Sci.* **2020**, *21*, 4866. [CrossRef] [PubMed]
82. Leo, R.; Pratico, D.; Iuliano, L.; Pulcinelli, F.M.; Ghiselli, A.; Pignatelli, P.; Colavita, A.R.; FitzGerald, G.A.; Violi, F. Platelet activation by superoxide anion and hydroxyl radicals intrinsically generated by platelets that had undergone anoxia and then reoxygenated. *Circulation* **1997**, *95*, 885–891. [CrossRef] [PubMed]

83. Praticò, D.; Iuliano, L.; Ghiselli, A.; Alessandri, C.; Violi, F. Hydrogen peroxide as trigger of platelet aggregation. *Haemostasis* **1991**, *21*, 169–174. [CrossRef] [PubMed]
84. Jang, J.Y.; Min, J.H.; Chae, Y.H.; Baek, J.Y.; Wang, S.B.; Park, S.J.; Oh, G.T.; Lee, S.H.; Ho, Y.S.; Chang, T.S. Reactive oxygen species play a critical role in collagen-induced platelet activation via shp-2 oxidation. *Antioxid. Redox Signal.* **2014**, *20*, 2528–2540. [CrossRef]
85. Jang, J.Y.; Wang, S.B.; Min, J.H.; Chae, Y.H.; Baek, J.Y.; Yu, D.Y.; Chang, T.S. Peroxiredoxin ii is an antioxidant enzyme that negatively regulates collagen-stimulated platelet function. *J. Biol. Chem.* **2015**, *290*, 11432–11442. [CrossRef]
86. Kumar, S.; Singh, R.K.; Bhardwaj, T.R. Therapeutic role of nitric oxide as emerging molecule. *Biomed. Pharmacother. Biomed. Pharmacother.* **2017**, *85*, 182–201. [CrossRef]
87. Essex, D.W.; Li, M.; Feinman, R.D.; Miller, A. Platelet surface glutathione reductase-like activity. *Blood* **2004**, *104*, 1383–1385. [CrossRef]
88. Essex, D.W.; Chen, K.; Swiatkowska, M. Localization of protein disulfide isomerase to the external surface of the platelet plasma membrane. *Blood* **1995**, *86*, 2168–2173. [CrossRef]
89. Manickam, N.; Ahmad, S.S.; Essex, D.W. Vicinal thiols are required for activation of the αiibβ3 platelet integrin. *J. Thromb. Haemost.* **2011**, *9*, 1207–1215. [CrossRef]
90. Manickam, N.; Sun, X.; Li, M.; Gazitt, Y.; Essex, D.W. Protein disulphide isomerase in platelet function. *Br. J. Haematol.* **2008**, *140*, 223–229. [CrossRef]
91. Lahav, J.; Wijnen, E.M.; Hess, O.; Hamaia, S.W.; Griffiths, D.; Makris, M.; Knight, C.G.; Essex, D.W.; Farndale, R.W. Enzymatically catalyzed disulfide exchange is required for platelet adhesion to collagen via integrin alpha2beta1. *Blood* **2003**, *102*, 2085–2092. [CrossRef] [PubMed]
92. Zhou, J.; Wu, Y.; Chen, F.; Wang, L.; Rauova, L.; Hayes, V.M.; Poncz, M.; Li, H.; Liu, T.; Liu, J.; et al. The disulfide isomerase erp72 supports arterial thrombosis in mice. *Blood* **2017**, *130*, 817–828. [CrossRef] [PubMed]
93. Essex, D.W. Redox control of platelet function. *Antioxid. Redox Signal.* **2009**, *11*, 1191–1225. [CrossRef] [PubMed]
94. Raijmakers, M.T.; Zusterzeel, P.L.; Roes, E.M.; Steegers, E.A.; Mulder, T.P.; Peters, W.H. Oxidized and free whole blood thiols in preeclampsia. *Obstet. Gynecol.* **2001**, *97*, 272–276.
95. Pimentel, A.M.; Pereira, N.R.; Costa, C.A.; Mann, G.E.; Cordeiro, V.S.; de Moura, R.S.; Brunini, T.M.; Mendes-Ribeiro, A.C.; Resende, A.C. L-arginine-nitric oxide pathway and oxidative stress in plasma and platelets of patients with pre-eclampsia. *Hypertens. Res.* **2013**, *36*, 783–788. [CrossRef]
96. Larson, M.C.; Hillery, C.A.; Hogg, N. Circulating membrane-derived microvesicles in redox biology. *Free Radic. Biol. Med.* **2014**, *73*, 214–228. [CrossRef]
97. Bhullar, J.; Bhopale, V.M.; Yang, M.; Sethuraman, K.; Thom, S.R. Microparticle formation by platelets exposed to high gas pressures—An oxidative stress response. *Free Radic. Biol. Med.* **2016**, *101*, 154–162. [CrossRef]
98. Oe, Y.; Ko, M.; Fushima, T.; Sato, E.; Karumanchi, S.A.; Sato, H.; Sugawara, J.; Ito, S.; Takahashi, N. Hepatic dysfunction and thrombocytopenia induced by excess sflt1 in mice lacking endothelial nitric oxide synthase. *Sci. Rep.* **2018**, *8*, 102. [CrossRef]
99. Koenig, M.; Roy, M.; Baccot, S.; Cuilleron, M.; de Filippis, J.P.; Cathebras, P. Thrombotic microangiopathy with liver, gut, and bone infarction (catastrophic antiphospholipid syndrome) associated with hellp syndrome. *Clin. Rheumatol.* **2005**, *24*, 166–168. [CrossRef]
100. Abraham, K.A.; Kennelly, M.; Dorman, A.M.; Walshe, J.J. Pathogenesis of acute renal failure associated with the hellp syndrome: A case report and review of the literature. *Eur. J. Obstet. Gynecol. Reprod. Biol.* **2003**, *108*, 99–102. [CrossRef]
101. Sadler, J.E. Biochemistry and genetics of von willebrand factor. *Annu. Rev. Biochem.* **1998**, *67*, 395–424. [CrossRef] [PubMed]
102. Mayadas, T.N.; Wagner, D.D. In vitro multimerization of von willebrand factor is triggered by low ph. Importance of the propolypeptide and free sulfhydryls. *J. Biol. Chem.* **1989**, *264*, 13497–13503. [PubMed]
103. Furlan, M. Von willebrand factor: Molecular size and functional activity. *Ann. Hematol.* **1996**, *72*, 341–348. [CrossRef] [PubMed]
104. Sporn, L.A.; Marder, V.J.; Wagner, D.D. Von willebrand factor released from weibel-palade bodies binds more avidly to extracellular matrix than that secreted constitutively. *Blood* **1987**, *69*, 1531–1534. [CrossRef] [PubMed]

105. Arya, M.; Anvari, B.; Romo, G.M.; Cruz, M.A.; Dong, J.F.; McIntire, L.V.; Moake, J.L.; Lopez, J.A. Ultralarge multimers of von willebrand factor form spontaneous high-strength bonds with the platelet glycoprotein ib-ix complex: Studies using optical tweezers. *Blood* **2002**, *99*, 3971–3977. [CrossRef]
106. Furlan, M.; Robles, R.; Lammle, B. Partial purification and characterization of a protease from human plasma cleaving von willebrand factor to fragments produced by in vivo proteolysis. *Blood* **1996**, *87*, 4223–4234. [CrossRef]
107. Tsai, H.M. Physiologic cleavage of von willebrand factor by a plasma protease is dependent on its conformation and requires calcium ion. *Blood* **1996**, *87*, 4235–4244. [CrossRef]
108. Levy, G.G.; Nichols, W.C.; Lian, E.C.; Foroud, T.; McClintick, J.N.; McGee, B.M.; Yang, A.Y.; Siemieniak, D.R.; Stark, K.R.; Gruppo, R.; et al. Mutations in a member of the adamts gene family cause thrombotic thrombocytopenic purpura. *Nature* **2001**, *413*, 488–494. [CrossRef]
109. Kroll, M.H.; Hellums, J.D.; McIntire, L.V.; Schafer, A.I.; Moake, J.L. Platelets and shear stress. *Blood* **1996**, *88*, 1525–1541. [CrossRef]
110. Chen, J.; Fu, X.; Wang, Y.; Ling, M.; McMullen, B.; Kulman, J.; Chung, D.W.; López, J.A. Oxidative modification of von willebrand factor by neutrophil oxidants inhibits its cleavage by adamts13. *Blood* **2010**, *115*, 706–712. [CrossRef]
111. Lancellotti, S.; De Filippis, V.; Pozzi, N.; Peyvandi, F.; Palla, R.; Rocca, B.; Rutella, S.; Pitocco, D.; Mannucci, P.M.; De Cristofaro, R. Formation of methionine sulfoxide by peroxynitrite at position 1606 of von willebrand factor inhibits its cleavage by adamts-13: A new prothrombotic mechanism in diseases associated with oxidative stress. *Free Radic. Biol. Med.* **2010**, *48*, 446–456. [PubMed]
112. Choi, H.; Aboulfatova, K.; Pownall, H.J.; Cook, R.; Dong, J.F. Shear-induced disulfide bond formation regulates adhesion activity of von willebrand factor. *J. Biol. Chem.* **2007**, *282*, 35604–35611. [CrossRef]
113. Li, Y.; Choi, H.; Zhou, Z.; Nolasco, L.; Pownall, H.J.; Voorberg, J.; Moake, J.L.; Dong, J.F. Covalent regulation of ulvwf string formation and elongation on endothelial cells under flow conditions. *J. Thromb. Haemost.* **2008**, *6*, 1135–1143. [CrossRef] [PubMed]
114. Butera, D.; Passam, F.; Ju, L.; Cook, K.M.; Woon, H.; Aponte-Santamaria, C.; Gardiner, E.; Davis, A.K.; Murphy, D.A.; Bronowska, A.; et al. Autoregulation of von willebrand factor function by a disulfide bond switch. *Sci. Adv.* **2018**, *4*, eaaq1477. [CrossRef] [PubMed]
115. Wang, Y.; Chen, J.; Ling, M.; Lopez, J.A.; Chung, D.W.; Fu, X. Hypochlorous acid generated by neutrophils inactivates adamts13: An oxidative mechanism for regulating adamts13 proteolytic activity during inflammation. *J. Biol. Chem.* **2015**, *290*, 1422–1431. [PubMed]
116. Yeh, H.C.; Zhou, Z.; Choi, H.; Tekeoglu, S.; May, W., III; Wang, C.; Turner, N.; Scheiflinger, F.; Moake, J.L.; Dong, J.F. Disulfide bond reduction of von willebrand factor by adamts-13. *J. Thromb. Haemost.* **2010**, *8*, 2778–2788. [CrossRef] [PubMed]
117. Aref, S.; Goda, H. Increased vwf antigen levels and decreased adamts13 activity in preeclampsia. *Hematology* **2013**, *18*, 237–241. [CrossRef]
118. Szpera-Gozdziewicz, A.; Gozdziewicz, T.; Boruczkowski, M.; Dworacki, G.; Breborowicz, G.H. Relationship between the von willebrand factor plasma concentration and ultrasonographic doppler findings in pregnancies complicated by hypertensive disorders: A pilot study. *Gynecol. Obstet. Investig.* **2018**, *83*, 252–258. [CrossRef]
119. Hulstein, J.J.; van Runnard Heimel, P.J.; Franx, A.; Lenting, P.J.; Bruinse, H.W.; Silence, K.; de Groot, P.G.; Fijnheer, R. Acute activation of the endothelium results in increased levels of active von willebrand factor in hemolysis, elevated liver enzymes and low platelets (hellp) syndrome. *J. Thromb. Haemost. JTH* **2006**, *4*, 2569–2575. [CrossRef]
120. Lattuada, A.; Rossi, E.; Calzarossa, C.; Candolfi, R.; Mannucci, P.M. Mild to moderate reduction of a von willebrand factor cleaving protease (adamts-13) in pregnant women with hellp microangiopathic syndrome. *Haematologica* **2003**, *88*, 1029–1034.
121. Palos, L.A. Oxidation of the coagulation factors. *Nature* **1949**, *164*, 926. [CrossRef] [PubMed]
122. Bayele, H.K.; Murdock, P.J.; Perry, D.J.; Pasi, K.J. Simple shifts in redox/thiol balance that perturb blood coagulation. *FEBS Lett.* **2002**, *510*, 67–70. [CrossRef]
123. Chen, V.M.; Ahamed, J.; Versteeg, H.H.; Berndt, M.C.; Ruf, W.; Hogg, P.J. Evidence for activation of tissue factor by an allosteric disulfide bond. *Biochemistry* **2006**, *45*, 12020–12028. [CrossRef] [PubMed]

124. Versteeg, H.H.; Ruf, W. Thiol pathways in the regulation of tissue factor prothrombotic activity. *Curr. Opin. Hematol.* **2011**, *18*, 343–348. [CrossRef]
125. Chen, V.M. Tissue factor de-encryption, thrombus formation, and thiol-disulfide exchange. *Semin. Thromb. Hemost.* **2013**, *39*, 40–47. [CrossRef]
126. Harutyunyan, H.A. Prothrombin and fibrinogen carbonylation: How that can affect the blood clotting. *Redox Rep. Commun. Free Radic. Res.* **2017**, *22*, 160–165. [CrossRef]
127. Roitman, E.V.; Azizova, O.A.; Morozov, Y.A.; Aseichev, A.V. Effect of oxidized fibrinogens on blood coagulation. *Bull. Exp. Biol. Med.* **2004**, *138*, 245–247. [CrossRef]
128. Weigandt, K.M.; White, N.; Chung, D.; Ellingson, E.; Wang, Y.; Fu, X.; Pozzo, D.C. Fibrin clot structure and mechanics associated with specific oxidation of methionine residues in fibrinogen. *Biophys. J.* **2012**, *103*, 2399–2407. [CrossRef]
129. White, N.J.; Wang, Y.; Fu, X.; Cardenas, J.C.; Martin, E.J.; Brophy, D.F.; Wade, C.E.; Wang, X.; St John, A.E.; Lim, E.B.; et al. Post-translational oxidative modification of fibrinogen is associated with coagulopathy after traumatic injury. *Free Radic. Biol. Med.* **2016**, *96*, 181–189. [CrossRef]
130. Mello, G.; Parretti, E.; Marozio, L.; Pizzi, C.; Lojacono, A.; Frusca, T.; Facchinetti, F.; Benedetto, C. Thrombophilia is significantly associated with severe preeclampsia: Results of a large-scale, case-controlled study. *Hypertension* **2005**, *46*, 1270–1274. [CrossRef]
131. Dudding, T.; Heron, J.; Thakkinstian, A.; Nurk, E.; Golding, J.; Pembrey, M.; Ring, S.M.; Attia, J.; Scott, R.J. Factor v leiden is associated with pre-eclampsia but not with fetal growth restriction: A genetic association study and meta-analysis. *J. Thromb. Haemost. JTH* **2008**, *6*, 1869–1875. [CrossRef] [PubMed]
132. Schlembach, D.; Beinder, E.; Zingsem, J.; Wunsiedler, U.; Beckmann, M.W.; Fischer, T. Association of maternal and/or fetal factor v leiden and g20210a prothrombin mutation with hellp syndrome and intrauterine growth restriction. *Clin. Sci.* **2003**, *105*, 279–285. [CrossRef] [PubMed]
133. Tranquilli, A.L.; Giannubilo, S.R.; Dell'Uomo, B.; Grandone, E. Adverse pregnancy outcomes are associated with multiple maternal thrombophilic factors. *Eur. J. Obstet. Gynecol. Reprod. Biol.* **2004**, *117*, 144–147. [CrossRef] [PubMed]
134. Kahn, S.R.; Platt, R.; McNamara, H.; Rozen, R.; Chen, M.F.; Genest, J., Jr.; Goulet, L.; Lydon, J.; Seguin, L.; Dassa, C.; et al. Inherited thrombophilia and preeclampsia within a multicenter cohort: The montreal preeclampsia study. *Am. J. Obstet. Gynecol.* **2009**, *200*, 151.e1–151.e9. [CrossRef] [PubMed]
135. Silver, R.M.; Zhao, Y.; Spong, C.Y.; Sibai, B.; Wendel, G., Jr.; Wenstrom, K.; Samuels, P.; Caritis, S.N.; Sorokin, Y.; Miodovnik, M.; et al. Prothrombin gene g20210a mutation and obstetric complications. *Obstet. Gynecol.* **2010**, *115*, 14–20. [CrossRef]
136. Hiltunen, L.M.; Laivuori, H.; Rautanen, A.; Kaaja, R.; Kere, J.; Krusius, T.; Paunio, M.; Rasi, V. Blood group ab and factor v leiden as risk factors for pre-eclampsia: A population-based nested case-control study. *Thromb. Res.* **2009**, *124*, 167–173. [CrossRef]
137. Stief, T.W. Oxidized fibrin stimulates the activation of pro-urokinase and is the preferential substrate of human plasmin. *Blood Coagul. Fibrinolysis Int. J. Haemost. Thromb.* **1993**, *4*, 117–121. [CrossRef]
138. Pretorius, E.; Bester, J.; Vermeulen, N.; Lipinski, B. Oxidation inhibits iron-induced blood coagulation. *Curr. Drug Targets* **2013**, *14*, 13–19. [CrossRef]
139. Stief, T.W.; Martin, E.; Jimenez, J.; Digon, J.; Rodriguez, J.M. Effect of oxidants on proteases of the fibrinolytic system: Possible role for methionine residues in the interaction between tissue type plasminogen activator and fibrin. *Thromb. Res.* **1991**, *61*, 191–200. [CrossRef]
140. Stief, T.W.; Kurz, J.; Doss, M.O.; Fareed, J. Singlet oxygen inactivates fibrinogen, factor v, factor viii, factor x, and platelet aggregation of human blood. *Thromb. Res.* **2000**, *97*, 473–480. [CrossRef]
141. Wang, Y.; Walsh, S.W. Placental mitochondria as a source of oxidative stress in pre-eclampsia. *Placenta* **1998**, *19*, 581–586. [CrossRef]
142. Afonso, C.B.; Spickett, C.M. Lipoproteins as targets and markers of lipoxidation. *Redox Biol.* **2019**, *23*, 101066. [CrossRef] [PubMed]
143. Slatter, D.A.; Percy, C.L.; Allen-Redpath, K.; Gajsiewicz, J.M.; Brooks, N.J.; Clayton, A.; Tyrrell, V.J.; Rosas, M.; Lauder, S.N.; Watson, A.; et al. Enzymatically oxidized phospholipids restore thrombin generation in coagulation factor deficiencies. *JCI Insight* **2018**, *3*. [CrossRef] [PubMed]
144. Belo, L.; Caslake, M.; Gaffney, D.; Santos-Silva, A.; Pereira-Leite, L.; Quintanilha, A.; Rebelo, I. Changes in ldl size and hdl concentration in normal and preeclamptic pregnancies. *Atherosclerosis* **2002**, *162*, 425–432. [CrossRef]

145. Witztum, J.L. Susceptibility of low-density lipoprotein to oxidative modification. *Am. J. Med.* **1993**, *94*, 347–349. [CrossRef]
146. Qiu, C.; Phung, T.T.; Vadachkoria, S.; Muy-Rivera, M.; Sanchez, S.E.; Williams, M.A. Oxidized low-density lipoprotein (oxidized ldl) and the risk of preeclampsia. *Physiol. Res.* **2006**, *55*, 491–500.
147. Ali Akbar, S.; Nicolaides, K.H.; Brown, P.R. Measurement of cu/zn sod in placenta, cultured cells, various fetal tissues, decidua and semen by elisa. *J. Obstet. Gynaecol. J. Inst. Obstet. Gynaecol.* **1998**, *18*, 331–335. [CrossRef]
148. Wang, Y.; Walsh, S.W. Antioxidant activities and mrna expression of superoxide dismutase, catalase, and glutathione peroxidase in normal and preeclamptic placentas. *J. Soc. Gynecol. Investig.* **1996**, *3*, 179–184. [CrossRef]
149. Mistry, H.D.; Kurlak, L.O.; Williams, P.J.; Ramsay, M.M.; Symonds, M.E.; Broughton Pipkin, F. Differential expression and distribution of placental glutathione peroxidases 1, 3 and 4 in normal and preeclamptic pregnancy. *Placenta* **2010**, *31*, 401–408. [CrossRef]
150. Sahay, A.S.; Sundrani, D.P.; Wagh, G.N.; Mehendale, S.S.; Joshi, S.R. Regional differences in the placental levels of oxidative stress markers in pre-eclampsia. *Int. J. Gynaecol. Obstet.* **2015**, *129*, 213–218. [CrossRef]
151. Nakamura, M.; Sekizawa, A.; Purwosunu, Y.; Okazaki, S.; Farina, A.; Wibowo, N.; Shimizu, H.; Okai, T. Cellular mrna expressions of anti-oxidant factors in the blood of preeclamptic women. *Prenat. Diagn.* **2009**, *29*, 691–696. [CrossRef] [PubMed]
152. Levytska, K.; Kingdom, J.; Baczyk, D.; Drewlo, S. Heme oxygenase-1 in placental development and pathology. *Placenta* **2013**, *34*, 291–298. [CrossRef] [PubMed]
153. Ehsanipoor, R.M.; Fortson, W.; Fitzmaurice, L.E.; Liao, W.X.; Wing, D.A.; Chen, D.B.; Chan, K. Nitric oxide and carbon monoxide production and metabolism in preeclampsia. *Reprod. Sci.* **2013**, *20*, 542–548. [CrossRef] [PubMed]
154. Wruck, C.J.; Huppertz, B.; Bose, P.; Brandenburg, L.O.; Pufe, T.; Kadyrov, M. Role of a fetal defence mechanism against oxidative stress in the aetiology of preeclampsia. *Histopathology* **2009**, *55*, 102–106. [CrossRef] [PubMed]
155. Xu, M.; Guo, D.; Gu, H.; Zhang, L.; Lv, S. Selenium and preeclampsia: A systematic review and meta-analysis. *Biol. Trace Elem. Res.* **2016**, *171*, 283–292. [CrossRef] [PubMed]
156. Zheng, Z.; Liu, L.; Zhou, K.; Ding, L.; Zeng, J.; Zhang, W. Anti-oxidant and anti-endothelial dysfunctional properties of nano-selenium in vitro and in vivo of hyperhomocysteinemic rats. *Int. J. Nanomed.* **2020**, *15*, 4501–4521. [CrossRef]
157. Khera, A.; Vanderlelie, J.J.; Perkins, A.V. Selenium supplementation protects trophoblast cells from mitochondrial oxidative stress. *Placenta* **2013**, *34*, 594–598. [CrossRef]
158. Khera, A.; Vanderlelie, J.J.; Holland, O.; Perkins, A.V. Overexpression of endogenous anti-oxidants with selenium supplementation protects trophoblast cells from reactive oxygen species-induced apoptosis in a bcl-2-dependent manner. *Biol. Trace Elem. Res.* **2017**, *177*, 394–403. [CrossRef]
159. Teran, E.; Hernandez, I.; Nieto, B.; Tavara, R.; Ocampo, J.E.; Calle, A. Coenzyme q10 supplementation during pregnancy reduces the risk of pre-eclampsia. *Int. J. Gynaecol. Obstet.* **2009**, *105*, 43–45. [CrossRef]
160. Gao, L.; Mao, Q.; Cao, J.; Wang, Y.; Zhou, X.; Fan, L. Effects of coenzyme q10 on vascular endothelial function in humans: A meta-analysis of randomized controlled trials. *Atherosclerosis* **2012**, *221*, 311–316. [CrossRef]
161. Kerley, R.N.; McCarthy, C.; Kell, D.B.; Kenny, L.C. The potential therapeutic effects of ergothioneine in pre-eclampsia. *Free Radic. Biol. Med.* **2018**, *117*, 145–157. [CrossRef] [PubMed]
162. Williamson, R.D.; McCarthy, F.P.; Manna, S.; Groarke, E.; Kell, D.B.; Kenny, L.C.; McCarthy, C.M. L-(+)-ergothioneine significantly improves the clinical characteristics of preeclampsia in the reduced uterine perfusion pressure rat model. *Hypertension* **2020**, *75*, 561–568. [CrossRef] [PubMed]
163. Roberge, S.; Demers, S.; Nicolaides, K.H.; Bureau, M.; Côté, S.; Bujold, E. Prevention of pre-eclampsia by low-molecular-weight heparin in addition to aspirin: A meta-analysis. *Ultrasound Obstet. Gynecol.* **2016**, *47*, 548–553. [CrossRef] [PubMed]
164. Taubert, D.; Berkels, R.; Grosser, N.; Schröder, H.; Gründemann, D.; Schömig, E. Aspirin induces nitric oxide release from vascular endothelium: A novel mechanism of action. *Br. J. Pharmacol.* **2004**, *143*, 159–165. [CrossRef] [PubMed]
165. Nascimento-Silva, V.; Arruda, M.A.; Barja-Fidalgo, C.; Fierro, I.M. Aspirin-triggered lipoxin a4 blocks reactive oxygen species generation in endothelial cells: A novel antioxidative mechanism. *Thromb. Haemost.* **2007**, *97*, 88–98.

166. Gil-Villa, A.M.; Alvarez, A.M. Role of aspirin-triggered lipoxin a4, aspirin, and salicylic acid in the modulation of the oxidative and inflammatory responses induced by plasma from women with pre-eclampsia. *Am. J. Reprod. Immunol.* **2020**, *83*, e13207. [CrossRef]
167. Dzieciuchowicz, Ł.; Checiński, P.; Krauss, H. Heparin reduces oxidative stress in the postoperative period. *Med. Sci. Monit. Int. Med. J. Exp. Clin. Res.* **2002**, *8*, Cr657–Cr660.
168. Manduteanu, I.; Dragomir, E.; Voinea, M.; Capraru, M.; Simionescu, M. Enoxaparin reduces h2o2-induced activation of human endothelial cells by a mechanism involving cell adhesion molecules and nuclear transcription factors. *Pharmacology* **2007**, *79*, 154–162. [CrossRef]
169. Chlopicki, S.; Olszanecki, R.; Janiszewski, M.; Laurindo, F.R.; Panz, T.; Miedzobrodzki, J. Functional role of nadph oxidase in activation of platelets. *Antioxid. Redox Signal.* **2004**, *6*, 691–698. [CrossRef]

Publisher's Note: MDPI stays neutral with regard to jurisdictional claims in published maps and institutional affiliations.

© 2020 by the authors. Licensee MDPI, Basel, Switzerland. This article is an open access article distributed under the terms and conditions of the Creative Commons Attribution (CC BY) license (http://creativecommons.org/licenses/by/4.0/).

Article

Pathways of 4-Hydroxy-2-Nonenal Detoxification in a Human Astrocytoma Cell Line

Eleonora Peroni [1], Viola Scali [2,†], Francesco Balestri [2,3], Mario Cappiello [2,3], Umberto Mura [2], Antonella Del Corso [2,3,*] and Roberta Moschini [2,3]

1 Ecoverde SpA, via IV Novembre, 55016 Porcari (LU), Italy; eleonoraperoni@hotmail.it

2 Department of Biology, Biochemistry Unit, University of Pisa, via S. Zeno, 51, 56123 Pisa, Italy; viola.scali@gmail.com (V.S.); francesco.balestri@unipi.it (F.B.); mario.cappiello@unipi.it (M.C.); umberto.mura@unipi.it (U.M.); roberta.moschini@unipi.it (R.M.)

3 Interdepartmental Research Center Nutrafood "Nutraceuticals and Food for Health", University of Pisa, 56124 Pisa, Italy

* Correspondence: antonella.delcorso@unipi.it

† PhD student at the Tuscany Region Pegaso PhD School in Biochemistry and Molecular Biology, Italy.

Received: 26 March 2020; Accepted: 4 May 2020; Published: 5 May 2020

Abstract: One of the consequences of the increased level of oxidative stress that often characterizes the cancer cell environment is the abnormal generation of lipid peroxidation products, above all 4-hydroxynonenal. The contribution of this aldehyde to the pathogenesis of several diseases is well known. In this study, we characterized the ADF astrocytoma cell line both in terms of its pattern of enzymatic activities devoted to 4-hydroxynonenal removal and its resistance to oxidative stress induced by exposure to hydrogen peroxide. A comparison with lens cell lines, which, due to the ocular function, are normally exposed to oxidative conditions is reported. Our results show that, overall, ADF cells counteract oxidative stress conditions better than normal cells, thus confirming the redox adaptation demonstrated for several cancer cells. In addition, the markedly high level of $NADP^+$-dependent dehydrogenase activity acting on the glutahionyl-hydroxynonanal adduct detected in ADF cells may promote, at the same time, the detoxification and recovery of cell-reducing power in these cells.

Keywords: 4-hydroxy-2-nonenal; 3-glutathionyl-4-hydroxynonanal; astrocytoma cells

---

1. Introduction

Cancer cells often have to deal with the generation of reactive oxygen species (ROS) and an altered redox status [1,2]. At the same time, an increased antioxidant capacity often enables cancer cells to adapt to oxidative stress. [3,4]. Thus, cancer cells can easily survive under oxidative stress conditions, and this redox adaptation increases their resistance to anticancer and radiation therapies [5–7].

The high rate of oxygen consumption, the high level of lipids, and the relatively low level of enzymes involved in the antioxidant defense make the central nervous system particularly susceptible to oxidative stress [8]. In fact, evidence of a marked increase in lipid peroxidation processes has been reported in patients with glioblastoma [9], the most common type of brain tumor. In addition, isocitrate dehydrogenase (IDH) mutations have been frequently observed in glioma cells [10]. ROS accumulation, disruption of the NADP/NADPH balance and abnormally high reduced glutathione (GSH)/glutathione disulfide (GSSG) ratios have been observed in several glioma cells and correlated with IDH mutations [11–13]. Astrocytoma is characterized by a high resistance to radiotherapy and chemotherapy treatments [14]. The alkylating agent temozolomide (TMZ) is currently used in standard chemotherapeutic treatment in addition to radiotherapy [15,16]. However, the sensitivity of glioblastomas to TMZ is variable and several mechanisms of drug resistance have been hypothesized, linked to different DNA repair pathways [17]. Recently, a role for oxidative stress and aldehyde dehydrogenase ALDH1A3 in the TMZ-induced therapeutic effects has been suggested [18,19]. It was shown that ALDH1A3, through its ability to detoxify lipid peroxidation products, confers to glioblastoma cells chemoresistance against TMZ.

Among lipid peroxidation products, 4-hydroxy-2-nonenal (HNE) is one of the most abundant and is considered a classical marker of oxidative stress in cells [20]. Due to its chemical reactivity, HNE easily covalently reacts with low molecular weight compounds, such as GSH, and with proteins and DNA, thus differentially affecting, depending on its intracellular concentration, cell cycle regulation towards proliferation or apoptosis [21].

The main metabolic pathways of HNE involve both NAD(P)-dependent oxidoreductases and GSH conjugation [20].

A set of specific enzymes presenting a high affinity for HNE are able to metabolize the free aldehyde (i.e., aldo-keto reductases AKR1B1 and AKR1B10 and aldehyde dehydrogenases) [22–27]. However, HNE is mainly metabolized by direct conjugation with GSH, forming 3-glutathionyl-4-hydroxynonanal (GSHNE) [28]. In addition to its possible occurrence through the chemical reactivity of the reagents, GSHNE formation is enhanced by the action of glutathione S-transferases (GSTs). Of the various GST isoforms, GSTA4 presents the highest HNE affinity [29].

In turn, the metabolic fate of GSHNE takes place both through an oxidative and a reductive pathway. Recent evidence indicates that GSHNE may undergo oxidation, at the level of its hemiacetal hydroxyl group, through an efficient $NADP^+$-dependent reaction catalysed by carbonyl reductase 1 (CBR1), leading to the corresponding 3-glutathionyl-nonanoic-γ-lactone (GSHNA-γ-lactone) [30]. On the other hand, GSHNE reduction, which is primarily driven by aldose reductase (AKR1B1) [26], generates glutathione-1,4-dihydroxynonane (GSDHN), a signalling molecule involved in the inflammation response through the NF-κB activation pathway. In addition, CBR1 can reduce GSHNE to GSDHN, thus contributing to the inflammatory signalling action associated with GSHNE reduction [31].

In this work, we report the occurrence in the human ADF astrocytoma cell line of a substantial enzymatic pattern that intervenes in HNE detoxification and is robust in oxidative stress conditions.

2. Materials and Methods

2.1. Materials

Cell culture media, fetal bovine serum (FBS), penicillin/streptomycin solution, gentamicin, and glutamine were purchased from Euroclone (Pero, Italy). HNE was synthesized as previously described [30]. GSHNE was synthesized as described [32]. $NADP^+$, NAD^+, NADPH and NADH were from Carbosynth (Compton, England). GSH, GSSG, bovine serum albumin, D,L-glyceraldehyde (GAL), propanal, 1,1,3,3-tetraetoxipropane (TEP), D,L-dithiothreitol (DTT), 1-(4,5-Dimethylthiazol-2-yl)-3,5-diphenylformazan (MTT), sorbinil, disulfiram and amphotericin B were from Merck Life Science (Milan, Italy). Hank's balanced salt solution (HBSS) composition was: 0.014% (*w/v*) $CaCl_2$, 0.01% (*w/v*) $MgSO_4$, 0.04% (*w/v*) KCl, 0.006% (*w/v*) KH_2PO_4, 0.8% (*w/v*) NaCl, 0.005% (*w/v*) Na_2HPO_4, 0.1% (*w/v*) D-glucose. All inorganic chemicals were of reagent grade from BDH (VWR International, Poole, Dorset, UK).

2.2. Cell Cultures

All the cell lines were cultured at 37 °C in a humidified atmosphere in the presence of 5% CO_2. Human astrocytoma ADF cells, established from a primary tumor (Grade IV astrocytoma) [33,34] were kindly provided by Dr. W. Malorni, Istituto Superiore di Sanità, Rome, Italy. Concerning genetic alteration regarding IDH mutations, ADF were classified, according to WHO, as NOS [35]. ADF cells were cultured in RPMI 1640 medium supplemented with 10% (*v/v*) FBS, 50 mU/mL penicillin/streptomycin and 2 mM glutamine. The human lens epithelial cells (HLEC) line B3, was obtained from American Type Culture Collection (Rockville, MD, USA) and cultured in Eagle's modified essential medium (MEM) supplemented with 20% (*v/v*) FBS, 50 mU/mL penicillin/streptomycin and 2 mM glutamine. Primary cultures of bovine lens epithelial cells (BLEC) were obtained from the dissection of bovine (*Bos taurus*) lenses and cultured in RPMI 1640 medium supplemented with 10% (*v/v*) FBS, 2 mM glutamine, 2.5 mg/mL amphotericin B and 50 mU/mL penicillin/streptomycin. The experiments were performed on BLEC that did not exceed three passages.

2.3. Cell Viability Assay

MTT assay was performed according to Mosman with minor modifications [36]. Briefly, an MTT solution (0.5 mg/mL final concentration) was added to the cells, which were incubated at 37 °C for 1 h in the case of the ADF cells and BLEC, or 30 min in the case of HLEC. Then, for the solubilization of formazan crystals, an equal volume of a solution of isopropanol containing 0.4 N HCl was added to the cell medium. Cells were maintained in agitation for 10 min at 37 °C. The absorbance of the solutions was then read at 563 nm with a EL-808 microplate reader (BioTek Instruments, Winooski, VT, USA).

2.4. Oxidative Treatment of Cells

Before the oxidative treatment, BLECs and ADF cells were incubated for 24 h in RPMI 1640 medium containing 2% (*v/v*) FBS, 50 mU/mL penicillin/streptomycin and 2 mM glutamine, while HLEC were incubated for 24 h in MEM containing 0.5% FBS (*v/v*), 50 µg/mL gentamicin and 2 mM glutamine.

Oxidative treatment with H_2O_2 was applied to ADF cells and BLEC maintained in HBSS supplemented with 2 mM glutamine, and to HLEC maintained in MEM containing 0.5% FBS (*v/v*), 50 µg/mL gentamicin and 2mM glutamine.

Measurement of H_2O_2 concentration was performed as described [37].

2.5. Enzyme Activity Assays

The enzymatic activities described below were measured on cell lysates obtained through a freezing and thawing protocol, followed by a 10,000 x *g* centrifugation at 4 °C for 30 min, and subjected to an over-night dialysis (dialysis tubing cut off: 10 kDa) against 10 mM sodium phosphate buffer pH 7. All the enzymatic activities were determined using a Libra Biochrom (Biochrom Ltd., Cambourne Cambridge, UK) spectrophotometer.

Glutathione reductase activity was determined at 30 °C following the decrease in absorbance at 340 nm due to the oxidation of 0.2 mM NADPH (ε_{340} = 6.22 mM^{-1} cm^{-1}) in a 0.2 M potassium phosphate pH 7 buffer, in the presence of 10 mM EDTA and 0.5 mM GSSG.

Glutathione peroxidase activity was determined at 37 °C following the decrease in absorbance at 340 nm due to the oxidation of 0.2 mM NADPH in a 0.1 M Tris-HCl pH 7 buffer, in the presence of 1.3 U/mL glutathione reductase, 0.5 mM EDTA, 0.2 mM GSH and 0.125 mM H_2O_2.

Glutathione S-transferase activity was evaluated at 25 °C following the increase in absorbance at 340 nm linked to the formation of the adduct between 1-chloro-2,4-dinitrobenzene (CDNB) and GSH (ε_{340} = 9.6 $mM^{-1}cm^{-1}$). The assay mixture contained 0.1 M potassium phosphate pH 6.5 buffer, 0.6 mM CDNB and 1 mM GSH.

Catalase activity was measured at 25 °C following the decrease in absorbance at 240 nm (ε_{240} = 0.04 mM^{-1} cm^{-1}) in a reaction mixture containing 0.05 M sodium phosphate pH 7.4 buffer and 10 mM H_2O_2.

The NADPH-dependent reductase activities were measured at 37 °C as described [38], using 5 mM of GAL or 0.1 mM of either HNE or GSHNE as substrates. The dehydrogenase activities were measured at 37 °C following the increase in absorbance at 340 nm due to the reduction of $NAD(P)^+$. The reaction mixtures contained 50 mM sodium phosphate buffer pH 8.4, 0.18 mM of either NAD^+ or $NADP^+$ and one of the following substrates: 5 mM GAL, 5 mM propanal, 0.1 mM HNE or 0.1 mM GSHNE.

For all the activities mentioned, one unit of enzyme activity refers to the amount of enzyme that catalyses the conversion of 1 µmol of substrate/min in the conditions described.

2.6. Malondialdehyde Determination

Malondialdehyde (MDA) levels were determined by the thiobarbituric acid (TBA) assay according to [39]. At the end of the treatment, harvested cells were immediately analyzed. Whole lysate was obtained by ultrasonic treatment in the presence of 0.375 mM butylated hydroxytoluene and 1 mM EDTA. The TBA solution (1.2 mL), composed of 15% (*w/v*) trichloroacetic acid, 2% (*v/v*) acetic acid and 0.375% (*w/v*) TBA, was added to the sample (0.25 mL). The mixture was placed in a boiling bath for 10 min and then subjected to a 3000 x *g* centrifugation at room temperature for 2 min. The fluorescence intensity of the supernatant obtained was measured with a Jasco FP6500 spectrofluorimeter (Jasco Europe, Lecco, Italy) with excitation and emission wavelengths of 515 nm and 550 nm, respectively. MDA content was evaluated referring to a calibration curve obtained using TEP as a standard.

2.7. Other Methods

Reduced, oxidized and total intracellular glutathione levels were measured as previously described [40]. Protein concentration was determined according to Bradford [41], using a Bio-Rad (Hercules, CA, USA) protein assay kit with a calibration curve obtained using bovine serum albumin as the standard. Statistical analysis was performed using one-way ANOVA and Tukey's post hoc test carried out with Graphpad 6.0.

3. Results and Discussion

3.1. HNE and GSHNE Detoxification Pathways in ADF Cells Crude Extracts

The incubation of ADF cell crude extracts with HNE and GSHNE in the presence of pyridine cofactors revealed that reduction took place on both substrates only through NADPH-dependent reactions. On the contrary, oxidation occurred only on GSHNE, through both NAD^+ and $NADP^+$-dependent reactions. The levels of the enzyme activities involved in the redox transformations of HNE and GSHNE in ADF cells are reported in Table 1, which also gives the results obtained on crude extracts of HLEC. These cells, because of the eye function, are constantly exposed to oxidative stress conditions, as a consequence of the daily exposure to sunlight and atmospheric oxygen [42–46]. As different eye components, also epithelial lens cells are well equipped with several antioxidant systems to counteract oxidative stress arising from the environmental exposure [47]. Thus, HLEC (as for BLEC, see below) were used as control non-tumor cells for the comparison with ADF to evaluate the antioxidant defense capability. The two cell extracts displayed similar NADPH-dependent activity patterns. In contrast, differences were observed in the GSHNE oxidation patterns. In fact, the NAD^+-dependent dehydrogenase activity on GSHNE measured in ADF cells was lacking in HLEC. Moreover, the $NADP^+$-dependent oxidative reaction on GSHNE was at least ten times higher in ADF cells than in HLEC. The specificity of NADPH as a cofactor for HNE transformation suggests the involvement of aldo-keto reductases (AKRs), pre-eminently AKR1B1. In fact, there is evidence that in astrocytomas, there is a simultaneous over-expression of AKR1B1 and under-expression of AKR1B10 [48]. In addition to the activity on HNE, in the ADF cell extracts, we measured an NADPH-dependent reduction of GAL, a typical AKR1B1 substrate (specific activity 7.5 ± 1.7 mU/mg).

Table 1. Levels of pyridine cofactor-dependent oxido/reductase activities acting on 4- hydroxynonenal (HNE) and glutathionylhydroxynonanal (GSHNE) in cultured ADF cells and human lens epithelial cells (HLEC).

Reaction		Specific Activity (mU/mg)	
Substrate	**Cofactor**	**ADF**	**HLEC**
HNE	NAD^+	n.d.	n.d.
	$NADP^+$	n.d	n.d.
	NADH	n.d.	n.d.
	NADPH	3.0 ± 0.3	2 ± 0.3
GSHNE	NAD^+	2.2 ± 0.1	n.d.
	$NADP^+$	57.0 ± 2.0 ****	5.7 ± 0.3
	NADH	n.d.	n.d.
	NADPH	2.5 ± 0.5	1.5 ± 0.2

n.d. stands for "not detectable". The values are the mean ± SEM from at least three independent biological repeats. See Section 2.5 for details on assays conditions. Statistical significance with respect to HLEC: **** $p < 0.0001$.

When 10 μM sorbinil (a classical AKR1B1 inhibitor) [49] was present in the assay mixture, both this reduction and the reduction of HNE were inhibited by 82% and 87%, respectively. No other possible pathway of HNE removal was considered to affect the NADPH-dependent reduction of the aldehyde. In fact, the α,β-double bond of HNE may undergo reduction catalysed by alkenal/one oxidase (also known as leucotriene B4 synthase) [50] generating 4-hydroxynonanal. Although this molecule theoretically has features that are the same as those in an AKR1B1 substrate [51], it is reduced by purified *h*AKR1B1 in the presence of NADPH at a rate of even less than 10% of that measured for GAL reduction (data not shown). This may be because the stabilization of 4-hydroxynonanal as γ-hemiacetal reduces the availability of the open aldehyde. In addition, 4-hydroxynonanoic acid (4HNA) may also be generated and further reduced. However, no NADPH-dependent activity was measured in the ADF extract in the presence of 100 μM 4HNA. Furthermore, given the $NAD(P)^+$-dependent activities present in the crude extract, we found that 4HNA was not an oxidable substrate. This evidence suggests that the main pyridine-dependent HNE reductive pathway is linked to a direct NADPH-dependent transformation. Table 1 highlights the markedly high potential of pyridine cofactor dependent activities of cell extracts toward GSHNE. This indirectly supports the relevance of a shift in the HNE metabolism toward the glutathionylated adduct pathway. Comparing ADF and HLEC enzymatic patterns shows that this metabolic strategy is highly developed in the cancer cell line.

The specific activities related to GSHNE transformation clearly underline the high value of the $NADP^+$-dependent dehydrogenase activity. In fact, it is approximately 26 times higher than the corresponding NAD^+-dependent dehydrogenase activity on the same substrate, and 23 times higher than the NADPH-dependent reductase.

Although it may be unwise to univocally ascribe activities detected in crude extracts to specific enzymes, the $NADP^+$-dependent dehydrogenase activity acting on GSHNE is conceivably related to CBR1, with the formation of GSHNA-γ-lactone [30].

It is well known that although different aldo-keto reductases (AKRs) are able to reduce the free aldehyde [27], the only NADPH-dependent activity reported to be able to reduce its glutathionylated adduct is represented by AKRB1. We tested this association in ADF cells by evaluating the inhibitory effect on the NADPH-dependent activities acting on GSHNE exerted by 10 μM sorbinil. This effect accounted for approximately 40% of inhibition. The failure of sorbinil to completely inhibit the GSHNE reducing activities may be due to the fact that also CBR1, which is insensitive to Sorbinil [52], may catalyse the NADPH-dependent reduction of GSHNE [30].

The catalytic action of CBR1 on GSHNE generates a disproportionation of the substrate that produces GSHNA-γ-lactone and GSDHN even in sub saturating conditions of the cofactor [53]. Thus, the HNE removal through glutathionylation may be more efficient than what appears from the direct measurement of NADPH oxidation in the presence of GSHNE. An NAD^+-dependent dehydrogenase activity associated with aldehyde dehydrogenases (ALDHs), which is able to act on GSHNE has already been hypothesised [54]. However, in our tests, the NAD^+ dependent oxidation of GSHNE measured in crude extracts of ADF cells was not affected at all when disulfiram, a classical inhibitor of ALDHs, was present in the assay at a final concentration of 10 μM. This suggests that disulfiram-insensitive NAD^+-dependent dehydrogenase activities may be involved in the HNE metabolism. These may include activities acting on oxidizable centres of GSHNE different than the aldehydic group, and that may generate, besides GSHNA, 3-glutathionyl-4-oxononanal or GSHNA-γ-lactone.

3.2. Antioxidant Enzymes' Pattern and Antioxidant Detoxification Ability of Human ADF Cells

ADF cells, in addition to the above described efficiency in the removal of GSHNE through both reductive and, even more efficiently, oxidative NADP-dependent steps, also appear to be characterized by an efficient antioxidant/detoxifying enzyme pattern devoted to antagonizing peroxidative phenomena. Table 2 compares the levels of antioxidant enzymes measured in ADF, HLEC and BLEC crude extracts.

Table 2. Levels of enzymatic activities involved in antioxidant defense.

Enzyme Activity	Specific Activity (mU/mg)		
	ADF	HLEC	BLEC
Glutathione reductase	50.7 ± 1.2 (****) (####)	13.7 ± 2.3	14.0 ± 0.98
Glutathione peroxidase	1.4 ± 0.4	1.77 ± 0.16	1.6 ± 0.58
Catalase	5800 ± 238 (****), (####)	1390 ± 142	348 ± 53.1
Glutathione S-transferase	580.0 ± 54.2 (****), (####)	34.7 ± 2.3	24 ± 5.4

The values are the mean ± SEM from at least three independent biological repeats. See Section 2.5 for details on the assays. Statistical significance of ADF with respect to HLEC (*) and of ADF with respect to bovine lens epithelial cells (BLEC) (#); **** and #### $p < 0.0001$.

Except for glutathione peroxidase, whose level was similar in the extracts of the three cell lines, the other detoxifying enzymes are far more represented in the tumor cells. Thus, glutathione reductase level in ADF cells is approximately 4-fold higher than those measured in the lens cell extracts, while catalase was 4- and 16-fold higher in ADF than in HLEC and BLEC, respectively. For glutathione S-transferase, we found an approximately 17- and 24-fold difference for ADF with respect to HLEC and BLEC, respectively. Glutathione S-transferase is considered to play a key role as a catalyst of the glutathionylation of alkenals. This is a crucial step in addressing HNE to its most relevant detoxification pathway.

The susceptibility of ADF cells to oxidative stress was assessed by measuring, under a prolonged treatment with hydrogen peroxide, both the ability of H_2O_2 removal from the medium and the cell viability by MTT assay (see Section 2.3). Thus, cells in HBSS medium were supplemented with different concentrations of H_2O_2 and incubated for 30 min at 37 °C. At the end of the incubation, the residual H_2O_2 was measured, consumed H_2O_2 was replenished, and the cells incubated again for 30 min before a third replenishment/incubation step. Sixty minutes after the last H_2O_2 supplementation (2 h of overall incubation), the cell viability was evaluated by MTT assay. The time course of H_2O_2 consumption as well as the results of the viability test are reported in Figure 1.

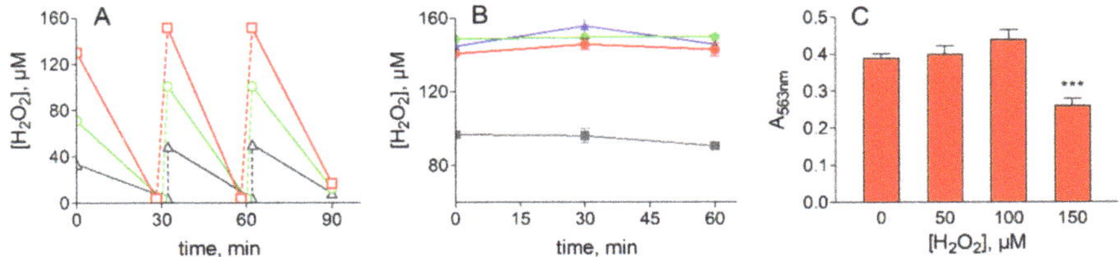

Figure 1. Treatment of ADF cells with H_2O_2. (**A**): removal of H_2O_2. ADF cells at 70% confluence (2.2×10^6 cells) were incubated in Hank's balanced salt solution (HBSS) (7 mL) at 37 °C; cells were supplemented with 50 (gray), 100 (green) or 150 (red) μM H_2O_2 at the times indicated and the concentration of the peroxide was measured as described in Section 2.4. (**B**): Stability of H_2O_2 in the free-cell medium. H_2O_2 at the concentration of 100 μM (gray squares) or 150 μM (red circles) was incubated at 37 °C in a fresh HBSS medium. Blue squares and green circles refer to the incubation of 150 μM H_2O_2 in cell free HBSS medium previously maintained for 60 min at 37 °C in the presence of ADF cells alone or supplemented with 100 μM H_2O_2, respectively. (**C**): Viability assay for ADF cells subjected to oxidative insult by H_2O_2 treatment. 1-(4,5-Dimethylthiazol-2-yl)-3,5-diphenylformazan (MTT) assay was performed at the end of the incubation in the presence of the indicated H_2O_2 concentrations under the conditions described in Panel (**A**). All the values are the mean ± SEM of at least three independent experiments. Significance was evaluated with respect to untreated cells; *** $p < 0.001$.

The figure highlights the marked ability of ADF cells to detoxify the peroxide. No significant differences were observed in the rate of H_2O_2 removal after repeated additions of the oxidant, which suggests a strong antioxidant capacity (Figure 1A). Given that H_2O_2 is stable in fresh HBSS medium or in the cell free medium withdrawn after 60 min of incubation of ADF cells (Figure 1B), oxidant removal appears to take place following cell permeation, likely exploiting the robust antioxidant enzyme pattern reported in Table 2.

The MTT assay performed on ADF cells at the end of the oxidative insult confirms the resistance of this cancer cell line to oxidative stress at least for up to 2 h of incubation with 100 μM H_2O_2 (Figure 1C).

In order to verify the intracellular impact of the oxidative insult on the activities acting on HNE and its glutathionyl adduct (Table 1), the levels of these activities were measured at the end of the most severe oxidative treatment described in Figure 1 (i.e., 150 μM H_2O_2). The results are reported in Figure 2. No significant changes were observed for the NADPH-dependent reductase activities acting on HNE and GSHNE, or for the $NAD(P)^+$-dependent dehydrogenase activities acting on GSHNE (Figure 2A, B and C). In addition, to evaluate the effect of peroxidative conditions on other reductase/dehydrogenase activities, different substrates were used (Figure 2D, E and F). No effect was observed on the NADPH-dependent reductase activity measured using GAL as substrate; this activity is likely associated with AKRs for which GAL, HNE and GSHNE are good substrates. On the other hand, there was a significant decrease in the NAD-dependent-dehydrogenase activity assayed with either GAL or propanal as substrates. The loss of these activities, which are probably linked to ALDH(s), which itself is indicative of the heterogeneity of the dehydrogenase pool present in the extract, could be easily reverted by DTT treatment.

These results show that the pattern of the pyridine cofactor dependent enzymes responsible for the HNE metabolic control is well preserved in ADF cells undergoing oxidative treatment. The efficient removal of H_2O_2 by these cells, despite the recurrent oxidative challenge, highlights the overall effectiveness of the detoxification. In addition, the possibility of GSHNE generation from HNE is also preserved, since GST activity results completely unaffected by the oxidative treatment (data not shown). The analysis of intracellular markers of oxidative stress indicate that no significant changes were observed for MDA levels, a common marker of lipid peroxidation (Figure 3A). In addition, the H_2O_2 treatment does not appear to affect the glutathione levels (Figure 3B); the tripeptide, besides being an antioxidant, is key for HNE metabolism and extrusion.

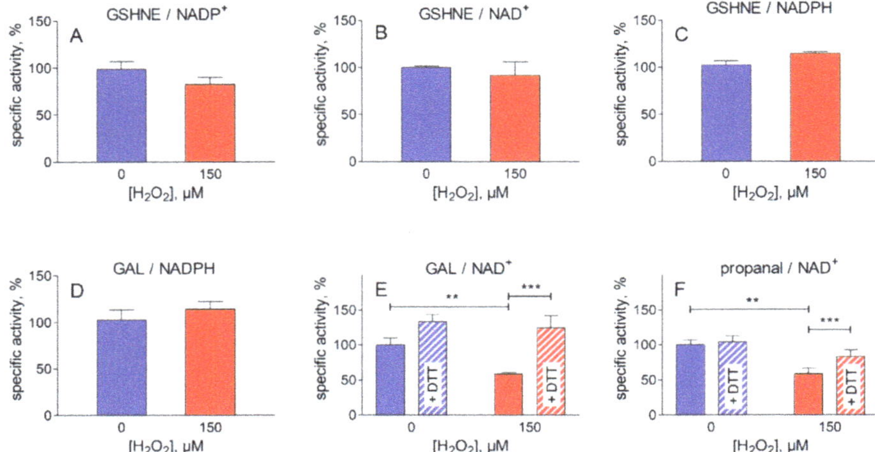

Figure 2. Effect of peroxidative treatment on oxidoreductase activities present in ADF cells. The cells were subjected to the treatment with H_2O_2 as described in Figure 1, using 150 μM of H_2O_2. The following activities were measured in ADF crude extracts as described in Section 2.5 before (blue bars) or after the oxidative treatment (red bars). $NADP^+$-dependent (**A**) and NAD^+-dependent (**B**) oxidation of glutathionylhydroxynonanal (GSHNE); NADPH-dependent reduction of GSHNE (**C**); NADPH-dependent reduction of glyceraldehyde (GAL) (**D**); NAD^+-dependent oxidation of GAL (**E**) or propanal (**F**). Dashed bars refer to the activity values measured after incubation of the indicated sample for 60 min at 37 °C in the presence of 2 mM dithiothreitol (DTT). The values are expressed as % of the specific activity measured in control cells (incubated in the absence of the peroxide) and are the mean ± SEM of at least three independent experiments. Significance was evaluated with respect to the treatment indicated. (** $p < 0.01$; *** $p < 0.001$).

In fact, although no significant changes were observed for the levels of oxidized glutathione (i.e., GSSG plus GS-protein mixed disulfides) upon oxidative treatment, total glutathione levels slightly increased, thus leaving the reduced active form of this metabolic defense tool essentially unaltered. The adaptation of the defense system to oxidative conditions by enhancing the GSH synthesis is likely to be underestimated, since the fraction of glutathione linked to targets through a non-disulfide bond was not evaluated by the assay.

On the bases of the modest antioxidant enzymes' pattern exhibited by HLEC (Table 2), these cells would be expected to be more sensitive to oxidative insult than ADF. However, the survival of these cells in the HBSS medium in the absence of peroxidative insult was too low (approximately 50% survival after 1 h of incubation) to enable a direct comparison with ADF cells. Such a comparison would have entailed using a more complete incubation medium, such as the MEM, in which HLEC are sufficiently stable (100% survival after 8 h of incubation) to be challenged with H_2O_2. In these conditions, the oxidative insult inflicted by H_2O_2 treatment, as described in ADF cells, highlights that HLEC have a significantly reduced resistance to oxidative insult, both in terms of the efficiency of H_2O_2 removal and cell viability (Figure 4A,B). In fact, a decrease in cell viability of up to 60% of the control value was observed along with an increase in the severity of the oxidative treatment.

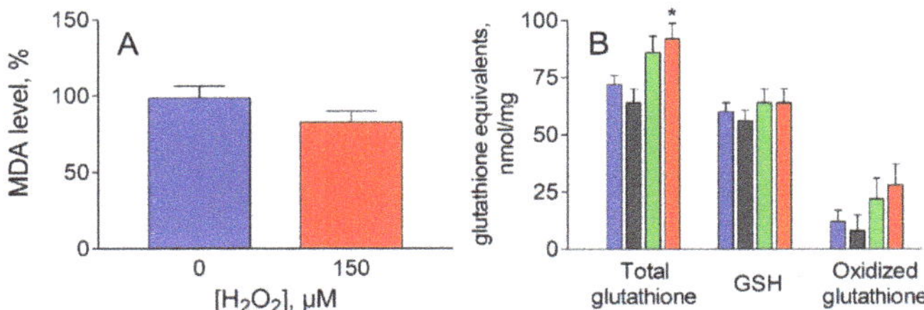

Figure 3. Oxidative stress markers in ADF cells following peroxidative insult. (**A**): Malondialdehyde (MDA) content in cells before (blue bar) and after (red bar) exposure to 150 μM H_2O_2 for 2 h at 37 °C as described in Figure 1. MDA values are expressed as % of MDA content measured in control cells ± SEM. (**B**): Total glutathione, reduced glutathione (GSH) and oxidized glutathione (i.e. glutathione disulfide plus GS-protein mixed disulfides) were measured after exposure of ADF cells, as described in Figure 1, at the following H_2O_2 concentrations: zero, blue bars; 50 μM, gray bars; 100 μM green bars; 150 μM red bars. Oxidized glutathione refers to the difference between total glutathione and GSH. The values are the mean ± SEM of at least three independent experiments. Significance was evaluated with respect to control cells (* $p < 0.05$).

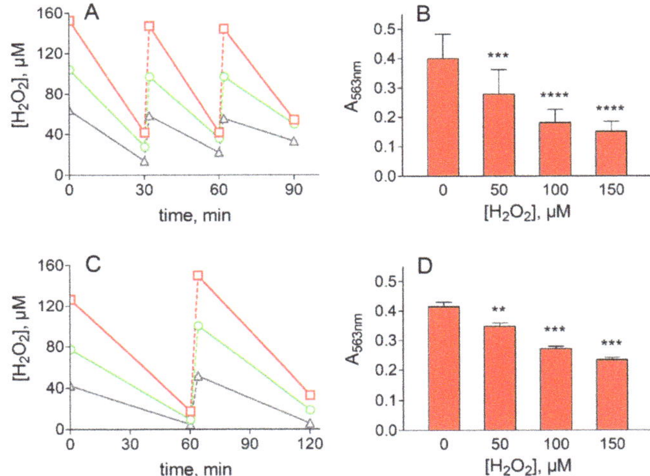

Figure 4. Treatment of lens epithelial cells with H_2O_2. (**A**) and (**B**) refer to H_2O_2 removal and viability assay in HLEC, respectively. (**C**) and (**D**) refer to H_2O_2 removal and viability assay in bovine lens epithelial cells (BLEC), respectively. Panel (**A**): HLEC at 70% confluence (2.5×10^6 cells) were incubated in Minimum Essential Medium (7 mL) at 37 °C; at the indicated times cells were supplemented with 50 (gray), 100 (green) or 150 (red) μM H_2O_2 and the concentration of the peroxide was measured as described in Section 2.4. C: BLEC at 70% confluence (1.0×10^6 cells) were incubated in HBSS (7 mL) at 37 °C; at the indicated times, cells were supplemented with 50 (gray), 100 (green) or 150 (red) μM H_2O_2 and the concentration of the peroxide was measured as described in Section 2.4. (**C**) and (**D**): Cell viability was evaluated by MTT assay (see Section 2.3) at the end of overall oxidative treatment at the indicated concentrations of H_2O_2. The values represent the mean ± SEM of at least three independent experiments. Significance was evaluated with respect to untreated cells (** $p < 0.01$, *** $p < 0.001$, **** $p < 0.0001$).

Similarly, primary cultures of BLEC, used as non-tumor cells normally highly exposed to oxidative conditions (as HLEC), were significantly less resistant to oxidative insult than ADF. BLEC, which are highly stable in HBSS medium (100% survival after 2 h of incubation in the absence of peroxidative insult) could be subjected to peroxidative stress in the same medium conditions as ADF cells by repeated exposure to H_2O_2. In this case, after several additions of H_2O_2, BLEC displayed both a lower ability to remove the peroxide from the medium (Figure 4C) and, similar to HLEC, a higher sensitivity to the oxidative treatment with respect to ADF cells. This is highlighted by a significant decrease in viability related to the severity in stress conditions from 17% to 40% of the control value (Figure 4D). As stated in Section 2.2, our experiments were performed at atmospheric oxygen pressure; oxygen pressure is known to differently affect the response to oxidative stress of cultured cells, depending not only on cell types, but also on cells being a primary line or an immortalized one [55]. The possible effect of oxygen pressure on the ADF response to oxidative stress conditions remains an aspect to be furthered in order to complete the characterization of this cell line.

4. Conclusions

We made a direct comparison of the astrocytoma cell line ADF with two non-tumor cell lines (HLEC and BLEC) that normally face in vivo radiating and chemical oxidative conditions. We thereby confirmed the resistance of tumor cells to oxidative stress conditions, which is considered a feature exhibited by several cancer cells. On the basis of the measured levels of antioxidant/detoxifying enzymes, the higher survival ability of ADF cells in peroxidative conditions shows that these cells are better equipped than normal cells to counteract oxidative stress. This aspect may be relevant in counteracting the oxidative stress induced by chemotherapy treatment, as specifically observed for ALDH1A3 in TMZ treatment [18,19].

In addition, in several cancer cells, an abnormally low $NADPH/NADP^+$ ratio has been reported [56–58]; this suggests the recovery of cell reducing power as a primary necessity for tumor cells, often linked to IDH mutations frequently observed in glioma cells. Despite the fact that ADF cells have not been characterized in terms of possible IDH mutations, the GSH/GSSG ratio measured in this cell line is far from the marked glutathione red/ox unbalance observed for other IDH1 mutated glioma cells. In any case, the markedly high level of an $NADP^+$ dependent dehydrogenase activity acting on GSHNE detected in ADF cells may help these cells to support cell detoxification, through the removal of the main HNE metabolite (i.e., the GSHNE adduct) and the recovery of cell reducing power. Whether the antioxidant pattern described for ADF is common in other astrocytoma cell lines remains a relevant aspect to be furthered in future work.

We believe that our results highlight the importance for ADF cells of NADP(H)-dependent oxidoreductases (conceivable CBR1 and AKR1B1) in antioxidant/detoxification processes. These enzymes may be useful additional inhibition targets to antagonize the survival of cancer ADF cells under stressful conditions.

Author Contributions: Conceptualization, A.D.C., R.M. and U.M.; Investigation, E.P., R.M. and V.S.; Resources, F.B.; Data Curation, M.C.; Writing—Original Draft Preparation, A.D.C and U.M.; Writing—Review & Editing, A.D.C., E.P., F.B., M.C., R.M., U.M., and V.S.; Funding Acquisition, A.D.C., M.C. and R.M. All authors have read and agreed to the published version of the manuscript.

Funding: This research was funded by Pisa University.

Acknowledgments: We are indebted to Dr. G. Pasqualetti and Dr. R. Di Sacco (veterinary staff of Consorzio Macelli S. Miniato, Pisa) for their valuable co-operation in the bovine lenses collection.

Conflicts of Interest: The authors declare no conflict of interest.

References

1. Szatrowski, T.P.; Nathan, C.F. Production of large amounts of hydrogen peroxide by human tumor cells. *Cancer Res.* **1991**, *51*, 794–798. [PubMed]
2. Kawanishi, S.; Hiraku, Y.; Pinlaor, S.; Ma, N. Oxidative and nitrative DNA damage in animals and patients with inflammatory diseases in relation to inflammation-related carcinogenesis. *Biol. Chem.* **2006**, *387*, 365–372. [CrossRef] [PubMed]
3. Hu, Y.; Rosen, D.G.; Zhou, Y.; Feng, L.; Yang, G.; Liu, J.; Huang, P. Mitochondrial manganese-superoxide dismutase expression in ovarian cancer: Role in cell proliferation and response to oxidative stress. *J. Biol. Chem.* **2005**, *280*, 39485–39492. [CrossRef] [PubMed]

4. Greenwood, H.E.; McCormick, P.N.; Gendron, T.; Glaser, M.; Pereira, R.; Maddocks, O.D.K.; Sander, K.; Zhang, T.; Koglin, N.; Lythgoe, M.F.; et al. Measurement of tumor antioxidant capacity and prediction of chemotherapy resistance in preclinical models of ovarian cancer by positron emission tomography. *Clin. Cancer Res.* **2019**, *25*, 2471–2482. [CrossRef]
5. Yang, H.; Villani, R.M.; Wang, H.; Simpson, M.J.; Roberts, M.S.; Tang, M.; Liang, X. The role of cellular reactive oxygen species in cancer chemotherapy. *J. Exp. Clin. Cancer Res.* **2018**, *37*, 266. [CrossRef]
6. Cetin, T.; Arpaci, F.; Yilmaz, M.I.; Saglam, K.; Ozturk, B.; Komurcu, S.; Gunay, M.; Ozet, A.; Akay, C.; Kilic, S.; et al. Oxidative stress in patients undergoing high-dose chemotherapy plus peripheral blood stem cell transplantation. *Biol. Trace Elem. Res* **2004**, *97*, 237–247. [CrossRef]
7. Toyokuni, S.; Okamoto, K.; Yodoi, J.; Hiai, H. Persistent oxidative stress in cancer. *FEBS Lett.* **1995**, *358*, 1–3. [CrossRef]
8. Coyle, J.T.; Puttfarcken, P. Oxidative stress, glutamate, and neurodegenerative disorders. *Science* **1993**, *262*, 689–695. [CrossRef]
9. Zajdel, A.; Wilczok, A.; Slowinski, J.; Orchelm, J.; Mazurek, U. Aldehydic lipid peroxidation products in human brain astrocytomas. *J. Neurooncol.* **2007**, *84*, 167–173. [CrossRef]
10. Yan, H.; Parsons, D.W.; Jin, G.; McLendon, R.; Rasheed, B.A.; Yuan, W.; Kos, I.; Batinic-Haberle, I.; Jones, S.; Riggins, G.J.; et al. IDH1 and IDH2 mutations in gliomas. *N. Engl. J. Med.* **2009**, *360*, 765–773. [CrossRef]
11. Tang, X.; Fu, X.; Liu, Y.; Yu, D.; Cai, S.J.; Yang, C. Blockade of glutathione metabolism in *IDH1*-mutated glioma. *Mol. Cancer Ther.* **2020**, *19*, 221–230. [CrossRef] [PubMed]
12. Liu, Y.; Lu, Y.; Celiku, O.; Li, A.; Wu, Q.; Zhou, Y.; Yang, C. Targeting IDH1-mutated malignancies with NRF2 blockade. *J. Natl. Cancer Inst.* **2019**, *111*, 1033–1041. [CrossRef]
13. Tiburcio, P.D.B.; Gillespie, D.L.; Jensen, R.L.; Huang, L.E. Extracellular glutamate and $IDH1^{R132H}$ inhibitor promote glioma growth by boosting redox potential. *J. Neurooncol.* **2020**, *146*, 427–437. [CrossRef]
14. Bryukhovetskiy, I.; Bryukhovetskiy, A.; Khotimchenko, Y.; Mischenko, P. Novel cellular and post-genomic technologies in the treatment of glioblastoma multiforme. *Oncol. Rep.* **2016**, *35*, 639–648. [CrossRef] [PubMed]
15. Stupp, R.; Mason, W.P.; van den Bent, M.J.; Weller, M.; Fisher, B.; Taphoorn, M.J.; Belanger, K.; Brandes, A.A.; Marosi, C.; Bogdahn, U. Radiotherapy plus concomitant and adjuvant temozolomide for glioblastoma. *N. Engl. J. Med.* **2005**, *352*, 987–996. [CrossRef] [PubMed]
16. Sulman, E.P.; Ismaila, N.; Chang, S.M. Radiation therapy for glioblastoma: American Society of Clinical Oncology clinical practice guideline endorsement of the American Society for Radiation Oncology guideline. *J. Oncol. Pract.* **2017**, *13*, 123–127. [CrossRef] [PubMed]
17. Nagel, Z.D.; Kitange, G.J.; Gupta, S.K.; Joughin, B.A.; Chaim, I.A.; Mazzucato, P.; Lauffenburger, D.A.; Sarkaria, J.N.; Samson, L.D. DNA Repair capacity in multiple pathways predicts chemoresistance in glioblastoma multiforme. *Cancer Res.* **2017**, *77*, 198–206. [CrossRef]
18. Wu, W.; Schecker, J.; Wurstle, S.; Schneider, F.; Schonfelder, M.; Schlegel, J. Aldehyde dehydrogenase 1A3 (ALDH1A3) is regulated by autophagy in human glioblastoma cells. *Cancer Lett.* **2018**, *417*, 112–123. [CrossRef]
19. Wu, W.; Wu, Y.; Mayer, K.; von Rosenstiel, C.; Schecker, J.; Baur, S.; Würstle, S.; Liesche-Starnecker, F.; Gempt, J.; Schlegel, J. Lipid peroxidation plays an important role in chemotherapeutic effects of temozolomide and the development of therapy resistance in human glioblastoma. *Transl. Oncol.* **2020**, *13*. [CrossRef]
20. Esterbauer, H.; Schaur, R.J.; Zollner, H. Chemistry and biochemistry of 4-hydroxynonenal, malondialdehyde and related aldehydes. *Free Radic. Biol. Med.* **1991**, *11*, 81–128. [CrossRef]
21. Poli, G.; Schaur, R.J.; Siems, W.G.; Leonarduzzi, G. 4-Hydroxynonenal: A membrane lipid oxidation product of medicinal interest. *Med. Res. Rev.* **2008**, *28*, 569–631. [CrossRef] [PubMed]
22. Pappa, A.; Estey, T.; Manzer, R.; Brown, D.; Vasiliou, V. Human aldehyde dehydrogenase 3A1 (ALDH3A1): Biochemical characterization and immunohistochemical localization in the cornea. *Biochem. J.* **2003**, *376*, 615–623. [CrossRef] [PubMed]
23. Murphy, T.C.; Amarnath, V.; Gibson, K.M.; Picklo, M.J., Sr. Oxidation of 4-hydroxy- 2-nonenal by succinic semialdehyde dehydrogenase (ALDH5A). *J. Neurochem.* **2003**, *86*, 298–305. [CrossRef] [PubMed]
24. Kong, D.; Kotraiah, V. Modulation of aldehyde dehydrogenase activity affects (±)-4-hydroxy-2E-nonenal (HNE) toxicity and HNE-protein adducts levels in PC12 cells. *J. Mol. Neurosci.* **2012**, *47*, 595–603. [CrossRef] [PubMed]

25. Vander Jagt, D.L.; Kolb, N.S.; Vander Jagt, T.J.; Chino, J.; Martinez, F.J.; Hunsaker, L.A.; Royer, R.E. Substrate specificity of human aldose reductase: Identification of 4-hydroxynonenal as an endogenous substrate. *Biochim. Biophys. Acta* **1995**, *1249*, 117–126. [CrossRef]
26. Srivastava, S.; Chandra, A.; Bhatnagar, A.; Srivastava, S.K.; Ansari, N.H. Lipid peroxidation product, 4-hydroxynonenal and its conjugate with GSH are excellent substrates of bovine lens aldose reductase. *Biochem. Biophys. Res. Commun.* **1995**, *217*, 741–746. [CrossRef]
27. Shen, Y.; Zhong, L.; Johnson, S.; Cao, D. Human aldo-keto reductases 1B1 and 1B10: A comparative study on their enzyme activity toward electrophilic carbonyl compounds. *Chem. Biol. Interact.* **2011**, *191*, 192–198. [CrossRef]
28. Hayes, J.D.; McLellan, L.I. Glutathione and glutathione-dependent enzymes represent a co-ordinately regulated defence against oxidative stress. *Free Radic. Res.* **1999**, *31*, 273–300. [CrossRef]
29. Balogh, L.M.; Atkins, W.M. Interactions of glutathione transferases with 4-hydroxynonenal. *Drug Metab. Rev.* **2011**, *43*, 165–178. [CrossRef]
30. Moschini, R.; Peroni, E.; Rotondo, R.; Renzone, G.; Melck, D.; Cappiello, M.; Srebot, M.; Napolitano, E.; Motta, A.; Scaloni, A.; et al. $NADP^+$-dependent dehydrogenase activity of carbonyl reductase on glutathionylhydroxynonanal as a new pathway for hydroxynonenal detoxification. *Free Radic. Biol. Med.* **2015**, *83*, 66–76. [CrossRef]
31. Rotondo, R.; Moschini, R.; Renzone, G.; Tuccinardi, T.; Balestri, F.; Cappiello, M.; Scaloni, A.; Mura, U.; Del-Corso, A. Human carbonyl reductase 1 as efficient catalyst for the reduction of glutathionylated aldehydes derived from lipid peroxidation. *Free Radic. Biol. Med.* **2016**, *99*, 323–332. [CrossRef]
32. Balestri, F.; Barracco, V.; Renzone, G.; Tuccinardi, T.; Pomelli, C.S.; Cappiello, M.; Lessi, M.; Rotondo, R.; Bellina, F.; Scaloni, A.; et al. Stereoselectivity of aldose reductase in the reduction of glutathionyl-hydroxynonanal adduct. *Antioxidants* **2019**, *8*. [CrossRef] [PubMed]
33. Fabrizi, C.; Colasanti, M.; Persichini, T.; Businaro, R.; Starace, G.; Lauro, G.M. Interferon gamma up-regulates alpha 2 macroglobulin expression in human astrocytoma cells. *J. Neuroimmunol.* **1994**, *53*, 31–37. [CrossRef]
34. Malorni, W.; Rainaldi, G.; Rivabene, R.; Santini, M.T. Different susceptibilities to cell death induced by t-butylhydroperoxide could depend upon cell histotype-associated growth features. *Cell Biol. Toxicol.* **1994**, *10*, 207–218. [CrossRef] [PubMed]
35. Louis, D.N.; Perry, A.; Reifenberger, G.; von Deimling, A.; Figarella-Branger, D.; Cavenee, W.K.; Ohgaki, H.; Wiestler, O.D.; Kleihues, P.; Ellison, D.W. The 2016 World Health Organization classification of tumors of the central nervous system: A summary. *Acta Neuropathol.* **2016**, *131*, 803–820. [CrossRef]
36. Mosmann, T. Rapid colorimetric assay for cellular growth and survival: Application to proliferation and cytotoxicity assays. *J. Immunol. Methods* **1983**, *65*, 55–63. [CrossRef]
37. Tinberg, C.E.; Song, W.J.; Izzo, V.; Lippard, S.J. Multiple roles of component proteins in bacterial multicomponent monooxygenases: Phenol hydroxylase and toluene/o-xylene monooxygenase from *Pseudomonas sp.* OX1. *Biochemistry* **2011**, *50*, 1788–1798. [CrossRef]
38. Balestri, F.; Rotondo, R.; Moschini, R.; Pellegrino, M.; Cappiello, M.; Barracco, V.; Misuri, L.; Sorce, C.; Andreucci, A.; Del Corso, A.; et al. Zolfino landrace (*Phaseolus vulgaris L.*) from Pratomagno: General and specific features of a functional food. *Food Nutr. Res.* **2016**, *60*, 31792. [CrossRef]
39. Yagi, K. Lipid peroxides and human disease. *Chem. Phys. Lipids* **1987**, *45*, 337–351. [CrossRef]
40. Cappiello, M.; Peroni, E.; Lepore, A.; Moschini, R.; Del Corso, A.; Balestri, F.; Mura, U. Rapid colorimetric determination of reduced and oxidized glutathione using an end point coupled enzymatic assay. *Anal. Bioanal. Chem.* **2013**, *405*, 1779–1785. [CrossRef]
41. Bradford, M.M. A rapid and sensitive method for the quantitation of microgram quantities of protein utilizing the principle of protein-dye binding. *Anal. Biochem.* **1976**, *72*, 248–254. [CrossRef]
42. Collman, G.W.; Shore, D.L.; Shy, C.M.; Checkoway, H.; Luria, A.S. Sunlight and other risk factors for cataracts: An epidemiological study. *Am. J. Public Health* **1988**, *78*, 1459–1462. [CrossRef] [PubMed]
43. Dolin, P.J.; Johnson, G.J. Solar ultraviolet radiation and ocular disease: A review of the epidemiological and experimental evidence. *Ophthalmic. Epidemiol.* **1994**, *1*, 155–164. [CrossRef] [PubMed]
44. Neale, R.E.; Purdie, J.L.; Hirst, L.W.; Green, A.C. Sun exposure as a risk factor for nuclear cataract. *Epidemiology* **2003**, *14*, 707–712. [CrossRef]
45. Truscott, R.J. Age-related nuclear cataract-oxidation is the key. *Exp. Eye Res.* **2005**, *80*, 709–725. [CrossRef]
46. West, S. Ocular ultraviolet B exposure and lens opacities: A review. *J. Epidemiol.* **1999**, *9*, S97–S101. [CrossRef]

47. Chen, Y.; Mehta, G.; Vasiliou, V. Antioxidant defenses in the ocular surface. *Ocul. Surf.* **2009**, *7*, 176–185. [CrossRef]
48. Laffin, B.; Petrash, J.M. Expression of the Aldo-Ketoreductases AKR1B1 and AKR1B10 in human cancers. *Front. Pharmacol.* **2012**, 3104. [CrossRef]
49. Huang, Q.; Liu, Q.; Ouyang, D. Sorbinil, an aldose reductase inhibitor, in fighting against diabetic complications. *Med. Chem.* **2019**, *15*, 3–7. [CrossRef]
50. Dick, R.A.; Kwak, M.K.; Sutter, T.R.; Kensler, T.W. Antioxidative function and substrate specificity of NAD(P)H-dependent alkenal/one oxidoreductase. A new role for leukotriene B4 12-hydroxydehydrogenase/15-oxoprostaglandin 13-reductase. *J. Biol. Chem.* **2001**, *276*, 40803–40810. [CrossRef]
51. Srivastava, S.; Watowich, S.J.; Petrash, J.M.; Srivastava, S.K.; Bhatnagar, A. Structural and kinetic determinants of aldehyde reduction by aldose reductase. *Biochemistry* **1999**, *38*, 42–54. [CrossRef] [PubMed]
52. Moschini, R.; Rotondo, R.; Renzone, G.; Balestri, F.; Cappiello, M.; Scaloni, A.; Mura, U.; Del-Corso, A. Kinetic features of carbonyl reductase 1 acting on glutathionylated aldehydes. *Chem. Biol. Interact.* **2017**, *276*, 127–132. [CrossRef]
53. Barracco, V.; Moschini, R.; Renzone, G.; Cappiello, M.; Balestri, F.; Scaloni, A.; Mura, U.; Del-Corso, A. Dehydrogenase/reductase activity of human carbonyl reductase 1 with NADP(H) acting as a prosthetic group. *Biochem. Biophys. Res. Commun.* **2020**, *522*, 259–263. [CrossRef] [PubMed]
54. Alary, J.; Fernandez, Y.; Debrauwer, L.; Perdu, E.; Gueraud, F. Identification of intermediate pathways of 4-hydroxynonenal metabolism in the rat. *Chem. Res. Toxicol.* **2003**, *16*, 320–327. [CrossRef] [PubMed]
55. Jagannathan, L.; Cuddapah, S.; Costa, M. Oxidative stress under ambient and physiological oxygen tension in tissue culture. *Curr. Pharmacol. Rep.* **2016**, *2*, 64–72. [CrossRef] [PubMed]
56. Ahmad, I.M.; Aykin-Burns, N.; Sim, J.E.; Walsh, S.A.; Higashikubo, R.; Buettner, G.R.; Venkataraman, S.; Mackey, M.A.; Flanagan, S.W.; Oberley, L.W.; et al. Mitochondrial O_2^- and H_2O_2 mediate glucose deprivation-induced stress in human cancer cells. *J. Biol. Chem.* **2005**, *280*, 4254–4263. [CrossRef] [PubMed]
57. Aykin-Burns, N.; Ahmad, I.M.; Zhu, Y.; Oberley, L.W.; Spitz, D.R. Increased levels of superoxide and H_2O_2 mediate the differential susceptibility of cancer cells versus normal cells to glucose deprivation. *Biochem. J.* **2009**, *418*, 29–37. [CrossRef]
58. Cairns, R.A.; Harris, I.S.; Mak, T.W. Regulation of cancer cell metabolism. *Nat. Rev. Cancer* **2011**, *11*, 85–95. [CrossRef]

© 2020 by the authors. Licensee MDPI, Basel, Switzerland. This article is an open access article distributed under the terms and conditions of the Creative Commons Attribution (CC BY) license (http://creativecommons.org/licenses/by/4.0/).

Review

Role of Phytochemicals in Perturbation of Redox Homeostasis in Cancer

Shreyas Gaikwad and Sanjay K. Srivastava *

Department of Immunotherapeutics and Biotechnology, Center for Tumor Immunology and Targeted Cancer Therapy, Texas Tech University Health Sciences Center, Abilene, TX 79601, USA; shreyas.gaikwad@ttuhsc.edu
* Correspondence: sanjay.srivastava@ttuhsc.edu; Tel.: +1-325-696-0464; Fax: +1-325-696-3875

Abstract: Over the past few decades, research on reactive oxygen species (ROS) has revealed their critical role in the initiation and progression of cancer by virtue of various transcription factors. At certain threshold values, ROS act as signaling molecules leading to activation of oncogenic pathways. However, if perturbated beyond the threshold values, ROS act in an anti-tumor manner leading to cellular death. ROS mediate cellular death through various programmed cell death (PCD) approaches such as apoptosis, autophagy, ferroptosis, etc. Thus, external stimulation of ROS beyond a threshold is considered a promising therapeutic strategy. Phytochemicals have been widely regarded as favorable therapeutic options in many diseased conditions. Over the past few decades, mechanistic studies on phytochemicals have revealed their effect on ROS homeostasis in cancer. Considering their favorable side effect profile, phytochemicals remain attractive treatment options in cancer. Herein, we review some of the most recent studies performed using phytochemicals and, we further delve into the mechanism of action enacted by individual phytochemicals for PCD in cancer.

Keywords: oxidative stress; cancer; reactive oxygen species (ROS); phytochemicals; dietary chemicals; natural compounds; programmed cell death; apoptosis; anoikis; autophagy; ferroptosis; pyroptosis

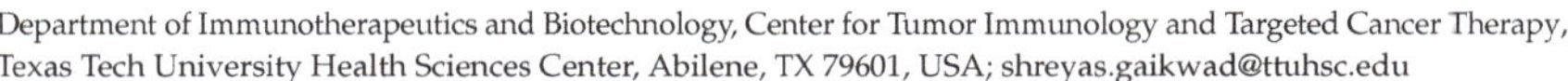

Citation: Gaikwad, S.; Srivastava, S.K. Role of Phytochemicals in Perturbation of Redox Homeostasis in Cancer. *Antioxidants* **2021**, *10*, 83. https://doi.org/10.3390/antiox10010083

Received: 25 November 2020
Accepted: 5 January 2021
Published: 9 January 2021

Publisher's Note: MDPI stays neutral with regard to jurisdictional claims in published maps and institutional affiliations.

Copyright: © 2021 by the authors. Licensee MDPI, Basel, Switzerland. This article is an open access article distributed under the terms and conditions of the Creative Commons Attribution (CC BY) license (https://creativecommons.org/licenses/by/4.0/).

1. Introduction

1.1. ROS and Its Physiological Role

Reactive oxygen species (ROS) is a blanket term which encompasses several reactive species derived from molecular oxygen [1]. ROS are short-lived highly reactive byproducts of aerobic metabolism. They are formed due to single-electron reduction of oxygen (O_2) which forms radical superoxide ($O_2^{\bullet-}$). The superoxide molecules are then converted to hydrogen peroxide (H_2O_2) in the presence of superoxide dismutase [2]. Reactions involving H_2O_2 and O_2^- result in further generation of other ROS molecules such as hydroxyl radicals (OH), singlet oxygen ($^1O_2$), and peroxynitrite ($ONOO^-$) [3]. Mitochondria is the biggest source of ROS where 80% of the ROS are generated. Once generated, ROS oxidize several cellular molecules and hence play diverse roles in various physiological processes of the cells [4]. ROS react with carbohydrates, lipids, proteins, and nucleic acids, resulting in functional alterations of pathways associated with cellular differentiation, proliferation, cytoskeletal regulation, and apoptosis [3,5]. Several protein structures such as ion channels, kinases, phosphatases are regulated by redox signaling [5,6]. In signal transduction, ROS transiently oxidizes the cysteine sulfhydryl group which is involved in the activation of most phosphatases [7]. ROS have been considered as a harmful by-product of aerobic metabolism, however, under balanced conditions, ROS play an important role as mediators of signaling pathways in various physiological systems [8]. The biological activity of ROS depends on the extent, location, and duration of action of the specific ROS molecule [9]. For example, in embryonic development, optimum levels of H_2O_2 are considered to be crucial for neuronal growth. A level of 1–10 nM H_2O_2 is required for neuronal growth as levels below 1 nM have shown spatial memory defects in mice studies, while levels above 10 nM result in axonal degeneration [10,11] In the cardiovascular system, mitochondrial ROS (mROS) are suggested to be involved in the dilation of human coronary resistance arteries in response to shear stress [12]. Additionally, in terms of optimal vascular angiogenesis, NADPH oxidase (NOX) is activated by vascular endothelial growth factor (VEGF) to produce ROS. The produced ROS and VEGF together then induce endothelial cell proliferation and migration [13]. For optimal blood vessel growth, a specific range of ROS called the "redox window" is required [14]. ROS have been observed to be involved in cognitive functioning wherein it was determined that mROS potentiates post-synaptic inhibitory signaling in cerebellar stellate cells via selective recruitment of α3-containing GABAA receptors [15]. Host-defense against pathogens was the first function of ROS to be discovered. Hypochlorous acid (HOCl) produced from chlorine and H_2O_2 generates chloramines which have cytotoxic effects [16]. In terms of host defense mechanism, after phagocytosis of the pathogen, NADPH oxidase (NOX) produces a high amount of ROS which in turn destructs the pathogen [17]. In the immune system, ROS coordinate the migration of polymorphonuclear leukocytes (PMNs) towards the pathogens, and then ROS also retain the PMNs at the site of infection [18]. T cell activation and proliferation are mediated through a signaling cascade mainly involving the MAPK/ERK pathway [19]. Protein tyrosine phosphatases (PTPs) negatively regulate the MAPK/ERK pathway and PTPs have a cysteine (Cys) residue in their active site which is regulated by ROS. ROS act as inhibitor of PTP by converting Cys residue in the active site to sulfenic acid (Cys-SOH), thus leading to activation of the MAPK/ERK pathway [20]. To summarize, ROS serve as important mediators in several physiological systems and the level of ROS molecules play a key role in its action.

1.2. Oxidative Stress in Cancer

A constant balance of ROS levels is required for appropriate homeostasis. Under normal conditions, our body's antioxidant defense system continuously counteracts the oxidative attack from ROS molecules [21]. However, excessive accumulation of ROS within the cell disrupts the balance, leading to the generation of oxidative stress [22,23]. The generated oxidative stress leads to irreversible damage of DNA, peroxidation of lipids, oxidation of proteins, and inhibition of enzymes [1]. A high level of ROS present in oxidative stress conditions leads to the activation of various oncogenic pathways [6]. The phosphoinositide 3-kinases (PI3K) pathway is highly activated in a number of cancers. ROS primarily inactivate phosphatase and tensin homolog (PTEN) in the PI3K pathway by oxidizing Cys124, an active site on PTEN, leading to a disulfide bond with another intraprotein Cys71. Consequently, the inactivation of PTEN leads to hyper-activation of the PI3K pathway [24]. Constant phosphorylation of PI3K subsequently results in the activation of protein kinase B (AKT) via downstream pyruvate dehydrogenase kinase 1 (PDK1), ultimately leading to the upregulation of cell cycle stimulating genes such as proliferating cell nuclear antigen (PCNA) and cyclin-dependent kinase 1 (CDK1) [25]. In many Kirsten rat sarcoma viral oncogene homolog (KRAS) mutated oncogenic cancers, ROS play an important role in anchorage-independent growth through regulation of the ERK MAPK signaling pathway [26]. In the initial stage of tumor formation, angiogenesis is absent or poorly developed, resulting in hypoxic conditions. The hypoxic condition affects the mitochondrial electron transport chain (ETC), leading to increased ROS levels. Hypoxia-driven ROS play a role in the activation of hypoxia inducing factor-1 (HIF-1), a transcription factor that contributes towards glucose to lactate conversion for tumor glucose metabolism and the induction of VEGF. ROS activate the HIF-1 subunit by inactivating a HIF-1 inhibitor, PHD (prolyl dydroxylase domain) [27]. Increased ROS levels have been linked to tumor proliferation via the c-Myc pathway. At increased ROS levels, HIF-1 α dependent activation of c-Myc occurs, resulting in tumor proliferation and DNA damage [28]. ROS also play a role in tumor stromal environment and activate invasive tumor progression. Increased ROS levels in stromal fibroblasts by transforming growth factor beta-1 (TGFß1) initiated the mesenchymal–mesenchymal transition (MMT) and secretion of vascular endothelial growth factor (VEGF), hepatocyte growth factor (HGF), and interleukin-6 which are biomarkers for invasive tumor cells [29]. ROS affect the matrix metalloproteinase (MMP)/tissue inhibitor of metalloproteinase (TIMPs) ratio by activating MMP synthesis. ROS activate MMP synthesis either via Ras and MAPK signaling cascades or through NFκB pathway [30]. Cancer cells survive on levels of ROS that are slightly greater than normal cell counterparts and activate various oncogenic pathways. The cells adapt themselves to the moderate redox environment and proliferate. However, if ROS levels rise beyond a certain threshold, the transformed cancer cells are longer able to adapt and turn to various cell death pathways [31].

2. ROS and Associated Programmed Cell Death (PCD) Pathways

2.1. Apoptosis

Among all the PCD pathways, apoptosis is the most studied pathway which is mediated via a group of proteins called cysteine-dependent aspartate-directed proteases known as caspases [32]. ROS play a role in both extrinsic and intrinsic apoptosis pathways. The intrinsic pathway is mediated through the mitochondrial permeability transition pore (mPTP). ROS cause cytoplasmic release of cytochrome c from the mitochondria via its action on PTP. Specifically, ROS govern the regulation of three proteins adenine nucleotide translocase (ANT), cyclophilin D, and voltage-dependent anion-selective channel (VDAC) involved in the opening of PTP. ROS oxidizes specific cysteines in their active sites [33,34]. The intrinsic pathway is a more common apoptosis mode mediated by ROS since mitochondria is very sensitive to the increased ROS levels as it lacks DNA repair enzymes. In hyperactivated oxidative stress, the shutdown of mitochondrial function contributes to apoptosis as cellular energy supply stops [35,36]. In the extrinsic pathway also known as the death receptor pathway, TNF receptors are involved which are located at the plasma membrane and contain an intracellular death domain. These receptors recruit pro-caspases and adaptor proteins, which results in the formation of death-inducing signaling complex (DISC) and activates the caspases [37]. In this process, the cellular FLICE-inhibitory protein (c-FLIP) impedes the formation of DISC [38]. ROS mediate the downregulation of FLIP by ubiquitination and subsequent degradation by the proteasome. Primarily, H_2O_2 and $O_2^{\bullet-}$ are responsible for FLIP down-regulation [39].

2.2. Autophagy

Autophagy is a catabolic process required for the regeneration of damaged organelles and cellular components under various stressful environments such as pathogen infection, growth factor deprivation, starvation, and intra-cellular stress [40]. Autophagy involves the generation of a double-membrane vesicle, called autophagosome, which engulfs the damaged components and merges with lysosome for hydrolytic degradation of the engulfed cargo [41,42]. A definite interplay exists between ROS and autophagy since both are induced by common oncogenic stimuli [43]. Under normal conditions, autophagy mediates cell survival by eliminating ROS and protecting the mitochondria. However, when stimulated, ROS induce excess autophagy resulting in cell death [44]. ROS regulates a negative regulator of autophagy, the mammalian target of rapamycin complex 1 (mTORC1). Inhibition of mTORC1 is mediated through cytoplasmic ataxia telangiectasia mutated (ATM). ATM activates the tuberous sclerosis complex 2 (TSC2) tumor suppressor which inhibits mTOR via stimulation of GTP hydrolysis of Ras family small GTPase Rheb (Ras homolog enriched in brain) [45,46]. Additionally, oxidative stress activates FOXO3 which stimulates the transcription of Bcl-2 nineteen kilodalton interacting protein (BNIP3) [47]. BNIP3 in turn competes with Beclin-1 for binding to Bcl-2. The unbound Beclin1 induce autophagic cell death [48]. However, autophagy acts in a context dependent manner and might play a pro-tumor role in certain conditions [49].

2.3. Ferroptosis

Ferroptosis, a novel form of programmed cell death, is characterized by the accumulation of iron and lipid peroxidation, leading to intracellular changes such as changes in mitochondrial structure and functioning [50]. The mechanism involved in ferroptosis is the suppression of system Xc, an amino acid antitransporter. System Xc is a part of an important antioxidant system in cells that provokes the synthesis of glutathione (GSH). GSH decreases ROS levels under the action of glutathione peroxidases (GPXs), thereby maintain a balanced oxidant-antioxidant environment. During ferroptosis, inhibiting system Xc causes decreased synthesis of GSH, leading to the accumulation of lipid ROS and ultimately, oxidative cell death [51]. Among GPXs, GPX4 is a pivotal player that counterbalances lipid peroxides and protects membrane integrity by using GSH as a cofactor [52]. In terms of cancer therapy, GPX4 is targeted as it is a specific and central regulator of ferroptotic cell death [53].

2.4. *Pyroptosis*

Pyroptosis is a caspase-dependent (caspase 1/4/5/11) mode of cell death, which is initiated by the formation of the inflammasome. The mechanism involves caspase-mediated cleavage of gasdermin D (GSDMD) and gasdermin E (GSDME). Following the cleavage, the N-terminal fragment of GSDMD is released and it forms pores in the cell membrane, leading to activation of inactive cytokines like IL-18 and IL-1β, water influx, cell swelling, and osmotic lysis [54]. ROS induce pyroptosis through the activation of NLR family pyrin domain containing 3 (NLRP3) inflammasome [55].

2.5. *Anoikis*

Anoikis is another apoptotic PCD wherein cells undergo death in the absence or improper attachment to the extracellular matrix (ECM). Anoikis is a Greek term for "homelessness" [56–58]. In cancer cells, the sonic hedgehog/Gli1 pathway is a well-known regulator of anoikis [59,60]. Additionally, ROS has been implicated to support anoikis resistance, however, depending upon the activation, ROS trigger anoikis-inducing cell death [61].

3. Phytochemicals: A Promising Role in ROS Mediated Cancer Cell Death

Considering the plethora of evidence generated from various studies, it is clear that beyond a threshold value, ROS play an anti-tumorigenic role. Therapeutic interventions to achieve the required threshold ROS level is a promising strategy for inducing PCD. Phytochemicals are naturally occurring compounds derived from plants and have been studied for their therapeutic effects in various physiological conditions. Multiple phytochemicals have proven their inheritance of cancer prevention and therapeutic properties [62,63]. The low side-effect profile of phytochemicals has made them a prominent arsenal against cancer. These naturally derived compounds are a valuable resource for cancer treatment and we hereupon discuss some of the phytochemicals studied in ROS-mediated anti-cancer activity.

3.1. *Phytochemicals Acting via ROS-Mediated Apoptosis*

3.1.1. Capsaicin

Capsaicin (*N*-vanillyl-8-methylnonenamide) is a homovanillic acid derivative obtained from hot chili pepper [64]. Capsaicin has been widely studied for its anti-cancer activity [65,66]. Its activity is proposed to be mediated through oxidative stress-induced apoptosis. In pancreatic cancer cell lines BXPC-3 and AsPC-1, 4 to 6-fold ROS generation was observed within 1 h of capsaicin treatment while ROS generation was absent in normal pancreatic cells. Capsaicin induced oxidation of cardiolipin, a component of mitochondria. Oxidation of cardiolipin subsequently led to apoptosis. ETC complex-I and complex-III were observed to be the ROS-producing entities following capsaicin treatment [67,68].

3.1.2. Sulforaphane

Sulforaphane (SFN; 1-isothiocyanato-4-(methyl-sulfinyl)-butane) is an isothiocyanate present in cruciferous vegetables like cabbage, broccoli, cauliflower, and kale [69]. Sulforaphane induced ROS mediated apoptosis in PC-3 and DU145 human prostate cancer cells. Specifically, caspase-8 activation and Fas protein induction following sulforaphane treatment confirmed apoptotic activity. ROS generation was suggested to be caused by both nonmitochondrial and mitochondrial mechanisms. GSH depletion was proposed to be the non-mitochondrial mechanism. ROS were postulated to be upstream activators of apoptosis based on caspase-8 attenuation following treatment with EUK-134 (SOD and catalase mimetic) [70].

3.1.3. α-Lipoic Acid

α-lipoic acid (1,2-Dithiolan-3-yl pentanoic acid) is a naturally occurring dithiol compound found in fruits and vegetables [71]. In hepatocellular carcinoma cell line HepG2, it induced mitochondrial apoptosis preceded by ROS generation. Precisely, α-lipoic acid-induced ER (endoplasmic reticulum) stress wherein increased levels of C/EBP homologous protein (CHOP) were detected. ER stress led to activation of PERK (phospho- extracellular signal-related kinase) pathway and IRE1 (inositol-requiring enzyme 1) pathway leading to apoptosis. α-lipoic acid also inhibited ATF6 (activating transcription factor 6)-mediated pro-survival pathway [72].

3.1.4. Benzyl Isothiocyanate (BITC)

BITC is an isothiocyanate present in cruciferous vegetables such as cauliflower, mustard, watercress, cabbage, and horseradish [73]. BITC has been studied for its anti-cancer effects in various studies previously [74–78]. BITC was studied for its anti-cancer potential in pancreatic cancer cell lines Capan-2 and MIAPaCa-2 cells. BITC treatment activated MAPK signaling with activation of ERK and JNK observed within as early as 1 h of treatment while ROS generation was observed within 30 min of treatment. Additionally, activation of MAPK family members was absent in normal HPDE-6 cell line. Thus, BITC treatment led to ROS generation followed by activation of MAPK family members ERK, JNK, and P38 and subsequently apoptosis [79].

3.1.5. Phenethyl Isothiocyanate (PEITC)

PEITC, another isothiocyanate present in cruciferous vegetables has been studied for its anti-cancer effects and ROS-generating properties [80–85]. In a study to investigate effects of PEITC on breast cancer cells, PEITC showed ROS-mediated apoptosis in MDA-MB-231 and MCF-7 cell lines. High levels of hydrogen peroxide were observed in the cancer cells following treatment with PEITC. An antioxidant pre-treatment blocked cellular death and percent apoptosis induced by PEITC. Thus, confirming high levels of ROS led to apoptotic cell death [86].

3.1.6. Piperine

Piperine [1-(5-[1,3-benzodioxol-5-yl]-1-oxo-2,4-pentadienyl) piperidine] is an alkaloid extracted from black pepper (*P. nigrum*) and long pepper (*P. longum*) [87]. Various small molecules derived from the *Piperaceae* family have been implicated in ROS-mediated apoptotic cell death of cancer cells [88,89]. Piperine has been implicated for ROS generation in SKMEL-28 and B16-F0 melanoma cell lines. ROS generation was observed within 30 min of treatment and the levels sustained until 24 h. Phosphorylation of Chk1 (checkpoint kinase 1) was observed in response to DNA damage which consequently led to G1 cell cycle arrest and apoptosis. Significant expression of phosphorylated ataxia telangiectasia and Rad3 related protein (ATR) at Ser 428 and Chk1 at Ser 296 was observed, suggesting DNA damage [90].

3.1.7. Curcumin

Curcumin (1,7-bis(4-hydroxy-3-methoxyphenyl)-1,6-heptadiene-3,5-dione) is a polyphenol derived from rhizome of Curcuma longa. It possesses pleiotropic therapeutic benefits including anti-cancer effects [91,92]. The apoptotic role of curcumin was studied in colorectal cancer and sequential dosing was considered important for ROS-mediated apoptotic induction. In the study, the sequential treatment caused lysosomal permeabilization and ROS generation, accompanied by autophagic dysregulation and ER stress. Lysosomal permeabilization resulted in BID-dependent mitochondrial membrane permeabilization and thereby caspase-dependent apoptosis. Cathepsin B was confirmed to be involved in apoptosis following sequential treatment [93]. Figure 1 depicts the ROS-mediated apoptotic action of phytochemicals.

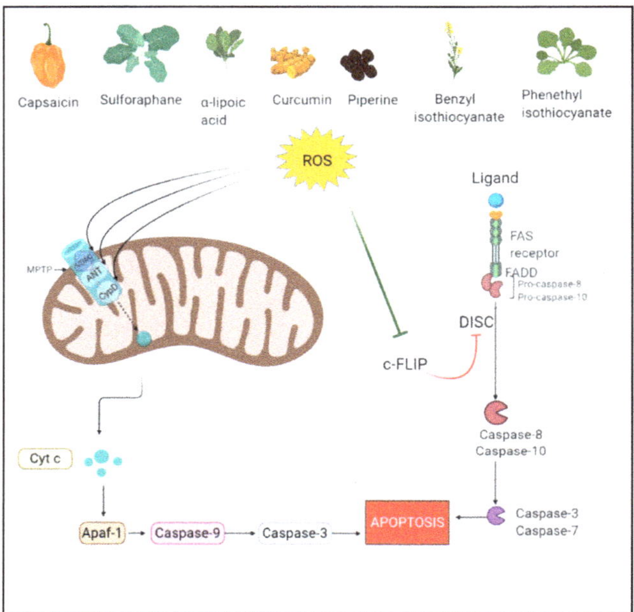

Figure 1. Reactive Oxygen Species (ROS)-mediated apoptotic action of phytochemicals. Figure created with BioRender.com.

3.2. *Phytochemicals Acting via ROS-Mediated Ferroptosis*

3.2.1. Withaferin A

Withaferin A (WA) is a steroidal lactone isolated from the plant *Withania somnifera*. It has been widely studied for anti-inflammatory and anti-tumorigenic properties [94]. Withaferin A induced ferroptotic cell death in neuroblastoma cells. WA targeted Keap-1 (kelch-like ECH-associated protein (1), a negative regulator of Nrf2 (nuclear factor erythroid 2-related factor (2). Keap-1 inhibition caused the release of Nrf2. The released Nrf2 upregulated HMOX1 [heme oxygenase (decycling) 1] which catalyzes heme degradation to generate ferrous iron (Fe^{2+}). A high concentration of Fe^{2+} led to ROS generation and ultimately cellular death [95].

3.2.2. Bromelain

Bromelain is a mixture of proteolytic enzymes derived from the stem of pineapple plants belonging to the *Bromeliaceae* family [96]. Bromelain was shown to induce ferroptotic cell death in KRAS-mutant colorectal cancer wherein bromelain increased the level of long-chain acyl-CoA synthetase-4 (ACSL4) an isozyme which plays role in lipid biosynthesis and fatty acid degradation. Specifically, ACSL4 accumulates oxidized cellular membrane phospholipids. ACSL4 mainly acts on phosphatidylethanolamine levels which act as an executioner of ferroptotic cell death [81,97].

3.2.3. Ruscogenin

Ruscogenin [(1β,3β,25*R*)-Spirost-5-ene-1,3-diol] is a steroidal sapogenin extracted from *Ruscus aculeatus* plant [98]. Ruscogenin was studied for ferroptotic effects in pancreatic cancer cell lines BxPC-3, SW1990, PANC-1, and AsPC-1. Ruscogenin treatment increased Fe^{2+} ion concentration by upregulating the expression of transferrin and downregulating the expression of ferroportin. To further confirm the mode of action, cell death by ruscogenin reduced following treatment with iron chelator and cell death increased when ferric ammonium citrate was added. To conclude, ruscogenin regulated iron transport resulted in overproduction of ROS, leading to cell death [99].

3.2.4. Oridonin

Oridonin (Ori) is a tetracyclic diterpenoid obtained from *Isodon Rubescens* [100]. The ferroptotic effect of Ori was studied in esophageal cancer cell line TE1 and inhibition of the gamma-glutamyl cycle was observed in oridonin treated cells. Precisely, Ori affected GGT1, a key enzyme in the gamma-glutamyl cycle, which is crucial for protecting cells from oxidative stress and maintaining cysteine homeostasis. Additionally, Ori inhibited GSH by a covalently binding with cysteine after entering the TE1 cells. The covalent binding reduces intracellular cysteine levels which is responsible for ferroptosis. A decrease in GPX4 level was also detected after Ori treatment [101,102]. Figure 2 depicts the ROS-mediated ferroptotic action of phytochemicals.

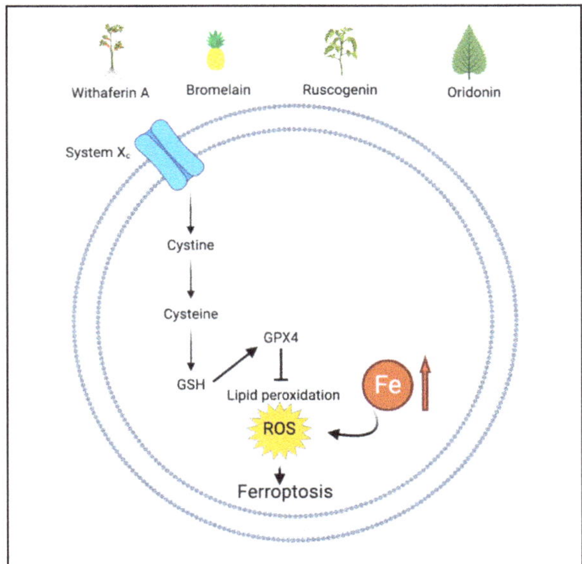

Figure 2. ROS-mediated ferroptotic action of phytochemicals. Figure created with BioRender.com.

3.3. *Phytochemicals Acting via ROS-Mediated Autophagy*

3.3.1. Cucurbitacin B

Cucurbitacin B (1,2-Dihydro-alpha-elaterin) is a tetracyclic triterpene compound found in the plants of *Cucurbitaceae* and other also a variety of other plant families. Some of the sources of cucurbitacin B are *Helicteres angustifolia, Licaniaintra petiolaris, Casearia arborea,* and *Cucumis prophetarum* [103]. Cucurbitacin B has been studied for its anticancer effects in human cervical cancer HeLa cells. Cucurbitacin B induced autophagic cell death and was observed to have mediated through mitochondrial ROS production. Treatment of HeLa cells with cucurbitacin-B resulted in accumulation of several lamellar structures having cytosolic autophagic vacuoles. Additionally, the conversion of non-autophagic soluble LC3 (LC3-I) to autophagic LC3 (LC3-II) was detected following treatment with cucurbitacin-B. Pretreatment with antioxidant NAC and Mito-TEMPO (mitochondria-targeting antioxidant) inhibited LC3-II conversion, cell death, and autophagosome formation. Thus, the results indicated that ROS plays an important role in promoting autophagic cell death following cucurbitacin treatment [104]. Another study confirmed cucurbitacin-B induced autophagic cell death via ROS in MCF-7 breast cancer cells. The role of ROS was confirmed by the use of antioxidant pretreatment which reduced protein expression of LC3 II [105].

3.3.2. Silibinin

Silibinin is a flavonoid derived from the milk thistle plant (*Silybum marianum*). The autophagic activity of silibinin was studied in human fibro-sarcoma HT1080 cells wherein silibinin was shown to activate p53 via the ROS-p38 pathway. Activation of p53 partially mediated autophagic cell death by inhibiting the MEK/ERK1/2 and PI3K/Akt pathway. An increased expression of Beclin1 (autophagosome marker) and conversion of LC3 I to LC3 II was observed following silibinin treatment [106]. Another study indicated that silibinin induces autophagic cell death mediated through mitochondrial dysfunction [107].

3.3.3. Allicin

Allicin (2-propene-1-sulfinothioic acid *S*-2-propenyl ester) is an organosulfur compound present in garlic initially studied for its anti-bacterial studies [108]. Allicin was shown to induce both autophagy and apoptosis via increased production of ROS in non-small cell lung cancer cell lines A549 (adenocarcinoma) and NCI-H460 (large cell carcinoma). The authors observed a dose-dependent autophagic action of allicin. At low doses, allicin induced autophagy associated with moderate ROS levels, while increased doses resulted in higher ROS production and lysosomal disruption leading to cell death. The lysosomal disruption resulted in intra-cellular hydrolytic enzyme release [109]. Similar autophagic action of allicin was studied in hepatocellular carcinoma cells. Interestingly, allicin caused autophagy induction in p53 normal human liver cancer cells while apoptosis in p53 deficient cell line showing the dependence of its autophagic action on p53 [110].

3.3.4. Carnosol

Carnosol is an ortho-diphenolic diterpene found in Mediterranean herbs rosemary, sage, and oregano [111]. Although it has been studied for its antioxidant activity, some evidence also shows the cytotoxic effects of carnosol via ROS overproduction. ROS production by carnosol is a dose and time-dependent process [112]. Carnosol study in MDA-MB 231 breast cancer cells showed ROS accumulation within 1 h of treatment while LC3II accumulation at 3 h post-treatment, suggesting ROS to be an upstream inducer of autophagy. It was observed that at lower concentration (<25 μM), carnosol-induced ROS-mediated autophagy only, while at higher concentration (>50 μM), ROS induced both autophagy as well as apoptosis. Interestingly, autophagy induction by carnosol was beclin-1 independent [113].

3.3.5. Quercetin

Quercetin is a flavonoid found in fruits such as grapes, red raspberry, apples, and vegetables like broccoli, shallots, tomatoes, onion, and tea [114]. Quercetin has been widely studied for its anti-cancer effects and exerts its effects through multiple pathways. A study of quercetin in osteosarcoma cells revealed it activates autophagic flux via its action on nuclear protein 1 (NUPR1) which is involved in autophagosome formation, cargo degradation, and autophagosome-lysosome fusion. It was observed that quercetin-induced overproduction of ROS led to increased expression and activation of NUPR1. Although a majority of the studies show an inverse relation between NUPR1 and ROS, a direct relation is hypothesized [115,116].

3.3.6. Berberine

Berberine (5,6-dihydro-9,10-dimethoxybenzo[g]-1,3-benzodioxolo[5,6-a] quinolizinium) is a quaternary benzylisoquinoline alkaloid derived from stem bark and roots of many plants and mainly the plants belonging to genus *Berberis* [117]. Recently, berberine was investigated for its anticancer role in the renal cell carcinoma model wherein it was combined with photodynamic therapy (PDT). The fluorescent nature of berberine was a key characteristic of the combinational effect. Berberine was used as a photosensitizer agent because the complementary action of a photosensitizer agent with a laser induces ROS generation, leading to cellular death. Berberine along with PDT showed significantly increased autophagy in comparison to berberine alone, suggesting photoactivation of berberine is required for ROS-mediated autophagy and ultimate cell death [118]. Figure 3 depicts the ROS-mediated autophagic action of phytochemicals.

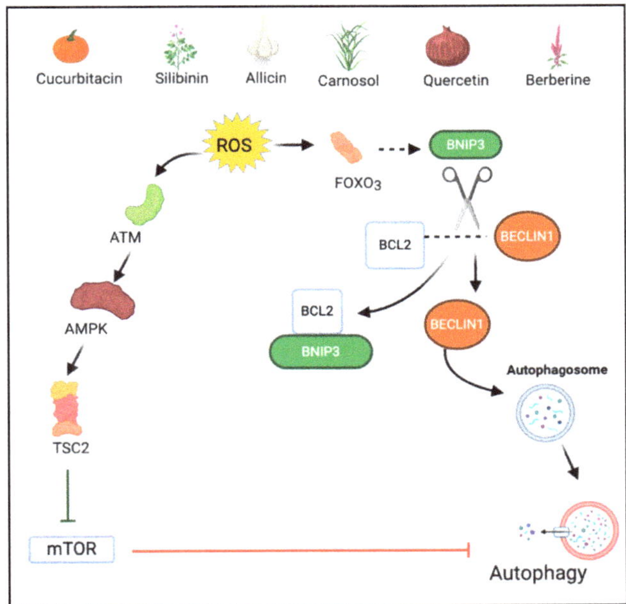

Figure 3. ROS-mediated autophagic action of phytochemicals. Figure created with BioRender.com.

3.4. *Phytochemicals Acting via ROS-Mediated Pyroptosis*

Nobiletin

Nobiletin (5,6,7,8,3′,4′-hexamethoxyflavone) is a polymethoxyflavonoid extracted from citrus fruits. Nobiletin has various pharmacological activities such as antimetabolic disorder, anticancer, neuroprotection, anti-inflammation, antioxidation, and cardiovascular protection [119]. Nobiletin was studied for its anti-cancer activity in human ovarian carcinoma cell lines A2780 and OVCAR3. In the study, nobiletin treatment resulted in ROS generation, exhibited by a significant decrease in mitochondrial membrane potential in a dose-dependent manner. Further, ROS contributed to the cleavage of GSDMD and GSDME, the two biomarkers of pyroptosis among which GSDMD is considered to be an executioner of pyroptosis [120,121]. Figure 4 depicts the ROS-mediated pyroptotic action of phytochemicals.

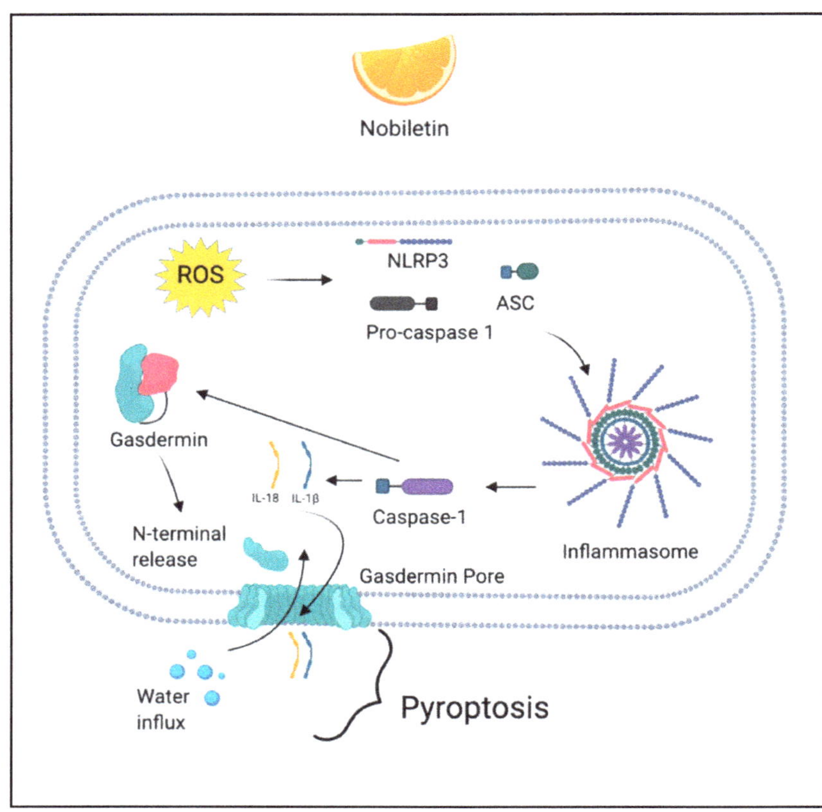

Figure 4. ROS-mediated pyroptotic action of phytochemicals. Figure created with BioRender.com.

3.5. Phytochemicals Acting via ROS-Mediated Anoikis

Emodin

Emodin (1,3,8-trihydroxy-6-methylanthraquinone) is an anthraquinone derivative obtained from rhubarb plant (*Rheum palmatum*) and *Aloe vera* [122]. Emodin was shown to induce ROS generation followed by anoikis in gastric cancer cells. Emodin induced oxidative stress and caused inactivation of RhoA, a crucial signaling molecule in the cytoskeletal rearrangement. Inactivation of RhoA led to the disruption of focal adhesion complex and ultimately anoikis [61]. Figure 5 depicts the ROS-mediated anoikis by phytochemicals.

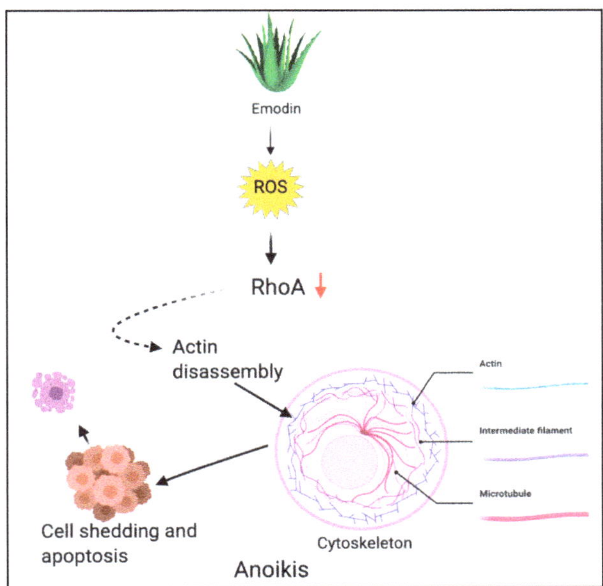

Figure 5. ROS-mediated anoikis by phytochemicals. Figure created with BioRender.com.

4. Conclusions

The ability of natural phytochemicals to modulate various signaling pathways make them promising therapeutic candidates in cancer. Many phytochemicals have progressed towards clinical trials against various malignancies such as breast, pancreatic, colon, and prostate cancers where promising results have been observed. Since cancer cells survive on levels of ROS that are slightly higher than normal cells, they are more sensitive to external perturbation. Fine-tuning of ROS levels using phytochemicals is a promising therapeutic strategy in cancer. Additionally, recent findings such as photooxidative stress using phytochemicals have provided selective targeting of cancer cells. However, there are a few challenges which need to be addressed in terms of the pharmacokinetic parameters of phytochemicals. In order to induce sufficient ROS levels for anti-cancer activity, specific plasma levels of the phytochemicals need to be attained. It has been observed that some of the phytochemicals do achieve the required plasma levels. A few such candidates having good pharmacokinetic parameters include BITC, PEITC, silibinin, and berberine. However, the majority of the phytochemicals fail to achieve required cytotoxic levels. This issue can be addressed by formulating the natural compound in advanced drug delivery systems such as nano drug delivery formulations. Various studies have been conducted in this aspect and nano drug delivery has proven to be an effective tool to achieve high plasma concentrations of the phytochemicals. Some of the successful candidates include sulforaphane and curcumin. A nano structured lipid formulation increased the bioavalability of sulforaphane in rat plasma while a nanoparticle formulation increased curcumin levels in human plasma by 10–15-fold. In the future, large cohort studies of ROS targeted phytochemicals are required, since these studies will provide data regarding efficacy, toxicity, bioavailability in clinical setup for their establishment as anti-cancer therapies in the market. An outline of pharmacokinetic data, selective action between normal and cancer cells by phytochemicals has been presented in Table 1.

Table 1. Phytochemicals are associated with ROS-mediated programmed cell death in cancer.

Sr. No.	Phytochemical	PCD Type	Effective Anti-Cancer Concentration (In Vitro)	Dose Administered (In Vivo)	Pharmacokinetic Data	TttCell Lines/Subjects	Outcome/Comments	References
1	Capsaicin	Apoptosis	150 µM	2.5 mg/kg	Not stated	Animal subject—Mice In vitro study Pancreatic cancer cells	Apoptosis was observed in AsPC-1 and $BXPC_3$ cell lines at 150 µM concentration, while no effect were observed in normal (HPDE-6) cells. In an animal study, an oral dose 2.5 mg/kg was effective in suppressing tumor growth. Human equivalent dose—0.202 mg/kg.	[67,68]
2	Capsaicin (Pharmacokinetic study)	-	-	5 g	C_{max}: 2.47 ± 0.13 ng/mL	Human subjects	The oral bioavailability of capsaicin is effective for anti-diabetic effect, but plasma level is less for its anti-cancer activity.	[123,124]
3	Sulforaphane	Apoptosis	5–10 µM	-	-	-	Various studies report anticancer activity in a range of 5–10 µM which is not attainable through dietary intake. A study in normal vs. cancerous prostate cells revealed selective cytotoxicity of sulforaphane in cancerous cells. No effect was observed in normal cells at a dose range of 0–15 µM.	[125,126]
4	Sulforaphane (Pharmacokinetic study)	-	-	30 mg/kg	C_{max}: 772.8 ± 54.36 ng/mL	Animal subject—Rats	Nano structured lipid formulation of sulforaphane increased its bioavailability in rat plasma wherein a dose close to effective anticancer level was achieved.	[127]
5	α-lipoic acid	Apoptosis	500 µM	-	-	In vitro study hepatoma cell line	ROS-mediated apoptosis was observed at 500 µM. However, in another study, no effect was observed in normal L02 liver cells at 5 mM concentration.	[72,128]

Table 1. *Cont.*

Sr. No.	Phytochemical	PCD Type	Effective Anti-Cancer Concentration (In Vitro)	Dose Administered (In Vivo)	Pharmacokinetic Data	TttCell Lines/Subjects	Outcome/Comments	References
6	α-lipoic acid (Pharmacokinetic study)	-	-	600 mg per day	C_{max}: 8–52 nM	Human subjects	Plasma levels achieved in plasma following oral administration of racemic α-lipoic acid is much less than required for its anti-cancer activity.	[129]
7	BITC	Apoptosis	-	-	-	In vitro study Pancreatic cancer cells	BITC caused ROS generation in a concentration dependent manner starting from 2.5 to 20 µM in MIAPaCa-2 and Capan-2 cell lines	[79,130]
8	BITC	Apoptosis	<2.5–5 µM	-	-	In vitro study Breast cancer cells and normal mammary epithelial cell line	BITC induced ROS-mediated apoptosis in MDA-MB-231 and MCF-7 at IC_{50} value of <2.5 µm and 5 µM respectively at 24-h time point. However, the IC_{50} was 20 µM in normal mammary epithelial cell line MCF-10A.	[131]
9	BITC	Apoptosis	8 µM	-	-	In vitro study Pancreatic cancer cells	An IC_{50} of 8 µM was observed when $BXPC_3$ cells were treated with BITC.	[132]
10	BITC (Pharmacokinetic study)	-	-	12 µM/day	C_{max}: 7.5 µM in tumor	Animal subject—Mice	BXPC3 cells were subcutaneously implanted and a 43% tumor inhibition was observed with a concentration near IC_{50} value achieved in plasma.	[74]
11	PEITC	Apoptosis	<5 µM	-	-	In vitro study Breast cancer cells	PEITC caused ROS-mediated apoptosis in MDA-MB-231 and MCF-7 cell lines.	[86]

Table 1. *Cont.*

Sr. No.	Phytochemical	PCD Type	Effective Anti-Cancer Concentration (In Vitro)	Dose Administered (In Vivo)	Pharmacokinetic Data	TttCell Lines/Subjects	Outcome/Comments	References
12	PEITC	Apoptosis	5–10 µM	-	-	In vitro study normal ovarian epithelial cell line	Ovarian epithelial cell line T72 was transfected to over express *Ras* oncogene to form oncogenic T72Ras cell line. Following PEITC treatment, the transformed cells showed higher sensitivity to ROS as compared to normal cells. Thus, showing selective activity of PEITC.	[133]
13	PEITC (Pharmacokinetic study)	-	-	10–100 µmol/kg	C_{max}: 9–42 µM	Animal subjects—Rats	A considerably high level of PEITC was achieved after oral administration of PEITC than required for its anti-cancer activity.	[134]
14	Piperine	Apoptosis	136–137 µM	-	-	In vitro study Melanoma cell lines	Piperine induced RPS mediated apoptosis in SKMEL-28 and B16-F0 cells at 136 µM and 137 µM respectively at 72-h time point.	[90]
15	Piperine	Apoptosis	132 µg/mL	-	-	In vitro study Lung cancer cells and normal lung fibroblasts	Piperine induced apoptotic cell death via p53 dependent mitochondrial pathway in A549 cancer cell line while no significant cytotoxicity was observed in normal WI38 human lung fibroblasts. ROS are considered to be downstream effectors of p53 mediated apoptosis.	[135,136]
16	Piperine (Pharmacokinetic study)	-	-	200 mg tablets (Benjakul formulation)	C_{max}: 1.078 µg/mL	Human subjects	Piperine is a major component of a traditional Thai medication called Benjakul. A dose of 200mg resulted in a Cmax level of 1.078 µg/mL which is significantly lower than that required for anticancer effect.	[137]

Table 1. *Cont.*

Sr. No.	Phytochemical	PCD Type	Effective Anti-Cancer Concentration (In Vitro)	Dose Administered (In Vivo)	Pharmacokinetic Data	TttCell Lines/Subjects	Outcome/Comments	References
17	Curcumin	Apoptosis	25 µM	5 mg/kg	-	In vitro study Colorectal cancer cells Animal subjects—Mice	Curcumin induced ROS-mediated apoptotic cell death in HCT-116 cell line at 25 µM concentration within 48hh. Curcumin showed considerable tumor inhibition in in vivo xenograft model following administration of 5 mg/kg I.P dose.	[93]
18	Curcumin	Apoptosis	20 µM	-	-	In vitro study Cervical cancer cells and normal cervical cells	Curcumin induced ROS-mediated apoptosis in cervical cancer cell lines C33A, CaSki, HeLa, and ME180 cells at approximately 20 µM in 48-h time period, while it did not induce significant toxicity in normal counterparts until 40 µM concentration.	[138]
19	Curcumin (Pharmacokinetic study)	-	-	Various doses (2–10 g)	C_{max}: 1–3200 ng/mL	Healthy human volunteers and patients (cancer, Alzheimer's disease etc.)	The serum levels of curcumin achieved by oral administration of crude curcumin are much lower than required for anti-cancer activity. However, a 10–15-fold increase in plasma levels was observed when curcumin was formulated as nanoparticle or combined with piperine, lecithin, etc.	[139]
20	Withaferin A	Ferroptosis	5–10 µM	4 mg/kg	Not stated	In vitro study Neuroblastoma cell lines Animal subjects—Mice	Crude Withaferin A and its nano particle formulation for tumor targeting showed ROS-mediated ferroptosis at 10 µM in neuroblastoma cell lines IMR-32, SK-N-SH, Kelly, NB69, and CHP-134 within 4–8 h. Tumor regression was mediated through lipid peroxidation.	[95]

Table 1. *Cont.*

Sr. No.	Phytochemical	PCD Type	Effective Anti-Cancer Concentration (In Vitro)	Dose Administered (In Vivo)	Pharmacokinetic Data	TttCell Lines/Subjects	Outcome/Comments	References
21	Withaferin A	Apoptotic	1–50 μM	4 mg/kg	C_{max}: 1.8 μM	In vitro study Breast cancer cell lines Animal subject—Mice	Withaferin A displayed cytotoxicity in breast carcinoma cell lines MDA-MB 231, H1299, T47D, MCF-7, LN686, as well as normal fibroblast cell line COS-7 in a wide range of 1–50 μM at 72-h time point. The IC_{50} was lesser than normal fibroblasts in majority of the cancer cell lines.	[140]
22	Bromelain	Ferroptosis	-	3 mg/kg	Not stated	In vitro study Colon carcinoma cell lines and normal colon cell line. Animal subjects—Mice	Bromelain inhibited proliferation of in Kras mutant colon cancer cell lines HCT-116 and DLD-1 at 50 μg/mL, while it induced significant ferroptosis in cancer cell lines at a concentration of 5 μm when combined with Erastin as compared to normal colon cells CCD1co. Bromelain increased survival rate in treatment group as compared to vehicle group.	[81]
23	Bromelain (Pharmacokinetic study)	-	-	143 mg/kg	C_{max}: Very low (specific value not stated)	Human subjects	Bromelain showed very low (ng/mL) plasma levels following oral administration at 143 mg/kg body weight.	[141]
24	Ruscogenin	Ferroptosis	7.3–28.19 μM	5 or 10 mg/kg	Not stated	In vitro study Pancreatic cancer cells Animal subjects—Mice	Ruscogenin induced significant ferroptotic cell death in pancreatic cancer cell lines BxPC-3, SW1990, PANC-1, and ASPC-1 cells, as compared to normal pancreatic cell line HPDE-6-C7 wherein not IC_{50} was detected at 72-h time point in HPDE6-C7 cells. Additionally, it inhibited pancreatic cancer growth in vivo.	[99]

Table 1. *Cont.*

Sr. No.	Phytochemical	PCD Type	Effective Anti-Cancer Concentration (In Vitro)	Dose Administered (In Vivo)	Pharmacokinetic Data	TttCell Lines/Subjects	Outcome/Comments	References
25	Ruscogenin (Pharmacokinetic study)	-	-	8 mg/kg	C_{max}: 504.50 ± 63.47 ng/mL (Mean value)	Animal subjects—Rats	Following IV administration of Ruscogenin in rats, a plasma level between 2–1000 ng/mL was observed which roughly translates to 2.35 µM, a level much lower than required for anticancer effect.	[142]
26	Oridonin	Ferroptosis	26.93 µM	-	-	In vitro study Esophageal cancer cells	Oridonin induced ferroptotic cell death in esophageal cancer cell line TE1 at 27 µM within 24-h time point.	[101]
27	Oridonin	Apoptosis	2.5–10 µM	-	-	In vitro study Esophageal cancer cell lines	Oridonin induced cellular death in human esophageal cancer cell lines (KYSE70, KYSE410, and KYSE450) at 2.5–10 µM concentration, while it showed 40% cell death in normal esophageal cell line SHEE at 40 µM, a significantly higher concentration.	[143]
28	Oridonin (Pharmacokinetic study)	-	-	20 mg/kg	C_{max}: 146.9 ± 10.17 ng/mL	Animal subjects—Rats	A concentration of approximately 150 ng/mL translating to 412 nM was achieved. Combination with verapamil increased the Cmax level of oridonin to 194 ± 10 ng/mL, still a lower level for anti-cancer efficacy.	[144]
29	Cucurbitacin B	Autophagy	Approximately 1 µM	-	-	In vitro study Cervical cancer cell line and breast cancer cell lines	Cucurbitacin B induced caspase independent autophagic cell death in HeLa cells at 1 µM. Additionally, it induced autophagy and increased ROS levels in MCF-7 at 200 nM. Autophagic cell death in normal counterparts was not studied.	[104]

Table 1. *Cont.*

Sr. No.	Phytochemical	PCD Type	Effective Anti-Cancer Concentration (In Vitro)	Dose Administered (In Vivo)	Pharmacokinetic Data	TttCell Lines/Subjects	Outcome/Comments	References
30	Cucurbitacin B	Apoptosis	0.2 µM	-	-	In vitro study Prostate cancer cell lines	Cucurbitacin B induced cell death and ROS production in human prostate cancer cell lines LNCaP and PC-3 at 0.2 µM while no significant cell death or ROS production was observed in normal prostate cell line PrEC.	[145]
31	Cucurbitacin B (Pharmacokinetic study)	-	-	2–4 mg/kg	C_{max}: 9–31 µg/L	Animal subjects—Rats	Following an oral administration of 2–4 mg/kg, a significantly low plasma level was achieved in rat plasma than required for anti-cancer activity.	[146]
32	Silibinin	Autophagy	40 µM	-	-	In vitro study Fibrosarcoma cells	Silibinin induced autophagy in human fibrosarcoma cell lines HT1080 at 40 µM within a period of 4 h. Cellular death was concluded to be an autophagy mediated apoptosis process. Effect on normal cells was not studied.	[106]
33	Silibinin (Pharmacokinetic study)	-	-	360, 720 and 1440 mg	C_{max}: 0.4, 1.4 and 4 ± 5.3 µM respectively	Human Subjects—Colorectal cancer patients	Silibinin formulated capsules showed considerably high levels required to exert its anti-tumor effect.	[147]
34	Allicin	Autophagy and apoptosis	15–30 µg/mL	-	-	In vitro study Non-small cell lung cancer cell lines	Allicin induced both autophagy and apoptosis in human lung cancer cell lines A549 and NCI-H460.	[109]
35	Carnosol	Autophagy	<25 µM	-	-	In vitro study Breast cancer cell lines	Carnosol induced ROS-mediated autophagy in triple negative breast cancer cell line MDA-MB-231. Effect on normal cell counterparts was not studied.	[113]

Table 1. *Cont.*

Sr. No.	Phytochemical	PCD Type	Effective Anti-Cancer Concentration (In Vitro)	Dose Administered (In Vivo)	Pharmacokinetic Data	TttCell Lines/Subjects	Outcome/Comments	References
36	Quercetin	Autophagy	200 µM	100 mg/kg	-	In vitro study Osteosarcoma cell line Animal subjects—Mice	Quercetin induced ROS-mediated autophagic cell death in human osteosarcoma cell line MG-63 and also inhibited tumor growth in mice model.	[116]
37	Berberine	Autophagy	-	-	-	In vitro study Renal carcinoma cell lines and normal kidney cell line.	Berberine induced autophagic cell death in 786-O, ACHN cell lines via its photosensitizer activity when combined with laser. Additionally, it induced cell death in normal kidney cell line HK-2 at when combined with laser. Treatment with Berberine alone failed to induce cell death >20% in cancer cells, while it induced cell death >20% in HK-2 normal cells at 40 µM within 48 h.	[118]
38	Berberine (Pharmacokinetic study)	-	-	200 mg/kg	C_{max}: 25.85 ± 7.34 µg/L	Animal subject—Rats	A plasma level of 25.85 ± 7.34 µg/L equivalent to 76 µM was achieved.	[148]
39	Nobiletin	Pyroptosis	34.85–35.31 µM	-	-	In vitro study Ovarian cancer cell lines	Nobiletin induced cytotoxicity at concentration of 35 µM but no data regarding its effect on normal cell lines were shown.	[120]
40	Nobiletin	Apoptosis	40–80 µM	100 mg/kg	-	In vitro study Ovarian cancer cell lines OVCAR-3 and A2780 Animal subject—Mice	Nobiletin exhibited cytotoxicity in ovarian cancer cell lines at 40–80 µM while the IC_{50} for normal ovarian cell line was around 160 µM. Additionally, significant tumor growth inhibition was observed in athymic nude mice model at a dose of 100 mg/kg.	[149]

Table 1. *Cont.*

Sr. No.	Phytochemical	PCD Type	Effective Anti-Cancer Concentration (In Vitro)	Dose Administered (In Vivo)	Pharmacokinetic Data	TttCell Lines/Subjects	Outcome/Comments	References
41	Nobiletin (Pharmacokinetic study)	-	-	50 mg/kg	C_{max}: 1.78 µg/mL in plasma	Animal subject—Rats	The plasma level achieved after oral administration is 1.78 µg/mL which correlates to approximately 4–5 µM.	[150]
42	Emodin	Anoikis	10 µM	-	-	In vitro study Gastric cancer cell lines	Significant difference observed in RhoA expression between cancer and normal cell lines.	[61]
43	Emodin	Apoptosis	70 µM	-	-	In vitro study Cancer cell lines isolated from breast, lung, colon, and cervix carcinomas, normal human fibroblasts, and normal human keratinocytes	No effect observed on normal cell lines after 48-h treatment while cytotoxicity was observed in cancer cell lines.	[151]
44	Emodin (Pharmacokinetic study)	-	-	4.5 mg/kg	C_{max}: 0.2 ± 0.1 µM	Animal subject—Rats	The concentration of free emodin achieved in rat plasma after oral administration of rhubarb extract was found to be much lower than required for its anti-cancer activity.	[152]

Author Contributions: Conception and design of the review, S.G.; writing—original and draft preparation, S.G.; editing, suggestions and supervision S.K.S. All authors have read and agreed to the published version of the manuscript.

Funding: Authors appreciate the financial support from Dodge Jones Foundation, Abilene, Texas.

Conflicts of Interest: The authors declare no conflict of interest.

References

1. Sies, H.; Jones, D.P. Reactive oxygen species (ROS) as pleiotropic physiological signalling agents. *Nat. Rev. Mol. Cell. Biol.* **2020**, *21*, 363–383. [CrossRef] [PubMed]
2. Pacher, P.; Beckman, J.S.; Liaudet, L. Nitric oxide and peroxynitrite in health and disease. *Physiol. Rev.* **2007**, *87*, 315–424. [CrossRef] [PubMed]
3. Patel, L.R.; Peng, J.; Chilian, W.M. Reactive Oxygen Species: The Good and the Bad. In *Reactive Oxygen Species (ROS) in Living Cells*; IntechOpen: London, UK, 2017. [CrossRef]
4. Milkovic, L.; Cipak Gasparovic, A.; Cindric, M.; Mouthuy, P.A.; Zarkovic, N. Short Overview of ROS as Cell Function Regulators and Their Implications in Therapy Concepts. *Cells* **2019**, *8*, 793. [CrossRef] [PubMed]
5. Winterbourn, C.C. Reconciling the chemistry and biology of reactive oxygen species. *Nat. Chem. Biol.* **2008**, *4*, 278–286. [CrossRef] [PubMed]
6. Schieber, M.; Chandel, N.S. ROS function in redox signaling and oxidative stress. *Curr. Biol.* **2014**, *24*, R453–R462. [CrossRef]
7. Miki, H.; Funato, Y. Regulation of intracellular signalling through cysteine oxidation by reactive oxygen species. *J. Biochem.* **2012**, *151*, 255–261. [CrossRef]
8. Droge, W. Free radicals in the physiological control of cell function. *Physiol. Rev.* **2002**, *82*, 47–95. [CrossRef]
9. Brieger, K.; Schiavone, S.; Miller, F.J., Jr.; Krause, K.H. Reactive oxygen species: From health to disease. *Swiss Med. Wkly.* **2012**, *142*, w13659. [CrossRef]
10. Oswald, M.C.W.; Garnham, N.; Sweeney, S.T.; Landgraf, M. Regulation of neuronal development and function by ROS. *FEBS Lett.* **2018**, *592*, 679–691. [CrossRef]
11. Wilson, C.; Gonzalez-Billault, C. Regulation of cytoskeletal dynamics by redox signaling and oxidative stress: Implications for neuronal development and trafficking. *Front. Cell. Neurosci.* **2015**, *9*, 381. [CrossRef]
12. Liu, Y.; Zhao, H.; Li, H.; Kalyanaraman, B.; Nicolosi, A.C.; Gutterman, D.D. Mitochondrial sources of H_2O_2 generation play a key role in flow-mediated dilation in human coronary resistance arteries. *Circ. Res.* **2003**, *93*, 573–580. [CrossRef] [PubMed]
13. Ushio-Fukai, M.; Alexander, R.W. Reactive oxygen species as mediators of angiogenesis signaling: Role of NAD(P)H oxidase. *Mol. Cell. Biochem.* **2004**, *264*, 85–97. [CrossRef] [PubMed]
14. Yun, J.; Rocic, P.; Pung, Y.F.; Belmadani, S.; Carrao, A.C.; Ohanyan, V.; Chilian, W.M. Redox-dependent mechanisms in coronary collateral growth: The "redox window" hypothesis. *Antioxid. Redox Signal.* **2009**, *11*, 1961–1974. [CrossRef] [PubMed]
15. Accardi, M.V.; Daniels, B.A.; Brown, P.M.; Fritschy, J.M.; Tyagarajan, S.K.; Bowie, D. Mitochondrial reactive oxygen species regulate the strength of inhibitory GABA-mediated synaptic transmission. *Nat. Commun.* **2014**, *5*, 3168. [CrossRef] [PubMed]
16. Fang, F.C. Antimicrobial actions of reactive oxygen species. *mBio* **2011**, *2*. [CrossRef] [PubMed]
17. Dupre-Crochet, S.; Erard, M.; Nubetae, O. ROS production in phagocytes: Why, when, and where? *J. Leukoc. Biol.* **2013**, *94*, 657–670. [CrossRef]
18. Kuiper, J.W.; Sun, C.; Magalhaes, M.A.; Glogauer, M. Rac regulates PtdInsP(3) signaling and the chemotactic compass through a redox-mediated feedback loop. *Blood* **2011**, *118*, 6164–6171. [CrossRef]
19. Tavassolifar, M.J.; Vodjgani, M.; Salehi, Z.; Izad, M. The Influence of Reactive Oxygen Species in the Immune System and Pathogenesis of Multiple Sclerosis. *Autoimmune Dis.* **2020**, *2020*, 5793817. [CrossRef]
20. Lee, K.; Esselman, W.J. cAMP potentiates H(2)O(2)-induced ERK1/2 phosphorylation without the requirement for MEK1/2 phosphorylation. *Cell. Signal.* **2001**, *13*, 645–652. [CrossRef]
21. Burton, G.J.; Jauniaux, E. Oxidative stress. *Best Pr. Res. Clin. Obs. Gynaecol.* **2011**, *25*, 287–299. [CrossRef]
22. Birben, E.; Sahiner, U.M.; Sackesen, C.; Erzurum, S.; Kalayci, O. Oxidative stress and antioxidant defense. *World Allergy Organ. J.* **2012**, *5*, 9–19. [CrossRef] [PubMed]
23. Zuo, L.; Zhou, T.; Pannell, B.K.; Ziegler, A.C.; Best, T.M. Biological and physiological role of reactive oxygen species—The good, the bad and the ugly. *Acta Physiol.* **2015**, *214*, 329–348. [CrossRef] [PubMed]
24. Leslie, N.R.; Bennett, D.; Lindsay, Y.E.; Stewart, H.; Gray, A.; Downes, C.P. Redox regulation of PI 3-kinase signalling via inactivation of PTEN. *EMBO J.* **2003**, *22*, 5501–5510. [CrossRef] [PubMed]
25. Okoh, V.O.; Felty, Q.; Parkash, J.; Poppiti, R.; Roy, D. Reactive oxygen species via redox signaling to PI3K/AKT pathway contribute to the malignant growth of 4-hydroxy estradiol-transformed mammary epithelial cells. *PLoS ONE* **2013**, *8*, e54206. [CrossRef]
26. Weinberg, F.; Hamanaka, R.; Wheaton, W.W.; Weinberg, S.; Joseph, J.; Lopez, M.; Kalyanaraman, B.; Mutlu, G.M.; Budinger, G.R.; Chandel, N.S. Mitochondrial metabolism and ROS generation are essential for Kras-mediated tumorigenicity. *Proc. Natl. Acad. Sci. USA* **2010**, *107*, 8788–8793. [CrossRef]
27. Gatenby, R.A.; Gillies, R.J. Why do cancers have high aerobic glycolysis? *Nat. Rev. Cancer* **2004**, *4*, 891–899. [CrossRef]

28. Vafa, O.; Wade, M.; Kern, S.; Beeche, M.; Pandita, T.K.; Hampton, G.M.; Wahl, G.M. c-Myc can induce DNA damage, increase reactive oxygen species, and mitigate p53 function: A mechanism for oncogene-induced genetic instability. *Mol. Cell* **2002**, *9*, 1031–1044. [CrossRef]
29. Cat, B.; Stuhlmann, D.; Steinbrenner, H.; Alili, L.; Holtkotter, O.; Sies, H.; Brenneisen, P. Enhancement of tumor invasion depends on transdifferentiation of skin fibroblasts mediated by reactive oxygen species. *J. Cell Sci.* **2006**, *119*, 2727–2738. [CrossRef]
30. Kar, S.; Subbaram, S.; Carrico, P.M.; Melendez, J.A. Redox-control of matrix metalloproteinase-1: A critical link between free radicals, matrix remodeling and degenerative disease. *Respir. Physiol. Neurobiol.* **2010**, *174*, 299–306. [CrossRef]
31. Ozben, T. Oxidative stress and apoptosis: Impact on cancer therapy. *J. Pharm. Sci.* **2007**, *96*, 2181–2196. [CrossRef]
32. Kim, B.; Srivastava, S.K.; Kim, S.H. Caspase-9 as a therapeutic target for treating cancer. *Expert Opin. Ther. Targets* **2015**, *19*, 113–127. [CrossRef] [PubMed]
33. Madesh, M.; Hajnoczky, G. VDAC-dependent permeabilization of the outer mitochondrial membrane by superoxide induces rapid and massive cytochrome c release. *J. Cell Biol.* **2001**, *155*, 1003–1015. [CrossRef] [PubMed]
34. Circu, M.L.; Aw, T.Y. Reactive oxygen species, cellular redox systems, and apoptosis. *Free Radic. Biol. Med.* **2010**, *48*, 749–762. [CrossRef] [PubMed]
35. Ott, M.; Gogvadze, V.; Orrenius, S.; Zhivotovsky, B. Mitochondria, oxidative stress and cell death. *Apoptosis* **2007**, *12*, 913–922. [CrossRef] [PubMed]
36. Wang, J.; Yi, J. Cancer cell killing via ROS: To increase or decrease, that is the question. *Cancer Biol. Ther.* **2008**, *7*, 1875–1884. [CrossRef]
37. Kumar, R.; Herbert, P.E.; Warrens, A.N. An introduction to death receptors in apoptosis. *Int. J. Surg.* **2005**, *3*, 268–277. [CrossRef]
38. Safa, A.R.; Day, T.W.; Wu, C.H. Cellular FLICE-like inhibitory protein (C-FLIP): A novel target for cancer therapy. *Curr. Cancer Drug Targets* **2008**, *8*, 37–46. [CrossRef]
39. Wang, L.; Azad, N.; Kongkaneramit, L.; Chen, F.; Lu, Y.; Jiang, B.H.; Rojanasakul, Y. The Fas death signaling pathway connecting reactive oxygen species generation and FLICE inhibitory protein down-regulation. *J. Immunol.* **2008**, *180*, 3072–3080. [CrossRef]
40. He, C.; Klionsky, D.J. Regulation mechanisms and signaling pathways of autophagy. *Annu. Rev. Genet.* **2009**, *43*, 67–93. [CrossRef]
41. Xie, Z.; Klionsky, D.J. Autophagosome formation: Core machinery and adaptations. *Nat. Cell. Biol* **2007**, *9*, 1102–1109. [CrossRef]
42. Ranjan, A.; Kaushik, I.; Srivastava, S.K. Pimozide Suppresses the Growth of Brain Tumors by Targeting STAT3-Mediated Autophagy. *Cells* **2020**, *9*, 2141. [CrossRef] [PubMed]
43. Cordani, M.; Donadelli, M.; Strippoli, R.; Bazhin, A.V.; Sanchez-Alvarez, M. Interplay between ROS and Autophagy in Cancer and Aging: From Molecular Mechanisms to Novel Therapeutic Approaches. *Oxid. Med. Cell. Longev.* **2019**, *2019*, 8794612. [CrossRef] [PubMed]
44. Chen, Y.F.; Liu, H.; Luo, X.J.; Zhao, Z.; Zou, Z.Y.; Li, J.; Lin, X.J.; Liang, Y. The roles of reactive oxygen species (ROS) and autophagy in the survival and death of leukemia cells. *Crit. Rev. Oncol. Hematol.* **2017**, *112*, 21–30. [CrossRef] [PubMed]
45. Li, Y.; Corradetti, M.N.; Inoki, K.; Guan, K.L. TSC2: Filling the GAP in the mTOR signaling pathway. *Trends Biochem. Sci.* **2004**, *29*, 32–38. [CrossRef] [PubMed]
46. Alexander, A.; Kim, J.; Walker, C.L. ATM engages the TSC2/mTORC1 signaling node to regulate autophagy. *Autophagy* **2010**, *6*, 672–673. [CrossRef] [PubMed]
47. Dobrowolny, G.; Aucello, M.; Rizzuto, E.; Beccafico, S.; Mammucari, C.; Boncompagni, S.; Belia, S.; Wannenes, F.; Nicoletti, C.; Del Prete, Z.; et al. Skeletal muscle is a primary target of SOD1G93A-mediated toxicity. *Cell Metab.* **2008**, *8*, 425–436. [CrossRef]
48. Burton, T.R.; Gibson, S.B. The role of Bcl-2 family member BNIP3 in cell death and disease: NIPping at the heels of cell death. *Cell Death Differ.* **2009**, *16*, 515–523. [CrossRef]
49. White, E.; DiPaola, R.S. The double-edged sword of autophagy modulation in cancer. *Clin. Cancer Res.* **2009**, *15*, 5308–5316. [CrossRef]
50. Dixon, S.J.; Lemberg, K.M.; Lamprecht, M.R.; Skouta, R.; Zaitsev, E.M.; Gleason, C.E.; Patel, D.N.; Bauer, A.J.; Cantley, A.M.; Yang, W.S.; et al. Ferroptosis: An iron-dependent form of nonapoptotic cell death. *Cell* **2012**, *149*, 1060–1072. [CrossRef]
51. Li, J.; Cao, F.; Yin, H.L.; Huang, Z.J.; Lin, Z.T.; Mao, N.; Sun, B.; Wang, G. Ferroptosis: Past, present and future. *Cell Death Dis.* **2020**, *11*, 1–13. [CrossRef]
52. Ribas, V.; Garcia-Ruiz, C.; Fernandez-Checa, J.C. Glutathione and mitochondria. *Front. Pharm.* **2014**, *5*, 151. [CrossRef] [PubMed]
53. Yang, W.S.; SriRamaratnam, R.; Welsch, M.E.; Shimada, K.; Skouta, R.; Viswanathan, V.S.; Cheah, J.H.; Clemons, P.A.; Shamji, A.F.; Clish, C.B.; et al. Regulation of ferroptotic cancer cell death by GPX4. *Cell* **2014**, *156*, 317–331. [CrossRef] [PubMed]
54. Fang, Y.; Tian, S.; Pan, Y.; Li, W.; Wang, Q.; Tang, Y.; Yu, T.; Wu, X.; Shi, Y.; Ma, P.; et al. Pyroptosis: A new frontier in cancer. *Biomed. Pharm.* **2020**, *121*, 109595. [CrossRef] [PubMed]
55. Yang, Y.; Liu, P.Y.; Bao, W.; Chen, S.J.; Wu, F.S.; Zhu, P.Y. Hydrogen inhibits endometrial cancer growth via a ROS/NLRP3/caspase-1/GSDMD-mediated pyroptotic pathway. *BMC Cancer* **2020**, *20*, 28. [CrossRef] [PubMed]
56. Gilmore, A.P. Anoikis. *Cell Death Differ.* **2005**, *12* (Suppl. 2), 1473–1477. [CrossRef]
57. Fofaria, N.M.; Srivastava, S.K. Critical role of STAT3 in melanoma metastasis through anoikis resistance. *Oncotarget* **2014**, *5*, 7051–7064. [CrossRef]
58. Fofaria, N.M.; Srivastava, S.K. STAT3 induces anoikis resistance, promotes cell invasion and metastatic potential in pancreatic cancer cells. *Carcinogenesis* **2015**, *36*, 142–150. [CrossRef]

59. Kandala, P.K.; Srivastava, S.K. Diindolylmethane-mediated Gli1 protein suppression induces anoikis in ovarian cancer cells in vitro and blocks tumor formation ability in vivo. *J. Biol. Chem.* **2012**, *287*, 28745–28754. [CrossRef]
60. Gupta, P.; Gupta, N.; Fofaria, N.M.; Ranjan, A.; Srivastava, S.K. HER2-mediated GLI2 stabilization promotes anoikis resistance and metastasis of breast cancer cells. *Cancer Lett.* **2019**, *442*, 68–81. [CrossRef]
61. Cai, J.; Niu, X.; Chen, Y.; Hu, Q.; Shi, G.; Wu, H.; Wang, J.; Yi, J. Emodin-induced generation of reactive oxygen species inhibits RhoA activation to sensitize gastric carcinoma cells to anoikis. *Neoplasia* **2008**, *10*, 41–51. [CrossRef]
62. Ranjan, A.; Ramachandran, S.; Gupta, N.; Kaushik, I.; Wright, S.; Srivastava, S.; Das, H.; Srivastava, S.; Prasad, S.; Srivastava, S.K. Role of Phytochemicals in Cancer Prevention. *Int. J. Mol. Sci.* **2019**, *20*, 4981. [CrossRef] [PubMed]
63. Boreddy, S.R.; Srivastava, S.K. Pancreatic cancer chemoprevention by phytochemicals. *Cancer Lett.* **2013**, *334*, 86–94. [CrossRef] [PubMed]
64. Cordell, G.A.; Araujo, O.E. Capsaicin: Identification, nomenclature, and pharmacotherapy. *Ann. Pharm.* **1993**, *27*, 330–336. [CrossRef] [PubMed]
65. Pramanik, K.C.; Fofaria, N.M.; Gupta, P.; Ranjan, A.; Kim, S.H.; Srivastava, S.K. Inhibition of beta-catenin signaling suppresses pancreatic tumor growth by disrupting nuclear beta-catenin/TCF-1 complex: Critical role of STAT-3. *Oncotarget* **2015**, *6*, 11561–11574. [CrossRef]
66. Pramanik, K.C.; Fofaria, N.M.; Gupta, P.; Srivastava, S.K. CBP-mediated FOXO-1 acetylation inhibits pancreatic tumor growth by targeting SirT. *Mol. Cancer Ther.* **2014**, *13*, 687–698. [CrossRef]
67. Pramanik, K.C.; Boreddy, S.R.; Srivastava, S.K. Role of mitochondrial electron transport chain complexes in capsaicin mediated oxidative stress leading to apoptosis in pancreatic cancer cells. *PLoS ONE* **2011**, *6*, e20151. [CrossRef]
68. Zhang, R.; Humphreys, I.; Sahu, R.P.; Shi, Y.; Srivastava, S.K. In vitro and in vivo induction of apoptosis by capsaicin in pancreatic cancer cells is mediated through ROS generation and mitochondrial death pathway. *Apoptosis* **2008**, *13*, 1465–1478. [CrossRef] [PubMed]
69. Houghton, C.A.; Fassett, R.G.; Coombes, J.S. Sulforaphane: Translational research from laboratory bench to clinic. *Nutr. Rev.* **2013**, *71*, 709–726. [CrossRef]
70. Singh, S.V.; Srivastava, S.K.; Choi, S.; Lew, K.L.; Antosiewicz, J.; Xiao, D.; Zeng, Y.; Watkins, S.C.; Johnson, C.S.; Trump, D.L.; et al. Sulforaphane-induced cell death in human prostate cancer cells is initiated by reactive oxygen species. *J. Biol. Chem.* **2005**, *280*, 19911–19924. [CrossRef]
71. Shay, K.P.; Moreau, R.F.; Smith, E.J.; Smith, A.R.; Hagen, T.M. Alpha-lipoic acid as a dietary supplement: Molecular mechanisms and therapeutic potential. *Biochim. Biophys. Acta* **2009**, *1790*, 1149–1160. [CrossRef]
72. Pibiri, M.; Sulas, P.; Camboni, T.; Leoni, V.P.; Simbula, G. alpha-Lipoic acid induces Endoplasmic Reticulum stress-mediated apoptosis in hepatoma cells. *Sci. Rep.* **2020**, *10*, 7139. [CrossRef] [PubMed]
73. Pramanik, K.C.; Srivastava, S.K.; Boreddy, S.R. *Molecular Targets of Benzyl Isothiocyanates in Pancreatic Cancer*; IntechOpen: London, UK, 2012.
74. Boreddy, S.R.; Pramanik, K.C.; Srivastava, S.K. Pancreatic tumor suppression by benzyl isothiocyanate is associated with inhibition of PI3K/AKT/FOXO pathway. *Clin. Cancer Res.* **2011**, *17*, 1784–1795. [CrossRef] [PubMed]
75. Boreddy, S.R.; Sahu, R.P.; Srivastava, S.K. Benzyl isothiocyanate suppresses pancreatic tumor angiogenesis and invasion by inhibiting HIF-alpha/VEGF/Rho-GTPases: Pivotal role of STAT-3. *PLoS ONE* **2011**, *6*, e25799. [CrossRef] [PubMed]
76. Batra, S.; Sahu, R.P.; Kandala, P.K.; Srivastava, S.K. Benzyl isothiocyanate-mediated inhibition of histone deacetylase leads to NF-kappaB turnoff in human pancreatic carcinoma cells. *Mol. Cancer Ther.* **2010**, *9*, 1596–1608. [CrossRef]
77. Sahu, R.P.; Epperly, M.W.; Srivastava, S.K. Benzyl isothiocyanate sensitizes human pancreatic cancer cells to radiation therapy. *Front. Biosci.* **2009**, *1*, 568–576. [CrossRef]
78. Wicker, C.A.; Sahu, R.P.; Kulkarni-Datar, K.; Srivastava, S.K.; Brown, T.L. BITC Sensitizes Pancreatic Adenocarcinomas to TRAIL-induced Apoptosis. *Cancer Growth Metastasis* **2010**, *2009*, 45–55. [CrossRef]
79. Sahu, R.P.; Zhang, R.; Batra, S.; Shi, Y.; Srivastava, S.K. Benzyl isothiocyanate-mediated generation of reactive oxygen species causes cell cycle arrest and induces apoptosis via activation of MAPK in human pancreatic cancer cells. *Carcinogenesis* **2009**, *30*, 1744–1753. [CrossRef]
80. Loganathan, S.; Kandala, P.K.; Gupta, P.; Srivastava, S.K. Inhibition of EGFR-AKT axis results in the suppression of ovarian tumors in vitro and in preclinical mouse model. *PLoS ONE* **2012**, *7*, e43577. [CrossRef]
81. Park, S.; Oh, J.; Kim, M.; Jin, E.J. Bromelain effectively suppresses Kras-mutant colorectal cancer by stimulating ferroptosis. *Anim. Cells Syst.* **2018**, *22*, 334–340. [CrossRef]
82. Gupta, P.; Wright, S.E.; Srivastava, S.K. PEITC treatment suppresses myeloid derived tumor suppressor cells to inhibit breast tumor growth. *Oncoimmunology* **2015**, *4*, e981449. [CrossRef]
83. Gupta, P.; Kim, B.; Kim, S.H.; Srivastava, S.K. Molecular targets of isothiocyanates in cancer: Recent advances. *Mol. Nutr. Food Res.* **2014**, *58*, 1685–1707. [CrossRef] [PubMed]
84. Gupta, P.; Wright, S.E.; Kim, S.H.; Srivastava, S.K. Phenethyl isothiocyanate: A comprehensive review of anti-cancer mechanisms. *Biochim. Biophys. Acta* **2014**, *1846*, 405–424. [CrossRef] [PubMed]
85. Gupta, P.; Adkins, C.; Lockman, P.; Srivastava, S.K. Metastasis of Breast Tumor Cells to Brain Is Suppressed by Phenethyl Isothiocyanate in a Novel In Vivo Metastasis Model. *PLoS ONE* **2013**, *8*, e67278. [CrossRef] [PubMed]

86. Gupta, P.; Srivastava, S.K. Antitumor activity of phenethyl isothiocyanate in HER2-positive breast cancer models. *BMC Med.* **2012**, *10*, 80. [CrossRef]
87. Gorgani, L.; Mohammadi, M.; Najafpour, G.D.; Nikzad, M. Piperine—The Bioactive Compound of Black Pepper: From Isolation to Medicinal Formulations. *Compr. Rev. Food Sci. Food* **2017**, *16*, 124–140. [CrossRef] [PubMed]
88. Dhillon, H.; Chikara, S.; Reindl, K.M. Piperlongumine induces pancreatic cancer cell death by enhancing reactive oxygen species and DNA damage. *Toxicol. Rep.* **2014**, *1*, 309–318. [CrossRef] [PubMed]
89. Fofaria, N.M.; Qhattal, H.S.; Liu, X.; Srivastava, S.K. Nanoemulsion formulations for anti-cancer agent piplartine—Characterization, toxicological, pharmacokinetics and efficacy studies. *Int. J. Pharm.* **2016**, *498*, 12–22. [CrossRef]
90. Fofaria, N.M.; Kim, S.H.; Srivastava, S.K. Piperine causes G1 phase cell cycle arrest and apoptosis in melanoma cells through checkpoint kinase-1 activation. *PLoS ONE* **2014**, *9*, e94298. [CrossRef]
91. Hewlings, S.J.; Kalman, D.S. Curcumin: A Review of Its Effects on Human Health. *Foods* **2017**, *6*, 92. [CrossRef]
92. Sahu, R.P.; Batra, S.; Srivastava, S.K. Activation of ATM/Chk1 by curcumin causes cell cycle arrest and apoptosis in human pancreatic cancer cells. *Br. J. Cancer* **2009**, *100*, 1425–1433. [CrossRef]
93. Khaket, T.P.; Singh, M.P.; Khan, I.; Kang, S.C. In vitro and in vivo studies on potentiation of curcumin-induced lysosomal-dependent apoptosis upon silencing of cathepsin C in colorectal cancer cells. *Pharm. Res.* **2020**, *161*, 105156. [CrossRef] [PubMed]
94. Vanden Berghe, W.; Sabbe, L.; Kaileh, M.; Haegeman, G.; Heyninck, K. Molecular insight in the multifunctional activities of Withaferin, A. *Biochem. Pharm.* **2012**, *84*, 1282–1291. [CrossRef] [PubMed]
95. Hassannia, B.; Wiernicki, B.; Ingold, I.; Qu, F.; Van Herck, S.; Tyurina, Y.Y.; Bayir, H.; Abhari, B.A.; Angeli, J.P.F.; Choi, S.M.; et al. Nano-targeted induction of dual ferroptotic mechanisms eradicates high-risk neuroblastoma. *J. Clin. Investig.* **2018**, *128*, 3341–3355. [CrossRef] [PubMed]
96. Pavan, R.; Jain, S.; Shraddha; Kumar, A. Properties and therapeutic application of bromelain: A review. *Biotechnol. Res. Int.* **2012**, *2012*, 976203. [CrossRef]
97. Doll, S.; Proneth, B.; Tyurina, Y.Y.; Panzilius, E.; Kobayashi, S.; Ingold, I.; Irmler, M.; Beckers, J.; Aichler, M.; Walch, A.; et al. ACSL4 dictates ferroptosis sensitivity by shaping cellular lipid composition. *Nat. Chem. Biol.* **2017**, *13*, 91–98. [CrossRef] [PubMed]
98. Masullo, M.; Pizza, C.; Piacente, S. Ruscus Genus: A Rich Source of Bioactive Steroidal Saponins. *Planta Med.* **2016**, *82*, 1513–1524. [CrossRef]
99. Song, Z.; Xiang, X.; Li, J.; Deng, J.; Fang, Z.; Zhang, L.; Xiong, J. Ruscogenin induces ferroptosis in pancreatic cancer cells. *Oncol. Rep.* **2020**, *43*, 516–524. [CrossRef]
100. Owona, B.A.; Schluesener, H.J. Molecular Insight in the Multifunctional Effects of Oridonin. *Drugs R D* **2015**, *15*, 233–244. [CrossRef]
101. Zhang, J.; Wang, N.; Zhou, Y.; Wang, K.; Sun, Y.; Yan, H.; Han, W.; Wang, X.; Wei, B.; Ke, Y.; et al. Oridonin induces ferroptosis by inhibiting gamma-glutamyl cycle in TE1 cells. *Phytother. Res.* **2020**. [CrossRef]
102. Fujii, J.; Homma, T.; Kobayashi, S. Ferroptosis caused by cysteine insufficiency and oxidative insult. *Free Radic. Res.* **2019**, *10*, 1–12. [CrossRef]
103. Garg, S.; Kaul, S.C.; Wadhwa, R. Cucurbitacin B and cancer intervention: Chemistry, biology and mechanisms (Review). *Int. J. Oncol.* **2018**, *52*, 19–37. [CrossRef]
104. Zhang, T.; Li, Y.; Park, K.A.; Byun, H.S.; Won, M.; Jeon, J.; Lee, Y.; Seok, J.H.; Choi, S.W.; Lee, S.H.; et al. Cucurbitacin induces autophagy through mitochondrial ROS production which counteracts to limit caspase-dependent apoptosis. *Autophagy* **2012**, *8*, 559–576. [CrossRef] [PubMed]
105. Ren, G.; Sha, T.; Guo, J.; Li, W.; Lu, J.; Chen, X. Cucurbitacin B induces DNA damage and autophagy mediated by reactive oxygen species (ROS) in MCF-7 breast cancer cells. *J. Nat. Med.* **2015**, *69*, 522–530. [CrossRef] [PubMed]
106. Duan, W.J.; Li, Q.S.; Xia, M.Y.; Tashiro, S.; Onodera, S.; Ikejima, T. Silibinin activated p53 and induced autophagic death in human fibrosarcoma HT1080 cells via reactive oxygen species-p38 and c-Jun N-terminal kinase pathways. *Biol. Pharm. Bull.* **2011**, *34*, 47–53. [CrossRef] [PubMed]
107. Jiang, K.; Wang, W.; Jin, X.; Wang, Z.; Ji, Z.; Meng, G. Silibinin, a natural flavonoid, induces autophagy via ROS-dependent mitochondrial dysfunction and loss of ATP involving BNIP3 in human MCF7 breast cancer cells. *Oncol Rep* **2015**, *33*, 2711–2718. [CrossRef] [PubMed]
108. Lawson, L.D.; Hunsaker, S.M. Allicin Bioavailability and Bioequivalence from Garlic Supplements and Garlic Foods. *Nutrients* **2018**, *10*, 812. [CrossRef]
109. Pandey, N.; Tyagi, G.; Kaur, P.; Pradhan, S.; Rajam, M.V.; Srivastava, T. Allicin Overcomes Hypoxia Mediated Cisplatin Resistance in Lung Cancer Cells through ROS Mediated Cell Death Pathway and by Suppressing Hypoxia Inducible Factors. *Cell Physiol. Biochem.* **2020**, *54*, 748–766. [CrossRef]
110. Chu, Y.L.; Ho, C.T.; Chung, J.G.; Raghu, R.; Lo, Y.C.; Sheen, L.Y. Allicin induces anti-human liver cancer cells through the p53 gene modulating apoptosis and autophagy. *J. Agric. Food Chem.* **2013**, *61*, 9839–9848. [CrossRef]
111. Johnson, J.J. Carnosol: A promising anti-cancer and anti-inflammatory agent. *Cancer Lett.* **2011**, *305*, 1–7. [CrossRef]
112. Choi, S.M.; Kim, D.H.; Chun, K.S. Carnosol induces apoptotic cell death through ROS-dependent inactivation of STAT3 in human melanoma G361 cells. *Appl. Biol. Chem.* **2019**, *62*, 1–11. [CrossRef]

113. Al Dhaheri, Y.; Attoub, S.; Ramadan, G.; Arafat, K.; Bajbouj, K.; Karuvantevida, N.; AbuQamar, S.; Eid, A.; Iratni, R. Carnosol induces ROS-mediated beclin1-independent autophagy and apoptosis in triple negative breast cancer. *PLoS ONE* **2014**, *9*, e109630. [CrossRef]
114. Mlcek, J.; Jurikova, T.; Skrovankova, S.; Sochor, J. Quercetin and Its Anti-Allergic Immune Response. *Molecules* **2016**, *21*, 623. [CrossRef] [PubMed]
115. Hamidi, T.; Cano, C.E.; Grasso, D.; Garcia, M.N.; Sandi, M.J.; Calvo, E.L.; Dagorn, J.C.; Lomberk, G.; Urrutia, R.; Goruppi, S.; et al. Nupr1-aurora kinase A pathway provides protection against metabolic stress-mediated autophagic-associated cell death. *Clin Cancer Res.* **2012**, *18*, 5234–5246. [CrossRef] [PubMed]
116. Wu, B.; Zeng, W.; Ouyang, W.; Xu, Q.; Chen, J.; Wang, B.; Zhang, X. Quercetin induced NUPR1-dependent autophagic cell death by disturbing reactive oxygen species homeostasis in osteosarcoma cells. *J. Clin. Biochem. Nutr.* **2020**, *67*, 137–145. [CrossRef] [PubMed]
117. Neag, M.A.; Mocan, A.; Echeverria, J.; Pop, R.M.; Bocsan, C.I.; Crisan, G.; Buzoianu, A.D. Berberine: Botanical Occurrence, Traditional Uses, Extraction Methods, and Relevance in Cardiovascular, Metabolic, Hepatic, and Renal Disorders. *Front. Pharmacol.* **2018**, *9*, 557. [CrossRef] [PubMed]
118. Lopes, T.Z.; de Moraes, F.R.; Tedesco, A.C.; Arni, R.K.; Rahal, P.; Calmon, M.F. Berberine associated photodynamic therapy promotes autophagy and apoptosis via ROS generation in renal carcinoma cells. *Biomed. Pharmacother.* **2020**, *123*, 109794. [CrossRef] [PubMed]
119. Huang, H.; Li, L.; Shi, W.; Liu, H.; Yang, J.; Yuan, X.; Wu, L. The Multifunctional Effects of Nobiletin and Its Metabolites In Vivo and In Vitro. *Evid Based Complement. Altern. Med.* **2016**, *2016*, 2918796. [CrossRef]
120. Zhang, R.; Chen, J.; Mao, L.; Guo, Y.; Hao, Y.; Deng, Y.; Han, X.; Li, Q.; Liao, W.; Yuan, M. Nobiletin Triggers Reactive Oxygen Species-Mediated Pyroptosis through Regulating Autophagy in Ovarian Cancer Cells. *J. Agric. Food Chem.* **2020**, *68*, 1326–1336. [CrossRef]
121. He, W.T.; Wan, H.; Hu, L.; Chen, P.; Wang, X.; Huang, Z.; Yang, Z.H.; Zhong, C.Q.; Han, J. Gasdermin D is an executor of pyroptosis and required for interleukin-1beta secretion. *Cell Res.* **2015**, *25*, 1285–1298. [CrossRef]
122. Huang, Q.; Lu, G.; Shen, H.M.; Chung, M.C.; Ong, C.N. Anti-cancer properties of anthraquinones from rhubarb. *Med. Res. Rev.* **2007**, *27*, 609–630. [CrossRef]
123. Suresh, D.; Srinivasan, K. Tissue distribution & elimination of capsaicin, piperine & curcumin following oral intake in rats. *Indian J. Med. Res.* **2010**, *131*, 682–691.
124. Chaiyasit, K.; Khovidhunkit, W.; Wittayalertpanya, S. Pharmacokinetic and the effect of capsaicin in Capsicum frutescens on decreasing plasma glucose level. *J. Med. Assoc. Thail.* **2009**, *92*, 108–113.
125. Sestili, P.; Fimognari, C. Cytotoxic and Antitumor Activity of Sulforaphane: The Role of Reactive Oxygen Species. *Biomed. Res. Int.* **2015**, *2015*, 402386. [CrossRef] [PubMed]
126. Clarke, J.D.; Hsu, A.; Yu, Z.; Dashwood, R.H.; Ho, E. Differential effects of sulforaphane on histone deacetylases, cell cycle arrest and apoptosis in normal prostate cells versus hyperplastic and cancerous prostate cells. *Mol. Nutr. Food Res.* **2011**, *55*, 999–1009. [CrossRef] [PubMed]
127. Soni, K.; Rizwanullah, M.; Kohli, K. Development and optimization of sulforaphane-loaded nanostructured lipid carriers by the Box-Behnken design for improved oral efficacy against cancer: In vitro, ex vivo and in vivo assessments. *Artif. Cells Nanomed. Biotechnol.* **2018**, *46*, 15–31. [CrossRef]
128. Shi, D.Y.; Liu, H.L.; Stern, J.S.; Yu, P.Z.; Liu, S.L. Alpha-lipoic acid induces apoptosis in hepatoma cells via the PTEN/Akt pathway. *FEBS Lett* **2008**, *582*, 1667–1671. [CrossRef]
129. Teichert, J.; Hermann, R.; Ruus, P.; Preiss, R. Plasma kinetics, metabolism, and urinary excretion of alpha-lipoic acid following oral administration in healthy volunteers. *J. Clin. Pharmacol.* **2003**, *43*, 1257–1267. [CrossRef]
130. Fofaria, N.M.; Ranjan, A.; Kim, S.H.; Srivastava, S.K. Mechanisms of the Anticancer Effects of Isothiocyanates. *Enzymes* **2015**, *37*, 111–137. [CrossRef]
131. Xiao, D.; Vogel, V.; Singh, S.V. Benzyl isothiocyanate-induced apoptosis in human breast cancer cells is initiated by reactive oxygen species and regulated by Bax and Bak. *Mol. Cancer Ther.* **2006**, *5*, 2931–2945. [CrossRef]
132. Srivastava, S.K.; Singh, S.V. Cell cycle arrest, apoptosis induction and inhibition of nuclear factor kappa B activation in anti-proliferative activity of benzyl isothiocyanate against human pancreatic cancer cells. *Carcinogenesis* **2004**, *25*, 1701–1709. [CrossRef]
133. Trachootham, D.; Zhou, Y.; Zhang, H.; Demizu, Y.; Chen, Z.; Pelicano, H.; Chiao, P.J.; Achanta, G.; Arlinghaus, R.B.; Liu, J.; et al. Selective killing of oncogenically transformed cells through a ROS-mediated mechanism by beta-phenylethyl isothiocyanate. *Cancer Cell* **2006**, *10*, 241–252. [CrossRef]
134. Ji, Y.; Kuo, Y.; Morris, M.E. Pharmacokinetics of dietary phenethyl isothiocyanate in rats. *Pharm. Res.* **2005**, *22*, 1658–1666. [CrossRef] [PubMed]
135. Lin, Y.; Xu, J.; Liao, H.; Li, L.; Pan, L. Piperine induces apoptosis of lung cancer A549 cells via p53-dependent mitochondrial signaling pathway. *Tumour Biol.* **2014**, *35*, 3305–3310. [CrossRef] [PubMed]
136. Johnson, T.M.; Yu, Z.X.; Ferrans, V.J.; Lowenstein, R.A.; Finkel, T. Reactive oxygen species are downstream mediators of p53-dependent apoptosis. *Proc. Natl. Acad. Sci. USA* **1996**, *93*, 11848–11852. [CrossRef] [PubMed]
137. Jumpa-ngern, P. Pharmacokinetics of piperine following single dose administration of benjakul formulation in healthy Thai subjects. *Afr. J. Pharm. Pharmacol.* **2013**, *7*, 560–566. [CrossRef]

138. Kim, B.; Kim, H.S.; Jung, E.J.; Lee, J.Y.; Tsang, B.K.; Lim, J.M.; Song, Y.S. Curcumin induces ER stress-mediated apoptosis through selective generation of reactive oxygen species in cervical cancer cells. *Mol. Carcinog.* **2016**, *55*, 918–928. [CrossRef]
139. Dei Cas, M.; Ghidoni, R. Dietary Curcumin: Correlation between Bioavailability and Health Potential. *Nutrients* **2019**, *11*, 2147. [CrossRef]
140. Thaiparambil, J.T.; Bender, L.; Ganesh, T.; Kline, E.; Patel, P.; Liu, Y.; Tighiouart, M.; Vertino, P.M.; Harvey, R.D.; Garcia, A.; et al. Withaferin A inhibits breast cancer invasion and metastasis at sub-cytotoxic doses by inducing vimentin disassembly and serine 56 phosphorylation. *Int. J. Cancer* **2011**, *129*, 2744–2755. [CrossRef]
141. Committee for Medicinal Products for Veterinary Use. *European Public MRL Assessment Report (EPMAR) Bromelain*; European Medicines Agency: Amsterdam, The Netherlands, 2017.
142. Ji, P.Y.; Li, Z.W.; Yang, Q.; Wu, R. Rapid determination of ruscogenin in rat plasma with application to pharmacokinetic study. *J. Chromatogr. B* **2015**, *985*, 71–74. [CrossRef]
143. Song, M.; Liu, X.; Liu, K.; Zhao, R.; Huang, H.; Shi, Y.; Zhang, M.; Zhou, S.; Xie, H.; Chen, H.; et al. Targeting AKT with Oridonin Inhibits Growth of Esophageal Squamous Cell Carcinoma In Vitro and Patient-Derived Xenografts In Vivo. *Mol. Cancer Ther.* **2018**, *17*, 1540–1553. [CrossRef]
144. Liu, J.; Zhang, N.; Li, N.; Fan, X.; Li, Y. Influence of verapamil on the pharmacokinetics of oridonin in rats. *Pharm. Biol.* **2019**, *57*, 787–791. [CrossRef]
145. Gao, Y.; Islam, M.S.; Tian, J.; Lui, V.W.; Xiao, D. Inactivation of ATP citrate lyase by Cucurbitacin B: A bioactive compound from cucumber, inhibits prostate cancer growth. *Cancer Lett.* **2014**, *349*, 15–25. [CrossRef] [PubMed]
146. Hunsakunachai, N.; Nuengchamnong, N.; Jiratchariyakul, W.; Kummalue, T.; Khemawoot, P. Pharmacokinetics of cucurbitacin B from Trichosanthes cucumerina L. in rats. *BMC Complement. Altern. Med.* **2019**, *19*, 157. [CrossRef] [PubMed]
147. Hoh, C.; Boocock, D.; Marczylo, T.; Singh, R.; Berry, D.P.; Dennison, A.R.; Hemingway, D.; Miller, A.; West, K.; Euden, S.; et al. Pilot study of oral silibinin, a putative chemopreventive agent, in colorectal cancer patients: Silibinin levels in plasma, colorectum, and liver and their pharmacodynamic consequences. *Clin. Cancer Res.* **2006**, *12*, 2944–2950. [CrossRef] [PubMed]
148. Tan, X.S.; Ma, J.Y.; Feng, R.; Ma, C.; Chen, W.J.; Sun, Y.P.; Fu, J.; Huang, M.; He, C.Y.; Shou, J.W.; et al. Tissue distribution of berberine and its metabolites after oral administration in rats. *PLoS ONE* **2013**, *8*, e77969. [CrossRef]
149. Chen, J.; Chen, A.Y.; Huang, H.; Ye, X.; Rollyson, W.D.; Perry, H.E.; Brown, K.C.; Rojanasakul, Y.; Rankin, G.O.; Dasgupta, P.; et al. The flavonoid nobiletin inhibits tumor growth and angiogenesis of ovarian cancers via the Akt pathway. *Int. J. Oncol.* **2015**, *46*, 2629–2638. [CrossRef]
150. Singh, S.P.; Wahajuddin; Tewari, D.; Patel, K.; Jain, G.K. Permeability determination and pharmacokinetic study of nobiletin in rat plasma and brain by validated high-performance liquid chromatography method. *Fitoterapia* **2011**, *82*, 1206–1214. [CrossRef]
151. Dumit, V.I.; Zerbes, R.M.; Kaeser-Pebernard, S.; Rackiewicz, M.; Wall, M.T.; Gretzmeier, C.; Kuttner, V.; van der Laan, M.; Braun, R.J.; Dengjel, J. Respiratory status determines the effect of emodin on cell viability. *Oncotarget* **2017**, *8*, 37478–37490. [CrossRef]
152. Wu, W.; Yan, R.; Yao, M.; Zhan, Y.; Wang, Y. Pharmacokinetics of anthraquinones in rat plasma after oral administration of a rhubarb extract. *Biomed. Chromatogr.* **2014**, *28*, 564–572. [CrossRef]

Article

NHERF1 Loss Upregulates Enzymes of the Pentose Phosphate Pathway in Kidney Cortex

Adrienne Bushau-Sprinkle [1], Michelle T. Barati [2], Kenneth B. Gagnon [2], Syed Jalal Khundmiri [3], Kathleen Kitterman [2], Bradford G. Hill [4], Amanda Sherwood [2], Michael Merchant [2], Shesh N. Rai [5,6], Sudhir Srivastava [6,7], Barbara Clark [8], Leah Siskind [1], Michael Brier [1,2], Jessica Hata [9,10] and Eleanor Lederer [2,11,*]

1 Department of Pharmacology and Toxicology, University of Louisville, Louisville, KY 40202, USA; ambush03@louisville.edu (A.B.-S.); leah.siskind@louisville.edu (L.S.); michael.brier@louisville.edu (M.B.)
2 Department of Medicine, Division of Nephrology, University of Louisville, Louisville, KY 40202, USA; michelle.barati@louisville.edu (M.T.B.); kenneth.gagnon@louisville.edu (K.B.G.); kathleen.kitterman@louisville.edu (K.K.); amanda.sherwood@louisville.edu (A.S.); michael.merchant@louisville.edu (M.M.)
3 Department of Physiology and Biophysics, Howard University, Washington, DC 20059, USA; syed.khundmiri@Howard.edu
4 Department of Medicine, Division of Environmental Medicine, University of Louisville, Louisville, KY 40202, USA; bradford.hill@louisville.edu
5 Department of Bioinformatics and Biostatistics, School of Public Health Information Sciences, University of Louisville, Louisville, KY 40202, USA; shesh.rai@louisville.edu
6 Biostatistics and Bioinformatics Facility, School of Medicine, University of Louisville, Louisville, KY 40202, USA; sudhir.srivastava@louisville.edu
7 Centre for Agricultural Bioinformatics, ICAR-Indian Agricultural Statistics Research Institute, New Delhi 110012, India
8 Department of Biochemistry and Molecular Genetics, University of Louisville, Louisville, KY 40202, USA; barbara.clark@louisville.edu
9 Department of Pathology, University of Louisville, Louisville, KY 40202, USA; jessica.hata@louisville.edu
10 Department of Pathology, Norton Children's Hospital, Louisville, KY 40202, USA
11 Robley Rex VA Medical Center, Louisville, KY 40206, USA
* Correspondence: eleanor.lederer@UTSouthwestern.edu; Tel.: +1-214-857-0908; Fax: +1-214-648-2071

Received: 7 August 2020; Accepted: 3 September 2020; Published: 14 September 2020

Abstract: (1) Background: We previously showed Na/H exchange regulatory factor 1 (NHERF1) loss resulted in increased susceptibility to cisplatin nephrotoxicity. NHERF1-deficient cultured proximal tubule cells and proximal tubules from NHERF1 knockout (KO) mice exhibit altered mitochondrial protein expression and poor survival. We hypothesized that NHERF1 loss results in changes in metabolic pathways and/or mitochondrial dysfunction, leading to increased sensitivity to cisplatin nephrotoxicity. (2) Methods: Two to 4-month-old male wildtype (WT) and KO mice were treated with vehicle or cisplatin (20 mg/kg dose IP). After 72 h, kidney cortex homogenates were utilized for metabolic enzyme activities. Non-treated kidneys were used to isolate mitochondria for mitochondrial respiration via the Seahorse XF24 analyzer. Non-treated kidneys were also used for LC-MS analysis to evaluate kidney ATP abundance, and electron microscopy (EM) was utilized to evaluate mitochondrial morphology and number. (3) Results: KO mouse kidneys exhibit significant increases in malic enzyme and glucose-6 phosphate dehydrogenase activity under baseline conditions but in no other gluconeogenic or glycolytic enzymes. NHERF1 loss does not decrease kidney ATP content. Mitochondrial morphology, number, and area appeared normal. Isolated mitochondria function was similar between WT and KO. Conclusions: KO kidneys experience a shift in metabolism to the pentose phosphate pathway, which may sensitize them to the oxidative stress imposed by cisplatin.

Keywords: cellular redox state; oxidative stress; mitochondrial function; cisplatin nephrotoxicity

1. Introduction

Although cisplatin is a widely used chemotherapeutic that treats a variety of solid malignant tumors (e.g., ovarian testicular, head and neck, and lung cancer), its nephrotoxicity limits its use [1]. Twenty percent to 30% of patients on cisplatin will develop cisplatin-induced acute kidney injury (AKI) with a single dose [1]. The mechanisms underlying cisplatin's nephrotoxicity remain perplexing. Furthermore, with no methods for the prevention or treatment of cisplatin nephrotoxicity, the identification of host factors (biomarkers) that confer susceptibility to cisplatin and new therapeutic targets for the prevention and/or treatment of cisplatin nephrotoxicity are promising areas of research.

The accumulation and bioactivation of cisplatin to a more nephrotoxic metabolite underlies the kidney's susceptibility to cisplatin-induced AKI. Cisplatin nephrotoxicity has been found to alter renal cell metabolism and induce cell death via both apoptosis and necrosis. However, the mechanism by which cisplatin induces these pathways remains unclear. Several studies have found that cisplatin treatment leads to renal tubular cell depletion of amino acids [2–5], influences lipid metabolism through the reduction of fatty acid oxidation resulting in the accumulation of fatty acids in kidney tissue [2,5,6], and decreases glycolytic enzymes and intermediates of the pentose phosphate pathway and the citric acid cycle [2,7]. Reactive oxygen species (ROS) and mitochondrial function may also contribute to cisplatin's mechanism of injury [8]. Mitochondria continuously produce ROS, such as superoxide, and scavenge these ROS using antioxidant enzymes (superoxide dismutase, glutathione peroxidase, catalase, and glutathione S-transferase) [9]. Cisplatin has been found to also accumulate in the mitochondria of renal epithelial cells [10,11], resulting in mitochondrial damage including decreased mitochondrial mass, disruption of cristae, and even mitochondrial swelling [12–14]. These morphological changes are associated with a significant reduction in mitochondrial activity and ATP production [12,13]. This structural damage to mitochondria leads to electron leakage and produces massive amounts of free radicals in the form of ROS, leading to cell injury and death.

We have recently published the novel finding that kidneys lacking the scaffolding protein Na/H exchange regulatory factor 1 (NHERF1) show increased susceptibility to cisplatin-induced AKI [15]. NHERF1 is a known monomeric membrane-associated protein that belongs to the NHERF family of post-synaptic density protein 9595/Drosophila Discs Large/ZonulaOcculens -1(PSD-95/DIg/ZO-1)homology (PDZ)-scaffold proteins [16]. NHERF1 is found in all epithelial cells and acts as a scaffold for multi-protein signaling complexes. In proximal tubule cells of the kidney, NHERF1 is anchored to the cytoskeleton in the subapical plasma membrane, where it acts as a key scaffolding protein of transport proteins and has critical roles in defining the renal proximal tubule brush border membrane (BBM) composition and in regulating ion transport [16]. Recent studies have uncovered increasing evidence that NHERF1 has a much broader role than as a scaffolding protein. In fact, alterations in NHERF1 expression, phosphorylation status, and/or localization have been associated with tumorigenesis [17–19], changes in cell structure and trafficking [19–21], inflammatory responses [19,22], and tissue injury [15,19,23,24]. These studies highlight the fact that changes in NHERF1 expression can affect cell proliferation through alterations in the WNT/beta-catenin pathway and cyclin D kinase pathways and that the absence of NHERF1 diminishes the inflammatory response by impairing the assembly of specific components of the response such as the intracellular adhesion molecule 1 (ICAM1) [23]. Previous studies from our laboratory comparing wild-type (WT) and NHERF1-deficient opossum kidney (OK) cells, a model of mammalian renal proximal tubule, demonstrated that NHERF1-deficient cells grew more slowly and were more likely to die in culture; however, they exhibited no overt morphological differences. The NHERF1-deficient cells did show decreased BBM expression of the type IIa sodium phosphate cotransporter (Npt2a), the sodium-dependent glucose cotransporter (SGLT1), and the enzyme γ-glutamyl transferase (GGTase) (unpublished data), which are defects that reversed when NHERF1 was re-introduced into the cells [25]. Likewise, NHERF1-deficient mice show normal kidney morphology but a marked decrease in the BBM expression of Npt2a, resulting in significant phosphate wasting, hypophosphatemia, hypercalciuria, and stone formation [26]. Formal studies to address metabolic changes in NHERF1-deficient kidneys have not been performed, but proteomic analysis from our laboratory comparing BBM from WT and NHERF1 KO kidneys revealed significant differences in the expression of mitochondrial proteins and enzymes from a variety of metabolic pathways. Notably, similar differences in BBM protein expression from the intestine of NHERF1 KO mice have been reported, but again, no specific studies to determine whether these changes in protein expression translate into changes in metabolism have been performed [27].

These observations suggest the hypothesis that alterations in renal cell metabolic pathways and/or mitochondrial dysfunction resulting from the loss of NHERF1 could prime these kidneys for a 'second hit' with cisplatin, therefore sensitizing these cells to cisplatin nephrotoxicity. The goals of this manuscript were twofold: (1) to determine if NHERF1 KO mice demonstrate changes in renal metabolic pathways, and (2) to determine if NHERF1 KO mice have altered mitochondrial function and/or structure that predisposes these animals to cisplatin nephrotoxicity.

2. Materials and Methods

2.1. Animals and Treatments

Two to 4-month-old male NHERF1$^{(-/-)}$ KO mice [28] and their WT littermates on C57BL/6J background were maintained on a 12:12 h light–dark cycle and were provided water and food ad libitum. At the time of sacrifice, animals were anesthetized with ketamine/xylazine (100/15 mg/kg, intraperitoneally (IP)). For enzyme kinetic assays, mice were given a single IP injection of 20 mg/kg cisplatin or vehicle (saline). Vehicle-treated and cisplatin-treated mice were euthanized after 72 h. All cisplatin studies were performed at the same time each day. Kidneys were removed and decapsulated; then, the cortex was separated from the medulla for homogenization. The kidney cortex was homogenized in 0.1 M Tris-HCl, pH 7.4 on ice. Then, tissue homogenates were sonicated on ice for 10 s/sample. Then, the samples were centrifuged for 15 min at 2500× *g* at 4 °C. For the mitochondrial studies, mice were not given a cisplatin or saline injection, and kidneys were removed and decapsulated for mitochondrial isolation or snap frozen in liquid nitrogen and stored at −80 °C for ATP analysis. All animal studies were approved by the Institutional Animal Care Use Committee (IACUC) at the University of Louisville and followed the guidelines of the American Veterinary Medical Association.

2.2. Fructose-1,6-Bisphosphatase (FBPase) Activity Assay

FBPase activity was measured as described previously [29]. Briefly, 20 µg protein (10 µL) was added to 170 µL assay buffer containing 0.1 M Tris-HCl, pH 8.6; 0.1 $MgCl_2$; 0.1 M cysteine HCl in a 96-well plate and incubated at 37 °C for 5 min. Then, 20 µL of freshly prepared 50 mM fructose-1,6-bisphosphate (final concentration 5 mM) was added. Samples along with blanks (protein and substrate) and freshly prepared standards (0, 10, 25, 50, 100, 150, 200, 250 nM KH_2PO_4) were incubated at 37 °C for 60 min. The reaction was stopped by the addition of 50 µL 10% trichloroacetic acid (TCA), and the samples were centrifuged at 10,000× *g* for 5 min. Then, 200 µL of supernatant from samples, standards, and blanks was carefully pipetted into a fresh 96-well plate. Then, 50 µL freshly prepared 5% $FeSO_4$ (0.5 g of FeSO4 dissolved in 1 mL of 10% ammonium molybdate in 0.2 N H_2SO_4 and the total volume was brought to 10 mL using distilled water) was added. The samples, standards, and blanks were read after 10 min at 820 nm wavelength. All samples, blanks, and standards were read in triplicate. The amount of inorganic phosphate released by the reaction of FBPase was determined by subtracting the substrate blank values from the sample values using the standards curve. All samples, blanks, and standards were read in triplicate. The average activity from the triplicate was considered as $n = 1$ and shown as activity per mg protein (WT vehicle $n = 3$), (NHERF1 KO vehicle $n = 4$), (WT cisplatin $n = 3$), (NHERF1 KO cisplatin $n = 5$).

2.3. Glucose-6-Phosphatase (G6Pase) Activity Assay

G6Pase activity was measured as described [29]. Briefly, 20 µg protein (10 µL) was added to 170 µL assay buffer containing 0.1 M Tris-HCl, pH 7.4, and 0.1 M $MgCl_2$ in a 96-well plate and incubated for 5 min at 37 °C. Then, 20 µL freshly prepared substrate (50 mM glucose-6-phosphate, final concentration 5 mM) was added to each well. Samples along with blanks (protein and substrate) and freshly prepared standards (0, 10, 25, 50, 100, 150, 200, 250 nM KH_2PO_4) were incubated at 37 °C for 60 min. The reaction was stopped by the addition of 50 µL 10% TCA, and the samples were centrifuged at 10,000× *g* for 5 min. Then, 200 µL of supernatant from samples, standards, and blanks was carefully pipetted into a fresh 96-well plate. Then, 50 µL freshly prepared 5% $FeSO_4$ (0.5 g of $FeSO_4$ dissolved in 1 mL of 10% ammonium molybdate in 0.2 N H_2SO_4 and the total volume was brought to 10 mL using distilled water) was added. The samples, standards, and blanks were read after 10 min at 820 nm wavelength. All samples, blanks, and standards were read in triplicate. The amount of inorganic phosphate released by the reaction of G6Pase was determined by subtracting the substrate blank values from the samples values using the standard curve. The average activity from triplicates was considered as $n = 1$ and shown as activity per mg protein (WT vehicle $n = 3$), (NHERF1 KO vehicle $n = 4$), (WT cisplatin $n = 3$), (NHERF1 KO cisplatin $n = 5$).

2.4. Lactate Dehydrogenase (LDH) Activity Assay

LDH activity was measured as described [29]. Briefly, samples (20 µg protein, 10 µL) were added to 170 µL assay buffer containing 0.1 M Tris-HCl, pH 7.4; 0.1 M $MgCl_2$ and 0.5 mM sodium pyruvate (substrate) and incubated at 37 °C for 5 min. Reaction was started by adding 20 µL 5 mM nicotinamide adenine dinucleotide and hydrogen (NADH) freshly prepared in 1 mM sodium bicarbonate, pH 9.0 solution (final concentration 0.5 mM). Samples and blanks (solution, protein, and NADH) were immediately read at 340 nm for 10 min. The optical density values were recorded every minute. The optical density measured before the addition of NADH was considered as time zero. Absorbance in the linear range was used for calculating activity. The activity was determined as a decrease in absorbance due to the oxidation of NADH per mg protein. The values of NADH blank were subtracted from samples to calculate the final activity. All samples were read in triplicate and the average value was considered $n = 1$ (WT vehicle $n = 3$), (NHERF1 KO vehicle $n = 4$), (WT cisplatin $n = 3$), (NHERF1 KO cisplatin $n = 5$).

2.5. Malate Dehydrogenase (MDH) Activity Assay

MDH activity is determined as described [29]. Briefly, a 10 µL sample (20 µg protein was added to assay buffer containing 0.1 M Tris-HCl, pH 7.4; 0.5 mM oxaloacetate; 0.1 M $MgCl_2$. Reaction was started by adding 20 µL 5 mM NADH freshly prepared in 1 mM sodium bicarbonate, pH 9.0 solution (final concentration 0.5 mM). Samples and blanks (solution, protein, and NADH) were immediately read at 340 nm for 10 min. The optical density values were recorded every minute. The optical density measured before the addition of NADH was considered as time zero. Absorbance in the linear range was used for calculating activity. The activity was determined as a decrease in absorbance due to the oxidation of NADH per mg of protein. The values of the NADH blank were subtracted from samples to calculate the final activity. All samples were read in triplicate, and the average value was considered $n = 1$ (WT vehicle $n = 3$), (NHERF1 KO vehicle $n = 4$), (WT cisplatin $n = 3$), (NHERF1 KO cisplatin $n = 5$).

2.6. Malic Enzyme (ME) Activity Assay

ME activity was measured as described [29]. Briefly, a 10 µL sample (20 µg protein) was added to 170 µL assay buffer containing 0.1 M Tris-HCl, pH 7.4; 0.5 mM L-malic acid; 0.1 M $MgCl_2$. Reaction was started by the addition of 20 µL freshly prepared 2 mg/mL of $NADP^+$ in 100 mM imidazole. Samples and blanks (solution, sample, and $NADP^+$ blank) were immediately read at 340 nm for 10 min. The optical density values were recorded every minute. The optical density measured before the addition of $NADP^+$ was considered as time zero. Absorbance in the linear range was used for calculating activity. The activity was determined as an increase in absorbance due to the reduction of $NADP^+$ to NADPH per mg protein. The values of $NADP^+$ blank were subtracted from samples to calculate the final activity. All samples were read in triplicate, and the average value was considered $n = 1$ (WT vehicle $n = 3$), (NHERF1 KO vehicle $n = 4$), (WT cisplatin $n = 3$), (NHERF1 KO cisplatin $n = 5$).

2.7. Glucose-6-Phosphate Dehydrogenase (G6PD) Activity Assay

G6PD activity was measured as described [29]. Briefly, a 10 µL sample (20 µg protein) was added to 170 µL assay buffer containing 0.1 M Tris-HCl, pH 7.4; 0.5 mM glucose-6-phosphate; 0.1 M $MgCl_2$. Reaction was started by the addition of 20 µL freshly prepared 2 mg/mL of $NADP^+$ in 100 mM imidazole. Samples and blanks (solution, sample, and $NADP^+$ blank) were immediately read at 340 nm for 10 min. The optical density values due to the conversion of $NADP^+$ to NADPH were recorded every minute. The optical density measured before the addition of $NADP^+$ was considered as time zero. Absorbance in the linear range was used for calculating activity. The activity was determined as increase in absorbance due to the reduction of $NADP^+$ to NADPH per mg protein. The values of $NADP^+$ blank were subtracted from samples to calculate the final activity. All samples were read in triplicate, and the average was considered $n = 1$ (WT vehicle $n = 3$), (NHERF1 KO vehicle $n = 4$), (WT cisplatin $n = 3$), (NHERF1 KO cisplatin $n = 5$).

2.8. Liquid Chromatography-Mass Spectrometry of Kidney Cortex for ATP Quantification

First, 10 mg of kidney cortex was snap-frozen and stored at −80 °C until the LC-MS procedure. The tissue was maintained in liquid nitrogen during the preparation for LC-MS. Tubes and beads were pre-cooled in liquid nitrogen and 2.5% kidney cortex homogenate was made using an extraction solution (70% acetonitrile (ACN) and 30% H_2O). Tissue was homogenized using a bead beater at 5 m/s for 15 s. Then, the homogenized tissue was transferred to another tube for centrifugation at 16,000× *g* for 10 min at 4 °C. The supernatant was saved for subsequent LC-MS analysis. To quantitate the ATP present in mouse kidney cortex tissue, a standard curve was prepared using 1 mM ATP diluted serially with the extraction buffer containing 20 µM $^{13}C_{10}$ATP. Each concentration point was diluted 50× with 60% ACN and 40% 15 mM ammonium acetate. Then, 5 µL was injected onto a SeQuant ZIC-cHILIC 100 × 2.1 mm metal-free HPLC column. Separation was performed by pumping 60% ACN and 40% 15 mM ammonium acetate through the column with a flow rate of 0.5 mL/min with no change in composition using a Waters Acquity UPLC. ATP was eluted from the column and flowed into a Waters Quattro Premier XE mass spectrometer and subsequently quantitated using optimized MRMs. A calibration curve was constructed using ATP concentration on the x-axis and the response of ATP over $^{13}C_{10}$ATP on the y-axis. Mouse kidney extracts were also diluted with 50x 60% ACN and 40% 15 mM ammonium acetate, 5 µL was injected, and the concentration in kidney tissue was interpolated using the calibration curve previously made for (WT $n = 5$) and (NHERF1 KO $n = 5$).

2.9. Perfusion Fixation of Total Kidney In Situ for Electron Microscopy

The abdominal cavity of the mouse was opened along the *linea alba* and intestines were moved aside to expose kidneys. A suture was loosely put around the right kidney for removal without glutaraldehyde exposure. Then, the right kidney was used for kidney cortex homogenates. Next, the chest cavity was opened to expose the heart for perfusion through the left ventricle, and the vena cava was cut, while the right kidney was tied off and removed. The left mouse kidney was perfused with 3% glutaraldehyde solution at a rate of 6 mL/min for approximately 1–3 min. Perfusion was stopped once kidneys had a change in color and consistency. The kidney was removed and placed in a petri dish of

3% glutaraldehyde where three to four 0.2 cm × 0.4 cm slices were cut and then stored in 3–4 mL of 3% glutaraldehyde at 4 °C. At this point, the tissue was sent to Norton Children's Hospital Pathology Department for a blind EM analysis of kidney tubule mitochondria. Images were taken by a renal pathologist at the Pathology Department. Sections with the highest concentration of mitochondria were randomly picked in a 4x field to be used to calculate mitochondria number and mitochondria area by Image J (WT $n = 6$) and (NHERF1 KO $n = 5$).

2.10. Seahorse XF24 Mitochondrial Respiration Analysis

Mice for this study had food taken away 6 h before sacrifice. Mitochondrial oxidative capacity was measured in isolated kidney mitochondria using a Seahorse Bioscience XF24 extracellular flux analyzer (Billerica, MA, USA). For measurements in isolated mitochondria, tissue from the kidney cortex of both kidneys (approximately 50 mg) was isolated and homogenized in 1 mL of isolation buffer (220 mM mannitol, 70 mM sucrose, 5 mM 3-(N-morpholino) propanesulfonic acid (MOPS), 1 mM ethylene glycol tetraacetic acid (EGTA), 0.3% fatty acid-free bovine serum albumin (BSA), pH 7.2). The homogenate was centrifuged at 500× *g* for five minutes at 4 °C. The supernatant containing mitochondria was centrifuged at 10,000× *g* for five minutes. Following two wash centrifugation steps in BSA-free isolation buffer, the mitochondria were suspended in respiration buffer (120 mM KCl, 25 mM sucrose, 10 mM 4-(2-hydroxyethyl)-1-piperazineethanesulfonic acid (HEPES), 1 mM $MgCl_2$, 5 mM KH_2PO_4, pH to 7.2). Protein in the mitochondrial suspension was estimated by the bicinchoninic acid method (Sigma) using BSA as a standard. Then, 10 μg of mitochondrial protein was sedimented in XF culture plates. Succinate (10 mM), rotenone (1 μM), and ADP (1 mM) were injected to assess state 3 respiratory activity (phosphorylating respiration). The oxygen consumption rate (OCR) of mitochondria after exposure to oligomycin (1 μg/mL) was used to estimate state 4 activity (non-phosphorylating respiration). Finally, Antimycin A (20 μM) was utilized to stop all respiration. Data are expressed as pmol O_2/min/μg protein (WT $n = 6$, NHERF1 KO $n = 6$).

2.11. Brush Border Membrane Isolation

Kidney cortex BBM from WT and NHERF1 KO mice were prepared at 4 °C using the $MgCl_2$ precipitation method as previously described [30]. Kidney cortex slices were minced and then briefly homogenized in 50 mM mannitol and 5 mM Tris-HEPES buffer, pH 7.0 (20 mL/g), in a glass teflon homogenizer with four complete strokes. Then, a polyron homogenizer was used to provide high-speed [20,500 revolutions/min (rpm)] homogenization for three strokes of 15 s each with an interval of 15 s between each stroke. $MgCl_2$ was added to the homogenate to a final concentration of 10 mM and slowly stirred for 20 min. The homogenate was spun at 2000× *g* in a Sorvall high-speed centrifuge using an SS-34 rotor. The supernatant was centrifuged at 35,000× *g* for 30 min using the SS-34 rotor. Then, the pellet was resuspended in 300 mM mannitol and 5 mM Tris-HEPES, pH 7.4, with four passes by a loose-fitting Dounce homogenizer (Wheaton, IL) and centrifuged at 35,000× *g* for 20 min in 15 mL Corex tubes using the SS-34 rotor. The outer final pellet was resuspended in a small volume of buffered 300 mM mannitol.

2.12. Label-Free Quantitative Liquid Chromatography-Mass Spectrometry

Protein samples were analyzed by 2D-LC-MS/MS using either a linear trap quadrupole (LTQ) ion trap mass spectrometer (Thermo Fisher Scientific, Waltham, MA, USA) and/or using a Thermo Scientific LTQ Orbitrap Elite hybrid FTMS system enabled with electron transfer dissociation (ETD) fragmentation. Routinely one-dimensional (1D) reversed phase (RP) or two-dimensional (2D) strong cation exchange (SCX)-(RP) HPLC experiments were conducted off line for the LC-MALDI approach or online for the LC-LTQ-XL-Orbitrap-ESI-MS/MS approach and protein assignments were made as previously described [31–34].

The acquired mass spectrometry data were searched against an appropriate REFSEQ protein database using the SEQUEST (version 27 revision 11) algorithm and Matrix Science Mascot v.1.4 using

Proteome Discoverer v1.3 to prepare and process RAW files as well as aggregate results. Separately for comparative LCMS analyses, the MS/MS database analysis was performed with SequestSorcerer (Sage-N Research, San Jose, CA) and high-probability peptide and protein identifications were assigned from the SEQUEST results using the Protein and PeptideProphet (tools.proteomecenter.org/software.php) and SageN Sorcerer statistical platforms. Scaffold 3 proteomic analysis software (ProteomeSoftware, Inc, Portland, OR) was used for quantitative comparison using a label-free spectral counting method. The qualitative comparison of protein expression patterns was performed by Ingenuity Pathways Analysis software (http://ingenuity.com).

2.13. Interpretation of Protein Assignment Results

Lists of identified proteins and expression patterns were submitted for pathways analysis using one of several publically available (David, http://david.abcc.ncifcrf.gov/; GO, http://www.geneontology.org/; KEGG Pathway database, http://genome.jp/kegg/pathway.html), GenMAPP (www.genmapp.org/) or commercially licensed tools (Ingenuity Pathways Analysis software, (Redwood City, California, USA) http://ingenuity.com).

2.14. Statistical Analysis

Data are shown as means ± standard error means (SEM) or standard deviation (SD). Two-way ANOVA was used to compare the different treatment groups for the enzyme kinetic assays. Student's *t*-test was performed for the LC-MS data measuring ATP content, mitochondrial number, mitochondrial area, and mitochondrial respiration via Seahorse XF24. All statistical analysis was conducted with SPSS version 24 software. p values of < 0.05 were considered statistically significant.

3. Results

3.1. Absence of NHERF1 Results in Extensive Changes in Kidney BBM Protein Expression

BBM were prepared from the kidney cortex of WT and NHERF1 KO mice, followed by comparative proteomic analysis. Pathway analysis of differentially expressed proteins revealed significant changes in proteins associated with mitochondrial function as well as protein components of signaling pathways, actin cytoskeleton, cell survival, and oxidative phosphorylation (Supplementary Materials Figure S1, Table S1). Based on these findings, we proceeded to determine if metabolic pathways of WT and NHERF1 KO mouse kidneys differed. Under normal conditions, gluconeogenesis is a signature metabolic pathway for kidney proximal tubules with little glycolysis [35]. Therefore, we examined enzymes of the pathways involved in carbohydrate metabolism.

3.2. Cisplatin Treatment Significantly Decreases FBPase and G6Pase Enzyme Activity in Both WT and NHERF1 KO Mice

Previous studies investigated the effect of cisplatin on gluconeogenesis [2,36,37]; however, the effect of NHERF1 protein loss on gluconeogenesis has not been investigated. FBPase is a critical regulatory enzyme in gluconeogenesis that catalyzes the hydrolysis of fructose-1,6-bisphosphate to fructose-6-phosphate and inorganic phosphate [38]. In order to determine if NHERF1 loss affected gluconeogenic enzyme activity alone or with cisplatin treatment, male 2–4-month-old mice were treated with vehicle or cisplatin to induce AKI and then sacrificed after 72 h. Kidney cortex tissue from these mice were used for the FBPase enzyme kinetic assay as described in the methods section. There were no significant differences between vehicle [(WT: 41.5 nmole/mg protein/min ± 6.5) (KO: 38.7 nmole/mg protein/min ± 4.1)] or cisplatin [(WT: 20.6 nmole/mg protein/min ± 0.4) (KO: 19.7 nmole/mg protein/min ± 1.2)] treated WT and NHERF KO kidneys (Figure 1A). However, cisplatin did decrease FBPase activity in both WT and NHERF1 KO kidneys ($p = 0.0001$) (Figure 1A).

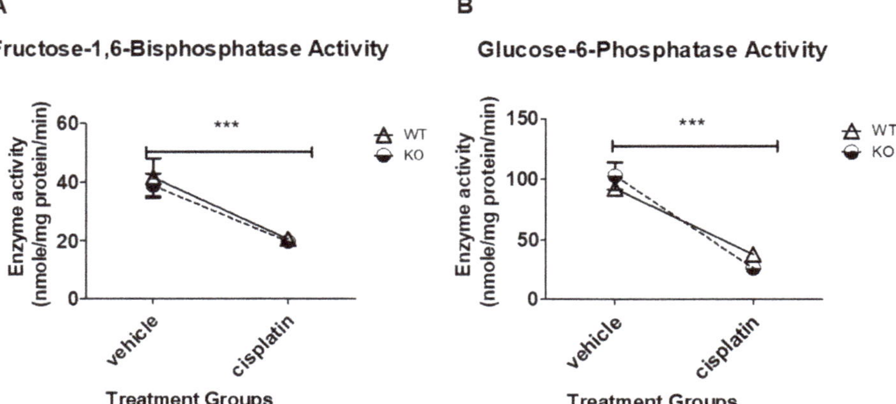

Figure 1. Effect of cisplatin treatment on fructose-1,6-bisphosphatase and glucose-6-phosphase enzyme activity in wild-type (WT) and Na/H exchange regulatory factor 1 (NHERF1) knockout (KO) mouse kidneys. Two to 4-month-old male C57BL/6J WT and NHERF1 KO mice were given cisplatin (20 mg/kg dose intraperitoneally (IP)) or vehicle (saline) and sacrificed after 72 h as described in the Methods section. (**A**) Fructose-1,6-bisphosphatase (FBPase) enzyme activity was determined from the kidney cortex tissue of these mice. Data are means ± SEM (WT vehicle $n = 3$), (KO vehicle $n = 4$), (WT cisplatin $n = 3$), and (KO cisplatin $n = 5$). *** $p = 0.001$ cisplatin-treated WT and NHERF1 KO mice compared to vehicle saline controls. (**B**) Glucose-6-phosphatase (G6Pase) enzyme activity was determined from the kidney cortex tissue of these mice. Data are means ± SEM (WT vehicle $n = 3$), (KO vehicle $n = 4$), (WT cisplatin $n = 3$), and (KO cisplatin $n = 5$). *** $p < 0.001$ cisplatin-treated WT and NHERF1 KO mice compared to vehicle saline controls.

G6Pase is the final step in gluconeogenesis, where it hydrolyzes glucose-6-phosphate to free glucose and a phosphate group [39]. Similarly to FBPase, G6Pase activity was comparable in WT and NHERF1 KO kidneys regardless of treatment [(WT vehicle: 92.3 nmole/mg protein/min ± 5.0), (KO vehicle: 103.0 nmole/mg protein/min ± 11.3), (WT cisplatin: 38.0 nmole/mg protein/min ± 5.1), and (KO cisplatin: 26.4 nmole/mg protein/min ± 3.8)]. Cisplatin led to a significant decrease in enzyme activity in both genotypes ($p < 0.0001$) (Figure 1B).

3.3. *NHERF1 Deficiency or Cisplatin Treatment Does Not Significantly Affect LDH or MDH Enzyme Activity in Mice*

Lactate dehydrogenase (LDH) catalyzes the interconversion of lactate and pyruvate, concomitantly with the interconversion of NADH and NAD^+ [40]. When oxygen is absent, LDH converts pyruvate, the final product of glycolysis, to lactate [40]. Thus, LDH activity was measured in vehicle and cisplatin mouse kidneys. NHERF1 loss ($p = 0.65$) or cisplatin treatment ($p = 0.71$) did not significantly affect lactate dehydrogenase activity in these mouse kidneys [(WT vehicle: 0.06 nmole/mg protein/min ± 0.02), (KO vehicle: 0.1 nmole/mg protein/min ± 0.05), (WT cisplatin: 0.1 nmole/mg protein/min ± 0.8), and (KO cisplatin: 0.02 nmole/mg protein/min ± 0.006)] (Figure 2A).

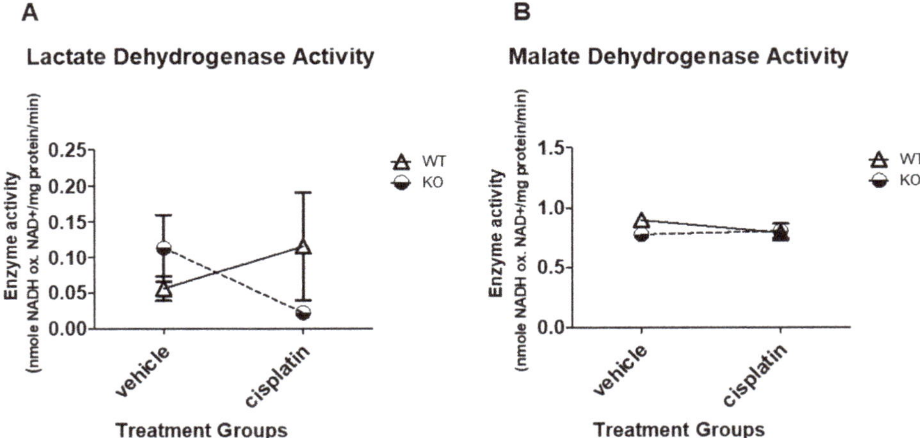

Figure 2. Lactate dehydrogenase and malate dehydrogenase enzyme activity in WT and NHERF1 KO mouse kidneys. Two to 4-month-old male C57BL/6J WT and NHERF1 KO mice were given cisplatin (20 mg/kg dose IP) or vehicle (saline) and sacrificed after 72 h as described in the Methods section. (**A**) Lactate dehydrogenase (LDH) enzyme activity was determined from kidney cortex tissue of these mice. Data are mean ± SEM (WT vehicle $n = 3$), (KO vehicle $n = 4$), (WT cisplatin $n = 3$), and (NHERF1 KO cisplatin $n = 5$). No significant differences were recorded. (**B**) Malate dehydrogenase (MDH) enzyme activity was determined from the kidney cortex tissue of these mice. Data are mean ± SEM (WT vehicle $n = 3$), (KO vehicle $n = 4$), (WT cisplatin $n = 3$), (NHERF1 KO cisplatin $n = 5$). No significant differences were reported.

Malate dehydrogenase (MDH) is an enzyme involved in many metabolic pathways including the citric acid cycle. MDH reversibly catalyzes the oxidation of malate to oxaloacetate with the reduction of NAD^+ to NADH [41]. In this study, we measured MDH activity as the conversion of oxaloacetate to malate and oxidation of NADH to NAD^+. The effect of NHERF1 loss and/or cisplatin treatment on MDH activity was analyzed. In a similar way to LDH, NHERF1 loss or cisplatin treatment did not significantly affect MDH activity in these mouse kidneys [(WT vehicle: 0.9 nmole/mg protein/min ± 0.06), (KO vehicle: 0.7 nmole/mg protein/min ± 0.02), (WT cisplatin: 0.8 nmole/mg protein/min ± 0.02), and (KO cisplatin: 0.81 nmole/mg protein/min ± 0.06)] (Figure 2B).

3.4. NHERF1 Deficiency Upregulates ME and G6PD Activity

Malic enzyme (ME) catalyzes the conversion of malic acid to pyruvate and produces NADPH [42]. ME serves as an additional source of NADPH for lipogenesis. In order to understand the effect that NHERF1 loss and/or cisplatin treatment may have on ME activity, kidney cortex tissue from vehicle or cisplatin-treated WT and NHERF1 KO were evaluated. Interestingly, there was a significant genotype effect on ME activity resulting in an increase in activity in NHERF1 KO kidneys ($P = 0.0065$) (Figure 3A). Additionally, a significant interaction was also noted between WT and NHERF1 KO kidneys after cisplatin treatment ($p = 0.0005$) [(WT vehicle: 0.07 nmole/mg protein/min ± 0.012), (KO vehicle: 0.21 nmole/mg protein/min ± 0.01), (WT cisplatin: 0.15 nmole/mg protein/min ± 0.012), and (KO cisplatin: 0.13 nmole/mg protein/min ± 0.02)] (Figure 3A).

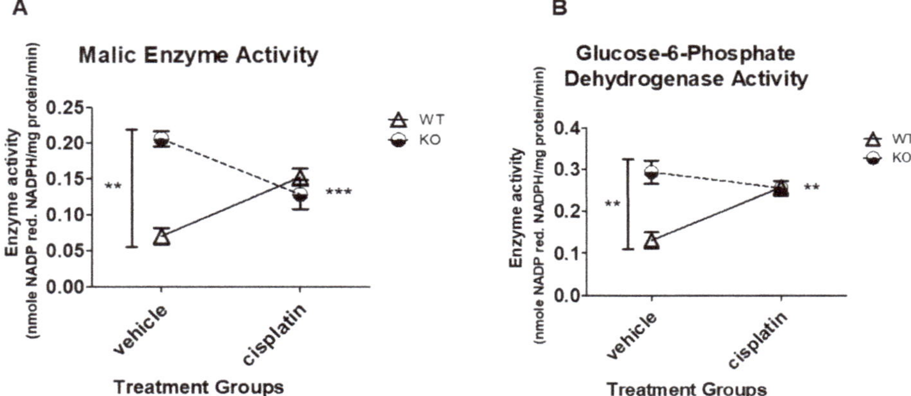

Figure 3. Effect of NHERF1 loss and cisplatin treatment on malic enzyme and glucose-6-phosphate dehydrogenase enzyme activity in WT and NHERF1 KO mouse kidneys. Two to 4-month-old male C57BL/6J WT and NHERF1 KO mice were given cisplatin (20 mg/kg dose IP) or vehicle (saline) and sacrificed after 72 h as described in the Methods section. (**A**) Malic enzyme (ME) activity was determined from the kidney cortex tissue of these mice. Data are mean ± SEM (WT vehicle $n = 3$), (KO vehicle $n = 4$), (WT cisplatin $n = 3$), and (KO cisplatin $n = 5$). ** $p = 0.0065$. Vehicle-treated NHERF1 KO mice compared to WT vehicle controls; *** $p = 0.0005$ interaction of cisplatin-treated NHERF1 KO mice to cisplatin-treated WT mice. (**B**) G6PD enzyme activity was determined from the kidney cortex tissue of these mice. Data are mean ± SEM (WT vehicle $n = 3$), (KO vehicle $n = 4$), (WT cisplatin $n = 3$), and (KO cisplatin $n = 5$). ** $p = 0.0033$ vehicle-treated NHERF1 KO mice compared to WT vehicle controls; ** $p = 0.00029$ interaction of cisplatin-treated NHERF1 KO mice to cisplatin-treated WT mice.

Cisplatin-induced AKI is known to decrease intermediates of the pentose phosphate pathway [2,7] in mice. Glucose-6-phosphate dehydrogenase (G6PD) is a cytosolic enzyme that participates in the pentose phosphate pathway, resulting in NADPH production [43]. This is accomplished when G6PD reduces $NADP^+$ to NADPH while oxidizing glucose-6-phosphate [43]. G6PD enzyme activity was analyzed in vehicle and cisplatin-treated WT and NHERF1 KO kidney cortex to elucidate if NHERF1 loss and/or cisplatin treatment affected the pentose phosphate pathway. Similarly, to ME, there was a significant genotype effect on G6PD activity, resulting in an increase in activity in NHERF1 KO kidneys ($p = 0.0033$) (Figure 3B). Additionally, a significant interaction was also noted between WT and NHERF1 KO kidneys after cisplatin treatment ($p = 0.00029$) [(WT vehicle: 0.13 nmole/mg protein/min ± 0.02), (KO vehicle: 0.3 nmole/mg protein/min ± 0.03), (WT cisplatin: 0.3 nmole/mg protein/min ± 0.007), and (KO cisplatin: 0.3 nmole/mg protein/min ± 0.02)] (Figure 3B).

3.5. NHERF1 Deficiency Does Not Affect ATP Abundance in Mouse Kidneys

ATP provides energy to drive many cellular processes and is consumed during many metabolic processes [44]. In eukaryotes, ATP is produced by three different metabolic pathways: [1] glycolysis, [2] the citric acid cycle or oxidative phosphorylation, and [3] beta-oxidation [44]. In order to determine if NHERF1 KO kidneys had differences in ATP content, kidneys were snap-frozen and processed while cold for LC-MS as described in the Methods section. LC-MS analysis revealed there were no significant differences in ATP amount in WT (3.4 nmoles/mg tissue ± 0.5) and NHERF1 KO (3.1 nmoles/mg tissue ± 0.5) kidneys ($p = 0.67$) (Figure 4).

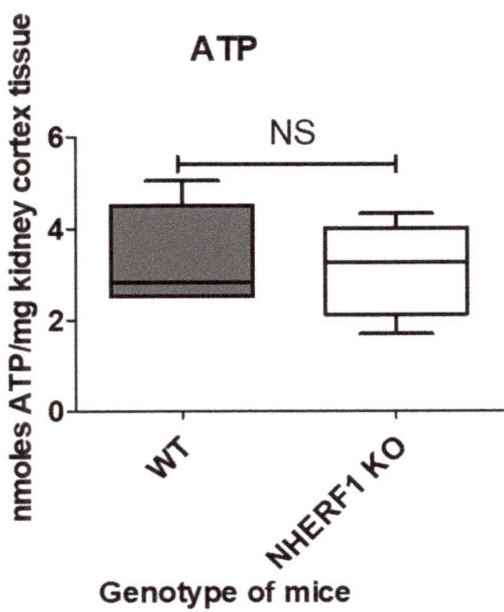

Figure 4. ATP content of WT and NHERF1 KO mouse kidneys.LC-MS was utilized to evaluate the amount of ATP in these tissues as described in the Methods section. Data are means ± SEM (WT $n = 5$) and (KO $n = 5$). No significant differences were reported in these kidneys.

3.6. NHERF1 Deficiency Does Not Affect Kidney Proximal Tubule Mitochondria Morphology, Number, or Area

The mitochondrial structure is essential for proper function; thus, EM images of WT and NHERF1 KO proximal tubule mitochondria were utilized to evaluate their morphology. These images were of 2–4-month-old male C57BL/6J WT and NHERF1 KO mice whose kidneys were perfused with 3% glutaraldehyde prior to EM analysis.

These were taken and evaluated by a renal pathologist for signs of injury, oxidative stress, and changes in cristae. There were no apparent changes in mitochondria morphology between WT and NHERF1 KO proximal tubules (Figure 5, panels A and B). The only injury reported was early ischemic changes most likely due to harvesting of the kidneys (Figure 5). Some endosomal swelling was noted but occurred across both genotypes. Additionally, the density and distribution of mitochondria within the tubules were alike, and no apparent signs of oxidative stress were found in either genotype (Figure 5).

Changes in mitochondrial number and a decrease in size have been associated with a decline in mitochondrial function [45]. Therefore, one goal of this study was to determine if mitochondrial number and/or size changed in NHERF1 KO proximal tubules when compared to WT. Images from electron microscopy (EM) of WT and NHERF1 KO kidney proximal tubules were utilized in order to count the number of mitochondria and to calculate the average area via Image J. There was not a significant difference between the average number of mitochondria between WT and NHERF1 KO tubules (WT average number: 128.8) and (NHERF1 KO average number: 115) ($p = 0.6$) (Figure 6A). In addition, there was not a significant difference between the average area of mitochondria in WT and NHERF1 KO tubules (WT average area: 580,540.9 μm^2) and (NHERF1 KO average area: 678,465.4 μm^2) ($p = 0.75$) (Figure 6B).

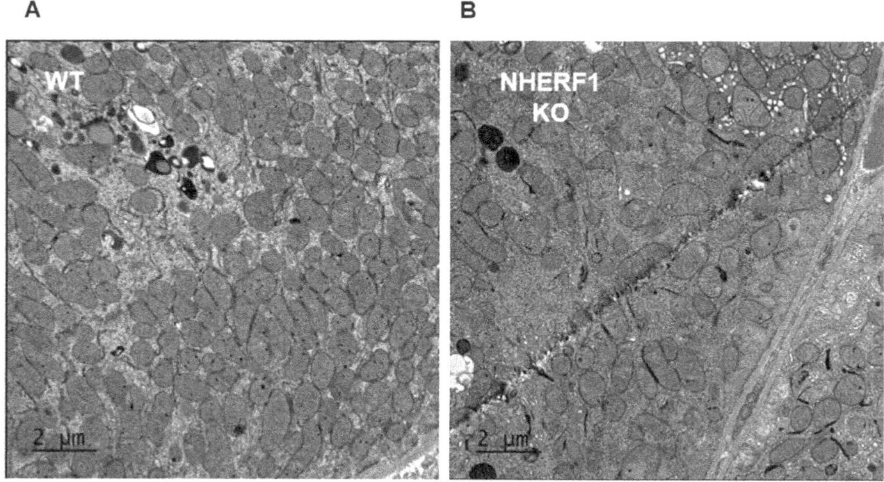

Figure 5. Electron microscopy of mitochondria in WT and NHERF1 KO proximal tubules.

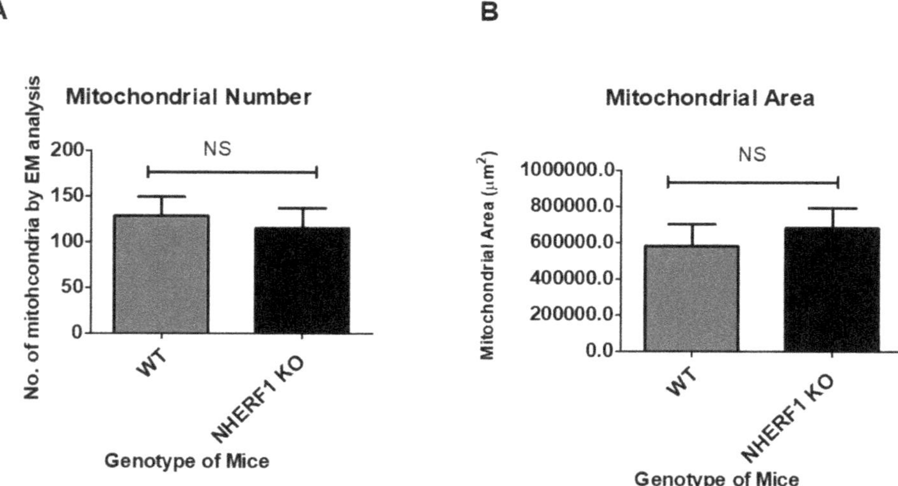

Figure 6. Evaluation of mitochondrial number and area of WT and NHERF1 KO proximal tubules. (**A**) Number of mitochondria were counted in random 4× visual fields with the highest density of mitochondria. Data are means ± SEM (WT $n = 6$) and (NHERF1 KO $n = 5$). The mitochondria number of NHERF1 KO proximal tubules was insignificant when compared to WT. (**B**) Mitochondria area was calculated using electron microscopy (EM) images and Image J. Data are means ± SEM (WT $n = 6$) and (NHERF1 KO $n = 5$). Mitochondria area of NHERF1 KO proximal tubules were insignificant when compared to WT.

3.7. WT and NHERF1 KO Mouse Kidney Mitochondria Have Similar Oxidative Capacities

The mitochondria's capacity to reduce oxygen is a critical aspect in the process of mitochondrial electron transport and ATP synthesis. Therefore, measuring mitochondrial oxygen consumption can provide a valuable method to assess mitochondrial function. One purpose of this work was to assess mitochondrial function by oxidative capacity in WT and NHERF1 KO kidneys using the Seahorse XF24 analyzer. In panel A of Figure 7, the oxygen consumption rate (OCR) of WT and NHERF1 KO kidney mitochondria are shown over time. Both WT and NHERF1 KO mitochondria exhibit a similar trend and response to added substrates and inhibitors. When adding the substrate Succinate/Rotenone plus ADP for the production of ATP, both genotypes exhibit a maximal increase in OCR. Moreover, both genotypes undergo a decrease in OCR after adding oligomycin, an inhibitor of complex V (formation of ATP from ADP via O_2 consumption). Lastly, antimycin A shuts down all respiration, where the OCR is close to the basal OCR. The difference between the basal OCR and OCR after antimycin A is the non-mitochondrial respiration.

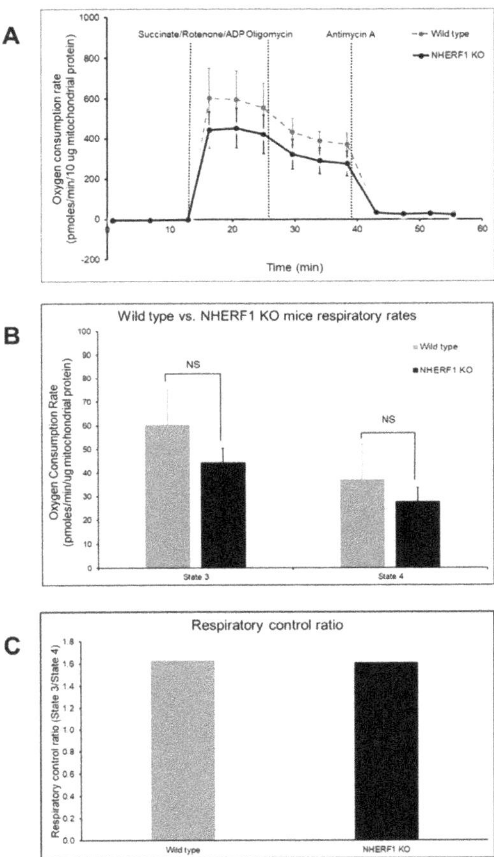

Figure 7. Mitochondrial function in isolated mitochondria of WT and NHERF1 KO kidneys. Mitochondria from two to 4-month-old male WT and NHERF1 KO mice were isolated and analyzed via the Seahorse XF24 for oxidative capacity as described in the Methods section. (**A**) Oxygen consumption rate (OCR) was recorded after the addition of both substrates (succinate/rotenone/ADP) and inhibitors (oligomycin and antimycin A) [(WT n = 6) and (NHERF1 KO n = 6)]. (**B**) State 3 and state 4 were calculated using the recorded OCRs of WT and NHERF1 KO mitochondria. Data are mean ± SD [(WT n = 6) and (NHERF1 KO n = 6)]. State 3 and state 4 respiration were considered insignificant between WT and NHERF1 KO mouse kidney mitochondria. (**C**) Respiratory control ratio (RCR) (state 3/state 4) was calculated between WT and NHERF1 KO kidney mitochondria. Data are represented as state 3/state 4 ratio [(WT n = 6) and (NHERF1 n = 6)]. RCR was insignificant between WT and NHERF1 KO mouse kidney mitochondria.

Changes in state 3 (conversion of ATP from ADP and consumption of O_2) and state 4 (non-phosphorylating or resting respiration) respiration are commonly used to evaluate mitochondria oxidative capacity. Panel B of Figure 7 shows both state 3 [(WT: 60 pmoles/min/μg protein ± 15) and (NHERF1 KO: 44 pmoles/min/μg protein ± 6)] and state 4 [(WT: 37 pmoles/min/μg protein ± 15) and (NHERF1 KO: 28 pmoles/min/μg protein ± 6)] of WT and NHERF1 KO kidney mitochondria, where state 3 ($p = 0.2$) and 4 ($p = 0.1$) was determined to not be significantly different between the groups.

The respiratory control ratio (RCR) is the best general measure of mitochondrial function in isolated mitochondria. RCR is measured by taking state 3/state 4 respiration and sums up the main function of mitochondria: the ability to respond to ADP from a resting state by making ATP at high rates. The RCR has no absolute value that is diagnostic of mitochondrial dysfunction [46]. Thus, RCR values are substrate and tissue-dependent, making the RCR advantageous when measuring mitochondrial function in isolated mitochondria [46]. A change in almost any aspect of oxidative phosphorylation will result in a change in the RCR when comparing isolated mitochondria [46].

Representative photomicrographs show a 4x field of WT and NHERF1 KO proximal tubule mitochondria. Panel A represents the proximal tubule mitochondria of WT mice, while panel B represents NHERF1 KO proximal tubule mitochondria (WT $n = 6$) and (NHERF1 KO $n = 5$). The scale bars were set at 2 μm.

Accordingly, the RCR was calculated between WT (1.63) and NHERF1 (1.61) KO kidney mitochondria and was also found to not be significantly different (Figure 7C).

4. Discussion

This work aimed to examine two aspects of proximal tubule cell function: metabolic enzymatic pathways and mitochondrial structure and function. We proposed that changes in kidney metabolism and/or mitochondrial structure or function could predispose NHERF1 KO mice to cisplatin nephrotoxicity. This is the first study to find changes in the kidney pentose phosphate pathway enzymes with NHERF1 loss and a novel proposed mechanism of susceptibility to cisplatin-induced AKI. Recent studies have found that cisplatin alters renal cell metabolism, contributing to injury and the secondary result of chronic kidney disease (CKD) development [2–6]. Cisplatin treatment results in the depletion of amino acids in the kidney [2–5], reduces fatty acid oxidation while concomitantly accumulating fatty acids in the kidney [2,5,6], and decreases renal glycolytic enzymes and intermediates of the pentose phosphate pathway [2,36]. In addition to affecting metabolic pathways cisplatin, nephrotoxicity has been established in inducing apoptotic and necrotic cell death. The mechanisms involved in cisplatin-induced nephrotoxic cell death remain unclear. However, there is increasing evidence that ROS and mitochondrial function have an important role in cisplatin's mechanism of injury. These observations combined with the increased susceptibility to cisplatin-induced AKI suggested the hypothesis that NHERF1 KO mice have metabolic alterations and/or mitochondrial dysfunction that predispose them to cisplatin nephrotoxicity.

NHERF1 KO mice did not exhibit changes in gluconeogenic or glycolytic enzyme activity. Indeed, cisplatin treatment resulted in a parallel decrease in FBPase and G6Pase activity in NHERF1 KO and WT mice. Additionally, there were no significant changes with LDH and MDH activity between non-treated and treated WT and NHERF1 KO mice. These results are in agreement with previous studies [2,7]. Interestingly, NHERF1 KO mouse kidneys exhibit increased activity of ME and G6PD under baseline conditions when compared to WT mouse kidneys. The significance of this shift in metabolism toward a greater utilization of the pentose phosphate pathway is not entirely clear. However, these findings suggest a potential compensatory mechanism for increased NADPH production as protection against oxidative stress. ME and G6PD activity provide necessary NADPH, a key cofactor in redox control and reductive biosynthesis. ME plays a role in the production of pyruvate and serves as an additional source of NADPH for lipogenesis. Additionally, there is recent evidence for direct cross-talk between ME and the pentose phosphate pathway [47], where G6PD is a rate-limiting enzyme. A study using a cell culture model of diabetes observed that the increased activity of G6PD restored redox balance in endothelial cells exposed to high glucose levels [48], where high glucose levels had previously decreased G6PD and increased levels of oxidative stress. A similar observation has been made in studies of liver cirrhosis in rats subject to oxidative stress, where an increase in ME and G6PD gene expression and activity [49] are also seen, presumably providing protection against the stress through an increased production of NADPH [49]. Multiple studies have noted the importance of cellular redox balance in both the development of and in protection from renal injury [50,51]. Additionally, one other investigator found that NHERF1 is a previously unidentified regulator of Nox1 (NADPH oxidase) and promotes Nox1 activity [52]. Taken together, these observations suggest that the kidneys of NHERF1 KO mice experience a greater degree of oxidative stress that is masked by increased NADPH production through the pentose phosphate pathway. If NHERF1 KO mouse kidneys are more reliant on the pentose phosphate pathway for maintenance of the cellular redox state, the decrease in the activity of the enzymes of the pentose phosphate pathway resulting from cisplatin toxicity could potentially result in more severe injury.

While proteomic data demonstrated changes in mitochondrial proteins in NHERF1 KO mice (Figure S1) (Table S1), we found no alterations in mitochondrial structure or function. The kidney ATP levels in NHERF1 KO mice are equivalent to WT. EM analysis revealed that NHERF1 KO mice have similar proximal tubule mitochondrial morphologies, size, distribution and number when compared to WT. This result is not surprising, as our proteomic analysis did not identify differences in any of the proteins associated with mitochondrial fission or fusion such as dynamin related protein 1 (Drp1), mitofusin-1 (Mfn1), or mitofusin-2 (Mfn2). Mitochondria from NHERF1 KO and WT kidneys were found to have similar oxidative capacities as demonstrated by the measurement of OCR and RCR (WT: 1.63 and KO: 1.61) (Figure 7C). However, these findings do not exclude completely a role for altered mitochondrial function as a contributor to enhanced susceptibility to cisplatin-induced AKI. Although isolated mitochondria from NHERF1 KO kidneys function normally, they may not do so in intact tissue. NHERF1 KO mice undergo phosphate wasting [28] due to the faulty trafficking of Npt2a to the apical membrane [21], which may result in intracellular phosphate deficiency. The loss of intracellular phosphate may create an intracellular environment where mitochondria cannot function properly. NHERF1 KO mouse kidneys may also sustain losses of other important nutrients. In addition to higher urine phosphate excretion, NHERF1 KO mice also demonstrate hypercalciuria and hyperuricosuria [26]. A full evaluation of the alterations in proximal tubule cell transport in the NHERF1 KO kidneys has not been examined. The absence of NHERF1 may also result in impaired signaling processes [53], alterations in intracellular or mitochondrial protein phosphorylation, loss of mitochondrial interaction with other organelles, or aberrant mitochondrial protein localization, as suggested by the proteomic data demonstrating changes in mitochondrial proteins associated with the renal cortical BBM (Figure S1) (Table S1). Thus, further studies to compare mitochondrial function of WT and NHERF1 KO in intact kidney tissue are warranted.

5. Conclusions

In conclusion, this study provides insight into metabolic and mitochondrial changes of NHERF1 KO mice and provides new avenues to explore regarding NHERF1 loss and susceptibility to cisplatin nephrotoxicity. We did not find changes in enzyme activity in gluconeogenesis or the citric acid cycle, ATP content, and mitochondrial morphology and function. However, we discovered that enzymes of the pentose phosphate pathway were found to be increased in NHERF1 KO mice and suggest these animals are expressing some differences in metabolic pathways and may be compensating for an underlying stress. The basis for these changes in the activity of this metabolic pathway and its significance for the increased susceptibility of NHERF1 KO mice to cisplatin nephrotoxicity remain unknown. These results provide another area to be explored in the future pertaining to NADPH levels in NHERF1 KO mouse kidneys. Further investigation into the bioenergetics of NHERF1 KO mouse kidneys may elucidate more insight into susceptibility to cisplatin injury and increase our understanding of the underlying mechanism of susceptibility to cisplatin-induced AKI. In the future, this information may provide novel therapeutic targets and/or biomarkers to use clinically for the prevention of cisplatin nephrotoxicity.

Supplementary Materials: The following are available online at http://www.mdpi.com/2076-3921/9/9/862/s1, Figure S1: Proteomic analysis of BBM from WT and NHERF1 KO mice, Table S1: Differential expression of BBM proteins from WT and NHERF1 KO mice.

Author Contributions: Conceptualization, A.B.-S. and E.L.; methodology, A.B.-S., E.L., M.T.B., S.J.K., B.G.H., K.K., J.H., A.S. and M.M.; Software, B.G.H., A.S., S.N.R., S.S., and M.M.; validation, A.B.-S., M.M., M.B., B.C., and E.L.; formal analysis, M.B., M.M., A.B.-S., S.N.R., and S.S.; investigation, A.B.-S., E.L., L.S., M.T.B., K.B.G., and K.K.; resources, B.G.H., L.S., M.M., M.B., M.T.B., and B.C.; data curation, A.B.-S., M.M., M.B., M.T.B., K.B.G. and J.H.; writing—original draft preparation, A.B.-S. and E.L.; writing—review and editing, M.M., B.G.H, S.J.K., K.K., B.C., M.T.B., and K.B.G.; visualization, A.B.-S., M.M., K.B.G., J.H., and E.L.; supervision, E.L., M.M., and M.T.B.; project administration, E.L.; funding acquisition, E.L. All authors have read and agreed to the published version of the manuscript.

Funding: This work was supported in part by Merit Review Award Number 1BX000610 from the United States Department of Veterans Affairs BLR&D Service. The views in the manuscript are those of the author and do not reflect the views of the Department of Veterans Affairs. This work was also supported by NIH P20GM113226 UofL Hepatobiology and Toxicology CoBRE, NIH P30ES P30ES030283 UofL Center for Integrative Environmental Health Sciences.

Acknowledgments: We are grateful to Susan Isaacs for expert technical support and David Hoetker for performing the LC-MS analysis in this manuscript. We would like to thank the University of Louisville Division of Nephrology for providing financial support.

Conflicts of Interest: Dr. Lederer is a consultant for WebMD and HQSI.

Data Sharing

Data files for acquired LCMS data (.RAW) and scaffold search results (.sf3) files along with a sample key and sequence database were deposited in MassIVE (http://massive.ucsd.edu/) data repository (MassIVE ID: MSV000085816) with the Center for Computational Mass Spectrometry at the University of California, San Diego and shared with the ProteomeXchange (www.proteomexchange.org) (Proteome Exchange ID: PXD020534).

References

1. Hanigan, M.H.; Devarajan, P. Cisplatin nephrotoxicity: Molecular mechanisms. *Cancer Ther.* **2003**, *1*, 47–61. [PubMed]
2. Lagies, S.; Pichler, R.; Kaminski, M.M.; Schlimpert, M.; Walz, G.; Lienkamp, S.S.; Kammerer, B. Metabolic characterization of directly reprogrammed renal tubular epithelial cells (iRECs). *Sci. Rep.* **2018**, *8*, 3878. [CrossRef] [PubMed]
3. Zhang, P.; Chen, J.-Q.; Huang, W.-Q.; Li, W.; Huang, Y.; Zhang, Z.-J.; Xu, F.-G. Renal Medulla is More Sensitive to Cisplatin than Cortex Revealed by Untargeted Mass Spectrometry-Based Metabolomics in Rats. *Sci. Rep.* **2017**, *7*, 44804. [CrossRef] [PubMed]
4. Boudonck, K.J.; Mitchell, M.W.; Német, L.; Keresztes, L.; Nyska, A.; Shinar, D.; Rosenstock, M. Discovery of Metabolomics Biomarkers for Early Detection of Nephrotoxicity. *Toxicol. Pathol.* **2009**, *37*, 280–292. [CrossRef]
5. Portilla, D.; Li, S.; Nagothu, K.K.; Megyesi, J.; Kaissling, B.; Schnackenberg, L.; Safirstein, R.L.; Beger, R.D. Metabolomic study of cisplatin-induced nephrotoxicity. *Kidney Int.* **2006**, *69*, 2194–2204. [CrossRef]
6. Wilmes, A.; Bielow, C.; Ranninger, C.; Bellwon, P.; Aschauer, L.; Limonciel, A.; Chassaigne, H.; Kristl, T.; Aiche, S.; Huber, C.G.; et al. Mechanism of cisplatin proximal tubule toxicity revealed by integrating transcriptomics, proteomics, metabolomics and biokinetics. *Toxicology In Vitro* **2015**, *30*, 117–127. [CrossRef]
7. Choi, Y.M.; Kim, H.K.; Shim, W.; Anwar, M.A.; Kwon, J.W.; Kwon, H.K.; Kim, H.J.; Jeong, H.; Kim, H.M.; Hwang, D.; et al. Mechanism of Cisplatin-Induced Cytotoxicity Is Correlated to Impaired Metabolism Due to Mitochondrial ROS Generation. *PLoS ONE* **2015**, *10*, e0135083. [CrossRef]
8. Ueda, N.; Kaushal, G.P.; Shah, S.V. Apoptotic mechanisms in acute renal failure. *Am. J. Med.* **2000**, *108*, 403–415. [CrossRef]

9. Richter, C.; Gogvadze, V.; Laffranchi, R.; Schlapbach, R.; Schweizer, M.; Suter, M.; Walter, P.; Yaffee, M. Oxidants in mitochondria: From physiology to diseases. *Biochim. Et Biophys. Acta* **1995**, *1271*, 67–74. [CrossRef]

10. Singh, G. A possible cellular mechanism of cisplatin-induced nephrotoxicity. *Toxicology* **1989**, *58*, 71–80. [CrossRef]

11. Gemba, M.; Fukuishi, N. Amelioration by ascorbic acid of cisplatin-induced injury in cultured renal epithelial cells. *Contrib. Nephrol.* **1991**, *95*, 138–142. [CrossRef] [PubMed]

12. Oh, G.-S.; Kim, H.-J.; Shen, A.; Lee, S.-B.; Yang, S.-H.; Shim, H.; Cho, E.-Y.; Kwon, K.-B.; Kwak, T.H.; So, H.-S. New Therapeutic Concept of NAD Redox Balance for Cisplatin Nephrotoxicity. *Biomed. Res. Int.* **2016**, *2016*, 4048390. [CrossRef] [PubMed]

13. Yang, Y.; Liu, H.; Liu, F.; Dong, Z. Mitochondrial dysregulation and protection in cisplatin nephrotoxicity. *Arch. Toxicol.* **2014**, *88*, 1249–1256. [CrossRef] [PubMed]

14. Zsengellér, Z.K.; Ellezian, L.; Brown, D.; Horváth, B.; Mukhopadhyay, P.; Kalyanaraman, B.; Parikh, S.M.; Karumanchi, S.A.; Stillman, I.E.; Pacher, P. Cisplatin nephrotoxicity involves mitochondrial injury with impaired tubular mitochondrial enzyme activity. *J. Histochem. Cytochem.* **2012**, *60*, 521–529. [CrossRef] [PubMed]

15. Bushau-Sprinkle, A.; Barati, M.; Conklin, C.; Dupre, T.; Gagnon, K.B.; Khundmiri, S.J.; Clark, B.; Siskind, L.; Doll, M.A.; Rane, M.; et al. Loss of the Na(+)/H(+) Exchange Regulatory Factor 1 Increases Susceptibility to Cisplatin-Induced Acute Kidney Injury. *Am. J. Pathol.* **2019**, *189*, 1190–1200. [CrossRef]

16. Khundmiri, S.J.; Weinman, E.J.; Steplock, D.; Cole, J.; Ahmad, A.; Baumann, P.D.; Barati, M.; Rane, M.J.; Lederer, E. Parathyroid hormone regulation of NA+,K+-ATPase requires the PDZ 1 domain of sodium hydrogen exchanger regulatory factor-1 in opossum kidney cells. *J. Am. Soc. Nephrol. Jasn* **2005**, *16*, 2598–2607. [CrossRef]

17. Dai, J.L.; Wang, L.; Sahin, A.A.; Broemeling, L.D.; Schutte, M.; Pan, Y. NHERF (Na+/H+ exchanger regulatory factor) gene mutations in human breast cancer. *Oncogene* **2004**, *23*, 8681–8687. [CrossRef]

18. Georgescu, M.M.; Morales, F.C.; Molina, J.R.; Hayashi, Y. Roles of NHERF1/EBP50 in cancer. *Curr. Mol. Med.* **2008**, *8*, 459–468. [CrossRef]

19. Bushau-Sprinkle, A.M.; Lederer, E.D. New roles of the Na(+)/H(+) exchange regulatory factor 1 scaffolding protein: A review. *Am. J. Physiology. Ren. Physiol.* **2020**, *318*, F804–F808. [CrossRef] [PubMed]

20. Sun, L.; Zheng, J.; Wang, Q.; Song, R.; Liu, H.; Meng, R.; Tao, T.; Si, Y.; Jiang, W.; He, J. NHERF1 regulates actin cytoskeleton organization through modulation of alpha-actinin-4 stability. *FASEB J. Off. Publ. Fed. Am. Soc. Exp. Biol.* **2016**, *30*, 578–589. [CrossRef]

21. Ketchem, C.J.; Khundmiri, S.J.; Gaweda, A.E.; Murray, R.; Clark, B.J.; Weinman, E.J.; Lederer, E.D. Role of Na+/H+ exchanger regulatory factor 1 in forward trafficking of the type IIa Na+-Pi cotransporter. *Am. J. Physiology. Ren. Physiol.* **2015**, *309*, F109-119. [CrossRef] [PubMed]

22. Jiang, Y.; Lu, G.; Trescott, L.R.; Hou, Y.; Guan, X.; Wang, S.; Stamenkovich, A.; Brunzelle, J.; Sirinupong, N.; Li, C.; et al. New conformational state of NHERF1-CXCR2 signaling complex captured by crystal lattice trapping. *PLoS ONE* **2013**, *8*, e81904. [CrossRef] [PubMed]

23. Li, M.; Mennone, A.; Soroka, C.J.; Hagey, L.R.; Ouyang, X.; Weinman, E.J.; Boyer, J.L. Na(+) /H(+) exchanger regulatory factor 1 knockout mice have an attenuated hepatic inflammatory response and are protected from cholestatic liver injury. *Hepatology* **2015**, *62*, 1227–1236. [CrossRef] [PubMed]

24. Leslie, K.L.; Song, G.J.; Barrick, S.; Wehbi, V.L.; Vilardaga, J.P.; Bauer, P.M.; Bisello, A. Ezrin-radixin-moesin-binding phosphoprotein 50 (EBP50) and nuclear factor-kappaB (NF-kappaB): A feed-forward loop for systemic and vascular inflammation. *J. Biol. Chem.* **2013**, *288*, 36426–36436. [CrossRef] [PubMed]

25. Khundmiri, S.J.; Ahmad, A.; Bennett, R.E.; Weinman, E.J.; Steplock, D.; Cole, J.; Baumann, P.D.; Lewis, J.; Singh, S.; Clark, B.J.; et al. Novel regulatory function for NHERF-1 in Npt2a transcription. *Am. J. Physiology. Ren. Physiol.* **2008**, *294*, F840-849. [CrossRef]

26. Weinman, E.J.; Mohanlal, V.; Stoycheff, N.; Wang, F.; Steplock, D.; Shenolikar, S.; Cunningham, R. Longitudinal study of urinary excretion of phosphate, calcium, and uric acid in mutant NHERF-1 null mice. *Am. J. Physiology. Ren. Physiol.* **2006**, *290*, F838-843. [CrossRef]

27. Donowitz, M.; Singh, S.; Singh, P.; Salahuddin, F.F.; Chen, Y.; Chakraborty, M.; Murtazina, R.; Gucek, M.; Cole, R.N.; Zachos, N.C.; et al. Alterations in the proteome of the NHERF1 knockout mouse jejunal brush border membrane vesicles. *Physiol. Genom.* **2010**, *42*, 200–210. [CrossRef]
28. Shenolikar, S.; Voltz, J.W.; minkoff, C.M.; Wade, J.B.; Weinman, E.J. Targeted disruption of the mouse NHERF-1 gene promotes internalization of proximal tubule sodium-phosphate cotransporter type IIa and renal phosphate wasting. *Proc. Natl. Acad. Sci. USA* **2002**, *99*, 11470–11475. [CrossRef]
29. Khundmiri, S.J.; Asghar, M.; Khan, F.; Salim, S.; Yusufi, A.N. Effect of ischemia and reperfusion on enzymes of carbohydrate metabolism in rat kidney. *J. Nephrol.* **2004**, *17*, 377–383.
30. Khundmiri, S.J.; Rane, M.J.; Lederer, E.D. Parathyroid hormone regulation of type II sodium-phosphate cotransporters is dependent on an A kinase anchoring protein. *J. Biol. Chem.* **2003**, *278*, 10134–10141. [CrossRef] [PubMed]
31. Merchant, M.L.; Powell, D.W.; Wilkey, D.W.; Cummins, T.D.; Deegens, J.K.; Rood, I.M.; McAfee, K.J.; Fleischer, C.; Klein, E.; Klein, J.B. Microfiltration isolation of human urinary exosomes for characterization by MS. *Proteom. Clin. Appl.* **2010**, *4*, 84–96. [CrossRef]
32. Luerman, G.C.; Powell, D.W.; Uriarte, S.M.; Cummins, T.D.; Merchant, M.L.; Ward, R.A.; McLeish, K.R. Identification of phosphoproteins associated with human neutrophil granules following chemotactic peptide stimulation. *Mol Cell Proteom.* **2011**, 10. [CrossRef]
33. Merchant, M.L.; Cummins, T.D.; Wilkey, D.W.; Salyer, S.A.; Powell, D.W.; Klein, J.B.; Lederer, E.D. Proteomic analysis of renal calculi indicates an important role for inflammatory processes in calcium stone formation. *Am. J. Physiol. Ren. Physiol.* **2008**, *295*, F1254–F1258. [CrossRef]
34. Uriarte, S.M.; Powell, D.W.; Luerman, G.C.; Merchant, M.L.; Cummins, T.D.; Jog, N.R.; Ward, R.A.; McLeish, K.R. Comparison of proteins expressed on secretory vesicle membranes and plasma membranes of human neutrophils. *J. Immunol.* **2008**, *180*, 5575–5581. [CrossRef] [PubMed]
35. Mandel, L.J. Metabolic substrates, cellular energy production, and the regulation of proximal tubular transport. *Annu. Rev. Physiol.* **1985**, *47*, 85–101. [CrossRef] [PubMed]
36. Zhou, R.; Vander Heiden, M.G.; Rudin, C.M. Genotoxic Exposure Is Associated with Alterations in Glucose Uptake and Metabolism. *Cancer Res.* **2002**, *62*, 3515–3520. [PubMed]
37. Hannemann, J.; Baumann, K. Cisplatin-induced lipid peroxidation and decrease of gluconeogenesis in rat kidney cortex: Different effects of antioxidants and radical scavengers. *Toxicology* **1988**, *51*, 119–132. [CrossRef]
38. Timson, D.J. Fructose 1,6-bisphosphatase: Getting the message across. *Biosci. Rep.* **2019**, *39*, BSR20190124. [CrossRef] [PubMed]
39. Van Schaftingen, E.; Gerin, I. The glucose-6-phosphatase system. *Biochem. J.* **2002**, *362*, 513–532. [CrossRef]
40. Valvona, C.J.; Fillmore, H.L.; Nunn, P.B.; Pilkington, G.J. The Regulation and Function of Lactate Dehydrogenase A: Therapeutic Potential in Brain Tumor. *Brain Pathol.* **2016**, *26*, 3–17. [CrossRef]
41. Gietl, C. Malate dehydrogenase isoenzymes: Cellular locations and role in the flow of metabolites between the cytoplasm and cell organelles. *Biochim. Et Biophys. Acta (BBA) Bioenerg.* **1992**, *1100*, 217–234. [CrossRef]
42. Frenkel, R. Regulation and Physiological Functions of Malic Enzymes. In *Current Topics in Cellular Regulation*; Horecker, B.L., Stadtman, E.R., Eds.; Academic Press: Cambridge, MA, USA, 1975; Volume 9, pp. 157–181.
43. Efferth, T.; Schwarzl, S.M.; Smith, J.; Osieka, R. Role of glucose-6-phosphate dehydrogenase for oxidative stress and apoptosis. *Cell Death Differ.* **2006**, *13*, 527–528. [CrossRef] [PubMed]
44. Bonora, M.; Patergnani, S.; Rimessi, A.; De Marchi, E.; Suski, J.M.; Bononi, A.; Giorgi, C.; Marchi, S.; Missiroli, S.; Poletti, F.; et al. ATP synthesis and storage. *Purinergic Signal.* **2012**, *8*, 343–357. [CrossRef]
45. Chistiakov, D.A.; Sobenin, I.A.; Revin, V.V.; Orekhov, A.N.; Bobryshev, Y.V. Mitochondrial aging and age-related dysfunction of mitochondria. *Biomed. Res. Int.* **2014**, *2014*, 238463. [CrossRef] [PubMed]
46. Brand, M.D.; Nicholls, D.G. Assessing mitochondrial dysfunction in cells. *Biochem. J.* **2011**, *435*, 297–312. [CrossRef] [PubMed]
47. Yao, P.; Sun, H.; Xu, C.; Chen, T.; Zou, B.; Jiang, P.; Du, W. Evidence for a direct cross-talk between malic enzyme and the pentose phosphate pathway via structural interactions. *J. Biol. Chem.* **2017**, *292*, 17113–17120. [CrossRef] [PubMed]
48. Zhang, Z.; Yang, Z.; Zhu, B.; Hu, J.; Liew, C.W.; Zhang, Y.; Leopold, J.A.; Handy, D.E.; Loscalzo, J.; Stanton, R.C. Increasing glucose 6-phosphate dehydrogenase activity restores redox balance in vascular endothelial cells exposed to high glucose. *PLoS ONE* **2012**, *7*, e49128. [CrossRef] [PubMed]

49. Sanz, N.; Díez-Fernández, C.; Valverde, A.M.; Lorenzo, M.; Benito, M.; Cascales, M. Malic enzyme and glucose 6-phosphate dehydrogenase gene expression increases in rat liver cirrhogenesis. *Br. J. Cancer* **1997**, *75*, 487–492. [CrossRef]
50. Zunino, F.; Pratesi, G.; Micheloni, A.; Cavalletti, E.; Sala, F.; Tofanetti, O. Protective effect of reduced glutathione against cisplatin-induced renal and systemic toxicity and its influence on the therapeutic activity of the antitumor drug. *Chem. Biol. Interact.* **1989**, *70*, 89–101. [CrossRef]
51. Shiraishi, F.; Curtis, L.M.; Truong, L.; Poss, K.; Visner, G.A.; Madsen, K.; Nick, H.S.; Agarwal, A. Heme oxygenase-1 gene ablation or expression modulates cisplatin-induced renal tubular apoptosis. *Am. J. Physiol. Ren. Physiol.* **2000**, *278*, F726–F736. [CrossRef]
52. Al Ghouleh, I.; Meijles, D.N.; Mutchler, S.; Zhang, Q.; Sahoo, S.; Gorelova, A.; Henrich Amaral, J.; Rodriguez, A.I.; Mamonova, T.; Song, G.J.; et al. Binding of EBP50 to Nox organizing subunit p47phox is pivotal to cellular reactive species generation and altered vascular phenotype. *Proc. Natl. Acad. Sci. USA* **2016**, *113*, E5308–E5317. [CrossRef] [PubMed]
53. Pera, T.; Tompkins, E.; Katz, M.; Wang, B.; Deshpande, D.A.; Weinman, E.J.; Penn, R.B. Specificity of NHERF1 regulation of GPCR signaling and function in human airway smooth muscle. *FASEB J. Off. Publ. Fed. Am. Soc. Exp. Biol.* **2019**, *33*, 9008–9016. [CrossRef] [PubMed]

© 2020 by the authors. Licensee MDPI, Basel, Switzerland. This article is an open access article distributed under the terms and conditions of the Creative Commons Attribution (CC BY) license (http://creativecommons.org/licenses/by/4.0/).

Article

Dietary and Lifestyle Factors Modulate the Activity of the Endogenous Antioxidant System in Patients with Age-Related Macular Degeneration: Correlations with Disease Severity

Zofia Ulańczyk [1], Aleksandra Grabowicz [2], Elżbieta Cecerska-Heryć [3], Daria Śleboda-Taront [3], Elżbieta Krytkowska [2], Katarzyna Mozolewska-Piotrowska [2], Krzysztof Safranow [4], Miłosz Piotr Kawa [1], Barbara Dołęgowska [3] and Anna Machalińska [2,*]

1 Department of General Pathology, Pomeranian Medical University, 70-111 Szczecin, Poland; zofia.litwinska@pum.edu.pl (Z.U.); kawamilosz@gmail.com (M.P.K.)
2 First Department of Ophthalmology, Pomeranian Medical University, 70-111 Szczecin, Poland; aleksandra.grabowicz@pum.edu.pl (A.G.); oko1@pum.edu.pl (E.K.); kmp@pum.edu.pl (K.M.-P.)
3 Department of Laboratory Medicine, Pomeranian Medical University, 70-111 Szczecin, Poland; cecerskaela@wp.pl (E.C.-H.); daria.sleboda@pum.edu.pl (D.Ś.-T.); barbara.dolegowska@pum.edu.pl (B.D.)
4 Department of Biochemistry and Medical Chemistry, Pomeranian Medical University, 70-111 Szczecin, Poland; chrissaf@mp.pl
* Correspondence: annam@pum.edu.pl

Received: 17 August 2020; Accepted: 2 October 2020; Published: 5 October 2020

Abstract: Age-related macular degeneration (AMD) is a common cause of blindness in the elderly population, but the pathogenesis of this disease remains largely unknown. Since oxidative stress is suggested to play a major role in AMD, we aimed to assess the activity levels of components of the antioxidant system in patients with AMD. We also investigated whether lifestyle and dietary factors modulate the activity of these endogenous antioxidants and clinical parameters of disease severity. We recruited 330 patients with AMD (39 with early, 100 with intermediate and 191 with late form of AMD) and 121 controls in this study. At enrolment, patients' dietary habits and physical activity were assessed, and each study participant underwent a thorough ophthalmologic examination. The activity of several components of the antioxidant system were measured in red blood cells and platelets using both kinetic and spectrophotometric methods. Patients with AMD consumed much lower levels of fatty fish and eggs than the control group ($p = 0.008$ and $p = 0.04$, respectively). In the nAMD group, visual acuity (VA) correlated positively with green vegetable consumption (Rs = +0.24, $p = 0.004$) and omega-3-rich oil intake (Rs = +0.17, $p = 0.03$). In the AMD group, the total physical activity MET score correlated positively with VA (Rs = +0.17, $p = 0.003$) and correlated negatively with the severity of AMD (Rs = −0.14, $p = 0.01$). A multivariate analysis of patients and controls adjusted for age, sex, and smoking status (pack-years) revealed that AMD was an independent variable associated with a lower RBC catalase ($\beta = -0.37$, $p < 0.001$) and higher PLT catalase ($\beta = +0.25$, $p < 0.001$), RBC GPx ($\beta = +0.26$, $p < 0.001$), PLT GPx ($\beta = +0.16$, $p = 0.001$), RBC R-GSSG ($\beta = +0.13$, $p = 0.009$), PLT R-GSSG ($\beta = +0.12$, $p = 0.02$) and RBC GSH transferase ($\beta = +0.23$, $p < 0.001$) activity. The activities of components of the antioxidant system were associated with disease severity and depended on dietary habits. The observed substantial increase in the activity of many critical endogenous antioxidants in patients with AMD further indicates that the required equilibrium in the antioxidant system is disturbed throughout the course of the disease. Our findings explicitly show that a diet rich in green vegetables, fish and omega-3-rich oils, supplemented by physical exercise, is beneficial for patients with AMD, as it might delay disease progression and help retain better visual function.

Keywords: age-related macular degeneration (AMD); antioxidants; diet

1. Introduction

Age-related macular degeneration (AMD) is an incurable ocular condition of the outer retina that affects approximately 8–10% of the elderly population worldwide [1]. AMD is characterised by a progressive visual impairment and remains the leading cause of visual disability in developed countries [2]. Because of the increasing life expectancy, the number of individuals with AMD is estimated to reach 288 million in 2040 [3,4]. This increasing prevalence of AMD represents a major financial challenge for healthcare systems and is expected to exert increasing socio-economic effects [5].

As the pathogenesis of AMD is still poorly understood, the treatment options remain limited and are available only for the advanced, neovascular form of AMD (nAMD) [6]. The antiangiogenic treatment targets the main pathophysiological feature of this subtype of AMD: the formation of new, largely malformed vessels (choroidal neovascularization, CNV) [7]. At present, no proven treatment is available for the earlier stages of AMD nor for GA, which are characterised by the accumulation of drusen and retinal atrophy [8]. Therefore, research has focused on the prevention and/or slowing the progression of AMD to its late stages by manipulating modifiable risk factors, among which nutrition and dietary habits are listed as the most important risk factors [9]. Based on research evidence, changes in a patient's dietary habits and the addition of supplements represent a simple and cost-effective method for modifying the risk of developing and progression of AMD [10]. Several observational and experimental studies have been conducted in humans to investigate associations between dietary antioxidants, the consumption of certain foods and AMD [11], including Age Related Eye Disease Study 1 (AREDS1) [12]. Two of the most notable studies, AREDS1 and most recently AREDS2, have contributed to the supplementation strategy currently used in clinical practice [13]. These studies have noted the effects of dietary factors such as omega-3s, carotenoids, lutein/zeaxanthin, vitamins A and D on eye health and suggested that they might affect the course of AMD. These findings resulted in the development of recommendations and clinical practice guidelines that have been used as a decision-making tool in clinical settings [14]. In general, the most cost-effective and seemingly achievable strategy for the prevention of progression of AMD to its later stages appears to be a general healthy lifestyle that is achieved by a healthy diet and exercise [15,16].

The pathogenesis of AMD has been attributed to both modifiable and unmodifiable factors, including age, genetics and active smoking [17]. The most well-known genetic factors associated with an increased risk of AMD are polymorphisms in the *CFH* Y402H (complement factor H) [18,19] and *ARMS2* (age-related maculopathy susceptibility 2) genes [20]. Recently, our team identified another variant associated with an increased risk for AMD that is located in the peripherin-2 (*PRPH2*) gene, which encodes a photoreceptor-specific protein vital for rod and cone cell formation and stability [21]. The detrimental effect of the major modifiable risk factor for AMD, cigarette smoking, which increases the risk of AMD 2–4 times [22–24], has been attributed to the induction of angiogenesis, impairments in choroidal circulation, activation of the immune system and generation of oxidative damage [22,25,26]. In fact, risk factors other than smoking that contribute to AMD development, such as light exposure, diet, and vitamin D levels, among others, also exert well-documented effects on oxidative stress, which corresponds to cellular damage caused by reactive oxygen species (ROS) [27].

At the molecular level, the retinal environment is particularly susceptible to oxidative stress, as it is constantly exposed to light and is characterised by increased oxygen consumption and a high proportion of polyunsaturated fatty acids [27,28]. ROS directly damage DNA (particularly in mitochondria) and lipids in the photoreceptors, leading to the deterioration of the retinal pigment epithelium (RPE) [29]. These oxidatively damaged molecules then accumulate in the macular area and become a continuous source of chronic oxidative stress [30]. The retina protects itself from oxidative damage by producing a considerable number of antioxidants in the photoreceptor and RPE cells, including the enzymes superoxide dismutase (SOD), glutathione peroxidase (GPx), catalase (CAT), glutathione transferase (GST) and glutathione reductase (R-GSSG) and nonenzymatic antioxidants glutathione (GSH), carotenoids, uric acid, albumin and many others [31]. Disturbances in the tight balance of antioxidant system might contribute to AMD pathogenesis [32].

In the present study, we aimed to assess the activity levels of several components of the antioxidant system in patients with AMD and controls and to explore the role of dietary habits in AMD development. In particular, we wanted to investigate whether lifestyle factors modulate the concentrations of these endogenous antioxidants and clinical parameters of disease severity. We also focused on possible associations of antioxidant activity with genetic risk factors for AMD.

2. Materials and Methods

2.1. Study Groups and Initial Management

For this study we recruited 330 patients with AMD (39 with early, 100 with intermediate and 191 with late form of AMD) from the outpatient population of the First Department of Ophthalmology of Pomeranian Medical University in Szczecin, Poland (ethical code is KB-0012/141/13). For the control group, we enrolled 121 age-matched participants with no symptoms or signs of macular degeneration (absence of drusen, neovascularization or pigmentary abnormalities). We excluded patients with significant chronic systemic conditions (diabetes mellitus, renal failure, collagen/neoplastic disease, hepatic dysfunction, etc.) or ongoing retinal disease except AMD (in AMD groups), i.e., glaucoma or intraocular inflammatory diseases from the participation in the study. A consent form was signed by all patients before enrolment in the trial, in accordance with the tenets of the Declaration of Helsinki.

Demographic characteristics (age, gender, time of symptom onset, time to presentation) and the following vascular risk factors present at the time of the enrolment were recorded: hypertension, hyperlipidaemia, diabetes mellitus, atrial fibrillation, ischaemic heart disease, cardiomyopathy, prior cerebrovascular events or renal artery stenosis (atherosclerosis). The following anthropometric and nutritional parameters were also assessed in all patients: waist circumference [cm], waist/hip ratio (WHR), and body mass index (BMI) [weight (kg)/height (m)^2]. Cumulative pack-years of smoking were determined from the recorded average number of smoked cigarettes per day and smoking years. The actual blood pressure (BP) was determined in all subjects prior to the ophthalmic examination.

2.2. Dietary Habits and Physical Activity Assessment

At the enrolment, each patient completed Food Frequency Questionnaire (FFQ) and International Physical Activity Questionnaire (IPAQ) with the help of the member of the research team.

We modified a quantitative Food Frequency Questionnaire (FFQ) to assess the intake of the following food groups rich in nutrients considered important in the AMD aetiology and oxidative processes to evaluate the participants' dietary habits: fatty fish, eggs, green vegetables, fruit and fruit juice, omega-3-rich oils, simple and complex carbohydrates, as recommended previously [11,13,33,34]. Three different frequencies in terms of portions per week were available for selection for each food type and alcoholic drink as follows: fatty fish: <1, 2–4, and >4; eggs: <1, 2–4, and >4; green vegetables: <2, 2–7, and >7; fruit and fruit juice: <2, 2–7, and >7; omega-3-rich oils: <2, 2–7, and >7; simple carbohydrates: <2, 2–7, and >7; complex carbohydrates: <2, 2–7, and >7; beer: 0, ≤1, and 2–7; wine: 0, <2, and 2–7; and vodka: 0, ≤1, and 2–3.

Each participant completed the International Physical Activity Questionnaire (IPAQ) with the assistance of the member of the research team due to participants' vision impairment, which comprises 7 questions regarding all types of physical activity associated with daily life, work and leisure performed in the last seven days. The duration of each activity included in the final data was 10 min or longer with no interruptions at any moment. The physical activity score was presented in MET-min per week units and was calculated by multiplying a factor specific for each activity by several days spent performing the activity and time in min spent on the activity per day. Weekly activity was measured by adding scores of each of the activities.

2.3. Ophthalmologic Examination

The patients were examined by an ophthalmologist with a comprehensive ophthalmologic evaluation, including best-corrected visual acuity using the Early Treatment Diabetic Retinopathy Study (ETDRS) chart, intraocular pressure measurement, fundus photography, autofluorescence imaging, spectral-domain OCT, and fluorescein or indocyanine green angiography (Spectralis, Heidelberg Engineering, Carlsbad, CA, USA). The severity of AMD was classified according to Ferris et al. [35]: patients with medium drusen (63–125 m) and without pigmentary abnormalities were classified as early AMD group, patients with large drusen or with pigmentary abnormalities associated with at least medium drusen were classified as intermediate AMD group, and patients with lesions associated with neovascular AMD or geographic atrophy were classified to have late AMD. The examinations were carried out in a blinded manner.

2.4. Blood Sample Collection, RBC and PLT Preparation

Peripheral venous blood (approx. 7.5 mL) was collected from the AMD group and controls into two types of tubes containing ethylenediaminetetraacetic acid (EDTA) or sodium citrate as an anticoagulant. The blood sample in the EDTA tube was centrifuged at 3000× *g* for 10 min to separate the plasma and buffy coat from red blood cells (RBCs). The plasma and buffy coat were removed from RBCs and 3 mL of deionised water were added to induce haemolysis. The sample was then centrifuged at 13,500× *g* for 5 min to separate the haemolysate from the pellet of red blood cell membranes. Haemoglobin levels were assayed using Drabkin's method. All results obtained for the activity of antioxidant enzymes were calculated per 1 g of haemoglobin in RBCs.

Platelets were obtained from venous blood collected in a tube containing 109 mM sodium citrate (3.2%, 9:1; *v*/*v*). Blood was centrifuged (10 min; 20 °C; 10,000 rpm) to obtain platelet-rich plasma (PRP), which was transferred to a new tube and centrifuged again (10 min; 20 °C; 3824 rpm). The resulting platelet-poor plasma (PPP) was placed in a fresh tube and stored at −80 °C until further analysis. The platelet pellet was washed twice with Tyrode's solution (pH 7.4), suspended in 1 mL of Tyrode's solution and the number of platelets was determined using spectrophotometry. The suspension was stored at −80 °C until further analysis. The platelet suspension was thawed (37 °C) and frozen (−80 °C) twice, and the obtained platelet lysate was centrifuged (10 min; 4 °C; 3824 rpm). All results obtained for the activity of antioxidant enzymes were calculated per 1 g of platelet lysate protein. Protein levels were assayed using the Lowry protein assay.

We used automated methods and commercially available assays to measure fasting glucose and lipid levels (including triglycerides, total cholesterol and high (HDL) and low-density (LDL) lipoproteins) in all patients with AMD and controls.

2.5. The Activity of Antioxidant Enzymes

A spectrophotometric method was used to establish the concentrations of reduced glutathione (GSH). The activities of superoxide dismutase (SOD), catalase (CAT), glutathione peroxidase (GPx), glutathione transferase (GST), glutathione reductase (R-GSSG) in red blood cells (RBCs) and platelets (PLT) were obtained using kinetic methods. The measurements in RBCs and PLT were performed using a UV/VIS Lambda 650 spectrophotometer (Perkin-Elmer, Waltham, MA, USA) and similar protocols (specified below). The extracellular haemoglobin concentration in plasma samples was determined using a spectrophotometric method [36,37] with the same spectrophotometer.

2.5.1. GSH Concentration

The (haemo)lysate was diluted, mixed with a precipitation solution (1.67 g of metaphosphoric acid, 0.2 g of EDTA-Na_2, 30 g of NaCl and 100 mL of H_2O; Sigma-Aldrich, St. Louis, MO, USA), incubated

(5 min, 4 °C) and centrifuged (550× *g* for 10 min). The supernatant was diluted with phosphoric buffer (pH 7.9), DTNB (5,5′-dithiobis-(2-nitrobenzoic acid), Sigma-Aldrich, St. Louis, MO, USA) was added, and then the mixture was incubated for 15 min at 25 °C. The detection wavelength was λ 412 nm. The GSH concentration was calculated using the molar absorption coefficient (e = 13,600 M^{-1} cm^{-1}).

2.5.2. SOD Activity

In a test tube, a mixture of (haemo)lysate, chloroform:ethanol (3:5; *v*/*v*) solution, and distilled water was combined. The mixtures were subsequently vortexed and centrifuged (5 min; 4 °C; 3824× *g*). The study material (Na_2CO_3/$NaHCO_3$ buffer, SOD extract and adrenaline solution; Sigma-Aldrich, St. Louis, MO, USA) was incubated for 3 min at 37 °C. The absorbance of the study material was recorded in 5 min at a wavelength of 320 nm (in 30 °C).

2.5.3. CAT Activity

The (haemo)lysate was diluted 500-fold (100-fold for PLT) with 50 mM phosphoric buffer. Absorbance measurements of the study sample ((haemo)lysate and 30 mM H_2O_2 solution; Sigma-Aldrich, St. Louis, MO, USA) were performed within 30 s at a wavelength of 240 nm (at 30 °C). CAT activity was determined from a calibration curve, which was obtained as a result of assays of several solutions with a known catalase activity pattern (Oxis Research, Portland, OR, USA).

2.5.4. GPx Activity

A reaction mixture was prepared (phosphoric buffer, glutathione reductase, GSH, NADPH + H^+, and (haemo)lysate with transforming reagent incubated for 5 min at room temperature beforehand; transforming reagent used only for RBC) and incubated for 10 min at 37 °C. After the incubation, the reaction was initiated by adding tert-butyl hydroxide (or hydrogen peroxide), and the decrease in the absorption at λ 340 nm was measured. The amount of enzyme that oxidised 1 µmol of GSH (0.5 µmol NADPH + H) in one minute was defined as a unit of enzyme activity.

2.5.5. GST Activity

A reaction mixture was combined (phosphoric buffer, GSH, CDNB (1-chloro-2,4-dinitrobenzene), and (haemo)lysate), and the increase in absorbance at λ 340 nm was measured. Glutathione transferase activity was determined using molar absorption coefficient of the synthesised conjugate (e = 9600 M^{-1} cm^{-1}).

2.5.6. R-GSSG Activity

The (haemo)lysate was diluted, mixed with 1 mL of a diluted RI working reagent (900µL EDTA and 100µL RI; RI: $NADPH^+$ + H^+ diluted in 0.01 M NaOH, Sigma Aldrich, St. Louis, MO, USA) and incubated (5 min, 30 °C). Then, RII reagent was added (GSSG (glutathione disulphide) diluted in EDTA) and extinction was measured at λ 340 nm over 3 min at 30 °C.

2.5.7. Statistical Analysis

The quantitative parameters measured in both eyes were averaged before further analysis. The distributions of all the analysed variables related to components of antioxidant system were right-skewed and significantly different from normal distribution ($p < 0.05$, Shapiro-Wilk test). Therefore, the nonparametric Kruskal-Wallis and Mann-Whitney tests were used to compare quantitative values between groups, whereas Spearman's rank correlation coefficient (Rs) was calculated to measure the strength of associations between these values. We used Fisher's exact test to compare qualitative variables between groups. A multivariate analysis of AMD as an independent variable was performed using a general linear model (GLM) adjusted for age, sex and smoking status (pack-years) by inclusion of the confounding factors as independent variables, with logarithmic transformation applied to the dependent variables to normalize their distributions. Standardised regression coefficients (β) were calculated to measure the strength of associations between independent and dependent variables. The interpretation of β and Rs coefficients is similar: values +1 and −1 indicate perfect positive and negative association, respectively, while 0 indicates complete lack of association. Quantitative variables were presented as mean ± standard deviation. $p < 0.05$ was considered statistically significant without correction for multiple testing other than included in the applied test itself (number of compared groups in Kruskal-Wallis test, number of variables in GLM). Statistica 13 software (Dell Inc., Round Rock, TX, USA) was used for statistical analyses.

3. Results

3.1. Characteristics of the Study Subjects

We enrolled 330 patients with AMD and 121 healthy controls in this study. The clinical characteristics of the studied groups are presented in Table 1. The AMD and control groups did not differ in age and well-known atherosclerotic risk factors, including serum lipid and glucose levels. Statistically significant differences in physical activity were not observed between the groups. On the other hand, the number of past smokers and the number of pack-years of smoking were considerably higher in the AMD group than in controls ($p < 0.001$). Importantly, a strong positive correlation between the number of pack-years of smoking and disease severity was identified (Rs = +0.23, $p < 0.001$), corroborating the well-documented relation between smoking and AMD. The positive correlation between the patient's age and clinical classification of AMD (Rs = +0.19, $p < 0.001$) indicates that age is one of the main factors affecting disease severity.

Table 1. Characteristics of the study groups. In bold, p-value < 0.05, which was considered statistically significant.

Parameter	AMD Group	Control Group	p-Value *
Number of patients	330	121	-
Sex (male/female)	135/219	32/89	**0.02**
Age [years] (mean ± SD)	73.4 ± 8.0	73.1 ± 6.0	0.41
Current smokers (%)	13.6%	6.3%	0.05
Former smokers (%)	51.4%	30.9%	**<0.001**
Smoking pack-years (mean ± SD)	13.6 ± 18.9	6.0 ± 13.1	**<0.001**
Period without smoking [years] (mean ± SD)	6.8 ± 10.9	5.3 ± 10.2	0.06
Iris colour (dark/light)	91/261	26/95	0.39
Outdoor/indor working conditions	40.1/59.9%	33.1/66.9%	0.19
MAP [mmHg] (mean ± SD)	98.3 ± 11.1	98.7 ± 9.7	0.86
Disease history:			
Hypertension (%)	64.7%	71.1%	0.27
Hypertension duration [years] (mean ± SD)	8.2 ± 9.5	9.2 ± 9.9	0.27
Ischemic heart disease (%)	16.2%	11.3%	0.33
Ischemic heart disease duration [years] (mean ± SD)	1.2 ± 4.2	0.8 ± 3.3	0.26
Aortic aneurysm (%)	1.6%	0.0%	0.59
Peripheral artery disease (%)	5.0%	6.2%	0.61
Cerebral stroke (%)	2.8%	3.1%	1.00
Cardiac infarction (%)	6.2%	6.1%	1.00
Currently taken medications:			
Hypotensive drugs/vasodilators	65.0%	70.1%	0.39
Cardiac/antiarrhythmic drugs	13.9%	14.4%	0.87
NSAIDs	20.2%	19.6%	1.00
Hormonal drugs	17.1%	20.6%	0.45
Thyroxine	13.7%	20.6%	0.11
Steroids	1.9%	1.0%	1.00
Other hormonal drugs	1.3%	0.0%	0.58
Statins	26.6%	36.1%	0.07
Antidepressants	4.7%	5.2%	0.79
Antiasthmatic drugs	7.4%	3.1%	0.16

Table 1. *Cont.*

Parameter	AMD Group	Control Group	*p*-Value *
BMI [kg/m^2] (mean ± SD)	26.9 ± 4.2	26.6 ± 3.7	0.43
WHR [cm/cm] (mean ± SD)	0.90 ± 0.10	0.88 ± 0.10	0.13
Waist circumference [cm] (mean ± SD)	103.3 ± 9.1	102.1 ± 7.3	0.33
Intensive physical activity (MET)	269.2 ± 824.2	173.5 ± 451.7	0.59
Average physical activity (MET)	490.3 ± 1151.9	433.8 ± 706.2	0.35
Walking (MET)	778.8 ± 914.3	785.6 ± 852.4	0.63
Total physical activity (MET)	1536.2 ± 2025.1	1382.5 ± 1493.5	0.97
Cholesterol [mg/dL] (mean ± SD)	204.4 ± 44.6	202.8 ± 43.3	0.99
HDL[mg/dL] (mean ± SD)	60.1 ± 14.0	59.8 ± 13.6	0.90
LDL[mg/dL] (mean ± SD)	119.9 ± 39.2	117.5 ± 37.5	0.71
Triglycerides [mg/dL] (mean ± SD)	106.2 ± 51.5	108.9 ± 52.4	0.38
Glucose [mg/dL] (mean ± SD)	104.8 ± 12.7	102.8 ± 11.0	0.12

* Mann—Whitney test/Fisher's exact test.

On the other hand, the HDL level correlated negatively with disease severity (Rs = −0.15, $p = 0.007$), suggesting its protective effect on AMD progression. Another negative correlation was observed between the education level and disease severity (Rs = −0.14, $p = 0.01$). Thus, a higher educational attainment might be associated with better health awareness and subsequently reduce AMD progression.

3.2. *Analysis of Lifestyle Habits*

Since diet is considered one of the potentially modifiable risk factors for AMD, we aimed to compare the dietary habits of patients with AMD and healthy controls (Table 2) and to analyse whether dietary habits differed between the early, intermediate and late AMD groups (Table 3).

Patients with AMD consumed much less fatty fish than the control group ($p = 0.008$). Similar findings were observed for egg consumption ($p = 0.04$). In contrast, the AMD group recorded higher consumption of fruits and fruit juices than the controls ($p = 0.01$). We did not observe significant differences in the consumption of green vegetables, omega-3 rich oils, and simple and complex carbohydrates between groups ($p = 0.39$, $p = 0.66$, $p = 0.18$, and $p = 0.26$, respectively). Accordingly, no differences in alcohol intake were observed between groups. Interestingly, we observed a significantly higher consumption of fatty fish in patients with intermediate AMD than in patients with the late form of the disease ($p = 0.01$). Significant differences in egg, green vegetable, fruit and fruit juice, omega-3-rich oil, simple and complex carbohydrate consumption and alcohol intake were not observed between the early, intermediate and late AMD groups.

We aimed to determine whether physical activity and the consumption of specific food groups or alcohol were associated with disease severity to better assess the roles of lifestyle and dietary habits in the progression of AMD. In the AMD group, we observed strong positive correlations between VA and the total physical activity MET score (Rs = +0.17, $p = 0.003$), but these correlations were not observed in controls. Positive correlations between VA and physical activity intensity were clearly detectable in the nAMD group, including the intense physical activity MET score (Rs = +0.17; $p = 0.04$), average physical activity MET score (Rs = +0.21; $p = 0.01$) and total physical activity MET score (Rs = +0.22, $p = 0.006$). Thus, patients with AMD who were more physically active displayed better visual function. Accordingly, the total physical activity MET score negatively correlated with the severity of AMD (Rs = −0.14, $p = 0.01$). Similarly, time (in min) spent sitting in the last 7 days was associated with more advanced stages of AMD (Rs = +0.20, $p = 0.0005$). Based on these findings, sedentary behaviour facilitates the progression of AMD.

Table 2. Dietary habits of the AMD group and control group. p-values < 0.05, which were considered statistically significant, are shown in bold.

Food Group	Portions Per Week (%)	AMD Group	Control Group	p-Value *
Fatty fish	<1	73.67%	59.57%	**0.008**
	2–4	26.02%	39.36%	
	>4	0.31%	1.06%	
Eggs	<1	23.82%	10.75%	**0.04**
	2–4	69.91%	84.95%	
	>4	6.27%	4.30%	
Green vegetables	<2	5.63%	1.08%	0.40
	2–7	63.44%	76.34%	
	>7	30.94%	22.58%	
Fruit and fruit juice	<2	6.94%	1.06%	**0.01**
	2–7	46.69%	72.34%	
	>7	46.37%	26.60%	
Omega-3 rich oils	<2	14.38%	4.21%	0.66
	2–7	64.38%	81.05%	
	>7	21.25%	14.74%	
Simple carbohydrates	<2	25.86%	11.58%	0.18
	2–7	55.45%	74.74%	
	>7	18.69%	13.68%	
Complex carbohydrates	<2	2.18%	3.16%	0.26
	2–7	77.26%	81.05%	
	>7	20.56%	15.79%	
Beer consumption	0	37.54%	36.08%	0.87
	≤1	58.68%	63.92%	
	2–7	3.79%	0.00%	
Wine consumption	0	37.78%	36.08%	0.88
	<2	60.95%	63.92%	
	2–7	1.27%	0.00%	
Vodka consumption	0	37.42%	36.08%	0.84
	≤1	62.26%	63.92%	
	2–3	0.31%	0.00%	

* Mann–Whitney test.

Regarding dietary factors, VA correlated positively with green vegetable consumption (Rs = +0.24, p = 0.004) and omega-3-rich oil intake (Rs = +0.17, p = 0.03) in the nAMD group. Therefore, the consumption of these food products might preserve visual function in patients with nAMD, although no differences in the consumption of those products were observed between the AMD and control groups. Accordingly, fatty fish consumption correlated positively with the retinal volume in the AMD group (Rs = +0.23, p = 0.003). We observed negative correlations between the drusen size and consumption of green vegetables (Rs = −0.19, p = 0.02), fruit and fruit juice (Rs = −0.28, p = 0.0004), omega-3-rich oils (Rs = −0.24, p = 0.002) and complex carbohydrates (Rs = −0.21, p = 0.01) in the AMD group. These correlations were not observed in the control group. Thus, the consumption of these products might exert beneficial effects on the disease course and reduce AMD progression.

Table 3. Dietary habits of the early, intermediate and late AMD groups. The bold font indicates p-values < 0.05, which were considered statistically significant.

Food Group	p-Value [1]	Portions Per Week (%)	Early AMD Group	Intermediate AMD Group	Late AMD Group	p-Value [2] Early AMD vs. Intermediate AMD	Early AMD vs. Late AMD	Intermediate AMD vs. Late AMD
Fatty fish	0.047	<1	76.47%	63.44%	77.46%	0.21	0.83	0.01
		2–4	20.59%	36.56%	22.54%			
		>4	2.94%	0.00%	0.00%			
Eggs	0.28	<1	11.76%	17.20%	28.90%	0.84	0.26	0.18
		2–4	88.24%	78.49%	63.01%			
		>4	0.00%	4.30%	8.09%			
Green vegetables	0.47	<2	0.00%	3.23%	7.47%	0.45	0.24	0.55
		2–7	64.71%	66.67%	63.22%			
		>7	35.29%	30.11%	29.31%			
Fruit and fruit juice	0.83	<2	2.94%	6.52%	8.72%	0.53	0.67	0.76
		2–7	50.00%	51.09%	45.35%			
		>7	47.06%	42.39%	45.93%			
Omega-3 rich oils	0.25	<2	5.88%	8.60%	18.39%	0.14	0.12	0.66
		2–7	64.71%	74.19%	58.62%			
		>7	29.41%	17.20%	22.99%			
Simple carbohydrates	0.36	<2	8.82%	22.34%	30.46%	0.29	0.17	0.59
		2–7	76.47%	61.70%	49.43%			
		>7	14.71%	15.96%	20.11%			
Complex carbohydrates	0.48	<2	0.00%	0.00%	4.02%	0.18	0.40	0.57
		2–7	73.53%	84.04%	73.56%			
		>7	26.47%	15.96%	22.41%			
Beer consumption	0.75	0	42.86%	36.17%	38.46%	0.46	0.64	0.64
		≤1	54.29%	59.57%	58.58%			
		2–7	2.86%	4.26%	2.96%			
Wine consumption	0.81	0	42.86%	36.56%	38.69%	0.56	0.76	0.63
		<2	54.29%	61.29%	60.71%			
		2–7	2.86%	2.15%	0.60%			
Vodka consumption	0.74	0	42.86%	36.17%	38.24%	0.45	0.61	0.66
		≤1	57.14%	62.77%	61.76%			
		2–3	0.00%	1.06%	0.00%			

[1] Kruskal-Wallis test, [2] Mann-Whitney test.

3.3. Components of the Antioxidant System

Excess oxidative stress coupled with overwhelmed antioxidant defence systems are thought to be the important contributors to the complex pathophysiology of AMD. We chose 6 factors to analyse the efficiency of the antioxidant system in patients with AMD and controls: the activities of five enzymes (SOD, CAT, GPx, R-GSSG and GSH transferase) and concentrations of reduced glutathione (GSH) in red blood cells (RBCs) and platelets (PLT). The AMD group presented higher values for 7 of the 12 tested factors compared with the control group (Table 4, Figure 1): GPx, R-GSSG, GSH transferase in RBCs and SOD, catalase, GPx and R-GSSG in PLT. A significant downregulation in catalase activity levels was observed in RBCs from patients with AMD (0.33 ± 0.21 U/mg Hb) compared to controls (0.53 ± 0.24; $p < 0.0001$). A multivariate analysis of patients and controls adjusted for age, sex, and smoking status (pack-years) revealed that AMD was an independent variable associated with a lower RBC catalase ($\beta = -0.37$, $p < 0.001$) and higher PLT catalase ($\beta = +0.25$, $p < 0.001$), RBC GPx ($\beta = +0.26$, $p < 0.001$), PLT GPx ($\beta = +0.16$, $p = 0.001$), RBC R-GSSG ($\beta = +0.13$, $p = 0.009$), PLT R-GSSG ($\beta = +0.12$, $p = 0.02$) and RBC GSH transferase ($\beta = +0.23$, $p < 0.001$) activity levels.

Table 4. Comparison of levels of components of the antioxidant system in patients with AMD and controls. The bold font indicates *p*-values < 0.05, which were considered statistically significant.

Antioxidant System Component	AMD Group	Control Group	*p*-Value *
	Mean ± SD	Mean ± SD	
SOD (RBC) [U/mg Hb]	0.31 ± 0.23	0.33 ± 0.19	0.56
Catalase (RBC) [U/mg Hb]	0.33 ± 0.21	0.53 ± 0.24	**<0.0001**
GPx (RBC) [U/g Hb]	0.05 ± 0.05	0.02 ± 0.01	**<0.0001**
R-GSSG (RBC) [U/g Hb]	7.28 ± 5.73	5.55 ± 2.62	**0.001**
GSH (RBC) [μmol/g Hb]	6.18 ± 2.7	6.49 ± 1.95	0.13
GSH transferase (RBC) [U/g Hb]	0.02 ± 0.02	0.01 ± 0.01	**<0.0001**
SOD (PLT) [U/mg of protein]	69.24 ± 145.07	29.07 ± 26.64	**0.03**
Catalase (PLT) [U/mg of protein]	28.31 ± 72.11	9.08 ± 17.91	**<0.0001**
GPx (PLT) [U/g of protein]	4.26 ± 16.083	1.45 ± 1.99	**0.002**
R-GSSG (PLT) [U/g of protein]	1604.07 ± 5624.53	515.63 ± 528.46	**0.04**
GSH (PLT) [μmol/g Hb]	1964.62 ± 3398.87	1126.1 ± 2141.95	0.44
GSH transferase (PLT) [U/g of protein]	1.83 ± 3.92	1.06 ± 1.12	0.26

* Mann-Whitney test.

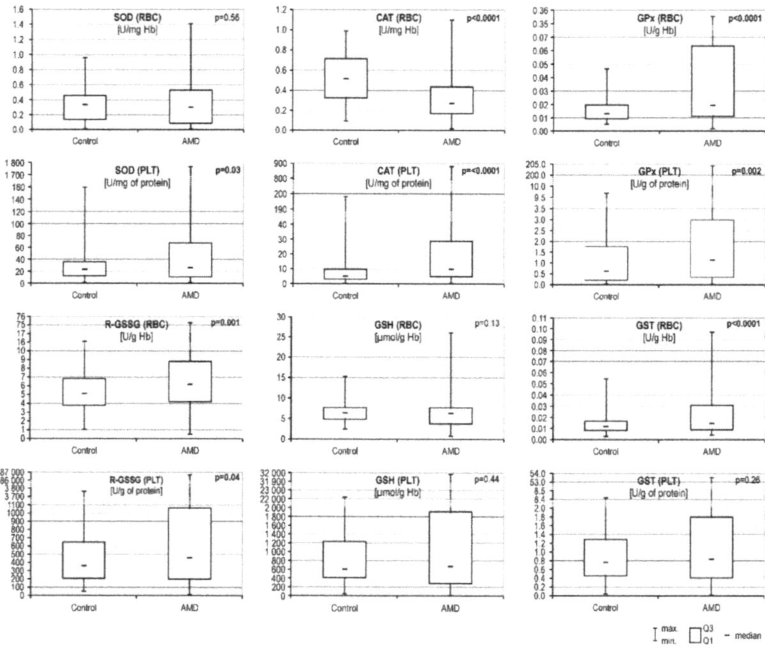

Figure 1. Comparison of the antioxidant system component levels in patients with AMD and controls. *p*-values < 0.05, which were considered statistically significant, are shown in bold.

An analysis stratified by AMD severity revealed that the early AMD group presented higher GSH (RBC) concentration and lower R-GSSG (PLT) activity than the late AMD group ($p = 0.03$ and $p = 0.04$, respectively) (Table 5). However, we should be rather careful in interpreting the results of the Mann–Whitney test, since the Kruskal-Wallis test showed no statistically significant differences between the early, intermediate and late AMD groups.

Table 5. Comparison of levels of components of the antioxidant system in patients with early, intermediate and late AMD. p-values < 0.05, which were considered statistically significant, are shown in bold.

Antioxidant System Component	p-Value [1]	Early AMD Group		Intermediate AMD Group		Late AMD Group		p-Value [2]		
		n	Mean ± SD Median (IQR)	n	Mean ± SD Median (IQR)	n	Mean ± SD Median (IQR)	Early AMD vs. Intermediate AMD	Early AMD vs. Late AMD	Intermediate AMD vs. Late AMD
SOD (RBC) [U/mg Hb]	0.27	39	0.27 ± 0.25 0.15 (0.38)	99	0.33 ± 0.23 0.33 (0.43)	184	0.30 ± 0.23 0.26 (0.42)	0.11	0.26	0.32
Catalase (RBC) [U/mg Hb]	0.12	39	0.38 ± 0.25 0.31 (0.31)	99	0.36 ± 0.22 0.30 (0.31)	184	0.32 ± 0.20 0.26 (0.24)	0.83	0.15	0.07
GPx (RBC) [U/g Hb]	0.17	39	0.04 ± 0.04 0.02 (0.01)	99	0.04 ± 0.04 0.02 (0.02)	183	0.05 ± 0.06 0.02 (0.07)	0.46	0.09	0.22
R-GSSG (RBC) [U/g Hb]	0.97	39	7.27 ± 4.54 6.18 (5.64)	99	6.59 ± 2.86 6.20 (3.23)	185	7.64 ± 7.12 6.07 (5.07)	0.94	0.82	0.86
GSH (RBC) [μmol/g Hb]	0.08	39	6.99 ± 2.38 6.89 (2.94)	99	6.56 ± 3.49 6.26 (3.59)	185	5.93 ± 2.25 6.36 (3.96)	0.09	**0.03**	0.47
GSH transferase (RBC) [U/g Hb]	0.38	39	0.02 ± 0.02 0.01 (0.01)	99	0.02 ± 0.02 0.01 (0.02)	185	0.02 ± 0.02 0.01 (0.02)	0.26	0.16	0.82
SOD (PLT) [U/mg of protein]	0.31	39	44.82 ± 55.41 24.35 (39.32)	98	57.93 ± 85.41 26.83 (43.76)	186	84.88 ± 184.59 27.49 (76.11)	0.48	0.14	0.40
Catalase (PLT) [U/mg of protein]	0.13	39	14.81 ± 19.31 6.37 (16.70)	98	21.68 ± 35.18 9.62 (13.72)	186	35.54 ± 93.80 9.95 (31.73)	0.14	0.05	0.51
GPx (PLT) [U/g of protein]	0.37	38	1.71 ± 1.77 1.28 (2.00)	98	3.98 ± 14.11 0.94 (2.94)	185	5.27 ± 19.33 1.20 (2.66)	0.80	0.32	0.23
R-GSSG (PLT) [U/g of protein]	0.12	39	1478.05 ± 5657.85 281.19 (750.52)	97	1914.16 ± 8866.18 472.54 (785.88)	185	1499.27 ± 3268.86 466.47 (1022.51)	0.15	**0.04**	0.52

n: number of observations, [1] Kruskal-Wallis test, [2] Mann-Whitney test.

Next, we investigated the associations between ophthalmic parameters and concentrations of the analysed antioxidants to more specifically evaluate the functions of antioxidants in the development of AMD. In the AMD group, RBC catalase activity and GSH concentrations negatively correlated with the disease severity (Rs = −0.11, $p = 0.04$; R = −0.11, $p = 0.05$, respectively). This relationship corresponds to lower RBC catalase activity in the AMD group than in controls. Similarly, we observed weak positive correlations between the clinical classification of AMD and RBC GPx (Rs = +0.10, $p = 0.07$), PLT catalase (Rs = +0.10, $p = 0.08$) and R-GSSG PLT (Rs = +0.10, $p = 0.08$) activities that corresponded to higher activities of these enzymes in RBCs and PLT from patients with AMD. Accordingly, the drusen size in patients with AMD correlated positively with SOD, GPx and GSH transferase activities in RBCs (Rs = 0.31, $p < 0.001$; Rs = 0.16, $p = 0.003$; and Rs = 0.18, $p < 0.001$, respectively) and negatively correlated with both RBC activity and PLT GSH concentration (Rs = −0.27, $p < 0.001$; and Rs = −0.24, $p < 0.001$, respectively). This association was similar to the higher RBC GPx and GSH transferase activity in the AMD group than in controls. Based on these results, the antioxidant system might be a major contributor to the clinical course of AMD.

3.4. Correlations between the Antioxidant System and Lifestyle Factors

Different nutritional factors are proposed to modulate the antioxidant potential of various cells [38,39]. We evaluated the possible associations between the activity levels of components of the antioxidant system with physical activity and diet to assess whether the activities of the investigated antioxidant enzymes in RBCs and PLT were associated with lifestyle factors and whether these correlations were specific for patients with AMD. Overall, GPx, R-GSSG and GSH transferase activities in RBCs from the AMD group correlated negatively with egg consumption (Rs = −0.22, $p < 0.001$; Rs = −0.17, $p = 0.003$; and Rs = −0.13, $p = 0.03$, respectively), whereas the RBC catalase activity was positively correlated with the amount of egg consumption (Rs = +0.11, $p = 0.05$). Similar correlations were not observed in controls. These relationships clearly correspond to higher GPx, R-GSSG and GSH transferase activities and lower catalase activity in the AMD group, as well as lower egg consumption, than in controls. Similarly, in the AMD group, positive correlations between RBC catalase activity and fatty fish consumption (Rs = +0.18, $p = 0.001$) paralleled the lower consumption of this dietary product and lower catalase activity in patients with AMD than in controls. Accordingly, we observed positive correlations between fruit and fruit juice consumption with two PLT enzymes: SOD and GPx (Rs = +0.12, $p = 0.03$ and Rs = +0.14, $p = 0.02$, respectively). This finding confirms the aforementioned observation, since the activities of both of these enzymes and fruit and fruit juice consumption were higher in the AMD group than in the control group. The identified relationships indicate the possible effect of dietary habits on antioxidant activity in patients with AMD.

In terms of physical activity, R-GSSG activity levels in RBCs correlated negatively with the total physical activity MET score (Rs = −0.17, $p = 0.04$) in patients with nAMD, but this correlation was not observed in controls.

3.5. Genotypes and Components of the Antioxidant System

We also explored the relationship between antioxidant activity and genetic risk factors for AMD to further elucidate the interactions between various risk factors for AMD and their contributions to disease pathogenesis. For this purpose, we investigated associations between the six selected antioxidants and polymorphisms in genes previously associated with AMD: *CFH* Y402H, *ARMS* A69S and a single nucleotide variant that our team recently linked to a higher AMD risk, *PRPH2* c.582-67T > A (rs3818086) (paper in press). However, when the correction for multiple testing was applied, we did not identify any statistically significant relationships between components of the antioxidant system and the genotypes of these genes.

4. Discussion

AMD remains the major cause of visual impairment among the elderly population, significantly reducing the quality of life of affected individuals [40]. Several genetic and environmental factors, including the *CFH* Y402H polymorphism, age and cigarette smoking, have been identified as contributing to the complex landscape of AMD, although the exact pathogenesis of the disease remains unclear [41,42]. At the molecular level, various risk factors for AMD share a common denominator, oxidative stress, which is thought to be the main component of AMD pathology [27,43]. In fact, manipulations of dietary and lifestyle habits, which are thought to contribute to the tight balance of the endogenous antioxidant system, might be beneficial in preventing and/or slowing the progression of AMD [11]. Thus, in the present study, we aimed to investigate the role of antioxidant components in AMD and to assess whether dietary and lifestyle factors modulate the levels of those endogenous antioxidants and clinical parameters of disease severity. We also assessed possible relationships between antioxidant activity and genetic risk factors for AMD.

First, we assessed the systemic levels of components of the antioxidant system in peripheral blood and found that the activity of the majority of tested substances were significantly increased in patients with AMD (GPx, R-GSSG, and GSH transferase levels in RBCs and SOD, CAT, GPx, and R-GSSG levels in PLT), whereas only the CAT activity in RBC was evidently reduced in patients with AMD compared with controls. Our observation of increased GPx activity is in contrast to the results reported by Mrowicka et al. [44] and Plestina-Borjan et al. [31], where significantly lower GPx (RBC) activity was observed in AMD patients in comparison with controls. On the other hand, in the large POLA study of a cohort of 2584 participants, the increased levels of plasma GPx concentration, which catalyses H_2O_2 degradation by GSH, correlated with a nine-fold increase in the prevalence of late AMD [45]. GPx not only protects RPE cells in models of oxidative damage-induced retinal degeneration but is also required for the maturation of photoreceptor cells [46]. As proposed by Tokarz et al., the increased activity of GPx in patients with AMD reflects the activity of RPE cells, which attempt to dispose of overwhelming amount of H_2O_2 formed during the disease course [47]. R-GSSG interacts with GPx to regulate the GSH concentration, as it converts glutathione disulfide (GSSG) to GSH [48]. In contrast to our results, the R-GSSG concentration and activity have been reported to be rather low in patients with AMD and was associated with a decrease in GSH levels [49,50], the product of enzymatic reaction catalysed by R-GSSG. In our study, we did not observe a significant association between R-GSSG activity and GSH concentration in any of the tested groups; thus, the increase in R-GSSG activity in patients with AMD might be an analogous indicator of increased antioxidant activity, similarly to GPx or increased GSH transferase activity. Interestingly, conflicting results on the involvement of *GSTM1* and *GSTM5* (glutathione s-transferase mu 1/5) polymorphisms and gene expression in AMD pathology exist [51,52], once again suggesting that the necessary balance of the antioxidant system is achieved through the proper activity of several enzymes, and not a single enzyme. The tight cooperation of endogenous antioxidants is reflected in SOD activity, which functions together with GPx and CAT to convert H_2O_2 to nontoxic products and by that protect the photoreceptors and RPE from oxidative damage [31,49]. SOD activity decreases in the RPE periphery with ageing and at the same time its immunoreactivity increases [53]. However, in the aforementioned POLA study, a high level of erythrocyte SOD activity was not associated with AMD [45]. This finding is consistent with our results, although we also observed a higher activity of SOD in PLT from the AMD group. On the contrary, Venza et al. reported lower SOD (both in plasma and RBC) activity in AMD patients compared to controls [54]. Interestingly, in vitro studies have shown a reduction in SOD activity in response to oxidative stress when ARPE-19 cells were treated with acrolein, a powerful initiator of oxidative stress and mitochondrial dysfunction [55], whereas the upregulation of *SOD1/2* expression resulted in oxidative damage in RPE cells [56]. In fact, excess SOD (in relation to the activities of GPx and CAT) may cause damage [57], which further suggests the need for a tight balance of the antioxidant system. Similar to SOD, CAT function decreases in the macular and peripheral RPE with ageing [58], but in contrast to SOD, CAT immunoreactivity is reduced in RPE cells in the eyes of patients with and without AMD [47,59]. Our finding of reduced CAT activity in patients with AMD is consistent with previous reports [44]. As proposed by Tate et al., treatment with ROS-generating compounds induces CAT expression in RPE cells, which protects against H_2O_2, even in the adjacent RPE cells without upregulated CAT expression. Overall, our findings of increased activity of several antioxidants in the AMD group suggest an enhanced response to oxidative damage that might contribute to AMD pathogenesis by disrupting the tight balance of the antioxidant system [32].

Dietary antioxidants aid the endogenous antioxidant defence system in protecting against oxidative damage and enhanced ROS production and consequently, prevent or slow related disorders, including AMD [60]. Several major clinical trials, in particular the Age-Related Eye Disease Study (AREDS) and AREDS2, have shown that nutrients with antioxidant properties, namely, lutein, zeaxanthin, polyunsaturated omega-3 fatty acids (PUFAs), zinc, vitamins C and E, delay the progression of advanced AMD in persons with intermediate AMD [13,61]. In our study, patients with AMD consumed much lower levels of fatty fish and eggs than controls, whereas greater consumption of green vegetables and omega-3-rich oils was correlated with favourable clinical outcomes (better visual acuity and a smaller drusen size in patients with AMD). These observations are consistent with previous studies, as patients with AMD are generally encouraged to increase their intake of green vegetables, eggs and fish [9,14,33]. Both green vegetables and eggs are rich sources of lutein and zeaxanthin, potent anti-inflammatory and antioxidant factors that exert beneficial effects on slowing the progression of AMD [15]. The antioxidant potential of these macular carotenoid pigments combined with their ability to filtrate blue-light may serve not only to protect the ocular tissue from oxidative damage, but also to improve visual acuity [62]. Our study further supports this notion, as we found a positive correlation between VA and consumption of green leafy vegetables, which are a well-known source of these potent macular carotenoid pigments [63]. According to Gopinath et al., eggs also contain large quantities of selenium, which directly protects cells from oxidative damage [64]. Additionally, eggs, similar to fatty fish, are a good source of omega-3 PUFAs, which may minimize retinal inflammation, oxidation, and degeneration [10,64,65]. Indeed, the Blue Mountains Eye Study (BMES) of a large Australian cohort has shown that a greater consumption of fish and omega-3s may slow AMD progression [66]. Interestingly, we did not observe any correlations between alcohol consumption and the clinical parameters in our patients, although at the molecular level, the toxicity of alcohol is associated with lipid peroxidation and oxidative stress and potentially represents another causative factor for AMD [67]. This finding is consistent with a study by Knudtson et al. [68], although some recent reports suggest a modest association between alcohol consumption and an increased AMD risk [69].

The preventive strategies incorporating a modification of the diet represent an attractive approach to slow AMD progression, but some debate exists in the literature regarding whether physical exercise is also recommended to protect against AMD [70]. Our study provides a clear indication that a sedentary lifestyle worsens the AMD course, as more physically active patients presented better visual acuity. Several previous studies have associated physical activity with a lower risk of AMD [71,72], but reports have also described the lack of significant relationships between exercise and AMD risk [73]. As shown in a recent study by Gopinath et al., the most physically active patients aged at least 75 years are 79% less likely to develop late AMD in 15 years [70]. However, when other confounding factors were considered, no significant association was observed in this group. Overall, the systemic benefits of physical activity (e.g., protective effects on obesity, diabetes, inflammation, etc.) make it a vital part of a healthy lifestyle [70,74], which should be recommended to patients with AMD, along with refraining from cigarette smoking and consuming a diet rich in vegetables, fish and eggs.

It is worth noting; however, that the correlations found in the present study do not necessarily indicate causation. Thus, further studies on larger cohorts are needed to provide more straightforward evidence of various dietary and lifestyle factors affecting AMD course.

5. Conclusions

In our study, we observed significant increases in the levels of several crucial endogenous antioxidants in blood samples from patients with AMD, which further suggests that the necessary balance of the antioxidant system is disrupted during the disease course. Since dietary habits potentially modulate AMD progression by contributing to the antioxidant status, we also investigated food intake in our cohort and identified several significant correlations. Based on our results, a diet rich in green vegetables, eggs, fish and omega-3-rich oils, accompanied by physical activity, is beneficial for patients with AMD and might slow disease progression and help maintain better visual functions. Further studies of larger cohorts might elucidate the potential effects of various other food components and nutritional factors on the severity and progression of AMD or even their roles in preventing this disease.

Author Contributions: Conceptualization, A.M.; ophthalmologic examination, K.M.-P., E.K., biochemical analysis, E.C.-H., D.Ś.-T.; database management, A.G.; statistical analysis, K.S.; writing—original draft preparation, Z.U.; writing—review and editing, M.P.K., A.M.; supervision, B.D., A.M. All authors have read and agreed to the published version of the manuscript.

Funding: This work was supported by Polish National Centre for Research and Development (grant number: STRATEGMED1/234261/2NCBR/2014).

Conflicts of Interest: The authors declare no conflict of interest.

Abbreviations

AMD	age-related macular degeneration
CAT	catalase
CNV	choroidal neovascularization
GA	geographic atrophy
GPx	glutathione peroxidase
R-GSSG	glutathione reductase
GST	glutathione transferase
nAMD	neovascular AMD
PLT	platelets
ROS	reactive oxygen species
RBCs	red blood cells
GSH	reduced glutathione
RPE	retinal pigment epithelium
SOD	superoxide dismutase

References

1. Al-Zamil, W.M.; Yassin, S.A. Recent developments in age-related macular degeneration: A review. *Clin. Interv. Aging* **2017**, *12*, 1313–1330. [CrossRef] [PubMed]
2. Alswailmi, F.K. Global prevalence and causes of visual impairment with special reference to the general population of Saudi Arabia. *Pak. J. Med Sci.* **2018**, *34*, 751–756. [CrossRef] [PubMed]
3. Jonas, J.B.; Cheung, C.M.G.; Panda-Jonas, S. Updates on the Epidemiology of Age-Related Macular Degeneration. *Asia Pac. J. Ophthalmol (Phila)* **2017**, *6*, 493–497. [CrossRef]
4. Pennington, K.L.; DeAngelis, M.M. Epidemiology of age-related macular degeneration (AMD): Associations with cardiovascular disease phenotypes and lipid factors. *Eye Vis. (Lond. Engl.)* **2016**, *3*, 34. [CrossRef] [PubMed]
5. Schmier, J.K.; Jones, M.L.; Halpern, M.T. The burden of age-related macular degeneration. *Pharmacoeconomics* **2006**, *24*, 319–334. [CrossRef]
6. Hernandez-Zimbron, L.F.; Zamora-Alvarado, R.; Ochoa-De la Paz, L.; Velez-Montoya, R.; Zenteno, E.; Gulias-Canizo, R.; Quiroz-Mercado, H.; Gonzalez-Salinas, R. Age-Related Macular Degeneration: New Paradigms for Treatment and Management of AMD. *Oxid. Med. Cell Longev.* **2018**, *2018*, 8374647. [CrossRef]

7. Morris, B.; Imrie, F.; Armbrecht, A.-M.; Dhillon, B. Age-related macular degeneration and recent developments: New hope for old eyes? *Postgrad. Med. J.* **2007**, *83*, 301–307. [CrossRef]
8. Gehrs, K.M.; Anderson, D.H.; Johnson, L.V.; Hageman, G.S. Age-related macular degeneration–emerging pathogenetic and therapeutic concepts. *Ann. Med.* **2006**, *38*, 450–471. [CrossRef]
9. Carneiro, Â.; Andrade, J.P. Nutritional and Lifestyle Interventions for Age-Related Macular Degeneration: A Review. *Oxidative Med. Cell. Longev.* **2017**, *2017*, 6469138. [CrossRef]
10. Pinazo-Durán, M.D.; Gómez-Ulla, F.; Arias, L.; Araiz, J.; Casaroli-Marano, R.; Gallego-Pinazo, R.; García-Medina, J.J.; López-Gálvez, M.I.; Manzanas, L.; Salas, A.; et al. Do nutritional supplements have a role in age macular degeneration prevention? *J. Ophthalmol.* **2014**, *2014*, 901686. [CrossRef]
11. Chapman, N.A.; Jacobs, R.J.; Braakhuis, A.J. Role of diet and food intake in age-related macular degeneration: A systematic review. *Clin. Exp. Ophthalmol.* **2019**, *47*, 106–127. [CrossRef] [PubMed]
12. Janko, N.-Z. A randomized, placebo-controlled, clinical trial of high-dose supplementation with vitamins C and E, beta carotene, and zinc for age-related macular degeneration and vision loss: AREDS report no. 8. *Arch. Ophthalmol.* **2001**, *119*, 1417–1436. [CrossRef]
13. Gorusupudi, A.; Nelson, K.; Bernstein, P.S. The Age-Related Eye Disease 2 Study: Micronutrients in the Treatment of Macular Degeneration. *Adv. Nutr. (BethesdaMd.)* **2017**, *8*, 40–53. [CrossRef] [PubMed]
14. Lawrenson, J.G.; Evans, J.R. Advice about diet and smoking for people with or at risk of age-related macular degeneration: A cross-sectional survey of eye care professionals in the UK. *BMC Public Health* **2013**, *13*, 564. [CrossRef] [PubMed]
15. Khoo, H.E.; Ng, H.S.; Yap, W.-S.; Goh, H.J.H.; Yim, H.S. Nutrients for Prevention of Macular Degeneration and Eye-Related Diseases. *Antioxidants* **2019**, *8*, 85. [CrossRef]
16. Vina, J.; Sanchis-Gomar, F.; Martinez-Bello, V.; Gomez-Cabrera, M.C. Exercise acts as a drug; the pharmacological benefits of exercise. *Br. J. Pharmacol.* **2012**, *167*, 1–12. [CrossRef]
17. Chen, Y.; Bedell, M.; Zhang, K. Age-related macular degeneration: Genetic and environmental factors of disease. *Mol. Interv.* **2010**, *10*, 271–281. [CrossRef]
18. Klein, R.J.; Zeiss, C.; Chew, E.Y.; Tsai, J.-Y.; Sackler, R.S.; Haynes, C.; Henning, A.K.; SanGiovanni, J.P.; Mane, S.M.; Mayne, S.T.; et al. Complement factor H polymorphism in age-related macular degeneration. *Science* **2005**, *308*, 385–389. [CrossRef]
19. Cascella, R.; Strafella, C.; Longo, G.; Ragazzo, M.; Manzo, L.; De Felici, C.; Errichiello, V.; Caputo, V.; Viola, F.; Eandi, C.M.; et al. Uncovering genetic and non-genetic biomarkers specific for exudative age-related macular degeneration: Significant association of twelve variants. *Oncotarget* **2017**, *9*, 7812–7821. [CrossRef]
20. Chakravarthy, U.; McKay, G.J.; de Jong, P.T.; Rahu, M.; Seland, J.; Soubrane, G.; Tomazzoli, L.; Topouzis, F.; Vingerling, J.R.; Vioque, J.; et al. ARMS2 increases the risk of early and late age-related macular degeneration in the European Eye Study. *Ophthalmology* **2013**, *120*, 342–348. [CrossRef]
21. Conley, S.M.; Naash, M.I. Gene therapy for PRPH2-associated ocular disease: Challenges and prospects. *Cold Spring Harb. Perspect. Med.* **2014**, *4*, a017376. [CrossRef] [PubMed]
22. Velilla, S.; García-Medina, J.J.; García-Layana, A.; Dolz-Marco, R.; Pons-Vázquez, S.; Pinazo-Durán, M.D.; Gómez-Ulla, F.; Arévalo, J.F.; Díaz-Llopis, M.; Gallego-Pinazo, R. Smoking and age-related macular degeneration: Review and update. *J. Ophthalmol.* **2013**, *2013*, 895147. [CrossRef] [PubMed]
23. Saunier, V.; Merle, B.M.J.; Delyfer, M.N.; Cougnard-Gregoire, A.; Rougier, M.B.; Amouyel, P.; Lambert, J.C.; Dartigues, J.F.; Korobelnik, J.F.; Delcourt, C. Incidence of and Risk Factors Associated with Age-Related Macular Degeneration: Four-Year Follow-up From the ALIENOR Study. *JAMA Ophthalmol.* **2018**, *136*, 473–481. [CrossRef] [PubMed]
24. Klein, R.; Myers, C.E.; Cruickshanks, K.J.; Gangnon, R.E.; Danforth, L.G.; Sivakumaran, T.A.; Iyengar, S.K.; Tsai, M.Y.; Klein, B.E. Markers of inflammation, oxidative stress, and endothelial dysfunction and the 20-year cumulative incidence of early age-related macular degeneration: The Beaver Dam Eye Study. *JAMA Ophthalmol.* **2014**, *132*, 446–455. [CrossRef]
25. Kauppinen, A.; Paterno, J.J.; Blasiak, J.; Salminen, A.; Kaarniranta, K. Inflammation and its role in age-related macular degeneration. *Cell Mol. Life Sci.* **2016**, *73*, 1765–1786. [CrossRef]
26. Lee, J.; Taneja, V.; Vassallo, R. Cigarette smoking and inflammation: Cellular and molecular mechanisms. *J. Dent. Res.* **2012**, *91*, 142–149. [CrossRef]
27. Shaw, P.X.; Stiles, T.; Douglas, C.; Ho, D.; Fan, W.; Du, H.; Xiao, X. Oxidative stress, innate immunity, and age-related macular degeneration. *AIMS Mol. Sci.* **2016**, *3*, 196–221. [CrossRef]

28. Masuda, T.; Shimazawa, M.; Hara, H. Retinal Diseases Associated with Oxidative Stress and the Effects of a Free Radical Scavenger (Edaravone). *Oxidative Med. Cell. Longev.* **2017**, *2017*, 9208489. [CrossRef]
29. Nita, M.; Grzybowski, A. The Role of the Reactive Oxygen Species and Oxidative Stress in the Pathomechanism of the Age-Related Ocular Diseases and Other Pathologies of the Anterior and Posterior Eye Segments in Adults. *Oxidative Med. Cell. Longev.* **2016**, *2016*, 3164734. [CrossRef]
30. Handa, J.T. How does the macula protect itself from oxidative stress? *Mol. Asp. Med.* **2012**, *33*, 418–435. [CrossRef]
31. Plestina-Borjan, I.; Katusic, D.; Medvidovic-Grubisic, M.; Supe-Domic, D.; Bucan, K.; Tandara, L.; Rogosic, V. Association of age-related macular degeneration with erythrocyte antioxidant enzymes activity and serum total antioxidant status. *Oxidative Med. Cell. Longev.* **2015**, *2015*, 804054. [CrossRef] [PubMed]
32. Yildirim, Z.; Ucgun, N.I.; Yildirim, F. The role of oxidative stress and antioxidants in the pathogenesis of age-related macular degeneration. *Clinics (Sao Paulo Braz.)* **2011**, *66*, 743–746. [CrossRef]
33. Nunes, S.; Alves, D.; Barreto, P.; Raimundo, M.; da Luz Cachulo, M.; Farinha, C.; Lains, I.; Rodrigues, J.; Almeida, C.; Ribeiro, L.; et al. Adherence to a Mediterranean diet and its association with age-related macular degeneration. The Coimbra Eye Study-Report 4. *Nutrition* **2018**, *51–52*, 6–12. [CrossRef] [PubMed]
34. Singh, N.; Srinivasan, S.; Muralidharan, V.; Roy, R.; Jayprakash, V.; Raman, R. Prevention of Age-Related Macular Degeneration. *Asia Pac. J. Ophthalmol. (Phila)* **2017**, *6*, 520–526. [CrossRef]
35. Ferris, F.L., III; Wilkinson, C.P.; Bird, A.; Chakravarthy, U.; Chew, E.; Csaky, K.; Sadda, S.R. Clinical classification of age-related macular degeneration. *Ophthalmology* **2013**, *120*, 844–851. [CrossRef] [PubMed]
36. Fleisch, H. The measure of hemolysis by the determination in the plasma of hemoglobin, methemalbumin and their derivatives. *Helv. Med. Acta* **1960**, *27*, 383–407.
37. Hunter, F.T.; Grove-Rasmussen, M.; Soutter, L. A spectrophotometric method for quantitating hemoglobin in plasma or serum. *Am. J. Clin. Pathol* **1950**, *20*, 429–433. [CrossRef] [PubMed]
38. Poljsak, B.; Šuput, D.; Milisav, I. Achieving the Balance between ROS and Antioxidants: When to Use the Synthetic Antioxidants. *Oxidative Med. Cell. Longev.* **2013**, *2013*, 956792. [CrossRef]
39. Schneider, C.D.; Bock, P.M.; Becker, G.F.; Moreira, J.C.F.; Bello-Klein, A.; Oliveira, A.R. Comparison of the effects of two antioxidant diets on oxidative stress markers in triathletes. *Biol. Sport* **2018**, *35*, 181–189. [CrossRef]
40. Ulanczyk, Z.; Sobus, A.; Luczkowska, K.; Grabowicz, A.; Mozolewska-Piotrowska, K.; Safranow, K.; Kawa, M.p.; Palucha, A.; Krawczyk, M.; Sikora, P.; et al. Associations of microRNAs, Angiogenesis-Regulating Factors and CFH Y402H Polymorphism-An Attempt to Search for Systemic Biomarkers in Age-Related Macular Degeneration. *Int. J. Mol. Sci.* **2019**, *20*, 5750. [CrossRef]
41. Sharma, N.K.; Gupta, A.; Prabhakar, S.; Singh, R.; Sharma, S.K.; Chen, W.; Anand, A. Association between CFH Y402H polymorphism and age related macular degeneration in North Indian cohort. *PLoS ONE* **2013**, *8*, e70193. [CrossRef] [PubMed]
42. Litwinska, Z.; Sobus, A.; Luczkowska, K.; Grabowicz, A.; Mozolewska-Piotrowska, K.; Safranow, K.; Kawa, M.p.; Machalinski, B.; Machalinska, A. The Interplay Between Systemic Inflammatory Factors and MicroRNAs in Age-Related Macular Degeneration. *Front. Aging Neurosci.* **2019**, *11*, 286. [CrossRef] [PubMed]
43. Liguori, I.; Russo, G.; Curcio, F.; Bulli, G.; Aran, L.; Della-Morte, D.; Gargiulo, G.; Testa, G.; Cacciatore, F.; Bonaduce, D.; et al. Oxidative stress, aging, and diseases. *Clin. Interv. Aging* **2018**, *13*, 757–772. [CrossRef] [PubMed]
44. Mrowicka, M.; Mrowicki, J.; Szaflik, J.p.; Szaflik, M.; Ulinska, M.; Szaflik, J.; Majsterek, I. Analysis of antioxidative factors related to AMD risk development in the polish patients. *Acta Ophthalmol.* **2017**, *95*, 530–536. [CrossRef]
45. Delcourt, C.; Cristol, J.p.; Leger, C.L.; Descomps, B.; Papoz, L. Associations of antioxidant enzymes with cataract and age-related macular degeneration. *Pola Study. Pathol. Ocul. Liees A L'age. Ophthalmol.* **1999**, *106*, 215–222. [CrossRef]
46. Ueta, T.; Inoue, T.; Furukawa, T.; Tamaki, Y.; Nakagawa, Y.; Imai, H.; Yanagi, Y. Glutathione peroxidase 4 is required for maturation of photoreceptor cells. *J. Biol. Chem.* **2012**, *287*, 7675–7682. [CrossRef]
47. Tokarz, P.; Kaarniranta, K.; Blasiak, J. Role of antioxidant enzymes and small molecular weight antioxidants in the pathogenesis of age-related macular degeneration (AMD). *Biogerontology* **2013**, *14*, 461–482. [CrossRef]

48. Tabatabaie, T.; Floyd, R.A. Susceptibility of glutathione peroxidase and glutathione reductase to oxidative damage and the protective effect of spin trapping agents. *Arch. Biochem. Biophys.* **1994**, *314*, 112–119. [CrossRef]
49. Cohen, S.M.; Olin, K.L.; Feuer, W.J.; Hjelmeland, L.; Keen, C.L.; Morse, L.S. Low glutathione reductase and peroxidase activity in age-related macular degeneration. *Br. J. Ophthalmol* **1994**, *78*, 791–794. [CrossRef]
50. Čolak, E.; Majkić-Singh, N.; Žoric, L.; Radosavljević, A.; Kosanović-Jaković, N. The impact of inflammation to the antioxidant defense parameters in AMD patients. *Aging Clin. Exp. Res.* **2012**, *24*, 588–594. [CrossRef]
51. Hunter, A.A., III; Smit-McBride, Z.; Anderson, R.; Bordbari, M.H.; Ying, G.S.; Kim, E.S.; Park, S.S.; Telander, D.G.; Dunaief, J.L.; Hjelmeland, L.M.; et al. GSTM1 and GSTM5 Genetic Polymorphisms and Expression in Age-Related Macular Degeneration. *Curr. Eye Res.* **2016**, *41*, 410–416. [CrossRef] [PubMed]
52. Oz, O.; Aras Ates, N.; Tamer, L.; Yildirim, O.; Adigüzel, U. Glutathione S-transferase M1, T1, and P1 gene polymorphism in exudative age-related macular degeneration: A preliminary report. *Eur. J. Ophthalmol.* **2006**, *16*, 105–110. [CrossRef] [PubMed]
53. Qin, S.; Rodrigues, G.A. Progress and perspectives on the role of RPE cell inflammatory responses in the development of age-related macular degeneration. *J. Inflamm. Res.* **2008**, *1*, 49–65. [CrossRef] [PubMed]
54. Venza, I.; Visalli, M.; Cucinotta, M.; Teti, D.; Venza, M. Association between oxidative stress and macromolecular damage in elderly patients with age-related macular degeneration. *Aging Clin. Exp. Res.* **2012**, *24*, 21–27. [CrossRef] [PubMed]
55. Liu, Z.; Sun, L.; Zhu, L.; Jia, X.; Li, X.; Jia, H.; Wang, Y.; Weber, P.; Long, J.; Liu, J. Hydroxytyrosol protects retinal pigment epithelial cells from acrolein-induced oxidative stress and mitochondrial dysfunction. *J. Neurochem.* **2007**, *103*, 2690–2700. [CrossRef]
56. Lu, L.; Oveson, B.C.; Jo, Y.J.; Lauer, T.W.; Usui, S.; Komeima, K.; Xie, B.; Campochiaro, P.A. Increased expression of glutathione peroxidase 4 strongly protects retina from oxidative damage. *Antioxid. Redox Signal.* **2009**, *11*, 715–724. [CrossRef]
57. Ighodaro, O.M.; Akinloye, O.A. First line defence antioxidants-superoxide dismutase (SOD), catalase (CAT) and glutathione peroxidase (GPX): Their fundamental role in the entire antioxidant defence grid. *Alex. J. Med.* **2018**, *54*, 287–293. [CrossRef]
58. Liles, M.R.; Newsome, D.A.; Oliver, P.D. Antioxidant enzymes in the aging human retinal pigment epithelium. *Arch. Ophthalmol.* **1991**, *109*, 1285–1288. [CrossRef]
59. Frank, R.N.; Amin, R.H.; Puklin, J.E. Antioxidant enzymes in the macular retinal pigment epithelium of eyes with neovascular age-related macular degeneration. *Am. J. Ophthalmol.* **1999**, *127*, 694–709. [CrossRef]
60. Tan, B.L.; Norhaizan, M.E.; Liew, W.-P.-P.; Sulaiman Rahman, H. Antioxidant and Oxidative Stress: A Mutual Interplay in Age-Related Diseases. *Front. Pharmacol.* **2018**, *9*, 1162. [CrossRef]
61. The Age-Related Eye Disease Study: A clinical trial of zinc and antioxidants–Age-Related Eye Disease Study Report No. 2. *J. Nutr.* **2000**, *130*, 1516s–1519s. [CrossRef]
62. Whitehead, A.J.; Mares, J.A.; Danis, R.P. Macular Pigment: A Review of Current Knowledge. *Arch. Ophthalmol.* **2006**, *124*, 1038–1045. [CrossRef] [PubMed]
63. Mares-Perlman, J.A.; Fisher, A.I.; Klein, R.; Palta, M.; Block, G.; Millen, A.E.; Wright, J.D. Lutein and zeaxanthin in the diet and serum and their relation to age-related maculopathy in the third national health and nutrition examination survey. *Am. J. Epidemiol.* **2001**, *153*, 424–432. [CrossRef] [PubMed]
64. Gopinath, B.; Liew, G.; Tang, D.; Burlutsky, G.; Flood, V.M.; Mitchell, P. Consumption of eggs and the 15-year incidence of age-related macular degeneration. *Clin. Nutr.* **2020**, *39*, 580–584. [CrossRef] [PubMed]
65. Smith, A.; Gray, J. Considering the benefits of egg consumption for older people at risk of sarcopenia. *Br. J. Community Nurs.* **2016**, *21*, 305–309. [CrossRef] [PubMed]
66. Tan, J.S.; Wang, J.J.; Flood, V.; Mitchell, P. Dietary fatty acids and the 10-year incidence of age-related macular degeneration: The Blue Mountains Eye Study. *Arch. Ophthalmol* **2009**, *127*, 656–665. [CrossRef]
67. Hernández, J.A.; López-Sánchez, R.C.; Rendón-Ramírez, A. Lipids and Oxidative Stress Associated with Ethanol-Induced Neurological Damage. *Oxidative Med. Cell. Longev.* **2016**, *2016*, 1543809. [CrossRef]
68. Knudtson, M.D.; Klein, R.; Klein, B.E. Alcohol consumption and the 15-year cumulative incidence of age-related macular degeneration. *Am. J. Ophthalmol.* **2007**, *143*, 1026–1029. [CrossRef]
69. Adams, M.K.M.; Chong, E.W.; Williamson, E.; Aung, K.Z.; Makeyeva, G.A.; Giles, G.G.; English, D.R.; Hopper, J.; Guymer, R.H.; Baird, P.N.; et al. 20/20—Alcohol and Age-related Macular Degeneration: The Melbourne Collaborative Cohort Study. *Am. J. Epidemiol.* **2012**, *176*, 289–298. [CrossRef]

70. Gopinath, B.; Liew, G.; Burlutsky, G.; Mitchell, P. Physical Activity and the 15-Year Incidence of Age-Related Macular Degeneration. *Investig. Ophthalmol. Vis. Sci.* **2014**, *55*, 7799–7803. [CrossRef]
71. Knudtson, M.D.; Klein, R.; Klein, B.E. Physical activity and the 15-year cumulative incidence of age-related macular degeneration: The Beaver Dam Eye Study. *Br. J. Ophthalmol.* **2006**, *90*, 1461–1463. [CrossRef] [PubMed]
72. McGuinness, M.B.; Le, J.; Mitchell, P.; Gopinath, B.; Cerin, E.; Saksens, N.T.M.; Schick, T.; Hoyng, C.B.; Guymer, R.H.; Finger, R.P. Physical Activity and Age-related Macular Degeneration: A Systematic Literature Review and Meta-analysis. *Am. J. Ophthalmol.* **2017**, *180*, 29–38. [CrossRef] [PubMed]
73. Risk factors for neovascular age-related macular degeneration. *Eye Dis. Case-Control Study Group. Arch. Ophthalmol.* **1992**, *110*, 1701–1708. [CrossRef]
74. Astrup, A. Healthy lifestyles in Europe: Prevention of obesity and type II diabetes by diet and physical activity. *Public Health Nutr.* **2001**, *4*, 499–515. [CrossRef]

© 2020 by the authors. Licensee MDPI, Basel, Switzerland. This article is an open access article distributed under the terms and conditions of the Creative Commons Attribution (CC BY) license (http://creativecommons.org/licenses/by/4.0/).

Article

Effects of A2E-Induced Oxidative Stress on Retinal Epithelial Cells: New Insights on Differential Gene Response and Retinal Dystrophies

Luigi Donato [1,2,*], Rosalia D'Angelo [1,2], Simona Alibrandi [1,3], Carmela Rinaldi [1], Antonina Sidoti [1,2,*] and Concetta Scimone [1,2]

1 Department of Biomedical and Dental Sciences and Morphofunctional Imaging, Division of Medical Biotechnologies and Preventive Medicine, University of Messina, 98125 Messina, Italy; rdangelo@unime.it (R.D.); simona.alibrandi@live.it (S.A.); crinaldi@unime.it (C.R.); cscimone@unime.it (C.S.)

2 Department of Biomolecular Strategies, Genetics and Avant-Garde Therapies, I.E.ME.S.T., 90139 Palermo, Italy

3 Department of Chemical, Biological, Pharmaceutical and Environmental Sciences, University of Messina, 98125 Messina, Italy

* Correspondence: ldonato@unime.it (L.D.); asidoti@unime.it (A.S.); Tel.: +39-090-221-3136 (L.D.); +39-090-221-3372 (A.S.)

Received: 20 March 2020; Accepted: 8 April 2020; Published: 10 April 2020

Abstract: Oxidative stress represents one of the principal inductors of lifestyle-related and genetic diseases. Among them, inherited retinal dystrophies, such as age-related macular degeneration and retinitis pigmentosa, are well known to be susceptible to oxidative stress. To better understand how high reactive oxygen species levels may be involved in retinal dystrophies onset and progression, we performed a whole RNA-Seq experiment. It consisted of a comparison of transcriptomes' profiles among human retinal pigment epithelium cells exposed to the oxidant agent N-retinylidene-N-retinylethanolamine (A2E), considering two time points (3h and 6h) after the basal one. The treatment with A2E determined relevant differences in gene expression and splicing events, involving several new pathways probably related to retinal degeneration. We found 10 different clusters of pathways involving differentially expressed and differentially alternative spliced genes and highlighted the sub- pathways which could depict a more detailed scenario determined by the oxidative-stress-induced condition. In particular, regulation and/or alterations of angiogenesis, extracellular matrix integrity, isoprenoid-mediated reactions, physiological or pathological autophagy, cell-death induction and retinal cell rescue represented the most dysregulated pathways. Our results could represent an important step towards discovery of unclear molecular mechanisms linking oxidative stress and etiopathogenesis of retinal dystrophies.

Keywords: RNA-Seq; RPE; Retinitis pigmentosa; A2E

1. Introduction

Oxidative stress, recently defined as "a state where oxidative forces exceed the antioxidant systems due to loss of the balance between them", represents one of the principal inductors of lifestyle-related and genetic diseases [1]. Among them, retinal dystrophies and, in particular, age-related macular degeneration (AMD) and the subgroup of retinitis pigmentosa (RP) are well known to be susceptible to oxidative stress [2,3]. Nowadays AMD is considered the principal cause of visual disability among patients over 50 years [4]. The typical cellular sign of early AMD is represented by drusen, characteristic macular pigmentary deposits, associated with intermediate vision loss [5]. A "dry" and a "wet" form of AMD are currently known, the first more diffused but the second responsible for 90% of acute blindness due to AMD [6]. As previously mentioned, the most relevant risk factor associated with AMD etiopathogenesis is represented by high levels of oxidative stress damaging the macula, generally induced by production of advanced glycation end products (AGE) and exposition to environmental factors [7]. The majority of such effects are also exerted by dysregulation of vascular endothelial growth factor (VEGF), impairment proteins and organelles clearance and glial cell dysfunctions [8]. RP consists of a very heterogeneous group of inherited eye disorders characterized by progressive vision loss [9]. RP has an incidence of 1:4000 people worldwide and represents the most prevalent form of photoreceptor-related diseases [10]. Primary symptoms can already occur in childhood or adolescence and, generally, consist of night blindness and gradual reduction of the visual field due to progressive death of rods. Total blindness, instead, is a feature typical of the late stage of the disease, following degeneration of macula photoreceptors [11]. Rods represent about 95% of all photoreceptors, and oxidative metabolism of fatty acids is their main energy source [12]. Main causes of rod death are genetic mutations, and more than 80 RP-causative genes have been already identified (https://sph.uth.edu/RetNet/sum-dis.htm#B-diseases), even if a relevant number of them are still unknown [13]. Conversely, cone degeneration is usually a late event frequently resulting from cytotoxic effects of high oxygen levels in the retina after rod reduction. Thus, oxidative damage is considered the first cause of cone apoptosis and progressive vision loss [14]. Interestingly, AMD and RP can also arise due to mutations in genes expressed in other retinal cell types, such as *MERTK* [15], *RLBP1* [16] and *RPE65* [17] expressed in retinal pigment epithelium (RPE). Originally, only a trophic function was hypothesized for RPE cells. Nowadays, it is well known that RPE is a monolayer of neural-crista-derived pigmented epithelial cells interacting with Bruch's membrane and choriocapillaris on the basolateral side and with the outer segments of the photoreceptors on the apical one [17]. RPE plays many vital roles for photoreceptor cells and the most fascinating certainly is the protection from oxidative stress [18]. Recent studies confirmed a high level of reactive oxygen species (ROS) in RPE, and fatty acids are one of their molecular targets. If oxidized, they may impair transduction pathways and gene expression [19]. Although fatty acid oxidation was already confirmed to cause macular degeneration, oxidative stress mechanism in RP development requires further clarifications [20]. Therefore, to better understand how high ROS levels may lead to retinal dystrophies onset and progression, we performed a comparison of transcriptomes' profiles among human RPE cells exposed to the oxidant agent N-retinylidene-N-retinylethanolamine (A2E). A2E is a toxic bis-retinoid that derives from the condensation and the oxidation of the trans-retinal [21]. Throughout life, A2E and other complex lipids accumulate and form lipofuscin in the RPE, ultimately determining photoreceptor death [22]. Additionally, A2E photo-oxidation is able to generate singlet oxygen, a highly reactive molecule that contributes to the increase of level of toxic metabolites such as epoxides and endoperoxides [23]. Moreover, lipid peroxidation is also responsible for A2E cleavage that releases cytotoxic reactive aldehydes. These reactive species could affect the lipid membranes fluidity and can damage the DNA and the cellular proteins [24]. On these bases, it was hypothesized that lipofuscin, A2E and its oxidized metabolites could accumulate if the cellular antioxidant system is unable to fight the oxidative-damaged lipids in rods and cones. Consequently, accumulation and photo-oxidation of A2E could lead to several retinal dystrophies, like already established for age-related macular degeneration (AMD) [25]. Various studies based on ARPE-19 cell line present pharmacological solution to retinal dystrophies. The same cell line is also used to identify new candidate compounds able to protect RPE against A2E oxidation [26]. As highlighted in the manuscript, the treatment with such a compound determines relevant differences in gene expression and splicing events, involving several new pathways probably related to retinal degeneration.

2. Materials and Methods

2.1. Cell Culture

Human RPE-derived Cells (H-RPE—Human Retinal Pigment Epithelial Cells, Clonetics™, Lonza, Walkersville, USA) were grown in T-75 flasks containing RtEGM™ Retinal Pigment Epithelial Cell Growth Medium BulletKit® (Clonetics™, Lonza, Walkersville, MD, USA) with 2% *v*/*v* fetal bovine serum (FBS), 1% of penicillin/streptomycin and incubated at 37 °C with 5% CO_2. H-RPE cells were then plated into 96-well plates (4×10^4 cells/well) and cultured for 24 h to reach confluence before treatment. Subsequently, A2E was added to a final concentration of 20 μM for 6 h before rinsing with medium. Control cell groups were incubated without A2E. Confluent cultures were transferred to PBS supplemented with calcium, magnesium and glucose (PBS–CMG) and then exposed to blue light emitted by a tungsten-halogen source (470 ± 20 nm; 0.4 mW/mm^2) for 30 min to induce phototoxicity of A2E and incubated at 37 °C. The 1–3 generation of subcultured RPE cells were used in this study.

2.2. MTT Assay

Cell viability was determined by mitochondrial-dependent reduction of methylthiazolyldiphenyl-tetrazolium bromide (MTT) (Sigma-Aldrich, St. Louis, MO, USA) to formazan-insoluble crystals. Briefly, 10 μL of 5 mg/mL of MTT in PBS was added to the cells following the A2E treatment. After incubation at 37 °C for 2 h, 100 μL of 10% SDS in 0.01 mol/L HCl was added to dissolve the crystals and incubated for 16 h. The absorbance was measured in a Dynatech microplate reader at 570 nm. Results were expressed as percentage of viable cells normalized with control conditions in the absence of A2E.

2.3. Total RNA Sequencing

Total RNA was extracted by TRIzol™ Reagent (Invitrogen™, ThermoFisher Scientific, Waltham, MA, USA), following manufacturer's protocol, and quantified at Qubit 2.0 fluorimeter by Qubit® RNA assay kit (Invitrogen™, ThermoFisher Scientific, Waltham, MA, USA). The RNA-seq samples consisted of 3 factor groups, represented by Human RPE cells, before the treatment with A2E and at 2 following different time points of 3 h and 6 h, respectively. For each group, 3 biological replicates were considered, for a total of 9 samples. The selection of 3 h and 6 h time points was based on previous experiments realized by our research group (unpublished data), confirmed by outcomes from MTT assay in this work. Such results highlighted that in wider time intervals, the death rate of oxidative stressed cells might be so high as to invalidate the following expression analysis. Libraries were generated using 1 μg of total RNA by the TruSeq Stranded Total RNA Sample Prep Kit with Ribo-Zero H/M/R (Illumina, San Diego, CA, USA), according to manufacturer's protocols. Sequencing runs were performed on an HiSeq 2500 Sequencer (Illumina, San Diego, CA, USA), using the HiSeq SBS Kit v4 (Illumina, San Diego, CA, USA). The experiment was repeated thrice.

2.4. Quality Validation and Read Mapping

Sequence reads were generated from RPE-specific cDNA libraries on the Illumina HiSeq 2500 Sequencer. Obtained raw sequences were filtered to remove low-quality reads (average per base Phred score <30) and adaptor sequences. The quality of analyzed data was checked using FastQC (v.0.11.9) (https://www.bioinformatics.babraham.ac.uk/projects/fastqc/) and QualiMap (v.2.2.1) [27], while trimming was realized by Trimmomatic (v.0.39). Filtered data were then mapped by CLC Genomics Workbench v.20.0 (https://digitalinsights.qiagen.com/products-overview/analysis-and-visualization/qiagen-clc-genomics-workbench/) against the Homo sapiens genome hg38 and the RNA database v.91, on Ensembl database. RNA-seq analysis was conducted using the following settings: quality trim limit = 0.01, ambiguity trim maximum value = 2. Map to annotated reference was as follows: mismatch cost = 2, insertion and deletion costs = 3, minimum length fraction and minimum similarity fraction = 0.8, maximum number of hits for a read = 10, strand-specific = both, expression value = TPM. Raw data are available upon request.

2.5. Gene Expression and Statistical Analysis

Mapped reads were quantified by alignment-dependent and alignment-independent methods. The first approach uses the expectation-maximization (EM) algorithm [28] in order to determine expressions even in cases where the majority of reads map equally well to multiple genes or transcripts. Once the algorithm has converged, every non-uniquely mapping read was assigned randomly to a particular transcript according to the abundances of the transcripts within the same mapping. The transcript per million reads (TPM) values were, then, computed from the counts assigned to each transcript. The second method has a higher accuracy for the point expression estimation and also allows the user to bootstrap the expression quantification to get an estimate of the technical variability. This approach was applied by the Salmon tool [29] using the transcript fasta files downloaded from GENCODE v.32 (gencode.v32.transcripts.fa). Salmon was run with the following settings for the RNA-seq data: quant –index/index –libType U –unmatedReads /single.fastq –incompatPrior '0.0' –biasSpeedSamp '5 ' –fldMax '1000' –fldMean '250' –fldSD '25' –forgettingFactor '0.65' –maxReadOcc '100' –numBiasSamples '2000000' –numAuxModelSamples '5000000' –numPreAuxModelSamples '5000' –numGibbsSamples '0' –numBootstraps '0' –thinningFactor '16' –sigDigits '3' –vbPrior '1e-05' -o/output. Once obtained, Salmon outputs were imported using tximport R package version 1.10.0 and lengthScaledTPM method [30] to generate read counts and Transcripts Per Million (TPMs). Low expressed transcripts and genes were filtered based on the data mean–variance trend analysis. The expected decreasing trend between data mean and variance was observed when expressed transcripts were determined as which had ≥1 of the 9 samples with count per million reads (CPM) ≥1, which provided an optimal filter of low expression. A gene was expressed if any of its transcripts with the above criteria were expressed. The trimmed mean of M-values (TMM) method was used to normalize the gene and transcript read counts to log_2-CPM [31]. The principal component analysis (PCA) plot showed that the RNA-seq data did not have distinct batch effects, so it was possible to proceed with downstream analyses.

2.6. DE, DAS and DTU Analysis

Limma R package was used for differential expression analyses [32]. General linear models were established to compare gene and transcript expression changes at the different conditions of experimental design, setting the contrast groups as 0 h.untreated versus 3 h.treated, 0 h.untreated versus 6 h.treated, 3 h.treated versus 6 h.treated, 0 h.untreated versus (3 h.treated + 6 h.treated)/2. For differentially expressed (DE) genes/transcripts, the log_2 fold change (L_2FC) of gene/transcript abundance were calculated based on contrast groups and significance of expression changes were determined using the t-test [33]. *P*-values of multiple testing were adjusted with BH to correct false discovery rate (FDR) [34]. A gene/transcript was significantly DE in a contrast group if it had adjusted *p*-value <0.01 and L_2FC ≥1. At the alternative splicing level, differential transcript usage (DTU) transcripts were determined by comparing the L_2FC of a transcript to the weighted average of L_2FCs (weights were based on their standard deviation) of all remaining transcripts in the same gene. A transcript was determined as significant DTU if it had adjusted *p*-value <0.01 and ΔPS ≥0.1. For differentially alternative spliced (DAS) genes, L_2FC of each individual transcript was compared to gene level L_2FC, which was calculated as the weighted average of L_2FCs of all transcripts of the gene. Then *p*-values of individual transcript comparisons were summarized to a single gene level *p*-value with F-test. A gene was significantly DAS in a contrast group if it had an adjusted *p*-value <0.01 and any of its transcript had a Δ Percent Spliced (ΔPS) ratio ≥0.1. Finally, time points (0 h, 3 h, 6 h) in groups (untreated, treated) were used for time-series trend analysis. Natural Cubic Spline method with degree of freedom was used to generate time-series trend (Figure S1).

2.7. Gene-Enrichment and Functional Pathway Analysis

The up- and down-regulated genes were analyzed by the Database for Annotation, Visualization and Integrated Discovery (DAVID) 6.8 [35]. This tool is based on more than 40 annotation categories, including GO terms to protein–protein interactions, from disease associations to gene functional summaries, and many others. In DAVID annotation system, EASE Score, a modified Fisher Exact P-Value, is adopted to measure the gene-enrichment in annotation terms. The EASE score provides a conservative adjustment to the Fisher exact probability that weights significance in favor of themes supported by more genes. The EASE score is calculated by penalizing (removing) one gene within the given category from the list and calculating the resulting Fisher exact probability for that category.

2.8. Selection of Single-Pathway "Master genes" and Selection of Retinitis Pigmentosa Candidate Genes by ToppGene Prioritization

In order to highlight new candidate genes involved into retinitis pigmentosa, based on oxidative-related candidate pathways, we chose the 15 most altered genes for each one. Firstly, we considered them for each time point; then, we chose the commons in all time points. Subsequently, chosen genes underwent prioritization by ToppGene (https://toppgene.cchmc.org), a web-based tool able to classify a selected group of candidate genes from a large set of genes correlated with a pathology, giving each one a score. The score is based on the intersection of data from various databases of annotations related to cellular and physiological functions, analyzing complex networks shared between genes already known to cause the disease (training genes) and candidate genes (test genes). Training genes were obtained from RetNet online database.

2.9. Data Validation by qRT-PCR.

Ten most dysregulated mRNAs from candidate genes previously identified were selected and validated by quantitative Real-Time-Polymerase Chain Reaction (qRT-PCR), in order to validate RNA-Seq data. Reverse transcription was carried out according to the manufacturer's protocol of GoScript™ Reverse Transcription System (Promega, Madison, WI, USA). The obtained cDNA was subjected to RT-PCR in the ABI 7500 fast sequence detection system (Applied Biosystems, Foster City, CA, USA), using BRYT-Green-based PCR reaction. PCR amplification was performed in a total reaction mixture of 20 μL containing 20 ng cDNA, 10 μL 2 × GoTaq1qPCR Master Mix (Promega, Madison, WI, USA) and 0.2 μM of each primer. PCR was run with the standard thermal cycle conditions using the two-step qRT-PCR method: an initial denaturation at 95 °C for 30 s, followed by 40 cycles of 30 s at 95 °C and 30 s at 60 °C. Each reaction was replicated six times, considering all analyzed time points (18 samples), and the average threshold cycle (Ct) was calculated for each replicate. The expression of mRNAs was calculated relative to expression level of endogenous control β-actin, and the relative expression of genes was calculated using the $2^{-\Delta\Delta Ct}$ method [36]. The results were shown as the mean ± SEM (Standard Error of Mean). Statistical significance was determined by analysis of variance between groups (ANOVA), followed by Bonferroni post-hoc test. Finally, a linear regression analysis was performed to check the correlation of the FC of the gene expression ratios between qRT-PCR and RNA-Seq. The statistical analyses were all performed using IBM SPSS 26.0 software (https://www.ibm.com/analytics/us/en/technology/spss/). Adjusted p-values <0.05 were considered statistically significant. The research was approved by the Scientific Ethics Committee of the Azienda Ospedaliera Universitaria—Policlinico "G. Martino" Messina.

3. Results

3.1. MTT Cell Viability Assay Results

The MTT cell viability assay showed a relevant and different trend in RPE-treated cells versus control. The addition of A2E to cultures led to a dose-dependent increase in cell death percentage (Figure 1).

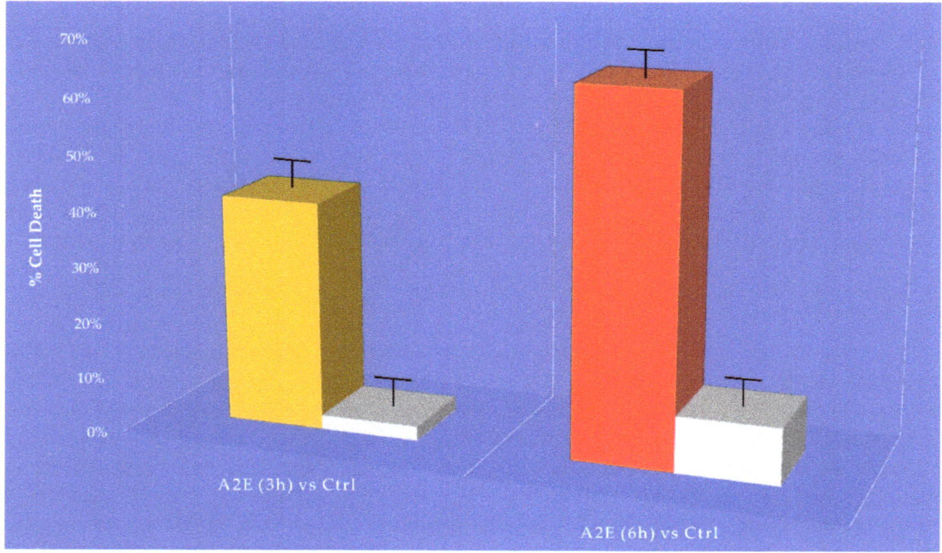

Figure 1. MTT determination of A2E treatment in retinal pigment epithelium (RPE) cells. Cell death was assessed at considered time points (3 h and 6 h) in A2E treated samples compared to basal untreated group. Results are shown as mean ± standard error of mean ($n = 3$). p-value < 0.05.

3.2. Sequencing Analysis and Mapping Statistics

RNA sequencing carried out on Illumina HiSeq 2500 yielded a total of 96,346,180 quality reads (mean mapping quality = 29) with a percentage of 67.8% uniquely mapped. A total of 16,173 genes and 69,653 transcripts were identified out of 60,609 and 227,462 reference counterparts, respectively, considering the whole human transcriptome. The annotated reference assembly v.32 (GRCh38.p13) was downloaded from GeneCode FTP server (ftp://ftp.ebi.ac.uk/pub/databases/gencode/Gencode_human/). All previous mapping statistics were based on average values calculated for all three replicates in each time point. Details are available in Figure S2.

3.3. Analysis of Gene Expression Profile of RPE Cells

As previously cited, in our transcriptome study, 16,173 genes (Table S1) and 69,653 transcripts (Table S2) were expressed with $\log_2$-CPM values ≥1 considered as the average value per sample. We first analyzed differential expression at the gene level (DE) (Figure 2a and Table S3). Considering the stringent criteria we have chosen, we identified 2432 genes that were significantly differentially expressed in response to A2E treatment. Of these, 59.7% resulted as up-regulated, while 40.3% down-regulated, showing a decreasing time-dependent trend for up-regulated ones and an increasing trend for down-regulated ones. Afterward, the analysis of transcript-level data allowed us to identify genes that were DAS between the contrast groups (Figure 2b and Table S4). We detected 5119 DAS genes, of which 1101 were also DE genes (regulated by both transcription and AS) and 4108 were regulated by AS only. Therefore, considering all 6540 genes that showed significantly altered levels of differential gene and/or transcript-level expression, 78.3% were differentially alternatively spliced, highlighting a consistent response to the induced oxidative stress. Furthermore, to identify the specific transcripts that characterize a gene as DAS, a DTU analysis was performed (Figure 2b and Table S5). Globally, around 12% (8587) of expressed transcripts were classified as DTU. The next step consisted in the analysis of early response to the induced oxidative stress in comparison with the response to late treatment with A2E. About 40% (858) and 55% (2457) of DE and DAS genes, respectively, resulted as common to both time point observations, while the residual was unique to 3 h and 6 h (Figure 3). Consequently, it is evident how changes in gene-level expression and alternative splicing occurred throughout the whole period, either transiently (occurring after 3 h and returning to initial level after 6 h), occurring later (only after 6 h from treatment) or enduring throughout the whole period. Such results probably suggest the different responses of RPE cells as a growing resistance to oxidative stress stimuli in a time-dependent manner.

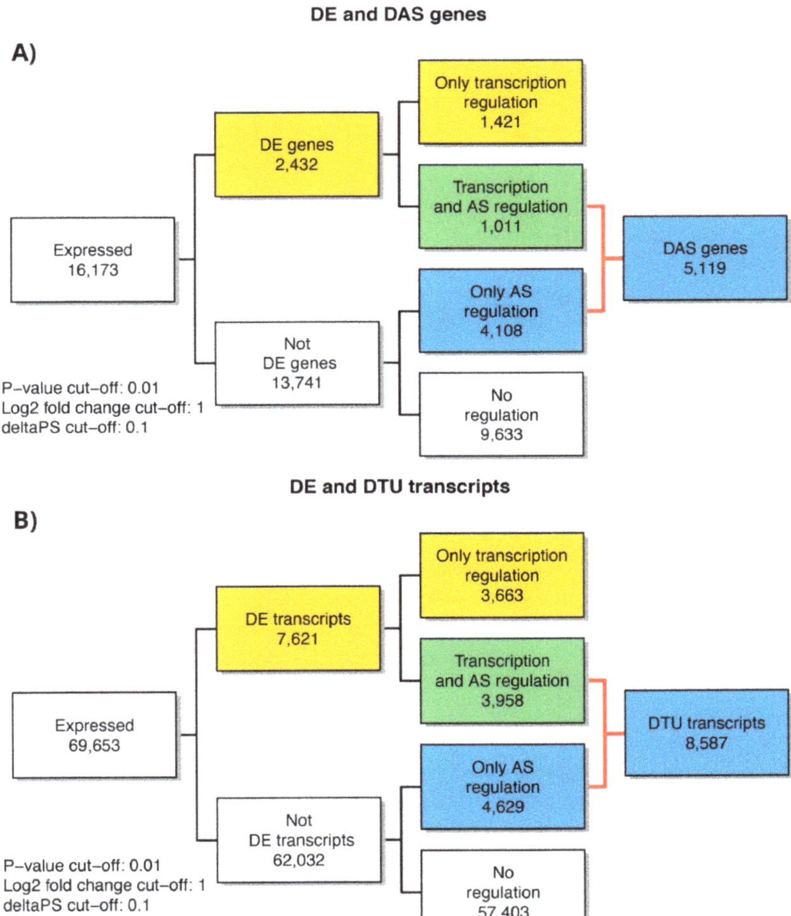

Figure 2. Summary Figure of expressed genes and significant DE, DE + DAS, DE + DTU, DAS and DTU genes from analysis of the considered time points of RPE cell data. (**A**) Number of genes regulated only by transcription (DE), only by alternative splicing (DAS) and by both transcription and alternative splicing (DE + DAS); (**B**) number of transcripts regulated only by transcription (DE), only by alternative splicing (DAS) and by both transcription and alternative splicing (DE + DAS). DTU = Differential Transcript Usage. AS = Alternative Splicing.

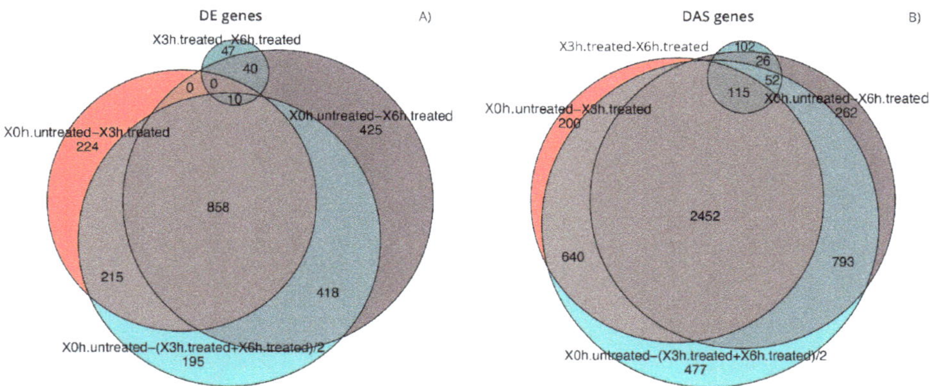

Figure 3. Comparison of the gene lists generated during differential expression analyses. Euler diagrams of DE (**A**) and DAS (**B**) genes identified during expression analyses in considered conditions of experimental design, setting the contrast groups as 0 h.untreated versus 3 h.treated, 0 h.untreated versus 6 h.treated, 3 h.treated versus 6 h.treated, 0 h.untreated versus (3 h.treated + 6 h.treated)/2.

3.4. DE and DAS Genes Highlighted Different Functionality Patterns

Functional enrichment analyses of DE and DAS genes showed relevant differences, as already supposed by the overlap of only 1011 genes (Figure 2). The most significantly enriched terms for DE genes were linked to vascular events and very heterogeneous responses to oxidative stress. As evidenced by hierarchical clustering of total gene expression levels of DE genes, adaptive, transient and late expression profiles followed the A2E induced stress, and an analog response was highlighted by transcript expression profiles of individual DE genes (Figure 4). Functional annotation of individual DE genes clusters was associated with methylation of RNA (cluster 1), various stress response (cluster 2), microtubule assembly and activity (cluster 3), angiogenesis (cluster 6), lipid biosynthesis (cluster 7), focal adhesions (cluster 8) and respiratory electron transport (cluster 10). For DAS genes, the most enriched functional terms deal with alteration of cellular proliferation and cell death. Hierarchical clustering of the DTU transcripts and expression profiles of individual DAS genes also evidenced heterogeneous response profiles. Functional annotation of DTU clusters of genes revealed enrichment of terms related to misfolded and damaged protein removal (clusters 1 and 3), autophagy (cluster 2), lipid biosynthesis (cluster 6), regulation of DNA damage response (cluster 7), induction of cell death (cluster 8) and translation regulation (cluster 10) (Figure 5).

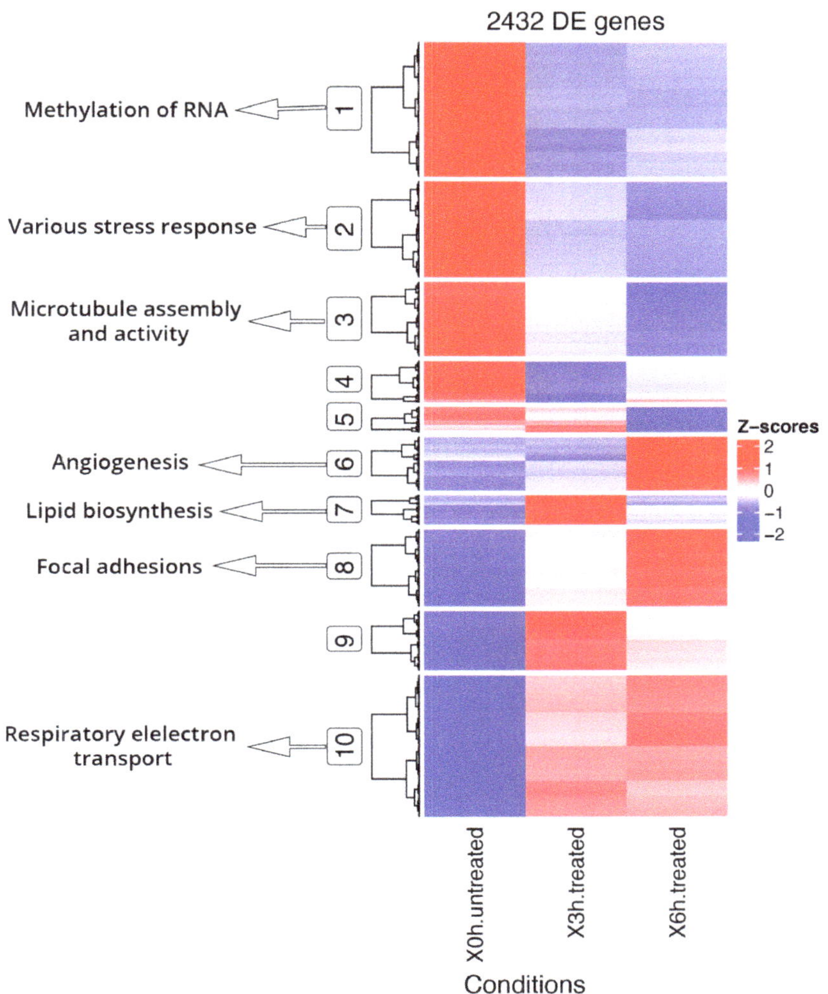

Figure 4. Hierarchical clustering and heatmap of DE genes and Key GO terms. DE genes show segregation into 10 coexpressed clusters, of which the main ones related to oxidative stress were highlighted (circled) and linked to GO specific terms. Full results of GO enrichment analyses are shown in Tables S6 and S7. The z-score scale represents mean-subtracted regularized log-transformed transcripts per million (TPMs).

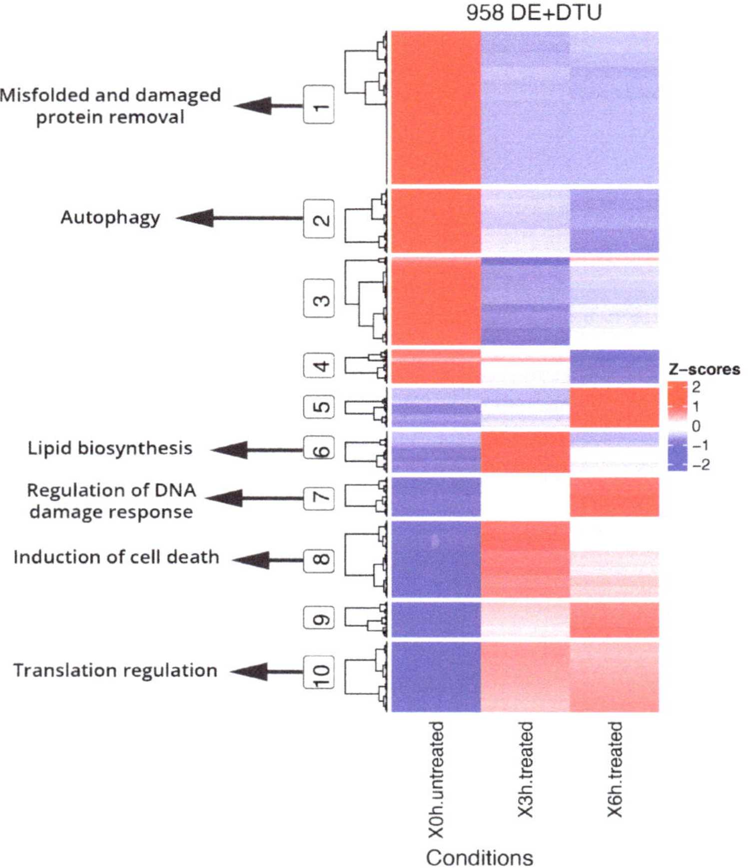

Figure 5. Hierarchical clustering and heatmap of DE genes + DTU transcripts from DAS genes and Key GO terms. DE genes and DTU transcripts from DAS genes show segregation into 10 coexpressed clusters, of which the main ones related to oxidative stress were highlighted (circled) and linked to GO specific terms. Full results of GO enrichment analyses are shown in Tables S6 and S7. The z-score scale represents mean-subtracted regularized log-transformed TPMs.

3.5. Early Cellular Response to Induced Stress Mainly Involves Pre-mRNA Splicing and Glycolysis-Related DE and DAS Genes

The applied statistical model used in our analyses permitted us to determine precisely at which time point each DE and DAS gene first showed a significant change, together with the magnitude and course of that change. The top 15 significant genes or transcripts belonging to the considered groups were deeply investigated, allowing the identification of new candidate genes and specific pathways possibly involved in retinal dystrophy etiopathogenesis (Figure 6). After the first 3 h of treatment, only one gene (*HNRNPA3P6*) reached a relevant DE status, and it is involved in cytoplasmic trafficking of RNA and pre-mRNA splicing. Main DAS genes that showed early important differences (*CTSH* and *GPI*), instead, were related to glycolysis, and the same biochemical pathway was the most correlated to initial DTU transcripts expression dysregulation (ENST00000615999.4, ENST00000588991.7, ENST00000586425.2, ENST00000525807.5 and ENST00000550050.5).

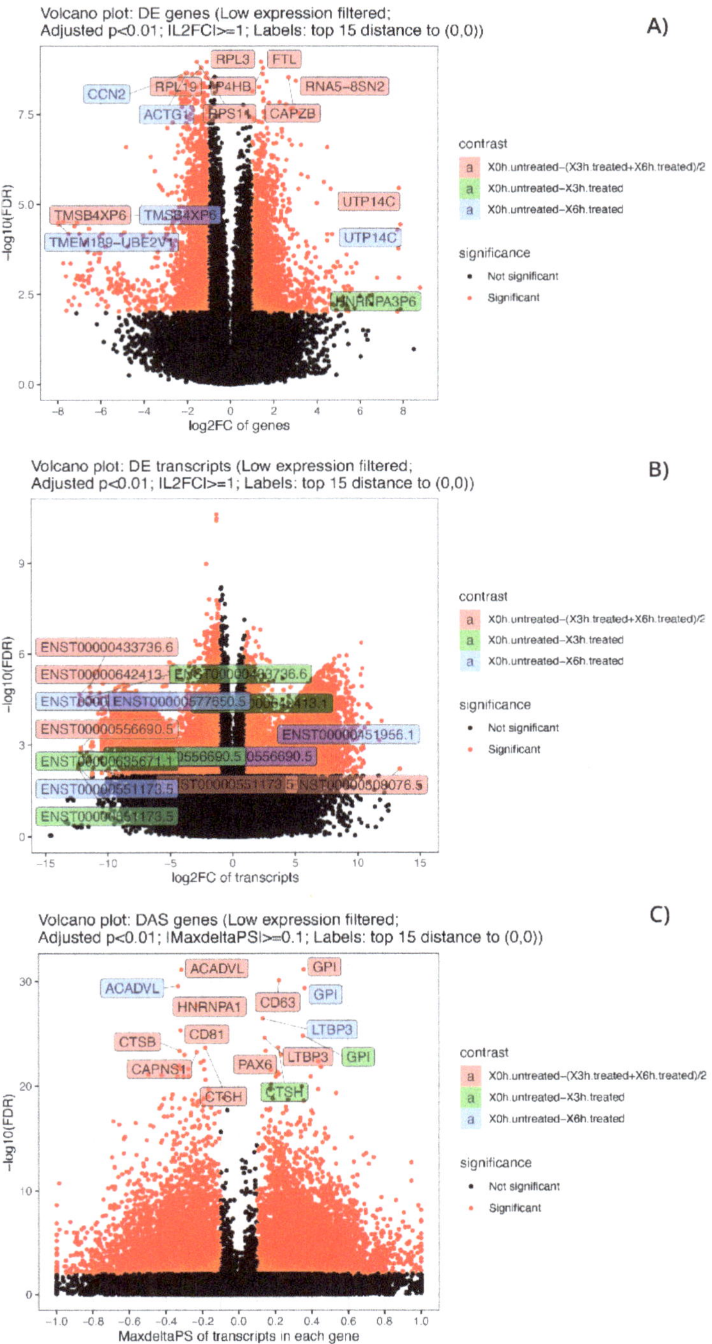

Figure 6. *Cont.*

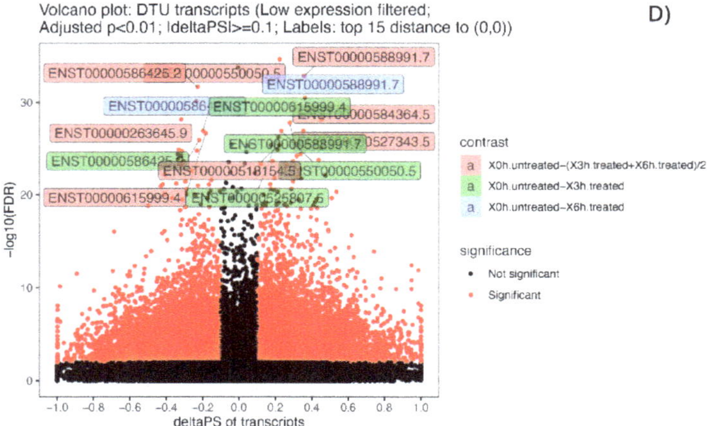

Figure 6. Volcano plots of significant DE and DAS genes and of DE and DTU transcripts. Volcano plots of significant (adjusted *p*-value < 0.01) DE genes (**A**), DE transcripts (**B**), DAS genes (**C**) and DTU transcripts (**D**). The low expressed genes and transcripts were filtered. The top 15 considered elements with the smallest corrected *p*-values and bigger fold-changes are highlighted, and different colors refer to different contrast groups. DE genes: log2FC vs. −log10(FDR) at gene level; DAS genes: maximum ΔPS of transcript in a gene vs. −log10(FDR) at gene level; DE transcripts: log2FC vs. −log10(FDR) at gene level at transcript level and DTU transcripts: ΔPS vs. −log10(FDR) at transcript level.

3.6. Late RPE Cell Response to A2E Treatment Could Impair Bioenergetic Specific Reactions, Extracellular Matrix Integrity and Neurotransmission-Related DE and DAS Genes

While the number of significant and relevant DE genes was limited in the early stage of treatment, it grew over time, reaching a climax at 6 h (Figure 6). At this time point, many of the DE genes (*ACTG1*, *CCN2*, *RPL19*, *RPL3*, *P4HB*, *RPS11*, *FTL*, *CAPZB* and *RNA5-8SN2*) (Figures 7 and 8) resulted as linked to new particular pathways, as iron metabolism, plasma lipoproteins assembly and F-actin capping. Furthermore, several over-expressed DE genes at 6 h (*TTC8*, *ARL3*, *REEP6*, *GUCA1B* and *PDE6G*) resulted as associated with alternative splicing of retina-preferred gene transcripts. About DAS genes, instead, the most significant ones (*ACADVL*, *GPI* and *LTBP3*) resulted as correlated to already-involved pathways (e.g., glycolysis). However, more interestingly, the same genes were together with DTU transcripts (ENST00000586425.2 and ENST00000588991.7), but even more interestingly, with other biochemical activities regarding oxidative processes in mitochondria (e.g., fatty acids reactions) and low conductance of potassium channels (as evidenced by DE transcripts ENST00000577650.5, ENST00000451956.1, ENST00000556690.5 and ENST00000551173.5).

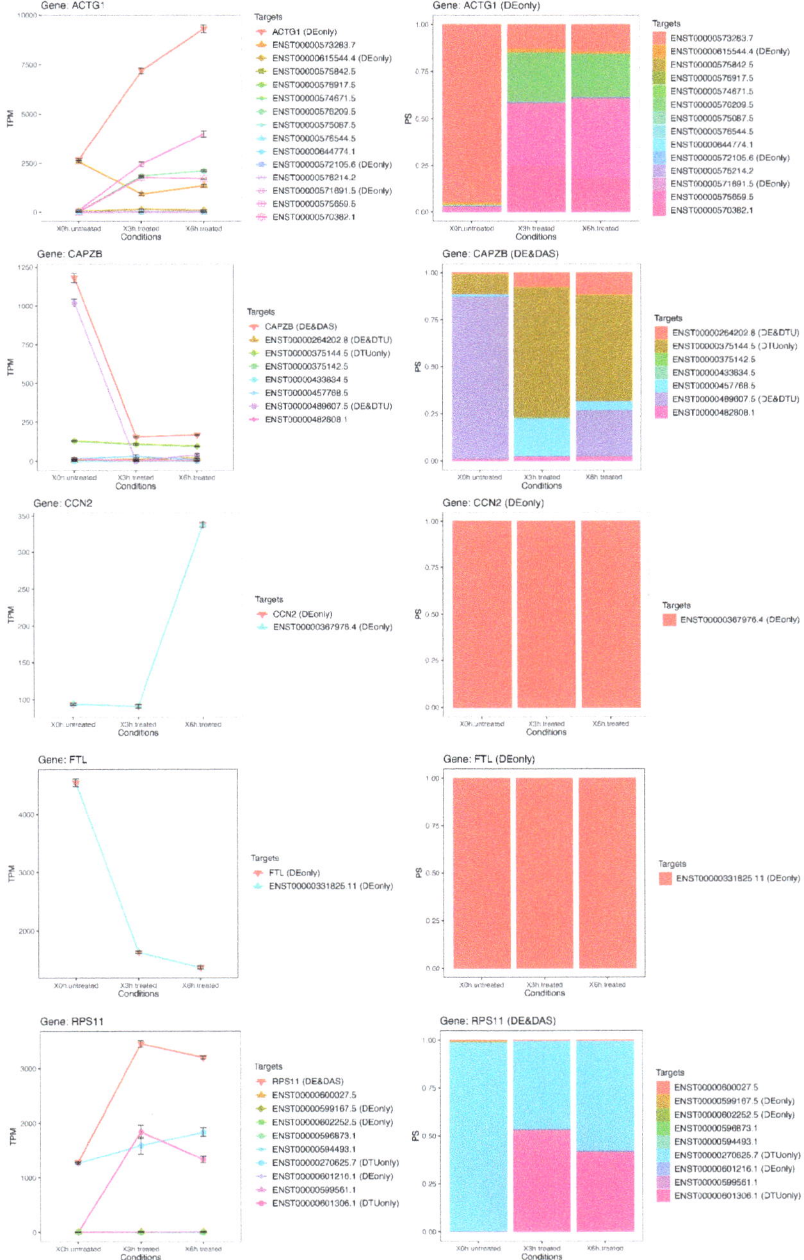

Figure 7. Expression profiles of most important DE genes/transcripts; first cluster associated with newly discovered pathways linked to oxidative stress. Detailed gene/transcript expression profiles across the time course of the first cluster of most important DE genes (ACTG1, CAPZB, CCN2, FTL and RPS11) linked to newly identified candidate pathways linked to oxidative stress.

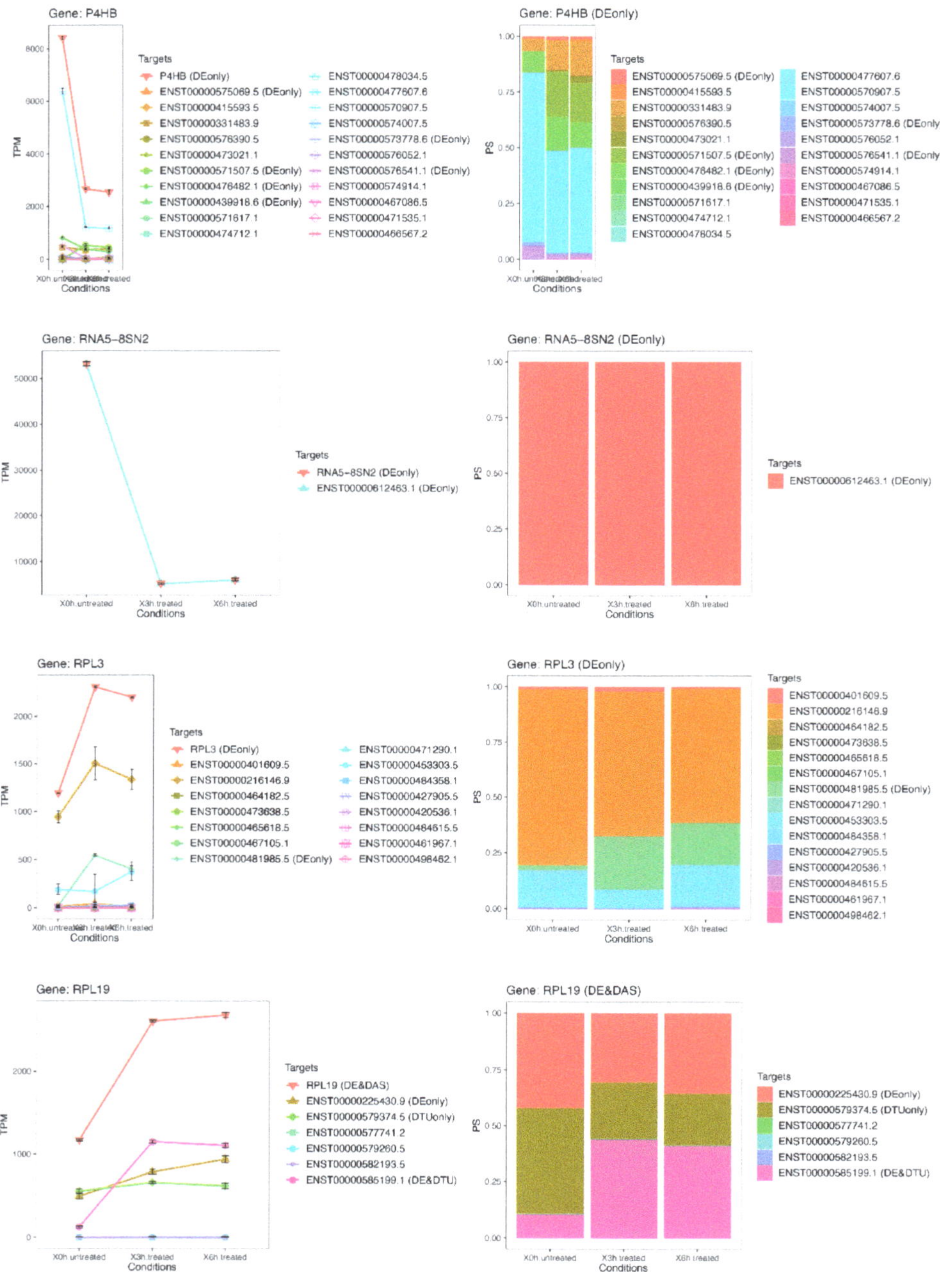

Figure 8. Expression profiles of most important DE genes/transcripts secondo cluster associated with newly discovered pathways linked to oxidative stress. Detailed gene/transcript expression profiles across the time course of the second cluster of most important DE genes (*P4HB*, *RNA5-8SN2*, *RPL3* and *RPL19)* linked to newly identified candidate pathways linked to oxidative stress.

3.7. The Transcriptome Comparison between Untreated (Time Zero) and Treated (3 h + 6 h) Rpe Cells Revealed the Possible Impairment of Retinal Cells Crosstalk and Synapses, Leading to Rescue or Cell Death

Many DE and DAS genes only showed significant differences only if analyzed during the whole considered treatment period. Among these DE genes, five (*CCN2, ACTG1, UTP14C, TMSB4XP6* and *TMEM189-UBE2V1*) were linked to extracellular matrix constituent secretion and cellular junctions, as well as to misfolded protein ubiquitination. Furthermore, the number of DAS genes and DTU transcripts changed during treatment was very high. The first ones (*ACADVL, GPI, HNRNPA1, CD81, CD63, CTSB, CTSH, LTBP3, CAPNS1* and *PAX6*) were enriched in terms related to neuropeptide catabolic processes and extracellular vesicles in the crosstalk of cells, while the second ones (ENST00000586425.2, ENST00000550050.5, ENST00000588991.7, ENST00000527343.5, ENST00000584364.5, ENST00000263645.9, ENST00000615999.4 and ENST00000518154.5) mainly regard dendrite regeneration. Therefore, the overall scenario could reveal a dual response by stressed cells: on one hand, the alteration of retinal cell crosstalk and synapses could lead to various forms of cell death (e.g., autophagy). On the other hand, RPE tries to survive, increasing regeneration of capable parts (e.g., dendrites), by synaptic plasticity. Detailed info on pathways enrichment are available in Table S6 and Table S7).

3.8. The Most Significant DAS Genes Represented the Main Retinal Dystrophy Candidate Genes

The ToppGene prioritization analysis on known causative retinal dystrophy genes that intersected with most significant DE and DAS genes previously described revealed 19 candidate genes (Bonferroni corrected p-value <0.05). Of these, seven showed a strong association with the training genes (Bonferroni corrected p-value <0.01). Most of the 19 significant obtained candidate genes, included three of most statistically associated (*PAX6, CTSH* and *HNRPA1*) belong to DAS genes, further highlighting the influence of oxidative stress on alternative splicing (Table 1). Details of ToppGene results are available in Table S8.

3.9. qRT-PCR Validation

To validate the authenticity and reproducibility of the RNA-Seq results, the 15 selected mRNAs were validated by qRT-PCR analysis, and obtained expression profiles were similar to the results of transcriptome analysis (Table S9). Moreover, the ANOVA statistics, conducted to compare the means among multiple groups, highlighted high significance (p-values < 0.05). The linear regression analysis showed a significantly positive correlation of the relationship between gene expression ratios of qRT-PCR and RNA-Seq for all selected time points (Figure S3), confirming our transcriptomic data validity.

Table 1. ToppGene prioritization analysis results. The ToppGene prioritization analysis on known causative retinal dystrophy genes that intersected with most significant DE and DAS genes obtained during the whole transcriptome analysis revealed 19 candidate genes (Bonferroni corrected p-value < 0.05), of which seven had a strong association with training genes (Bonferroni-corrected p-value < 0.01). All shown p-values are Bonferroni-corrected.

Rank	Gene Symbol	GeneId	GO: Mol. Func. Score	GO: Mol. Func. p-Value	GO: Bio. Proc. Score	GO: Bio. Proc. p-Value	GO: Cell. Comp. Score	GO: Cell. Comp. p-Value	Human Pheno. Score	Human Pheno. p-Value	Pathway Score	Pathway p-Value	Pubmed Score	Pubmed p-Value	Disease Score	Disease p-Value	Average Score	Overall p-Value
1	PAX6	5080	0.0422	0.0931	1.0	0.0102	0.2814	0.0807	1.0	0.0135	0.0	0.503	1.0	0.001	0.999	7.75×10^{-10}	0.6176	1.88×10^{-12}
2	ACTG1	71	0.8143	0.0106	0.996	0.0278	0.9686	0.0127	0.999	0.0135	0.0	0.503	0.330	0.032	0.730	0.006	0.7267	5.15×10^{-11}
3	TGFBI	7045	0.6329	0.0288	0.982	0.0462	0.4683	0.0534	0.988	0.0331	0.0	0.503	0.204	0.073	0.819	0.003	0.6226	0.002
4	CCN2	1490	0.4318	0.0491	0.999	0.0209	0.4362	0.0544	0.999	0.0253	0.0	0.503	0.204	0.073	0.551	0.011	0.5529	0.003
5	CTSH	1512	0.2796	0.0677	0.999	0.0185	0.9986	0.0036	0.976	0.0358	0.0	0.503	0.0	0.536	0.445	0.016	0.5302	0.003
6	GNAI2	2771	0.9263	0.0029	0.962	0.0573	0.9500	0.0166	0.999	0.0195	0.0	0.503	0.095	0.073	0.0	0.512	0.5951	0.004
7	LTBP2	4053	0.5918	0.0375	0.813	0.1049	0.0350	0.2254	0.999	0.0135	0.0	0.503	0.204	0.073	0.636	0.009	0.5168	0.007
8	HNRNPA1	3178	0.0421	0.0931	0.910	0.0772	0.5481	0.0499	0.999	0.0214	0.0	0.503	0.801	0.006	0.0	0.512	0.4913	0.011
9	GPI	2821	0.0421	0.0931	0.965	0.0563	0.9563	0.0156	0.986	0.0335	0.0	0.503	0.490	0.019	0.0	0.512	0.5249	0.011
10	CD81	975	0.6702	0.0213	0.999	0.0209	0.3057	0.0782	0.999	0.0232	0.0	0.503	0.095	0.073	0.0	0.512	0.4691	0.011
11	FTL	2512	0.2622	0.0766	0.587	0.1403	0.2370	0.0979	0.999	0.0205	0.0	0.503	0.095	0.073	0.331	0.021	0.4151	0.012
12	ITGAV	3685	0.4732	0.0476	0.999	0.0156	0.9885	0.0094	−1.0	0.0	0.0	0.503	0.076	0.073	0.0	0.512	0.4563	0.014
13	CAPZB	832	0.5061	0.0426	0.998	0.0242	0.9934	0.0079	−1.0	0.0	0.0	0.503	0.205	0.073	0.0	0.512	0.4919	0.015
14	LTBP3	4054	0.0421	0.0931	0.982	0.0458	0.0350	0.2254	0.999	0.0178	0.0	0.503	0.0	0.536	0.636	0.009	0.3864	0.018
15	CTSB	1508	0.2996	0.0642	0.915	0.0753	0.3862	0.0600	0.183	0.0553	0.0	0.503	0.343	0.032	0.0	0.512	0.3794	0.026
16	P4HB	5034	0.2996	0.0642	0.851	0.0960	0.2370	0.0979	0.941	0.0400	0.0	0.503	0.351	0.032	0.0	0.512	0.4317	0.031
17	MATN2	4147	0.4747	0.0464	0.987	0.0408	0.3796	0.0606	−1.0	0.0	0.0	0.503	0.204	0.073	0.0	0.512	0.3834	0.042
18	TMEM189-UBE2V1	387522	−1.0	0.0	−1.0	0.0	−1.0	0.0	−1.0	0.0	0.0	0.503	0.059	0.073	0.579	0.011	0.2746	0.050
19	ACADVL	37	0.6272	0.0290	0.903	0.0792	0.2422	0.0857	0.980	0.0348	0.0	0.503	0.0	0.536	0.0	0.512	0.4014	0.053
20	CD151	977	0.3015	0.0639	0.776	0.1113	0.4489	0.0542	0.994	0.0313	0.0	0.503	0.0	0.536	0.0	0.512	0.3755	0.065
21	CAPNS1	826	0.0421	0.0931	0.819	0.1039	0.1428	0.1432	−1.0	0.0	0.0	0.503	0.076	0.073	0.0	0.512	0.2277	0.103
22	CD63	967	0.0	0.5652	0.986	0.0416	0.3390	0.0613	−1.0	0.0	0.0	0.503	0.095	0.073	0.0	0.512	0.2735	0.104
23	RNA5-8SN2	109864281	0.3250	0.0559	−1.0	0.0	0.2466	0.0846	−1.0	0.0	−1.0	0.0	0.0	0.536	−1.0	0.0	0.1905	0.112
24	RPL3	6122	0.3250	0.0559	0.952	0.0609	0.2814	0.0807	−1.0	0.0	0.0	0.503	0.0	0.536	0.0	0.512	0.2637	0.117
25	RPS11	6205	0.3250	0.0559	0.877	0.0861	0.2814	0.0807	−1.0	0.0	0.0	0.503	0.0	0.536	0.0	0.512	0.2571	0.132
26	RPL19	6143	0.3250	0.0559	0.877	0.0861	0.2814	0.0807	−1.0	0.0	0.0	0.503	0.0	0.536	0.0	0.512	0.2571	0.132
27	SLC16A3	9123	0.0	0.5652	0.212	0.2039	0.3057	0.0782	−1.0	0.0	0.0	0.503	0.065	0.073	0.0	0.512	0.1703	0.192
28	UTP14C	9724	0.0	0.5652	0.220	0.2037	0.2814	0.0807	−1.0	0.0	0.0	0.503	0.0	0.536	0.0	0.512	0.1251	0.353
29	TMSB4XP6	7120	−1.0	0.0	−1.0	0.0	−1.0	0.0	−1.0	0.0	−1.0	0.0	0.0	0.536	−1.0	0.0	0.0	1.0

4. Discussion

Retinal dystrophies like age-related macular degeneration and, particularly, retinitis pigmentosa represent a very heterogeneous group of ocular pathologies characterized by a very complex pattern of environmental and genetic causes. One of the most challenging aspects regards the incomplete knowledge of all causative genes and their involved biochemical and molecular pathways, leading to a huge group of orphan forms [37]. Gene mutations or dysfunctional processes not only in the retina but also in RPE could cause inherited retinal degeneration, age-related macular degeneration and other retinal diseases [38]. Such a feature highlights the relevant role of RPE, a high metabolic demand monolayer of pigmented cells that plays fundamental functions for both rods and cones, such as metabolite transport and photoreceptor excitability, regulation of visual cycle, secretion of growth factors, phagocytosis of photoreceptor outer segments (POSs) and oxidative stress defense. Regarding the latter point, oxidative stress represents one of the major lethal mechanisms responsible for age-related RPE damages [39]. Many studies have demonstrated that accumulation of lipid deposit called lipofuscin generates reactive oxygen species through phototoxicity in RPE cells [40–42]. Oxidative stress triggered by photo-oxidation of bis-retinoid A2E, a lipofuscin constituent, is well known to be a progression factor of age-related macular degeneration and also in genetic macular degeneration syndromes such as Stargardt disease [43], but very little is known about A2E involvement in retinitis pigmentosa.

In this study, we treated RPE cells with A2E during a follow-up of two time points (3 h and 6 h) after exposure and compared them to untreated time zero controls. The main purpose of our experiment was the discovery of new pathways potentially involved in retinal dystrophies development, with the further detection of new candidate genes that could be associated or causative of such ocular diseases, emerging from the expression analysis in such altered conditions.

Starting from an initial average value of 16,173 detected genes per sample, about 2432 showed changes in their expression level and 119 were differentially alternative spliced with transiently, late or enduring fluctuations. Selected altered DE and DAS genes were then functionally and statistically analyzed and clustered into final 10 candidate "macro-pathways", showing a very variable time of exposure-related trends. Considering the average fold-change of each constituting genes and their reciprocal connections, we revealed a more detailed functional network. Such connections could help to depict several causative/associative clusters, underlining a more complex pattern of possible retinal dystrophies etiologies.

From analyses of DE genes related pathways, it emerged that, after an early increase of apoptosis processes, the programmed cell death decreased in both considered time points following A2E exposure, probably due to activation of rescue systems and to a limited percentage of survived cells. An opposite trend was shown by steroid receptor and nucleoside transport activities, which evidenced a huge up-regulation of involved genes after 3 h and 6 h. Such results could reflect late alterations in RPE antioxidant and anti-inflammatory abilities [44], as well as inhibition of photoreceptor outer segment (POS) phagocytosis and impairment of ion currents in retinal cells [45,46]. Furthermore, a very interesting result was the sinusoidal trend involving isoprenoid pathway, related to cholesterol-dependent homeostasis of POS [47] and angiogenesis [48]. After 3 h from treatment, it could be possible that disc bulk membranes increased trying to improve phototransduction by residual photoreceptors, despite the decreased choriocapillaris viability due to the reduced vascular endothelial growth factor (VEGF) [49]. The situation reversed at 6 h observation, when the discs turnover was drastically decreased and the VEGF level increased, which could contribute to the subretinal neovascularization already characterized in wet age macular degeneration [50].

Additionally, the extensive alternative splicing information identified from DAS-related pathway analysis highlighted a much higher degree of complexity of regulation in response to A2E-induced oxidative stress, which has been significantly underestimated by analysis of DE genes only. In particular, speed and extent of the oxidative stress-induced AS suggested that AS, together with the transcriptional response, is a major driver of transcriptome reprogramming for RPE cell death and their attempts to survive. From 3 h and up to 6 h from treatment, an impairment of intracellular traffic related to Rab proteins, already reported in choroideremia [51], was observed along with the alteration of autophagy and accumulation of proteins and damaged organelles. These events are typical of AMD [52] and are also enforced by inactivation of chaperone genes [53,54]. This scenario could reflect a strong reduction of macroautophagy (a catabolic cell survival system) and of a hybrid autophagy–phagocytosis-degradative pathway called LC3-associated phagocytosis (LAP) [55], which plays a critical role in visual pigment regeneration, as well as the complete degradation of phagosomes [56]. The other resulted pathway "big cluster", instead, highlighted a global up-regulation of DAS-involved genes up to the end of the last considered time point. In particular, we demonstrated the dynamic contribution of AS by changes in multiple different mechanisms of transcription and translation. Photosensitization of A2E stimulates oxidative DNA damage, such as the production of 8-oxo-guanines in telomeres, leading to their possible damage [57]. Thus, the resulted alteration of telomerase RNA localization to Cajal body could accelerate the RPE senescence [58], and this process could be further increased by the reduction of FGFR1 signaling and the consequent POS phagocytosis decrease [59]. Moreover, DNA damage response could increase the activity of miRNAs involved in [60,61] and cell death genes transcription by TP53 [62], determining a possible role in retina degeneration. In the meantime, as already discussed, retinal cells could try to fight against induced stress, and one resistance mechanism could be represented by the improved maturation of the large ribosomal unit (LSU) [63] and by the increased polyribosome activity, showing a fundamental role in translation regulation of many retinal genes [64,65]. Detailed analysis of DE and DAS gene-involved pathways, with their possible impact on retinal dystrophies etiopathogenesis, is reported in Table 2.

Table 2. Detailed analysis of DE and DAS gene-involved pathways, with their possible impact on retinal dystrophies etiopathogenesis. DE and DAS genes were dysregulated during the whole analysis and showed several fluctuations during observed time points, suggesting that changes in gene-level expression and alternative splicing occurred throughout the whole period, either transiently (occurring after 3 h and returning to initial level after 6 h), occurring later (only after 6 h from treatment) or enduring throughout the whole period.

De Gene-Involved Pathways	Expression Changes	Das Gene-Involved Pathways
RNA methyltransferase	3 h and 6 h = DOWN-REGULATED	Endosomal sorting complex required for transport (ESCRT) and RAB geranylgeranylation
TRAF6 mediated NF-kB activation and activation of IKK by MEKK1		Phagopore assembly site membrane; C-terminal protein lipidation; protein localization to microtubule cytoskeleton; regulation of TNFR1 signaling; TNF signaling
Condensed chromosome outer kinetochore and kinesin complex		Methylation; activation of chaperone genes; nucleotide-sugar biosynthetic process
Transport of nucleoside and free purine and pyrimidine; histone pre-mRNA DCP binding; formation of AT-AC complex; respiratory electron transport	3 h and 6 h = UP-REGULATED	Negative regulation of FGFR1 signaling
		Regulation of telomerase RNA localization to Cajal body; maturation of LSU-rRNA; miRNAs involved in DNA damagv/ve response
		Mithocondrial intermembrane space and TP53 regulates transcription of cell death genes
		Polysomal ribosome

Table 2. *Cont.*

De Gene-Involved Pathways	Expression Changes	Das Gene-Involved Pathways
Mevalonate pathway and cholesterol biosynthesis	3 h = UP-REGULATED 6 h = DOWN-REGULATED	Cholesterol biosynthesis
CBL binds and ubiquinates Sprouty and MAP kinase phosphatase activity	3 h = DOWN-REGULATED 6 h = UP-REGULATED	/
PTK2/SRC-1 phosphorylates BCAR1		

5. Conclusions

We realized a whole RNA-seq experiment on RPE cells treated with A2E, considering two time points (3 h and 6 h) after the basal one. We found 10 different clusters of pathways involving DE and DAS genes, with many highlighted sub-pathways, which could depict a more detailed scenario determined by induced oxidative stress. Regulation and/or alterations of angiogenesis, extracellular matrix integrity, isoprenoid-mediated reactions, physiological or pathological autophagy, cell death induction and retinal cell rescue represented the most dysregulated pathway, probably involved in retinal degeneration. Assembly of splicing and transcriptional networks from analyzed data will further define the contribution of AS, as an extra level of regulation, and the interplay and coordination of the transcriptional and AS responses. However, it is fundamental to highlight several limitations of our study: RPE-cultured cells were not in contact with photoreceptors' outer segments and did not perform any phagocytosis. Moreover, the short-term response (at 3 h and 6 h) detected in vitro do not surely reflect what happens in vivo. Finally, even if it was important to underline the importance to realize an in vivo experiment to confirm what observed in RPE cells, our results could represent an important step towards discovery of unclear molecular mechanisms involved in etiopathogenesis of retinal dystrophies.

Supplementary Materials: The following are available online at http://www.mdpi.com/2076-3921/9/4/307/s1, Figure S1: Time-series trend analysis, Figure S2: Gene expression data statistics, Figure S3: qRT-PCR validation of ten most differentially expressed genes, Table S1: Gene expressed in RPE cells transcriptome analysis, Table S2: Gene transcripts expressed in RPE cells transcriptome analysis, Table S3: DE genes profile of RPE Cells, Table S4: DAS genes profile of RPE Cells, Table S5: DTU transcripts profile of RPE Cells, Table S6: Significant DE vs DAS gene enrichment by DAVID, Table S7: Significant DE vs DAS gene enrichment by ClueGO, Table S8: ToppGene prioritization detailed results, Table S9: Expression profiles of 10 among most DE genes.

Author Contributions: Conceptualization, L.D.; methodology, L.D.; software, L.D. and C.S.; validation, C.S. and S.A.; formal analysis, L.D.; investigation, L.D.; resources, C.S.; data curation, L.D. and C.S.; writing—original draft preparation, L.D.; writing—review and editing, C.S., R.D. and C.R.; visualization, C.R. and R.D.; supervision, A.S.; project administration, A.S. All authors have read and agreed to the published version of the manuscript.

Funding: This research received no external funding.

Conflicts of Interest: The authors declare no conflict of interest.

References

1. Sies, H. Oxidative stress: A concept in redox biology and medicine. *Redox Biol.* **2015**, *4*, 180–183. [CrossRef]
2. Donato, L.; Scimone, C.; Nicocia, G.; D'Angelo, R.; Sidoti, A. Role of oxidative stress in Retinitis pigmentosa: New involved pathways by an RNA-Seq analysis. *Cell Cycle* **2019**, *18*, 84–104. [CrossRef]
3. Datta, S.; Cano, M.; Ebrahimi, K.; Wang, L.; Handa, J.T. The impact of oxidative stress and inflammation on RPE degeneration in non-neovascular AMD. *Prog. Retin. Eye Res.* **2017**, *60*, 201–218. [CrossRef] [PubMed]
4. Randolph, S.A. Age-related macular degeneration. *Workplace Health Saf.* **2014**, *62*, 352. [CrossRef] [PubMed]
5. Gheorghe, A.; Mahdi, L.; Musat, O. Age-Related Macular Degeneration. *Rom. J. Ophthalmol.* **2015**, *59*, 74–77. [PubMed]
6. Lim, L.S.; Mitchell, P.; Seddon, J.M.; Holz, F.G.; Wong, T.Y. Age-related macular degeneration. *Lancet* **2012**, *379*, 1728–1738. [CrossRef]

7. Kivinen, N.; Koskela, A.; Kauppinen, A.; Kaarniranta, K. Pathogenesis of age-related macular degeneration—dialogue between autophagy and inflammasomes. *Duodecim* **2017**, *133*, 641–646. [PubMed]
8. Xu, B.; Zhang, H.; Zhu, M.; Le, Y.Z. Critical Role of Trophic Factors in Protecting Muller Glia: Implications to Neuroprotection in Age-Related Macular Degeneration, Diabetic Retinopathy, and Anti-VEGF Therapies. *Adv. Exp. Med. Biol.* **2019**, *1185*, 469–473. [CrossRef]
9. Natarajan, S. Decoding retinitis pigmentosa. *Indian J. Ophthalmol.* **2013**, *61*, 91–94. [CrossRef]
10. Verbakel, S.K.; van Huet, R.A.C.; Boon, C.J.F.; den Hollander, A.I.; Collin, R.W.J.; Klaver, C.C.W.; Hoyng, C.B.; Roepman, R.; Klevering, B.J. Non-syndromic retinitis pigmentosa. *Prog. Retin. Eye Res.* **2018**, *66*, 157–186. [CrossRef]
11. Fahim, A. Retinitis pigmentosa: Recent advances and future directions in diagnosis and management. *Curr. Opin. Pediatr.* **2018**, *30*, 725–733. [CrossRef] [PubMed]
12. Agbaga, M.P.; Merriman, D.K.; Brush, R.S.; Lydic, T.A.; Conley, S.M.; Naash, M.I.; Jackson, S.; Woods, A.S.; Reid, G.E.; Busik, J.V.; et al. Differential composition of DHA and very-long-chain PUFAs in rod and cone photoreceptors. *J. Lipid Res.* **2018**, *59*, 1586–1596. [CrossRef] [PubMed]
13. Birtel, J.; Gliem, M.; Oishi, A.; Muller, P.L.; Herrmann, P.; Holz, F.G.; Mangold, E.; Knapp, M.; Bolz, H.J.; Charbel Issa, P. Genetic testing in patients with retinitis pigmentosa: Features of unsolved cases. *Clin. Exp. Ophthalmol.* **2019**, *47*, 779–786. [CrossRef] [PubMed]
14. Campochiaro, P.A.; Mir, T.A. The mechanism of cone cell death in Retinitis Pigmentosa. *Prog. Retin. Eye Res.* **2018**, *62*, 24–37. [CrossRef]
15. Audo, I.; Mohand-Said, S.; Boulanger-Scemama, E.; Zanlonghi, X.; Condroyer, C.; Demontant, V.; Boyard, F.; Antonio, A.; Mejecase, C.; El Shamieh, S.; et al. MERTK mutation update in inherited retinal diseases. *Hum. Mutat.* **2018**, *39*, 887–913. [CrossRef]
16. Scimone, C.; Donato, L.; Esposito, T.; Rinaldi, C.; D'Angelo, R.; Sidoti, A. A novel RLBP1 gene geographical area-related mutation present in a young patient with retinitis punctata albescens. *Hum. Genom.* **2017**, *11*, 18. [CrossRef]
17. Miraldi Utz, V.; Coussa, R.G.; Antaki, F.; Traboulsi, E.I. Gene therapy for RPE65-related retinal disease. *Ophthalmic Genet.* **2018**, *39*, 671–677. [CrossRef]
18. Sparrow, J.R.; Hicks, D.; Hamel, C.P. The retinal pigment epithelium in health and disease. *Curr. Mol. Med.* **2010**, *10*, 802–823. [CrossRef]
19. Nowak, J.Z. Oxidative stress, polyunsaturated fatty acids-derived oxidation products and bisretinoids as potential inducers of CNS diseases: Focus on age-related macular degeneration. *Pharmacol. Rep.* **2013**, *65*, 288–304. [CrossRef]
20. Donato, L.; Scimone, C.; Rinaldi, C.; D'Angelo, R.; Sidoti, A. Non-coding RNAome of RPE cells under oxidative stress suggests unknown regulative aspects of Retinitis pigmentosa etiopathogenesis. *Sci. Rep.* **2018**, *8*, 16638. [CrossRef]
21. Crouch, R.K.; Koutalos, Y.; Kono, M.; Schey, K.; Ablonczy, Z. A2E and Lipofuscin. *Prog. Mol. Biol. Transl. Sci.* **2015**, *134*, 449–463. [CrossRef] [PubMed]
22. Sparrow, J.R.; Fishkin, N.; Zhou, J.; Cai, B.; Jang, Y.P.; Krane, S.; Itagaki, Y.; Nakanishi, K. A2E, a byproduct of the visual cycle. *Vis. Res.* **2003**, *43*, 2983–2990. [CrossRef]
23. Parmar, V.M.; Parmar, T.; Arai, E.; Perusek, L.; Maeda, A. A2E-associated cell death and inflammation in retinal pigmented epithelial cells from human induced pluripotent stem cells. *Stem Cell Res.* **2018**, *27*, 95–104. [CrossRef] [PubMed]
24. Dontsov, A.; Koromyslova, A.; Ostrovsky, M.; Sakina, N. Lipofuscins prepared by modification of photoreceptor cells via glycation or lipid peroxidation show the similar phototoxicity. *World J. Exp. Med.* **2016**, *6*, 63–71. [CrossRef]
25. Alaimo, A.; Linares, G.G.; Bujjamer, J.M.; Gorojod, R.M.; Alcon, S.P.; Martinez, J.H.; Baldessari, A.; Grecco, H.E.; Kotler, M.L. Toxicity of blue led light and A2E is associated to mitochondrial dynamics impairment in ARPE-19 cells: Implications for age-related macular degeneration. *Arch. Toxicol.* **2019**, *93*, 1401–1415. [CrossRef] [PubMed]
26. Alaimo, A.; Di Santo, M.C.; Dominguez Rubio, A.P.; Chaufan, G.; Garcia Linares, G.; Perez, O.E. Toxic effects of A2E in human ARPE-19 cells were prevented by resveratrol: A potential nutritional bioactive for age-related macular degeneration treatment. *Arch. Toxicol.* **2019**, *94*, 553–572. [CrossRef]

27. Okonechnikov, K.; Conesa, A.; Garcia-Alcalde, F. Qualimap 2: Advanced multi-sample quality control for high-throughput sequencing data. *Bioinformatics* **2016**, *32*, 292–294. [CrossRef]
28. Li, B.; Ruotti, V.; Stewart, R.M.; Thomson, J.A.; Dewey, C.N. RNA-Seq gene expression estimation with read mapping uncertainty. *Bioinformatics* **2010**, *26*, 493–500. [CrossRef]
29. Patro, R.; Duggal, G.; Love, M.I.; Irizarry, R.A.; Kingsford, C. Salmon provides fast and bias-aware quantification of transcript expression. *Nat. Methods* **2017**, *14*, 417–419. [CrossRef]
30. Soneson, C.; Matthes, K.L.; Nowicka, M.; Law, C.W.; Robinson, M.D. Isoform prefiltering improves performance of count-based methods for analysis of differential transcript usage. *Genome Biol.* **2016**, *17*, 12. [CrossRef]
31. Pereira, M.B.; Wallroth, M.; Jonsson, V.; Kristiansson, E. Comparison of normalization methods for the analysis of metagenomic gene abundance data. *BMC Genom.* **2018**, *19*, 274. [CrossRef]
32. Ritchie, M.E.; Phipson, B.; Wu, D.; Hu, Y.; Law, C.W.; Shi, W.; Smyth, G.K. limma powers differential expression analyses for RNA-sequencing and microarray studies. *Nucleic Acids Res.* **2015**, *43*, e47. [CrossRef] [PubMed]
33. Guo, W.T.N.; Stephen, G.; Milne, I.; Calixto, C.; Waugh, R.; Brown, J.W.; Zhang, R. 3D RNA-seq—A powerful and flexible tool for rapid and accurate differential expression and alternative splicing analysis of RNA-seq data for biologists. *bioRxiv* **2019**, 656686. [CrossRef]
34. Ge, Y.; Sealfon, S.C.; Speed, T.P. Some Step-down Procedures Controlling the False Discovery Rate under Dependence. *Stat. Sin.* **2008**, *18*, 881–904. [PubMed]
35. Jiao, X.; Sherman, B.T.; Huang da, W.; Stephens, R.; Baseler, M.W.; Lane, H.C.; Lempicki, R.A. DAVID-WS: A stateful web service to facilitate gene/protein list analysis. *Bioinformatics* **2012**, *28*, 1805–1806. [CrossRef]
36. Livak, K.J.; Schmittgen, T.D. Analysis of relative gene expression data using real-time quantitative PCR and the 2(-Delta Delta C(T)) Method. *Methods* **2001**, *25*, 402–408. [CrossRef] [PubMed]
37. Nivison-Smith, L.; Milston, R.; Madigan, M.; Kalloniatis, M. Age-related macular degeneration: Linking clinical presentation to pathology. *Optom. Vis. Sci.* **2014**, *91*, 832–848. [CrossRef] [PubMed]
38. Vancura, P.; Csicsely, E.; Leiser, A.; Iuvone, P.M.; Spessert, R. Rhythmic Regulation of Photoreceptor and RPE Genes Important for Vision and Genetically Associated With Severe Retinal Diseases. *Investig. Ophthalmol. Vis. Sci.* **2018**, *59*, 3789–3799. [CrossRef] [PubMed]
39. Fuhrmann, S.; Zou, C.; Levine, E.M. Retinal pigment epithelium development, plasticity, and tissue homeostasis. *Exp. Eye Res.* **2014**, *123*, 141–150. [CrossRef]
40. D'Angelo, R.; Donato, L.; Venza, I.; Scimone, C.; Aragona, P.; Sidoti, A. Possible protective role of the ABCA4 gene c.1268A>G missense variant in Stargardt disease and syndromic retinitis pigmentosa in a Sicilian family: Preliminary data. *Int. J. Mol. Med.* **2017**, *39*, 1011–1020. [CrossRef]
41. Donato, L.; Scimone, C.; Rinaldi, C.; Aragona, P.; Briuglia, S.; D'Ascola, A.; D'Angelo, R.; Sidoti, A. Stargardt Phenotype Associated With Two ELOVL4 Promoter Variants and ELOVL4 Downregulation: New Possible Perspective to Etiopathogenesis? *Investig. Ophthalmol. Vis. Sci.* **2018**, *59*, 843–857. [CrossRef] [PubMed]
42. Wiktor, A.; Sarna, M.; Wnuk, D.; Sarna, T. Lipofuscin-mediated photodynamic stress induces adverse changes in nanomechanical properties of retinal pigment epithelium cells. *Sci. Rep.* **2018**, *8*, 17929. [CrossRef] [PubMed]
43. Moiseyev, G.; Nikolaeva, O.; Chen, Y.; Farjo, K.; Takahashi, Y.; Ma, J.X. Inhibition of the visual cycle by A2E through direct interaction with RPE65 and implications in Stargardt disease. *Proc. Natl. Acad. Sci. USA* **2010**, *107*, 17551–17556. [CrossRef] [PubMed]
44. Kaarniranta, K.; Machalinska, A.; Vereb, Z.; Salminen, A.; Petrovski, G.; Kauppinen, A. Estrogen signalling in the pathogenesis of age-related macular degeneration. *Curr. Eye Res.* **2015**, *40*, 226–233. [CrossRef]
45. Akanuma, S.; Soutome, T.; Hisada, E.; Tachikawa, M.; Kubo, Y.; Hosoya, K. Na+-independent nucleoside transporters regulate adenosine and hypoxanthine levels in Muller cells and the inner blood-retinal barrier. *Investig. Ophthalmol. Vis. Sci.* **2013**, *54*, 1469–1477. [CrossRef]
46. Dos Santos-Rodrigues, A.; Pereira, M.R.; Brito, R.; de Oliveira, N.A.; Paes-de-Carvalho, R. Adenosine transporters and receptors: Key elements for retinal function and neuroprotection. *Vitam. Horm.* **2015**, *98*, 487–523. [CrossRef]
47. Fliesler, S.J.; Keller, R.K. Isoprenoid metabolism in the vertebrate retina. *Int. J. Biochem. Cell Biol.* **1997**, *29*, 877–894. [CrossRef]

48. Roysommuti, S.; Thaeomor, A.; Khimsuksri, S.; Lerdweeraphon, W.; Wyss, J.M. Perinatal taurine imbalance alters the interplay of renin-angiotensin system and estrogen on glucose-insulin regulation in adult female rats. *Adv. Exp. Med. Biol.* **2013**, *776*, 67–80. [CrossRef]
49. Okabe, K.; Kobayashi, S.; Yamada, T.; Kurihara, T.; Tai-Nagara, I.; Miyamoto, T.; Mukouyama, Y.S.; Sato, T.N.; Suda, T.; Ema, M.; et al. Neurons limit angiogenesis by titrating VEGF in retina. *Cell* **2014**, *159*, 584–596. [CrossRef]
50. Joyal, J.S.; Gantner, M.L.; Smith, L.E.H. Retinal energy demands control vascular supply of the retina in development and disease: The role of neuronal lipid and glucose metabolism. *Prog. Retin. Eye Res.* **2018**, *64*, 131–156. [CrossRef]
51. Preising, M.; Ayuso, C. Rab escort protein 1 (REP1) in intracellular traffic: A functional and pathophysiological overview. *Ophthalmic Genet.* **2004**, *25*, 101–110. [CrossRef] [PubMed]
52. Kauppinen, A.; Paterno, J.J.; Blasiak, J.; Salminen, A.; Kaarniranta, K. Inflammation and its role in age-related macular degeneration. *Cell Mol. Life Sci.* **2016**, *73*, 1765–1786. [CrossRef] [PubMed]
53. Piippo, N.; Korhonen, E.; Hytti, M.; Skottman, H.; Kinnunen, K.; Josifovska, N.; Petrovski, G.; Kaarniranta, K.; Kauppinen, A. Hsp90 inhibition as a means to inhibit activation of the NLRP3 inflammasome. *Sci. Rep.* **2018**, *8*, 6720. [CrossRef] [PubMed]
54. Subrizi, A.; Toropainen, E.; Ramsay, E.; Airaksinen, A.J.; Kaarniranta, K.; Urtti, A. Oxidative stress protection by exogenous delivery of rhHsp70 chaperone to the retinal pigment epithelium (RPE), a possible therapeutic strategy against RPE degeneration. *Pharm. Res.* **2015**, *32*, 211–221. [CrossRef]
55. Dhingra, A.; Alexander, D.; Reyes-Reveles, J.; Sharp, R.; Boesze-Battaglia, K. Microtubule-Associated Protein 1 Light Chain 3 (LC3) Isoforms in RPE and Retina. *Adv. Exp. Med. Biol.* **2018**, *1074*, 609–616. [CrossRef]
56. Frost, L.S.; Lopes, V.S.; Bragin, A.; Reyes-Reveles, J.; Brancato, J.; Cohen, A.; Mitchell, C.H.; Williams, D.S.; Boesze-Battaglia, K. The Contribution of Melanoregulin to Microtubule-Associated Protein 1 Light Chain 3 (LC3) Associated Phagocytosis in Retinal Pigment Epithelium. *Mol. Neurobiol.* **2015**, *52*, 1135–1151. [CrossRef]
57. Sparrow, J.R.; Zhou, J.; Cai, B. DNA is a target of the photodynamic effects elicited in A2E-laden RPE by blue-light illumination. *Investig. Ophthalmol. Vis. Sci.* **2003**, *44*, 2245–2251. [CrossRef]
58. Wang, J.; Feng, Y.; Han, P.; Wang, F.; Luo, X.; Liang, J.; Sun, X.; Ye, J.; Lu, Y.; Sun, X. Photosensitization of A2E triggers telomere dysfunction and accelerates retinal pigment epithelium senescence. *Cell Death Dis.* **2018**, *9*, 178. [CrossRef]
59. Cornish, E.E.; Natoli, R.C.; Hendrickson, A.; Provis, J.M. Differential distribution of fibroblast growth factor receptors (FGFRs) on foveal cones: FGFR-4 is an early marker of cone photoreceptors. *Mol. Vis.* **2004**, *10*, 1–14.
60. Donato, L.; Bramanti, P.; Scimone, C.; Rinaldi, C.; Giorgianni, F.; Beranova-Giorgianni, S.; Koirala, D.; D'Angelo, R.; Sidoti, A. miRNAexpression profile of retinal pigment epithelial cells under oxidative stress conditions. *FEBS Open Bio* **2018**, *8*, 219–233. [CrossRef]
61. Olejniczak, M.; Kotowska-Zimmer, A.; Krzyzosiak, W. Stress-induced changes in miRNA biogenesis and functioning. *Cell Mol. Life Sci.* **2018**, *75*, 177–191. [CrossRef] [PubMed]
62. Westlund, B.S.; Cai, B.; Zhou, J.; Sparrow, J.R. Involvement of c-Abl, p53 and the MAP kinase JNK in the cell death program initiated in A2E-laden ARPE-19 cells by exposure to blue light. *Apoptosis* **2009**, *14*, 31–41. [CrossRef] [PubMed]
63. McCann, K.L.; Charette, J.M.; Vincent, N.G.; Baserga, S.J. A protein interaction map of the LSU processome. *Genes Dev.* **2015**, *29*, 862–875. [CrossRef] [PubMed]
64. Marchesi, N.; Thongon, N.; Pascale, A.; Provenzani, A.; Koskela, A.; Korhonen, E.; Smedowski, A.; Govoni, S.; Kauppinen, A.; Kaarniranta, K.; et al. Autophagy Stimulus Promotes Early HuR Protein Activation and p62/SQSTM1 Protein Synthesis in ARPE-19 Cells by Triggering Erk1/2, p38(MAPK), and JNK Kinase Pathways. *Oxid. Med. Cell Longev.* **2018**, *2018*, 4956080. [CrossRef]
65. McGrew, D.A.; Hedstrom, L. Towards a pathological mechanism for IMPDH1-linked retinitis pigmentosa. *Adv. Exp. Med. Biol.* **2012**, *723*, 539–545. [CrossRef]

© 2020 by the authors. Licensee MDPI, Basel, Switzerland. This article is an open access article distributed under the terms and conditions of the Creative Commons Attribution (CC BY) license (http://creativecommons.org/licenses/by/4.0/).

Article

Substance-P Restores Cellular Activity of ADSC Impaired by Oxidative Stress

Jeong Seop Park [1], Jiyuan Piao [2], Gabee Park [2] and Hyun Sook Hong [1,3,*]

[1] Department of Biomedical Science and Technology, Graduate School, Kyung Hee University, Seoul 02447, Korea; godjs@khu.ac.kr

[2] Department of Genetic Engineering, College of Life Science and Graduate School of Biotechnology, Kyung Hee University, Yong In 17104, Korea; fredericpark210@gmail.com (J.P.); yuiop2690@gmail.com (G.P.)

[3] East-West Medical Research Institute, Kyung Hee University, Seoul 02447, Korea

* Correspondence: hshong@khu.ac.kr; Tel.: +82-2-958-1828

Received: 15 September 2020; Accepted: 11 October 2020; Published: 12 October 2020

Abstract: Oxidative stress induces cellular damage, which accelerates aging and promotes the development of serious illnesses. Adipose-derived stem cells (ADSCs) are novel cellular therapeutic tools and have been applied for tissue regeneration. However, ADSCs from aged and diseased individuals may be affected in vivo by the accumulation of free radicals, which can impair their therapeutic efficacy. Substance-P (SP) is a neuropeptide that is known to rescue stem cells from senescence and inflammatory attack, and this study explored the restorative effect of SP on ADSCs under oxidative stress. ADSCs were transiently exposed to H_2O_2, and then treated with SP. H_2O_2 treatment decreased ADSC cell viability, proliferation, and cytokine production and this activity was not recovered even after the removal of H_2O_2. However, the addition of SP increased cell viability and restored paracrine potential, leading to the accelerated repopulation of ADSCs injured by H_2O_2. Furthermore, SP was capable of activating Akt/GSK-3β signaling, which was found to be downregulated following H_2O_2 treatment. This might contribute to the restorative effect of SP on injured ADSCs. Collectively, SP can protect ADSCs from oxidant-induced cell damage, possibly by activating Akt/GSK-3β signaling in ADSCs. This study supports the possibility that SP can recover cell activity from oxidative stress-induced dysfunction.

Keywords: Substance-P; adipose-derived stem cells; oxidative stress; paracrine factors

---

1. Introduction

Oxidative stress is an imbalance of free radicals and antioxidants in the body, which can lead to cell and tissue damage. There are several factors that cause oxidative stress and excess free radical production, including obesity, smoking, alcohol consumption, pollution, and chemicals. These risk factors can activate immune cells, which subsequently produce free radicals and impair normal cells. Consistent accumulation of free radicals and cellular damage causes inflammation, contributing to the development of serious diseases including cancer, diabetes, Alzheimer's, and cardiovascular diseases.

Aging is the progressive loss of tissue and organ function over time. Oxidative stress is believed to be closely related to the aging process because age-associated functional losses are caused by the accumulation of oxidative damage to macromolecules. Although the exact mechanism of oxidative stress-induced aging is unclear, increased free radical production obviously leads to cellular senescence, decreased cell proliferation, and increased inflammatory cytokine production [1]. In addition, several studies have found that with aging, endogenous antioxidant levels, antioxidant enzyme activity, gene expression, and protein levels decrease. This alteration in the antioxidant defense system worsens ROS imbalances and contributes to oxidative-stress-induced aging [1–4].

Oxidative stress and cellular senescence are involved in several acute and chronic pathological processes such as cardiovascular diseases (CVDs), acute and chronic kidney disease (CKD), neurodegenerative diseases (NDs), macular degeneration (MD), biliary diseases, and cancer. Cardiovascular (CV) risk factors are associated with the inflammatory pathway mediated by interleukin (IL)-1α, IL-6, IL-8, and with increased cellular senescence [5]. The associations between oxidative stress, inflammation, and aging produce a vicious cycle whereby chronic ROS production and inflammation feed each other and accelerate aging and age-related morbidity [6]. Thus, modulating oxidative stress-mediated cellular injury may be a fundamental solution to prevent the progression of lethal diseases in aged individuals.

Stem cells facilitate tissue repair in vivo, and many, such as bone marrow, adipose-derived, and umbilical cord blood stem cells are currently being used as novel treatments for various diseases. The application of mesenchymal stem/stromal cells (MSCs) in regenerative medicine has been intensively studied in many clinical trials, as these cells represent a promising source of multipotent adult stem cells for cell therapy and tissue engineering [7–12]. The therapeutic effects of MSCs are generally mediated by various secreted cytokines, growth factors, extracellular matrix proteins, and factors involved in matrix remodeling, as well as different types of extracellular vesicles.

Currently, as alternative of bone marrow stem cells, adipose-derived stem cells (ADSCs) are being applied extensively in the clinic because they can be easily isolated and cause lower donor-site morbidity [12–14]. Aging, oxidative stress, and/or inflammation can affect tissue-resident cells as well as circulating cells, which suggests that stem cells may also be influenced by oxidative stress in vivo. Considering that the aged population is the main population to be treated by stem cell therapy, the restoration of the cellular function of stem cells injured by oxidative stress is likely important.

Substance-P (SP) is an endogenous neuropeptide that interacts with neurokinin receptor 1 (NK-1R). SP has been reported to stimulate cell proliferation and prevent apoptosis under inflammatory or oxidative stress by activating the extracellular signal-regulated kinases 1/2 (ERK 1/2) or Akt and by translocating β-catenin to cell nuclei [15–18]. SP could suppress inflammation to promote tissue repair in severe diseases by modulating the immune cell profile in circulation-associated and lymphoid organs [19–22]. Moreover, SP can recover the cellular activity of senescent stem cells [23,24] and stimulate stem cell mobilization from the bone marrow to the peripheral blood [17].

Considering these functions of SP, we hypothesized that SP would be able to restore stem cells from oxidative stress-induced injury. To explore the potential recovery role of SP in injured stem cells due to oxidative stress, ADSCs were exposed to H_2O_2 at various concentrations. Subsequently, SP was added to the damaged ADSCs and the effect of SP was assessed by evaluating cell viability, cell proliferation, paracrine factors, and early signaling molecules.

2. Materials and Methods

2.1. Materials

SP and phenylmethylsulfonyl fluoride (PMSF) were purchased from Sigma-Aldrich (St. Louis, MO, USA). Penicillin/streptomycin, 0.25% trypsin-EDTA solution, and phosphate-buffered saline (PBS) were provided by Welgene (Daegu, Korea). Fetal bovine serum and alpha-MEM were purchased from Gibco (Grand Island, NY, USA). An anti-GAPDH antibody (Abcam, Cambridge, MA, USA), and anti-WST-1 antibody (Roche; Indianapolis, IN, USA) were used in this study. Cell lysis buffer, anti-Akt antibody, anti-phospho-Akt antibody, anti-GSK-3β antibody, and anti-phospho-GSK-3β antibody were purchased from Cell Signaling Technology (Danvers, MA, USA).

2.2. Cell Culture

Healthy adipose tissues were provided by the Kyung Hee University Hospital Institutional Review Board (eight donors, M6/F2; Seoul, Korea; (IRB# 2016-12-022)). All consents were informed. Fat tissue (two fat tissues (1 × 1 cm) from one donor; total 16 tissues) washed twice with PBS, and treated with collagenase (GMP grade, Vivagen, Los Angeles, CA, USA) for 1 h at 37 °C. Red blood cells and debris were removed using a cell strainer (SPL Life, Pocheon, Korea). The vascular fraction was seeded in α-MEM supplemented with 10% FBS at 37 °C with 5% CO_2. ADSCs were characterized by analyzing the expression of CD29, CD73, CD105, and CD90 using a fluorescence-activated cell sorting (FACS) Calibur flow cytometer and CELLQuest software (Becton Dickinson, San Jose, CA, USA) (Figure S1).

2.3. *Hydrogen Peroxide Exposure Procedure and SP Treatment*

Cells were seeded in a 96-well plate at a density of 1×10^4 cells/well or in a 6-well plate with a density of 3×10^4 cells/well. These cells were allowed to adhere to the bottom of the well. Twenty-four hours later, different concentrations of H_2O_2 (50, 100, 200, 300, and 400 μM) were added to the wells for 2 h and then removed by changing the culture media. After 24 h, SP was added to each well at a final concentration of 100 nM, and this was repeated 24 h later.

2.4. *Wst-1 Assay*

Ten microliters of water-soluble tetrazolium salt (WST-1; Roche) solution was added to each well at 10% the total volume of the medium, and the 96-well-plate was incubated for 1 h at 37 °C in 5% CO_2. After incubation, the optical density values were measured at a wavelength of 450 nm using an Enzyme Linked Immunosorbent Assay (ELISA) microplate reader (Molecular Devices, Sunnyvale, CA, USA).

2.5. *Enzyme Linked Immunosorbent Assay (ELISA)*

The total TGF-β1 and VEGF levels in the supernatants were quantified using ELISA kits, according to the manufacturer's instructions. In brief, standards and samples were added to the wells of anti-TGF-β1 or anti-VEGF antibody-coated 96-well plates and incubated for 2 h at room temperature. After discarding the supernatant, a horseradish peroxidase-conjugated secondary antibody was added to each well and incubated again for 2 h at room temperature. After rinsing with washing solution three times, 100 μL of substrate solution was added, followed by the addition of 100 μL of stop solution. The optical density was measured at 450 nm using an ELISA microplate reader (Molecular Devices, Sunnyvale, CA, USA).

2.6. *Preparation of Cell Extracts and Western Blot Analysis*

Cells were rapidly washed with chilled 1× PBS and lysed with 1× lysis buffer/1 mM phenylmethylsulfonyl fluoride (PMSF) solution. Cells were then scraped and supernatants were collected by centrifugation (Rotor radius: 70 mm) at 12,000 rpm for 10 min at 4 °C. Protein concentrations of lysates were determined using the bicinchoninic acid (BCA) assay (Thermo Fisher, Rockford, IL, USA). Ten micrograms of lysates were denatured and electrophoresed using sodium dodecyl sulfate-polyacrylamide gel electrophoresis (SDS-PAGE) and transferred to a nitrocellulose membrane. After blocking with 5% skim milk, membranes were incubated with primary anti-Akt, anti phospho-Akt, anti-GSK-3β, anti-phospho-GSK-3β, or anti-GAPDH antibodies, followed by an anti-IgG horseradish peroxidase-conjugated secondary antibody. The blots were processed using enhanced chemiluminescence (GE Healthcare, Buckinghamshire, UK).

2.7. *Statistical Analysis*

All data are presented as the mean ± standard deviation (SD) of more than three independent experiments. p values of less than 0.05 were considered statistically significant. Statistical analysis of the data was carried out using an unpaired, two-tailed Student's t-test.

3. Results

3.1. H_2O_2 Impairs the Viability and Morphology of ADSCs

Previous reports have induced oxidative stress in vitro by treating cells with H_2O_2 for 24 h [25,26]. However, ADSCs appear to be particularly susceptible to H_2O_2-dependent oxidative stress, with 1 or 2 h of exposure being sufficient to impair ADSC activity [27,28]. Thus, in this study, we induced oxidative stress in ADSCs by exposing these cells to different concentrations of H_2O_2 for 2 h (Figure 1A).

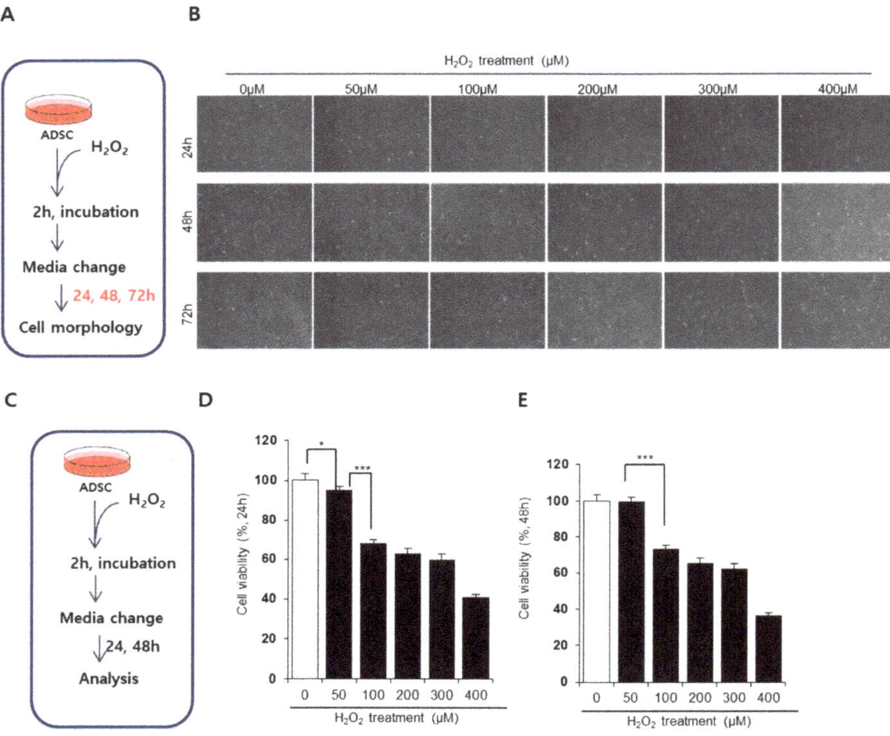

Figure 1. H_2O_2 treatment impairs cellular activity of adipose derived stem cells (ADSCs). (**A**,**C**) Experimental scheme for oxidative stress induction in ADSCs. (**B**) Alterations in ADSC cellular morphology depending on the concentration of H_2O_2. (**D–E**) The viability of ADSCs treated with H_2O_2 at 24 h (**D**) and 48 h (**E**) after the removal of H_2O_2. The untreated control was set as 100%, and cell viability was expressed as the percentage relative to the activity of the control group. Values of $p < 0.05$ were interpreted as statistically significant (* $p < 0.05$, *** $p < 0.001$). The data are expressed as the mean ± SD of three independent experiments.

The transient treatment of ADSCs with H_2O_2 primarily affected cell proliferation and morphology (Figure 1B). At 24 h after the removal of H_2O_2, cell density was significantly different between nontreated and H_2O_2-treated conditions. While nontreated cells started to proliferate and became confluent at 72 h, H_2O_2-treated ADSCs displayed suppressed cell proliferation. Notably, a distinct difference in cellular density and shape was observed at H_2O_2 concentrations above 100 μM. High concentrations of H_2O_2 (i.e., 200, 300, and 400 μM) severely impaired the cellular activity of ADSCs, significantly reducing density at 72 h. H_2O_2 at 50 μM seemed to damage ADSCs, but these cells appeared to recover as time passed. To compare cellular impairment quantitatively, we determined cell viability at 24 and 48 h after oxidative stress (Figure 1C). H_2O_2 treatment decreased cell viability in a dose-dependent manner (Figure 1D,E). Consistent with our results on cell shape and density, ADSCs treated with 50 μM H_2O_2 spontaneously restored their activity within 48 h (Figure 1E). However, concentrations above 100 μM H_2O_2 reduced cell viability and this did not return to normal levels. Based on these data, we determined that 100 μM H_2O_2 induced substantial oxidative damage in ADSCs in vitro. This indicates that transient H_2O_2 treatment impairs the cellular viability of ADSCs, which might affect cell survival and the function of ADSCs.

3.2. Substance-P Restores Cell Viability of ADSCs Injured by Oxidative Stress

In order to determine the effect of SP on damaged ADSCs, cells were treated with 100 μM H_2O_2 for 2 h and then provided with fresh media. After 24 h, SP was added to the damaged ADSCs at a concentration of 100 nM (Figure 2A). This dose of SP was determined based on previous reports [15–17].

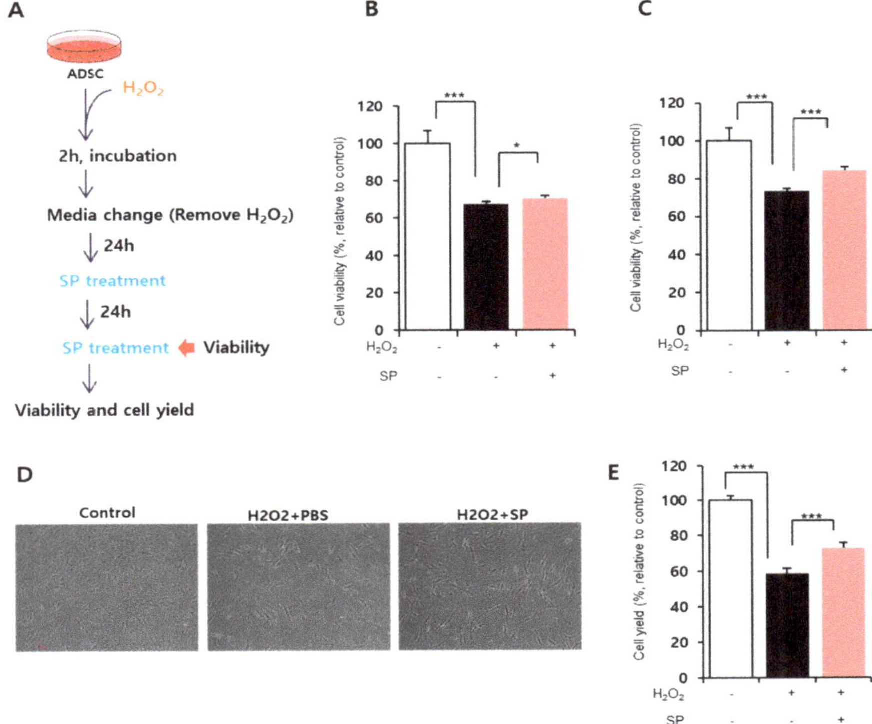

Figure 2. Substance-P improves the viability of ADSCs damaged by oxidative stress. (**A**) Experimental scheme for inducing oxidative stress in ADSCs and the subsequent SP treatment. (**B**,**C**) Cell viability was measured by WST assay at 24 (**B**) and 48 h (**C**) after SP treatment. The untreated control was set as 100%, and cell viability was expressed as the percentage relative to the activity of the control group. (**D**) The representative image of cellular morphology. (**E**) Final cell yield was measured by counting total cell number. Values of $p < 0.05$ were interpreted as statistically significant (* $p < 0.05$, *** $p < 0.001$). The data are expressed as the mean ± SD of three independent experiments.

Substance-P treatment started to improve ADSC viability within 24 h (Figure 2B; H_2O_2 treated: 67.47 ± 1.29%, H_2O_2 + SP-treated: 70.4 ± 1.4%, $p < 0.05$) and fully restored cell viability at 48 h (Figure 2C; H_2O_2 treated: 73.16 ± 1.83%, H_2O_2 + SP-treated: 84.7 ± 3.6%, $p < 0.001$). Cell viability is directly related to cell repopulation. Therefore, we examined the cell yield by counting the total number of cells and comparing it with that of the control. The analysis of cell yield confirmed the restorative effect of SP on damaged ADSCs (Figure 2E, H_2O_2 treated: 58.79 ± 1.96%, H_2O_2 + SP-treated: 72.97 ± 4.9%, $p < 0.001$, relative to control). SP treatment did not influence cell morphology, but enhanced cell proliferation (Figure 2D).

This revealed that SP treatment enhances the viability of ADSCs impaired by oxidative stress. This effect of SP finally led to the reduced cellular senescence of ADSCs (Figure S2).

3.3. Effect of SP on Paracrine Potential of ADSCs Exposed to Oxidative Stress

Oxidative stress is well known to decrease cell viability and negatively affect cell function. Stem cells exert their function via paracrine factors, and thus, it is critical to evaluate ADSC cytokine production when investigating the effects of oxidative stress (Figure 3A). VEGF and TGF-β are constitutively produced from MSCs, and their levels are typically reduced by aging or cellular damage [29]. Therefore, VEGF and TGF-β were selected as surrogate markers to represent the paracrine action of ADSCs in this experiment.

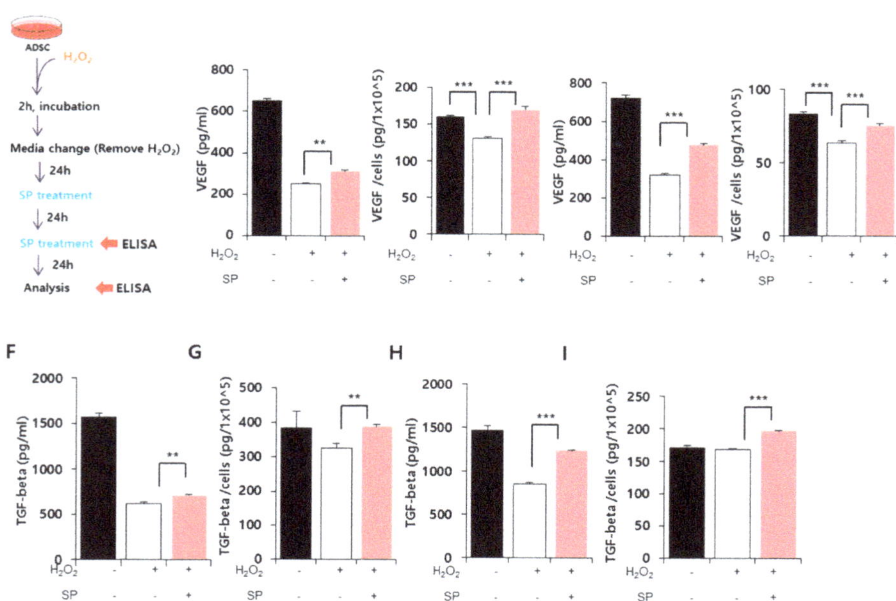

Figure 3. Substance-P recovers the paracrine potential of oxidatively damaged ADSCs. (**A**) Experimental scheme for ADSC with oxidative stress and SP treatment. (**B–E**) VEGF in the conditioned medium of ADSCs was quantified using ELISA at 24 (**B**,**C**) and 48 h (**D**,**E**) after the first SP treatment. The absolute concentration of VEGF (**B**,**D**) and secreted amount per 1×10^5 ADSCs (**C**,**E**) were evaluated. (**F–I**) TGF-β in the conditioned medium of ADSCs was quantified by ELISA at 24 h (**F**,**G**) and 48 h (**H**,**I**) after the first SP treatment. The absolute concentration of TGF-β (**F**,**H**) and the secreted amount per 1×10^5 ADSCs (**G**,**I**) were evaluated. Values of $p < 0.05$ were interpreted as statistically significant (** $p < 0.01$, *** $p < 0.001$). The data are expressed as the mean ± SD of three independent experiments.

VEGF levels from ADSCs significantly decreased after oxidative stress (Figure 3B, 48 h after oxidative stress, Control: 651.8 ± 8.7 pg/mL, H_2O_2 treated: 251.4 ± 4.7 pg/mL), and this reduction was sustained by 72 h after oxidative stress (Figure 3D; 72 h after oxidative stress, Control: 721.8 ± 12.9 pg/mL, H_2O_2 treated: 321.74 ± 9.77 pg/mL). This indicates that VEGF secretion did not increase significantly between 48 and 72 h, and that impairment of cytokine secretion occurred early. However, SP treatment elevated VEGF production in ADSCs (Figure 3B; 48 h after oxidative stress, H_2O_2 + SP-treated: 312.6 ± 8.1 pg/mL; Figure 3D; 72 h after oxidative stress, H_2O_2 + SP-treated: 477 ± 5.4 pg/mL).

In this experiment, oxidative stress decreased the total cell number (Figure 2), and thus, it could be inferred that the reduction in cytokine secretion was due to the low cell number. To clarify this, the amount of cytokines was assessed per cell. It was found that oxidative stress disabled the paracrine function of ADSCs (Figure 3C,E). Interestingly, SP treatment increased the concentration of VEGF in the conditioned medium of ADSCs, which might be attributed to their improved ability to produce VEGF as well as the increased cell number. This phenomenon was also observed for TGF-β secretion (Figure 3F,H), with TGF-β levels decreasing following oxidative stress and then recovering after SP treatment (Figure 3G,I).

This suggests that SP is able to improve the paracrine action of ADSCs under oxidative stress, and that repeated SP treatment can intensify the restorative function of SP in damaged ADSCs.

3.4. SP Activates Akt Signaling in ADSCs Injured by Oxidative Stress

Typically, when cells are under oxidative stress, signaling associated with cell survival is activated, allowing the cells to survive. The phosphoinositide 3-kinase (PI3K)-Akt pathway is a pro-survival pathway regulated by ROS. When oxidative stress is exerted on cells, Akt is phosphorylated in a PI3K-dependent manner, which induces the phosphorylation and subsequent inactivation of pro-apoptotic factors, including glycogen synthase kinase (GSK)-3 [30,31]. To examine whether the increase in ADSCs viability by SP was accompanied by the activation of Akt signaling, we determined the phosphorylation state of Akt and GSK-3β following ADSCs treatment with H_2O_2 and then with SP for 20 min (Figure 4A). ADSCs treated with H_2O_2 failed to maintain phosphorylated Akt levels, whereas SP treatment promoted Akt phosphorylation (Figure 4B). Additionally, GSK-3β, a downstream effector of Akt signaling and a pro-apoptotic molecule [14], was phosphorylated and inactivated following SP treatment. The expression levels of phospho-Akt and phospho-GSK-3β were quantified relative to the levels of total Akt and GSK-3β (Figure 4B). Taken together, these results demonstrate that SP can activate Akt/GSK-3β signaling, which contributes to the SP-induced recovery of oxidatively damaged ADSCs.

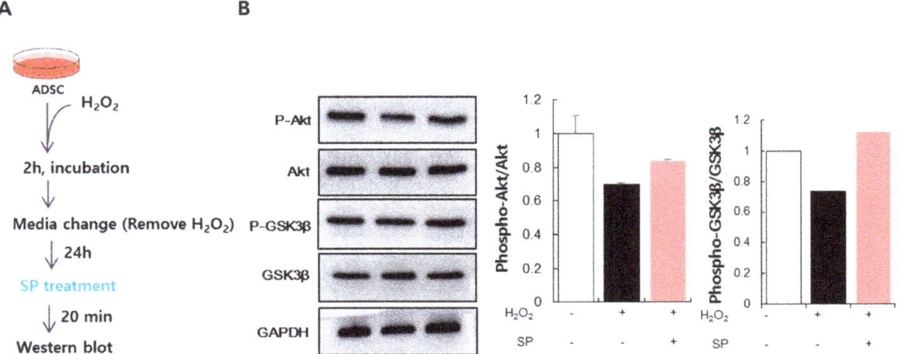

Figure 4. (**A**) Experimental scheme for ADSC oxidative stress and SP treatment. (**B**) Level of phospho-Akt (**B**) and phospho-GSK-3β as detected by Western blotting. Phospho-Akt and phospho-GSK-3β protein expression levels relative to total Akt and GSK-3β were quantified using the Image J program. Expression levels were represented relative to that of the untreated control. The data are expressed as the mean ± SD of three independent experiments.

4. Discussion

Aging and oxidative stress are highly associated with inflammation and the development of mortal diseases [1,21]. To combat oxidative stress-related diseases, antioxidants are typically applied from natural compounds or various medicines; however, their effects were equivocal, and are often accompanied by unwanted side effects. Therefore, additional treatment options with increased efficacy are urgently needed.

Most studies deal with oxidative stress to study retinal disease [6,16]. Stem cell therapies have emerged as an exciting option in the treatment of a variety of diseases including those that are related to oxidative stress. ADSCs are among the most popular stem cells to be used in novel therapies [12,32], as they have a high repopulation potential and their tissue of origin (adipose) is easily accessible. Ideally, therapies involving ADSCs would involve their autologous transplantation; however, ADSCs from diseased or aged patients may not be fully functional given the heightened levels of free radicals and inflammation in the individual. Indeed, we found that ADSCs from diseased animals have low repopulation rates and decreased cytokine secretion with a lack of differentiation potential even in early passages [33]. Moreover, a recent study corroborated the impairment of ADSCs by oxidative stress. Several studies have also indicated that, compared to other cells, stem cells are more susceptible to damage due to free radicals [27]. Thus, the effect of oxidative stress on stem cells should be taken into consideration for the application of stem cell therapy and endogenous regeneration.

In this study, oxidative stress was induced by treating ADSCs with H_2O_2 for 2 h. We found that this transient exposure to H_2O_2 was sufficient to affect ADSC activity and function, but was not completely detrimental to the ADSC population. H_2O_2 treatment altered cell morphology, reduced cell viability, inhibited the proliferation of ADSCs, and decreased their paracrine potential. This impairment could not be restored by removing the oxidant, which might lead to cellular senescence and death.

In an attempt to rescue ADSCs from oxidative stress, SP was employed. SP treatment of injured ADSCs enhanced cell viability and restored the paracrine potential of ADSCs. Moreover, at an early time point, SP activated the Akt/Gsk-3β pathway, which was downregulated by oxidative stress, and might contribute to the improvement of cell survival. Differentiation potential was also improved by SP but its effect was so slight, comparing to that of cell viability and cytokine secretion. Therefore, it was inferred that SP can augment cell survival and secretome production (rather than stemness), which might contribute to enhanced differentiation potential, to some extent. Notably, paracrine factors including VEGF and TGF-beta are deeply involved in osteogenesis [34–36]. SP could restore VEGF and TGF-beta production from ADSC with H_2O_2. This might suggest the possibility for the correlation between paracrine potential and osteogenesis (Figure S3)

In conclusion, this study demonstrated that SP could stimulate the recovery of ADSCs under oxidative stress, possibly by promoting cell proliferation through the activation of Akt/GSK-3β signaling. SP is anticipated to enhance the activity of ADSCs from aged or diseased individuals. Constant treatment of SP is anticipated to further augment the restoration of impaired stem cells.

Supplementary Materials: The following are available online at http://www.mdpi.com/2076-3921/9/10/978/s1, Figure S1: The analysis for marker expression of ADSC; Figure S2: Beta-galatoxidase staining of ADSC with H_2O_2 and SP; Figure S3: The differentiation ability of ADSC with H_2O_2 and SP.

Author Contributions: Conceptualization: H.S.H., J.S.P.; Methodology: H.S.H., J.S.P., J.P., G.P., Validation: H.S.H., J.S.P.; Formal analysis: H.S.H., J.S.P., J.P.; Investigation: H.S.H., J.S.P., G.P.; Data curation: H.S.H., J.S.P., J.P., G.P.; writing—original draft preparation: H.S.H., J.S.P., supervision: H.S.H.; project administration: H.S.H.; funding acquisition: H.S.H. All authors have read and agreed to the published version of the manuscript.

Funding: This study was supported by Basic Science Research Program through the National Research Foundation of Korea (NRF) funded by the Ministry of Education (2018R1D1A1B0704104813) and by a Korean Health Technology R&D Project grant [HI18C1492] from the Ministry of Health and Welfare (Sejong, Republic of Korea).

Conflicts of Interest: The authors declare no conflict of interest.

References

1. Liguori, I.; Russo, G.; Curcio, F.; Bulli, G.; Aran, L.; Della-Morte, D.; Gargiulo, G.; Testa, G.; Cacciatore, F.; Bonaduce, D.; et al. Oxidative stress, aging, and diseases. *Clin. Interv. Aging* **2018**, *13*, 757. [CrossRef]
2. Okoduwa, S.I.; Umar, I.A.; Ibrahim, S.; Bello, F.; Habila, N. Age-dependent alteration of antioxidant defensesystem in hypertensive and type-2 diabetes patients. *J. Diabetes Metab. Disord.* **2015**, *14*, 32. [CrossRef] [PubMed]
3. Luo, L.; Chen, H.; Trush, M.A.; Show, M.D.; Anway, M.D.; Zirkin, B.R. Aging and the brown Norway ratleydig cell antioxidant defense system. *J. Androl.* **2006**, *27*, 240–247. [CrossRef] [PubMed]
4. Oh, S.; Kim, Y.J.; Lee, E.K.; Park, S.W.; Yu, H.G. Antioxidative Eects of Ascorbic Acid and Astaxanthin on ARPE-19 Cells in an Oxidative Stress Model. *Antioxidants* **2020**, *9*, 833. [CrossRef]
5. Chandrasekaran, A.; Idelchik, M.; Melendez, J.A. Redox control of senescence and age-related disease. *Redox Biol.* **2017**, *11*, 91–102. [CrossRef] [PubMed]
6. Fuente, M.D.; Miquel, J. An update of the oxidation-inflammation theory of aging: The involvement of the immune system in oxi-inflamm-aging. *Curr. Pharm. Des.* **2009**, *15*, 3003–3026. [CrossRef] [PubMed]
7. Robb, K.P.; Fitzgerald, J.C.; Barry, F.; Viswanathan, S. Mesenchymal stromal cell therapy: Progress in manufacturing and assessments of potency. *Cytotherapy* **2018**, *21*, 1–18. [CrossRef] [PubMed]
8. Fitzsimmons, R.E.B.; Mazurek, M.S.; Soos, A.; Simmons, C.A. Mesenchymal Stromal/Stem Cells in Regenerative Medicine and Tissue Engineering. *Stem Cells Int.* **2018**, *2018*, 8031718. [CrossRef] [PubMed]
9. Xiaorong, F.; Ge, L.; Alexander, H.; Yang, J.; Qing, L.; Guanbin, S. Mesenchymal Stem Cell Migration and Tissue Repair. *Cells* **2019**, *8*, 784.
10. Bhansali, S.; Dutta, P.; Kumar, V.; Yadav, M.K.; Jain, A.; Mudaliar, S.; Bhansali, S.; Sharma, R.R.; Jha, V.; Marwaha, N.; et al. Efficacy of Autologous Bone Marrow-Derived Mesenchymal Stem Cell and Mononuclear Cell Transplantation in Type 2 Diabetes Mellitus: A Randomized, Placebo-Controlled Comparative Study. *Stem Cell Dev.* **2017**, *26*, 471–481. [CrossRef]
11. Lina, B.; Judit, C. Adipose tissue depots and inflammation: Effects on plasticity and resident mesenchymal stem cell function. *Cardiovasc. Res.* **2017**, *113*, 1064–1073.
12. Lucie, B.; Jana, Z.; Martina, T.; Jana, M.; Julia, P.; Petr, S.; Nikola, S.K.; Vaclav, S.; Zdenka, K.; Hooman, M.; et al. Stem cells: Their source, potency and use in regenerative therapies with focus on adipose-derived stem cells—A review. *Biotechnol. Adv.* **2018**, *36*, 1111–1126.
13. Gronthos, S.; Franklin, D.M.; Leddy, H.A.; Robey, P.G.; Storms, R.W.; Gimble, J.M. Surface protein characterization of human adipose tissue-derived stromal cells. *J. Cell Physiol.* **2011**, *189*, 54–63. [CrossRef]
14. Zimmerlin, L.; Donnenberg, A.D.; Rubin, J.P.; Basse, P.; Landreneau, R.J.; Donnenberg, V.S. Regenerative therapy and cancer: In vitro and in vivo studies of the interaction between adipose-derived stem cells and breast cancer cells from clinical isolates. *Tissue Eng. Part A* **2011**, *17*, 93–106. [CrossRef] [PubMed]
15. Piao, J.; Hong, H.S.; Son, Y. Substance P ameliorates tumor necrosis factor-alpha-induced endothelial cell dysfunction by regulating eNOS expression in vitro. *Microcirculation* **2018**, *25*, e12443. [CrossRef] [PubMed]
16. Baek, S.M.; Yu, S.Y.; Son, Y.; Hong, H.S. Substance P promotes the recovery of oxidative stress-damaged retinal pigmented epithelial cells by modulating Akt/GSK-3β signaling. *Mol. Vis.* **2016**, *22*, 1015–1023.
17. Hong, H.S.; Lee, J.; Lee, E.; Kwon, Y.S.; Lee, E.; Ahn, W.; Jiang, M.H.; Kim, J.C.; Son, Y. A new role of substance P as an injury-inducible messenger for mobilization of CD29 + stromal-like cells. *Nat. Med.* **2009**, *15*, 425–435. [CrossRef]
18. Amadoro, G.; Pieri, M.; Ciotti, M.T.; Carunchio, I.; Canu, N.; Calissano, P.; Zona, C.; Severini, C. Substance P provides neuroprotection in cerebellar granule cells through Akt and MAPK/Erk activation: Evidence for the involvement of the delayed rectifier potassium current. *Neuropharmacology* **2007**, *52*, 1366–1377. [CrossRef]
19. Hong, H.S.; Kim, S.; Lee, S.; Woo, J.S.; Lee, K.H.; Cheng, X.W.; Son, Y.; Kim, W. Substance-P prevents cardiac ischemia-reperfusion injury by modulating stem cell mobilization and causing early suppression of injury-mediated inflammation. *Cell Physiol. Biochem.* **2019**, *52*, 40–56.
20. Kim, S.; Piao, J.; Hwang, D.Y.; Park, J.S.; Son, Y.; Hong, H.S. Substance P accelerates wound repair by promoting neovascularization and preventing inflammation in an ischemia mouse model. *Life Sci.* **2019**, *225*, 98–106. [CrossRef]

21. Park, H.J.; Kim, S.; Jeon, E.J.; Song, I.T.; Lee, H.; Son, Y.; Hong, H.S.; Cho, S.W. PEGylated substance P augments therapeutic angiogenesis in diabetic critical limb ischemia. *J. Ind. Eng. Chem.* **2019**, *78*, 396–409. [CrossRef]
22. Lim, J.E.; Chung, E.; Son, Y. A neuropeptide, Substance-P, directly induces tissue-repairing M2 like macrophages by activating the PI3K/Akt/mTOR pathway even in the presence of IFN. *Sci. Rep.* **2017**, *7*, 1–19. [CrossRef] [PubMed]
23. Jin, Y.; Hong, H.S.; Son, Y. Substance P enhances mesenchymal stem cells-mediated immune Modulation. *Cytokine* **2015**, *71*, 145–153. [CrossRef] [PubMed]
24. Kang, M.H.; Kim, D.H.; Yi, J.Y.; Son, Y. Substance P accelerates intestinal tissue regeneration after gamma-irradiation-induced damage. *Wound Repair Regen.* **2009**, *17*, 216–223. [CrossRef]
25. Spector, A. Oxidative stress-induced cataract: Mechanism of action. *FASEB J.* **1995**, *9*, 1173–1182. [CrossRef]
26. Zheng, Y.; Liu, Y.; Ge, J.; Wang, X.; Liu, L.; Bu, Z.; Liu, P. Resveratrol protects human lens epithelial cells against H_2O_2-induced oxidative stress by increasing catalase, SOD-1, and HO-1 expression. *Mol. Vis.* **2010**, *16*, 1467–1474.
27. Facchin, F.; Bianconi1, E.; Romano, M.; Impellizzeri, A.; Alviano, F.; Maioli, M.; Canaider, S.; Ventura, C. Comparison of Oxidative Stress Effects on Senescence Patterning of Human Adult and Perinatal Tissue-Derived Stem Cells in Short and Long-term Cultures. *Int. J. Med. Sci.* **2018**, *15*, 1486–1501. [CrossRef]
28. Maioli, M.; Rinaldi, S.; Santaniello, S.; Castagna, A.; Pigliaru, G.; Delitala, A.; Margotti, M.L.; Bagella, L.; Fontani, V.; Ventura, C. Anti-senescence efficacy of radio-electric asymmetric conveyer technology. *Age* **2014**, *36*, 9–20. [CrossRef]
29. Legzdina, D.; Romanauska, A.; Nikulshin, S.; Kozlovska, T.; Berzins, U. Characterization of senescence of culture-expanded human adipose-derived mesenchymal stem cells. *Int. J. Stem Cells* **2016**, *9*, 124–136. [CrossRef]
30. Osaki, M.; Oshimura, M.; Ito, H. PI3K-Akt pathway: Its functions and alterations in human cancer. *Apoptosis* **2004**, *9*, 667–676. [CrossRef]
31. Chang, F.; Lee, J.T.; Navolanic, P.M.; Steelman, L.S.; Shelton, J.G.; Blalock, W.L.; Franklin, R.A.; McCubrey, J.A. Involvement of PI3K/Akt pathway in cell cycle progression, apoptosis, and neoplastic transformation: A target for cancer chemotherapy. *Leukemia* **2003**, *17*, 590–603. [CrossRef]
32. Zuk, P.A.; Zhu, M.; Mizuno, H.; Huang, J.; Futrell, J.W.; Katz, A.J.; Benhaim, P.; Lorenz, H.P.; Hedrick, M.H. Multilineage cells from human adipose tissue: Implications for cell based therapies. *Tissue Eng.* **2001**, *7*, 211–228. [CrossRef] [PubMed]
33. Park, J.S.; Piao, J.; Park, G.; Yoo, K.S.; Hong, H.S. Osteoporotic Conditions Influence the Activity of Adipose-Derived stem cells. *Tissue Eng. Regen. Med.* **2020**, 1–11. [CrossRef]
34. Chen, G.; Deng, C.; Li, Y. TGF-b and BMP signaling in osteoblast differentiation and bone formation. *Int. J. Biol. Sci.* **2012**, *8*, 272–288. [CrossRef]
35. Hu, K.; Olsen, B.R. The role of vascular endothelial growth factor in bone repair and regeneration. *Bone* **2016**, *91*, 30–38. [CrossRef] [PubMed]
36. Rumney, R.M.H.; Lanham, S.A.; Kanczler, J.M.; Kao, A.P.; Thiagarajan, L.; Dixon, J.E.; Tozzi, G.; Oreffo, R.O.C. In vivo delivery of VEGF RNA and protein to increase osteogenesis and intraosseous angiogenesis. *Sci. Rep.* **2019**, *9*, 1–10. [CrossRef] [PubMed]

© 2020 by the authors. Licensee MDPI, Basel, Switzerland. This article is an open access article distributed under the terms and conditions of the Creative Commons Attribution (CC BY) license (http://creativecommons.org/licenses/by/4.0/).

Article

Antioxidants N-Acetylcysteine and Vitamin C Improve T Cell Commitment to Memory and Long-Term Maintenance of Immunological Memory in Old Mice

Andreas Meryk [1,2], Marco Grasse [3], Luigi Balasco [4], Werner Kapferer [1], Beatrix Grubeck-Loebenstein [1] and Luca Pangrazzi [1,4,*]

1 Immunology group, Institute for Biomedical Aging Research, University of Innsbruck, 6020 Innsbruck, Austria; andreas.meryk@i-med.ac.at (A.M.); Werner_Kapferer@hotmail.com (W.K.); beatrix.grubeck@uibk.ac.at (B.G.-L.)
2 Department of Pediatrics I, Medical University of Innsbruck, 6020 Innsbruck, Austria
3 Division of Hygiene and Medical Microbiology, Medical University of Innsbruck, 6020 Innsbruck, Austria; marco.grasse@i-med.ac.at
4 Center for Mind/Brain Sciences (CIMeC), University of Trento, 38068 Rovereto, Italy; luigi.balasco@unitn.it
* Correspondence: luca.pangrazzi-1@unitn.it

Received: 27 October 2020; Accepted: 18 November 2020; Published: 19 November 2020

Abstract: Aging is characterized by reduced immune responses, a process known as immunosenescence. Shortly after their generation, antigen-experienced adaptive immune cells, such as CD8$^+$ and CD4$^+$ T cells, migrate into the bone marrow (BM), in which they can be maintained for long periods of time within survival niches. Interestingly, we recently observed how oxidative stress may negatively support the maintenance of immunological memory in the BM in old age. To assess whether the generation and maintenance of immunological memory could be improved by scavenging oxygen radicals, we vaccinated 18-months (old) and 3-weeks (young) mice with alum-OVA, in the presence/absence of antioxidants vitamin C (Vc) and/or N-acetylcysteine (NAC). To monitor the phenotype of the immune cell population, blood was withdrawn at several time-points, and BM and spleen were harvested 91 days after the first alum-OVA dose. Only in old mice, memory T cell commitment was boosted with some antioxidant treatments. In addition, oxidative stress and the expression of pro-inflammatory molecules decreased in old mice. Finally, changes in the phenotype of dendritic cells, important regulators of T cell activation, were additionally observed. Taken together, our data show that the generation and maintenance of memory T cells in old age may be improved by targeting oxidative stress.

Keywords: antioxidants; vitamin C; NAC; immunosenescence; T cells; vaccination; aging

---

1. Introduction

One of the most dramatic changes in the aging immune system is the involution of the thymus, which leads to a consistent decline in the generation of new naïve T cells [1,2]. As the amounts of naïve T cells are reduced in old age, adaptive immunity is mainly supported by antigen-experienced cells, the maintenance of which is fundamental to fight infections [3]. One typical hallmark of immunosenescence is the accumulation of highly differentiated T cells, known to be detrimental for elderly persons as they are associated with oxidative stress, inflammation, and senescence, which overall supports an increased risk of age-related diseases and mortality [4,5].

Several studies described that the bone marrow (BM) plays an important role in the long-term survival of effector/memory CD8$^+$ and CD4$^+$ T cells, as well as long-lived plasma cells [6,7].

After clearance of antigens, some newly generated adaptive cells migrate to the BM, where they can be maintained for long periods of time within survival niches [6,8]. In particular, while all antigen-experienced T cells require the cytokine IL-15 for their survival, both memory CD8$^+$ and CD4$^+$ T cells are maintained by IL-7 [9,10]. Specifically, the survival of terminally differentiated, senescent-like T cells is almost totally based on IL-15, as this subset show high levels of IL-2/IL-15Rβ and low expression of IL-7Rα. In addition, pro-inflammatory cytokine IL-6 is required for the maintenance of highly differentiated T cells.

Recently, we documented that the expression of IL-15 increases, while IL-7 decreases in the human BM in old age, which overall suggests that the survival of bona fide memory T cells may be impaired in the elderly [11,12]. Interestingly, oxidative stress and age-related inflammation ("inflammaging") were shown to be involved in these changes. In addition, highly differentiated CD8$^+$ T cells increased in the aged BM and supported the over-production of oxygen radicals and pro-inflammatory molecules, leading to impaired maintenance of immunological memory as a result [11,13].

In the current study, we investigated whether the phenotype of adaptive immune cells and dendritic cells (DCs), an important regulator of T cell activation, may change in an environment with reduced oxidative stress, using antioxidants N-acetylcysteine (NAC) and vitamin C (Vc). Furthermore, we assessed whether inflammatory parameters in the BM and in the spleen, organ regulating the maintenance and the generation of immunological memory, respectively, may change after administering antioxidants. To achieve this aim, we took advantage of a vaccination protocol using the alum-ovalbumin (OVA) vaccination model, optimal for studying both CD4$^+$ and CD8$^+$ T cell responses [14]. Young and old mice were vaccinated with three doses of alum-OVA and treated with NAC and Vc, alone or in combination. Only in old mice and in certain conditions, effector/senescent-like CD8$^+$ T cells decreased, while memory and memory precursors CD8$^+$ T cells increased with antioxidant treatments, in all peripheral blood (PB), BM, and spleen. In addition, oxidative stress in both BM and spleen, and the expression of pro-inflammatory cytokine IL-6 in the spleen decreased in old mice after the treatments. Finally, changes in the phenotype of DCs in the spleen were additionally observed after antioxidant administration.

Taken together, our results show for the first time that the generation and maintenance of memory T cells in old age may be improved after the administration of antioxidants. This may help in planning novel strategies to counteract immunosenescence in the elderly.

2. Materials and Methods

2.1. Mice

Three-week- and 18-month-old female mice on a C57BL/6J genetic background were maintained under specific pathogen-free conditions at the Institute for Biomedical Aging Research, University of Innsbruck, Austria.

Information about vaccination protocol, antioxidant treatment, bleedings, and organ harvesting are reported in Figure 1. Mice were intraperitoneally (i.p.) injected with Endofit OVA (Invivogen, San Diego, CA, USA; 10 µg/mouse) dissolved in a 1:1 solution of alhydrogel adjuvant 2% (Invivogen) and PBS. Three doses of vaccine were administered to the mice (days 0, 14, and 77). From day −7 (7 days before the administration of the first alum-OVA dose) until day 21, NAC was administered in the drinking water at a concentration of 1 g/L (as reported by Lehmann and coworkers [15]). Vc was i.p. injected, once per day, from day −7 until day 21. Optimal Vc concentration to use in the treatments was chosen after a preliminary experiment (Supplementary Materials Figure S1). In this titration experiment, the vaccination protocol was performed using a small group of old mice, and the lowest concentration significantly reducing ROS levels in the spleen (10 mg/kg) was chosen.

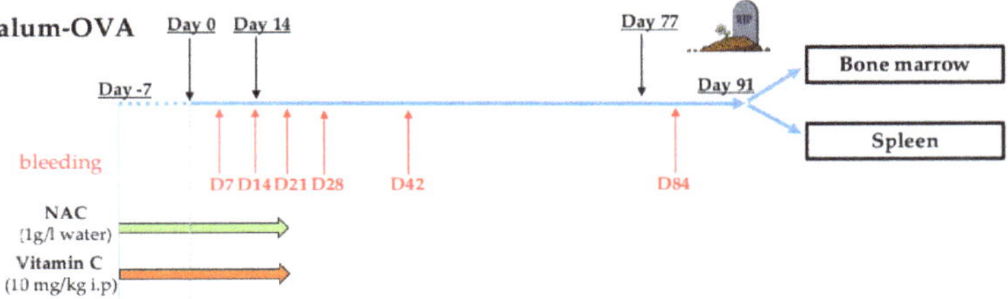

Figure 1. Vaccination protocol (see Materials and Methods). Mice were intraperitoneally (i.p.) injected with three doses of alum-OVA on days 0, 14, and 77. From day −7 until day 21, NAC was administered in the drinking water, and Vc and PBS (control group) were i.p. injected once per day. Blood was withdrawn from each mouse on days 7, 14, 21, 28, 42, and 84, and BM and spleen were harvested on day 91.

PB was withdrawn from each mouse on days 7, 14, 21, 28, 42, and 84, and BM and spleen were harvested on day 91. The following experimental groups were included in the experiments, each for 3-week-old and 18-month-old mice: PBS i.p. (control group), NAC, Vc, and NAC+Vc. To ensure reproducibility, the vaccination protocol was repeated twice, in both young and old mice. In both independent experiments, three mice were included in each group, and results from both experiments were pooled, and statistical analysis was performed. One mouse from the groups PBS old (included in experiment 2) and NAC old (experiment 1), and two mice from the group Vc old (one from experiment 1 and one from experiment 2) died before the end of the experiment. No signs of lymphoma or other age-related malignancies were found in the mice reaching the end of the vaccination protocol.

2.2. Material Processing and Cell Culture

200 µL blood was harvested from the facial vein and collected in a heparinized tube. Erythrocytes were lysed, incubating the blood with 10 mL lysis buffer (155 mM NH_4CL, 10 mM $KHCO_3$, 0.1 mM EDTA pH 7.4; all Merck KGaA) for 5 min in at RT. After the lysis, blood cells were washed once with RPMI 1640 (Sigma-Aldrich, St. Louis, MO, USA) and resuspended in complete medium (RPMI 1640 supplemented with 10% FCS, 100 U/mL penicillin, and 100 µg/mL streptomycin; Sigma-Aldrich and Invitrogen, Carlsbad, CA, USA, respectively). BM cells were obtained from mice by flushing the femur and tibia with PBS. Spleen cells were digested with Liberase™ Research Grade (Roche, Basel, Switzerland) and DNAse I (Roche) for 30 min, smashed through a Falcon 70 µm cell strainer (Corning, Corning, NY, USA), and washed with complete medium. Isolated cells from the spleen underwent erythrocyte lysis by incubating them with lysis buffer. After the isolation, both BM and spleen cells were washed once with RPMI and resuspended in complete medium.

2.3. Flow Cytometry

Immunofluorescence surface staining was performed by adding a panel of directly conjugated antibodies to freshly prepared blood, BM, and spleen cells. To analyze the expression of IL-6, cells were incubated with 10 mg/mL brefeldin A (BFA) for 15 h at 37 °C. To assess the expression of IFNγ, TNF, and IL-2, cells were incubated with 30 ng/mL PMA and 500 ng/mL ionomycin in the presence of 10 mg/mL BFA for 4 h at 37 °C. After surface staining, cells were permeabilized using the Cytofix/Cytoperm kit (BD Biosciences, San Jose, CA, USA), and incubated with the intracellular antibody. Dead cells were excluded from the analysis using 7-AAD or fixable viability dye (FVD) BV421 (both BD Biosciences). Labeled cells were measured using a Fluorescence activated cell sorter (FACS) Canto II (BD Biosciences). Data were analyzed using Flowjo software. The antibodies used in the experiments are shown in Table 1.

Table 1. Antibodies used in the FACS stainings.

Antigen	Fluorocrome	Company	Clone
KLRG-1	BV421	Biolegend	2F1/KLRG1
CD44	FITC	Biolegend	IM7
IL-7Rα	PE	Biolegend	A7R34
CD8	PeCy7	Biolegend	53-6.7
CD8	PerCp	Biolegend	53-6.7
CD62L	APC	Biolegend	MEL-14
CD4	BV510	Biolegend	GK1.5
CD4	Pecy7	Biolegend	GK1.5
CD3	APC-Vio770	Miltenyi	REA641
PD-1	PE	Biolegend	29F.1A12
IL-6	APC	Biolegend	MP5-20F3
IFNg	PE	Biolegend	XMG1.2
TNF	FITC	Biolegend	MP6-XT22
IL-2	APC	Biolegend	JES6-5H4
CD3	BV510	Biolegend	17A2
CD11c	BV510	Biolegend	N418
MHC II	FITC	Biolegend	M5/114.15.2
CD40	PE	Biolegend	3/23
NKp46	PerCp	Biolegend	29A1.4
CD19	PerCp	Biolegend	6D5
CD3	PerCp	Biolegend	145-2C11
CD80	APC	Biolegend	16-10A1
CD11b	APC-Cy7	Miltenyi	M1/70.15.11.5

2.4. ROS Measurement

ROS levels were measured after incubation of BM and spleen cells with the fluorescent dye dihydroethidium (Sigma-Aldrich) at a concentration of 1:250 in complete medium for 20 min at 37 °C.

2.5. Statistical Analysis

Statistical significance was assessed after two-way ANOVA and Tuckey post-hoc test, as indicated in the figure legends. A p value less than 0.05 was considered significant.

2.6. Study Approval

Animal protocols were approved by the Federal Ministry of Science, Research, and Economy of Austria, and carried out in accordance with the Austrian law for animal protection and the institutional guidelines at the University of Innsbruck.

3. Results

3.1. Effector/Memory T Cell Subsets Change in the Peripheral Blood (PB) after Treatment with Antioxidants

Using the markers IL-7Rα and KLRG-1, the populations IL-7Rα$^-$KLRG-1$^+$ short-living effector cells (SLEC), and IL-7Rα$^+$KLRG-1$^-$ memory progenitor effector cells (MPEC) can be identified within CD8$^+$ T cells [16,17]. While the first subset may either die soon after the resolution of infections or alternatively accumulate as senescent-like cells, MPEC differentiate into long-living memory CD8$^+$ T cells. We investigated the levels of SLEC in the peripheral blood (PB) of untreated and NAC, vitamin C (Vc), and NAC+Vc treated old mice 7, 14, 21, 28, 42, and 84 days after the first alum-OVA injection (Figure 2A). A significant reduction in the frequency of SLEC was observed in Vc treated mice already on day 7, although trends were present also for the other treatments. These differences were stable over the following time-points, and they were significant for all NAC, Vc, and NAC+Vc treatments on day 84. Despite this, no differences between the different treatments were observed. Complete gating strategy is reported in . Representative FACS plots are shown in Figure 2C.

No differences could be observed between treated and untreated young mice in the levels of SLEC in the PB (Figure S3A).

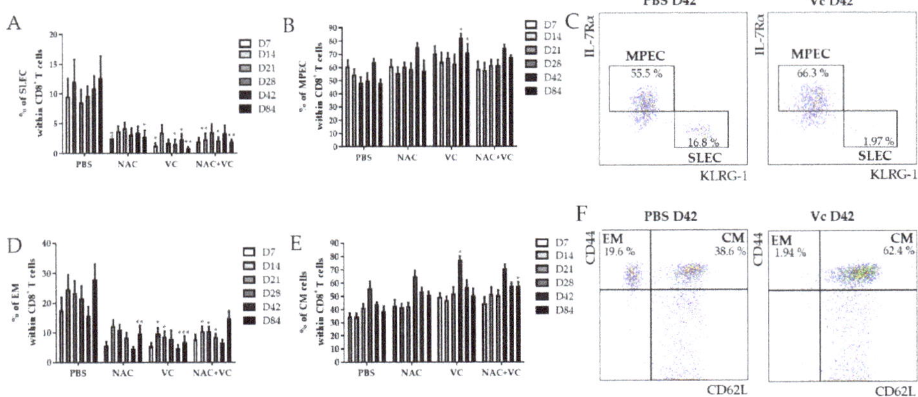

Figure 2. Effector/memory CD8$^+$ T cell subsets in the PB of old mice. Frequency of (**A**) IL-7Rα^-KLRG-1$^+$ SLEC and (**B**) IL-7Rα^+KLRG-1$^-$ MPEC in the blood of old mice treated with PBS, NAC, Vc, or NAC+Vc. (**C**) Representative FACS plots of SLEC and MPEC in a control and in a Vc-treated old mouse on day 42. Frequency of (**D**) CD44^{hi}CD62L$^-$ EM and (**E**) CD44^{hi}CD62L$^+$ CM CD8$^+$ T cells in the blood of old mice treated with PBS, NAC, Vc, or NAC+Vc. (**F**) Representative FACS plots of SLEC and MPEC in a control and in a Vc-treated old mouse on day 42. Blood was harvested on days 7, 14, 21, 28, 42, and 84. Data are shown as mean ± SEM. Two-way ANOVA, Tukey post-hoc test. * $p < 0.05$; ** $p < 0.01$; *** $p < 0.001$.

We next assessed whether the treatments may additionally change the frequency of MPEC in the PB of old mice (Figure 2B). Only when Vc alone was administered to the mice, a significant increase of MPEC levels was found on days 42 and 84. Again, no differences were identified in young mice (Figure S3B). In addition, the frequency of CD44^{hi}CD62L$^-$ effector memory (EM) CD8$^+$ T cells decreased in the PB of old mice (Figure 2D). Specifically, while NAC treatment reduced EM CD8$^+$ T cell levels only on day 84, the treatments with Vc and NAC+Vc were effective in inducing a significant downregulation in the levels of this subset already from day 14. On day 84, NAC and Vc administered alone, but not in combination, induced a consistent reduction in the frequency of EM CD8$^+$ T cells. No differences between the treated and untreated groups were observed for EM cells within CD4$^+$ T cells in old mice and for EM CD8$^+$ T cells in young mice (Figure S3C,D). In addition, the frequency of CD44^{hi}CD62L$^+$ central memory (CM) CD8$^+$ T cells increased in the PB of old mice treated with Vc on day 28, and in mice treated with NAC+Vc on day 84, in comparison with the untreated control (Figure 2D). Increased levels of CM CD4$^+$ T cells were additionally found on day 28 in the PB of old mice treated with NAC (Figure S3E). No differences were found in young mice, neither for CM CD8$^+$ T cells (Figure S3F) nor for CM CD4$^+$ T cells (data not shown).

Taken together, these results indicate that antioxidant treatments affect the levels of effector/memory T cell subsets in the PB of old mice, particularly within CD8$^+$ T cells.

3.2. *Effector/Memory T Cell Subsets Change in the BM and in the Spleen after Treatment with Antioxidants*

We next assessed whether the levels of effector/memory T cell subsets may additionally change in the BM and in the spleen after treatments with NAC, Vc, or NAC+Vc, in comparison with untreated control mice (Figure 3). Similar to the situation in the periphery, the frequency of SLEC in the BM decreased in old mice treated with Vc and NAC+Vc, in comparison with the controls (Figure 3A). While no differences between young and old mice were present in the control groups, the levels of SLEC were significantly lower in the old Vc group, in comparison to young mice treated with Vc.

In young mice, no differences were observed after any of the treatments. A similar situation was found in the spleen (Figure 3B). In this case, the levels of SLEC were always higher in old compared to young mice. Again, only in old mice, treatment with Vc and NAC+Vc like in the BM, but also with NAC alone, could significantly reduce the frequency of SLEC. In addition, MPEC increased in old mice treated with NAC and NAC+Vc (both BM and spleen) or Vc alone (only spleen), in comparison with the untreated controls (Figure 3C,D). Interestingly, levels of MPEC in untreated old mice were lower than their younger counterparts. After any of the three treatments, the frequency of this subset was similar between young and old mice. These results indicate that antioxidants may boost MPEC commitment in old mice.

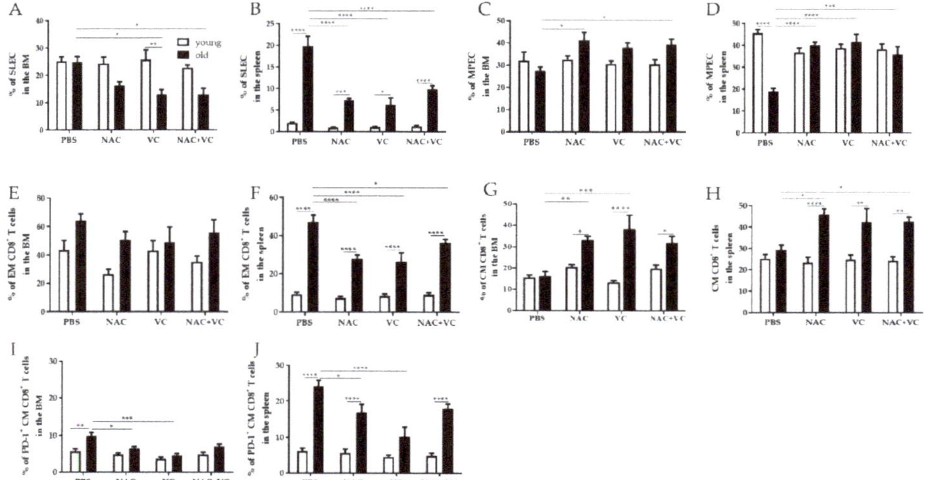

Figure 3. Effector/memory CD8$^+$ T cell subsets in the BM and in the spleen. Frequency of (**A**) SLEC in the BM, (**B**) SLEC in the spleen, (**C**) MPEC in the BM, (**D**) MPEC in the spleen, (**E**) EM CD8$^+$ T cells in the BM, (**F**) EM CD8$^+$ T cells in the spleen, (**G**) CM CD8$^+$ T cells in the BM, (**H**) CM CD8$^+$ T cells in the spleen, (**I**) PD-1$^+$CM CD8$^+$ T cells in the BM, (**J**) PD-1$^+$CM CD8$^+$ T cells in the spleen of young (white columns) and old (black columns) mice treated with PBS, NAC, Vc, or NAC+Vc. Data are shown as mean ± SEM. Two-way ANOVA, Tukey post-hoc test. * $p < 0.05$; ** $p < 0.01$; *** $p < 0.001$; **** $p < 0.0001$.

When the frequency of EM CD8$^+$ T cells was assessed in the BM, no significant differences were found between all the groups, neither in young nor in old mice (Figure 3E). As reported for SLEC, EM CD8$^+$ T cells in the spleen were higher in old compared to young mice and were reduced after NAC, Vc, and NAC+Vc treatments. While no differences were observed for EM CD4$^+$ T cells in the BM, this subset was more present in the spleen of old mice, in comparison with any groups of young mice (Figure S4A,B). Despite this, no differences between the treatments were identified. In both BM and spleen, the frequency of CM CD8$^+$ T cells was similar in untreated young and old mice (Figure 3G,H). Only in old mice, treatments with NAC or Vc alone (BM), and NAC alone or NAC+Vc (spleen) could increase the levels of this subset, in relationship to untreated controls. Again, no differences were found in young mice. In addition, similar frequencies of CM CD4$^+$ T cells were found in all the groups, in both BM and spleen (Figure S4C,D). The expression of activation/exhaustion marker PD-1 was quite low within BM CM CD8$^+$ T cells (Figure 3I). Despite this, higher PD-1 levels were found in old compared to young untreated mice, in both BM and spleen (Figure 3I,J). After treatments with NAC and Vc alone, the frequency of PD-1$^+$ CM CD8$^+$ T cells in old mice was significantly reduced. This was particularly evident in the spleen of old mice treated with Vc, which showed similar PD-1 levels when compared with young-Vc treated mice. No differences were found for co-inhibitory molecules CTLA-4 and TIM-3 (data not shown).

In summary, our data indicate that antioxidants may change the frequency of effector/memory T cell subsets in the BM and in the spleen of old mice.

3.3. Antioxidant Treatments Affect ROS Levels and Pro-Inflammatory Molecules in the BM and Spleen

We next investigated whether antioxidant treatments may change the levels of oxygen radicals and pro-inflammatory molecule IL-6 in the BM and in the spleen. As expected, in the untreated groups, ROS levels were higher in old in comparison with young mice, in both BM and spleen (Figure 4A,B). In old mice, after the treatment with NAC alone and NAC+Vc (BM + spleen) and additionally with Vc alone (BM only), oxygen radicals were reduced in comparison with the untreated control group. After any treatments, in both BM and spleen, ROS levels were similar between old and young mice, although no differences were found between treated and untreated young mice. In addition, the expression of IL-6 in the BM was similar between young and old mice, and it did not change after the treatments (Figure 4C, Figure S5A). In all experimental groups, IL-6 levels in the spleen were higher in old compared to young mice (Figure 4D, Figure S5B). Despite this, treatment with NAC could reduce IL-6 expression in old mice, in relationship to the untreated group.

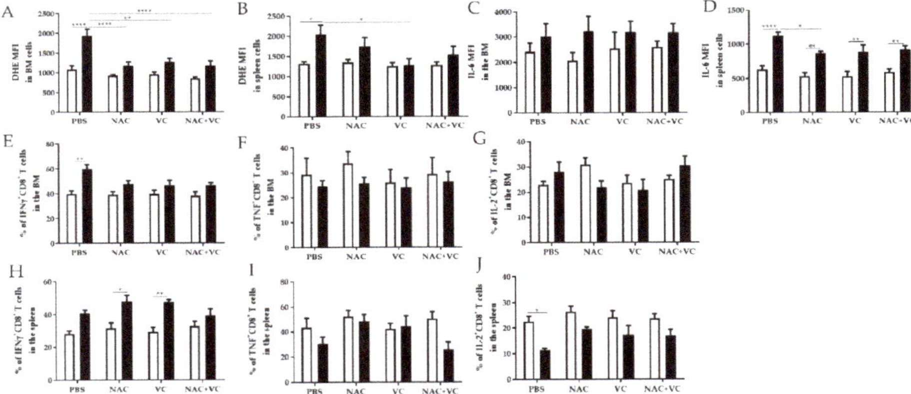

Figure 4. ROS levels and pro-inflammatory molecules in the BM and spleen. (**A**) ROS levels (=DHE Mean Fluorescence Intensity, MFI) in the BM, (**B**) ROS levels in the spleen, (**C**) IL-6 MFI in the BM, (**D**) IL-6 MFI in the spleen, (**E**) IFNγ$^+$CD8$^+$ T cells in the BM, (**F**) TNF$^+$CD8$^+$ T cells in the BM, (**G**) IL-2$^+$ CD8$^+$ T cells in the BM, (**H**) IFNγ$^+$CD8$^+$ T cells in the spleen, (**I**) TNF$^+$CD8$^+$ T cells in the spleen, (**J**) IL-2$^+$ CD8$^+$ T cells in the BM of young (white columns) and old (black columns) mice treated with PBS, NAC, Vc, or NAC+Vc. Data are shown as mean ± SEM. Two-way ANOVA, Tukey post-hoc test. * $p < 0.05$; ** $p < 0.01$; **** $p < 0.0001$.

We then measured the production of IFNγ and TNF, two additional pro-inflammatory and T cell effector molecules, in the BM and in the spleen (Figure 4E–J, Figure S6). In the BM of untreated old mice, more CD8$^+$ T cells produced IFNγ in comparison with young mice (Figure 4E). After treatments with NAC, Vc and NAC+Vc, no differences could be found between young and old mice. No age-related differences and no significant differences between the treatments were observed for TNF and IL-2 production in CD8$^+$ T cells within the BM (Figure 4F,G). In parallel, more IFNγ, TNF, and IL-2 was produced by BM CD4$^+$ T cells from old compared with young mice (Figure S6A–C). Although the treatments did not significantly change the production of IFNγ and TNF, more IL-2$^+$CD4$^+$ T cells were present in old mice injected with Vc, in comparison with old control mice (Figure S6C).

Similar to the situation in the BM, the frequency of IFNγ^{+}CD8^{+} T cells in the spleen was higher in old mice, particularly in the groups treated with NAC and Vc alone (Figure 4H). Despite this, no changes were induced by the treatments. When we assessed the production of IFNγ in CD4^{+} T cells from the spleen, higher levels were present in the old group (Figure S6D). After Vc treatment, IFNγ production was similar between old and young mice. No differences were observed for the expression of TNF within CD8^{+} T cells (Figure 4I) and CD4^{+} T cells (Figure S6E). Reduced IL-2 production was found in the spleen of untreated old mice, in comparison to the young control group. After any of the three treatments, no differences were present between old and young mice. No differences in the expression of IL-2 were present in CD4^{+} T cells from the spleen (Figure S6F). Representative FACS plots of IFNγ, TNF, and IL-2 expression within CD8^{+} T cells from BM and spleen are shown in Figure S7.

In summary, our results suggest that, in some conditions, but only in old mice, treatment with NAC and/or Vc could reduce the levels of ROS and affect the production of several T cell cytokines, in the BM and in the spleen.

3.4. Dendritic Cell Subsets Change in the Spleen after Antioxidant Treatments

Dendritic cells (DC) are antigen-presenting cells fundamental for the activation of adaptive immune responses. Thus, we investigated whether OVA vaccination in the presence of antioxidants NAC and/or Vc could affect the frequency of certain DC subsets and the phenotype of these cells (Figure 5). The gating strategy reporting the subsets considered in this part of the study is shown in Figure 5A. Classical DC (cDC) are known to express both MHC class II (MHC II) and CD11c, and can additionally be divided into the CD11b^{-} (cDC1) and CD11b^{+} (cDC2) subsets [18].

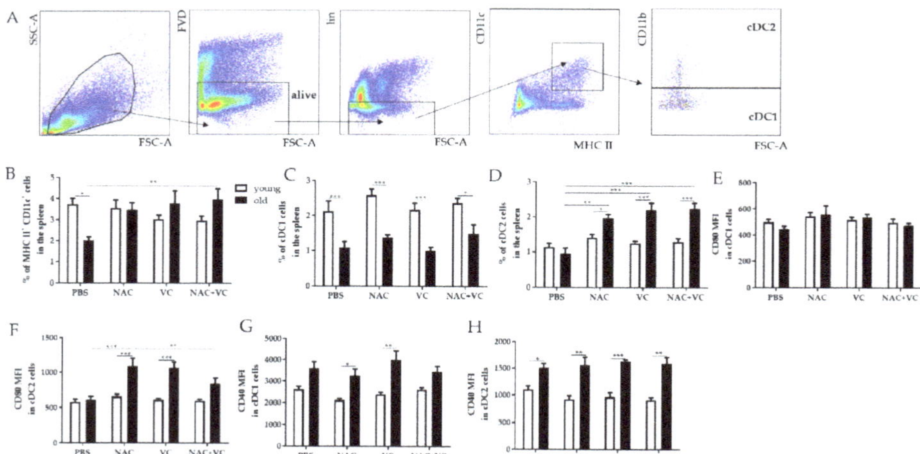

Figure 5. Dendritic cell subsets in the spleen. (**A**) Gating strategy for DC subsets. After excluding dead cells using the FVD Zombie Violet, lineage negative (lin-) CD3^{-}NKp46^{-}CD19^{-} cells were considered. FACS plots showing the gating for CD11c^{+}MHC II^{+} DCs, CD11c^{+}MHC II^{+}CD11b^{-} classical DCs (cDC1) and CD11c^{+}MHC II^{+}CD11b^{+} DCs (cDC2) are reported. Frequency of (**B**) CD11c^{+}MHC II^{+} DCs, (**C**) cDC1, (**D**) cDC2, and MFI of (**E**) CD80 in cDC1, (**F**) CD80 in cDC2, (**G**) CD40 in cDC1, and (**H**) CD40 in cDC2 in the spleen of young (white columns) and old (black columns) mice treated with PBS, NAC, Vc, or NAC+Vc. Data are shown as mean ± SEM. Two-way ANOVA, Tukey post-hoc test. * $p < 0.05$; ** $p < 0.01$; ***$p < 0.0001$.

When we measured all CD11c^{+}MHC II^{+} cDCs in the spleen, the frequency was lower in old untreated mice in comparison to the young control group (Figure 5B). NAC+Vc lead to a significant increase in cDC frequency in old mice, and thus, no age-related differences were present after the treatment. In addition, cDC1 cells were higher in young mice in all experimental group, but no significant differences were found after administering the antioxidants (Figure 5C). Furthermore, all NAC, VC, and NAC+Vc treatments lead to an increased frequency of the cDC2 subset, in comparison with old and young untreated groups. Again, no effects were observed when young mice were treated.

Costimulatory molecule CD80 is known to bind its ligands CD28 and CTLA-4 expressed by T cells and regulate the activation of these cells [19]. In cDC1 cells, similar CD80 levels were found in young and old mice, and no differences were present after the treatments (Figure 5E and Figure S8A). Only in cDC2 the administration of NAC and Vc, alone or in combination, increased CD80 expression in old mice, in comparison to the untreated group (Figure 5F and Figure S8B). Furthermore, the levels of CD40, a costimulatory protein necessary for DC activation [20], were higher in old mice in both cDC1 and cDC2 subsets, particularly in the groups treated with NAC and Vc alone for cDC1 and in all groups for cDC2 (Figure 5G,H and Figure S8C,D). Despite this, no significant changes were induced by the treatments.

Overall, antioxidant treatments may increase the frequency and the activation of cDC subsets in old mice, which may therefore be more efficient in antigen presentation and in activating T cells.

4. Discussion

Aging negatively affects the quality of the immune system, which leads to increased frequency and severity of infectious diseases and to a reduced efficiency of vaccinations [21]. Although both innate and adaptive immunity are impaired in old age, T lymphocytes are most severely affected, as they lose the organ involved in their maturation, the thymus [1,2]. Although reduced amounts of antigen-inexperienced, naïve T cells can be observed in the elderly, homeostatic proliferation mechanisms guarantee the presence of a certain amount of naïve T cells, necessary to fight against new infections. Despite this, in old age, naïve T cells may preferentially be committed to an effector, senescent-like T cells, with reduced capability to differentiate into memory cells [22]. In addition, chronic infections, such as cytomegalovirus (CMV), support the accumulation of senescent-like T cells [23]. These cells may accumulate in the BM, therefore reducing the space otherwise available for "true" memory T cells [13,24]. For this reason, finding strategies is helpful—not only to improve the generation, but also the maintenance of memory cells in the BM is fundamental to counteract immunosenescence, and boost immunity against infections in old age.

We recently observed how oxidative stress may impair the long-term maintenance of memory T cells and long-lived plasma cells in the BM [11,13]. In particular, we showed that accumulation of ROS in the BM leads to increased expression of IFNγ and TNF, which induce the production of IL-15 and IL-6, a survival factor for senescent-like T cells. In addition, a negative relationship between Diphtheria antibody concentration in the PB and ROS levels in the BM was identified [24]. Thus, oxidative stress may represent an important feature to target, to test novel strategies aiming at boosting immunological memory.

In our approach, we took advantage of NAC and Vc, well-known antioxidants, already available on the market, and with rare side effects. NAC is known to boost endogenous glutathione levels, as it is a source of amino acid cysteine, which is the rate-limiting step of glutathione synthesis [25,26]. In addition, NAC has shown glutathione-independent antioxidant properties [27]. Vc and glutathione can support each other in the neutralization of ROS. Indeed, reduced Vc (dehidroascorbate, DHA) can rapidly neutralize oxidants, and after this step, it is converted into its oxidized form (ascorbate). Reduced glutathione (GSH) can convert ascorbate into DHA, and thus, after this step Vc is able to react again against ROS. At moderate/high doses, less than 50% of orally administered Vc can reach the bloodstream, while the rest is excreted in the urine [28]. For this reason, we decided to administer this antioxidant i.p. to the mice. To counteract this problem, liposomal Vc formulation has recently been developed, which shows increased absorption rates in comparison to water-soluble Vc [29]. This situation does not happen for NAC, which is very well absorbed when administered orally. Although NAC and Vc were shown to improve immune function in several experimental settings [30–32], no studies reported how these molecules affect the phenotype of subsets of immune cells during a vaccination protocol.

We, therefore, established a vaccination schedule in which mice were pre-treated with antioxidants for 7 days, before administering the first alum-OVA dose. In this way, ROS levels may already be lower at the beginning of the vaccination. The treatments with NAC and Vc were stopped on day 21, as mice received two doses of vaccine, and therefore, the generation of memory T cells already started. Finally, a third alum-OVA dose was administered on day 77. This choice has two specific aims: (1) Studying the generation of memory T cells and the phenotype of DC in the spleen two weeks after; (2) investigating the phenotype of T cells migrated into the BM during the first part of the experiment; as memory T cells require 3–8 weeks for the migration to the BM [33], adaptive immune cells generated after the third dose of vaccine can preferentially be found in the spleen. In this way, both the generation and maintenance of immunological memory can be studied with one protocol.

We first assessed the levels of effector/memory T cell subsets in the PB after the injection of alum-OVA. In particular, we investigated whether antioxidants may affect the commitment of naïve T cells into SLEC/EM $CD8^+$ T cells, which include short-living cells and/or senescent-like T cells, and MPEC/CM, subsets of memory T cell precursors/memory T cells [16,34].

While the expression of CCR7 and CD62L on CM cells facilitate the homing to secondary lymphoid organs, EM cells preferentially migrate to tissues, are more cytolytic, and express receptors necessary for localization to areas of inflammation [35]. Thus, while SLEC and EM cells show rapid effector functions, and thus, are fundamental when pathogens are present in tissues, MPEC and CM differentiate into long-living memory T cells, and therefore, represent the primary aim of a vaccination in which immunological memory must be triggered. Interestingly, a reduction in the frequency of SLEC/EM and an increased in MPEC/CM $CD8^+$ T cells was observed in the PB of old mice treated with antioxidants. While the reduction of SLEC and EM was evident from day 7, the upregulation of MPEC/CM was present only at later time-points. On day 28, an increased frequency of CM $CD8^+$ T cells was present in every experimental group. This may be due to administering the second alum-OVA dose, which triggers the generation of new CM cells. In addition, an overall increase of MPEC cells was observed on day 42. CM $CD4^+$ T cells additionally increased after NAC treatment, but no differences were observed for EM $CD4^+$ T cells. The treatments did not lead to any differences in young mice, which overall showed reduced levels of SLEC and EM $CD8^+$ T cells in comparison to untreated old mice. After administering NAC and Vc, similar amounts of circulating SLEC and EM $CD8^+$ T cells were found in old and young mice, indicating that the treatment rescued the differences between the two groups. In the spleen of old mice, a consistent decline in both subsets could be observed.

In parallel, MPEC and CM $CD8^+$ T cells showed a strong increase. Thus, while without the treatments, old mice produced many more effector cells and less memory $CD8^+$ T cell precursors in comparison to young mice, a partial rescue was observed in the treated groups. We can, therefore, speculate that oxidative stress may affect $CD8^+$ T cell commitment to memory in old mice. Similar trends were obtained in the BM, in which the differences were less significant in comparison to the spleen. This may be attributable to the fact that effector/memory T cells present in the BM were generated after two doses of vaccine and not after three doses like the ones present in the spleen, and therefore, differences may be attenuated. In addition, the expression of inhibitory receptor PD-1 [36] on CM $CD8^+$ T cells, higher in old untreated mice in comparison to young mice, was reduced at the level of young animals after the treatments, in both BM and spleen. This indicates that, in old mice, CM $CD8^+$ T cells may be more "ready to react" after NAC and Vc administration.

A key question is whether antioxidant treatment may affect oxidative stress and pro-inflammatory parameters in the BM and in the spleen. In both organs, ROS levels were higher in old untreated mice, which for the BM is in accordance with our data in humans [11]. Interestingly, all antioxidant treatments (BM) and Vc administration (spleen) could rescue this condition, as no differences were present between old and young mice after the treatments. According to the mechanism of action of NAC and Vc, the effects of NAC and Vc together may be additive. Despite this, in our settings, we saw that administering a single antioxidant (either NAC or Vc) may be enough to significantly reduce ROS. Furthermore, we can speculate that the effects of antioxidants may be more stable in the BM in comparison to the spleen, as in this organ, they are still present 70 days after interrupting the treatments. Despite this, we believe that NAC and Vc may act on the spleen environment from the very beginning of their administration, and thus, in the PB, differences can be observed already from day 7. In the spleen, decreased ROS levels were paralleled by a reduction of IL-6 expression. As this cytokine is not only a T cell survival molecule, but also a pro-inflammatory cytokine, we expect that inflammatory processes may be reduced in the spleen of treated mice. No relationship between ROS and IL-6 levels was found in the BM. This is different from the situation in humans and in a small group of $SOD1^{-/-}$ and WT mice, in which a strong correlation between ROS levels and IL-6 MFI in the BM was observed [11].

Furthermore, in the BM, the differences in the expression of IFNγ in $CD8^+$ and $CD4^+$ T cells between young and old mice disappeared after some of the antioxidant treatments. Thus, we can speculate that inflammation within the BM environment, known to be detrimental for the maintenance of immunological memory in old age [11,13], may partially be attenuated by NAC and/or Vc. Although not completely "rescued" age-related inflammation, commonly known as "inflammaging" and known to support the onset and the severity of age-related diseases, among which immunosenescence [37], may be reduced after antioxidant treatment.

As the last step, we assessed whether the phenotype of DC subsets may be affected by ROS. Overall, in NAC and Vc treated old mice, MHC II^+ $CD11c^+$ cDCs, fundamental for delivering antigens to T cells and regulating the activation of these cells, increased in old mice after antioxidant treatment. In addition, in these mice, administering NAC and Vc lead to increased frequency of cDC2, a subset of migratory DC known to induce potent T follicular helper (Tfh) responses [38]. After the treatment, cDC2 in old mice expressed higher levels of CD80, a costimulatory molecule necessary for the complete activation of T cells. Thus, these results indicate that antioxidants may improve cDC2 functions, which may therefore be more efficient in activating T cells in lymph nodes. Furthermore, activation molecule CD40 was higher in old mice in both cDC1 and cDC2 subsets. Thus, it is important to underline that in old mice, after NAC and Vc administration, the expression of both CD80 and CD40 are higher in comparison to young mice. Overall, this suggests that the capability of priming T cells by DCs may be boosted, reducing oxidative stress in the elderly.

5. Conclusions

In summary, our work showed for the first time that commitment of naïve T cells to long-living memory T cells may be boosted by targeting oxidative stress. In addition, the maintenance of immunological memory in the BM may be supported. To achieve this aim, administering one antioxidant (either NAC or Vc) is generally sufficient. Indeed, a certain amount of ROS is required for immune functions [39]. In addition, it is important to underline that, in some conditions, an excessive administration of antioxidants may be even detrimental [40]. Furthermore, our results suggest that the effects of antioxidants are very stable in old mice, as 28 days of treatment led to sustained shifts in T cell subsets and in the production of pro-inflammatory molecules two months later. The administration of NAC and Vc did not lead to any changes in young mice. As these animals showed lower ROS levels in comparison to old mice, antioxidant treatments may be totally ineffective at this age. An optimal age should be identified, to start the treatments to target ROS and boost the functionality of the immune system. Thus, ROS scavengers administered at the right timepoint may represent an optimal tool to support not only the generation of new memory T cells, but also the maintenance of adaptive immune cells generated years before and maintained in the BM niches.

As NAC and Vc seem to boost immune responses in old age, future studies must assess the impact of antioxidants on a vaccination protocol performed using an antigen from a pathogen relevant for aging, such as a mouse-adapted influenza virus strain. Indeed, it is known that the efficacy of influenza vaccination is reduced in the elderly [41], and therefore, it will be important to understand whether antioxidants may be effective also in this context. The results obtained in our study using alum-OVA will certainly help to define optimal settings for future vaccination protocols.

In addition, the effects of NAC and Vc on levels of pro-inflammatory cytokines in the plasma will be analyzed in future studies. Furthermore, it must be assessed whether the long-term exposure to antioxidants may change molecular pathways controlling T cell differentiation, such as JAK/STAT, MAPK signaling, and HIF1α stabilization.

Finally, nutritional intervention, including natural antioxidants and/or nutraceutical compounds, must be tested to assess whether a healthy lifestyle may additionally boost the generation and maintenance of immunological memory in the elderly.

Supplementary Materials: The following are available online at http://www.mdpi.com/2076-3921/9/11/1152/s1. Figure S1: Titration of optimal Vc concentration to use in the study, Figure S2: T cell subsets in the PB, Figure S3: Effector/memory T cell subsets in the PB of young and old mice, Figure S4: Effector/memory T cell subsets in the BM and spleen of young and old mice, Figure S5: Frequency of IL-6+ cells; Figure S6: Expression of pro-inflammatory molecules within CD4+ T cells in the BM and spleen, Figure S7: Representative FACS plots, Figure S8: Frequency of CD80 and CD40 in DC subsets.

Author Contributions: Conceptualization, L.P., A.M. and B.G.-L.; methodology, L.P. and A.M.; validation, L.P., A.M. and M.G.; formal analysis, L.P.; investigation, L.P., A.M., M.G. and W.K.; resources, L.P. and B.G.-L.; data curation, L.B. and L.P.; writing—original draft preparation, L.P.; writing—review and editing, L.P., A.M., M.G., L.B., W.K., B.G.-L.; visualization, L.P.; supervision, L.P.; project administration, L.P.; funding acquisition, L.P., B.G.-L. All authors have read and agreed to the published version of the manuscript.

Funding: The research leading to these results received funding from Tyrolean Science Funds (TWF, grant number ZAP746006) and by the EU H2020 project "An integrated approach to dissect determinants, risk factors, and pathways of ageing of the immune system" (ImmunoAgeing, H2020-PHC-2014 grant agreement no. 633964).

Acknowledgments: The authors thank Michael Keller and Brigitte Jenewein for the excellent technical support.

Conflicts of Interest: The authors declare no conflict of interest. The funders had no role in the design of the study; in the collection, analyses, or interpretation of data; in the writing of the manuscript, or in the decision to publish the results.

References

1. George, A.J.T.; Ritter, M.A. Thymic involution with ageing. Obsolescence or good housekeeping? *Immunol. Today* **1996**, *17*, 267–272. [CrossRef]
2. Aspinall, R.; Andrew, D. Thymic involution in aging. *J. Clin. Immunol.* **2000**, *20*, 250–256. [CrossRef] [PubMed]
3. Boyman, O.; Letourneau, S.; Krieg, C.; Sprent, J. Homeostatic proliferation and survival of naive and memory T cells. *Eur. J. Immunol.* **2009**, *39*, 2088–2094. [CrossRef] [PubMed]
4. Wikby, A.; Nilsson, B.O.; Forsey, R.; Thompson, J.; Strindhall, J.; Lofgren, S.; Ernerudh, J.; Pawelec, G.; Ferguson, F.; Johansson, B. The immune risk phenotype is associated with IL-6 in the terminal decline stage: Findings from the Swedish NONA immune longitudinal study of very late life functioning. *Mech. Ageing Dev.* **2006**, *127*, 695–704. [CrossRef] [PubMed]
5. Dumitriu, I.E. The life (and death) of CD4 + CD28(null) T cells in inflammatory diseases. *Immunology* **2015**, *146*, 185–193. [CrossRef] [PubMed]
6. Tokoyoda, K.; Hauser, A.E.; Nakayama, T.; Radbruch, A. Organization of immunological memory by bone marrow stroma. *Nat. Rev. Immunol.* **2010**, *10*, 193–200. [CrossRef]
7. Okhrimenko, A.; Grün, J.R.; Westendorf, K.; Fang, Z.; Reinke, S.; von Roth, P.; Wassilew, G.; Kuhl, A.A.; Kudernatsch, R.; Demski, S.; et al. Human memory T cells from the bone marrow are resting and maintain long-lasting systemic memory. *Proc. Natl. Acad. Sci. USA* **2014**, *111*, 9229–9234. [CrossRef]
8. Herndler-Brandstetter, D.; Landgraf, K.; Tzankov, A.; Jenewein, B.; Brunauer, R.; Laschober, G.T.; Parson, W.; Kloss, F.; Gassner, R.; Lepperdinger, G.; et al. The impact of aging on memory T cell phenotype and function in the human bone marrow. *J. Leukoc Biol.* **2012**, *91*, 197–205. [CrossRef]
9. Li, J.; Valentin, A.; Ng, S.; Beach, R.K.; Alicea, C.; Bergamaschi, C.; Felber, B.K.; Pavlakis, G.N. Differential effects of IL-15 on the generation, maintenance and cytotoxic potential of adaptive cellular responses induced by DNA vaccination. *Vaccine* **2015**, *33*, 1188–1196. [CrossRef]

10. Osborne, L.C.; Dhanji, S.; Snow, J.W.; Priatel, J.J.; Ma, M.C.; Miners, M.J.; Teh, H.; Goldsmith, M.A.; Abraham, N. Impaired CD8 T cell memory and CD4 T cell primary responses in IL-7R alpha mutant mice. *J. Exp. Med.* **2007**, *204*, 619–631. [CrossRef]
11. Pangrazzi, L.; Meryk, A.; Naismith, E.; Koziel, R.; Lair, J.; Krismer, M.; Trieb, K.; Grubeck-Loebenstein, B. "Inflamm-aging" influences immune cell survival factors in human bone marrow. *Eur. J. Immunol.* **2017**, *47*, 481–492. [CrossRef] [PubMed]
12. Pangrazzi, L.; Naismith, E.; Meryk, A.; Keller, M.; Jenewein, B.; Trieb, K.; Grubeck-Loebenstein, B. Increased IL-15 Production and Accumulation of Highly Differentiated CD8+ Effector/Memory T Cells in the Bone Marrow of Persons with Cytomegalovirus. *Front. Immunol.* **2017**, *8*, 715. [CrossRef] [PubMed]
13. Naismith, E.; Pangrazzi, L. The impact of oxidative stress, inflammation, and senescence on the maintenance of immunological memory in the bone marrow in old age. *Biosci. Rep.* **2019**, *39*, BSR20190371. [CrossRef] [PubMed]
14. McKee, A.S.; Munks, M.W.; MacLeod, M.K.; Fleenor, C.J.; Van Rooijen, N.; Kappler, J.W.; Marrack, P. Alum induces innate immune responses through macrophage and mast cell sensors, but these sensors are not required for alum to act as an adjuvant for specific immunity. *J. Immunol.* **2009**, *183*, 4403–4414. [CrossRef]
15. Lehmann, D.; Karussis, D.; Misrachi-Koll, R.; Shezen, E.; Ovadia, H.; Abramsky, O. Oral administration of the oxidant-scavenger N-acetyl–cysteine inhibits acute experimental autoimmune encephalomyelitis. *J. Neuroimmunol.* **1994**, *50*, 35–42. [CrossRef]
16. Joshi, N.S.; Cui, W.; Chandele, A.; Lee, H.K.; Urso, D.R.; Hagman, J.; Gapin, L.; Kaech, S.M. Inflammation directs memory precursor and short-lived effector CD8(+) T cell fates via the graded expression of T-bet transcription factor. *Immunity* **2007**, *27*, 281–295. [CrossRef]
17. Remmerswaal, E.; Hombrink, P.; Nota, B.; Pircher, H.; Ten Berge, I.; van Lier, R.; van Aalderen, M.C. Expression of IL-7Rα and KLRG1 defines functionally distinct CD8+ T-cell populations in humans. *Eur. J. Immunol.* **2019**, *49*, 694–708. [CrossRef]
18. Guilliams, M. Dendritic cells, monocytes and macrophages: A unified nomenclature based on ontogeny. *Nat. Rev. Immunol.* **2014**, *14*, 571–578. [CrossRef]
19. Van der Merwe, P.A.; Bodian, D.L.; Daenke, S.; Linsley, P.; Davis, S.J. CD80 (B7–1) binds both CD28 and CTLA-4 with a low affinity and very fast kinetics. *J. Exp.* **1997**, *185*, 393–403. [CrossRef]
20. Ma, D.Y.; Clark, E.A. The role of CD40 and CD154/CD40L in dendritic cells. *Semin. Immunol.* **2009**, *21*, 265–272. [CrossRef]
21. Boraschi, D.; Aguado, M.T.; Dutel, C.; Goronzy, J.; Louis, J.; Grubeck-Loebenstein, B.; Rappuoli, R.; del Giudice, G. The gracefully aging immune system. *Sci. Transl. Med.* **2013**, *5*, 185. [CrossRef] [PubMed]
22. Pangrazzi, L.; Reidla, J.; Carmona Arana, J.A.; Naismith, E.; Miggitsch, C.; Meryk, A.; Keller, M.; Krause, A.A.N.; Melzer, F.L.; Trieb, K. CD28 and CD57 define four populations with distinct phenotypic properties within human CD8 + T cells. *Eur. J. Immunol.* **2020**, *50*, 363–379. [CrossRef] [PubMed]
23. Northfield, J.; Lucas, M.; Jones, H.; Young, N.T.; Klenerman, P. Does memory improve with age? CD85j (ILT-2/LIR-1) expression on CD8 T cells correlates with 'memory inflation' in human cytomegalovirus infection. *Immunol. Cell Biol.* **2005**, *83*, 182–188. [CrossRef]
24. Naismith, E.; Pangrazzi, L.; Grasse, M.; Keller, M.; Miggitsch, C.; Weinberger, B.; Trieb, K.; Grubeck-Loebenstein, B. Peripheral antibody concentrations are associated with highly differentiated T cells and inflammatory processes in the human bone marrow. *Immun. Ageing* **2019**, *16*, 21. [CrossRef] [PubMed]
25. Rushworth, G.F.; Megson, I.L. Existing and potential therapeutic uses for N-acetylcysteine: The need for conversion to intracellular glutathione for antioxidant benefits. *Pharmacol. Ther.* **2014**, *141*, 150–159. [CrossRef] [PubMed]
26. Elbini Dhouib, I.; Jallouli, M.; Annabi, A.; Gharbi, N.; Elfazaa, S.; Lasram, M.M. A minireview on N-acetylcysteine: An old drug with new approaches. *Life Sci.* **2016**, *15*, 359–363. [CrossRef] [PubMed]
27. Sun, S.Y. N-acetylcysteine, reactive oxygen species and beyond. *Cancer Biol. Ther.* **2010**, *9*, 109–110. [CrossRef]
28. Jacob, R.A.; Sotoudeh, G. Vitamin C function and status in chronic disease. *Nutr. Clin. Care* **2002**, *5*, 66–74. [CrossRef]
29. Łukawski, M.; Dałek, P.; Borowik, T.; Foryś, A.; Langner, M.; Witkiewicz, W.; Przybylo, M. New oral liposomal vitamin C formulation: Properties and bioavailability. *J. Liposome Res.* **2020**, *30*, 227–234. [CrossRef]

30. De Flora, S.; Grassi, C.; Carati, L. Attenuation of influenza-like symptomatology and improvement of cell-mediated immunity with long-term N-acetylcysteine treatment. *Eur. Respir. J.* **1997**, *10*, 1535–1541. [CrossRef]
31. Arranz, L.; Fernández, C.; Rodríguez, A.; Ribera, J.M.; De la Fuente, M. The glutathione precursor N-acetylcysteine improves immune function in postmenopausal women. *Free Radic. Biol. Med.* **2008**, *45*, 1252–1262. [CrossRef] [PubMed]
32. Hemilä, H. Vitamin C and Infections. *Nutrients* **2017**, *9*, 339. [CrossRef] [PubMed]
33. Tokoyoda, K.; Zehentmeier, S.; Hegazy, A.N.; Albrecht, I.; Grun, J.R.; Lohning, M.; Radbruch, A. Professional memory CD4+ T lymphocytes preferentially reside and rest in the bone marrow. *Immunity* **2009**, *30*, 721–730. [CrossRef] [PubMed]
34. Crauste, F.; Mafille, J.; Boucinha, L.; Djebali, S.; Gandrillon, O.; Marvel, J.; Arpin, C. Identification of Nascent Memory CD8 T Cells and Modeling of Their Ontogeny. *Cell Syst.* **2017**, *4*, 306–317. [CrossRef] [PubMed]
35. Sallusto, F.; Lenig, D.; Förster, R.; Lipp, M.; Lanzavecchia, A. Two subsets of memory T lymphocytes with distinct homing potentials and effector functions. *Nature* **1999**, *401*, 708–712. [CrossRef]
36. Francisco, L.M.; Sage, P.T.; Sharpe, A.H. The PD-1 pathway in tolerance and autoimmunity. *Immunol. Rev.* **2010**, *236*, 219–242. [CrossRef]
37. Franceschi, C.; Bonafe, M.; Valensin, S.; Olivieri, F.; De Luca, M.; Ottaviani, E.; De Benedictis, G. Inflamm-aging. An evolutionary perspective on immunosenescence. *Ann. N. Y. Acad. Sci.* **2000**, *908*, 244–254. [CrossRef]
38. Krishnaswamy, J.K.; Gowthaman, U.; Zhang, B.; Mattsson, J.; Szeponik, L.; Liu, D.; Wu, R.; White, T.; Calabro, S.; Xu, L.; et al. Migratory CD11b+ conventional dendritic cells induce T follicular helper cell-dependent antibody responses. *Sci. Immunol.* **2017**, *2*. [CrossRef]
39. Yang, Y.; Bazhin, A.V.; Werner, J.; Karakhanova, S. Reactive oxygen species in the immune system. *Int. Rev. Immunol.* **2013**, *32*, 249–270. [CrossRef]
40. Seifirad, S.; Ghaffari, A.; Amoli, M.M. The antioxidants dilemma: Are they potentially immunosuppressants and carcinogens? *Front. Physiol.* **2014**, *5*, 245. [CrossRef]
41. Smetana, J.; Chlibek, R.; Shaw, J.; Splino, M.; Prymula, R. Influenza vaccination in the elderly. *Hum. Vaccin. Immunother.* **2018**, *4*, 540–549. [CrossRef] [PubMed]

Publisher's Note: MDPI stays neutral with regard to jurisdictional claims in published maps and institutional affiliations.

© 2020 by the authors. Licensee MDPI, Basel, Switzerland. This article is an open access article distributed under the terms and conditions of the Creative Commons Attribution (CC BY) license (http://creativecommons.org/licenses/by/4.0/).

Article

Cigarette Smoke Extract Stimulates MMP-2 Production in Nasal Fibroblasts via ROS/PI3K, Akt, and NF-κB Signaling Pathways

Joo-Hoo Park [1], Jae-Min Shin [1,2], Hyun-Woo Yang [1], Tae Hoon Kim [2], Seung Hoon Lee [2], Heung-Man Lee [2], Jae-Gu Cho [2,*] and Il-Ho Park [1,2,*]

1 Upper Airway Chronic inflammatory Diseases Laboratory, Korea University College of Medicine, Seoul 05505, Korea; pjh52763@naver.com (J.-H.P.); shinjm0601@hanmail.net (J.-M.S.); yhw444@gmail.com (H.-W.Y.)

2 Department of Otorhinolaryngology-Head and Neck Surgery, Korea University College of medicine, Seoul 05505, Korea; doctorth@korea.ac.kr (T.H.K.); shleeent@korea.ac.kr (S.H.L.); lhman@korea.ac.kr (H.-M.L.)

* Correspondence: jgcho@korea.ac.kr (J.-G.C.); parkil5@korea.ac.kr (I.-H.P.); Tel.: +82-2-2626-1298 (J.-G.C.); +82-2-2626-1298 (I.-H.P.); Fax: +82-2-868-0475 (J.-G.C.); +82-2-868-0475 (I.-H.P.)

Received: 14 July 2020; Accepted: 3 August 2020; Published: 12 August 2020

Abstract: Cigarette smoke exposure has been shown to be associated with chronic rhinosinusitis and tissue remodeling. The present study aimed to investigate the effects of cigarette smoke extract (CSE) on matrix metalloproteinase (MMP) and tissue inhibitor of metalloproteinase (TIMP) production in nasal fibroblasts and to determine the underlying molecular mechanisms. Primary nasal fibroblasts from six patients were isolated and cultured. After the exposure of fibroblasts to CSE, the expression levels of MMP-2, MMP-9, TIMP-1, and TIMP-2 were measured by real-time PCR, ELISA, and immunofluorescence staining. The enzymatic activities of MMP-2 and MMP-9 were measured by gelatin zymography. Reactive oxygen species (ROS) production was analyzed using dichloro-dihydro-fluorescein diacetate and Amplex Red assays. PI3K/Akt phosphorylation and NF-κB activation were determined by Western blotting and luciferase assay. CSE significantly increased MMP-2 expression and inhibited TIMP-2 expression but did not affect MMP-9 and TIMP-1 expression. Furthermore, CSE significantly induced ROS production. However, treatment with ROS scavengers, specific PI3K/Akt inhibitors, NF-κB inhibitor, and glucocorticosteroids significantly decreased MMP-2 expression and increased TIMP-2 expression. Our results suggest that steroids inhibit CSE-regulated MMP-2 and TIMP-2 production and activation through the ROS/ PI3K, Akt, and NF-κB signaling pathways in nasal fibroblasts. CSE may contribute to the pathogenesis of chronic rhinosinusitis by regulating MMP-2 and TIMP-2 expression.

Keywords: cigarette smoke extract; nasal fibroblasts; tissue inhibitor of metalloproteinases; matrix metalloproteinase; steroids

---

1. Introduction

Chronic rhinosinusitis (CRS) is a nasal inflammatory disease with symptoms of nasal discharge/postnasal drip, nasal congestion, sinus pain/pressure, and anosmia/hyposmia lasting for at least 12 weeks. CRS is one of the most frequent chronic diseases in humans and consequently has an important socio-economic impact. Its influence on patient quality of life is even more detrimental than that of congestive heart failure, chronic obstructive pulmonary disease, and back pain [1]. CRS is a multifactorial disorder, and its pathogenesis involves interactions between environmental insults, infectious loading, and genetic predisposition. Among environmental factors, inhaled pollutants, including cigarette smoke, may play a significant role in CRS, which is a characteristic of chronic inflammatory upper airway disease [2].

Tissue remodeling is an energetic process that results in both production and degradation of the extracellular matrix (ECM) and is an important aspect in the pathogenesis of chronic inflammatory diseases in many organs. In airways, tissue remodeling is characterized by loss of epithelial integrity, goblet cell metaplasia, excessive ECM deposition, hyperplasia of mucosal cells, and basement membrane thickening. The differentiation of fibroblasts to myofibroblast is a key event in physiological and pathological tissue remodeling. Myofibroblasts are a source of matrix metalloproteinases (MMPs) and tissue inhibitors of MMPs (TIMPs) [3]. Like other chronic inflammatory diseases of the airway, tissue remodeling is present in CRS, and obvious remodeling features differentiate the various CRS subgroups [4]. Several factors, such as MMPs, ECM, and TGF-β, are related with remodeling. Specifically, MMP-2 and MMP-9 have been associated with airway inflammatory diseases [5]. MMPs are a large family of enzymes with zinc-binding catalytic domains and are involved in the degradation of ECM components. Their extracellular activities are regulated by TIMPs. MMPs play a crucial role in various physiological processes, including tissue remodeling [6]. MMP-2 (gelatinase A) and MMP-9 (gelatinase B) are involved in ECM remodeling, which is associated with upper airway remodeling [7]. MMP-2 and MMP-9 are the principal fibroblast-derived proteinases capable of degrading substrates, including collagen, gelatin, elastin, and fibronectin.

Chronic exposure to cigarette smoke is known to be the cause of several inflammatory diseases including asthma, chronic obstructive pulmonary disease, and CRS. Cigarette smoke extract (CSE) exerts harmful effects on processes, including cell viability, adhesion, migration, and myofibroblast differentiation, in the airway. Tissue remodeling is a pathologic process believed to be related with cigarette smoke. It was previously shown that smoke can directly induce remodeling without any need for exogenous inflammatory cells in the airway [8].

CSE contains many toxic and carcinogenic chemicals, as well as unsuitable free radical that enhance reactive oxygen species (ROS) production leading to oxidative stress. ROS is known to be higher expressed in patients with CRS was higher than in control subject [9]. It damages proteins, lipids, and DNA that plays an important roles in cellular process involved in the generation and development processes of nasal polyps [10]. Airway inflammation, airway hyper-responsiveness, tissue injury, and remodeling can be induced by excessive ROS production in epithelial cells, fibroblasts and severer inflammatory cells [11,12]. Increased ROS can mediate activation of AKT and NF-κB [13]. PI3K and Akt proteins in CRS were higher than those in the control subjects [14]. The PI3K/Akt signaling pathway is an important cellular signaling that is involved in cell growth, proliferation, apoptosis, metabolism, angiogenesis, metastasis and the cellular defiance against inflammatory stimuli. Akt cascade is known to mediate ECM proteolysis [15]. NF-κB is a transcriptional factor which plays a central role in diverse cellular processes, including inflammation and immune response in airway diseases. CSE modulates variety molecular mechanisms such as ROS, PI3K/AKT and NF-κB. However, the precise mechanisms by which cigarette smoke affects airway structure and function are still under investigation.

We hypothesized that CSE-induced tissue remodeling in the upper airway is related with an imbalance between MMPs and TIMPs from nasal fibroblasts. In the present study, we aimed to determine the effects of CSE on the expression of MMPs and TIMPS and the underlying molecular mechanisms in nasal fibroblasts. We also examined the involvement of ROS, AKT, and NF-κB signaling pathways, which are known to be closely related with CSE-induced inflammation in several diseases, in these mechanisms. Additionally, we aimed to determine the effect of glucocorticoids, the first line of treatment for CRS, on the CSE-induced imbalance between MMPs/TIMPs in nasal fibroblasts.

2. Materials and Methods

2.1. Preparation of CSE

CSE was obtained from 3R4F research cigarettes (University of Kentucky, Lexington, KY, USA), with each cigarette containing 0.60 mg of nicotine and 8.0 mg of tar, for use in all experiments. CSE was prepared by bubbling smoke from cigarettes into 15 mL of serum-free Dulbecco's modified Eagle's medium (DMEM; Invitrogen, Grand Island, NY, USA) at a rate of 1 cigarette/min using a modification of the method developed by Carp and Janoff [16]. The CSE solution is defined as 100% CSE and diluted with DMEM in the following experiments. CSE was standardized by measuring of the absorbance at 320 nm confirmed that the prepared CSE was reproducible, and various CSE preparations showed few differences. Freshly prepared CSE was used immediately in all the experiments.

2.2. Patients and Tissue Collection

Sinus tissue explants were collected during surgery from patients with blowout fracture (n = 6) at the Department of Otorhinolaryngology at Korea University Guro Hospital, Korea. All patients had no history of smoking, allergies, asthma, or aspirin sensitivity, and they were not treated with any antibiotics or oral or topical steroids for at least 4 weeks before surgery. Written consent was obtained from all patients, and the study was approved by the Ethics Committee of the Faculty of Medicine, Korea University, Korea.

2.3. Nasal Fibroblast Cultures

To obtain nasal fibroblasts, cells were cultured in DMEM with 10% heat-inactivated fetal bovine serum (FBS), 1% (*v/v*) 10,000 U/mL penicillin, and 10,000 μg/mL streptomycin (Invitrogen) in a humidified incubator under 5% CO_2 at 37 °C. Experiments were performed using cells at 80% confluence. Before treatment of agents, the cells were starved in serum-free media for 12 h. The purity of obtained human nasal fibroblasts was confirmed microscopically based on the characteristic spindle-like cell phenotype. Approximately 95% of cells in cultured nasal fibroblasts were positive for vimentin and Thy-1, which were used as fibroblast markers, and negative for E-cadherin, which was used as an epithelial cell marker. The nasal fibroblasts were cultured for four passages [17].

2.4. Real-Time PCR

The nasal fibroblasts were incubated with CSE for 12 h after pre-treatment of ROS scavengers (NAC, ebselen and DPI) PI3K/Akt inhibitor (LY294002) and NF-κB inhibitor (BAY11-7082) for 1 h. RNAs were extracted from nasal fibroblasts using TRIzol reagent (Invitrogen, Carlsbad, CA, USA). The total amount of RNA was determined using NanoDrop 2000 (Thermo Fisher Scientific Inc., Wilmington, DE, USA). Synthesis of cDNA was performed with 1 μg of total RNA using the Maxime RT PreMix cDNA Kit (iNtRON Biotechnology, Sungnam, Korea). The expression levels of mRNAs were analyzed using Quantstudio3 (Applied Biosystems, Foster City, CA, USA) and Power SYBR Green PCR Master Mix (Applied Biosystems). Real-time PCR analysis was performed to evaluate the expression levels of *MMP-2, MMP-9, TIMP-1,* and *TMIP-2* and that of *GAPDH,* which was used as a housekeeping gene. The primer sequences are shown in Table 1. The results were normalized by *GAPDH* mRNA expression and are shown as the fold ratio over the expression of the control group.

Table 1. Sequences of real-time PCR oligonucleotide primers.

Primer		Sequence
MMP-2	Forward	5′- AGA TCT TCT TCT TCA AGG AAC CGT T -3′
	Reverse	5′- GGC TGG TCA GTG GCT TGG GGT A -3′
MMP-9	Forward	5′- GCG GAG ATT GGG AAC CAG CTG TA -3′
	Reverse	5′- GAC GCG CCT GTG TAC ACC CAC A -3′
TIMP-1	Forward	5′- ACC ACC TTA TAC CAG CGT TAT GA -3′
	Reverse	5′- GGT GTA GAC GAA CCG GAT GTC -3′
TIMP-2	Forward	5′- GCT GCG AGT GCA AGA TCA C -3′
	Reverse	5′- TGG TGC CCG TTG ATG TTC TTC
TLR4	Forward	5′- TGA GCA GTC GTG CTG GTA TC -3′
	Reverse	5′- CAG GGC TTT TCT GAG TCG TC -3′

2.5. Gelatin Zymography

Nasal fibroblasts were exposed to CSE for 72 h after pretreatment with ROS scavengers, PI3K/Akt inhibitor and NF-κB inhibitor for 1 h. Aliquots of fibroblast-conditioned medium (10 μL) were analyzed using gelatin zymography for MMP-2 and MMP-9 in 1 mg/mL gelatin-10% polyacrylamide gels. Following electrophoresis, the gels were washed twice with 2.5% Triton X-100 for 30 min while shaking to remove sodium dodecyl sulfate and renature MMP-2 and MMP-9 in the gels. Renatured gels were incubated in developing buffer containing 50 mM Tris-HCl (pH 7.5), 200 mM NaCl, 5 mM $CaCl_2$, and 0.02% Brij-35 overnight at 37 °C. Gels were stained with 0.25% Coomassie brilliant blue G-250 (50% methanol, 10% acetic acid) and destained using destaining solution (50% methanol, 10% acetic acid). Proteinase activity was observed as cleared (unstained) regions on the gels. Finally, the gels were dried for 2 h using a gel dryer (Bio-Rad, Hercules, CA, USA).

2.6. Enzyme-Linked Immunosorbent Assay (ELISA)

MMP-2, MMP-9, TIMP-1, and TMIP-2 concentrations in the culture media were determined using ELISA (R&D systems, Minneapolis, MN, USA). Nasal fibroblasts were exposed to CSE for 72 h after pretreatment with ROS scavengers, PI3K/Akt inhibitor and NF-κB inhibitor for 1 h. Standards and samples were added and incubated at room temperature for 2 h. After 3 washes, MMP-2, MMP-9, TIMP-1, or TMIP-2 conjugate was added to the wells for 2 h at room temperature. The reaction was stopped with a stop solution, and the product was quantified at 450 nm using a microplate reader (Bio-Rad).

2.7. Western Blotting Analysis

Nasal fibroblasts were treated with CSE for 72 h. A total of 5×10^5 fibroblasts were lysed in PRO-PREP™ protein extraction solution (iNtRON Biotechnology) and stored overnight at −20 °C. Cell debris was removed from the lysates by centrifugation at 13,000× *g* for 30 min at 4 °C. Total protein concentration was determined using the Bradford assay (Bio-Rad). An equal quantity of protein from samples (30 μg) were separated using 12% sodium dodecyl sulfate polyacrylamide gel electrophoresis, transferred to 0.45 μm polyvinyl difluoride membranes, (Millipore Inc., Billerica, MA, USA), and analyzed separately. Membranes were blocked with 5% skim milk at room temperature for 60 min, rinsed three times with Tris-buffered saline containing Tween-20, and treated with the following primary antibodies: polyclonal p-PI3K (1:1000, #4228, Cell Signaling Technology, Danvers, MA, USA), total-PI3K (1:1000, #4292, Cell Signaling Technology), p-Akt (1:1000, #9271, Cell Signaling Technology), total-Akt (1:1000, #9272, Cell Signaling Technology), p-p65 (1:1000, sc-136548, Santa Cruz Biotechnology Inc., Santa Cruz, CA, USA), total-p65 (1:1000, sc-8008, Santa Cruz Biotechnology Inc.) and β-actin (1:10000, sc-47778, Santa Cruz Biotechnology Inc.). Bands were visualized using horseradish peroxidase-conjugated secondary antibodies and an enhanced chemiluminescence system (Pierce, Rockford, IL, USA).

2.8. Immunofluorescent Staining

Nasal fibroblasts were placed onto coverslips and treated with CSE for 72 h. Fibroblasts were fixed with 4% paraformaldehyde and then permeated with 0.2% Triton X-100 in 1% FBS for 10 min at room temperature. After treating coverslips with 5% BSA to block for 1 h at room temperature, fibroblasts were incubated overnight at 4 °C with anti-MMP-2, anti-MMP-9, anti-TIMP-1, or anti-TIMP-2 antibodies (Santa Cruz Biotechnology, Inc. CA, USA). Goat anti-mouse Alexa 488 (Invitrogen) secondary antibody was also added to fibroblasts and incubated. Lastly, 4′-6-diamidino-2-phenylindole (DAPI) was applied for counterstaining. The stained normal fibroblasts were subsequently maintained on a confocal laser scanning microscope (LSM700, Zeiss, Oberkochen, Germany).

2.9. Measurement of ROS Produced

Intracellular ROS production was determined using a fluorescent probe, 2, 7-dichlorodihydrofluorescein diacetate (Molecular Probes, Inc., Eugene, OR, USA). Cells were pretreated with N-acetyl-L-cysteine (NAC, 5 mM), ebselen (10 μM), or diphenyleneiodonium (DPI, 2 μM) for 1 h and then stimulated with CSE for 12 h. After that, CSE-stimulated cells were suspended in serum-free culture medium with H2DCF-DA (10 μM) for 30 min. Cellular fluorescence was measured using fluorescence microscopy (IX71; Olympus, Life Science Solutions, Tokyo, Japan) to observe ROS production. The H_2O_2 production was measured using an Amplex Red Hydrogen Peroxide/Peroxidase Assay Kit (Thermo Fisher Scientific Inc.), according to the manufacturer's instructions. Nasal fibroblasts were pretreated with NAC, ebselen, or DPI. Prior to CSE stimulation, the cell lysate (50 μL) was treated with a working solution of 100 μM Amplex Red reagent (Thermo Fisher Scientific Inc.) and 0.2 U/mL horseradish peroxidase. After incubation for 30 min at 37 °C, fluorescence was measured at 560 nm on a microplate reader (Bio-Rad).

2.10. Statistical Data Analysis

For all outcomes, data were obtained in triplicate from at least three separate experiments. Data are shown as mean ± SEM of three different experiments in triplicate. All significant differences between controls and examined samples were analyzed by one-way analysis of variance followed by Tukey's test (GraphPad Prism version 5; GraphPad Software, San Diego, CA, USA). Significance was considered at a 95% confidence level. P-values below 0.05 were considered statistically significant.

3. Results

3.1. Effects of CSE on MMP and TIMP Production in Nasal Fibroblasts

To determine whether CSE regulates the production of MMPs and TIMPs, we treated fibroblasts with CSE at various concentrations (0–5%). CSE increased MMP-2 expression dose-dependently but did not affect the expression of MMP-9 mRNA and protein (Figure 1A,B). Gelatin zymography, used to evaluate enzymatic activity, showed that only MMP-2 activity was increased by CSE treatment (Figure 1C). The stimulatory effect of CSE on MMP-2 protein was also confirmed by immunofluorescence staining (Figure 1D). CSE decreased TIMP-2 expression but did not affect the expression of TIMP-1 (Figure 1E,F). These results indicate that CSE induced MMP-2 production and decreased TIMP-2 production in nasal fibroblasts.

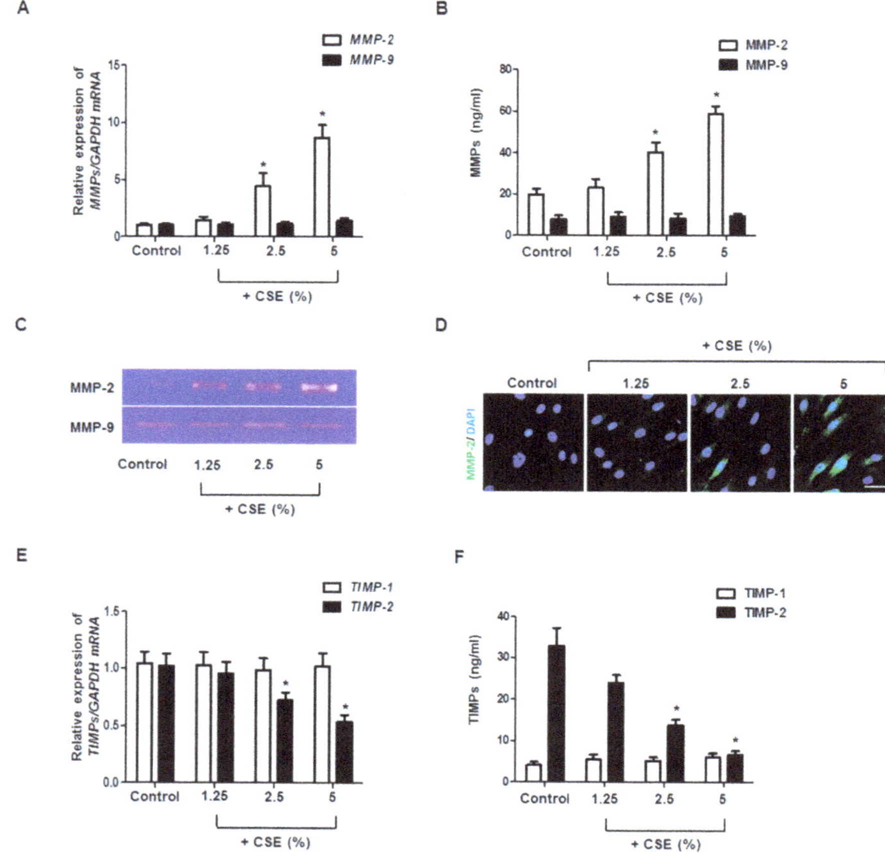

Figure 1. Effect of CSE on MMP and TIMP expression in nasal fibroblasts. After treatment of nasal fibroblasts with 5% CSE, the mRNA levels of *MMPs* and *TIMPs* were measured using real-time PCR (**A**,**E**), and MMP and TIMP protein expression levels were determined using ELISA (**B**,**F**). The enzymatic activities of MMP-2 and MMP-9 were measured by gelatin zymography (**C**). The expression and localization of MMP-2 protein (green) were observed using immunofluorescence staining (**D**). * $p < 0.05$ vs. control. Scale bar = 100 μm.

3.2. Role of ROS in MMP and TIMP Production in CSE-Stimulated Nasal Fibroblasts

To investigate the role of ROS in MMP-2 and TIMP-2 production, nasal fibroblasts were pretreated with ROS scavengers 1 h before CSE treatment. In DCFH-DA and Amplex Red staining assays, CSE treatment induced the production of ROS and hydrogen peroxide; however, such production was blocked by the addition of NAC, ebselen, or DPI (Figure 2A,B). Next, the effects of ROS scavengers on mRNA and protein expression were measured using real-time PCR and western blotting. Pretreatment with ROS scavengers blocked the stimulatory effects of CSE on MMP-2 expression (Figure 2C,D). This result was also observed through gelatin zymography (Figure 2E). Inversely, pretreatment with antioxidants blocked the inhibitory effect of CSE on TIMP-2 expression (Figure 2F,G). MMP-9 and TIMP-1 production was not affected by ROS scavengers in nasal fibroblasts.

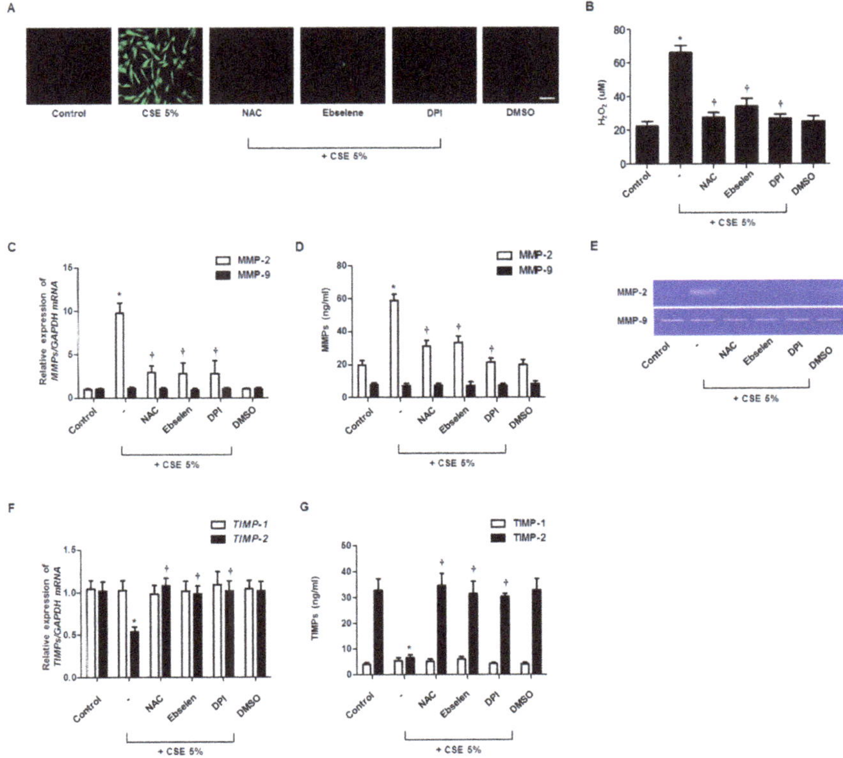

Figure 2. Effects of ROS on CSE-regulated MMP and TIMP expression in nasal fibroblasts. Total ROS and mitochondrial superoxides were quantified using the 2,7-dichlorofluorescein diacetate assay and Amplex Red assay (**A**,**B**). Fibroblasts were pretreated with NAC (1 mM), ebselen (10 μM), and DPI (2 μM) before being treated with CSE (5%). The expression levels of MMP and TIMP mRNAs were determined by real-time PCR (**C**,**F**). MMP and TIMP protein levels were determined using ELISA (**D**,**G**). The enzymatic activities of MMP-2 and MMP-9 were measured by gelatin zymography (**E**). * $p < 0.05$ vs. control; † $p < 0.05$ vs. CSE only. Scale bar = 50 μm.

3.3. Involvement of the PI3K/Akt Cascade in MMP and TIMP Production in CSE-Stimulated Nasal Fibroblasts

The fibroblasts were treated with CSE and a PI3K/Akt inhibitor to confirm whether a PI3K/Akt pathway is involved in the expression of MMP-2 and TIMP-2. CSE induced PI3K/Akt phosphorylation, which was inhibited by the PI3K/Akt inhibitor (LY294002) (Figure 3A). Next, we confirmed that the PI3K/Akt inhibitor suppressed MMP-2 mRNA and protein expression and enzymatic activation, which are induced by CSE in nasal fibroblasts (Figure 3B–D). On the contrary, TIMP-2 mRNA and protein expression was significantly increased by treatment with the PI3K/Akt inhibitor in CSE-stimulated nasal fibroblasts (Figure 3E,F). Additionally, we confirmed that ROS inhibition suppressed the activation of PI3K/AKT in nasal fibroblasts (data not shown). Therefore, we could assume that the signaling pathway associated with fibroblast activation under oxidative stress is attributable in part to the activation of the PI3K and AKT signaling pathways.

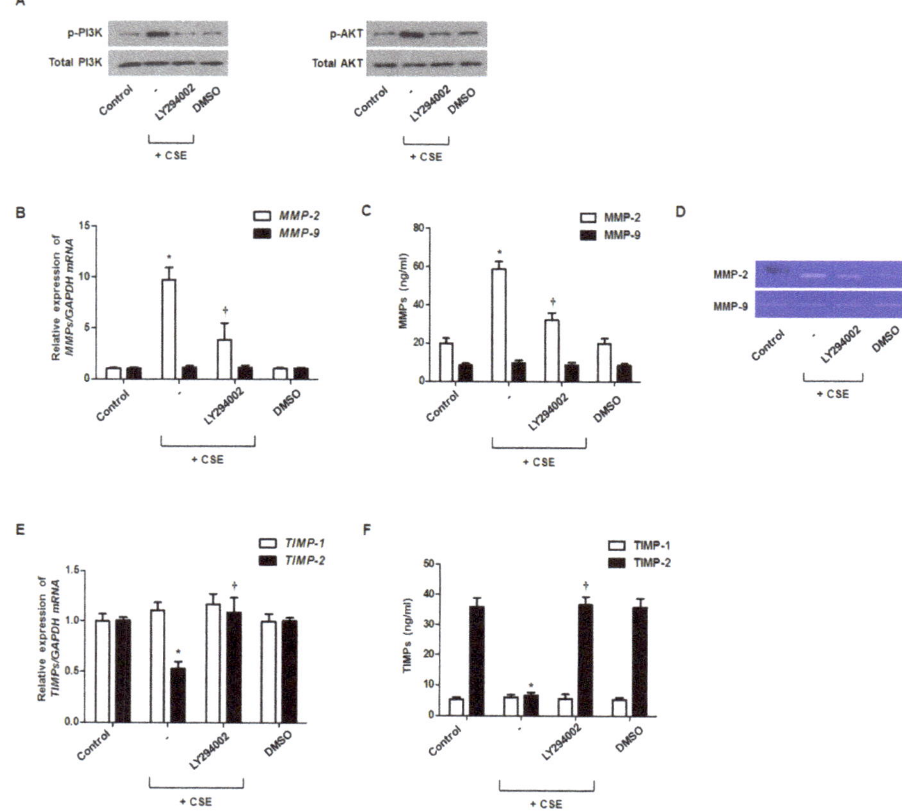

Figure 3. Regulation of PI3K/Akt signaling pathways with CSE regulated MMP and TIMP expression. Nasal fibroblasts were pretreated with LY294002 (PI3K/Akt inhibitor) before treatment with 5% CSE. Levels of phosphorylated (p)-PI3K and p-Akt were determined using western blotting (**A**). The expression levels of MMP and TIMP mRNAs were determined by real-time PCR (**B**,**E**). MMP and TIMP protein levels were determined using ELISA (**C**,**F**). The enzymatic activities of MMP-2 and MMP-9 were measured by gelatin zymography (**D**). * $p < 0.05$ vs. control; † $p < 0.05$ vs. CSE only.

3.4. Effect of CSE on NF-κB Activation for MMP and TIMP Production

The activated NF-κB enhanced MMP-2 expression in airway. To determine whether NF-κB activation is involved in MMP-2 and TIMP-2 expression, fibroblasts were pretreated with an NF-κB inhibitor (BAY11-7082) and then stimulated with CSE. Phosphorylated p65, a subunit of NF-κB, was induced by CSE treatment and inhibited by treatment with the NF-κB inhibitor (Figure 4A). NF-κB transcriptional activity, assessed by a luciferase reporter, was increased by CSE and then inhibited by the NF-κB inhibitor (Figure 4B). Immunocytochemical staining showed that CSE induced the translocation of p-p65 to the nucleus and that this was blocked by the NF-κB inhibitor (Figure 4C,D). Treatment with the NF-κB inhibitor inhibited MMP-2 mRNA and protein expression and activation of enzymatic capacity, which were stimulated by CSE (Figure 4E–G), and reversed the change in TIMP-2 mRNA and protein expression that had been inhibited by CSE treatment (Figure 4H,I). MMP-9 and TIMP-1 production was not affected by the NF-κB inhibitor in nasal fibroblasts.

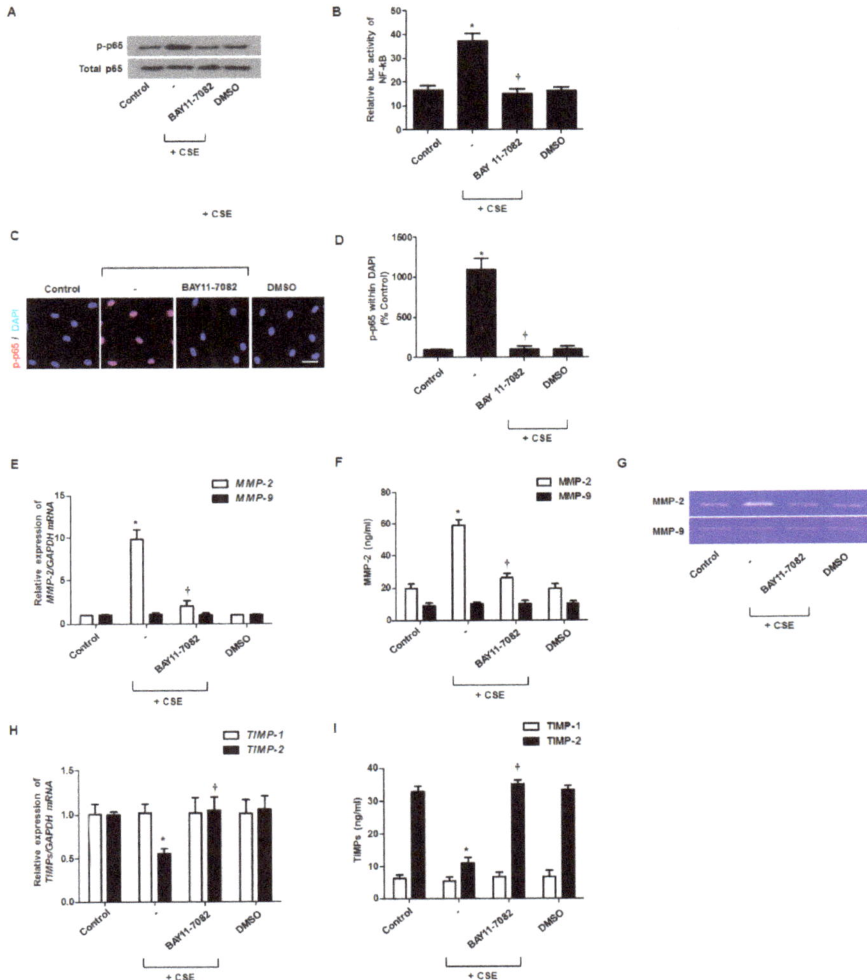

Figure 4. Effects of NF-κB activation on CSE-regulated MMP and TIMP levels in nasal fibroblasts. Nasal fibroblasts were pretreated with an NF-κB inhibitor (BAY 11-7082) before treatment with 5% CSE. Phospho (p)-p65 was measured using western blotting to determine NF-κB activation (**A**). NF-κB transcriptional activation was measured using the luciferase assay (**B**). After stimulation with CSE, the translocation of p-p65 protein (red) was observed by immunofluorescent staining. Magnification, ×400 (**C**,**D**). Nuclei were stained using DAPI (blue). The expression levels of MMP and TIMP mRNAs were determined by real-time PCR (**E**,**F**). The secretion levels of MMP and TIMP proteins were determined using ELISA (**H**,**I**). The enzymatic activities of MMP-2 and MMP-9 were measured by gelatin zymography (**G**). * $p < 0.05$ vs. control; † $p < 0.05$ vs. CSE only. Each experiment was performed three biological replicates. Scale bar = 100 μm.

3.5. Effect of Steroids on CSE-Regulated MMP and TIMP Production

Dexamethasone (Dex) and fluticasone propionate (FP) are potent synthetic corticosteroids that are widely used as anti-inflammatory agents to treat respiratory diseases [18]. To assess whether steroids inhibited CSE-regulated MMP and TIMP production, fibroblasts were pretreated with dexamethasone or fluticasone propionate and then stimulated with CSE. MMP-2 expression was increased by CSE treatment and suppressed by steroids, whereas the opposite pattern was observed for TIMP-2 mRNA (Figure 5A,D) and protein levels (Figure 5B,E). Gelatin zymography showed similar patterns for MMP-2 (Figure 5C). Steroids did not affect MMP-9 and TIMP1 expression. Steroids significantly decreased ROS production, PI3K/Akt phosphorylation, and NF-κB activation in CSE-stimulated nasal fibroblasts (Figure 5F–H). These results suggested that steroids could regulate MMP-2 and TIMP-2 expression by blocking the ROS/PI3K/Akt and NF-κB signaling pathways in fibroblasts.

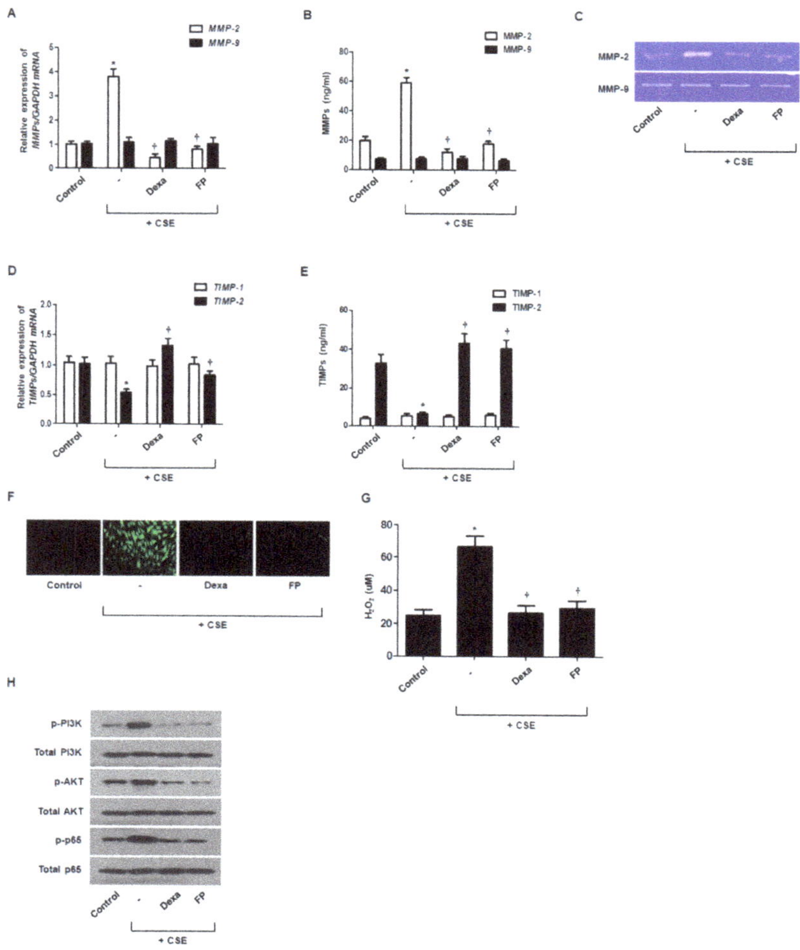

Figure 5. Effects of steroids on CSE-regulated MMP and TIMP levels in nasal fibroblasts. Nasal fibroblasts were pretreated with or without dexamethasone (Dexa, 2.5 μM) and fluticasone propionate (FP, 2.5 μM) before treatment with 5% CSE. The expression levels of MMP and TIMP mRNAs were determined by real-time PCR (**A**,**D**). MMP and TIMP protein secretion levels were determined using ELISA (**B**,**E**). The enzymatic activities of MMP-2 and MMP-9 were measured by gelatin zymography (**C**). Total ROS production and mitochondrial superoxides were quantified using the 2,7-dichlorofluorescein diacetate assay and Amplex Red assay (**F**,**G**). Levels of phosphorylated p-PI3K, p-Akt, and p-p65 were determined using western blotting (**H**). * $p < 0.05$ vs. control; † $p < 0.05$ vs. CSE only. Scale bar = 50 μm.

4. Discussion

The present study showed that CSE induced MMP-2 expression and decreased TIMP-2 expression but did not affect MMP-9 and TIMP-1 expression in nasal fibroblasts. CSE exposure induced ROS production. Treatment with ROS scavengers, such as NAC, ebselen, and DPI, suppressed CSE-regulated MMP-2 and TIMP-2 expression. Additionally, CSE induced PI3K/AKT phosphorylation and NF-κB activation. When CSE-stimulated nasal fibroblasts were treated with PI3K/AKT and NF-κB inhibitors, the CSE-mediated regulation of MMP-2 and TIMP-2 expression was suppressed in nasal fibroblasts. These data indicated that CSE induces MMP-2 expression and inhibits TIMP-2 expression via ROS, PI3K/AKT, and NF-κB signaling pathways in nasal fibroblasts.

Histopathological changes in the nasal mucosa of smokers are reported to differ from those in the nasal mucosa in nonsmokers [19]. In several studies, CSE exposure in airways increased inflammatory responses and tissue remodeling to aggravate chronic upper respiratory inflammation and diseases, such as CRS [2,20,21]. In CRS, histomorphological changes, including epithelial cell hyperplasia, basement membrane thickening, and ECM accumulation, occur in the respiratory system, leading to tissue remodeling [22]. An imbalance in MMPs and TIMPs, which are important factors involved in ECM homeostasis, leads to tissue remodeling [23]. Particularly, gelatinases, MMP-2, and MMP-9 are known to degrade almost all basement membrane components, including collagens, laminins, and gelatins, in tissue remodeling [24]. Bachert et al. showed that MMP-2 and MMP-9 expression was significantly enhanced in CRS patients compared to that in healthy controls [25]. We have previously reported that MMP-2 expression was increased in TGF-β1-stimulated nasal polyp-derived fibroblasts [26]. The present study demonstrated that CSE induces MMP-2 expression and inhibits TIMP-2 expression in nasal fibroblasts.

CSE contains high concentrations of oxidants that induce ROS, which play a significant role in the pathogenesis of diseases, such as CRS [27]. In particular, CSE-induced ROS may promote tissue remodeling. Fordham et al. showed that ROS were increased in the epithelia of patients with CRS, and exposure to CSE increased ROS in nasal tissues [27]. Thus, we showed that CSE induced an imbalance in MMP-2 and TIMP-2 through ROS in nasal fibroblasts. Additionally, ROS scavengers, including NAC, ebselen, and DPI, inhibited not only the imbalance in MMP-2 and TIMP-2 production, but also PI3K/Akt phosphorylation and NF-κB activation in CSE-stimulated nasal fibroblasts. PI3K/Akt phosphorylation and NF-κB activation are known to occur in inflammatory response progression [28]. Previous studies suggested that the PI3K/Akt pathway regulates various signaling pathways that lead to NF-κB activation [29]. CSE also caused PI3K/Akt phosphorylation and NF-κB activation in these signaling pathways in lung fibroblasts [30,31]. Intracellular signaling pathways, such as PI3K/Akt and NF-κB, have been shown to modulate MMP-2 and MMP-9 expression in lung tissues [32]. In our study, the inhibition of PI3K/Akt and NF-κB significantly suppressed CSE-enhanced MMP-2 expression and inhibited TIMP-2 expression in nasal fibroblasts, which might be associated with CRS aggravation. These findings correspond well with those of earlier studies, which indicated that the PI3K/Akt and NF-κB signaling pathways play a crucial role in MMP-2 and TIMP-2 expression.

Although, steroids have severe side effects, they are still one of the most important drugs that can treat various diseases with anti-inflammatory and antioxidant properties. Steroids, such as dexamethasone and fluticasone propionate, have been successfully used in the therapy of various human inflammatory diseases, such as CRS. Steroids may also reduce mucosal inflammation and edema in paranasal sinuses and improve symptoms associated with CRS. However, recent evidence showed that glucocorticoids may affect not only the prevention of inflammation, but also the inhibition of tissue remodeling in CRS. It was shown that inhaled steroids inhibited TGF-β1-induced MMP expression in bronchial fibroblasts [33]. Steroids have been known to inhibit inflammation response and tissue remodeling by inhibiting various signaling pathway such as ROS and AKT signaling pathways [34,35]. Our study demonstrated that steroids down-regulated CSE-induced MMP-2 expression through inhibiting the ROS/PI3K/Akt and NF-κB signaling pathways. These results suggested that steroids contributed to the inhibition of MMP-2 expression in nasal fibroblasts.

Our study has some limitations. CRS is a multifactorial disorder, and its pathogenesis involves interactions between environmental insults, infectious loading, and genetic predisposition, it is not possible to directly translate the obtained results into the CRS model. However, indirectly, it may be thought that the regulation of MMP may influence tissue remodeling and thus contribute to the development of CRS. In addition to, we did not check the effect of CSE in CRS mouse model. We will check whether the exposure of CSE can aggravate symptoms of CRS.

5. Conclusions

We have provided here the first evidence that CSE increases MMP-2 production and inhibits TIMP-2 expression in nasal fibroblasts. However, CSE did not affect MMP-9 and TIMP-1 expression in nasal fibroblasts. Additionally, our study demonstrated the role of the ROS/PI3K/Akt and NF-κB signaling pathways in mediating the CSE-regulated MMP-2/TIMP-2 imbalance in nasal fibroblasts, which might contribute to tissue remodeling in CRS. The present study did not demonstrate the effect of CSE on MMP and TIMP expression, which was only examined in one type of submerged cell culture. Additional experiments are needed to clarify the results of the present study.

Author Contributions: Conceptualization, J.-H.P., J.-M.S., and H.-M.L.; Data curation, J.-H.P. and J.-G.C.; Funding acquisition, J.-G.C. and I.-H.P.; Investigation, J.-H.P., J.-M.S., and H.-M.L.; Methodology, J.-H.P. and I.-H.P.; Validation, J.-H.P. and H.-W.Y.; Writing—original draft, J.-H.P. and I.-H.P.; Writing—review & editing, T.H.K., S.H.L., J.-M.S., and I.-H.P. All authors have read and agreed to the published version of the manuscript.

Funding: This work was supported by the National Research Foundation of Korea (NRF) grant funded by the Korea Government (MSIT) (2020R1F1A1071985). This research was supported and by a Korea University grant (K1912871).

Conflicts of Interest: The authors declare no conflict of interest.

References

1. Lange, B.; Holst, R.; Thilsing, T.; Baelum, J.; Kjeldsen, A. Quality of life and associated factors in persons with chronic rhinosinusitis in the general population: A prospective questionnaire and clinical cross-sectional study. *Clin. Otolaryngol.* **2013**, *38*, 474–480. [CrossRef] [PubMed]
2. Reh, D.D.; Higgins, T.S.; Smith, T.L. Impact of tobacco smoke on chronic rhinosinusitis: A review of the literature. *Int. Forum Allergy Rhinol.* **2012**, *2*, 362–369. [CrossRef]
3. Gueders, M.M.; Foidart, J.M.; Noel, A.; Cataldo, D.D. Matrix metalloproteinases (MMPs) and tissue inhibitors of MMPs in the respiratory tract: Potential implications in asthma and other lung diseases. *Eur. J. Pharmacol.* **2006**, *533*, 133–144. [CrossRef] [PubMed]
4. Van Bruaene, N.; Bachert, C. Tissue remodeling in chronic rhinosinusitis. *Curr. Opin. Allergy Clin. Immunol.* **2011**, *11*, 8–11. [CrossRef] [PubMed]
5. Caley, M.P.; Martins, V.L.; O'Toole, E.A. Metalloproteinases and Wound Healing. *Adv. Wound Care (New Rochelle)* **2015**, *4*, 225–234. [CrossRef]
6. Kessenbrock, K.; Plaks, V.; Werb, Z. Matrix Metalloproteinases: Regulators of the Tumor Microenvironment. *Cell* **2010**, *141*, 52–67. [CrossRef]
7. Chen, Y.S.; Langhammer, T.; Westhofen, M.; Lorenzen, J. Relationship between matrix metalloproteinases MMP-2, MMP-9, tissue inhibitor of matrix metalloproteinases-1 and IL-5, IL-8 in nasal polyps. *Allergy* **2007**, *62*, 66–72. [CrossRef]
8. Wang, R.D.; Wright, J.L.; Churg, A. Transforming growth factor-beta (1) drives airway remodeling in cigarette smoke-exposed tracheal explants. *Am. J. Resp. Cell. Mol.* **2005**, *33*, 387–393. [CrossRef]
9. Jia, M.; Chen, X.; Liu, J.; Chen, J. PTEN promotes apoptosis of H2O2-injured rat nasal epithelial cells through PI3K/Akt and other pathways. *Mol. Med. Rep.* **2018**, *17*, 571–579. [CrossRef]
10. Dagli, M.; Eryilmaz, A.; Besler, T.; Akmansu, H.; Acar, A.; Korkmaz, H. Role of free radicals and antioxidants in nasal polyps. *Laryngoscope* **2004**, *114*, 1200–1203. [CrossRef]
11. Zeng, H.; Wang, Y.; Gu, Y.; Wang, J.; Zhang, H.; Gao, H.; Jin, Q.; Zhao, L. Polydatin attenuates reactive oxygen species-induced airway remodeling by promoting Nrf2-mediated antioxidant signaling in asthma mouse model. *Life Sci.* **2019**, *218*, 25–30. [CrossRef] [PubMed]

12. Al-Azzam, N.; Teegala, L.R.; Pokhrel, S.; Ghebreigziabher, S.; Chachkovskyy, T.; Thodeti, S.; Gavilanes, I.; Covington, K.; Thodeti, C.K.; Paruchuri, S. Transient Receptor Potential Vanilloid channel regulates fibroblast differentiation and airway remodeling by modulating redox signals through NADPH Oxidase 4. *Sci. Rep.* **2020**, *10*, 9827. [CrossRef] [PubMed]
13. Wang, K.S.; Lv, Y.; Wang, Z.; Ma, J.; Mi, C.; Li, X.; Xu, G.H.; Piao, L.X.; Zheng, S.Z.; Jin, X. Imperatorin efficiently blocks TNF-α-mediated activation of ROS/PI3K/Akt/NF-κB pathway. *Oncol. Rep.* **2017**, *37*, 3397–3404. [CrossRef] [PubMed]
14. Tian, T.F.; Yuan, Q.; Ye, J. Relationship and significance among IL-6,PI3K/Akt and GSK 3β in chronic rhinosinusitis. *Lin Chuang Er Bi Yan Hou Tou Jing Wai Ke Za Zhi J. Clin. Otorhinolaryngol. Head Neck Surg.* **2016**, *30*, 1859–1864. [CrossRef]
15. Jung, C.H.; Kim, E.M.; Park, J.K.; Hwang, S.G.; Moon, S.K.; Kim, W.J.; Um, H.D. Bmal1 suppresses cancer cell invasion by blocking the phosphoinositide 3-kinase-Akt-MMP-2 signaling pathway. *Oncol. Rep.* **2013**, *29*, 2109–2113. [CrossRef]
16. Yang, S.R.; Chida, A.S.; Bauter, M.R.; Shafiq, N.; Seweryniak, K.; Maggirwar, S.B.; Kilty, I.; Rahman, I. Cigarette smoke induces proinflammatory cytokine release by activation of NF-kappa B and posttranslational modifications of histone deacetylase in macrophages. *Am. J. Physiol. Lung C.* **2006**, *291*, L46–L57. [CrossRef]
17. Cho, J.S.; Kang, J.H.; Um, J.Y.; Han, I.H.; Park, I.H.; Lee, H.M. Lipopolysaccharide Induces Pro-Inflammatory Cytokines and MMP Production via TLR4 in Nasal Polyp-Derived Fibroblast and Organ Culture. *PLoS ONE* **2014**, *9*. [CrossRef]
18. Léguillette, R.; Tohver, T.; Bond, S.L.; Nicol, J.A.; McDonald, K.J. Effect of Dexamethasone and Fluticasone on Airway Hyperresponsiveness in Horses with Inflammatory Airway Disease. *J. Vet. Intern. Med.* **2017**, *31*, 1193–1201. [CrossRef]
19. Hadar, T.; Yaniv, E.; Shvili, Y.; Koren, R.; Shvero, J. Histopathological changes of the nasal mucosa induced by smoking. *Inhal. Toxicol.* **2009**, *21*, 1119–1122. [CrossRef]
20. Amin, K.A.M. Allergic Respiratory Inflammation and Remodeling. *Turk. Thorac. J.* **2015**, *16*, 133–140. [CrossRef]
21. Christensen, D.N.; Franks, Z.G.; McCrary, H.C.; Saleh, A.A.; Chang, E.H. A Systematic Review of the Association between Cigarette Smoke Exposure and Chronic Rhinosinusitis. *Otolaryng. Head Neck* **2018**, *158*, 801–816. [CrossRef] [PubMed]
22. Vent, J.; Robinson, A.M.; Gentry-Nielsen, M.J.; Conley, D.B.; Hallworth, R.; Leopold, D.A.; Kern, R.C. Pathology of the olfactory epithelium: Smoking and ethanol exposure. *Laryngoscope* **2004**, *114*, 1383–1388. [CrossRef] [PubMed]
23. Bonnans, C.; Chou, J.; Werb, Z. Remodelling the extracellular matrix in development and disease. *Nat. Rev. Mol. Cell Biol.* **2014**, *15*, 786–801. [CrossRef] [PubMed]
24. Chao, W.; Deng, J.S.; Huang, S.S.; Li, P.Y.; Liang, Y.C.; Huang, G.J. 3, 4-dihydroxybenzalacetone attenuates lipopolysaccharide-induced inflammation in acute lung injury via down-regulation of MMP-2 and MMP-9 activities through suppressing ROS-mediated MAPK and PI3K/AKT signaling pathways. *Int. Immunopharmacol.* **2017**, *50*, 77–86. [CrossRef] [PubMed]
25. Li, X.Y.; Meng, J.; Qiao, X.M.; Liu, Y.F.; Liu, F.; Zhang, N.; Zhang, J.; Holtappels, G.; Luo, B.; Zhou, P.; et al. Expression of TGF, matrix metalloproteinases, and tissue inhibitors in Chinese chronic rhinosinusitis. *J. Allergy Clin. Immun.* **2010**, *125*, 1061–1068. [CrossRef]
26. Shin, J.M.; Park, J.H.; Kang, B.; Lee, S.A.; Park, I.H.; Lee, H.M. Effect of doxycycline on transforming growth factor-beta-1-induced matrix metalloproteinase 2 expression, migration, and collagen contraction in nasal polyp-derived fibroblasts. *Am. J. Rhinol. Allergy* **2016**, *30*, 385–390. [CrossRef]
27. Fordham, M.T.; Mulligan, J.K.; Casey, S.E.; Mulligan, R.M.; Wang, E.W.; Sansoni, E.R.; Schlosser, R.J. Reactive Oxygen Species in Chronic Rhinosinusitis and Secondhand Smoke Exposure. *Otolaryng. Head Neck* **2013**, *149*, 633–638. [CrossRef]
28. Platt, M.P.; Soler, Z.; Metson, R.; Stankovic, K.M. Pathways analysis of molecular markers in chronic sinusitis with polyps. *Otolaryngol. Head Neck Surg.* **2011**, *144*, 802–808. [CrossRef]
29. Liu, H.J.; Dai, X.N.; Cheng, Y.S.; Fang, S.C.; Zhang, Y.M.; Wang, X.G.; Zhang, W.; Liao, H.; Yao, H.H.; Chao, J. MCPIP1 mediates silica-induced cell migration in human pulmonary fibroblasts. *Am. J. Physiol. Lung C.* **2016**, *310*, L121–L132. [CrossRef]

30. Park, J.W.; Ryter, S.W.; Kyung, S.Y.; Lee, S.P.; Jeong, S.H. The phosphodiesterase 4 inhibitor rolipram protects against cigarette smoke extract-induced apoptosis in human lung fibroblasts. *Eur. J. Pharm.* **2013**, *706*, 76–83. [CrossRef]
31. Krimmer, D.I.; Burgess, J.K.; Wooi, T.K.; Black, J.L.; Oliver, B.G.G. Matrix Proteins from Smoke-Exposed Fibroblasts Are Pro-proliferative. *Am. J. Resp. Cell Mol.* **2012**, *46*, 34–39. [CrossRef]
32. Hou, G.; Yin, Y.; Han, D.; Wang, Q.Y.; Kang, J. Rosiglitazone attenuates the metalloprotease/anti-metalloprotease imbalance in emphysema induced by cigarette smoke: Involvement of extracellular signal-regulated kinase and NF kappa B signaling. *Int. J. Chronic Obs.* **2015**, *10*, 715–724. [CrossRef]
33. Todorova, L.; Bjermer, L.; Westergren-Thorsson, G.; Miller-Larsson, A. TGFbeta-induced matrix production by bronchial fibroblasts in asthma: Budesonide and formoterol effects. *Respir. Med.* **2011**, *105*, 1296–1307. [CrossRef] [PubMed]
34. Gero, D.; Szabo, C. Glucocorticoids Suppress Mitochondrial Oxidant Production via Upregulation of Uncoupling Protein 2 in Hyperglycemic Endothelial Cells. *PLoS ONE* **2016**, *11*, e154813. [CrossRef] [PubMed]
35. Okoh, V.O.; Felty, Q.; Parkash, J.; Poppiti, R.; Roy, D. Reactive oxygen species via redox signaling to PI3K/AKT pathway contribute to the malignant growth of 4-hydroxy estradiol-transformed mammary epithelial cells. *PLoS ONE* **2013**, *8*, e54206. [CrossRef] [PubMed]

© 2020 by the authors. Licensee MDPI, Basel, Switzerland. This article is an open access article distributed under the terms and conditions of the Creative Commons Attribution (CC BY) license (http://creativecommons.org/licenses/by/4.0/).

Article

Oxidative Chemical Stressors Alter the Physiological State of the Nasal Olfactory Mucosa of Atlantic Salmon

Carlo C. Lazado [1,*], Vibeke Voldvik [1], Mette W. Breiland [2], João Osório [1,3], Marianne H. S. Hansen [1] and Aleksei Krasnov [1]

[1] Nofima, The Norwegian Institute of Food, Fisheries and Aquaculture Research, 1433 Ås, Norway; vibeke.voldvik@nofima.no (V.V.); joao.osorio96@gmail.com (J.O.); marianne.h.s.hansen@nofima.no (M.H.S.H.); Aleksei.Krasnov@Nofima.no (A.K.)

[2] Nofima, The Norwegian Institute of Food, Fisheries and Aquaculture Research, 9019 Tromsø, Norway; Mette.W.Breiland@Nofima.no

[3] CIISA, Faculty of Veterinary Medicine, University of Lisbon, 1300-477 Lisbon, Portugal

* Correspondence: carlo.lazado@nofima.no

Received: 5 September 2020; Accepted: 17 November 2020; Published: 18 November 2020 

Abstract: The olfactory organs of fish have vital functions for chemosensory and defence. Though there have been some ground-breaking discoveries of their involvement in immunity against pathogens in recent years, little is known about how they respond to non-infectious agents, such as exogenous oxidants, which fish encounter regularly. To this end, we employed Atlantic salmon (*Salmo salar*) as a model to study the molecular responses at the nasal olfactory mucosa of a teleost fish when challenged with oxidants. Microarray analysis was employed to unravel the transcriptional changes at the nasal olfactory mucosa following two types of in vivo exposure to peracetic acid (PAA), a highly potent oxidative agent commonly used in aquaculture: Trial 1: periodic and low dose (1 ppm, every 3 days over 45 days) to simulate a routine disinfection; and Trial 2: less frequent and high dose (10 ppm for 30 min, every 15 days, 3 times) to mimic a bath treatment. Furthermore, leukocytes from the olfactory organ were isolated and exposed to PAA, as well as to hydrogen peroxide (H_2O_2) and acetic acid (AA)—the two other components of PAA trade products—to perform targeted cellular and molecular response profiling. In the first trial, microarrays identified 32 differentially expressed genes (DEG) after a 45-day oxidant exposure. Erythrocyte-specific genes were overly represented and substantially upregulated following exogenous oxidant exposure. In Trial 2, in which a higher dose was administered, 62 DEGs were identified, over 80% of which were significantly upregulated after exposure. Genes involved in immune response, redox balance and stress, maintenance of cellular integrity and extracellular matrix were markedly affected by the oxidant. All chemical stimuli (i.e., PAA, H_2O_2, AA) significantly affected the proliferation of nasal leukocytes, with indications of recovery observed in PAA- and H_2O_2-exposed cells. The migration of nasal leukocytes was promoted by H_2O_2, but not much by PAA and AA. The three chemical oxidative stressors triggered oxidative stress in nasal leukocytes as indicated by an increase in the intracellular reactive oxygen species level. This resulted in the mobilisation of antioxidant defences in the nasal leukocytes as shown by the upregulation of crucial genes for this response network. Though qPCR revealed changes in the expression of selected cytokines and heat shock protein genes following in vitro challenge, the responses were stochastic. The results from the study advance our understanding of the role that the nasal olfactory mucosa plays in host defence, particularly towards oxidative chemical stressors.

Keywords: fish; mucosal immunity; nasal immunity; oxidative stress; peroxide

1. Introduction

Oxidative stress is a physiological state in an organism in which the redox balance is altered, as characterised by an increase in the levels of reactive oxygen species (ROS) but normal or low amounts of antioxidants, which may be due to compromised neutralisation property and/or scavenging potential [1,2]. Fish, like many other organisms, have an extensive repertoire to counteract oxidative stress [2,3]. The integrated antioxidant systems, which include enzymatic and nonenzymatic antioxidants, are at the forefront of blocking the harmful effects of ROS [1]. Redox imbalance associated with oxidative stress promotes genetic instability, changes in gene expression patterns, alterations in cellular signalling cascades/cell metabolism, and disruption of the cell cycle, leading to several pathophysiological conditions [4,5].

Oxidants can be endogenously produced or derived from external sources. Endogenous ROS are produced from molecular oxygen as a result of normal cellular metabolism [1], and ROS are constantly produced in all living cells in which roughly up to 1% of an animal's total oxygen consumption may be attributed to ROS generation and detoxification [6]. Exogenous ROS may come from various sources, and their impacts on redox status have consequences on cell viability, activation, proliferation, and organ function. Farmed fish encounter an increased flow of exogenous ROS several times during a lifetime, as many husbandry practices employ ROS-generating compounds either as a form of disinfectant or water treatment, or as a chemotherapeutant [7–11]. The antimicrobial activity of ROS towards opportunistic and pathogenic microorganisms underlines their use in providing fish with a favourable rearing environment [12]. Nonetheless, our knowledge of the physiological alterations associated with exogenous ROS, mainly from ROS-generating agents being used in fish farming, is fragmentary.

Mucosal organs of fish are multifunctional; besides their role in defence, they carry a multitude of other physiological functions [13,14]. They are often considered the first line of defence because these structures interact with the water matrix where several biological and chemical challenges present themselves regularly. In recent years, there has been a dramatic development in the study of the physiology and immunology of mucosal surfaces in fish, driven mostly by their warranted importance in maintaining the health of farmed fish [13,15].

The nasal olfactory system plays a role not only in chemoreception but also in immune defence, as it is considered an ancient component of the mucosal immune system of vertebrates [16]. It is a highly specialised sensory organ for the detection and identification of minute quantities of chemicals in the environment [17,18], and because water constantly circulates through the nasal cavities, they are continuously prompted with environmental challenges [19]. The mucosal regions of the fish olfactory lamellae have different cellular elements such as goblet cells, sustentacular cells, olfactory sensory neurons, and, most importantly, a rich assemblage of immune cells [19–21]. Vertebrate olfactory sensory neurons rapidly sense chemical stimuli in the environment and transduce signals to the central nervous system [18]. The nasopharynx-associated lymphoid tissue (NALT) protects the teleost olfactory organ from water-borne pathogens, just as for airborne pathogens in terrestrial animals [16]. Several recently published studies have demonstrated how viral and bacterial stimulations affect the immunological repertoire of the nasal mucosa in fish. They reveal a very distinct microenvironment that can mount a localised immunity and, at the same time, influence distant immune functions [14,16,18–20,22]. Non-infectious agents such as exogenous oxidants are delivered via water and are expected to pass through the nasal cavity of fish. In mammalian models, oxidative stressors are highly potent modulators of the nasal epithelium, and the interaction could induce morphological and pathological alterations [23–25]. However, in fish, the influence of exogenous oxidants on the nasal olfactory mucosa is barely explored, despite its common use.

This study explored the impacts of oxidative chemical stressors on the nasal olfactory mucosa of Atlantic salmon (*Salmo salar*). We employed both in vivo and in vitro strategies to unravel the molecular changes in the nasal olfactory mucosa when challenged with exogenous oxidants relevant in fish farming. In this study, we employed peracetic acid (PAA) as the main oxidant, as it is currently being developed as a chemotherapeutant (i.e., for amoebic gill disease—AGD) and disinfectant in recirculating aquaculture systems for salmon [4,26], and the results here are expected to help underline its potential for use. Both in vivo trials were designed to simulate the prospective use of PAA as either a routine disinfectant (Trial 1) or a treatment for a parasitic infection (Trial 2). In addition, in vitro trials were conceived to understand the physiological state of a specific cell type at the mucosa in response to not only PAA but also hydrogen peroxide (H_2O_2) and acetic acid (AA). These two compounds are present in equilibrium with PAA in its trade product.

2. Materials and Methods

2.1. Oxidant Exposure Experiment

All fish handling procedures described in this paper followed the Guidelines of the European Union (2010/63/EU) and the in vivo exposure trials received approvals from the Norwegian Food Safety Authority (FOTS, Forsøksdyrforvatningen tilsyns- og søknadssystem IDs 20831, 19321). All key personnel have a FELASA (Federation of European Laboratory Animal Science Associations) C certificate to conduct experimentation on live animals. Two independent in vivo trials (Figure 1) were performed in which the application of the exogenous oxidant was based on its proposed use for that particular stage of Atlantic salmon production. It was ensured that all fish used in the experiments (both in vivo and in vitro) did not have a history of oxidant exposure. Peracetic acid is available under different trade products and two commercially available PAA-based disinfectants were used in this study. All three major components of PAA trade (i.e., PAA, hydrogen peroxide and acetic acid) products have disinfection power, though PAA is the most potent contributor to the disinfection property of PAA-based disinfectants.

Trial 1 was conducted at Nofima Centre for Recirculation in Aquaculture (NCRA; Sunndalsøra, Norway) and was aimed at evaluating the impacts of periodic low-dose oxidant exposure on the transcriptome of the nasal olfactory mucosa. This experiment was designed to mimic the use of the oxidant as a routine water disinfectant in a recirculating aquaculture system [27]. Briefly, each of the four 3.2 m^3 octagonal tanks in a recirculation system was stocked with 735 smolts with an average weight of around 90 g. After four weeks of acclimatisation, the oxidant in the form of a peracetic acid-based disinfectant (Perfectoxid, Novadan ApS, Kolding, Denmark) was directly applied to each tank at a nominal concentration of 1 ppm every 3 days for 45 days, making a total of 15 applications in the duration of the trial. This mode of application was patterned on a previous PAA experiment conducted in rainbow trout, a closely related species of salmon [26]. Moreover, the concentration is within the range safe for use in salmon [4,28]. It was ensured that the application of PAA on each occasion was between 0900 and 1000 to avoid temporal effects. The following parameters were maintained during the trial: water flow rate at 100 L min^{-1}, salinity at 11.6 ± 0.5 ‰, temperature at 12.8 ± 0.6 °C, pH at 7.5, dissolved oxygen > 90% saturation, photoperiod at 24 L: 0 D, and a continuous feeding regime (Nutra Olympic 3 mm, Skretting, Averøy, Norway).

Trial 2 was performed at Havbruksstasjonen i Tromsø (HiT; Tromsø, Norway) and was designed to unravel the changes in the nasal transcriptome after repeated but less frequent exposure to higher doses of oxidant. PAA is currently being explored as a potential treatment for amoebic gill disease (AGD), a gill health issue affecting mostly seawater-adapted salmon [4,7]. The trial was designed to simulate an oxidant exposure as a treatment protocol for AGD [26]. Forty fish with an average weight of 80–90 g were stocked into a 500 L circular tank in a flow-through system. There were six tanks in total: three for the control group and three for the oxidant-exposed group. Fish were allowed to acclimatise for a week before the first oxidant treatment was performed. Fish were fasted for 24 h prior to each treatment occasion. Oxidant treatment was performed as follows: Water flow in the tank was stopped. Thereafter, the oxidant (Divosan Forte™, Lilleborg AS, Olso, Norway) was added to the water column to achieve a final concentration of 10 ppm. This concentration was 2x higher than the concentration we earlier tested and reported [4]. Aeration was supplied to enable mixing and to maintain the oxygen level. After 30 min, the water flow was opened, and over 90% of the water was replaced within 10 min. Feeding was continued a day after the exposure. This exposure protocol was performed every 15 days and there were three exposure occasions in the whole trial. The following parameters were maintained during the trial: water flow rate at 6–7 L min^{-1}, salinity at 35‰, temperature at 12.0 ± 1 °C, dissolved oxygen > 90% saturation, photoperiod at 24 L: 0 D, and a continuous feeding regime (Nutra Olympic 3 mm, Skretting, Averøy, Norway).

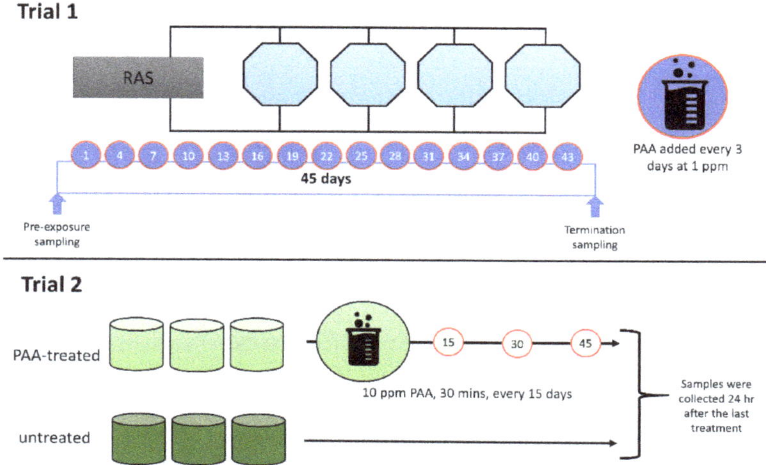

Figure 1. Diagrammatic summary of the in vivo trials. Trial 1 aimed to profile the impacts of periodic and low dose peracetic acid (PAA) application (1 ppm, every 3 days over 45 days), while Trial 2 was designed to investigate the nasal responses following less frequent and high dose PAA treatment (10 ppm for 30 min, every 15 days, 3 times). Details of each trial are described in Section 2.1.

2.2. Olfactory Organ Collection

All fish for sampling were humanely euthanised with an overdose of either Tricaine methanesulfonate (Trial 1; MS222, PHARMAQ Ltd., Ås, Norway) or Benzocaine (Trial 2; Benzoak®, ACD Pharmaceuticals AS, Oslo, Norway). Percussive stunning was avoided because it triggered the influx of blood to the head, including the olfactory epithelium. Olfactory organ was collected by opening the nostrils to expose the outermost section of the nasal mucosa. The rosette from the left side was then dissected out, immediately suspended in RNA*later*™ (Thermo Fisher Scientific, Sacramento, CA, USA), kept at room temperature overnight for penetration, and then stored at −70 °C until RNA isolation. In Trial 1, olfactory rosettes (n = 8 per time-point) were collected before and 45 days (i.e., 24 h after the last PAA application) after the start of periodic oxidant exposure. In Trial 2, the tissue samples (n = 9, per group) were collected 24 h after the 3rd exposure.

2.3. Isolation of Leukocytes from Olfactory Organ

Leukocytes from olfactory organ were isolated from 15 freshwater-adapted salmon (ca. 80–90 g) with similar genetic background, following a previously published method [22], with slight modifications. Briefly, fish were humanely euthanised with an anaesthetic overdose (Aqui-S, MSD Animal Health, Drammen, Norway). The olfactory organs from both sides were dissected and immediately placed in a modified L-15 (supplemented with 5% foetal bovine serum, 1% Penstrep, 1% HEPES, 4-(2-hydroxyethyl)-1-piperazineethanesulfonic acid) on ice. Rosettes from all fish were combined, and the tissues were cut into small pieces (0.5–1 cm) and mechanically dissociated by incubating the tissue suspension at 4 °C for 30 min with constant agitation. The cell supernatant was collected and temporarily stored at 4 °C. The remaining tissue fragments were suspended in modified L-15 medium and the process of mechanical dissociation was repeated four times. The collected supernatant from the four recurrences was combined and stored at 4 °C. The remaining tissue fragments were suspended in phosphate-buffered saline (PBS) with EDTA (1 mM) and DTT (0.9 mM) and incubated at 4 °C for 30 min with constant gentle agitation. The PBS supernatant was thereafter discarded. Enzymatic digestion was carried out by incubating the remaining fragments in collagenase solution (0.15 mg/mL in L-15, with 1% Penstrep) for 2 h at room temperature (20 °C) with agitation. The supernatants from mechanical dissociations and enzymatic digestion were combined, gently passed through a 100 µm filter, and spun down at 300× *g* for 10 min. The cell pellet was washed and resuspended in modified L-15, laid over a 34%/51% isotonic Percoll® (Sigma-Aldrich, Oslo, Norway) gradient. The tubes were then centrifuged for 30 min at 400× *g* at 4 °C. The cell layer between the gradients was carefully transferred to a tube with modified L-15 medium, centrifuged for 10 min at 400× *g* at 4 °C, and suspended in new modified L-15 medium. Cell viability and number were determined by CellCountess™ II (Thermo Fisher Scientific, Boston, MA, USA). The cells were seeded out onto a 12-well plate (Corning® CellBIND® Surface, Sigma-Aldrich, Oslo, Norway) at a density of 2×10^5 cells per well and incubated at 13 °C.

2.4. *In Vitro Exposure Trial*

The cells were allowed to adhere for 48 h before the exposure was performed. The leukocytes were exposed to three chemical stressors at physiological concentrations—100 µM PAA, 100 µM H_2O_2, and 100 µM AA—for 30 min. The concentrations were based on several preliminary in vitro trials and the concentrations were selected because they were able to trigger significant increase in intracellular ROS, thereby providing an indication that the internal redox balance had been altered by the tested oxidants. Each treatment group had four independent wells. Untreated cells served as controls and were handled similarly to the treatment groups, though no chemical stressor was added. After 30 min, the media were removed, cells were washed gently with modified L-15, and 300 µL of the same media was added to each well. After 24 h, the media were removed. The cells were suspended in lysis buffer (ZYMO Quick-RNA™ Microprep kit, Sacramento, CA, USA) and scraped, and the cell suspension was stored at 70 °C until RNA isolation.

2.5. *Proliferation and Migration Assays*

Nasal leukocytes were isolated from 12 freshwater-adapted salmon (ca. 80–90 g) following the method described in Section 2.3 and seeded onto a 96-well plate (Corning®, CellBIND® Surface, Sigma-Aldrich, Oslo, Norway) at a density of 10^5 cells per well. After the cells were allowed to settle and adhere for 24 h, they were exposed to PAA, H_2O_2, and AA at a physiological concentration of 100 µM for 30 min and washed. New media were added and the exposed cells were allowed to recover in the incubator. Unexposed cells, serving as controls, were likewise washed, and new media were added. Cell proliferation (proxy for cytotoxicity) was measured using the CyQUANT Direct Proliferation Assay (Thermo Fisher Scientific, Boston, MA, USA) 24 and 48 h after the challenge. Each treatment group, including the unexposed control group, had six replicate wells. Rate of proliferation was expressed relative to the control group of that time-point.

The effects of the chemical stressors on cellular migration were determined by the CytoSelect™ Cell Migration Assay kit (Cell Biolabs, Inc., Sacramento, CA, USA). Freshly isolated nasal leukocytes were suspended in modified L-15 medium without serum. The lower receptacle of the migration chamber was added with L-15 media containing either PAA, H_2O_2, or AA in a final concentration of 100 µM. Wells with L-15 medium containing 10% FBS served as the positive control for chemotaxis while L-15 medium alone was designated as a negative control. Each treatment, including the controls, had been assigned three wells. Thereafter, the cells were added to the upper receptacle of the migration chamber at a density of 2×10^5 cells. The migration chamber was incubated for 24 h at 13 °C before the migratory cells were dislodged from the membrane, lysed and labelled with CyQuant® GR dye solution (Sacramento, CA, USA), and fluorescence was read at 480 nm/520 nm.

2.6. Intracellular ROS Quantification

Nasal leukocytes were isolated, cultured, and treated with the chemical stressors as described in detail in Sections 2.4 and 2.5. The intracellular level of reactive oxygen species (ROS) in the treated cells, including untreated control, was quantified using the OxiSelect™ Intracellular ROS Assay Kit (Cell Biolabs, Inc., San Diego, CA, USA) at 24 and 48 h after challenge. The level of ROS is given as a proportion of fluorescent dichlorodihydrofluorescein (DCF).

2.7. RNA Isolation, cDNA Synthesis, and Quantitative Real-Time PCR

The RNA from both olfactory tissues and nasal leukocytes were isolated using Quick-RNA™ Microprep kit (Zymo Research, CA, USA). RNA concentration was measured in a NanoDrop 1000 Spectrophotometer (Thermo Fisher Scientific, DE, USA), and the quality of the samples for microarray was further assessed using an Agilent® 2100 Bioanalyzer™ RNA 6000 Nano kit (Agilent Technology Inc., Santa Clara, CA, USA). All samples had an RNA Integrity Value higher than 8.8.

A High Capacity RNA-to-cDNA Reverse Transcription kit (Applied Biosystems, Sacramento, CA, USA) was used to prepare the complementary DNA from the nasal leukocyte samples using 300 ng RNA input following a synthesis protocol of 25 °C for 10 min, followed by 37 °C for 120 min and then 5 min at 85 °C. The expression of selected antioxidant defence, cytokines, and heat shock protein genes (Supplementary File 1) was quantified using the PowerUp™ SYBR™ Green master chemistry (Applied Biosystems, Sacramento, CA, USA) in a QuantStudio5 real-time quantitative PCR system (Applied Biosystems, Sacramento, CA, USA). The qPCR reaction mixture included 4 μL of diluted cDNA, 5 μL SYBR™ Green Master (Thermo Fisher Scientific, Boston, MA, USA), and 1 μL of the forward and reverse primer. All samples were run in duplicate, including minus reverse transcriptase and no template controls. The thermocycling protocol included pre-incubation at 95 °C for 2 min, amplification with 40 cycles at 95 °C for 1 s and at 60 °C for 30 s, and a dissociation step series of 95 °C for 15 s, 60 °C for 1 min, and 95 °C for 15 s. Amplification efficiency was calculated from a five-point standard curve of 2-fold dilution series of pooled cDNA. The expression of the target genes was normalised using three reference genes, namely *elongation factor 1a* (*eef1a*), *acidic ribosomal protein* (*arp*), and *β-actin* (*actb*) [29].

2.8. Microarray Analysis

Olfactory rosettes from Trials 1 and 2 were subjected to microarray analysis using Nofima's Atlantic salmon DNA oligonucleotide microarray SIQ-6 (custom design, GPL16555, Sacramento, CA, USA), which contains 15 K probes for protein-coding genes involved in immunity, tissue structure, integrity and functions, cell communication and junctions, and extracellular matrix, amongst many others [30]. Agilent Technologies manufactured and supplied the microarrays, reagents, and equipment used in the analysis. Using 110 ng of total RNA template per reaction, RNA was amplified using a One-Color Quick Amp Labeling Kit, and thereafter Cy3 was labelled. Subsequently, fragmentation of the labelled RNA was carried out using a Gene Expression Hybridization Kit and hybridisation followed in an oven thermostatted at 65 °C with a constant rotation speed of 10 rpm for 17 h. The arrays were washed in sequence with Gene Expression Wash Buffers 1 and 2 and were scanned through an Agilent SureScan Microarray scanner. Data processing was carried out in Nofima's bioinformatics package STARS.

2.9. Data Handling and Statistics

The significant difference in the transcript level of the target marker genes between before and after periodic oxidant exposure in Trial 1 and between the unexposed-control and oxidant-exposed groups in Trial 2 was determined by Student's t-test for independent samples; the threshold of differential expression in microarray analyses was 1.75-fold. The level of significance was set at 5% ($p < 0.05$).

A Shapiro–Wilk test was used to evaluate the normal distribution and a Brown–Forsyth test to check for equal variance of the proliferation assay and gene expression data set. Two-way ANOVA was then employed to investigate significant differences amongst treatment groups over time. In addition, the Holm–Sidak test was used to identify pairwise differences. One-way ANOVA was used for migration assay data. All statistical tests were performed using SigmaPlot 14.0 Statistical Software (Systat Software Inc., London, UK), with a level of significance set at $p < 0.05$.

3. Results

3.1. Transcriptomic Changes in the Olfactory Rosettes from Trial 1

After 45 days of periodic low-dose oxidant exposure, microarray analysis identified 32 differentially regulated genes (DEG) in the nasal olfactory rosette (Table 1, Supplementary File 2). The numbers of upregulated (16/32) and downregulated genes were equal (16/32). Erythrocyte-specific genes were the most represented group, with nine transcripts/variants identified as being upregulated in response to the exogenous oxidant. Genes involved in immune response such as *c-c motif chemokine 28*, *interleukin 13 receptor alpha-2*, *defensin beta 4*, *ig heavy chain*, and *mannose-specific lectin-like*, were all downregulated after 45 days of oxidant exposure. Two genes encoding cytokeratins were downregulated. Three out of four genes with metabolic functions were likewise downregulated.

Table 1. Some of the differentially expressed genes in the olfactory rosette of salmon from Trial 1. Transcripts are annotated for their known or predicted function. Expression data are ratios of means of 45 days after the exposure to pre-exposure.

Annotation	Name	Fold
Cytoskeleton	*Keratin, type I cytoskeletal 20*	−2.69
Cytoskeleton	*Keratin, type I cytoskeletal*	−3.16
DNA replication	*DNA replication licensing factor MCM6*	2.34
Chemokine	*C-C motif chemokine 28*	−4.33
Cytokine receptor	*Interleukin 13 receptor alpha-2*	−1.83
Antibacterial	*Defensin beta 4*	−2.60
B cell	*Ig heavy chain*	−3.69
Lectin	*Mannose-specific lectin-like*	−2.20
Lipid metabolism	*Phosphoethanolamine/phosphocholine phosphatase*	−2.43
Protease	*Duodenase-1*	−2.20
Protease	*Serine protease 23-like*	1.76
Energy metabolism	*Pyruvate dehydrogenase kinase isozyme 2*	−2.12
Tissue ECM mucus	*Mucin-2*	2.14
Erythrocyte	*Hemoglobin subunit beta-1*	2.58
Erythrocyte	*Hemoglobin subunit alpha (5)*	2.80
Erythrocyte	*Hemoglobin subunit alpha-4-like (2)*	5.40
Erythrocyte	*Hemoglobin subunit beta*	1.76

For genes with several variants (number enclosed in parentheses), mean values are presented. NB: The complete list of differentially expressed genes (DEGs) is given in Supplementary File 2.

3.2. Transcriptomic Changes in the Olfactory Rosettes from Trial 2

Sixty-two DEGs were identified in the olfactory rosettes from fish exposed to an oxidant on three occasions, 56 of which, accounting for 82% of the DEGs, were upregulated (Table 2, Supplementary File 2). From this group, genes related to immunity, including cytokines and effectors, were largely represented with 14 upregulated transcripts. Genes with innate immune functions constitute a considerable number in the DEG panel. Genes with known involvement in cellular structural integrity such as *keratin* and *plekstrin* were likewise upregulated. A similar effect was observed on genes encoding extracellular proteins (e.g., *fibronectin*, *mucin5b*). Several genes involved in various metabolic pathways were represented in the DEGs panel, such as those involved in amine, amino acid, calcium, and xenobiotic metabolism. A total of 3/4 DEGs of lipid metabolism were downregulated following oxidant exposure.

Exogenous oxidant exposure upregulated the expression of genes with function in cellular processes such as DNA replication, signalling, and protein folding/modification including the *heat shock proteins*. Oxidant-induced changes in cellular redox balance were likewise manifested with two DEGs.

Table 2. Some of the differentially expressed genes in the olfactory rosette of salmon from Trial 2 are annotated for their known or predicted function. Expression data are ratios of means of the 10 ppm PAA-treated group to the unexposed/control treated group. Samples were collected 24 h after the 3rd exposure.

Annotation	Name	Fold
Cytoskeleton	*Keratin 14*	1.83
Cytoskeleton	*Keratin 4*	2.05
Cytoskeleton	*Keratin cytoskeletal 17*	3.48
Cytoskeleton	*Pleckstrin 2*	2.11
DNA replication	*DNA replication licensing factor MCM6*	1.75
Protein folding	*14-3-3 protein alpha*	1.82
Protein folding	*Heat shock cognate 70 (2)*	3.63
Protein folding	*Heat shock protein 90, alpha (2)*	3.53
Signaling	*Guanine nucleotide binding protein*	2.06
Redox homeostasis	*Glutathione reductase, mitochondrial*	2.26
Redox homeostasis	*Redox-regulatory protein FAM213A*	−2.67
Signaling	*Tyrosine phosphatase type IVA, member 1 (2)*	1.97
Cell Surface	*Vacuole membrane protein 1*	2.12
B cell	*IgH-locus*	1.90
Cytokine receptor	*Interferon-alpha/beta receptor alpha chain*	1.93
Cytokine receptor	*Interleukin 13 receptor alpha-2*	1.96
Cytokine receptor	*Interleukin-1 receptor type II (2)*	2.37
Effector	*Differentially regulated trout protein 1*	2.85
Effector	*Ornithine decarboxylase (3)*	2.28
Effector	*Thrombin-like enzyme cerastocytin*	2.85
Lymphocyte	*T-lymphocite maturation associated protein*	2.72
T cell	*CD276 antigen-like*	2.66
TNF	*TNF decoy receptor*	1.90
Effector	*Spermidine/spermine N1-acetyltransferase 1*	4.07
Amino acid metabolism	*Methionine adenosyltransferase II, alpha b*	1.97
Calcium metabolism	*Protein S100-A1*	2.22
Iron metabolism	*Ferritin, heavy polypeptide 1b (5)*	2.34
Lipid metabolism	*Mid1-interacting protein 1-like*	−1.93
Lipid metabolism	*Phospholipid transfer protein*	−1.92
Lipid metabolism	*Short-chain dehydrogenase/reductase 3*	−2.18
Mitochondria	*Malic enzyme 3, NADP*	−1.87
Protease	*Calpain 9 (2)*	2.24
Steroid metabolism	*Cholesterol 25-hydroxylase-like protein A*	1.92
Sugar metabolism	*Glycogen debranching enzyme*	−1.90
Xenobiotic metabolism	*Glutathione S-transferase P-like*	2.44
Tissue ECM	*Fibronection*	1.74
Collagen	*Collagen alpha-2(VI) chain*	−3.01
Mucus	*GMP Giant mucus protein*	1.85
Mucus	*Mucin-5B (2)*	1.91
Mucus	*Arylalkylamine N-acetyltransferase*	−1.92
Epithelium	*Epithelial membrane protein 2-like*	1.75
Secretory	*Gastrotropin-like*	2.83
Lipid metabolism	*Globoside alpha-1,3-N-acetylgalactosaminyltransferase 1-like*	2.62

For genes with several variants (number enclosed in parentheses), mean values are presented. NB: The complete list of DEGs is given in Supplementary File 2.

3.3. *Effects of Oxidative Chemical Stressors on Leukocyte Proliferation and Migration*

The proliferation of nasal leukocytes 24 h after the challenge was significantly affected by PAA, as well as by its two main components, H_2O_2 and AA (Figure 2A). Cellular proliferation reduced by at least 0.5-fold in all treatment groups and no inter-treatment differences were observed. After 48 h, nasal leukocytes exposed to PAA and H_2O_2 slightly recovered, and the proliferation rate did not significantly differ from that of the control group. However, the effect of AA on proliferation was still persistent after 48 h, when the proliferation index in the group was 0.6-fold lower compared to control. Moreover, a significant difference was observed between the AA-exposed and two other treatment groups.

PAA and AA did not significantly affect the migration potential of the nasal leukocytes (Figure 2B). On the other hand, H_2O_2 promoted the migration of nasal leukocytes with a significant increase compared to control. A comparison of treatments revealed that H_2O_2-induced migration was significantly different from PAA but not from AA.

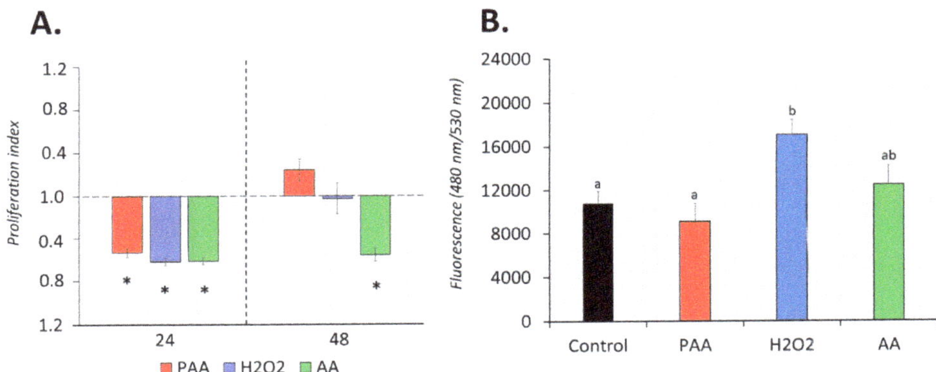

Figure 2. Effects of oxidative chemical stressors on the (**A**) proliferation and (**B**) migration of nasal olfactory leukocytes. For the proliferation assay, cells were isolated and cultured for 2 days before they were treated with 100 µM of PAA (peracetic acid), hydrogen peroxide (H_2O_2), and acetic acid (AA) for 30 min. Proliferation was quantified 24 and 48 h after challenge. Asterisk (*) denotes that proliferation was significantly affected relative to the unstimulated group. Results are presented as index of proliferation, where we expressed the proliferation relative to the control, unstimulated group. Two-way ANOVA followed by the Holm–Sidak test identified statistical difference amongst treatment groups over time. For migration assay, cells were allowed to migrate to a chamber containing the chemical stressor at a 100 µM concentration for 24 h. The positive control group has foetal bovine serum (FBS) as a chemoattractant. One-way ANOVA was used for migration assay data. Different letters a,b indicate significant difference at $p < 0.05$. Results are presented as mean ± SD of 5 (proliferation)/3 (migration) wells, with cells from 12 fish.

3.4. Level of ROS in Chemically Stressed Cells

The three chemical stressors significantly increased the intracellular ROS level of nasal olfactory leukocytes 24 h after challenge (Figure 3). The highest increment was identified in H_2O_2-exposed cells, followed by PAA and AA. In addition, the intracellular ROS of PAA- and H_2O_2-exposed cells were significantly higher than in the AA-exposed group at this time-point. After 48 h, the ROS level in PAA and AA-exposed cells displayed no significant difference from the control. The intracellular ROS in H_2O_2-exposed cells remained elevated 48 h after challenge. Moreover, the level was significantly higher compared to the control and AA-exposed cells but not to the PAA-exposed cells. Though not statistically significant, the level in PAA- and H_2O_2-exposed cells was apparently lower than the measured level at 24 h post-challenge.

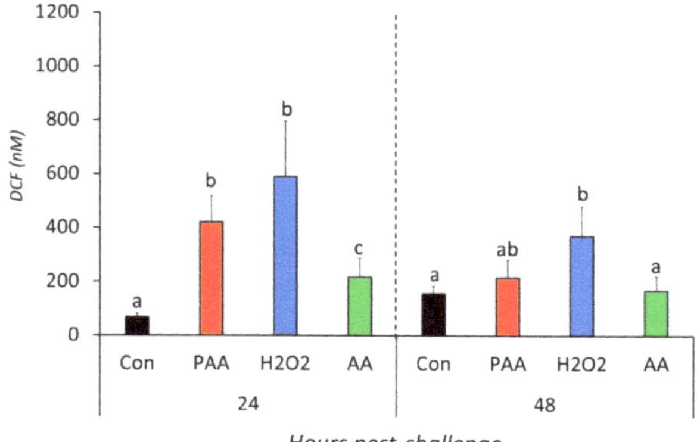

Figure 3. Levels of intracellular reactive oxygen species in nasal olfactory leukocytes exposed to different oxidative chemical stressors. The level was measured at 24 and 48 h after exposure to 100 μM of PAA (peracetic acid), H_2O_2, and AA (acetic acid) for 30 min. The control group was handled similarly, but without any chemical stimulation. Two-way ANOVA followed by the Holm–Sidak test identified statistical difference amongst the treatment groups over time. Different letters a,b indicate a significant difference at $p < 0.05$ between groups within a time-point. There was no time-related difference within a treatment group. Results are presented as mean ± SD of 5 wells, with cells from 12 fish.

3.5. *Changes in the Expression of Antioxidant Defence Genes in the Nasal Olfactory Leukocytes*

The expression of six genes with known function in antioxidant defence was differentially affected by the three chemical stressors (Figure 4). The transcript level of *gpx* was significantly higher in PAA- and H_2O_2-exposed cells compared to control at both time-points (Figure 4A). Such an increase was only identified in AA-exposed cells 48 h after challenge and the expression was significantly higher than the level at 24 h post-challenge. Chemically induced stress resulted in quite the opposite temporal expression profiles for *gr* and *gsta*. An elevated level of *gr* transcripts was observed in PAA- and H_2O_2-exposed groups at 24 h post-challenge while *gsta* expression was significantly altered after 48 h (Figure 4B,C). In both cases, the expression relative to the other time-point was significantly higher. The transcription of both genes in AA-exposed cells remained unaltered at both time-points. *Cat* expression was significantly upregulated in H_2O_2-exposed cells at both time-points, whereas for the PAA-exposed group, a significant increase was observed only 48 h post-challenge (Figure 4D). There was also a significant difference in *cat* expression in the PAA-exposed group between the two time-points. AA did not elicit a significant transcriptional change from *cat*. *Cu/zn sod* was significantly downregulated in the AA-exposed group as compared to control at 24 h, but not at 48 h after challenge. While *cu/zn sod* expression was unaltered in the PAA-exposed group at both time-points, a significant increase was observed in the H_2O_2-exposed group but only after 48 h. *Cu/zn sod* expression in H_2O_2-exposed cells at 24 h was likewise higher compared to the level at 48 h. The overall expression of *mnsod* displayed no significant alterations, except in PAA-exposed cells 48 h after challenge, where the expression was at least three times higher than the control and other treatment groups.

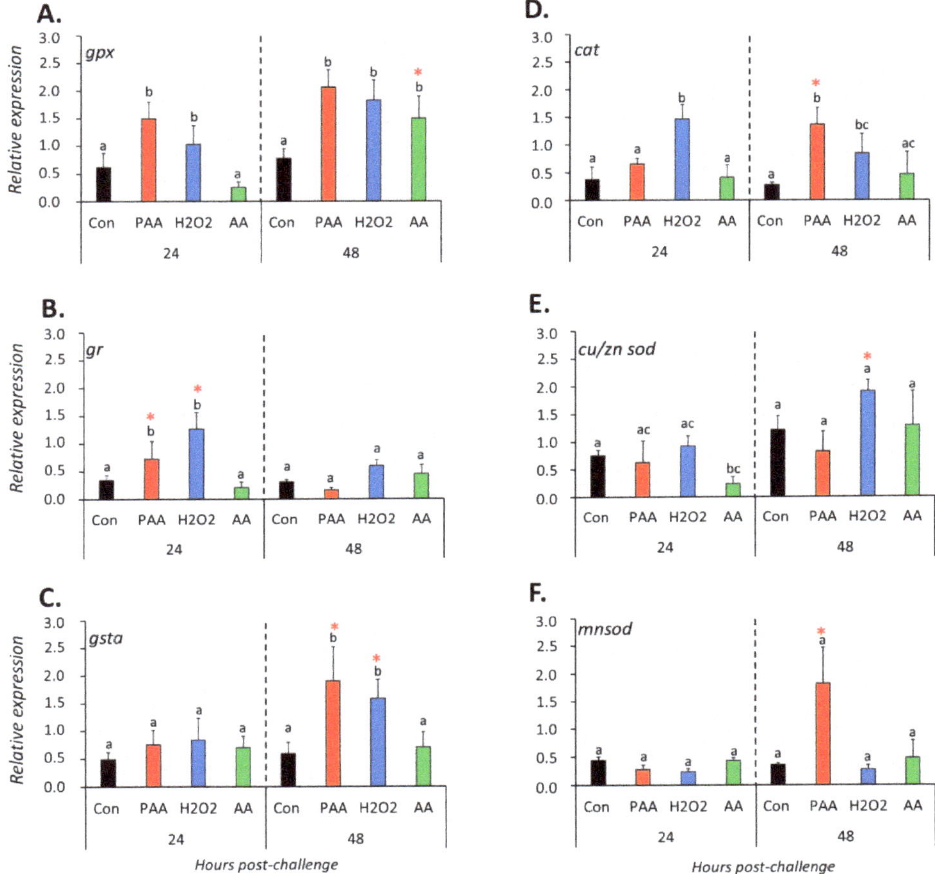

Figure 4. Changes in the expression of antioxidant defence genes following exposure to oxidative chemical stressors (**A**–**F**). The transcript level of six selected genes (i.e., *gpx*, *gr*, *gsta*, *cat*, *cu/zn sod*, *mnsod*) was quantified by RT-qPCR at 24 and 48 h after exposure to 100 µM of PAA, H_2O_2, and AA for 30 min. The control group was handled similarly, but without any chemical stimulation. Two-way ANOVA followed by the Holm–Sidak test identified statistical difference amongst treatment groups over time. Different letters indicate a significant difference at $p < 0.05$ amongst the treatment groups within a time-point. Asterisk (*) denotes that the level in a treatment group differs between the two time-points. Results are presented as mean ± SD of four wells, with ca. 10^5 cells from 15 fish.

3.6. Changes in the Gene Expression of Cytokine and Heat Shock Proteins in the Nasal Olfactory Leukocytes

The expression of *il1β* increased significantly in PAA- and H_2O_2-exposed groups compared to controls 24 h after challenge, but such a change was no longer observed at 48 h (Figure 5A). On the other hand, AA exposure did not alter *il1β* expression at both time-points. Generally, *il10* expression was not significantly affected by the oxidative chemical stressors, except in AA-exposed cells at 24 h post-challenge, where a significant downregulation was observed. The transcript level of *il13r1a* was significantly lower in PAA- and AA-exposed cells compared to control 24 h after challenge, though the change did not persist until 48 h (Figure 5C). The expression of *ifn* was unaltered in most treatment scenarios at both time-points (Figure 5D).

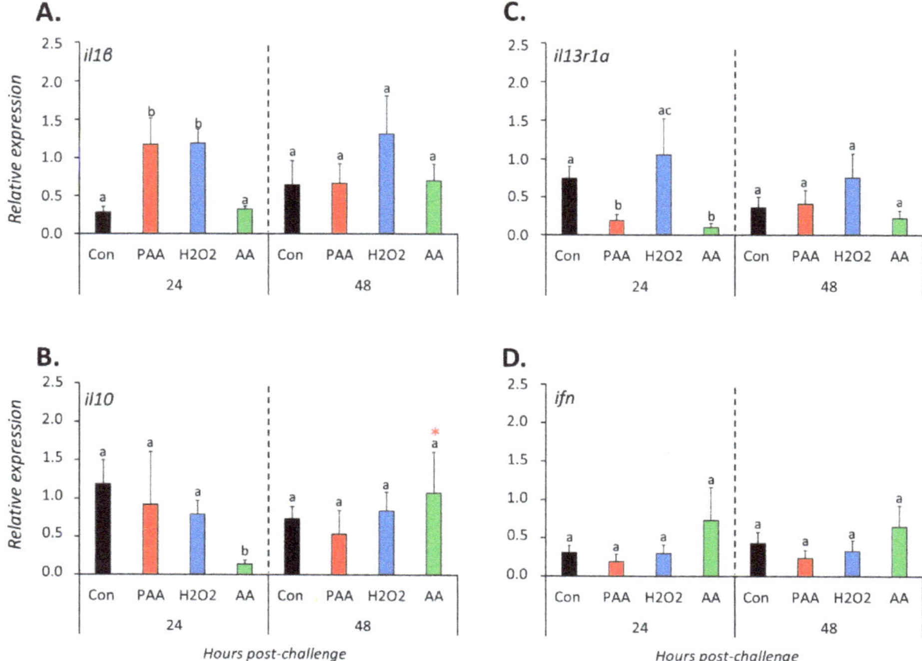

Figure 5. Changes in the expression of cytokine genes following exposure to oxidative chemical stressors (**A–D**). The transcript level of 4 selected genes (i.e., *il1β*, *il10*, *il13r1a*, *ifn*) was quantified by RT-qPCR at 24 and 48 h after exposure to 100 μM of PAA, H_2O_2, and AA for 30 min. The control group was handled similarly, but without any chemical stimulation. For the explanation on statistics and notations, please refer to Figure 4.

The three chemical stressors did not significantly change the expression of *hsp70* 24 h post-challenge (Figure 6A). After 48 h, however, *hsp70* expression was significantly elevated in all treatment groups compared to control, and the highest increment was identified in H_2O_2-exposed cells. *Hsp90* expression was not significantly altered in PAA- and H_2O_2-exposed cells at both time-points (Figure 6B). A significant downregulation was detected in AA-exposed cells at 24 but not at 48 h post-challenge, and the expression was higher in the latter than with the former time-point.

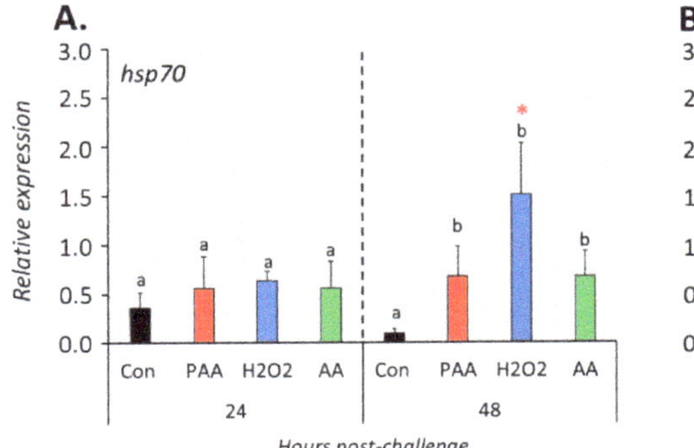

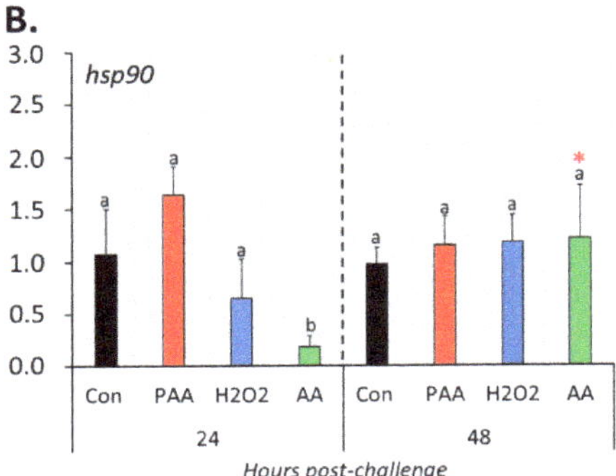

Figure 6. Changes in the expression of gene coding for heat shock proteins following oxidative chemical challenge (**A**,**B**). The transcript level of two selected genes (i.e., *hsp70*, *hsp90*) was quantified by RT-qPCR at 24 and 48 h after exposure to 100 μM of PAA, H_2O_2, and AA for 30 min. The control group was handled similarly, but without any chemical stimulation. For the explanation on statistics and notations, please refer to Figure 4.

4. Discussion

The present study reveals, through in vivo and in vitro exposure trials, how the nasal olfactory mucosa of Atlantic salmon mobilised its defence repertoires when challenged with oxidative chemical stressors. To our knowledge, this is the first report that demonstrates the molecular changes initiated by chemically induced oxidative stress in the nasal mucosa of a teleost fish. Using a nasal leukocyte model, it was further shown how the constituent oxidants of the tested therapeutics alter the cellular redox balance and the associated response mounted by a specific group of cells at the nasal olfactory mucosa to these challenges.

The two in vivo exposure studies uncovered the molecular repertoire of the nasal olfactory mucosa when challenged with either a periodic-low dose or less frequent-high dose of the oxidative agent PAA. In the first trial, we addressed the nasal consequences in the application of oxidant as a routine disinfectant in salmon. No conclusive implications can be drawn as to whether the exogenous oxidant was a stimulator or an inhibitor of nasal mucosal physiology as the ratios of upregulated and downregulated genes were equal after 45 days of exposure. However, two groups—haemoglobins and immune genes—displayed a trend in response to the exogenous oxidant. Haemoglobin (Hb) is a predominant component in erythrocytes responsible for oxygen transport to the different tissues in vertebrates [31]. The Hb transcripts were upregulated in the nasal mucosa following periodic low-dose oxidant exposure, and both α and β subunits were represented. In murine models, it has been demonstrated that overexpression of Hb affected a network of genes involved in O_2 homeostasis, which subsequently resulted in the suppression of oxidative stress [32,33]. Exposing HepG2 cells to H_2O_2 induced the expression of both haemoglobin α and β, and their overexpression likewise resulted in cellular protection against oxidative stress [34]. The upregulation of several haemoglobin transcripts in response to periodic low-dose oxidant challenge was probably a protective mechanism of the nasal mucosa against oxidative stress. Despite the limited number, there was an indication that intermittent oxidant exposure may negatively regulate nasal immunity. The constitutive presence of oxidant in the environment may trigger several responses from an organism – continual mounting of protective action, accommodation, or habituation (i.e., decreasing response through time). It was reported earlier that periodic application of oxidant (i.e., PAA) in trout—a species closely related to salmon – somehow resulted in a dampening response, which could be indicative of habituation [27,35]. Such a consequence was likewise implicated in salmon post-smolts [28]. This partly sheds light on the downregulation of these immune genes after several weeks of exposure. We cannot disregard the potential oxidant-mediated immunosuppression in the nasal mucosa, as some of these transcripts have earlier been implicated in compromised immunity under oxidative stress in higher animal models [36,37].

Trial 2 provided a relatively clearer picture of how an oxidant administered at a higher dose, but less frequently, altered the nasal transcriptome as shown with a higher number of DEGs and a prominent regulatory profile. It was evident that oxidant treatment resulted in dysregulation of nasal redox balance, and hence triggered mucosal oxidative stress. In earlier publications, we have demonstrated that antioxidant defences in mucosal tissues (i.e., gills and skin) were remarkably altered by a similar oxidant though delivered at a much lower concentration [4,7]. The mobilisation of these antioxidant defences following oxidant exposure highlighted the capability of mucosal surfaces to muster physiological responses to increased environmental ROS, thereby protecting the mucosa from eventual oxidative damage, as supported by in vitro cell works demonstrated here as well. A substantial upregulation in immune-related genes was also observed, which likely offered insights into the complementarity of the immune and antioxidant defence mechanism at the mucosa during oxidative stress. There is a tight relationship between oxidative stress and immunity, and often the co-regulation of these two defence mechanisms provides robust responses during oxidative challenges [3,38]. One of the immune effector molecules that were markedly regulated was *ornithine decarboxylase*, a gene coding for an enzyme responsible for catalysing the conversion of ornithine to putrescine, the first and rate-limiting step in the synthesis of putrescine and the polyamines spermidine and spermine [39]. It is important to highlight that *spermidine/spermine N1-acetyltransferase 1*, a key molecule in amine metabolism, and with a counteractive function against oxidative stress, was significantly upregulated too. Increased expression of *ornithine decarboxylase* had been demonstrated in human hepatoma HUH7 cells subjected to chemically induced oxidative stress [40]. The upregulation observed in the present study is indicative of a similar function for protection against oxidative damage. Heat shock proteins play a role in several cellular processes that occur during and after exposure to oxidative stress [41]. There are indications, both in this trial as well as from previous studies using this oxidant, that it can induce transient oxidative stress [4,7,27,28]. The upregulation of both *hsps* in the nasal mucosa following oxidant exposure suggests intervention in oxidative stress-triggered changes by correction of conformation or selecting and directing aberrant proteins to the proteasome or lysosomes for degradation, in which these molecules are known to be key regulators. Besides the activation of several effector molecules that provide an interconnected response to oxidant-triggered oxidative stress, it is compelling to observe that several genes for molecules for cytostructural and extracellular matrix were represented in the list of upregulated genes. Previous studies using this oxidant found histostructural changes, though minimal, in the gill and skin mucosa [4,27,42]. We can speculate that the upregulation observed here may play a role in maintaining the structural integrity and barrier functionality under oxidative challenge. The upregulation of two genes responsible for extracellular matrix, *GMP giant mucus protein* and *mucin 5b*, implies that mucus physiology is affected by the oxidant and that the changes in these two genes underline their function in providing a layer of protective defence at the mucosa. Two related genes have been shown to participate in modulating the mucus layer of the olfactory epithelium in mammalian models [43,44], and they likely exert a similar function in the nasal mucosa of salmon. However, this must be functionally ascertained in the future. DEGs with known functions in lipid metabolism were downregulated in the oxidant-challenged nasal mucosa. Though it is difficult to conclude whether oxidative stress triggered an imbalanced lipid metabolism in the nasal mucosa because of a limited panel of DEGs under this category, it is interesting to note that such interaction has been reported in other animals [45,46]. The downregulation observed in these lipid metabolic mediators indicates that oxidant exposure may interfere with this process, though the magnitude remains an open question.

We then focused on one specific cell type at the nasal mucosa to investigate how the cells respond to the oxidant, as well as to the two other constituents of PAA trade products. The proliferation of nasal olfactory leukocytes was affected by the three chemical stressors 24 h after exposure at an almost similar rate. However, such an influence was no longer observable 48 h after exposure in

PAA- and H_2O_2-exposed cells. This indicates that the effects on cellular proliferation following PAA and H_2O_2 stimulation could be transitory, and the cells were able to recover quickly. AA was more potent in inhibiting cellular proliferation, as the effects persisted until 48 h. Nasal olfactory leukocytes exhibited migratory potential, as all factors resulted in migration of the cells, though at varying levels. Cell migration plays an important role in many normal biological and pathophysiological processes, and oxidants can either promote or inhibit migration [47]. H_2O_2 administered individually and not in mixture with PAA and AA modulated the migration of the nasal olfactory leukocytes, suggesting its potent chemoattractant function, as demonstrated by earlier studies in other animal models [47,48]. The migratory potential of nasal olfactory leukocytes is vital in orchestrating a cellular response when environmental ROS levels reach a concerning concentration or when pathophysiological response had been caused by oxidative stress.

We further asked: *If the oxidants triggered changes in the cellular phenotypic response* (i.e., proliferation and migration), *can they also induce oxidative stress*? The oxidant used had been shown to prompt oxidative stress at the systemic [7] and tissue [4,7] levels, but this has yet to be demonstrated at the cellular level. Exposure to the three chemical oxidative stressors resulted in an increase in the intracellular concentration of ROS, which indicates that it induced redox imbalance; hence, oxidative stress occurred. Both H_2O_2 and AA have long been identified as inducers of oxidative stress [11,48,49]. Here, we have shown that PAA and H_2O_2 were far more potent inducers of intracellular ROS production than AA in nasal leukocytes. Interestingly, for both PAA and AA, induction of intracellular level was transient because, after an elevated level 24 h after exposure, the concentration returned to the normal/control level. This was not observed in H_2O_2-exposed cells; their intracellular ROS was still at an elevated level 48 h after exposure, though the level was slightly lower compared to that at 24 h. This indicates that cells exposed to PAA and AA have a faster capability to abate an increased intracellular ROS level than H_2O_2-exposed cells, which further suggests that H_2O_2 has a higher and more persisting impact as an oxidative stressor on the nasal leukocytes. Different oxidants could trigger different mechanisms of oxidative stress induction [1,50], and somehow the observations in the present study are in agreement with this differential regulatory impact.

The increased intracellular ROS incited by the three chemical stressors initiated a response from the antioxidant system of the nasal olfactory leukocytes. The expression of enzymatic antioxidants was predominantly upregulated, indicating their key role towards the threats of the oxidative stressors. The overall expression profile of these antioxidant genes shows that PAA and H_2O_2 were more potent triggers than AA, which, to some extent, is in agreement with the results in terms of how the oxidative stressors affected the intracellular ROS level. Antioxidant markers *gpx*, *gr*, *gsta*, and *cat* were perhaps the key response molecules, as their expression was significantly elevated in at least one time-point in PAA- and H_2O_2-exposed cells. The response profile can be divided into two arbitrary groups based on their elevated temporal expression – early (i.e., *gr*) and late (i.e., *gsta* and *cat*) oxidant responders. *Glutathione peroxidase* metabolises H_2O_2 to H_2O, and the reduced glutathione provides an antioxidant function by resetting the redox status in tissues [51]. *Gpx* has been implicated in the responses of fish to environmental toxicants that can trigger oxidative stress [52], and this mechanism may also be working in nasal leukocytes. We have previously documented oxidant-induced alteration in *gpx* expression in the gills and skin of salmon; hence, the results provide further evidence that it is a vital molecule for the mucosal antioxidant system in this fish species [7]. It is interesting to note that *glutathione reductase gr* was upregulated both in the cells and as one of the DEGs identified in Trial 2. *Gr*, as an antioxidant, is responsible for the regeneration of reduced glutathione during detoxification of peroxides and free radicals formed in mitochondria, thus maintaining the redox potential of the cell [53]. The upregulation of *gr* in both instances, as well as in previous oxidant studies in salmon [7,10], provides strong support for its canonical function in mucosal antioxidant defence, that is likely ubiquitously regulated by the oxidative chemical stressor. There was no apparent tendency for AA-induced changes in the antioxidant repertoire in the nasal olfactory leukocytes, though both *gpx* and *cu/zn sod* were responsive.

The overall profile in the expression of cytokine markers and heat shock protein genes could not be conclusively established, though the stochastic changes provide insights into how oxidative stressors may likely influence these molecules. The relationship between oxidants and inflammatory response is well-established in mammalian systems [54], though such interaction is less understood in fish, especially concerning exogenous oxidative stressors. We have shown that *il1b* expression was modulated by PAA and H_2O_2 24 h after exposure. An earlier publication reported that increased IL-1β stimulated glutathione production, thereby protecting neurons from oxidative damage [55]. The present data could not decisively ascertain whether such a directional effect was also initiated in the nasal leukocytes; however, the upregulation of *il1b* and the two genes of glutathione metabolism offer a potential link. In Trial 2, we have identified several cytokines genes that were upregulated following the oxidant challenge. It is possible that this increase in expression facilitated the migration of inflammatory cells to the epithelial surface where the oxidant was in close contact. One of the areas that must be explored in the future is the early phase of the inflammatory response, which was not captured by the current data. Both *hsp70* and *hsp90* were significantly upregulated in the olfactory rosette in Trial 2, but a similar change in the cell experiment could not be observed. The mode of oxidant application may play in part in this apparent difference.

5. Conclusions

Fish encounter environmental oxidants during production, as several husbandry practices rely on the use of oxidative chemical compounds, such as during water disinfection [27] or disease treatment [9]. Application frequency may vary, from continuous to periodic, each depending on its intended use (i.e., disinfectant vs. therapy). PAA is a potent oxidant, though the window of safe dose is limited [56]. The present study contributes to a better understanding of how the nasal olfactory mucosa of Atlantic salmon, one of the least explored mucosal tissues with regards to redox physiology, mount physiological and immunological responses when prompted with exogenously generated oxidative challenges. Oxidative stress is a physiological imbalance that requires a coordinated response to protect the organism from oxidative damage and, eventually, facilitate repair and recovery. The nasal mucosa of salmon can activate different molecules that may likely participate in the adaptive responses to oxidative stress. Nasal immunology is one of the emerging fields in fish immunology research [20], and the several oxidant-responsive genes identified in the paper are potential molecules for in-depth functional characterisation for their role in the nasal microenvironment, mainly towards non-infectious agents. One area that is interesting for future experiment is on whether exogenous antioxidants (e.g., in-feed antioxidants) can mitigate the effects of the chemical oxidative stressors by augmenting the innate antioxidant system of fish.

Supplementary Materials: The following are available online at http://www.mdpi.com/2076-3921/9/11/1144/s1, Supplementary File 1. List of primers used in qPCR analysis. Supplementary File 2. List of differentially expressed genes identified in Trials 1,2.

Author Contributions: C.C.L. conceived the research idea. C.C.L. and M.W.B. designed the in vivo fish trial. C.C.L., J.O. and M.W.B. performed the in vivo fish trial. C.C.L., J.O. and M.W.B. collected the samples. C.C.L. and V.V. performed the cell experiment. C.C.L., J.O., V.V. and M.H.S.H. conducted the lab analyses. M.H.S.H. and A.K. performed the microarray. C.C.L. and A.K. handled and processed the data. All authors contributed to the writing and review of the final version of the manuscript. All authors have read and agreed to the published version of the manuscript.

Funding: The study was funded by grants from the Research Council of Norway (237856 and 194050) and the Norwegian Seafood Research Fund (901472).

Acknowledgments: J. Osorio wishes to acknowledge the ERASMUS+ Student Mobility Programme for financing his 5-month research stay at Nofima. Lars-Flemming Pedersen of DTU Aqua is thanked for his insights on the trial conducted at HiT. The assistance of the technical staff at NCRA (Britt-Kristin Megård Reiten, Bernhard Eckel, Yuriy Marchenko, Jascha Gerwins) and HiT (Mads Korfitz Mortensen) during the fish trials is also acknowledged. We would like to thank Fiskefjøset at NMBU for providing the fish used in the cell experiment and Lisbeth Rørmark of Lilleborg AS for the PAA product. We would like to acknowledge the 4 independent reviewers for their comments and suggestions that significantly improved the scientific value of this paper.

Conflicts of Interest: The authors declare no competing interests. Mention of trade names in this manuscript does not imply any recommendation or endorsement by Nofima or the University of Lisbon.

References

1. Birben, E.; Sahiner, U.M.; Sackesen, C.; Erzurum, S.; Kalayci, O. Oxidative stress and antioxidant defense. *World Allergy Organ. J.* **2012**, *5*, 9–19. [CrossRef]
2. Lackner, R. "Oxidative stress" in fish by environmental pollutants. In *Fish Ecotoxicology*; Braunbeck, T., Hinton, D.E., Streit, B., Eds.; Birkhäuser: Basel, Switzerland, 1998; pp. 203–224. [CrossRef]
3. Biller, J.D.; Takahashi, L.S. Oxidative stress and fish immune system: Phagocytosis and leukocyte respiratory burst activity. *An. Acad. Bras. Ciênc.* **2018**, *90*, 3403–3414. [CrossRef]
4. Lazado, C.C.; Sveen, L.R.; Soleng, M.; Pedersen, L.-F.; Timmerhaus, G. Crowding reshapes the mucosal but not the systemic response repertoires of Atlantic salmon to peracetic acid. *Aquaculture* **2020**, *531*, 735830. [CrossRef]
5. Miller, J.K.; Brzezinska-Slebodzinska, E.; Madsen, F.C. Oxidative stress, antioxidants, and animal function. *J. Dairy Sci.* **1993**, *76*, 2812–2823. [CrossRef]
6. Yoneyama, M.; Kawada, K.; Gotoh, Y.; Shiba, T.; Ogita, K. Endogenous reactive oxygen species are essential for proliferation of neural stem/progenitor cells. *Neurochem. Int.* **2010**, *56*, 740–746. [CrossRef] [PubMed]
7. Soleng, M.; Johansen, L.-H.; Johnsen, H.; Johansson, G.S.; Breiland, M.W.; Rørmark, L.; Pittman, K.; Pedersen, L.-F.; Lazado, C.C. Atlantic salmon (*Salmo salar*) mounts systemic and mucosal stress responses to peracetic acid. *Fish Shellfish Immunol.* **2019**, *93*, 895–903. [CrossRef] [PubMed]
8. Spiliotopoulou, A.; Rojas-Tirado, P.; Chhetri, R.K.; Kaarsholm, K.M.S.; Martin, R.; Pedersen, P.B.; Pedersen, L.-F.; Andersen, H.R. Ozonation control and effects of ozone on water quality in recirculating aquaculture systems. *Water Res.* **2018**, *133*, 289–298. [CrossRef] [PubMed]
9. Bechmann, R.K.; Arnberg, M.; Gomiero, A.; Westerlund, S.; Lyng, E.; Berry, M.; Agustsson, T.; Jager, T.; Burridge, L.E. Gill damage and delayed mortality of Northern shrimp (*Pandalus borealis*) after short time exposure to anti-parasitic veterinary medicine containing hydrogen peroxide. *Ecotoxicol. Environ. Saf.* **2019**, *180*, 473–482. [CrossRef] [PubMed]
10. Stiller, K.T.; Kolarevic, J.; Lazado, C.C.; Gerwins, J.; Good, C.; Summerfelt, S.T.; Mota, V.C.; Espmark, Å.M. The effects of ozone on Atlantic salmon post-smolt in brackish water—Establishing welfare indicators and thresholds. *Int. J. Mol. Sci.* **2020**, *21*, 5109. [CrossRef] [PubMed]
11. Vera, L.M.; Migaud, H. Hydrogen peroxide treatment in Atlantic salmon induces stress and detoxification response in a daily manner. *Chronobiol. Int.* **2016**, *33*, 530–542. [CrossRef] [PubMed]
12. Vatansever, F.; de Melo, W.C.M.A.; Avci, P.; Vecchio, D.; Sadasivam, M.; Gupta, A.; Chandran, R.; Karimi, M.; Parizotto, N.A.; Yin, R.; et al. Antimicrobial strategies centered around reactive oxygen species—Bactericidal antibiotics, photodynamic therapy, and beyond. *FEMS Microbiol. Rev.* **2013**, *37*, 955–989. [CrossRef] [PubMed]
13. Cabillon, N.A.R.; Lazado, C.C. Mucosal barrier functions of fish under changing environmental conditions. *Fishes* **2019**, *4*, 2. [CrossRef]
14. Sepahi, A.; Casadei, E.; Tacchi, L.; Muñoz, P.; LaPatra, S.E.; Salinas, I. Tissue microenvironments in the nasal epithelium of rainbow trout (*Oncorhynchus mykiss*) define two distinct CD8α^+ cell populations and establish regional immunity. *J. Immunol.* **2016**, *197*, 4453–4463. [CrossRef] [PubMed]
15. Peatman, E.; Beck, B.H. 1—Why mucosal health? In *Mucosal Health in Aquaculture*; Beck, B.H., Peatman, E., Eds.; Academic Press: San Diego, CA, USA, 2015; pp. 1–2. [CrossRef]
16. Tacchi, L.; Musharrafieh, R.; Larragoite, E.T.; Crossey, K.; Erhardt, E.B.; Martin, S.A.M.; LaPatra, S.E.; Salinas, I. Nasal immunity is an ancient arm of the mucosal immune system of vertebrates. *Nat. Commun.* **2014**, *5*, 5205. [CrossRef]
17. Døving, K.B. Functional properties of the fish olfactory system. In *Progress in Sensory Physiology 6*; Autrum, H., Ottoson, D., Perl, E.R., Schmidt, R.F., Shimazu, H., Willis, W.D., Eds.; Springer: Berlin/Heidelberg, Germany, 1986; pp. 39–104. [CrossRef]
18. Sepahi, A.; Kraus, A.; Casadei, E.; Johnston, C.A.; Galindo-Villegas, J.; Kelly, C.; García-Moreno, D.; Muñoz, P.; Mulero, V.; Huertas, M.; et al. Olfactory sensory neurons mediate ultrarapid antiviral immune responses in a TrkA-dependent manner. *Proc. Natl. Acad. Sci. USA* **2019**, *116*, 12428–12436. [CrossRef] [PubMed]

19. Das, P.K.; Salinas, I. Fish nasal immunity: From mucosal vaccines to neuroimmunology. *Fish Shellfish Immunol.* **2020**, *104*, 165–171. [CrossRef] [PubMed]
20. Sepahi, A.; Salinas, I. The evolution of nasal immune systems in vertebrates. *Mol. Immunol* **2016**, *69*, 131–138. [CrossRef]
21. Hansen, A.; Zielinski, B.S. Diversity in the olfactory epithelium of bony fishes: Development, lamellar arrangement, sensory neuron cell types and transduction components. *J. Neurocytol.* **2005**, *34*, 183–208. [CrossRef]
22. Zhang, Y.-A.; Salinas, I.; Li, J.; Parra, D.; Bjork, S.; Xu, Z.; LaPatra, S.E.; Bartholomew, J.; Sunyer, J.O. IgT, a primitive immunoglobulin class specialised in mucosal immunity. *Nat. Immunol.* **2010**, *11*, 827–835. [CrossRef]
23. Pace, E.; Ferraro, M.; Di Vincenzo, S.; Gerbino, S.; Bruno, A.; Lanata, L.; Gjomarkaj, M. Oxidative stress and innate immunity responses in cigarette smoke stimulated nasal epithelial cells. *Toxicol. In Vitro* **2014**, *28*, 292–299. [CrossRef]
24. Dokic, D.; Trajkovska-Dokic, E.; Howarth, H.P. Effects of ozone on nasal mucosa (endothelial cells). *Prilozi* **2011**, *32*, 87–99.
25. Wagner, J.G.; Hotchkiss, J.A.; Harkema, J.R. Enhancement of nasal inflammatory and epithelial responses after ozone and allergen coexposure in brown norway rats. *Toxicol. Sci.* **2002**, *67*, 284–294. [CrossRef]
26. Pedersen, L.F.; Lazado, C.C. Decay of peracetic acid in seawater and implications for its chemotherapeutic potential in aquaculture. *Aquac. Environ. Interact.* **2020**, *12*, 153–165. [CrossRef]
27. Liu, D.; Lazado, C.C.; Pedersen, L.-F.; Straus, D.L.; Meinelt, T. Antioxidative, histological and immunological responses of rainbow trout after periodic and continuous exposures to a peracetic acid-based disinfectant. *Aquaculture* **2020**, *520*, 734956. [CrossRef]
28. Lazado, C.C.; Pedersen, L.-F.; Kirste, K.H.; Soleng, M.; Breiland, M.W.; Timmerhaus, G. Oxidant-induced modifications in the mucosal transcriptome and circulating metabolome of Atlantic salmon. *Aquat. Toxicol.* **2020**, *227*, 105625. [CrossRef]
29. Nagasawa, K.; Lazado, C.; Fernandes, J.M. Validation of endogenous reference genes for qPCR quantification of muscle transcripts in Atlantic cod subjected to different photoperiod regimes. In *Aquaculture*; IntechOpen: London, UK, 2012.
30. Krasnov, A.; Timmerhaus, G.; Afanasyev, S.; Jørgensen, S.M. Development and assessment of oligonucleotide microarrays for Atlantic salmon (*Salmo salar* L.). *Comp. Biochem. Physiol. Part D* **2011**, *6*, 31–38. [CrossRef] [PubMed]
31. Quinn, N.L.; Boroevich, K.A.; Lubieniecki, K.P.; Chow, W.; Davidson, E.A.; Phillips, R.B.; Koop, B.F.; Davidson, W.S. Genomic organisation and evolution of the Atlantic salmon hemoglobin repertoire. *BMC Genom.* **2010**, *11*, 539. [CrossRef] [PubMed]
32. Biagioli, M.; Pinto, M.; Cesselli, D.; Zaninello, M.; Lazarevic, D.; Roncaglia, P.; Simone, R.; Vlachouli, C.; Plessy, C.; Bertin, N.; et al. Unexpected expression of alpha- and beta-globin in mesencephalic dopaminergic neurons and glial cells. *Proc. Natl. Acad. Sci. USA* **2009**, *106*, 15454–15459. [CrossRef]
33. Nishi, H.; Inagi, R.; Kato, H.; Tanemoto, M.; Kojima, I.; Son, D.; Fujita, T.; Nangaku, M. Hemoglobin is expressed by mesangial cells and reduces oxidant stress. *J. Am. Soc. Nephrol.* **2008**, *19*, 1500–1508. [CrossRef]
34. Liu, W.; Baker, S.S.; Baker, R.D.; Nowak, N.J.; Zhu, L. Upregulation of hemoglobin expression by oxidative stress in hepatocytes and its implication in nonalcoholic steatohepatitis. *PLoS ONE* **2011**, *6*, e24363. [CrossRef]
35. Liu, D.; Pedersen, L.-F.; Straus, D.L.; Kloas, W.; Meinelt, T. Alternative prophylaxis/disinfection in aquaculture—Adaptable stress induced by peracetic acid at low concentration and its application strategy in RAS. *Aquaculture* **2017**, *474*, 82–85. [CrossRef]
36. Tran, H.B.; Ahern, J.; Hodge, G.; Holt, P.; Dean, M.M.; Reynolds, P.N.; Hodge, S. Oxidative stress decreases functional airway mannose binding lectin in COPD. *PLoS ONE* **2014**, *9*, e98571. [CrossRef] [PubMed]
37. Kawashima, R.; Shimizu, T.; To, M.; Saruta, J.; Jinbu, Y.; Kusama, M.; Tsukinoki, K. Effects of stress on mouse β-defensin-3 expression in the upper digestive mucosa. *Yonsei Med. J.* **2014**, *55*, 387–394. [CrossRef] [PubMed]
38. Biller-Takahashi, J.D.; Takahashi, L.S.; Mingatto, F.E.; Urbinati, E.C. The immune system is limited by oxidative stress: Dietary selenium promotes optimal antioxidative status and greatest immune defense in pacu Piaractus mesopotamicus. *Fish Shellfish Immunol.* **2015**, *47*, 360–367. [CrossRef]

39. Bacchi, C.; Nathan, H.; Hutner, S.; McCann, P.; Sjoerdsma, A. Polyamine metabolism: A potential therapeutic target in trypanosomes. *Science* **1980**, *210*, 332–334. [CrossRef]
40. Smirnova, O.A.; Isaguliants, M.G.; Hyvonen, M.T.; Keinanen, T.A.; Tunitskaya, V.L.; Vepsalainen, J.; Alhonen, L.; Kochetkov, S.N.; Ivanov, A.V. Chemically induced oxidative stress increases polyamine levels by activating the transcription of ornithine decarboxylase and spermidine/spermine-N1-acetyltransferase in human hepatoma HUH7 cells. *Biochimie* **2012**, *94*, 1876–1883. [CrossRef] [PubMed]
41. Kalmar, B.; Greensmith, L. Induction of heat shock proteins for protection against oxidative stress. *Adv. Drug Deliv. Rev.* **2009**, *61*, 310–318. [CrossRef]
42. Lazado, C.C.; Haddeland, S.; Timmerhaus, G.; Berg, R.S.; Merkin, G.; Pittman, K.; Pedersen, L.-F. Morphomolecular alterations in the skin mucosa of Atlantic salmon (*Salmo salar*) after exposure to peracetic acid-based disinfectant. *Aquac. Rep.* **2020**, *17*, 100368. [CrossRef]
43. Débat, H.; Eloit, C.; Blon, F.; Sarazin, B.; Henry, C.; Huet, J.-C.; Trotier, D.; Pernollet, J.-C. Identification of Human Olfactory Cleft Mucus Proteins Using Proteomic Analysis. *J. Proteome Res.* **2007**, *6*, 1985–1996. [CrossRef]
44. Kennel, C.; Gould, E.A.; Larson, E.D.; Salcedo, E.; Vickery, T.; Restrepo, D.; Ramakrishnan, V.R. Differential expression of mucins in murine olfactory versus respiratory epithelium. *Chem. Senses* **2019**, *44*, 511–521. [CrossRef]
45. Chen, Q.; Tang, L.; Xin, G.; Li, S.; Ma, L.; Xu, Y.; Zhuang, M.; Xiong, Q.; Wei, Z.; Xing, Z.; et al. Oxidative stress mediated by lipid metabolism contributes to high glucose-induced senescence in retinal pigment epithelium. *Free Radic. Biol. Med.* **2019**, *130*, 48–58. [CrossRef] [PubMed]
46. Seo, E.; Kang, H.; Choi, H.; Choi, W.; Jun, H.-S. Reactive oxygen species-induced changes in glucose and lipid metabolism contribute to the accumulation of cholesterol in the liver during aging. *Aging Cell* **2019**, *18*, e12895. [CrossRef] [PubMed]
47. Hurd, T.R.; DeGennaro, M.; Lehmann, R. Redox regulation of cell migration and adhesion. *Trends Cell Biol.* **2012**, *22*, 107–115. [CrossRef] [PubMed]
48. Klyubin, I.; Kirpichnikova, K.; Gamaley, I. Hydrogen peroxide-induced chemotaxis of mouse peritoneal neutrophils. *Eur. J. Cell Biol.* **1996**, *70*, 347–351. [PubMed]
49. Terasaki, M.; Ito, H.; Kurokawa, H.; Tamura, M.; Okabe, S.; Matsui, H.; Hyodo, I. Acetic acid is an oxidative stressor in gastric cancer cells. *J. Clin. Biochem. Nutr.* **2018**, *63*, 36–41. [CrossRef]
50. Costantini, D. Understanding diversity in oxidative status and oxidative stress: The opportunities and challenges ahead. *J. Exp. Biol.* **2019**, *222*, jeb194688. [CrossRef]
51. Parihar, M.S.; Javeri, T.; Hemnani, T.; Dubey, A.K.; Prakash, P. Responses of superoxide dismutase, glutathione peroxidase and reduced glutathione antioxidant defenses in gills of the freshwater catfish (*Heteropneustes fossilis*) to short-term elevated temperature. *J. Therm. Biol.* **1997**, *22*, 151–156. [CrossRef]
52. Sukhovskaya, I.V.; Borvinskaya, E.V.; Smirnov, L.P.; Kochneva, A.A. Role of glutathione in functioning of the system of antioxidant protection in fish (review). *Inland Water Biol.* **2017**, *10*, 97–102. [CrossRef]
53. Čolak, E.; Žorić, L. 6—Antioxidants and Age-related macular degeneration. In *Handbook of Nutrition, Diet, and the Eye*, 2nd ed.; Preedy, V.R., Watson, R.R., Eds.; Academic Press: Cambridge, MA, USA, 2019; pp. 85–106. [CrossRef]
54. Lugrin, J.; Rosenblatt-Velin, N.; Parapanov, R.; Liaudet, L. The role of oxidative stress during inflammatory processes. *Biol. Chem.* **2013**, *395*, 241. [CrossRef]
55. Chowdhury, T.; Allen, M.F.; Thorn, T.L.; He, Y.; Hewett, S.J. Interleukin-1β protects neurons against oxidant-induced injury via the promotion of astrocyte glutathione production. *Antioxidants* **2018**, *7*, 100. [CrossRef]
56. Acosta, F.; Montero, D.; Izquierdo, M.; Galindo-Villegas, J. High-level biocidal products effectively eradicate pathogenic γ-proteobacteria biofilms from aquaculture facilities. *Aquaculture* **2021**, *532*, 736004. [CrossRef]

Publisher's Note: MDPI stays neutral with regard to jurisdictional claims in published maps and institutional affiliations.

© 2020 by the authors. Licensee MDPI, Basel, Switzerland. This article is an open access article distributed under the terms and conditions of the Creative Commons Attribution (CC BY) license (http://creativecommons.org/licenses/by/4.0/).

Review

An Update of Palmitoylethanolamide and Luteolin Effects in Preclinical and Clinical Studies of Neuroinflammatory Events

Marika Cordaro [1,†], Salvatore Cuzzocrea [2,3,*] and Rosalia Crupi [2,†]

1 Department of Biomedical and Dental Sciences and Morphofunctional Imaging, University of Messina, Via Consolare Valeria 1, 98100 Messina, Italy; cordarom@unime.it
2 Department of Chemical, Biological, Pharmaceutical and Environmental Sciences, University of Messina, Via F. Stagno D'Alcontres 31, 98166 Messina, Italy; rcrupi@unime.it
3 Department of Pharmacology and Physiology, Saint Louis University, St. Louis, MO 63103, USA
* Correspondence: salvator@unime.it; Tel.: +390-906-765-208
† The authors equally contributed to this work.

Received: 29 January 2020; Accepted: 26 February 2020; Published: 5 March 2020

Abstract: The inflammation process represents of a dynamic series of phenomena that manifest themselves with an intense vascular reaction. Neuroinflammation is a reply from the central nervous system (CNS) and the peripheral nervous system (PNS) to a changed homeostasis. There are two cell systems that mediate this process: the glia of the CNS and the lymphocites, monocytes, and macrophages of the hematopoietic system. In both the peripheral and central nervous systems, neuroinflammation plays an important role in the pathogenesis of neurodegenerative diseases, such as Parkinson's and Alzheimer's diseases, and in neuropsychiatric illnesses, such as depression and autism spectrum disorders. The resolution of neuroinflammation is a process that allows for inflamed tissues to return to homeostasis. In this process the important players are represented by lipid mediators. Among the naturally occurring lipid signaling molecules, a prominent role is played by the N-acylethanolamines, namely N-arachidonoylethanolamine and its congener N-palmitoylethanolamine, which is also named palmitoylethanolamide or PEA. PEA possesses a powerful neuroprotective and anti-inflammatory power but has no antioxidant effects per se. For this reason, its co-ultramicronization with the flavonoid luteolin is more efficacious than either molecule alone. Inhibiting or modulating the enzymatic breakdown of PEA represents a complementary therapeutic approach to treating neuroinflammation. The aim of this review is to discuss the role of ultramicronized PEA and co-ultramicronized PEA with luteolin in several neurological diseases using preclinical and clinical approaches.

Keywords: neuroinflammation; clinical; palmitoylethanolamide; luteolin; co-ultramicronization; CNS pathology; adaptive immune response; cell homeostasis

1. Introduction

Pain, redness, increased heat, and swelling are the four cardinal signs of an inflammatory response; loss of function is occasionally added with these four as the fifth mark of an inflammatory response [1]. The "inflammation process" represents the body's response to different tissue lesions, and as such, involves the recruitment of immune cells and the release of mediators. Consequently, innate and adaptive immune cells, as well as vascular cells and neurons initiate a defense process in order to maintain or restore tissue integrity. Inflammatory events of the central nervous system (CNS) occur at different levels from those of other tissues and involve several types of cells [2,3].

In particular, the first difference relies in the absence of resident dendritic cells in the CNS parenchyma, where perivascular macrophages and vascular pericytes take over the functions of mature dendritic cells in the CNS [4]. As a second feature, the activation of the innate immune cells of the CNS parenchyma, such as astrocytes, microglia, and in some regions, mast cells, may be increase in pathological conditions, such as stroke, trauma, growth of a tumor, or neurodegenerative disease [5–7]. In addition, for the body to respond adequately during an inflammatory event, extravasation of the immune cells and inflammatory molecules must take place. These events are indispensable for triggering the immune response and activating the complementary cascade. Nevertheless, in the CNS, the blood–CNS barrier' reduces the permeability of microvessels, thus making the whole inflammatory reaction more difficult. Only activated T cells may penetrate the barrier, but this does not elicit an efficient reaction to inflammation when compared with that observed in peripheral tissues, where dendritic cells are responsible for the adaptive immune response [8]. Due to these features, it is interesting to point out that the CNS reacts to inflammatory events when these exert a direct effect on the CNS, i.e., in the case of pathogens and tissue damage, and when the inflammatory events are so severe that infiltrating T cells are involved. These observations lead to the introduction of the term "neuroinflammation," which distinguishes the inflammatory reaction in the CNS from inflammation in different tissues. From this perspective, neuroinflammation is a response of the CNS to a changed homeostasis. There are two cell systems that are able to mediate this response: glia of the CNS, and lymphocytes, monocytes, and macrophages of the hematopoietic system [9]. The neuroinflammation can be triggered by infection, autoimmunity, and toxins, which are defined not just by classical factors, but also by noxious stimuli or psychological stress, such as neurogenic factors. Consequently, the actions promoted by the neuroinflammations are classified as: homeostatic (vasodilation and release of cytokines and neurotrophic factors); maladaptive (release of pro-inflammatory factors); neurotoxic (release of pro-inflammatory factors and breakdown of blood–CNS barrier); and anti-inflammatory (release of pro-inflammatory cytokines, neurotrophic factors, neurotransmitters, and cell adhesion molecules). Neuroinflammation after damage is actively controlled by a complex network of regulatory mechanisms that restrict the potentially harmful effects of persistent inflammation. In particular, chronic, uncontrolled inflammation is characterized by the overexpression of cytokines, such as TNF and IL, reactive oxygen species (ROS), and other inflammatory mediators (such as inducible nitric oxide synthase). All of these signals are identified during injuries to the CNS and are transferred to the injury site by attracting and transporting peripheral macrophages and neutrophils. However, when neuroinflammation is prolonged and macrophage iperactivation is extended, it overwhelms the limits of physiological regulation and causes deleterious effects, including pro-inflammatory signaling pathways, elevated oxidative stress, and death of adjacent neurons that relate to chronic pain pathogenesis, such as neuropathic pain, contributing to neurodegeneration [10,11]. Neuroinflammation is a common mechanism that influences the severity and progression of neurodegenerative disease; for this reason, it is an important target for neuroprotective therapies (Table 1).

Table 1. Publications in 2019 about the relationship between neuroinflammation and neurodegenerative disorders.

Pathology	References
Vascular Dementia	[12–18]
Depression	[19–48]
Alzheimer's Disease	[49–65]
Parkinson's Disease	[53,63,65–82]
Schizophrenia	[21,37,72,83–90]
Epilepsy	[21,37,72,83–90]

2. Microglia

Crosstalk between the glia and the neurons, which is necessary for the preservation of homeostasis, is exemplified by the coordination between the immune system and the central nervous system. The glia family is composed of three distinct cell types: microglia, astrocytes, and oligodendrocytes [91].

Both microglia and macrophages, located in the brain and spinal cord parenchyma, represent the key players in the regulation of immunity in the CNS. The production of cytokines and chemokines, as well as the production of free radicals, such as reactive oxygen species (ROS) and nitric oxide (NO) known to contribute to neuronal and tissue damage, are implicated in the regulation of neuroinflammation.

Furthermore, microglia activation, which is important in the development of neurogenesis, plays a crucial role in both synaptic pruning and damage restoration by removing apoptotic cells and secreting growth factors [92]. It is then clear that glia cells are not just cells that fill "the space not occupied by neurons," as Virchow suggested in the late nineteenth century, but dynamically play a role in neuronal support and dysfunction [93]. Microglia comprise only 10% of the total cell population of the brain, but represent the major resident immune cells that survey the microenvironment looking for cellular debris and pathogens, and produce factors, such as cytokines, that influence the surrounding astrocytes and neurons. Meanwhile, microglia cells are classified as good players because remove cellular debris via phagocytosis, release neurotrophic factors and anti-inflammatory cytokines, prevent neuronal injury, restore tissue integrity in the injured brain, maintain the plasticity in the neuronal circuits, and encourage the protection and remodeling of synapses [94,95]. Structurally, microglia show a dynamic and active phenotype with ongoing retraction and extension of processes in the brain's structure [96]. Following insults that are not due to pathogenic agents, tissue damage, or exposure to neurotoxic substances, microglia are activated by stimulating the inflammatory response, which further involves the immune system.

Microglia can act via specialized pro-resolving lipid mediators (SPMs) receptors, cannabinoid receptors 2 (CB2), and aryl hydrocarbon receptors (AHR) in response to their ligands, such as SPMs; cannabinoids; and gut-derived metabolites, such as tryptophan (TRP). Activation of these microglial receptors enhances the phagocytic activity, reduces the expression of pro-inflammatory mediators, and increases the production of anti-inflammatory mediators and SPMs, thus promoting the resolution of neuroinflammation and pathological pain [11].

Following this activation, pro- and anti-inflammatory cytokines are released in the cerebral parenchyma. There are two phenotypes that describe this dichotomy at the level of the microglial phenotype: M1 and M2. The microglial phenotype changes in relation to the insult, the extent of the damage, and the post-injury recovery [97,98].

For example, lipopolysaccharide (LPS) and the pro-inflammatory cytokine interferon γ (IFNγ) promote a "classically activated" M1 phenotype, which produces high levels of pro-inflammatory cytokines and oxidative metabolites that are essential for host defense and phagocytic activity, but that also cause damage to healthy cells and tissue. Conversely, activating microglia in the presence of cytokines, such as IL-4 or IL-13, promote an "alternatively activated" M2 phenotype [99,100]. Although there is limited data on resident M2 microglia in the brain, it is thought that much like the M2 macrophages, these cells can promote angiogenesis, wound healing and tissue repair, extracellular matrix remodeling, and suppress destructive immune responses [101]. M2 microglia express specific antigens, such as arginase 1 (Arg1); mannose receptor (MRC), found in the inflammatory zone 1 (FIZZ1); and chitinase 3-like 3 (Ym1), among others. Furthermore, cultured microglia exposed to IL-4 or IL-13 develop the M2 phenotype, which can result in extensive neurite elongation and outgrowth across inhibitory surfaces in in vitro co-culture systems [99,102–104].

In the human brain, microglial activation and neuroinflammation have been associated with viral or bacterial infection, autoimmune diseases, head trauma, vascular system damage, neuropsychiatric disorders, and neurodegenerative illness.

3. Astroglia

Astrocytes are classified as immunocompetent cells that are good regulators of brain inflammation and can be divided into two major groups: protoplasmic astrocytes (gray matter, type-1) and fibrous astrocytes (white matter, type-2). Protoplasmic astrocytes, which are widely distributed in gray matter, have a larger size (≈50 μm) and more organelles than fibrous astrocytes, and at least one process contacts blood vessels through perivascular endfeet, as well as forming multiple contacts with neurons. These astrocytes regulate local blood flow and neuronal homeostasis [105,106]. Fibrous astrocytes originate from radial glial cells that are capable of differentiating neurons, astrocytes, and oligodendrocytes during brain development, and these astrocytes highly express glial fibrillary acidic protein (GFAP), nestin, and vimentin [107,108]. Although the specific function remains to be characterized, fibrous astrocytes also contact vessel capillaries like the protoplasmic astrocytes [109]. Generally, elevated GFAP is a common feature of the activation state of astrocytes.

Like microglia, astrocytes can become activated through a process known as astrogliosis. This process is characterized by altered gene expression, hypertrophy, and proliferation [110]. Because of their structure, astrocytes are defined as "territorial cells" with an essential role in the encouraging of cells to provide neurons for homeostasis [111]. Astrocytes strictly cooperate with the surrounding structures in the nervous system and provide the regulation of their functions. One of the most vital roles of astrocytes at the synapse is the clearance of neurotransmitters. For example, the astrocytic processes associated with excitatory synapses are enclosed with glutamate transporters, which maintain a low ambient glutamate level in the CNS and allow for glutamate receptor activation at synapses [112]. Nonetheless, astrocytes form the glia limitans around blood vessels, preventing the entry of immune cells via the blood–brain barrier (BBB) into the CNS parenchyma. They are characterized by the augmented expression of glial fibrillary acidic protein (GFAP) and signs of functional deficiency [113]. Astrocytes are able to release interleukins, NO, cytokines, and other cytotoxic molecules that are able to exert either neuroprotective or neurotoxic effects [114]. In particular, recent studies have shown two different reactive astroglia populations in the adult CNS: A1 and A2. A1 astrocytes can aggravate disease pathogenesis and destroy both neurons and oligodendrocytes. In fact, in several neurodegenerative diseases, A1 astroglia are indeed present in mouse and human, particularly around areas of disease pathology. Vice versa, A2 astrocytes seem to upregulate neurotrophic genes that encourage neuronal survival. Additionally, activation of both A1 and A2 astroglia is a result of crosstalk between activated microglia and astroglia in diseases [115–117]. It is very likely that the astrocyte immune activation is the result of a very complex interaction between the pro- and anti-inflammatory process, and the neurotoxic and neurotrophic processes. The two main characteristics of astrocytes are their elaborated intracellular calcium signaling, also termed calcium waves, and their high degree of intercellular communication, which is mediated by gap junctional channels (GJCs) [118,119]. Connexins (Cx) are the molecular constituent of GJC, where Cx43 is the major astrocytic connexin identified. However, this does not exclude the possibility that other astrocytic connexins could be expressed, either in specific brain regions or in pathological situations [118,119].

4. Oligodendroglia

In the central nervous system, myelin is made up of cells called oligodendrocytes. Such cells are important in propagating action potentials along axons; an important duck function they perform is the support they give to neurons by producing neurotrophic factors. However, these cells are very vulnerable to both oxidative stress and the toxicity caused by an excess of glutamate. The oligodendroglia cells are characterized by the presence of a large pool of iron, but despite this, they have a low capacity for anti-oxidative mechanisms, which render the cells especially sensitive to oxidative stress [120]. The crosstalk between microglia and oligodendrocytes regulates cerebral homeostasis. After being activated at the microglia, it produces the pro-inflammatory mediators [121]; they are essential for killing pathogens, but on the other hand, these mediators can damage both the glia and adjacent neurons. Under these conditions, oligodendrocytes are particularly susceptible to microglial factors due to their high metabolic activity and their energy needs.

Its response is represented by the production of poor-quality myelin, which may stimulate the pathology detected in many neurological diseases. Oligodendrocytes, in particular oligodendrocyte progenitor cells (OPCs), express receptors for fibroblast growth factor (FGF), epidermal growth factor (EGF), platelet-derived growth factor (PDGF), insulin-like growth factor 1 (IGF-1), and vascular endothelial growth factor (VEGF). Recent studies demonstrate that VEGF plays a key role in the remyelination process by promoting the migration of OPCs, despite being considered one of the most important factors in the regulation of angiogenesis [122]. Moreover, oligodendrocytes cell surfaces have receptors for a wide-ranging of mediators, such as IL-4, IL-6, IL-7, IL-10, IL-11, IL-12, and IL-18 [123]. Recent evidence also demonstrates that oligodendrocyte also express antigen-presenting molecules and co-stimulatory molecules, complement and complement receptor molecules, complement regulatory molecules, tetraspanins, neuroimmune regulatory proteins, extracellular matrix proteins, and many others [124]. Necrosis and apoptosis are events that can activate oligodendrocytes in various ways in the central nervous system. In relation to the activated path, oligodendrocytes can also release factors that stimulate microglia; it is still possible that stressed cells can trigger the protective pro-repair responses of microglia.

5. Mast Cells

Mast cells, also called effector cells, are first responders that are able to intervene as catalysts and recruiters in initiating, amplifying, and prolonging all the immune and nervous responses that arise from their activation. They originate from a hematopoietic lineage and are involved in a number of normal physiologic functions, such as immunity [125], angiogenesis, and tissue remodeling, as well as being implicated in multiple pathologic conditions [126]. Their purpose is to produce various mediators, such as cytokines, lipid metabolites, enzymes, biogenic amines, growth factors, NO, ATP, heparin, and neuropeptides [127]. It is important to consider that the release of the mediator through the degranulation process is very rapid, whereas the activation is more long lasting, allowing the release of newly formed mediators [128]. Under basal conditions, without disease, trauma, or stressful events, the number of mast cells is lower than that of neurons, microglia, and other cells residing in the brain. Even if in a limited number of the activated mast cells are able to influence the BBB, neurons, astrocytes, and microglia, there is a considerable increase in mast cell degranulation after stroke in the immature brain, as well as after a transient global ischemia in adult rats, or even after the deprivation of in vitro oxygen glucose; all these events imply that mast cells really play a role in determining neuronal damage [129–131]. There are numerous molecular mechanisms that have been identified as determining the potential interactions between mast cells and microglia [5]. It has been shown that the activation of P2 receptors on microglia by ATP stimulates the release of IL-33, which after binding to mast cell receptors, triggers the release of IL-6, IL-13, and chemo-attractive monocytes 1; in turn, these molecules are able to regulate the activity of the microglia. Likewise, mast cell tryptide activates microglia-activated receptor 2 (PAR2) receptors, and promotes the release of pro-inflammatory mediators, such as TNF-α, IL-6, and ROS, which in turn upregulate the expression of PAR2 receptors on mast cells [132]. Activated mast cells upregulate P2X purinergic receptors, ligand-ions, PR2X4 microglia receptors, promoting brain-derived neurotrophic factor (BDNF) [133]. Mast cells react dynamically with astrocytes like microglia. Both mast cells and astrocytes are co-located in both perivascularization and in the thalamus [134]. In vitro studies have shown that mast cells activate astrocytes through activation of CD40–CD40 ligand interactions, and are stimulated to release cytokines, leukotrienes, and histamine [134]. Interestingly, astrocytes also have histamine receptors (H1R and H2R) [135], and cytokines released from astrocytes can induce mast cell degranulation [136].

6. Mediators during Neuroinflammation: Inflammasome, Cytokines, and Chemokines

Inflammasome, cytokines, and chemokines play an important role in mediating neuroinflammation; for this reason, they have received a considerable amount of attention as possible therapeutic targets [137,138]. Inflammasomes are cytosolic multiprotein complexes that upon assembly, activate the pro-inflammatory caspase-1, which is responsible for the maturation and secretion of the inflammatory cytokines IL-1β, IL-18, and IL-33, as well as the induction of the pyroptosis process [138,139]. These pro-inflammatory cytokines have been shown to promote a variety of innate immune processes linked to infection, inflammation, and autoimmunity, and thus play an instrumental role in instigating neuroinflammation during old age and the eventual occurrence of cognitive impairment, neurodegenerative diseases, and dementia [139]. Everything we know about the inflammasome activation and its role in CNS inflammation is still incomplete and is principally based on in vitro studies with primary microglia and microglial cell lines, and in vivo studies with transgenic KO mice that lack the expression of specific inflammasome components throughout the body [138].

In addition to microglia and astrocytes, cytokines are large proteins (15–25 kDa) that are mainly released from immune cells, such as monocytes, macrophages, and lymphocytes. Cytokines are stimulated in circumstances where inflammation, infection, and/or immunological changes occur, and are specifically involved in tissue repair and homeostasis restoration [140,141]. Cytokines are an exceptionally large and diverse group of pro- or anti-inflammatory factors that are grouped into families based upon their structural homology or that of their receptors [137]. Interleukin-1β (IL-1β), interleukin-6 (IL-6), and tumor necrosis factor-α (TNF-α) are among the most widely investigated pro-inflammatory cytokines, whereas interleukin-4 (IL-4) and interleukin-10 (IL-10) are well-known anti-inflammatory cytokines.

Chemokines are a category of secreted proteins in the family of cytokines whose basic role is to cause cell migration [142,143]. These "chemotactic cytokines" include the chemoattraction of leukocytes and the flow of immune cells to locations throughout the body [137]. The two principal chemokine ligand superfamiles are named according to the arrangement of the (typically four) cytokines within them: in the CC family, the first two cysteines near the amino terminus are adjacent, whereas in the CXC family, there is one amino acid between them [144].

In addition to being essential for the synchronization of immune responses throughout the body, cytokines and chemokines are implicated in the control of CNS–immune system interactions. They are produced primarily not only by white blood cells or leukocytes, but also by a variety of other cells as a response to various stimuli under both pathological and physiological conditions. In the nervous system, cytokines and chemokines function as neuromodulators and regulate neurodevelopment, neuroinflammation, and synaptic transmission [137]. Cytokines and chemokines are essential to the immune function of the brain, helping to maintain immune surveillance, promote leukocyte traffic, and employ other inflammatory factors [145]. During neuroinflammation, immune cells and cells of the nervous system can release cytokines and chemokines, as well as respond to them by way of cytokine and chemokine receptors [146–148].

7. Resolution of Neuroinflammation

Inflammatory states may activate a program of resolution that involves the production of lipid mediators with the capacity to switch off inflammatory processes [149]. Several studies have shown that neuroinflammatory processes are noticeably dangerous for the body; as such, numerous studies have been performed, and others are ongoing, to develop new therapeutic strategies. However, to do this, it is necessary to know not only about the neuronal and non-neuronal cells, but also their mechanism of action. The resolution of neuroinflammation is a process that allows for inflamed tissues to return to homeostasis. In this process the important players are the lipid mediators (LM).

Lipid mediators are widely appreciated for their important roles in initiating the leukocyte traffic required in the host defense [150]. These include the classic eicosanoids, prostaglandins (PGs), and leukotrienes (LTs) [151,152] that stimulate blood flow changes, edema, and neutrophil influx to tissues [153]. Novel resolution-phase mediators that possess potent proresolving actions were identified and named resolvins, protectins, and maresins.

Biochemically, lipid mediators represent a diverse family of endogenous bioactive molecules that are enzymatically derived from fatty acid substrates. Prostaglandins, a family of extensively studied lipid mediators, are synthesized from arachidonic acid and are elevated after an acquired neurological injury, such as a traumatic brain injury (TBI). The first LMs synthesized are prostaglandins and leukotrienes, which are generated to activate and amplify the cardinal signs of inflammation. In particular, prostaglandins E2 and D2 induce production of mediators (lipoxins [154], resolvins, and protectins [155]) with both anti-inflammatory and pro-resolution activities. These mediators are classified as endogenous pro-resolution molecules that are not immunosuppressive. SPMs are able to stimulate and accelerate resolution via mechanisms that are multi-factorial. On this basis, resolution is not a passive process, but conversely is an active and dynamic process that is orchestrated by distinct cellular events and endogenous chemical mediators. In this context, prostaglandins are considered LMs that can promote the neuroinflammation process. In an acute inflammatory response, prostaglandins regulate local changes in blood flow and pain sensitization.

On the other hand, SPMs also show beneficial effects in different neuroinflammation models, including stroke, Alzheimer's disease (AD), spinal cord injury, and neuropathic pain [156–160]. In order to neutralize the chronic inflammatory processes, a resolution program based on the production of lipid mediators is necessary and therefore able to block inflammation. In the presence of chronic inflammatory conditions, several mechanisms of resolution are triggered, and among these, the production of lipid mediators is known to extinguish inflammation [161]. The existence of molecules involved in endogenous protective mechanisms that are activated in the body due to severe tissue damage or the stimulation of inflammatory responses and nociceptive fibers is well recognized. In this context, we believe the N-acylethanolamines, a class of naturally occurring lipidic mediators, are constituted of a fatty acid and ethanolamine, namely the fatty acid ethanolamines (FAEs). The main FAE family members include the endocannabinoid N-arachidonoylethanolamine (anandamide or 5Z,8Z,11Z,14Z)-N-(2-hydroxyethyl)icosa-5,8,11,14-tetraenamide), and its congeners N-stearoylethanolamine (N-(2-hydroxyethyl)stearamide), N-oleoylethanolamine (N-2-hydroxyethyl -9(Z)-octadecenamide), and palmitoylethanolamide (PEA) (Figure 1).

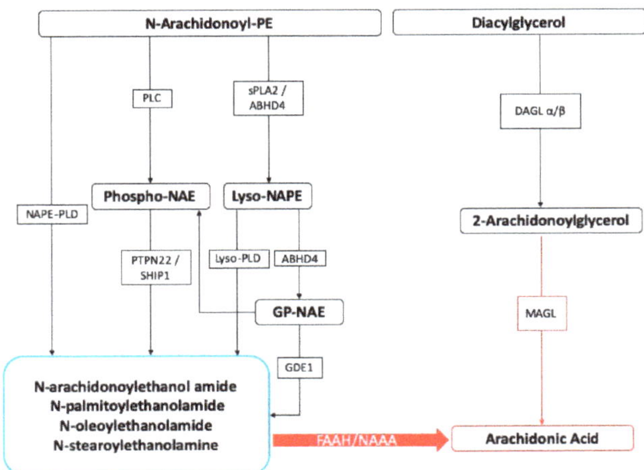

Figure 1. Fatty acid ethanolamines' metabolism and catabolism. Abbreviations: ABHD: α/β-Hydrolase domain containing, DAGL: diacylglycerol lipase, FAAH: fatty acid amide hydrolase, GDE: glycerophosphodiesterase, MAGL: monoacylglycerol lipase, NAAH: N-Acyl-ethanolamine-hydrolyzing acid amidase, NAPE: N-acyl-phosphatidylethanolamine, PLC: phospholipase C, PLD: phospholipase D, PTPN22: tyrosine phosphatase, SHIP1: inositol 5′-phosphatase, sPLA2: secretory phospholipase A2.

8. PEA: Mechanism of Action

N-acylethanolamines are classified as naturally occurring lipid mediator molecules composed of a fatty acid and ethanolamine, and are collectively identified as "fatty acid ethanolamines" (FAEs). These are endogenous molecules that have the characteristic of being involved in various mechanisms of endogenous protection, which are activated in the body by different types of tissue damage or stimulation of inflammatory responses and nociceptive fibers. The members of the FAE family are the endocannabinoid N-arachidonoylethanolamine (anandamide, or 5Z,8Z,11Z,14Z)-N-(2-hydroxyethyl)icosa-5,8,11,14-tetraenamide) and its congeners N-stearoylethanolamine (N-(2-hydroxyethyl)-stearamide), N-oleoylethanolamine (N-2-hydroxyethyl-9(Z)-octadecenamide), and PEA.

PEA and its congeners are constituted from N-acylated phosphatidylethanolamine (NAPE) via different enzymatic pathways [162], the principal one producing a membrane-associated NAPE-phospholipase D, which further produces the respective NAE and phosphatidic acid. This enzyme is able to convert N-palmitoyl-phosphatidyl-ethanolamine into PEA. In the mammalian brain, NAEs are hydrolyzed by (1) fatty acid amide hydrolase in the endoplasmic reticulum, which breaks down NAEs into the corresponding fatty acid and ethanolamine, and (2) lysosomal NAE-hydrolyzing acid amidase (NAAA) [163]. NAAA is found mainly in macrophages, as well as in the brain, where it hydrolyses NAEs with less than 18 carbon atoms, i.e., PEA, but not N-oleoylethanolamine and N-stearoylethanolamine [16,164].

In contrast, fatty acid amide hydrolase hydrolyses all three NAEs. PEA is abundant in mammals; there is evidence for the presence of PEA, as well as other FAEs, in marine species and sea urchin ovaries [165]. The regional distribution in the rat brain of orally administered PEA ($\approx$100 mg$\cdot$kg^{-1}) has been investigated through the use of N-[9-3H]-PEA by Artamonov and Gabrielsson [166,167]. The authors found that N-[9,10-3H]-PEA mainly accumulated in the hypothalamus, pituitary, and adrenal glands 20 min after oral administration [166]. The presence of the labelled PEA in the brain ($\approx$98 ng$\cdot$mg^{-1} of brain tissue) demonstrated the ability of the compound to penetrate, although in small amounts, through the blood–brain barrier [166].

Mechanistically, PEA may be a ligand for peroxisome proliferator activated receptor α (PPARα), one of a group of nuclear receptor proteins that function as transcription factors regulating the expression of genes. In particular, the α- and γ-isoforms of PPAR are associated with pro-inflammatory effects. Moreover, in PPARα null mice or mice with blocking PPARα antagonists, the anti-inflammatory, anti-nociceptive/anti-neuropathic, and neuroprotective effect of PEA was not detected [168]. PEA is produced through an "on-demand" synthesis within the lipid bilayer, where N-phosphatidylethanolamine-specific phospholipase D (NAPE-PLD) releases it from its membrane precursor, N-palmitoylphosphatidylethanolamine, but it is also present in "homeostatic" conditions [169].

An "entourage effect" was also proposed to explain PEA's pharmacological activities regarding improving the anti-inflammatory and anti-nociceptive function of other endogenous substances through potentiating their receptor binding or inhibiting metabolic degradation. Both anandamide and its congeners, such as PEA, have the type 1 transient vanilloid receptor (TRPV1) in common, whose function can be stimulated by harmful heat, low pH, and capsaicin. Moreover, anandamide itself is a TRPV1 receptor agonist, and PEA can enhance anandamide stimulation of the human TRPV1 receptor in a cannabinoid CB2 receptor antagonist-sensitive fashion, which could be considered as PEA acting indirectly by potentiating the actions of anandamide. Mast cells and microglia reportedly express TRPV1 receptors [170].

PEA also has an affinity to cannabinoid-like G-coupled receptors GPR55 and GPR119 [171]. The GPR55 receptor is widely expressed, and therefore its activity is correlated with multiple physiological processes. In particular, GPR55 is expressed in the frontal cortex, striatum, hippocampus, hypothalamus, cerebellum, brainstem, and microglia [172–174]. Balenga and colleagues showed that GPR55 activity modulates RhoA-dependent neutrophil migration, and it may prevent oxidative damage.

GPRl19 is expressed mainly in the pancreas and gut, in particular in β cells, where its activity modulates glucose-dependent insulin secretion, as well as in enteroendocrine L-cells, where it regulates the secretion of glucagon-like peptide 1 [175,176]. In normal-weight and healthy patients, it was observed that gut GPR119 expression rapidly increased following acute fat exposure, thus suggesting a potential involvement of GPR119 in type 2 diabetes, metabolic disorders, and obesity.

The main endogenous ligands of GPR119 are OEA, PEA, and AEA [177–179].

9. PEA: Role in Pathological and Physiological Conditions

Several reports showed a role of PEA in maintaining cellular homeostasis by acting as a mediator of the resolution of inflammatory processes. PEA is synthesized/metabolized by both microglia and mast cells. Moreover, it down-modulates mast cell and microglia activation, and very interesting tissue levels of PEA are augmented in brain regions that are implicated in nociception and in the spinal cord following the induction of neuropathic pain and other conditions related to strokes (Figure 2).

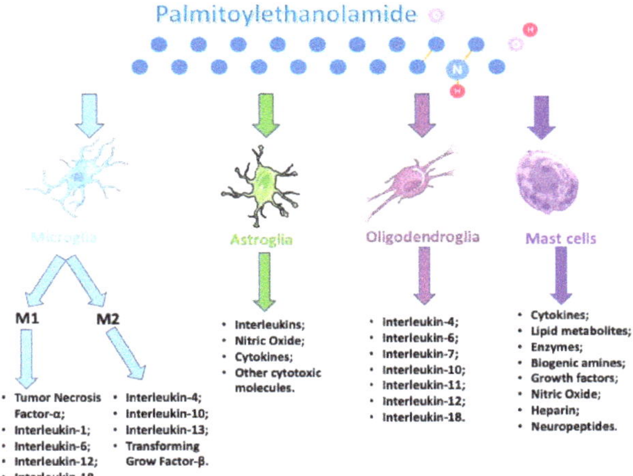

Figure 2. Palmitoylethanolamide (PEA) acts on several types of cells that are involved during an neuroinflammation event.

In particular, endogenous levels of PEA are altered in pathological conditions accompanied by cell damage and inflammatory processes. In inflamed tissues, for example following tissue ischemia, the levels of PEA increase significantly [180,181]. Increased PEA levels have also been observed in the brain tissues of animals in a state similar to multiple sclerosis and following a spinal cord injury [182,183]. At the cutaneous level, increases in PEA levels have been observed in animals affected by diabetic neuropathy, after exposure to irritants, and in the case of atopic dermatitis [184–186]. In a pain model, PEA inhibits peripheral inflammation and mast cell degranulation, as well as exerts neuroprotective and antinociceptive effects; these actions are accompanied by a decrease in NO production, neutrophil influx, and the expression of proinflammatory proteins, such as iNOS [187]. Moreover, PEA protects endothelial function from oxidative and inflammatory injuries. PEA administration improves neurological, emotional, and biochemical outcomes following TBI by ameliorating the secondary injury components of TBI, reducing both the infiltration and activation of mast cells [188]. It was demonstrated that in mice subjected to a compression model of spinal cord trauma, PEA administered systemically at 6- and 12-h post-injury considerably diminished the severity of spinal cord trauma via the reduction both mast cell activation and infiltration [189]. Furthermore, another study reported that intraperitoneal administration of PEA is able to reduce tissue injury and spinal cord inflammation, nitrotyrosine formation, neutrophil infiltration, pro-inflammatory cytokine expression, nuclear transcription factor kB activation, and inducible nitric oxide synthase expression and apoptosis, as well as ameliorate the recovery of motor limb function [190]. In several preclinical studies on intestinal inflammation, it was reported that PEA improves the symptoms of colitis in mice [191,192]

All these data indicate that the alterations of the levels of PEA in the conditions accompanied by inflammatory processes suggest that the exogenous contribution of the molecule may favor or accelerate the process of resolution of inflammation and the restoration of tissue homeostasis (Table 2).

Table 2. Preclinical studies reporting efficacy of PEA.

Model	Animals	Effects	References
Visceral inflammatory pain	Rats	Decrease in NO production and neutrophil influx	[187]
Traumatic brain injury	Mice	Improves neurological, emotional, and biochemical outcomes	[188]
Spinal cord injury	Mice	Reduce mast cell activation and infiltration	[189]
Spinal cord injury	Mice	Reduce inflammatory markers	[190]
Parkinson's disease	Mice	Reduce neuroinflammation	[193]
Colitis	Mice	Normalize the functional post-inflammatory accelerated intestinal transit	[192]
Colitis	Mice	Improve symptoms of colitis	[191]
Dermatitis	Dogs	Reduce pruritus and skin lesions	[194]
Contrast-agent-induced nephropathy	Rats	Prevent nephropathy in and alteration of biochemical parameters	[195]
Idiopathic pulmonary fibrosis	Mice	Reduces lung inflammation	[196]
Tibia fracture model	Mice	Decrease mast cell density, nerve growth factor, matrix metalloproteinase 9, and cytokines expression	[197]
Alzheimer's disease	Mice	Normalize astrocytic function, rebalance glutamatergic transmission, and restrain neuroinflammation	[198]
Joint pain	Rats	Reduce igeminal nerve sensitization	[199]

In neurological diseases, increases in PEA levels were observed in the patients liquor with schizophrenia, chronic migraine, and multiple sclerosis [200–202]. In multiple sclerosis, the levels of PEA also increased in plasma [203]. Orefice et al. reported a significant increase of PEA, anandamide, and oleoylethanolamide plasma levels in patients [204].

At the abdomino-pelvic level, PEA levels increase in the duodenum of celiac patients and in the colon of patients with colitis ulcerative [184,205].

On the other hand, there exist some pathological situations in which the levels of PEA are dramatically decreased. This happens, for example, in the synovial fluid of patients with arthritis or rheumatoid arthritis, in which the levels of PEA compared to subjects without joint pathologies are about a thousand times lower [206].

Also some clinical studies demonstrated the effect of PEA on the pain condition; in particular, Andresen et al. reported on a ultramicronized PEA that has good safety and tolerability in the treatment of patients with spinal cord injury [207]. In these studies, PEA represents a promising supplement to therapeutic management for neuropathic pain, with the potential for good tolerability and a low susceptibility for side effects [208–214].

10. PEA: Ultramicronization Process

The therapeutic use of PEA is strictly dependent on its lipophilic nature; on this basis, PEA is insoluble in water and poorly soluble in most other aqueous solvents, with the logarithm of its partition coefficient (log P) being greater than 5 [215].

PEA is usually administered via a parenteral route, but in particular in humans, oral treatment represents the goal standard due to patient compliance, versatility, and ease. Structurally, PEA is formed using a heterogeneous particle size, and for these reasons, presents some limitations in bioavailability and solubility. Due to these considerations, the oral absorption of PEA is strongly limited by the dissolution rate, with the amount absorbed conceivably showing an inverse relation to particle size [216]. A possible approach to bypass this problem is represented using micronization. The micronization process is applied both to reduce the particle size and to improve the bioavailability and effectiveness of particularly low-water-soluble molecules, thereby increasing the dissolution rate [217–219].

Both micronized and ultramicronized PEA oral administration showed superior pharmacological action in inflammatory pain induced by carrageenan, in contrast to a non-micronized PEA. Based on this, several studies were performed with these formulations; in fact, micronized pharmaceutical-grade formulations of PEA acquired via a jet milling process are now employed in both human and veterinary medicine to study inflammatory, hyperalgesic, and allergic disorders [220–222].

Similar to naïve PEA (particle size profile ranging between 100 and 700 μm), micronized and ultramicronized PEA showed a different particle size profile (2–10 μm and 0.8–6 μm at most, respectively). Micronization and ultramicronization processes yield a different crystalline structure with a higher energy content and smaller particle size, which result in better diffusion and distribution of these molecules compared to the naïve form, and thus superior biological efficacy [220,222]. The pharmacological profile of PEA and the pre-clinical results have inspired a series of clinical studies that have used PEA in the micronized and ultramicronized forms, where they predominantly focused on the pathological conditions associated with neuroinflammation and chronic and/or neuropathic pain [220,223]. In particular, a multicenter, double-blind, randomized study controlled with two doses of PEA versus placebo showed the analgesic dose-dependent effect of micronized PEA (PEA-m) in patients affected by lombosciatalgia with a compressive origin [211]. In the same conditions, therapy with PEA-m allows for a considerable reduction of the use of nonsteroidal anti-inflammatory drugs NSAIDs, suggesting the possibility of simultaneous use in order to reduce the use of NSAIDs, which if taken for long periods, can induce significant side effects [210]. The results obtained in patients with low back pain have also been confirmed on a large number of patients with different chronic pain pathological conditions, such as radiculopathy, osteoarthritis, joint pain, tibia fracture, spinal surgical failure, post herpetic neuropathy, diabetic neuropathy, and oncological pain [197,199,209,212,213] (Table 3).

Table 3. Clinical studies reporting efficacy of PEA.

Pathology	Effects	References
Lombosciatalgia	Analgesic effect	[210]
Lombosciatalgia	Analgesic effect	[211]
Chronic pain	Analgesic effect	[209]
Temporomandibular joint pain	Analgesic effect	[212]
Peripheral diabetic neuropathy	Reduce pain symptoms characteristic of diabetic neuropathy	[213]
Carpal tunnel syndrome	Improve sleep quality, confirming a correlation between sleep disorders and pain intensity	[208]

Moreover, Evangelista et al. reported that in patients affected by carpal tunnel syndrome, the oral administration of ultramicronized PEA improved sleep quality, verifying a correlation between sleep disorders and pain intensity [208]. PEA-m, in a preclinical study conducted by Crupi et al., used for the first time as a pre-treatment, showed a significantly neuroprotection after 1-methyl-4-phenyl-1,2,3,6-tetrahydropyridine (MPTP) intoxication in mice by protecting not only against loss of TH+ neurons, but also the alterations of the dopamine active transporter

(DAT), α-synuclein, and β3-tubulin in the substantia nigra. Moreover, pre-treatment with PEA-m showed an important reduction in the MPTP-induced proinflammatory cytokine expression and showed a pro-neurogenic effect in the hippocampus. These data support the idea of PEA-m as a potential therapeutic candidate to prevent dopamine neuronal cell degeneration in the pathogenesis of Parkinson's disease [193,224].

The data, obtained with ultramicronized PEA, showed that it was also able to decrease all the inflammatory pathways that were activated during pulmonary fibrosis, as well as dermatitis, myocardial ischemia, or contrast-agent-induced nephropathy [194–197]. Furthermore, Scuderi et al. showed that PEA-um was a novel and effective promising treatment for AD with the potential to be integrated into a multitargeted treatment strategy in combination with other drugs [198]. Finally, the association of PEA-um with standard neuropathic pain medications, such as pregabalin and oxycodone, allowed for highlighting analgesic pharmacological effects with very low doses of these drugs, suggesting an additive or synergistic effect among neuropathic pain medications acting on neuronal cells with PEA, which acts predominantly on non-neuronal cells [225,226] (Table 3).

11. Luteolin

Several studies support the association between inflammation; neurodegenerative diseases; oxidative stress; neuropsychological disorders, such as anxiety and depression; and mild cognitive disorder [227–230]. Flavonoids have been shown to display many neuroprotective and inflammatory actions [208,231]. Among the large family of flavonoids, luteolin has a good spectrum of action. Luteolin (3′,4′,5,7-tetrahydroxyflavone) is a common flavonoid in many fruits, vegetables, and medicinal herbs [232]. Chemical data show that luteolin has a C6-C3-C6 structure with two benzene rings and one oxygen-containing ring with a C2-C3 carbon double bond [233,234]. Structure–activity studies have revealed that the presence of hydroxyl moieties at carbons 5, 7, 3′, and 4′ positions of the luteolin structure, and the presence of the 2–3 double bond, are responsible for its multiple pharmacological effects [233]. Luteolin represents one of the most powerful and effective polyphenols, which has displayed numerous biological properties, such as antioxidant, anticancer, anti-inflammatory, and neuroprotective properties in in vitro and in vivo models [233,235–241]. It was formerly supposed that the oral bioavailability of flavonoids was very low. Nevertheless, Zhou et al. studied the bioavailability of luteolin in peanut hull extracts (PHEs) [242]. In particular, the research reported that the efficient permeability (P_{eff}) and absorption rate constant (k_a) of pure luteolin (5.0 μg/mL) were not significantly dissimilar in both the duodenum and jejunum, but notably greater in the colon and ileum. Pharmacokinetics analysis after subsequent oral administration of a single dose of pure luteolin (14.3 mg/kg) or PHE (containing 14.3 mg/kg of luteolin) in rats displayed that the peak concentration of luteolin in plasma was 1.97 ± 0.15 μg/mL for luteolin and 8.34 ± 0.98 μg/mL for PHE. Luteolin seems to be inertly absorbed in the rat intestine [242].

Moreover, luteolin has been revealed to inhibit cytokine expression, nuclear factor kappa B (NF-kB) signaling, and TLR4 signaling at micromolar concentrations in immune cells, such as mast cells [243–245]. Additionally, the ability of luteolin to inhibit the Keap1-Nrf2-ARE pathway in PC12 cells was demonstrated [246]. Numerous studies have shown that luteolin can improve the neurotoxic effect of Aβ fragment 25–35 ($A\beta_{25-35}$) in murine cortical neurons [247], as well as increase locomotor and muscular alteration in mice exposed to MPTP [248]. Moreover, luteolin has been used for the treatment of multiple sclerosis (MS) due to its inhibition of activated peripheral blood leukocytes that have been isolated from MS patients [249,250].

Luteolin can reduce experimental allergic encephalomyelitis (EAE), which is now employed as a murine model to analyze MS [251]. Luteolin treatment has also displayed additive effects in regulating cell proliferation and the production of pro-inflammatory cytokines, as well as the ratio of the cell migration mediator matrix metallopeptidase-9 (MMP-9) and its inhibitor called tissue inhibitor of metalloproteinase (TIMP-1) [249]. In this context, this suggests that it is reasonable to consider luteolin as a valuable adjuvant to IFN-β for the management of MS.

Another work by Xu et al. showed that treatment with luteolin can decrease secondary brain injury provoked by traumatic brain injury (TBI), involving brain water content, neuronal apoptosis, and neurological deficits [252].

A clinical study containing fifty children (4–10 years old; 42 boys and 8 girls) with autism spectrum disorder (ASD) described that there was an important improvement in adaptive functioning (as measured using the Vineland Adaptive Behavioral Scales domains) following a combination of luteolin (100 mg/capsule) and quercetin (70 mg/capsule) for 26 weeks in dietary supplementation [253]. Likewise, a significant reduction was also reported after stratification for age, sex, and history of allergies in the Aberrant Behaviour Checklist subscale scores.

12. Co-Ultra PEA + Luteolin

The pharmacodynamic properties of PEA and those of luteolin appear complementary, suggesting that if administered in combination, they can fight the two main conspirators of chronic diseases: low-grade inflammation and oxidative stress. Confirming this hypothesis, many studies have shown that joint treatment using PEA plus luteolin has a superior effect compared to the molecules used alone by stimulating both hippocampal neurogenesis and dendritic spine maturation [254,255]. These studies have also shown that the pharmacological properties of PEA + luteolin (PEALut) become synergistic when PEA and luteolin were simultaneously submitted to the micronization process with a jet mill technique in a mass ratio of 10:1. This is probably due to the fact that PEA is poorly soluble in aqueous media and is difficult to formulate using traditional approaches; for this reason, the micronized and ultramicronized forms have a reduced particle size compared to the native molecule, along with a greater bioavailability. Additionally, it seems that the association with flavones stabilizes the two molecules and enhances their pharmacological activities [256].

In particular, the association of these two molecules has been evaluated in a lot of different experimental models, such as in Alzheimer's and Parkinson's disease, vascular dementia, anxiety and depression, brain and spinal cord injury, and arthritis [254–263]. In these studies, it was proved that the association of PEA with luteolin was able to significantly decrease the neuroinflammatory and apoptotic pathways, modulating the cytokines release and the activation of astrocytes and microglia, decreasing the oxidative and nitrosative stress, and was able to enhance the expression of the neurotrophic factors promoting neuronal regeneration, as well as demonstrate that this association possesses the ability to modulate the autophagic process [254–263] (Table 4).

Table 4. Preclinical studies reporting the efficacy of PEA + luteolin (PEALut).

Model	Animals	Effects	References
Sci	Mice	Stimulate hippocampal neurogenesis and dendritic spine maturation	[257]
Paw edema	Rats	Reduce inflammation and pain	[256]
Alzheimer's disease	Cell culture	Reduce inducible nitric oxide synthase and GFAP expression, restore neuronal NO synthase and BDNF, and reduce the apoptosis	[261]
Parkinson's disease	Mice	Modulate the neuroinflammatory process and the autophagic pathway	[259]
Anxiety and depression	Mice	Antidepressant effect	[254]
Spinal Cord Injury	Mice and cell culture	Improve motor activity, reduced cyclooxygenase-2 (cox-2), and inducible nitric oxide synthase (inos)	[262]
Autism	Mice	Ameliorate social and nonsocial behaviors	[256]
Stroke	Rats	Reduce neuroinflammation	[259]
Spinal Cord Injury	Mice	Promote neuronal regeneration	[257]

Sci: Spinal cord injury.

The use of co-ultra PEALut was also evaluated in patients with autism disorders and in patients with stroke undergoing rehabilitation therapy [257,259]. Co-ultra PEALut treatment was tolerated well by the patients. Evaluation using the Autism Treatment Evaluation Checklist (ATEC) test revealed a decrease of scores, indicative of an increased behavioral outcome of about 23%. Co-ultra PEALut also reduced most indices of hyperactivity and improved the ability to understand simple commands and execute them. Unfortunately, no significant progress in speech profile has been observed yet [256].

In patients with stroke treated with co-ultra PEALut, the observations demonstrated a positive outcome in cognitive function, as well as in muscle spasticity and in mobility in daily living activities [258] (Table 5).

Table 5. Clinical studies reporting the efficacy of PEALut.

Pathology	Effects	References
Autism	Reduction in stereotypes	[256]
Stroke	Improve clinical outcome	[258]

13. Future Directions

Neuroinflammatory processes represent an important component in several diseases. Understanding the mechanisms that relate to systemic inflammation, even at a low-grade level, regarding neuroinflammation is an interesting area for research. The very early, possibly preclinical, stages of many pathologies could be the best window for therapeutic approaches or preventive interventions, which act on modifiable factors, such as systemic inflammation. Several studies have classified neuroinflammation as one of the driving forces for neurodegeneration. For example, glia activation, with the consequent release of pro-inflammatory mediators, is an important goal for diagnostic and therapeutic approaches in research. It would be fascinating to know whether activation of microglia revealed in the earliest phases of the disease is due to the anti-inflammatory M2 phenotype, later switching to the M1 phenotype, leading to further neuron damage. Once recognized, a clear panel of pro-inflammatory mediators could be added to the existing biomarker pool and used in the diagnostic process.

The influence of systemic inflammation on the CNS has been evident in acute episodes when the inflammatory status determines sickness behavior. A chronic inflammatory status appears to be responsible for the activation of glial cells, triggering a more persistent inflammatory status, leading to neuronal damage and neurodegeneration. Further studies are needed to understand the communication between the periphery and the CNS better, and to direct diagnostic evaluations, follow-up programs, and therapeutic and preventative approaches.

The growing understanding that central immune mediators are important in controlling physiological activities of the CNS has updated the field of neuroimmunology. Both microglia and cytokines have been connected to the regulation of neurodevelopment, neuronal wiring, and synaptic plasticity. The functional significance and underlying mechanisms of these non-immunological functions persist unknown and await additional study. It is well-defined that the reductive conception of microglia as basically central immune cells is too simplistic. Reasonably, they appear as a different but heterogeneous cell population of the CNS with a high degree of functional diversity and complexity. Implying that modifications in microglia activity profiles and/or inflammatory factors with ongoing neuroinflammation may therefore be too simplistic and could result in misconceptions. In contrast, changes in neuroimmune systems should be deduced in relation to the functional complexity of immune cells and molecules in physiological brain processes.

14. Conclusions

As summarized in this review, fitting the endogenous neuroinflammatory regulators may represent a valid therapeutic strategy against the disorders of the central nervous system by targeting non-neuronal cells. In this context, the capacity of PEA to modulate several protective responses in inflammatory and pain conditions show that endogenously produced PEA may be an element of the homeostatic system that is able to control the inflammatory process. The findings obtained show that the treatment of disorders of the nervous system must be based on endogenous control of inflammatory processes. In this context, the role of the PEA that modulates the activity of both mast cells and glial cells contributes to the resolution of neuroinflammatory processes.

Clinical studies demonstrate the ability of PEA in micronized and ultramicronized forms to counter the painful symptoms of the electrophysiological alterations present in different pathologies that involve the peripheral and/or central nervous system, which are sustained due to excessive mast cell activation and almost always associated with microglia. Moreover, PEA can improve the alterations that characterize many neurological diseases. Several studies also confirmed the excellent tolerability of products containing co-ultra PEALut, PEA-m, and PEA-um, even when administered in combination with anti-inflammatory and analgesic/anticonvulsant drugs.

These results suggest the use of PEA in micronized and ultramicronized forms as an innovative therapeutic tool for the treatment of all conditions characterized by the presence of neuroinflammatory processes and chronic painful states.

Author Contributions: R.C. and M.C. drafted the manuscript and S.C. critically revised the manuscript. All authors declare contribution to this paper. All authors have read and agreed to the published version of the manuscript.

Funding: This research received no external funding.

Acknowledgments: The authors would like to thank Valentina Malvagni for editorial support with the manuscript.

Conflicts of Interest: Salvatore Cuzzocrea is a co-inventor on patent WO2013121449 A8 (Epitech Group Srl), which deals with methods and compositions for the modulation of amidases capable of hydrolyzing N-acylethanolamines that are employable in the treatment of inflammatory diseases. This invention is wholly unrelated to the present study. Moreover, Cuzzocrea is also, with the Epitech Group, a co-inventor on the following patents: EP 2 821 083; MI2014 A001495; 102015000067344, that are unrelated to the study. The remaining authors report no conflict of interest.

References

1. Gibbins, I. The five cardinal signs of inflammation. *Med. J. Aust.* **2018**, *208*, 295. [CrossRef]
2. Ellis, A.; Bennett, D.L. Neuroinflammation and the generation of neuropathic pain. *Br. J. Anaesth* **2013**, *111*, 26–37. [CrossRef] [PubMed]
3. Ji, R.R.; Xu, Z.Z.; Gao, Y.J. Emerging targets in neuroinflammation-driven chronic pain. *Nat. Rev. Drug Discov.* **2014**, *13*, 533–548. [CrossRef] [PubMed]
4. Hickey, W.F.; Kimura, H. Perivascular microglial cells of the CNS are bone marrow-derived and present antigen in vivo. *Science* **1988**, *239*, 290–292. [CrossRef]
5. Skaper, S.D.; Giusti, P.; Facci, L. Microglia and mast cells: Two tracks on the road to neuroinflammation. *FASEB J. Off. Publ. Fed. Am. Soc. Exp. Biol.* **2012**, *26*, 3103–3117. [CrossRef]
6. Pekny, M.; Nilsson, M. Astrocyte activation and reactive gliosis. *Glia* **2005**, *50*, 427–434. [CrossRef]
7. Streit, W.J.; Mrak, R.E.; Griffin, W.S. Microglia and neuroinflammation: A pathological perspective. *J. Neuroinflamm.* **2004**, *1*, 14. [CrossRef]
8. Melchior, B.; Puntambekar, S.S.; Carson, M.J. Microglia and the control of autoreactive T cell responses. *Neurochem. Int.* **2006**, *49*, 145–153. [CrossRef]
9. Stoll, G.; Jander, S. The role of microglia and macrophages in the pathophysiology of the CNS. *Prog. Neurobiol.* **1999**, *58*, 233–247. [CrossRef]
10. Ji, R.R.; Chamessian, A.; Zhang, Y.Q. Pain regulation by non-neuronal cells and inflammation. *Science* **2016**, *354*, 572–577. [CrossRef]
11. Chen, G.; Zhang, Y.Q.; Qadri, Y.J.; Serhan, C.N.; Ji, R.R. Microglia in Pain: Detrimental and Protective Roles in Pathogenesis and Resolution of Pain. *Neuron* **2018**, *100*, 1292–1311. [CrossRef] [PubMed]
12. Cipollini, V.; Troili, F.; Giubilei, F. Emerging Biomarkers in Vascular Cognitive Impairment and Dementia: From Pathophysiological Pathways to Clinical Application. *Int. J. Mol. Sci.* **2019**, *20*. [CrossRef] [PubMed]

13. Braun, D.J.; Bachstetter, A.D.; Sudduth, T.L.; Wilcock, D.M.; Watterson, D.M.; Van Eldik, L.J. Genetic knockout of myosin light chain kinase (MLCK210) prevents cerebral microhemorrhages and attenuates neuroinflammation in a mouse model of vascular cognitive impairment and dementia. *Geroscience* **2019**. [CrossRef] [PubMed]
14. Fulop, G.A.; Tarantini, S.; Yabluchanskiy, A.; Molnar, A.; Prodan, C.I.; Kiss, T.; Csipo, T.; Lipecz, A.; Balasubramanian, P.; Farkas, E.; et al. Role of age-related alterations of the cerebral venous circulation in the pathogenesis of vascular cognitive impairment. *Am. J. Physiol. Heart Circ. Physiol.* **2019**, *316*, H1124–H1140. [CrossRef] [PubMed]
15. Sohn, E.; Kim, Y.J.; Lim, H.S.; Kim, B.Y.; Jeong, S.J. Hwangryunhaedok-Tang Exerts Neuropreventive Effect on Memory Impairment by Reducing Cholinergic System Dysfunction and Inflammatory Response in a Vascular Dementia Rat Model. *Molecules* **2019**, *24*. [CrossRef] [PubMed]
16. Impellizzeri, D.; Siracusa, R.; Cordaro, M.; Crupi, R.; Peritore, A.F.; Gugliandolo, E.; D'Amico, R.; Petrosino, S.; Evangelista, M.; Di Paola, R.; et al. N-Palmitoylethanolamine-oxazoline (PEA-OXA): A new therapeutic strategy to reduce neuroinflammation, oxidative stress associated to vascular dementia in an experimental model of repeated bilateral common carotid arteries occlusion. *Neurobiol. Dis.* **2019**, *125*, 77–91. [CrossRef]
17. Brook, E.; Mamo, J.; Wong, R.; Al-Salami, H.; Falasca, M.; Lam, V.; Takechi, R. Blood-brain barrier disturbances in diabetes-associated dementia: Therapeutic potential for cannabinoids. *Pharm. Res.* **2019**, *141*, 291–297. [CrossRef]
18. Hu, Y.; Zhang, M.; Chen, Y.; Yang, Y.; Zhang, J.J. Postoperative intermittent fasting prevents hippocampal oxidative stress and memory deficits in a rat model of chronic cerebral hypoperfusion. *Eur. J. Nutr.* **2019**, *58*, 423–432. [CrossRef]
19. Albrecht, D.S.; Kim, M.; Akeju, O.; Torrado-Carvajal, A.; Edwards, R.R.; Zhang, Y.; Bergan, C.; Protsenko, E.; Kucyi, A.; Wasan, A.D.; et al. The neuroinflammatory component of negative affect in patients with chronic pain. *Mol. Psychiatry* **2019**. [CrossRef]
20. Ashraf, A.; Mahmoud, P.A.; Reda, H.; Mansour, S.; Helal, M.H.; Michel, H.E.; Nasr, M. Silymarin and silymarin nanoparticles guard against chronic unpredictable mild stress induced depressive-like behavior in mice: Involvement of neurogenesis and NLRP3 inflammasome. *J. Psychopharmacol.* **2019**, *33*, 615–631. [CrossRef]
21. Atique-Ur-Rehman, H.; Neill, J.C. Cognitive dysfunction in major depression: From assessment to novel therapies. *Pharmacol. Ther.* **2019**. [CrossRef] [PubMed]
22. Banagozar Mohammadi, A.; Torbati, M.; Farajdokht, F.; Sadigh-Eteghad, S.; Fazljou, S.M.B.; Vatandoust, S.M.; Golzari, S.E.J.; Mahmoudi, J. Sericin alleviates restraint stress induced depressive- and anxiety-like behaviors via modulation of oxidative stress, neuroinflammation and apoptosis in the prefrontal cortex and hippocampus. *Brain Res.* **2019**, *1715*, 47–56. [CrossRef] [PubMed]
23. Cheng, L.; Huang, C.; Chen, Z. Tauroursodeoxycholic Acid Ameliorates Lipopolysaccharide-Induced Depression Like Behavior in Mice via the Inhibition of Neuroinflammation and Oxido-Nitrosative Stress. *Pharmacology* **2019**, *103*, 93–100. [CrossRef] [PubMed]
24. Fan, C.; Song, Q.; Wang, P.; Li, Y.; Yang, M.; Yu, S.Y. Neuroprotective Effects of Curcumin on IL-1beta-Induced Neuronal Apoptosis and Depression-Like Behaviors Caused by Chronic Stress in Rats. *Front. Cell Neurosci.* **2018**, *12*, 516. [CrossRef] [PubMed]
25. Fang, M.; Zhong, L.; Jin, X.; Cui, R.; Yang, W.; Gao, S.; Lv, J.; Li, B.; Liu, T. Effect of Inflammation on the Process of Stroke Rehabilitation and Poststroke Depression. *Front. Psychiatry* **2019**, *10*, 184. [CrossRef] [PubMed]
26. Fernandes, J.; Gupta, G.L. N-acetylcysteine attenuates neuroinflammation associated depressive behavior induced by chronic unpredictable mild stress in rat. *Behav. Brain Res.* **2019**, *364*, 356–365. [CrossRef]
27. Fourrier, C.; Singhal, G.; Baune, B.T. Neuroinflammation and cognition across psychiatric conditions. *CNS Spectr.* **2019**, *24*, 4–15. [CrossRef]
28. Gong, X.; Hu, H.; Qiao, Y.; Xu, P.; Yang, M.; Dang, R.; Han, W.; Guo, Y.; Chen, D.; Jiang, P. The Involvement of Renin-Angiotensin System in Lipopolysaccharide-Induced Behavioral Changes, Neuroinflammation, and Disturbed Insulin Signaling. *Front. Pharm.* **2019**, *10*, 318. [CrossRef]
29. Guo, L.T.; Wang, S.Q.; Su, J.; Xu, L.X.; Ji, Z.Y.; Zhang, R.Y.; Zhao, Q.W.; Ma, Z.Q.; Deng, X.Y.; Ma, S.P. Baicalin ameliorates neuroinflammation-induced depressive-like behavior through inhibition of toll-like receptor 4 expression via the PI3K/AKT/FoxO1 pathway. *J. Neuroinflamm.* **2019**, *16*, 95. [CrossRef]

30. Gupta, G.L.; Fernandes, J. Protective effect of Convolvulus pluricaulis against neuroinflammation associated depressive behavior induced by chronic unpredictable mild stress in rat. *Biomed. Pharm.* **2019**, *109*, 1698–1708. [CrossRef]
31. He, M.C.; Shi, Z.; Sha, N.N.; Chen, N.; Peng, S.Y.; Liao, D.F.; Wong, M.S.; Dong, X.L.; Wang, Y.J.; Yuan, T.F.; et al. Paricalcitol alleviates lipopolysaccharide-induced depressive-like behavior by suppressing hypothalamic microglia activation and neuroinflammation. *Biochem. Pharm.* **2019**, *163*, 1–8. [CrossRef] [PubMed]
32. Khan, A.; Shal, B.; Naveed, M.; Shah, F.A.; Atiq, A.; Khan, N.U.; Kim, Y.S.; Khan, S. Matrine ameliorates anxiety and depression-like behaviour by targeting hyperammonemia-induced neuroinflammation and oxidative stress in CCl4 model of liver injury. *Neurotoxicology* **2019**, *72*, 38–50. [CrossRef] [PubMed]
33. Lenart, L.; Balogh, D.B.; Lenart, N.; Barczi, A.; Hosszu, A.; Farkas, T.; Hodrea, J.; Szabo, A.J.; Szigeti, K.; Denes, A.; et al. Novel therapeutic potential of angiotensin receptor 1 blockade in a rat model of diabetes-associated depression parallels altered BDNF signalling. *Diabetologia* **2019**. [CrossRef] [PubMed]
34. Li, Q.; Liu, S.; Zhu, X.; Mi, W.; Maoying, Q.; Wang, J.; Yu, J.; Wang, Y. Hippocampal PKR/NLRP1 Inflammasome Pathway Is Required for the Depression-Like Behaviors in Rats with Neuropathic Pain. *Neuroscience* **2019**. [CrossRef]
35. Liu, L.L.; Li, J.M.; Su, W.J.; Wang, B.; Jiang, C.L. Sex differences in depressive-like behaviour may relate to imbalance of microglia activation in the hippocampus. *Brain Behav. Immun.* **2019**. [CrossRef]
36. Mattei, D.; Notter, T. Basic Concept of Microglia Biology and Neuroinflammation in Relation to Psychiatry. *Curr. Top. Behav. Neurosci.* **2019**. [CrossRef]
37. Milenkovic, V.M.; Stanton, E.H.; Nothdurfter, C.; Rupprecht, R.; Wetzel, C.H. The Role of Chemokines in the Pathophysiology of Major Depressive Disorder. *Int. J. Mol. Sci.* **2019**, *20*. [CrossRef]
38. Mindt, S.; Neumaier, M.; Hoyer, C.; Sartorius, A.; Kranaster, L. Cytokine-mediated cellular immune activation in electroconvulsive therapy: A CSF study in patients with treatment-resistant depression. *World J. Biol. Psychiatry Off. J. World Fed. Soc. Biol. Psychiatry* **2019**, 1–9. [CrossRef]
39. Morganti-Kossmann, M.C.; Semple, B.D.; Hellewell, S.C.; Bye, N.; Ziebell, J.M. The complexity of neuroinflammation consequent to traumatic brain injury: From research evidence to potential treatments. *Acta Neuropathol.* **2019**, *137*, 731–755. [CrossRef]
40. Pellegrini, C.; Fornai, M.; Antonioli, L.; Blandizzi, C.; Calderone, V. Phytochemicals as Novel Therapeutic Strategies for NLRP3 Inflammasome-Related Neurological, Metabolic, and Inflammatory Diseases. *Int. J. Mol. Sci.* **2019**, *20*. [CrossRef]
41. Poletti, S.; Leone, G.; Hoogenboezem, T.A.; Ghiglino, D.; Vai, B.; de Wit, H.; Wijkhuijs, A.J.M.; Locatelli, C.; Colombo, C.; Drexhage, H.A.; et al. Markers of neuroinflammation influence measures of cortical thickness in bipolar depression. *Psychiatry Res. Neuroimaging* **2019**, *285*, 64–66. [CrossRef] [PubMed]
42. Sahu, P.; Mudgal, J.; Arora, D.; Kinra, M.; Mallik, S.B.; Rao, C.M.; Pai, K.S.R.; Nampoothiri, M. Cannabinoid receptor 2 activation mitigates lipopolysaccharide-induced neuroinflammation and sickness behavior in mice. *Psychopharmacology* **2019**. [CrossRef] [PubMed]
43. Shal, B.; Khan, A.; Naveed, M.; Khan, N.U.; AlSharari, S.D.; Kim, Y.S.; Khan, S. Effect of 25-methoxy hispidol A isolated from Poncirus trifoliate against bacteria-induced anxiety and depression by targeting neuroinflammation, oxidative stress and apoptosis in mice. *Biomed. Pharm.* **2019**, *111*, 209–223. [CrossRef] [PubMed]
44. Su, K.P.; Lai, H.C.; Peng, C.Y.; Su, W.P.; Chang, J.P.; Pariante, C.M. Interferon-alpha-induced depression: Comparisons between early- and late-onset subgroups and with patients with major depressive disorder. *Brain Behav. Immun.* **2019**. [CrossRef] [PubMed]
45. Yu, X.B.; Zhang, H.N.; Dai, Y.; Zhou, Z.Y.; Xu, R.A.; Hu, L.F.; Zhang, C.H.; Xu, H.Q.; An, Y.Q.; Tang, C.R.; et al. Simvastatin prevents and ameliorates depressive behaviors via neuroinflammatory regulation in mice. *J. Affect. Disord.* **2019**, *245*, 939–949. [CrossRef]
46. Zhang, C.; Kalueff, A.V.; Song, C. Minocycline ameliorates anxiety-related self-grooming behaviors and alters hippocampal neuroinflammation, GABA and serum cholesterol levels in female Sprague-Dawley rats subjected to chronic unpredictable mild stress. *Behav. Brain Res.* **2019**, *363*, 109–117. [CrossRef]
47. Zhao, J.; Gao, X.; Wang, A.; Wang, Y.; Du, Y.; Li, L.; Li, M.; Li, C.; Jin, X.; Zhao, M. Depression comorbid with hyperalgesia: Different roles of neuroinflammation induced by chronic stress and hypercortisolism. *J. Affect. Disord.* **2019**, *256*, 117–124. [CrossRef]

48. Zhu, Y.; Klomparens, E.A.; Guo, S.; Geng, X. Neuroinflammation caused by mental stress: The effect of chronic restraint stress and acute repeated social defeat stress in mice. *Neurol. Res.* **2019**, 1–8. [CrossRef]
49. Ashraf, G.M.; Tarasov, V.V.; Makhmutovsmall a, C.A.; Chubarev, V.N.; Avila-Rodriguez, M.; Bachurin, S.O.; Aliev, G. The Possibility of an Infectious Etiology of Alzheimer Disease. *Mol. Neurobiol.* **2019**, *56*, 4479–4491. [CrossRef]
50. Bostancikliogl u, M. An update on the interactions between Alzheimer's disease, autophagy and inflammation. *Gene* **2019**, *705*, 157–166. [CrossRef]
51. Chaney, A.; Williams, S.R.; Boutin, H. In vivo molecular imaging of neuroinflammation in Alzheimer's disease. *J. Neurochem.* **2019**, *149*, 438–451. [CrossRef] [PubMed]
52. Dhiman, K.; Blennow, K.; Zetterberg, H.; Martins, R.N.; Gupta, V.B. Cerebrospinal fluid biomarkers for understanding multiple aspects of Alzheimer's disease pathogenesis. *Cell. Mol. Life Sci.* **2019**, *76*, 1833–1863. [CrossRef] [PubMed]
53. Dinda, B.; Dinda, M.; Kulsi, G.; Chakraborty, A.; Dinda, S. Therapeutic potentials of plant iridoids in Alzheimer's and Parkinson's diseases: A review. *Eur. J. Med. Chem.* **2019**, *169*, 185–199. [CrossRef] [PubMed]
54. Dong, Y.; Li, X.; Cheng, J.; Hou, L. Drug Development for Alzheimer's Disease: Microglia Induced Neuroinflammation as a Target? *Int. J. Mol. Sci.* **2019**, *20*. [CrossRef]
55. Jha, N.K.; Jha, S.K.; Kar, R.; Nand, P.; Swati, K.; Goswami, V.K. Nuclear factor-kappa beta as a therapeutic target for Alzheimer's disease. *J. Neurochem.* **2019**. [CrossRef]
56. Kowalski, K.; Mulak, A. Brain-Gut-Microbiota Axis in Alzheimer's Disease. *J. Neurogastroenterol. Motil.* **2019**, *25*, 48–60. [CrossRef]
57. McQuade, A.; Blurton-Jones, M. Microglia in Alzheimer's Disease: Exploring How Genetics and Phenotype Influence Risk. *J. Mol. Biol.* **2019**, *431*, 1805–1817. [CrossRef]
58. Menta, B.W.; Swerdlow, R.H. An Integrative Overview of Non-Amyloid and Non-Tau Pathologies in Alzheimer's Disease. *Neurochem. Res.* **2019**, *44*, 12–21. [CrossRef]
59. Naseri, N.N.; Wang, H.; Guo, J.; Sharma, M.; Luo, W. The complexity of tau in Alzheimer's disease. *Neurosci. Lett.* **2019**, *705*, 183–194. [CrossRef]
60. Nichols, M.R.; St-Pierre, M.K.; Wendeln, A.C.; Makoni, N.J.; Gouwens, L.K.; Garrad, E.C.; Sohrabi, M.; Neher, J.J.; Tremblay, M.E.; Combs, C.K. Inflammatory mechanisms in neurodegeneration. *J. Neurochem.* **2019**, *149*, 562–581. [CrossRef]
61. Qin, Q.; Li, Y. Herpesviral infections and antimicrobial protection for Alzheimer's disease: Implications for prevention and treatment. *J. Med. Virol.* **2019**. [CrossRef] [PubMed]
62. Toffa, D.H.; Magnerou, M.A.; Kassab, A.; Hassane Djibo, F.; Sow, A.D. Can magnesium reduce central neurodegeneration in Alzheimer's disease? Basic evidences and research needs. *Neurochem. Int.* **2019**, *126*, 195–202. [CrossRef] [PubMed]
63. Van Bulck, M.; Sierra-Magro, A.; Alarcon-Gil, J.; Perez-Castillo, A.; Morales-Garcia, J.A. Novel Approaches for the Treatment of Alzheimer's and Parkinson's Disease. *Int. J. Mol. Sci.* **2019**, *20*. [CrossRef] [PubMed]
64. Wang, S.; Colonna, M. Microglia in Alzheimer's disease: A target for immunotherapy. *J. Leukoc. Biol.* **2019**. [CrossRef]
65. Yimer, E.M.; Hishe, H.Z.; Tuem, K.B. Repurposing of the beta-Lactam Antibiotic, Ceftriaxone for Neurological Disorders: A Review. *Front. Neurosci.* **2019**, *13*, 236. [CrossRef]
66. Baird, J.K.; Bourdette, D.; Meshul, C.K.; Quinn, J.F. The key role of T cells in Parkinson's disease pathogenesis and therapy. *Parkinsonism Relat. Disord.* **2019**, *60*, 25–31. [CrossRef]
67. Caggiu, E.; Arru, G.; Hosseini, S.; Niegowska, M.; Sechi, G.; Zarbo, I.R.; Sechi, L.A. Inflammation, Infectious Triggers, and Parkinson's Disease. *Front. Neurol.* **2019**, *10*, 122. [CrossRef]
68. Campos-Acuna, J.; Elgueta, D.; Pacheco, R. T-Cell-Driven Inflammation as a Mediator of the Gut-Brain Axis Involved in Parkinson's Disease. *Front. Immunol.* **2019**, *10*, 239. [CrossRef]
69. Cuenca, L.; Gil-Martinez, A.L.; Cano-Fernandez, L.; Sanchez-Rodrigo, C.; Estrada, C.; Fernandez-Villalba, E.; Herrero, M.T. Parkinson's disease: A short story of 200 years. *Histol. Histopathol.* **2019**, *34*, 573–591. [CrossRef]
70. Deb, S.; Phukan, B.C.; Mazumder, M.K.; Dutta, A.; Paul, R.; Bhattacharya, P.; Sandhir, R.; Borah, A. Garcinol, a multifaceted sword for the treatment of Parkinson's disease. *Neurochem. Int.* **2019**, *128*, 50–57. [CrossRef]
71. Garretti, F.; Agalliu, D.; Lindestam Arlehamn, C.S.; Sette, A.; Sulzer, D. Autoimmunity in Parkinson's Disease: The Role of alpha-Synuclein-Specific T Cells. *Front. Immunol.* **2019**, *10*, 303. [CrossRef] [PubMed]

72. Heiss, C.N.; Olofsson, L.E. The role of the gut microbiota in development, function and disorders of the central nervous system and the enteric nervous system. *J. Neuroendocr.* **2019**, *31*, e12684. [CrossRef] [PubMed]
73. Kaur, R.; Mehan, S.; Singh, S. Understanding multifactorial architecture of Parkinson's disease: Pathophysiology to management. *Neurol. Sci.Off. J. Ital. Neurol. Soc. Ital. Soc. Clin. Neurophysiol.* **2019**, *40*, 13–23. [CrossRef] [PubMed]
74. Kodani, S.D.; Morisseau, C. Role of epoxy-fatty acids and epoxide hydrolases in the pathology of neuro-inflammation. *Biochimie* **2019**, *159*, 59–65. [CrossRef]
75. Norwitz, N.G.; Hu, M.T.; Clarke, K. The Mechanisms by Which the Ketone Body D-beta-Hydroxybutyrate May Improve the Multiple Cellular Pathologies of Parkinson's Disease. *Front. Nutr.* **2019**, *6*, 63. [CrossRef]
76. Patrick, K.L.; Bell, S.L.; Weindel, C.G.; Watson, R.O. Exploring the "Multiple-Hit Hypothesis" of Neurodegenerative Disease: Bacterial Infection Comes Up to Bat. *Front. Cell. Infect. Microbiol.* **2019**, *9*, 138. [CrossRef]
77. Silva, A.R.; Grosso, C.; Delerue-Matos, C.; Rocha, J.M. Comprehensive review on the interaction between natural compounds and brain receptors: Benefits and toxicity. *Eur. J. Med. Chem.* **2019**, *174*, 87–115. [CrossRef]
78. Staff, N.P.; Jones, D.T.; Singer, W. Mesenchymal Stromal Cell Therapies for Neurodegenerative Diseases. *Mayo Clin. Proc.* **2019**, *94*, 892–905. [CrossRef]
79. Storelli, E.; Cassina, N.; Rasini, E.; Marino, F.; Cosentino, M. Do Th17 Lymphocytes and IL-17 Contribute to Parkinson's Disease? A Systematic Review of Available Evidence. *Front. Neurol.* **2019**, *10*, 13. [CrossRef]
80. Wang, S.; Yuan, Y.H.; Chen, N.H.; Wang, H.B. The mechanisms of NLRP3 inflammasome/pyroptosis activation and their role in Parkinson's disease. *Int. Immunopharmacol.* **2019**, *67*, 458–464. [CrossRef]
81. Yuan, J.; Amin, P.; Ofengeim, D. Necroptosis and RIPK1-mediated neuroinflammation in CNS diseases. *Nat. Rev. Neurosci.* **2019**, *20*, 19–33. [CrossRef] [PubMed]
82. Zella, S.M.A.; Metzdorf, J.; Ciftci, E.; Ostendorf, F.; Muhlack, S.; Gold, R.; Tonges, L. Emerging Immunotherapies for Parkinson Disease. *Neurol. Ther.* **2019**, *8*, 29–44. [CrossRef]
83. Beroun, A.; Mitra, S.; Michaluk, P.; Pijet, B.; Stefaniuk, M.; Kaczmarek, L. MMPs in learning and memory and neuropsychiatric disorders. *Cell. Mol. Life Sci.* **2019**. [CrossRef] [PubMed]
84. Chen, A.T.; Nasrallah, H.A. Neuroprotective effects of the second generation antipsychotics. *Schizophr. Res.* **2019**, *208*, 1–7. [CrossRef]
85. Guilarte, T.R. TSPO in diverse CNS pathologies and psychiatric disease: A critical review and a way forward. *Pharm. Ther.* **2019**, *194*, 44–58. [CrossRef] [PubMed]
86. Huang, L.; Otrokocsi, L.; Sperlagh, B. Role of P2 receptors in normal brain development and in neurodevelopmental psychiatric disorders. *Brain Res. Bull.* **2019**. [CrossRef]
87. Kroken, R.A.; Sommer, I.E.; Steen, V.M.; Dieset, I.; Johnsen, E. Constructing the Immune Signature of Schizophrenia for Clinical Use and Research; An Integrative Review Translating Descriptives Into Diagnostics. *Front. Psychiatry* **2018**, *9*, 753. [CrossRef]
88. Michetti, F.; D'Ambrosi, N.; Toesca, A.; Puglisi, M.A.; Serrano, A.; Marchese, E.; Corvino, V.; Geloso, M.C. The S100B story: From biomarker to active factor in neural injury. *J. Neurochem.* **2019**, *148*, 168–187. [CrossRef]
89. Nicoletti, F.; Orlando, R.; Di Menna, L.; Cannella, M.; Notartomaso, S.; Mascio, G.; Iacovelli, L.; Matrisciano, F.; Fazio, F.; Caraci, F.; et al. Targeting mGlu Receptors for Optimization of Antipsychotic Activity and Disease-Modifying Effect in Schizophrenia. *Front. Psychiatry* **2019**, *10*, 49. [CrossRef]
90. Shattuck, E.C. A biocultural approach to psychiatric illnesses. *Psychopharmacology* **2019**. [CrossRef]
91. Teismann, P.; Schulz, J.B. Cellular pathology of Parkinson's disease: Astrocytes, microglia and inflammation. *Cell Tissue Res.* **2004**, *318*, 149–161. [CrossRef] [PubMed]
92. Wake, H.; Moorhouse, A.J.; Nabekura, J. Functions of microglia in the central nervous system–beyond the immune response. *Neuron Glia Biol.* **2011**, *7*, 47–53. [CrossRef] [PubMed]
93. Kettenmann, H.; Verkhratsky, A. Neuroglia: The 150 years after. *Trends Neurosci.* **2008**, *31*, 653–659. [CrossRef] [PubMed]
94. Ji, K.; Akgul, G.; Wollmuth, L.P.; Tsirka, S.E. Microglia actively regulate the number of functional synapses. *PLoS ONE* **2013**, *8*, e56293. [CrossRef] [PubMed]
95. Loane, D.J.; Kumar, A. Microglia in the TBI brain: The good, the bad, and the dysregulated. *Exp. Neurol.* **2016**, *275*, 316–327. [CrossRef] [PubMed]

96. Raivich, G. Like cops on the beat: The active role of resting microglia. *Trends Neurosci.* **2005**, *28*, 571–573. [CrossRef]
97. Saijo, K.; Crotti, A.; Glass, C.K. Regulation of microglia activation and deactivation by nuclear receptors. *Glia* **2013**, *61*, 104–111. [CrossRef]
98. Takeda, M.; Tanaka, T.; Cacabelos, R. (Eds.) *Molecular Neurobiology of Alzheimer Disease and Related Disorders*; Karger Medical and Scientific Publishers: Basel, Switzerland, 2004.
99. Colton, C.A.; Mott, R.T.; Sharpe, H.; Xu, Q.; Van Nostrand, W.E.; Vitek, M.P. Expression profiles for macrophage alternative activation genes in AD and in mouse models of AD. *J. Neuroinflamm.* **2006**, *3*, 27. [CrossRef]
100. Ponomarev, E.D.; Maresz, K.; Tan, Y.; Dittel, B.N. CNS-derived interleukin-4 is essential for the regulation of autoimmune inflammation and induces a state of alternative activation in microglial cells. *J. Neurosci.* **2007**, *27*, 10714–10721. [CrossRef]
101. Colton, C.A. Heterogeneity of microglial activation in the innate immune response in the brain. *J. Neuroimmune. Pharm.* **2009**, *4*, 399–418. [CrossRef]
102. Butovsky, O.; Talpalar, A.E.; Ben-Yaakov, K.; Schwartz, M. Activation of microglia by aggregated beta-amyloid or lipopolysaccharide impairs MHC-II expression and renders them cytotoxic whereas IFN-gamma and IL-4 render them protective. *Mol. Cell Neurosci.* **2005**, *29*, 381–393. [CrossRef] [PubMed]
103. Kigerl, K.A.; Gensel, J.C.; Ankeny, D.P.; Alexander, J.K.; Donnelly, D.J.; Popovich, P.G. Identification of two distinct macrophage subsets with divergent effects causing either neurotoxicity or regeneration in the injured mouse spinal cord. *J. Neurosci.* **2009**, *29*, 13435–13444. [CrossRef] [PubMed]
104. Kumar, A.; Loane, D.J. Neuroinflammation after traumatic brain injury: Opportunities for therapeutic intervention. *Brain Behav. Immun.* **2012**, *26*, 1191–1201. [CrossRef] [PubMed]
105. Takano, T.; Tian, G.F.; Peng, W.; Lou, N.; Libionka, W.; Han, X.; Nedergaard, M. Astrocyte-mediated control of cerebral blood flow. *Nat. Neurosci.* **2006**, *9*, 260–267. [CrossRef] [PubMed]
106. Verkhratsky, A.; Nedergaard, M. The homeostatic astroglia emerges from evolutionary specialization of neural cells. *Philos. Trans. R Soc. Lond. B Biol. Sci.* **2016**, *371*. [CrossRef] [PubMed]
107. Noctor, S.C.; Flint, A.C.; Weissman, T.A.; Dammerman, R.S.; Kriegstein, A.R. Neurons derived from radial glial cells establish radial units in neocortex. *Nature* **2001**, *409*, 714–720. [CrossRef]
108. Tsai, H.H.; Li, H.; Fuentealba, L.C.; Molofsky, A.V.; Taveira-Marques, R.; Zhuang, H.; Tenney, A.; Murnen, A.T.; Fancy, S.P.; Merkle, F.; et al. Regional astrocyte allocation regulates CNS synaptogenesis and repair. *Science* **2012**, *337*, 358–362. [CrossRef]
109. Marin-Padilla, M. Prenatal development of fibrous (white matter), protoplasmic (gray matter), and layer I astrocytes in the human cerebral cortex: A Golgi study. *J. Comp. Neurol.* **1995**, *357*, 554–572. [CrossRef]
110. Ridet, J.L.; Malhotra, S.K.; Privat, A.; Gage, F.H. Reactive astrocytes: Cellular and molecular cues to biological function. *Trends Neurosci.* **1997**, *20*, 570–577. [CrossRef]
111. Haydon, P.G.; Carmignoto, G. Astrocyte control of synaptic transmission and neurovascular coupling. *Physiol. Rev.* **2006**, *86*, 1009–1031. [CrossRef]
112. Chung, W.S.; Allen, N.J.; Eroglu, C. Astrocytes Control Synapse Formation, Function, and Elimination. *Cold Spring Harb. Perspect. Biol.* **2015**, *7*, a020370. [CrossRef] [PubMed]
113. Brown, A.M.; Sickmann, H.M.; Fosgerau, K.; Lund, T.M.; Schousboe, A.; Waagepetersen, H.S.; Ransom, B.R. Astrocyte glycogen metabolism is required for neural activity during aglycemia or intense stimulation in mouse white matter. *J. Neurosci. Res.* **2005**, *79*, 74–80. [CrossRef] [PubMed]
114. Farina, C.; Aloisi, F.; Meinl, E. Astrocytes are active players in cerebral innate immunity. *Trends Immunol.* **2007**, *28*, 138–145. [CrossRef] [PubMed]
115. Liddelow, S.A.; Guttenplan, K.A.; Clarke, L.E.; Bennett, F.C.; Bohlen, C.J.; Schirmer, L.; Bennett, M.L.; Munch, A.E.; Chung, W.S.; Peterson, T.C.; et al. Neurotoxic reactive astrocytes are induced by activated microglia. *Nature* **2017**, *541*, 481–487. [CrossRef]
116. Miller, S.J. Astrocyte Heterogeneity in the Adult Central Nervous System. *Front. Cell Neurosci.* **2018**, *12*, 401. [CrossRef]
117. Stogsdill, J.A.; Ramirez, J.; Liu, D.; Kim, Y.H.; Baldwin, K.T.; Enustun, E.; Ejikeme, T.; Ji, R.R.; Eroglu, C. Astrocytic neuroligins control astrocyte morphogenesis and synaptogenesis. *Nature* **2017**, *551*, 192–197. [CrossRef]

118. Micevych, P.E.; Abelson, L. Distribution of mRNAs coding for liver and heart gap junction proteins in the rat central nervous system. *J. Comp. Neurol.* **1991**, *305*, 96–118. [CrossRef]
119. Giaume, C.; Venance, L. Intercellular calcium signaling and gap junctional communication in astrocytes. *Glia* **1998**, *24*, 50–64. [CrossRef]
120. Benarroch, E.E. Oligodendrocytes: Susceptibility to injury and involvement in neurologic disease. *Neurology* **2009**, *72*, 1779–1785. [CrossRef]
121. Benveniste, E.N. Role of macrophages/microglia in multiple sclerosis and experimental allergic encephalomyelitis. *J. Mol. Med.* **1997**, *75*, 165–173. [CrossRef]
122. Hayakawa, K.; Pham, L.D.; Som, A.T.; Lee, B.J.; Guo, S.; Lo, E.H.; Arai, K. Vascular endothelial growth factor regulates the migration of oligodendrocyte precursor cells. *J. Neurosci. Off. J. Soc. Neurosci.* **2011**, *31*, 10666–10670. [CrossRef] [PubMed]
123. Zhang, Y.; Taveggia, C.; Melendez-Vasquez, C.; Einheber, S.; Raine, C.S.; Salzer, J.L.; Brosnan, C.F.; John, G.R. Interleukin-11 potentiates oligodendrocyte survival and maturation, and myelin formation. *J. Neurosci. Off. J. Soc. Neurosci.* **2006**, *26*, 12174–12185. [CrossRef] [PubMed]
124. Enz, L.S.; Zeis, T.; Hauck, A.; Linington, C.; Schaeren-Wiemers, N. Combinatory Multifactor Treatment Effects on Primary Nanofiber Oligodendrocyte Cultures. *Cells* **2019**, *8*. [CrossRef] [PubMed]
125. Collington, S.J.; Williams, T.J.; Weller, C.L. Mechanisms underlying the localisation of mast cells in tissues. *Trends Immunol.* **2011**, *32*, 478–485. [CrossRef]
126. Hong, G.U.; Kim, N.G.; Jeoung, D.; Ro, J.Y. Anti-CD40 Ab- or 8-oxo-dG-enhanced Treg cells reduce development of experimental autoimmune encephalomyelitis via down-regulating migration and activation of mast cells. *J. Neuroimmunol.* **2013**, *260*, 60–73. [CrossRef]
127. Johnson, D.; Krenger, W. Interactions of mast cells with the nervous system–recent advances. *Neurochem. Res.* **1992**, *17*, 939–951. [CrossRef]
128. Galli, S.J.; Nakae, S.; Tsai, M. Mast cells in the development of adaptive immune responses. *Nat. Immunol.* **2005**, *6*, 135–142. [CrossRef]
129. Biran, V.; Cochois, V.; Karroubi, A.; Arrang, J.M.; Charriaut-Marlangue, C.; Heron, A. Stroke induces histamine accumulation and mast cell degranulation in the neonatal rat brain. *Brain Pathol.* **2008**, *18*, 1–9. [CrossRef]
130. Hu, W.W.; Chen, Z.; Xu, L.S.; Du, X.F.; Xu, C.F.; Wei, E.Q. [Changes of brain mast cells after transient global ischemia in rats]. *Zhejiang Da Xue Xue Bao Yi Xue Ban* **2004**, *33*, 193–196, 200.
131. Hu, W.; Shen, Y.; Fu, Q.; Dai, H.; Tu, H.; Wei, E.; Luo, J.; Chen, Z. Effect of oxygen-glucose deprivation on degranulation and histamine release of mast cells. *Cell Tissue Res.* **2005**, *322*, 437–441. [CrossRef]
132. Zhang, S.; Zeng, X.; Yang, H.; Hu, G.; He, S. Mast cell tryptase induces microglia activation via protease-activated receptor 2 signaling. *Cell. Physiol. Biochem. Int. J. Exp. Cell. Physiol. Biochem. Pharm.* **2012**, *29*, 931–940. [CrossRef] [PubMed]
133. Yuan, H.; Zhu, X.; Zhou, S.; Chen, Q.; Zhu, X.; Ma, X.; He, X.; Tian, M.; Shi, X. Role of mast cell activation in inducing microglial cells to release neurotrophin. *J. Neurosci. Res.* **2010**, *88*, 1348–1354. [CrossRef] [PubMed]
134. Kim, D.Y.; Jeoung, D.; Ro, J.Y. Signaling pathways in the activation of mast cells cocultured with astrocytes and colocalization of both cells in experimental allergic encephalomyelitis. *J. Immunol.* **2010**, *185*, 273–283. [CrossRef] [PubMed]
135. Hosli, L.; Hosli, E.; Schneider, U.; Wiget, W. Evidence for the existence of histamine H1- and H2-receptors on astrocytes of cultured rat central nervous system. *Neurosci. Lett.* **1984**, *48*, 287–291. [CrossRef]
136. Dong, Y.; Benveniste, E.N. Immune function of astrocytes. *Glia* **2001**, *36*, 180–190. [CrossRef]
137. Ramesh, G.; MacLean, A.G.; Philipp, M.T. Cytokines and chemokines at the crossroads of neuroinflammation, neurodegeneration, and neuropathic pain. *Mediat. Inflamm.* **2013**, *2013*, 480739. [CrossRef]
138. Voet, S.; Srinivasan, S.; Lamkanfi, M.; van Loo, G. Inflammasomes in neuroinflammatory and neurodegenerative diseases. *EMBO Mol. Med.* **2019**, *11*. [CrossRef]
139. Singhal, G.; Jaehne, E.J.; Corrigan, F.; Toben, C.; Baune, B.T. Inflammasomes in neuroinflammation and changes in brain function: A focused review. *Front. Neurosci.* **2014**, *8*, 315. [CrossRef]
140. Nathan, C. Points of control in inflammation. *Nature* **2002**, *420*, 846–852. [CrossRef]
141. Woodroofe, M.N. Cytokine production in the central nervous system. *Neurology* **1995**, *45*, S6–S10. [CrossRef]

142. Walz, A.; Peveri, P.; Aschauer, H.; Baggiolini, M. Purification and amino acid sequencing of NAF, a novel neutrophil-activating factor produced by monocytes. *Biochem. Biophys. Res. Commun.* **1987**, *149*, 755–761. [CrossRef]
143. Yoshimura, T.; Matsushima, K.; Oppenheim, J.J.; Leonard, E.J. Neutrophil chemotactic factor produced by lipopolysaccharide (LPS)-stimulated human blood mononuclear leukocytes: Partial characterization and separation from interleukin 1 (IL 1). *J. Immunol.* **2005**, *175*, 5569–5574. [PubMed]
144. Zlotnik, A.; Yoshie, O.; Nomiyama, H. The chemokine and chemokine receptor superfamilies and their molecular evolution. *Genome Biol.* **2006**, *7*, 243. [CrossRef] [PubMed]
145. Takeshita, Y.; Ransohoff, R.M. Inflammatory cell trafficking across the blood-brain barrier: Chemokine regulation and in vitro models. *Immunol. Rev.* **2012**, *248*, 228–239. [CrossRef]
146. Benveniste, E.N. Inflammatory Cytokines within the Central-Nervous-System—Sources, Function, and Mechanism of Action. *Am. J. Physiol.* **1992**, *263*, C1–C16. [CrossRef]
147. Camara-Lemarroy, C.R.; Guzman-de la Garza, F.J.; Fernandez-Garza, N.E. Molecular Inflammatory Mediators in Peripheral Nerve Degeneration and Regeneration. *Neuroimmunomodulation* **2010**, *17*, 314–324. [CrossRef]
148. Watkins, L.R.; Maier, S.F. Beyond neurons: Evidence that immune and glial cells contribute to pathological pain states. *Physiol. Rev.* **2002**, *82*, 981–1011. [CrossRef]
149. Buckley, C.D.; Gilroy, D.W.; Serhan, C.N.; Stockinger, B.; Tak, P.P. The resolution of inflammation. *Nat. Rev. Immunol.* **2013**, *13*, 59–66. [CrossRef]
150. Cotran, R.S.; Kumar, V.; Collins, T.; Robbins, S.L. *Robbins Pathologic Basis of Disease*; WB Saunders CompHny: Philadelphia, PA, USA, 2004.
151. Samuelsson, B.; Dahlen, S.E.; Lindgren, J.A.; Rouzer, C.A.; Serhan, C.N. Leukotrienes and Lipoxins—Structures, Biosynthesis, and Biological Effects. *Science* **1987**, *237*, 1171–1176. [CrossRef]
152. Samuelsson, B. Role of Basic Science in the Development of New Medicines: Examples from the Eicosanoid Field. *J. Biol. Chem.* **2012**, *287*, 10070–10080. [CrossRef]
153. Flower, R.J. Prostaglandins, bioassay and inflammation. *Br. J. Pharm.* **2006**, *147*, S182–S192. [CrossRef] [PubMed]
154. Levy, B.D.; Clish, C.B.; Schmidt, B.; Gronert, K.; Serhan, C.N. Lipid mediator class switching during acute inflammation: Signals in resolution. *Nat. Immunol.* **2001**, *2*, 612–619. [CrossRef] [PubMed]
155. Serhan, C.N.; Clish, C.B.; Brannon, J.; Colgan, S.P.; Chiang, N.; Gronert, K. Novel functional sets of lipid-derived mediators with antiinflammatory actions generated from omega-3 fatty acids via cyclooxygenase 2-nonsteroidal antiinflammatory drugs and transcellular processing. *J. Exp. Med.* **2000**, *192*, 1197–1204. [CrossRef] [PubMed]
156. Aid, S.; Bosetti, F. Targeting cyclooxygenases-1 and -2 in neuroinflammation: Therapeutic implications. *Biochimie* **2011**, *93*, 46–51. [CrossRef] [PubMed]
157. Serhan, C.N.; Chiang, N.; Dalli, J. The resolution code of acute inflammation: Novel pro-resolving lipid mediators in resolution. *Semin. Immunol.* **2015**, *27*, 200–215. [CrossRef]
158. Serhan, C.N.; Chiang, N. Resolution phase lipid mediators of inflammation: Agonists of resolution. *Curr. Opin. Pharm.* **2013**, *13*, 632–640. [CrossRef]
159. Serhan, C.N.; Chiang, N. Endogenous pro-resolving and anti-inflammatory lipid mediators: A new pharmacologic genus. *Br. J. Pharm.* **2008**, *153*, S200–S215. [CrossRef]
160. Serhan, C.N. Pro-resolving lipid mediators are leads for resolution physiology. *Nature* **2014**, *510*, 92–101. [CrossRef]
161. Leung, D.; Saghatelian, A.; Simon, G.M.; Cravatt, B.F. Inactivation of N-acyl phosphatidylethanolamine phospholipase D reveals multiple mechanisms for the biosynthesis of endocannabinoids. *Biochemistry* **2006**, *45*, 4720–4726. [CrossRef]
162. Ueda, N.; Tsuboi, K.; Uyama, T. Metabolism of endocannabinoids and related N-acylethanolamines: Canonical and alternative pathways. *FEBS J.* **2013**, *280*, 1874–1894. [CrossRef]
163. Tsuboi, K.; Takezaki, N.; Ueda, N. The N-acylethanolamine-hydrolyzing acid amidase (NAAA). *Chem. Biodivers.* **2007**, *4*, 1914–1925. [CrossRef] [PubMed]
164. Skaper, S.D.; Facci, L.; Giusti, P. Mast cells, glia and neuroinflammation: Partners in crime? *Immunology* **2014**, *141*, 314–327. [CrossRef] [PubMed]

165. Bisogno, T.; Ventriglia, M.; Milone, A.; Mosca, M.; Cimino, G.; Di Marzo, V. Occurrence and metabolism of anandamide and related acyl-ethanolamides in ovaries of the sea urchin Paracentrotus lividus. *Biochim. Biophys. Acta* **1997**, *1345*, 338–348. [CrossRef]
166. Artamonov, M.; Zhukov, O.; Shuba, I.; Storozhuk, L.; Khmel, T.; Klimashevsky, V.; Mikosha, A.; Gula, N. Incorporation of labelled N-acylethanolamine (NAE) into rat brain regions in vivo and adaptive properties of saturated NAE under x-ray irradiation. *Ukr. Biokhim. Zh.* **2005**, *77*, 51–62.
167. Gabrielsson, L.; Mattsson, S.; Fowler, C.J. Palmitoylethanolamide for the treatment of pain: Pharmacokinetics, safety and efficacy. *Br. J. Clin. Pharm.* **2016**, *82*, 932–942. [CrossRef]
168. de Novellis, V.; Luongo, L.; Guida, F.; Cristino, L.; Palazzo, E.; Russo, R.; Marabese, I.; D'Agostino, G.; Calignano, A.; Rossi, F.; et al. Effects of intra-ventrolateral periaqueductal grey palmitoylethanolamide on thermoceptive threshold and rostral ventromedial medulla cell activity. *Eur. J. Pharm.* **2012**, *676*, 41–50. [CrossRef]
169. Okamoto, Y.; Morishita, J.; Tsuboi, K.; Tonai, T.; Ueda, N. Molecular characterization of a phospholipase D generating anandamide and its congeners. *J. Biol. Chem.* **2004**, *279*, 5298–5305. [CrossRef]
170. Kim, S.R.; Kim, S.U.; Oh, U.; Jin, B.K. Transient receptor potential vanilloid subtype 1 mediates microglial cell death in vivo and in vitro via Ca2+-mediated mitochondrial damage and cytochrome c release. *J. Immunol.* **2006**, *177*, 4322–4329. [CrossRef]
171. Godlewski, G.; Offertaler, L.; Wagner, J.A.; Kunos, G. Receptors for acylethanolamides-GPR55 and GPR119. *Prostaglandins Other Lipid Mediat.* **2009**, *89*, 105–111. [CrossRef]
172. Sawzdargo, M.; Nguyen, T.; Lee, D.K.; Lynch, K.R.; Cheng, R.; Heng, H.H.; George, S.R.; O'Dowd, B.F. Identification and cloning of three novel human G protein-coupled receptor genes GPR52, PsiGPR53 and GPR55: GPR55 is extensively expressed in human brain. *Brain Res. Mol. Brain Res.* **1999**, *64*, 193–198. [CrossRef]
173. Ryberg, E.; Larsson, N.; Sjogren, S.; Hjorth, S.; Hermansson, N.O.; Leonova, J.; Elebring, T.; Nilsson, K.; Drmota, T.; Greasley, P.J. The orphan receptor GPR55 is a novel cannabinoid receptor. *Br. J. Pharm.* **2007**, *152*, 1092–1101. [CrossRef] [PubMed]
174. Pietr, M.; Kozela, E.; Levy, R.; Rimmerman, N.; Lin, Y.H.; Stella, N.; Vogel, Z.; Juknat, A. Differential changes in GPR55 during microglial cell activation. *FEBS Lett.* **2009**, *583*, 2071–2076. [CrossRef] [PubMed]
175. Soga, T.; Ohishi, T.; Matsui, T.; Saito, T.; Matsumoto, M.; Takasaki, J.; Matsumoto, S.; Kamohara, M.; Hiyama, H.; Yoshida, S.; et al. Lysophosphatidylcholine enhances glucose-dependent insulin secretion via an orphan G-protein-coupled receptor. *Biochem. Biophys. Res. Commun.* **2005**, *326*, 744–751. [CrossRef] [PubMed]
176. Sakamoto, Y.; Inoue, H.; Kawakami, S.; Miyawaki, K.; Miyamoto, T.; Mizuta, K.; Itakura, M. Expression and distribution of Gpr119 in the pancreatic islets of mice and rats: Predominant localization in pancreatic polypeptide-secreting PP-cells. *Biochem. Biophys. Res. Commun.* **2006**, *351*, 474–480. [CrossRef] [PubMed]
177. Cvijanovic, N.; Isaacs, N.J.; Rayner, C.K.; Feinle-Bisset, C.; Young, R.L.; Little, T.J. Duodenal fatty acid sensor and transporter expression following acute fat exposure in healthy lean humans. *Clin. Nutr.* **2017**, *36*, 564–569. [CrossRef] [PubMed]
178. Overton, H.A.; Babbs, A.J.; Doel, S.M.; Fyfe, M.C.; Gardner, L.S.; Griffin, G.; Jackson, H.C.; Procter, M.J.; Rasamison, C.M.; Tang-Christensen, M.; et al. Deorphanization of a G protein-coupled receptor for oleoylethanolamide and its use in the discovery of small-molecule hypophagic agents. *Cell Metab.* **2006**, *3*, 167–175. [CrossRef]
179. Chu, Z.L.; Carroll, C.; Chen, R.; Alfonso, J.; Gutierrez, V.; He, H.; Lucman, A.; Xing, C.; Sebring, K.; Zhou, J.; et al. N-oleoyldopamine enhances glucose homeostasis through the activation of GPR119. *Mol. Endocrinol.* **2010**, *24*, 161–170. [CrossRef]
180. Kondo, S.; Sugiura, T.; Kodaka, T.; Kudo, N.; Waku, K.; Tokumura, A. Accumulation of various N-acylethanolamines including N-arachidonoylethanolamine (anandamide) in cadmium chloride-administered rat testis. *Arch. Biochem. Biophys.* **1998**, *354*, 303–310. [CrossRef]
181. Epps, D.E.; Schmid, P.C.; Natarajan, V.; Schmid, H.H. N-Acylethanolamine accumulation in infarcted myocardium. *Biochem. Biophys. Res. Commun.* **1979**, *90*, 628–633. [CrossRef]
182. Loria, F.; Petrosino, S.; Mestre, L.; Spagnolo, A.; Correa, F.; Hernangomez, M.; Guaza, C.; Di Marzo, V.; Docagne, F. Study of the regulation of the endocannabinoid system in a virus model of multiple sclerosis reveals a therapeutic effect of palmitoylethanolamide. *Eur. J. Neurosci.* **2008**, *28*, 633–641. [CrossRef]

183. Garcia-Ovejero, D.; Arevalo-Martin, A.; Petrosino, S.; Docagne, F.; Hagen, C.; Bisogno, T.; Watanabe, M.; Guaza, C.; Di Marzo, V.; Molina-Holgado, E. The endocannabinoid system is modulated in response to spinal cord injury in rats. *Neurobiol. Dis.* **2009**, *33*, 57–71. [CrossRef] [PubMed]
184. Darmani, N.A.; Izzo, A.A.; Degenhardt, B.; Valenti, M.; Scaglione, G.; Capasso, R.; Sorrentini, I.; Di Marzo, V. Involvement of the cannabimimetic compound, N-palmitoyl-ethanolamine, in inflammatory and neuropathic conditions: Review of the available pre-clinical data, and first human studies. *Neuropharmacology* **2005**, *48*, 1154–1163. [CrossRef]
185. Petrosino, S.; Iuvone, T.; Di Marzo, V. N-palmitoyl-ethanolamine: Biochemistry and new therapeutic opportunities. *Biochimie* **2010**, *92*, 724–727. [CrossRef] [PubMed]
186. Abramo, F.; Campora, L.; Albanese, F.; della Valle, M.F.; Cristino, L.; Petrosino, S.; Di Marzo, V.; Miragliotta, V. Increased levels of palmitoylethanolamide and other bioactive lipid mediators and enhanced local mast cell proliferation in canine atopic dermatitis. *BMC Vet. Res.* **2014**, *10*, 21. [CrossRef] [PubMed]
187. Farquhar-Smith, W.P.; Rice, A.S. Administration of endocannabinoids prevents a referred hyperalgesia associated with inflammation of the urinary bladder. *Anesthesiology* **2001**, *94*, 507–513. [CrossRef]
188. Ahmad, A.; Crupi, R.; Impellizzeri, D.; Campolo, M.; Marino, A.; Esposito, E.; Cuzzocrea, S. Administration of palmitoylethanolamide (PEA) protects the neurovascular unit and reduces secondary injury after traumatic brain injury in mice. *Brain Behav. Immun.* **2012**, *26*, 1310–1321. [CrossRef]
189. Esposito, E.; Paterniti, I.; Mazzon, E.; Genovese, T.; Di Paola, R.; Galuppo, M.; Cuzzocrea, S. Effects of palmitoylethanolamide on release of mast cell peptidases and neurotrophic factors after spinal cord injury. *Brainbehav. Immun.* **2011**, *25*, 1099–1112. [CrossRef]
190. Genovese, T.; Esposito, E.; Mazzon, E.; Di Paola, R.; Meli, R.; Bramanti, P.; Piomelli, D.; Calignano, A.; Cuzzocrea, S. Effects of palmitoylethanolamide on signaling pathways implicated in the development of spinal cord injury. *J. Pharm. Exp.* **2008**, *326*, 12–23. [CrossRef]
191. Borrelli, F.; Romano, B.; Petrosino, S.; Pagano, E.; Capasso, R.; Coppola, D.; Battista, G.; Orlando, P.; Di Marzo, V.; Izzo, A.A. Palmitoylethanolamide, a naturally occurring lipid, is an orally effective intestinal anti-inflammatory agent. *Br. J. Pharm.* **2015**, *172*, 142–158. [CrossRef]
192. Capasso, R.; Orlando, P.; Pagano, E.; Aveta, T.; Buono, L.; Borrelli, F.; Di Marzo, V.; Izzo, A.A. Palmitoylethanolamide normalizes intestinal motility in a model of post-inflammatory accelerated transit: Involvement of CB(1) receptors and TRPV1 channels. *Br. J. Pharm.* **2014**, *171*, 4026–4037. [CrossRef]
193. Crupi, R.; Impellizzeri, D.; Cordaro, M.; Siracusa, R.; Casili, G.; Evangelista, M.; Cuzzocrea, S. N-palmitoylethanolamide Prevents Parkinsonian Phenotypes in Aged Mice. *Mol. Neurobiol.* **2018**, *55*, 8455–8472. [CrossRef] [PubMed]
194. Noli, C.; Della Valle, M.F.; Miolo, A.; Medori, C.; Schievano, C.; Skinalia Clinical Research, G. Efficacy of ultra-micronized palmitoylethanolamide in canine atopic dermatitis: An open-label multi-centre study. *Vet. Derm.* **2015**, *26*, 432–440. [CrossRef] [PubMed]
195. Cordaro, M.; Impellizzeri, D.; Bruschetta, G.; Siracusa, R.; Crupi, R.; Di Paola, R.; Esposito, E.; Cuzzocrea, S. A novel protective formulation of Palmitoylethanolamide in experimental model of contrast agent induced nephropathy. *Toxicol. Lett.* **2016**, *240*, 10–21. [CrossRef] [PubMed]
196. Di Paola, R.; Impellizzeri, D.; Fusco, R.; Cordaro, M.; Siracusa, R.; Crupi, R.; Esposito, E.; Cuzzocrea, S. Ultramicronized palmitoylethanolamide (PEA-um((R))) in the treatment of idiopathic pulmonary fibrosis. *Pharm. Res.* **2016**, *111*, 405–412. [CrossRef] [PubMed]
197. Fusco, R.; Gugliandolo, E.; Campolo, M.; Evangelista, M.; Di Paola, R.; Cuzzocrea, S. Effect of a new formulation of micronized and ultramicronized N-palmitoylethanolamine in a tibia fracture mouse model of complex regional pain syndrome. *PLoS ONE* **2017**, *12*, e0178553. [CrossRef] [PubMed]
198. Scuderi, C.; Bronzuoli, M.R.; Facchinetti, R.; Pace, L.; Ferraro, L.; Broad, K.D.; Serviddio, G.; Bellanti, F.; Palombelli, G.; Carpinelli, G.; et al. Ultramicronized palmitoylethanolamide rescues learning and memory impairments in a triple transgenic mouse model of Alzheimer's disease by exerting anti-inflammatory and neuroprotective effects. *Transl. Psychiatry* **2018**, *8*, 32. [CrossRef]
199. Bartolucci, M.L.; Marini, I.; Bortolotti, F.; Impellizzeri, D.; Di Paola, R.; Bruschetta, G.; Crupi, R.; Portelli, M.; Militi, A.; Oteri, G.; et al. Micronized palmitoylethanolamide reduces joint pain and glial cell activation. *Inflamm. Res.* **2018**, *67*, 891–901. [CrossRef]
200. Leweke, F.M.; Giuffrida, A.; Wurster, U.; Emrich, H.M.; Piomelli, D. Elevated endogenous cannabinoids in schizophrenia. *Neuroreport* **1999**, *10*, 1665–1669. [CrossRef]

201. Sarchielli, P.; Pini, L.A.; Coppola, F.; Rossi, C.; Baldi, A.; Mancini, M.L.; Calabresi, P. Endocannabinoids in chronic migraine: CSF findings suggest a system failure. *Neuropsychopharmacology* **2007**, *32*, 1384–1390. [CrossRef]
202. Di Filippo, M.; Pini, L.A.; Pelliccioli, G.P.; Calabresi, P.; Sarchielli, P. Abnormalities in the cerebrospinal fluid levels of endocannabinoids in multiple sclerosis. *J. Neurol. Neurosurg. Psychiatry* **2008**, *79*, 1224–1229. [CrossRef]
203. Jean-Gilles, L.; Feng, S.; Tench, C.R.; Chapman, V.; Kendall, D.A.; Barrett, D.A.; Constantinescu, C.S. Plasma endocannabinoid levels in multiple sclerosis. *J. Neurol. Sci.* **2009**, *287*, 212–215. [CrossRef] [PubMed]
204. Orefice, N.S.; Alhouayek, M.; Carotenuto, A.; Montella, S.; Barbato, F.; Comelli, A.; Calignano, A.; Muccioli, G.G.; Orefice, G. Oral Palmitoylethanolamide Treatment Is Associated with Reduced Cutaneous Adverse Effects of Interferon-beta1a and Circulating Proinflammatory Cytokines in Relapsing-Remitting Multiple Sclerosis. *Neurotherapeutics* **2016**, *13*, 428–438. [CrossRef] [PubMed]
205. D'Argenio, G.; Petrosino, S.; Gianfrani, C.; Valenti, M.; Scaglione, G.; Grandone, I.; Nigam, S.; Sorrentini, I.; Mazzarella, G.; Di Marzo, V. Overactivity of the intestinal endocannabinoid system in celiac disease and in methotrexate-treated rats. *J. Mol. Med.* **2007**, *85*, 523–530. [CrossRef] [PubMed]
206. Richardson, D.; Pearson, R.G.; Kurian, N.; Latif, M.L.; Garle, M.J.; Barrett, D.A.; Kendall, D.A.; Scammell, B.E.; Reeve, A.J.; Chapman, V. Characterisation of the cannabinoid receptor system in synovial tissue and fluid in patients with osteoarthritis and rheumatoid arthritis. *Arthritis Res.* **2008**, *10*, R43. [CrossRef] [PubMed]
207. Andresen, S.R.; Bing, J.; Hansen, R.M.; Biering-Sorensen, F.; Johannesen, I.L.; Hagen, E.M.; Rice, A.S.; Nielsen, J.F.; Bach, F.W.; Finnerup, N.B. Ultramicronized palmitoylethanolamide in spinal cord injury neuropathic pain: A randomized, double-blind, placebo-controlled trial. *Pain* **2016**, *157*, 2097–2103. [CrossRef]
208. Evangelista, M.; Cilli, V.; De Vitis, R.; Militerno, A.; Fanfani, F. Ultra-micronized Palmitoylethanolamide Effects on Sleep-wake Rhythm and Neuropathic Pain Phenotypes in Patients with Carpal Tunnel Syndrome: An Open-label, Randomized Controlled Study. *CNS Neurol. Disord. Drug Targets* **2018**, *17*, 291–298. [CrossRef]
209. Gatti, A.; Lazzari, M.; Gianfelice, V.; Di Paolo, A.; Sabato, E.; Sabato, A.F. Palmitoylethanolamide in the Treatment of Chronic Pain Caused by Different Etiopathogenesis. *Pain Med.* **2012**, *13*, 1121–1130. [CrossRef]
210. Guida, G.; Cantieri, L.; Petrosino, S. Reducción del consumo de antiinflamatorios y analgésicos en el tratamiento del dolor neuropático crónico en pacientes afectados por lumbociatalgia de tipo compresivo y en tratamiento con Normast® 300 mg. *Dolor. Investig. Clínica Ter.* **2010**, *25*, 227–234.
211. Guida, G.; De Martino, M.; De Fabiani, A.; Cantieri, L.; Alexandre, A.; Vassallo, G.; Rogai, M.; Lanaia, F.; Petrosino, S. La palmitoiletanolamida (Normast®) en el dolor neuropático crónico por lumbociatalgia de tipo compresivo: Estudio clínico multicéntrico. *Dolor. Investig. Clínica Ter.* **2010**, *25*, 35–42.
212. Marini, I.; Bartolucci, M.L.; Bortolotti, F.; Gatto, M.R.; Bonetti, G.A. Palmitoylethanolamide Versus a Nonsteroidal Anti-Inflammatory Drug in the Treatment of Temporomandibular Joint Inflammatory Pain. *J. Orofac. Pain* **2012**, *26*, 99–104.
213. Schifilliti, C.; Cucinotta, L.; Fedele, V.; Ingegnosi, C.; Luca, S.; Leotta, C. Micronized palmitoylethanolamide reduces the symptoms of neuropathic pain in diabetic patients. *Pain Res. Treat.* **2014**, *2014*, 849623. [CrossRef] [PubMed]
214. Hesselink, J.M.; Hekker, T.A. Therapeutic utility of palmitoylethanolamide in the treatment of neuropathic pain associated with various pathological conditions: A case series. *J. Pain Res.* **2012**, *5*, 437–442. [CrossRef] [PubMed]
215. Lambert, D.M.; Vandevoorde, S.; Diependaele, G.; Govaerts, S.J.; Robert, A.R. Anticonvulsant activity of N-palmitoylethanolamide, a putative endocannabinoid, in mice. *Epilepsia* **2001**, *42*, 321–327. [CrossRef] [PubMed]
216. Takano, R.; Furumoto, K.; Shiraki, K.; Takata, N.; Hayashi, Y.; Aso, Y.; Yamashita, S. Rate-limiting steps of oral absorption for poorly water-soluble drugs in dogs; Prediction from a miniscale dissolution test and a physiologically-based computer simulation. *Pharm. Res.* **2008**, *25*, 2334–2344. [CrossRef] [PubMed]
217. Leleux, J.; Williams, R.O. Recent advancements in mechanical reduction methods: Particulate systems. *Drug Dev. Ind. Pharm.* **2014**, *40*, 289–300. [CrossRef] [PubMed]
218. Campardelli, R.; Oleandro, E.; Scognamiglio, M.; Della Porta, G.; Reverchon, E. Palmitoylethanolamide sub-micronization using fast precipitation followed by supercritical fluids extraction. *Powder Technol.* **2017**, *305*, 217–225. [CrossRef]

219. Petrosino, S.; Cordaro, M.; Verde, R.; Moriello, A.S.; Marcolongo, G.; Schievano, C.; Siracusa, R.; Piscitelli, F.; Peritore, A.F.; Crupi, R.; et al. Oral Ultramicronized Palmitoylethanolamide:Plasma and Tissue Levels and Spinal Anti-hyperalgesic Effect. *Front. Pharm.* **2018**, *9*. [CrossRef]
220. Skaper, S.D.; Facci, L.; Fusco, M.; della Valle, M.F.; Zusso, M.; Costa, B.; Giusti, P. Palmitoylethanolamide, a naturally occurring disease-modifying agent in neuropathic pain. *Inflammopharmacology* **2014**, *22*, 79–94. [CrossRef]
221. Petrosino, S.; Di Marzo, V. The pharmacology of palmitoylethanolamide and first data on the therapeutic efficacy of some of its new formulations. *Brit. J. Pharm.* **2017**, *174*, 1349–1365. [CrossRef]
222. Impellizzeri, D.; Bruschetta, G.; Cordaro, M.; Crupi, R.; Siracusa, R.; Esposito, E.; Cuzzocrea, S. Micronized/ultramicronized palmitoylethanolamide displays superior oral efficacy compared to nonmicronized palmitoylethanolamide in a rat model of inflammatory pain. *J. Neuroinflamm.* **2014**, *11*, 136. [CrossRef]
223. Freitag, C.M.; Miller, R.J. Peroxisome proliferator-activated receptor agonists modulate neuropathic pain: A link to chemokines? *Front. Cell Neurosci.* **2014**, *8*, 238. [CrossRef] [PubMed]
224. Emamzadeh, F.N.; Surguchov, A. Parkinson's Disease: Biomarkers, Treatment, and Risk Factors. *Front. Neurosci.* **2018**, *12*, 612. [CrossRef] [PubMed]
225. Desio, P. Associazione tra pregabalin e palmitoiletanolamide per il trattamento del dolore neuropatico. *Pathos* **2010**, *7*, 9–14.
226. Desio, P. Associazione dell'ossicodone a lenta titolazione con Palmitoiletanolamide per il trattamento del low back pain. *Anest E Med. Crit.* **2011**, *1*, 63–71.
227. Cervellati, C.; Cremonini, E.; Bosi, C.; Magon, S.; Zurlo, A.; Bergamini, C.M.; Zuliani, G. Systemic oxidative stress in older patients with mild cognitive impairment or late onset Alzheimer's disease. *Curr. Alzheimer Res.* **2013**, *10*, 365–372. [CrossRef] [PubMed]
228. Radi, E.; Formichi, P.; Battisti, C.; Federico, A. Apoptosis and oxidative stress in neurodegenerative diseases. *J. Alzheimers Dis.* **2014**, *42*, S125–S152. [CrossRef] [PubMed]
229. Salim, S. Oxidative stress and psychological disorders. *Curr. Neuropharmacol.* **2014**, *12*, 140–147. [CrossRef]
230. Bruschetta, G.; Impellizzeri, D.; Morabito, R.; Marino, A.; Ahmad, A.; Spano, N.; Spada, G.L.; Cuzzocrea, S.; Esposito, E. Pelagia noctiluca (Scyphozoa) crude venom injection elicits oxidative stress and inflammatory response in rats. *Mar. Drugs* **2014**, *12*, 2182–2204. [CrossRef]
231. Grosso, C.; Valentao, P.; Ferreres, F.; Andrade, P.B. The use of flavonoids in central nervous system disorders. *Curr. Med. Chem.* **2013**, *20*, 4694–4719. [CrossRef]
232. Nabavi, S.F.; Braidy, N.; Gortzi, O.; Sobarzo-Sanchez, E.; Daglia, M.; Skalicka-Wozniak, K.; Nabavi, S.M. Luteolin as an anti-inflammatory and neuroprotective agent: A brief review. *Brain Res. Bull.* **2015**, *119*, 1–11. [CrossRef]
233. Lin, Y.; Shi, R.X.; Wang, X.; Shen, H.M. Luteolin, a Flavonoid with Potential for Cancer Prevention and Therapy. *Curr. Cancer Drug Tar.* **2008**, *8*, 634–646. [CrossRef]
234. Bravo, L. Polyphenols: Chemistry, dietary sources, metabolism, and nutritional significance. *Nutr. Rev.* **1998**, *56*, 317–333. [CrossRef] [PubMed]
235. Chen, H.Q.; Jin, Z.Y.; Wang, X.J.; Xua, X.M.; Deng, L.; Zhao, J.W. Luteolin protects dopaminergic neurons from inflammation-induced injury through inhibition of microglial activation. *Neurosci. Lett.* **2008**, *448*, 175–179. [CrossRef] [PubMed]
236. Cheng, H.Y.; Hsieh, M.T.; Tsai, F.S.; Wu, C.R.; Chiu, C.S.; Lee, M.M.; Xu, H.X.; Zhao, Z.Z.; Peng, W.H. Neuroprotective Effect of Luteolin on Amyloid beta Protein (25-35)-Induced Toxicity in Cultured Rat Cortical Neurons. *Phytother. Res.* **2010**, *24*, S102–S108. [CrossRef] [PubMed]
237. Dirscherl, K.; Karlstetter, M.; Ebert, S.; Kraus, D.; Hlawatsch, J.; Walczak, Y.; Moehle, C.; Fuchshofer, R.; Langmann, T. Luteolin triggers global changes in the microglial transcriptome leading to a unique anti-inflammatory and neuroprotective phenotype (vol 7, pg 3, 2010). *J. Neuroinflamm.* **2012**, *9*. [CrossRef]
238. Dirscherl, K.; Karlstetter, M.; Ebert, S.; Kraus, D.; Hlawatsch, J.; Walczak, Y.; Moehle, C.; Fuchshofer, R.; Langmann, T. Luteolin triggers global changes in the microglial transcriptome leading to a unique anti-inflammatory and neuroprotective phenotype. *J. Neuroinflamm.* **2010**, *7*. [CrossRef]
239. Kang, S.S.; Lee, J.Y.; Choi, Y.K.; Kim, G.S.; Han, B.H. Neuroprotective effects of flavones on hydrogen peroxide-induced apoptosis in SH-SY5Y neuroblostoma cells. *Bioorg. Med. Chem. Lett.* **2004**, *14*, 2261–2264. [CrossRef]

240. Pandurangan, A.K.; Esa, N.M. Luteolin, a Bioflavonoid Inhibits Colorectal Cancer through Modulation of Multiple Signaling Pathways: A Review. *Asian Pac. J. Cancer Prev.* **2014**, *15*, 5501–5508. [CrossRef]
241. Zhang, Y.C.; Gan, F.F.; Shelar, S.B.; Ng, K.Y.; Chew, E.H. Antioxidant and Nrf2 inducing activities of luteolin, a flavonoid constituent in Ixeris sonchifolia Hance, provide neuroprotective effects against ischemia-induced cellular injury. *Food Chem. Toxicol.* **2013**, *59*, 272–280. [CrossRef]
242. Zhou, P.; Li, L.P.; Luo, S.Q.; Jiang, H.D.; Zeng, S. Intestinal absorption of luteolin from peanut hull extract is more efficient than that from individual pure luteolin. *J. Agric. Food Chem.* **2008**, *56*, 296–300. [CrossRef]
243. Kim, J.S.; Jobin, C. The flavonoid luteolin prevents lipopolysaccharide-induced NF-kappa B signalling and gene expression by blocking I kappa B kinase activity in intestinal epithelial cells and bone-marrow derived dendritic cells. *Immunology* **2005**, *115*, 375–387. [CrossRef]
244. Lee, J.K.; Kim, S.Y.; Kim, Y.S.; Lee, W.H.; Hwang, D.H.; Lee, J.Y. Suppression of the TRIF-dependent signaling pathway of Toll-like receptors by luteolin. *Biochem. Pharm.* **2009**, *77*, 1391–1400. [CrossRef] [PubMed]
245. Weng, Z.; Patel, A.B.; Panagiotidou, S.; Theoharides, T.C. The novel flavone tetramethoxyluteolin is a potent inhibitor of human mast cells. *J. Allergy Clin. Immun.* **2015**, *135*, 1044. [CrossRef] [PubMed]
246. Lin, C.W.; Wu, M.J.; Liu, I.Y.C.; Su, J.D.; Yen, J.H. Neurotrophic and Cytoprotective Action of Luteolin in PC12 Cells through ERK-Dependent Induction of Nrf2-Driven HO-1 Expression. *J. Agric. Food Chem.* **2010**, *58*, 4477–4486. [CrossRef] [PubMed]
247. Choi, S.M.; Kim, B.C.; Cho, Y.H.; Choi, K.H.; Chang, J.; Park, M.S.; Kim, M.K.; Cho, K.H.; Kim, J.K. Effects of Flavonoid Compounds on beta-amyloid-peptide-induced Neuronal Death in Cultured Mouse Cortical Neurons. *Chonnam Med. J.* **2014**, *50*, 45–51. [CrossRef] [PubMed]
248. Patil, S.P.; Jain, P.D.; Sancheti, J.S.; Ghumatkar, P.J.; Tambe, R.; Sathaye, S. Neuroprotective and neurotrophic effects of Apigenin and Luteolin in MPTP induced parkinsonism in mice. *Neuropharmacology* **2014**, *86*, 192–202. [CrossRef] [PubMed]
249. Sternberg, Z.; Chadha, K.; Lieberman, A.; Drake, A.; Hojnacki, D.; Weinstock-Guttman, B.; Munschauer, F. Immunomodulatory responses of peripheral blood mononuclear cells from multiple sclerosis patients upon in vitro incubation with the flavonoid luteolin: Additive effects of IFN-beta. *J. Neuroinflamm.* **2009**, *6*. [CrossRef]
250. Theoharides, T.C. Luteolin as a therapeutic option for multiple sclerosis. *J. Neuroinflamm.* **2009**, *6*. [CrossRef]
251. Verbeek, R.; van Tol, E.A.F.; van Noort, J.M. Oral flavonoids delay recovery from experimental autoimmune encephalomyelitis in SJL mice. *Biochem. Pharm.* **2005**, *70*, 220–228. [CrossRef]
252. Xu, J.; Wang, H.; Ding, K.; Zhang, L.; Wang, C.; Li, T.; Wei, W.; Lu, X. Luteolin provides neuroprotection in models of traumatic brain injury via the Nrf2-ARE pathway. *Free Radic. Biol. Med.* **2014**, *71*, 186–195. [CrossRef]
253. Taliou, A.; Zintzaras, E.; Lykouras, L.; Francis, K. An Open-Label Pilot Study of a Formulation Containing the Anti-Inflammatory Flavonoid Luteolin and Its Effects on Behavior in Children With Autism Spectrum Disorders. *Clin. Ther.* **2013**, *35*, 592–602. [CrossRef]
254. Crupi, R.; Paterniti, I.; Ahmad, A.; Campolo, M.; Esposito, E.; Cuzzocrea, S. Effects of palmitoylethanolamide and luteolin in an animal model of anxiety/depression. *CNS Neurol. Disord. Drug Targets* **2013**, *12*, 989–1001. [CrossRef] [PubMed]
255. Impellizzeri, D.; Esposito, E.; Di Paola, R.; Ahmad, A.; Campolo, M.; Peli, A.; Morittu, V.M.; Britti, D.; Cuzzocrea, S. Palmitoylethanolamide and luteolin ameliorate development of arthritis caused by injection of collagen type II in mice. *Arthritis Res.* **2013**, *15*, R192. [CrossRef] [PubMed]
256. Bertolino, B.; Crupi, R.; Impellizzeri, D.; Bruschetta, G.; Cordaro, M.; Siracusa, R.; Esposito, E.; Cuzzocrea, S. Beneficial Effects of Co-Ultramicronized Palmitoylethanolamide/Luteolin in a Mouse Model of Autism and in a Case Report of Autism. *CNS Neurosci. Ther.* **2017**, *23*, 87–98. [CrossRef] [PubMed]
257. Crupi, R.; Impellizzeri, D.; Bruschetta, G.; Cordaro, M.; Paterniti, I.; Siracusa, R.; Cuzzocrea, S.; Esposito, E. Co-Ultramicronized Palmitoylethanolamide/Luteolin Promotes Neuronal Regeneration after Spinal Cord Injury. *Front. Pharm.* **2016**, *7*, 47. [CrossRef]
258. Caltagirone, C.; Cisari, C.; Schievano, C.; Di Paola, R.; Cordaro, M.; Bruschetta, G.; Esposito, E.; Cuzzocrea, S.; Stroke Study, G. Co-ultramicronized Palmitoylethanolamide/Luteolin in the Treatment of Cerebral Ischemia: From Rodent to Man. *Transl. Stroke Res.* **2016**, *7*, 54–69. [CrossRef]

259. Siracusa, R.; Paterniti, I.; Impellizzeri, D.; Cordaro, M.; Crupi, R.; Navarra, M.; Cuzzocrea, S.; Esposito, E. The Association of Palmitoylethanolamide with Luteolin Decreases Neuroinflammation and Stimulates Autophagy in Parkinson's Disease Model. *CNS Neurol. Disord. Drug Targets* **2015**, *14*, 1350–1365. [CrossRef]
260. Siracusa, R.; Paterniti, I.; Bruschetta, G.; Cordaro, M.; Impellizzeri, D.; Crupi, R.; Cuzzocrea, S.; Esposito, E. The Association of Palmitoylethanolamide with Luteolin Decreases Autophagy in Spinal Cord Injury. *Mol. Neurobiol.* **2016**, *53*, 3783–3792. [CrossRef]
261. Paterniti, I.; Cordaro, M.; Campolo, M.; Siracusa, R.; Cornelius, C.; Navarra, M.; Cuzzocrea, S.; Esposito, E. Neuroprotection by association of palmitoylethanolamide with luteolin in experimental Alzheimer's disease models: The control of neuroinflammation. *CNS Neurol. Disord. Drug Targets* **2014**, *13*, 1530–1541. [CrossRef]
262. Paterniti, I.; Impellizzeri, D.; Di Paola, R.; Navarra, M.; Cuzzocrea, S.; Esposito, E. A new co-ultramicronized composite including palmitoylethanolamide and luteolin to prevent neuroinflammation in spinal cord injury. *J. Neuroinflamm.* **2013**, *10*, 91. [CrossRef]
263. Cordaro, M.; Cuzzocrea, S.; Di Paola, R. Vascular dementia and aliamides: A new approach for the future. *J. Transl. Sci.* **2018**, *5*, 1–4. [CrossRef]

© 2020 by the authors. Licensee MDPI, Basel, Switzerland. This article is an open access article distributed under the terms and conditions of the Creative Commons Attribution (CC BY) license (http://creativecommons.org/licenses/by/4.0/).

Article

Redox Regulation of Microvascular Permeability: IL-1β Potentiation of Bradykinin-Induced Permeability Is Prevented by Simvastatin

Felipe Freitas [1,†], Eduardo Tibiriçá [2], Mita Singh [1], Paul A. Fraser [1,*] and Giovanni E. Mann [1,*]

1 Centre of Research Excellence, King's College London British Heart Foundation, School of Cardiovascular Medicine & Sciences, Faculty of Life Sciences & Medicine, King's College London, 150 Stamford Street, London SE1 9NH, UK.; f.freitas@ucl.ac.uk (F.F.); mita.singh@kcl.ac.uk (M.S.)

2 National Institute of Cardiology, Ministry of Health, Rio de Janeiro 22240-006, Brazil; etibi@uol.com.br

* Correspondence: paul.fraser@kcl.ac.uk (P.A.F.); giovanni.mann@kcl.ac.uk (G.E.M.); Tel.: +44-(0)20-78484306 (G.E.M.)

† Present Address: Department of Neuroscience, Physiology, and Pharmacology, University College London, London WC1E 6BT, UK.

Received: 9 November 2020; Accepted: 9 December 2020; Published: 14 December 2020

Abstract: Antioxidant effects of statins have been implicated in the reduction in microvascular permeability and edema formation in experimental and clinical studies. Bradykinin (Bk)-induced increases in microvascular permeability are potentiated by IL-1β; however, no studies have examined the protection afforded by statins against microvascular hyperpermeability. We investigated the effects of simvastatin pretreatment on albumin–fluorescein isothiocyanate conjugate (FITC-albumin) permeability in post-capillary venules in rat cremaster muscle. Inhibition of nitric oxide synthase with L-NAME (10μM) increased basal permeability to FITC-albumin, which was abrogated by superoxide dismutase and catalase. Histamine-induced (1 μM) permeability was blocked by L-NAME but unaffected by scavenging reactive oxygen species with superoxide dismutase (SOD) and catalase. In contrast, bradykinin-induced (1–100 nM) permeability increases were unaffected by L-NAME but abrogated by SOD and catalase. Acute superfusion of the cremaster muscle with IL-1β (30 pM, 10 min) resulted in a leftward shift of the bradykinin concentration–response curve. Potentiation by IL-1β of bradykinin-induced microvascular permeability was prevented by the nicotinamide adenine dinucleotide phosphate oxidase (NADPH oxidase) inhibitor apocynin (1 μM). Pretreatment of rats with simvastatin (5 mg·kg^{-1}, i.p.) 24 h before permeability measurements prevented the potentiation of bradykinin permeability responses by IL-1β, which was not reversed by inhibition of heme oxygenase-1 with tin protoporphyrin IX (SnPP). This study highlights a novel mechanism by which simvastatin prevents the potentiation of bradykinin-induced permeability by IL-1β, possibly by targeting the assembly of NADPH oxidase subunits. Our findings highlight the therapeutic potential of statins in the prevention and treatment of patients predisposed to inflammatory diseases.

Keywords: microvascular permeability; bradykinin; interleukin 1β; NADPH oxidase; reactive oxygen species; simvastatin

1. Introduction

Microvascular endothelial barrier disruption occurs in a large number of disease states, such as stroke, sepsis, diabetes, hereditary and acquired angioedema, commonly induced by a variety of endogenous inflammatory mediators such as bradykinin [1–6]. Novel therapeutic approaches to prevent or reduce microvascular permeability are paramount to avoid tissue edema and to maintain sufficient blood supply to target organs. In this context, statins have been described to reduce vascular permeability and edema formation in different animal and clinical studies [7–10], yet the underlying mechanisms have not been investigated in an intact muscle microvasculature.

Bradykinin has several pathophysiological functions and activates the B2 receptor, which is constitutively expressed on the vasculature and increases vascular permeability in post-capillary venules [11]. Moreover, bradykinin is an important mediator in stroke, sepsis, diabetes, hereditary and acquired angioedema [1–6]. Bradykinin may also play a key role in the vascular leakage and pulmonary edema in patients with COVID-19 [12–14]. Angiotensin converting enzyme 2 (ACE2) has been implicated as the cellular receptor of SARS-CoV-2 virus [15,16], and reduced ACE2 activity may indirectly activate the kallikrein–bradykinin pathway to increase vascular permeability [17].

In vitro and in vivo studies have shown that the increase in vascular permeability induced by bradykinin depends on the generation of reactive oxygen species [18,19]. We previously reported that bradykinin-induced microvascular permeability in the brain pial microvasculature in vivo is directly associated with the release of reactive oxygen species following bradykinin receptor activation [20]. The pro-inflammatory cytokine IL-1β has been shown potentiate the actions of bradykinin and to increase microvascular permeability and edema formation after experimental cerebral ischemia reperfusion injury [19,21,22]. Under ischemic conditions, IL-1β is rapidly released from brain tissue, leading to NADPH oxidase assembly and activation, which then rapidly potentiates the permeability response to bradykinin [19]. Notably, potentiation of bradykinin-induced increases in cerebral microvascular permeability are blocked by the IL-1 receptor antagonist, IL-1ra [19]. Moreover, acute release of IL-1β has been described as a key inflammatory event in patients with COVID-19 [23–25] that could also potentiate bradykinin-induced vascular permeability.

Clinical and experimental studies indicate several beneficial effects of statins independent of their cholesterol-lowering action [26–28]. Statins may have the potential to reduce oxidative stress by modulating Nrf2-regulated antioxidant genes [29,30], such as heme oxygenase 1 (HO-1) known to afford protection in rodent models of ischemia in vivo [31,32] and in vascular cells in vitro [29,33]. Further evidence suggests that simvastatin may upregulate HO-1 independently of Nrf2 [34].

To date there are no studies focused on the protective actions of statins against IL-1β mediated potentiation of bradykinin-induced microvascular permeability. In this study, we investigate for the first time the effects of pretreatment of rats with simvastatin on bradykinin- and IL-1β-induced microvascular permeability using intravital microscopy in an intact cremaster muscle preparation that to date has not been reported. Our findings suggest that simvastatin prevents microvascular hyperpermeability induced by IL-1β and bradykinin via inhibition of NADPH oxidase and inhibition of reactive oxygen species generation.

2. Materials and Methods

2.1. Animals and Isolation of the Cremaster Skeletal Muscle Preparation

This study conforms with the Guide for the Care and Use of Laboratory Animals published by the US National Institutes of Health (NIH Publication No. 85–23, revised 1996) and is in accordance with UK Home Office regulations (Animals Scientific Procedures) Act, 1986. Approved by UK Home Office Animal Project License (PPL Number: 70/8934).

Male Wistar rats (Charles River, UK), 4–6 weeks old and weighing 80–100 g), were killed by exposure to a rising concentration of CO_2 followed by cervical dislocation. A longitudinal midline incision (1–2 cm) was made along the abdomen to expose the underlying organs. All the small branches of the aorta except the common iliac arteries leading branches that did not supply the chosen cremaster were tied off and the vena cava was then punctured to create an outlet for the blood that was flushed out of the circulation. The aorta was cannulated orthogradely with a polythene tubing (outside diameter 0.61 mm). The left common iliac and the right femoral and internal iliac arteries were ligated to ensure that perfusion was directed to the right external iliac artery supplying the cremaster artery and the cremaster muscle microvasculature. The tissue was perfused with a modified St. Thomas' cardioplegic solution (mM: 10 $MgCl_2$, 110 NaCl, 8 KCl, 1 $CaCl_2$, 10 HEPES) [35] containing heparin (30 U/mL) and isoproterenol 10 μM buffered to pH 7.0 ± 0.05 for 10 min.

2.2. Superfusion of Cremaster Muscle Preparation

A longitudinal incision was made along the length of the ventral aspect of the scrotum and the overlying fascia and connective tissue were carefully removed. The cremaster was pulled out with the testicle using a pair of blunt forceps, and the distal end of the muscle was secured on a SylGard block using histology pins (Watkins and Doncaster, Kent, England). The intact cremaster preparation was then transferred to a modified stage of an intravital microscope (ACM, Zeiss, Oberkochen, Germany) and continuously superfused (2 $mL{\cdot}min^{-1}$) with an albumin-free Krebs solution (pH 7.4) gassed with 5% CO_2 in air and maintained at 37 °C. The superfusate contained the Na^+ channel blocker lidocaine (20 $mg{\cdot}L^{-1}$) to block neural activity and to minimize cremaster muscle contractions.

2.3. Measurement of Post-Capillary Venule Permeability to FITC-Albumin

The stabilizing solution perfusing the cremaster vasculature was replaced with Krebs solution (mM: 118 NaCl; 4.7 KCl; 2.52 $CaCl_2$; 1.18 $MgSO_4.7H_2O$; 1.18 KH_2PO_4; 25 $NaHCO_3$, 5 glucose and buffered to pH 7.4 ± 0.05) containing bovine albumin (10 $mg{\cdot}mL^{-1}$) delivered by a gravity controlled reservoir at 0.5 $mL{\cdot}min^{-1}$. After 30 min, Krebs perfusion of the vasculature was stopped and a bolus of Krebs solution containing FITC-albumin (5 $mg{\cdot}mL^{-1}$) was injected into the perfusion line. Post-capillary venules were identified by noting the direction of flow, as the microvasculature was filled with the fluorescent dye, using a 10× water immersion objective (numerical aperture 0.5). Images were captured using an FITC filter cube (Chroma Technology Bellows Falls, VT, USA) via image-intensified CCD camera (Photonic Sciences, Robertsbridge, E. Sussex, UK) for subsequent analysis (ImageHopper; Samsara Research, Dorking, Surrey, UK).

Perfusion pressure was lowered to atmospheric, and pressure differences in the vasculature were allowed to dissipate over the course of 3 min. In previous studies, we have demonstrated a linear correlation between the light collected with the dye concentration and as well as with the square of the diameter of the microvessel up to a 60 μm limit [36]. Permeability measurements were obtained from an image sequence acquired at 1 s intervals over 100 s. The dye concentration difference across a vessel was calculated from the difference between the regions of interest positioned on an image stack (see Figure 1A,B). Permeability was determined from the rate of decrease in that difference, obtained by fitting an exponential to the data (Figure 1C) such that P = kD/4, where k is the rate constant and r is the vessel diameter. The lack of axial flow under the experimental conditions was confirmed by viewing fluorescent microspheres (1 μm diameter) within the vasculature (data not shown). It was possible to generate a permeability map on a few occasions when the venule was on the exposed surface of the cremaster preparation, so that there was no overlying tissue and that any escaped dye dissipated rapidly. Under these circumstances, the rate constants could be calculated on a pixel by pixel basis during the exponential fall of dye (see Figure 1D) by taking linear regression of the log (V-I), where V and I are the pixel values within the vessel and the interstitium, respectively.

2.4. Role of Nitric Oxide and Reactive Oxygen Species in Microvascular Permeability

To determine the role of nitric oxide and reactive oxygen species on basal permeability, the cremaster preparation was superfused for 5 min with a nitric oxide synthase (NOS) inhibitor, N-ω-nitro-L-arginine methyl ester (L-NAME; 10 μM) and/or the free radical scavengers superoxide dismutase (SOD, 100 $U{\cdot}mL^{-1}$) and catalase (CAT, 100 $U{\cdot}mL^{-1}$). Further experiments examined the effects of the vasoactive mediators histamine (1 μM) and bradykinin (100 nM) on permeability in the absence or presence of L-NAME or SOD and CAT.

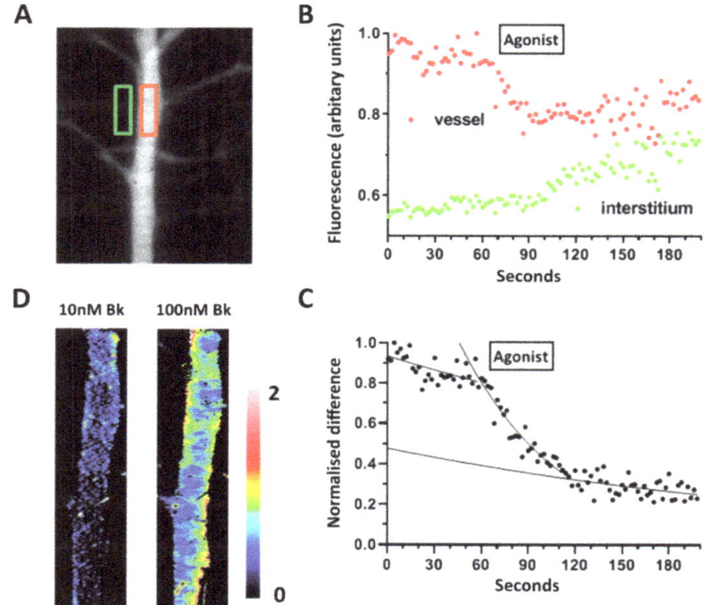

Figure 1. Basal and agonist stimulated permeability measurements in single venules in a rat cremaster muscle preparation. (**A**) Representative fluorescence image of the cremaster microvasculature following arterial FITC-albumin infusion. A sequence of images was captured at 1 s intervals after all axial flow had ceased, during which time (70–110 s, panel **B**) either histamine or bradykinin was applied topically. (**B**) Image stack was analyzed by placing regions of interest (ROIs) over the 33 μm diameter venule (red) and the neighboring interstitium (green). (**C**) Difference between these values for the two ROI yields the albumin concentration gradient across the microvessel. The rate constant (k) for the fitted monoexponential and the diameter gives the permeability value P = kD/4, assuming a circular diameter. (**D**) A few venules, such as the one illustrated in panel A, were on the surface of the cremaster, not overlaid with skeletal muscle fibers, which allowed a color-coded permeability map to be generated: the scale values are expressed as $cm{\cdot}s^{-1} \times 10^{-6}$. The left-hand image was generated following application of 10 nM bradykinin and the right-hand image after 100 nM bradykinin.

2.5. Bradykinin- and IL-1β-Induced Increases in Microvascular Permeability

Increasing concentrations of bradykinin (10^{-9}, 10^{-8} and 10^{-7} M) were applied abluminally to the cremaster muscle to elicit dose dependent increases in FITC-albumin permeability. After a dose–response curve to bradykinin, the cremaster muscle was rapidly superfused and IL-1β (30 pM) applied abluminally for 10 min. The preparation was then superfused to remove IL-1β, and a new dose–response curve to bradykinin (10^{-9}, 10^{-8} and 10^{-7} M) performed. To reduce variability between drug applications, the same region of the cremaster microvasculature was observed throughout an entire experiment, paired permeability measurements obtained in single post-capillary venules.

2.6. Inhibition of NADPH Oxidase Assembly

To determine the involvement of NADPH oxidase in the microvascular hyperpermeability induced by IL-1β and bradykinin, the cremaster preparation was superfused for 10 min with IL-1β (30 pM) in the presence of apocynin (Apo, 1 μM), a specific inhibitor of NADPH oxidase in control rats as well as in simvastatin pretreated rats. The preparation was then rapidly superfused to remove IL-1β and apocynin and bradykinin (100 nM) applied abluminally.

2.7. Pretreatment of Animals with Simvastatin

Simvastatin (5 mg·kg^{-1}) was administered to rats intraperitoneally 24 h before isolation of the cremaster muscle preparation.

2.8. Inhibition of Heme Oxygenase-1 with Tin Protophoryrin IX (SnPP)

The HO-1 inhibitor, tin protoporphyrin IX (SnPP) (5 μM), was applied abluminally for 10 min. The preparation was then superfused to remove the (SnPP), and bradykinin (100 nM) was applied. The cremaster muscle was then rapidly superfused (washed) and IL-1β (30 pM) was co-applied with (SnPP) (5 μM) abluminally for 10 min. The preparation was then superfused to remove IL-1β and (SnPP) and bradykinin (100 nM) applied abluminally.

2.9. Reagents

All chemicals were purchased from Sigma-Aldrich (Dorset, UK).

2.10. Statistical Analysis

Experimental data represent paired permeability measurements in single venules from different animals or are expressed as mean ± SEM of measurements in single venules from n = 4–10 animals. Data were analyzed using a paired or unpaired Student's t-test and ANOVA in GraphPad Prism 6.0 (La Jolla, CA, USA), with p <0.05 considered statistically significant.

3. Results

3.1. Role of NO and Reactive Oxygen Species in Modulating Basal Microvascular Permeability

Application of the nitric oxide synthase inhibitor L-NAME (10 μM) increased permeability (0.69 ± 0.26 cm·s^{-1} × 10^{-6}, p < 0.05) above basal levels (0.33 ± 0.23 cm·s^{-1} × 10^{-6}, Figure 2). Notably, co-application of superoxide dismutase (SOD, 100 U·mL^{-1}) and catalase (CAT, 100 U·mL^{-1}) with L-NAME abrogated the permeability increase (0.15 ± 0.09 cm·s^{-1} × 10^{-6} vs. 0.25 ±0.05 cm·s^{-1} × 10^{-6}, Figure 2) evoked by L-NAME.

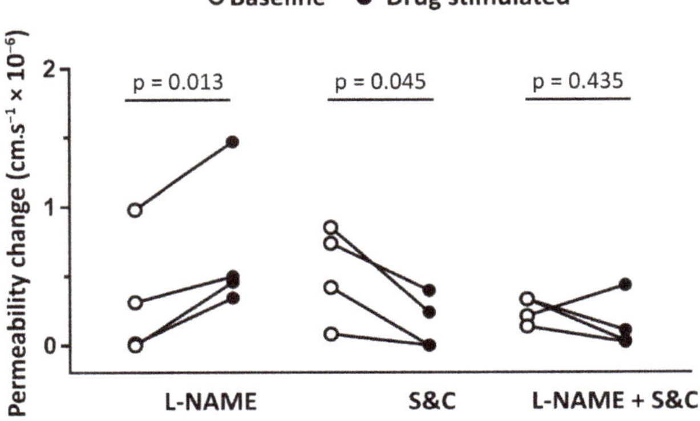

Figure 2. Constitutive nitric oxide reduces basal post-capillary venule permeability. Inhibiting constitutive eNOS with L-NAME (10 μM) resulted in a significant permeability increase, while scavenging reactive oxygen species with a combination superoxide dismutase and catalase (100 U·mL^{-1} each) reduced basal permeability. When superoxide dismutase and catalase were co-applied with L-NAME, there was no permeability change. Data from paired measurements in 4 venules from 4 different animals. Data were analyzed using a paired Student's t-test.

3.2. Histamine- and Bradykinin-Induced Microvascular Permeability Is Mediated by Different Signaling Pathways

Histamine-induced (1 μM) permeability increases (3.4 ± 1.0 cm·s^{-1} × 10^{-6}) were blocked by L-NAME (−0.1 ± 0.1 cm·s^{-1} × 10^{-6}) but unaffected by the free radical scavengers superoxide dismutase and catalase (Figure 3A). In contrast, as shown in Figure 3B, the permeability increase induced by bradykinin (100 nM, 2.2 ± 0.2 cm·s^{-1} × 10^{-6}) was unaffected by L-NAME (1.6 ± 0.4 cm·s^{-1} × 10^{-6}) but blocked by co-application of superoxide dismutase and catalase (100 U·mL^{-1}; −0.1 ± 0.1 cm·s^{-1} × 10^{-6}).

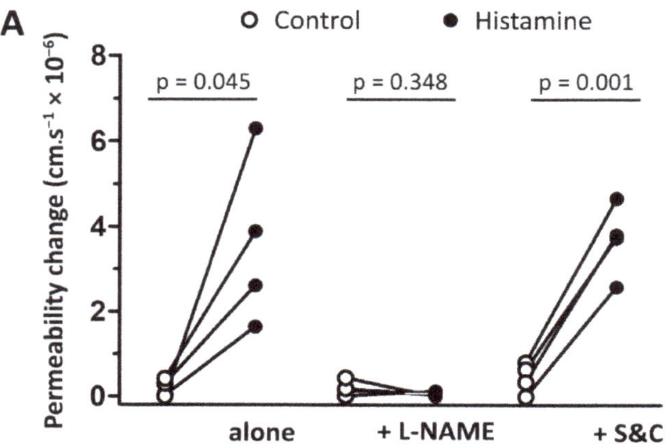

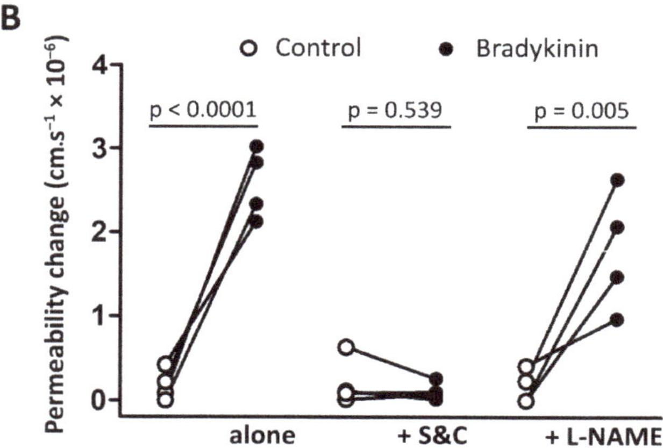

Figure 3. Involvement of reactive oxygen species and nitric oxide in histamine- and bradykinin-induced permeability. Changes in venular permeability following application of histamine (**A**) or bradykinin (**B**) at 1 μM in the absence or presence of superoxide dismutase (SOD, 100 U·mL^{-1}) and catalase (CAT, 100 U·mL^{-1}) or L-NAME (10 μM). Data from paired measurements in 4 venules from 4 different animals. Data were analyzed using a paired Student's *t*-test.

3.3. Bradykinin-Induced Microvascular Permeability Is Potentiated by IL-1β

Bradykinin applied abluminally to the cremaster microcirculation induced a dose-dependent increase in permeability to FITC-albumin (Figure 4A,B). Acute treatment with IL-1β (30 pM) for 10 min, followed by wash-off of IL-1β, resulted in a significant potentiation of bradykinin-induced permeability responses (Figure 4B).

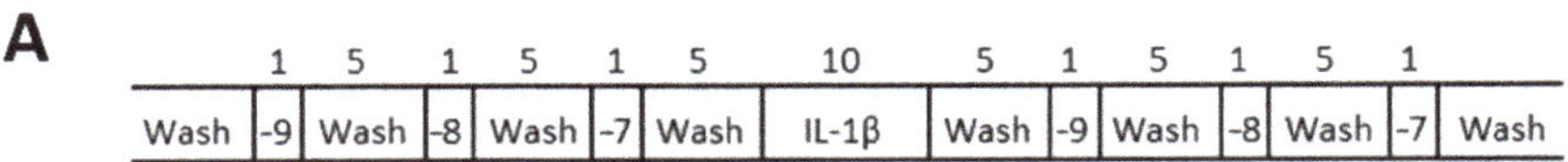

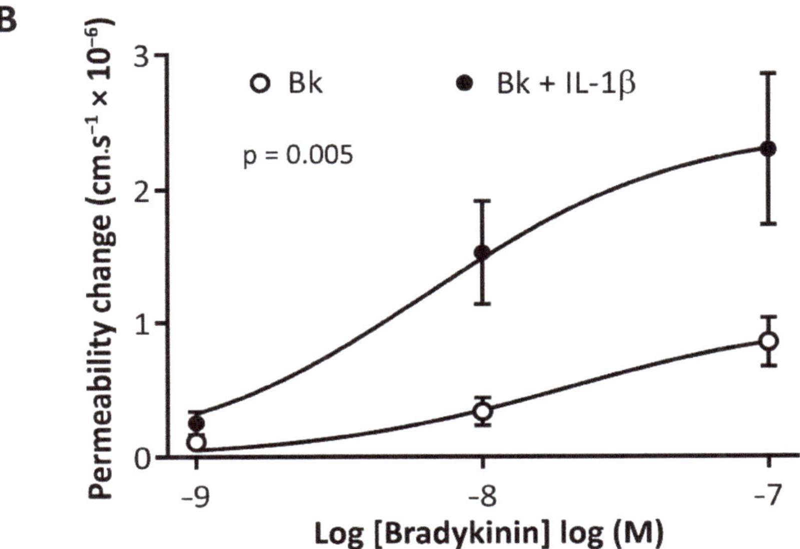

Figure 4. Acute treatment with IL-1β potentiates bradykinin-induced microvascular permeability. (**A**) Experimental protocol for dose–response curves to bradykinin (Bk) in the presence of kininase inhibitors. Bradykinin dose–response curves were obtained in the absence of IL-1β and after acute application of IL-1β (30 pM) for 10 min, followed by wash-off and reapplication of bradykinin applications to the same post-capillary venule. The numbers in panel A indicate the time in minutes for each phase of the protocol. (**B**) Bradykinin permeability response curve following IL-1β preapplication was significantly greater than all other responses. Data denote mean ± SEM of measurements in 8 vessels from 8 different animals, repeated measures, analysis of co-variance.

3.4. A Role for NADPH Oxidase and Reactive Oxygen Species in the Potentiation of Bradykinin-Induced Microvascular Permeability by IL-1β

Figure 5 summarizes changes in permeability obtained in single post-capillary venules in response to bradykinin (100 nM), IL-1β (30 pM), or bradykinin (100 nM) following 10 min treatment with IL-1β (30 pM). Apocynin, co-applied with IL-1β, effectively prevented the potentiation of bradykinin-induced permeability (Figure 5). Free radical scavenging by a mixture of by superoxide dismutase and catalase completely blocked the permeability response to bradykinin following IL-1β.

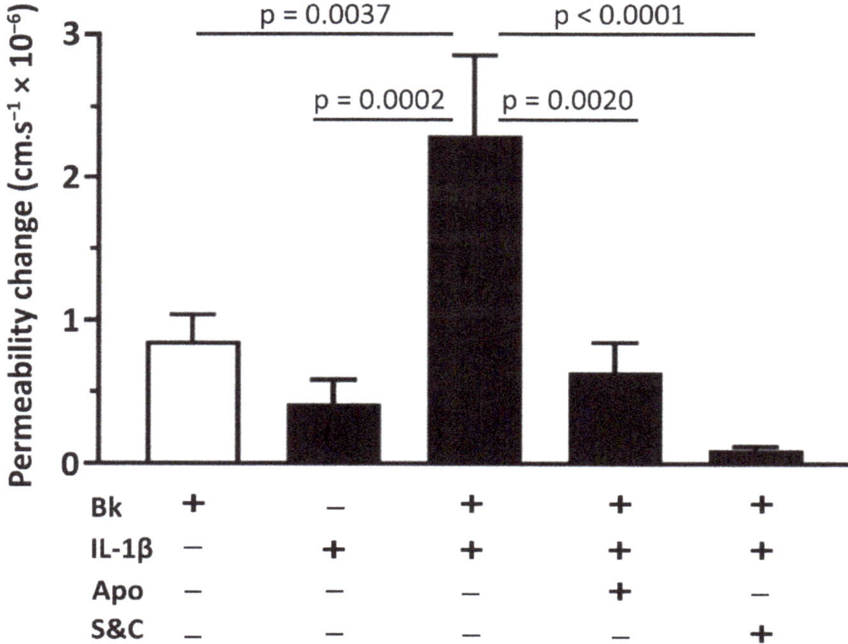

Figure 5. IL-1β potentiates bradykinin-induced microvascular permeability and involves NADPH oxidase and reactive oxygen species. The potentiated response to bradykinin (100 nM) application, following application of IL-1β (30 pM, 10 min), was prevented by co-application of apocynin (1 μM) with IL-1β (30 pM). Scavenging reactive oxygen species with superoxide dismutase (100 U/mL) and catalase (100 U/mL) completely blocked the permeability response to bradykinin. Data denote mean ± SEM, n = 10 venules from 10 animals. One-way ANOVA with Tukey's multiple comparison test.

3.5. Pretreatment of Animals with Simvastatin

Figure 6A demonstrates that in non-treated animals, application of IL-1β (30 pM, 10 min) in the absence of bradykinin resulted in a small permeability increase, which was prevented by the pretreatment of animals with simvastatin (5 mg·mL^{-1}) 24 h before. In subsequent experiments, animals were pretreated with simvastatin (5 mg·mL^{-1}) 24 h before an acute application of IL-1β (30 pM), followed by wash-off of IL-1β and application of bradykinin (100 nM). Pretreatment with simvastatin did not alter hyperpermeability induced by bradykinin alone (p = 0.411; Figure 6B). As shown in Figure 6B, pretreatment with simvastatin abolished the potentiation of bradykinin-induced microvascular permeability by IL-1β, with no significant effect on the permeability response to bradykinin alone. To examine whether the simvastatin induced loss of IL-1β potentiation of the bradykinin permeability response was due to an upregulation of HO-1, the cremaster preparation was treated with tin protophoryrin IX (SnPP), a known inhibitor of HO-1 [37,38]. Notably, inhibition of HO-1 with SnPP could not restore IL-1β mediated potentiation of bradykinin-induced permeability (Figure 6C). Apocynin (1 μM), a specific inhibitor of NADPH oxidase, also had no effect in simvastatin pretreated animals, suggesting that pretreatment with simvastatin was sufficient to prevent the assembly of NADPH oxidase induced by IL-1β (Figure 6D).

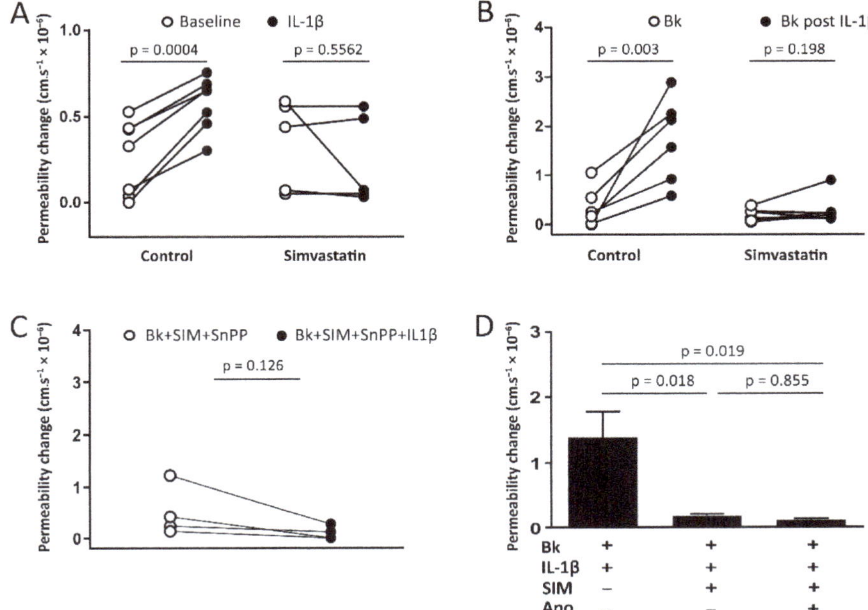

Figure 6. Pretreatment with simvastatin abolishes potentiation of bradykinin-induced microvascular permeability by IL-1β. (**A**) IL-1β (30 pM) application itself for 10 min resulted in a small permeability increase in non-treated rats (control), which was abrogated by the pretreatment with simvastatin (5 mg·mL^{-1}) 24 h before. (**B**) Potentiation of bradykinin-induced (100 nM) permeability by IL-1β (30 pM) was compared in cremaster muscle post-capillary venules from control and simvastatin pretreated (5 mg/kg; i.p.) animals. (**C**) Inhibition of HO-1 with SnPP (5 μM) did not restore IL-1β potentiation of bradykinin-induced permeability in simvastatin pretreated (5 mg/kg; i.p.) animals. Data were analyzed using a paired Student's *t*-test. (**D**) Apocynin had no effect on simvastatin-treated animals, suggesting that the pretreatment with simvastatin was sufficient to prevent the assembly of NADPH oxidase induced by IL-1β. One-way ANOVA with Tukey's multiple comparison test. Data denote mean ± SEM of paired measurements in 3–6 venules from 4–6 different animals in each group.

4. Discussion

The present study in an intact skeletal muscle microvasculature provides the first evidence that simvastatin prevents small permeability increases induced by IL-1β alone, as well as IL-1β mediated potentiation of bradykinin-induced microvascular permeability, highlighting the importance of pleiotropic effects of statins. Importantly, inhibition of Nox2 assembly by apocynin [37] or scavenging of reactive oxygen species with superoxide dismutase and catalase abolished the microvascular hyperpermeability induced by IL-β and bradykinin, strongly implicating Nox2 mediated free radical generation in increased microvascular permeability.

Our study confirms our previous findings in cerebral microvessels in vivo that acute bradykinin application results in a reactive oxygen species mediated increase in microvascular permeability. We report here that basal skeletal muscle microvascular permeability is reduced by scavenging reactive oxygen species, and that an increased permeability observed following inhibition of nitric oxide generation is abrogated by superoxide dismutase and catalase (Figure 2). This finding indicates that constitutive NO generation effectively scavenges basal formation of reactive oxygen species. There are numerous indications in the literature that NOS inhibition exacerbates inflammatory conditions, and this may provide an explanation for this.

Bradykinin-induced microvascular permeability has been associated with increased NO production and vasodilation [39,40], and a key role for reactive oxygen species generated following bradykinin receptor activation has been reported in cultured endothelial cells in vitro [18,41] and in rat cerebral microvessels in vivo [19]. Further studies in vivo, using scavengers of reactive oxygen species, confirmed these findings and showed that superoxide generation contributed to the vasodilation [42] and increased permeability following bradykinin application [19,20]. Similar to these findings, we have shown that bradykinin-induced permeability in rat cremaster muscle post-capillary venules was inhibited by superfusion with superoxide dismutase and catalase (Figure 3B). In addition, the fact that L-NAME did not inhibit bradykinin-induced permeability in cremaster muscle venules argues against a role for NO and supports findings in rat mesentery [43] and brain [20].

Histamine has been shown to increase cGMP production in endothelial cells via endothelial derived NO production, with increased vascular permeability and vasodilation mediated via activation of soluble guanylyl cyclase [44,45]. In this context, treatment of the cremaster muscle preparation with L-NAME allowed us to establish that histamine-induced permeability increases were NO-dependent but unaffected by scavenging of reactive oxygen species.

Although intracellular signaling pathways underlying reactive oxygen species mediated permeability increases were not studied, it is likely that bradykinin may induce permeability changes via the generation of free radicals during arachidonic acid metabolism leading to Ca^{2+} entry through areas of lipid peroxidation, as we previously reported for brain pial microvessels in vivo [20]. The attenuation of bradykinin-induced permeability responses in the presence of superoxide dismutase and catalase suggests that bradykinin-induced permeability increases are linked to free radical generation in rat cremaster muscle. This finding is consistent with previous reports from our laboratory that permeability responses to bradykinin in the brain microvasculature in vivo involve the generation of reactive oxygen species [19,20].

Statins have been described to improve endothelial function, reduce vascular permeability and edema formation in different experimental and clinical studies [9,46–50]. A clinical study with hypercholesterolemic patients assessed transcapillary albumin escape rate as an index of macromolecular permeability, and notably simvastatin treatment over 1 month normalized increases in transvascular albumin leakage independently of lipid levels in these patients [51]. Using an Evans blue dye exclusion test, simvastatin treatment for 1 month reduced vascular leakage in the aorta of hyperlipidemic rabbits [52]. Moreover, simvastatin treatment for 5 weeks improves endothelial barrier permeability changes in the brain, retina and myocardium of streptozotocin-induced diabetes rats [53].

Notably, administration of simvastatin 24 h before and along with intratracheal injection of lipopolysaccharide (LPS) attenuates vascular leak and inflammation in a murine inflammatory model of acute lung injury [7]. Simvastatin reduced approximately 50% of albumin levels in the bronchoalveolar lavage, and leakage of albumin conjugated with Evans blue dye into the pulmonary parenchyma in a murine inflammatory lung injury model [7]. Additionally, acute oral administration of simvastatin reduces brain edema formation and blood–brain barrier permeability after traumatic brain injury in rats [9]. In a model of experimental intracerebral hemorrhage in rats, simvastatin treatment increases cerebral blood flow in the injured region of the brain and reduces blood-brain barrier (BBB) permeability and cerebral edema [10]. Simvastatin also acutely protects the neurovascular unit, reducing blood–brain barrier permeability, when administered subcutaneously 30 min after transient cerebral ischemia induced by middle cerebral artery occlusion [8]. It is important, however, to highlight that most of these previous studies evaluated permeability changes using indirect methods, such as the Evans blue dye test. Our findings establish that simvastatin has the potential to protect the endothelial barrier and reduce vascular permeability; however, further studies are necessary to elucidate the mechanisms involved in these processes and measuring permeability coefficients.

It has been reported that lovastatin induces expression of bradykinin type 2 receptors in cultured human coronary artery endothelial cells [54]. However, in order to confirm these in vitro findings, additional in vivo studies with statin treatment in humans and in animal models are required.

Simvastatin was chosen in the present study based on its potency and pharmacokinetic properties. The potency rank order for HMG-CoA reductase inhibition among the second-generation statins is simvastatin > pravastatin > lovastatin ≅ mevastatin [55]. Furthermore, lipophilic statins, such as simvastatin, are considered more likely to enter endothelial cells by passive diffusion in contrast to hydrophilic statins, such as pravastatin and rosuvastatin, which are primarily targeted to the liver [56]. Hydrophilic statins have been described to exert similar effects on the vasculature to lipophilic statins suggesting that specific mechanisms may exist for the uptake of the former; however, this may take longer than the lipophilic statins [57].

Bradykinin has been shown to play a key role in blood–brain barrier disruption and edema formation in different pathophysiological processes, including stroke [58,59]. IL-1β is rapidly released from the brain parenchyma after an ischemic event, triggering NADPH activation and thereby potentiating bradykinin-induced microvascular permeability [60]. Moreover, the release of bradykinin and IL-1β contribute to reactive oxygen species generation in the early stages of cerebral ischemia and reperfusion injury [19]. IL-1β application increases superoxide anion release from human endothelial cells and increases reactive oxygen species generation from mitochondria and NADPH oxidase in cultured retinal epithelial cells [61]. Additionally, bradykinin may act as a potential mediator of vascular leakage and pulmonary edema in patients with COVID-19 [12–14]. In this context, IL-1β release has been proposed as one of the key inflammatory mediators in COVID-19 [23–25] and could potentially exacerbate bradykinin-induced vascular permeability in these patients. Thus, employing drugs already in clinical use, such as simvastatin, could offer a therapeutic strategy for decreasing bradykinin- and/or IL-1β-induced pulmonary edema in patients with COVID-19.

In accordance with previous studies [19,62], we observed that concomitant application of IL-1β with apocynin, a specific inhibitor of NADPH oxidase, abolished the potentiation of bradykinin-induced microvascular permeability by IL-1β (Figure 5). Apocynin rapidly prevents the assembly of NADPH oxidase, by blocking the cytosolic subunit p47phox translocation to the cell membrane [37]. Furthermore, apocynin had no effect on simvastatin pretreated rats, suggesting that simvastatin pretreatment was sufficient to prevent the assembly of NADPH oxidase induced by IL-1β (Figure 6C). Pretreatment with simvastatin was effective in inhibiting IL-1β actions on bradykinin-induced permeability, suggesting that protection afforded by simvastatin against microvascular hyperpermeability may in part be due to inhibition of Nox2. Furthermore, it has been reported that IL-1β alone rapidly (within 10 to 15 min of its application) increases superoxide release in both cultured endothelial cells [63] and retinal epithelial cells, with the latter study suggesting that NADPH oxidase activation was involved [61]. Similarly, we have also demonstrated that IL-1β itself results in a small permeability increase (see Figure 6A), which was abrogated by simvastatin. These findings strengthen the proposition that simvastatin pretreatment prevents IL-1β stimulation of ROS generation via Nox2 assembly. Nevertheless, additional studies are necessary to investigate whether other pro-inflammatory cytokines, such as IL-2 and IL-6, could also increase bradykinin-induced microvascular permeability and whether statins could modulate the profile of these cytokines.

By inhibiting reactive oxygen species generation and reducing the NAD+/NADH ratio, statins will reduce cellular oxidative stress [64–66]. Thus, protective cardiovascular effects of statins may be directly associated with their cellular antioxidant properties, independent of the cholesterol-lowering effects of these agents. As statins have been reported to activate the redox sensitive transcription factor Nrf2 and upregulate the cytoprotective antioxidant enzyme HO-1 [29–33], we postulated that loss of IL-1β potentiation of bradykinin-induced permeability may be a consequence of enhanced HO-1 activity. Notably, inhibition of HO-1 with SnPP did not restore the IL-1β-induced potentiation (see Figure 6B), suggesting that simvastatin probably acts via reducing NADPH oxidase activity. Statins have been reported to reduce NADPH oxidase activity by inhibiting isoprenylation of the protein Rac1 [28,66–68].

Isoprenylated Rac1 is essential for assembly of the NADPH oxidase enzymatic complex on the cell membrane [69]. In patients with heart failure, statin treatment reduces Rac1 function,

NADPH oxidase activity and levels of reactive oxygen species [70], a finding consistent with our observation that simvastatin pretreatment reduces IL-1β/bradykinin mediated microvascular hyperpermeability. Reactive oxygen species have been reported to negatively regulate cell–cell adhesion controlled by intercellular adhesion molecules, such as VE-cadherin and β-catenin, which are linked to transmembrane molecules and the actin cytoskeleton. In addition to a role for reactive oxygen species, RhoA activation is important for bradykinin-induced permeability [71]. RhoA-GTP activation leads to actin cytoskeleton contraction, resulting in the breakdown of the endothelial barrier [72]. In this context, statins protect the endothelial barrier, reduce oxidative stress and inhibit isoprenylation and activation of RhoA and Rac1 [52,66].

In the present study, protection afforded by simvastatin against increased microvascular permeability in cremaster muscle venules in response to IL-1β and bradykinin may be associated with inhibitory effects on the assembly of NADPH oxidase subunit, leading to diminished NADPH oxidase mediated superoxide release. Although not investigated in the present study, other cytokines such as such as IL-6, TNF-α and IL17 may similarly potentiate bradykinin-induced microvascular permeability. It has been reported that simvastatin inhibits IL-6, IL-8 and IL-1β production in vitro [73,74], which may contribute to its protective role in cardiovascular diseases. We have now demonstrated that a key anti-inflammatory action of simvastatin is to prevent IL-1β mediated potentiation of bradykinin-induced permeability in skeletal muscle microvasculature. This study highlights a novel action by which simvastatin prevents the potentiation of bradykinin-induced permeability by IL-1β, possibly by targeting the assembly of NADPH oxidase subunits. The approach undertaken in this study was functional, and future studies focusing on the molecular pathways are needed to elucidate the exact mechanism by which simvastatin reduces NADPH oxidase assembly.

5. Conclusions

Simvastatin could play an important role in the prevention and/or treatment of patients with a high predisposition to microvascular hyperpermeability mediated by pro-inflammatory cytokines potentiating the actions of bradykinin, with implications perhaps for vascular leakage and pulmonary edema.

Author Contributions: F.F., P.A.F. and G.E.M. designed the experiments and critically discussed and analyzed experimental data. F.F. and M.S. conducted all experiments. F.F. drafted the manuscript and P.A.F. and G.E.M. revised the manuscript. F.F., E.T., P.A.F. and G.E.M. read and approved the final manuscript and all authors agree to accept accountability for all aspects of the work. All authors have read and agreed to the published version of the manuscript.

Funding: This study was supported by CNPq, Brazil (205398/2014-6/SWE, to F.F) and British Heart Foundation (FS/15/6/31298, to G.E.M.).

Conflicts of Interest: The authors declare no conflict of interest.

References

1. Nussberger, J.; Cugno, M.; Cicardi, M. Bradykinin-mediated angioedema. *N. Engl. J. Med.* **2002**, *347*, 621–622. [CrossRef]
2. Bas, M.; Adams, V.; Suvorava, T.; Niehues, T.; Hoffmann, T.K.; Kojda, G. Nonallergic angioedema: Role of bradykinin. *Allergy* **2007**, *62*, 842–856. [CrossRef]
3. Obtulowicz, K. Bradykinin-mediated angioedema. *Pol. Arch. Med. Wewn.* **2016**, *126*, 76–85. [CrossRef]
4. Zausinger, S.; Lumenta, D.B.; Pruneau, D.; Schmid-Elsaesser, R.; Plesnila, N.; Baethmann, A. Effects of LF 16-0687 Ms, a bradykinin B(2) receptor antagonist, on brain edema formation and tissue damage in a rat model of temporary focal cerebral ischemia. *Brain Res.* **2002**, *950*, 268–278. [CrossRef]
5. Ruiz, S.; Vardon-Bounes, F.; Buleon, M.; Guilbeau-Frugier, C.; Seguelas, M.H.; Conil, J.M.; Girolami, J.P.; Tack, I.; Minville, V. Kinin B1 receptor: A potential therapeutic target in sepsis-induced vascular hyperpermeability. *J. Transl. Med.* **2020**, *18*, 174. [CrossRef]

6. Othman, R.; Vaucher, E.; Couture, R. Bradykinin Type 1 Receptor—Inducible Nitric Oxide Synthase: A New Axis Implicated in Diabetic Retinopathy. *Front. Pharmacol.* **2019**, *10*, 300. [CrossRef]
7. Jacobson, J.R.; Barnard, J.W.; Grigoryev, D.N.; Ma, S.F.; Tuder, R.M.; Garcia, J.G. Simvastatin attenuates vascular leak and inflammation in murine inflammatory lung injury. *Am. J. Physiol. Lung Cell. Mol. Physiol.* **2005**, *288*, L1026–L1032. [CrossRef]
8. Nagaraja, T.N.; Knight, R.A.; Croxen, R.L.; Konda, K.P.; Fenstermacher, J.D. Acute neurovascular unit protection by simvastatin in transient cerebral ischemia. *Neurol. Res.* **2006**, *28*, 826–830. [CrossRef]
9. Beziaud, T.; Ru Chen, X.; El Shafey, N.; Frechou, M.; Teng, F.; Palmier, B.; Beray-Berthat, V.; Soustrat, M.; Margaill, I.; Plotkine, M.; et al. Simvastatin in traumatic brain injury: Effect on brain edema mechanisms. *Crit. Care Med.* **2011**, *39*, 2300–2307. [CrossRef]
10. Yang, D.; Knight, R.A.; Han, Y.; Karki, K.; Zhang, J.; Ding, C.; Chopp, M.; Seyfried, D.M. Vascular recovery promoted by atorvastatin and simvastatin after experimental intracerebral hemorrhage: Magnetic resonance imaging and histological study. *J. Neurosurg.* **2011**, *114*, 1135–1142. [CrossRef] [PubMed]
11. Regoli, D. Neurohumoral regulation of precapillary vessels: The kallikrein-kinin system. *J. Cardiovasc. Pharmacol.* **1984**, *6*, S401–S412. [CrossRef]
12. Teuwen, L.A.; Geldhof, V.; Pasut, A.; Carmeliet, P. COVID-19: The vasculature unleashed. *Nat. Rev. Immunol.* **2020**, *20*, 389–391. [CrossRef]
13. Roche, J.A.; Roche, R. A hypothesized role for dysregulated bradykinin signaling in COVID-19 respiratory complications. *FASEB J. Off. Publ. Fed. Am. Soc. Exp. Biol.* **2020**, *34*, 7265–7269. [CrossRef]
14. Van de Veerdonk, F.L.; Netea, M.G.; van Deuren, M.; van der Meer, J.W.; de Mast, Q.; Bruggemann, R.J.; van der Hoeven, H. Kallikrein-kinin blockade in patients with COVID-19 to prevent acute respiratory distress syndrome. *eLife* **2020**, *9*, e57555. [CrossRef]
15. Yan, R.; Zhang, Y.; Li, Y.; Xia, L.; Guo, Y.; Zhou, Q. Structural basis for the recognition of SARS-CoV-2 by full-length human ACE2. *Science* **2020**, *367*, 1444–1448. [CrossRef]
16. Hoffmann, M.; Kleine-Weber, H.; Schroeder, S.; Kruger, N.; Herrler, T.; Erichsen, S.; Schiergens, T.S.; Herrler, G.; Wu, N.H.; Nitsche, A.; et al. SARS-CoV-2 Cell Entry Depends on ACE2 and TMPRSS2 and Is Blocked by a Clinically Proven Protease Inhibitor. *Cell* **2020**, *181*, 271–280. [CrossRef]
17. Imai, Y.; Kuba, K.; Penninger, J.M. The discovery of angiotensin-converting enzyme 2 and its role in acute lung injury in mice. *Exp. Physiol.* **2008**, *93*, 543–548. [CrossRef]
18. Shimizu, S.; Ishii, M.; Yamamoto, T.; Kawanishi, T.; Momose, K.; Kuroiwa, Y. Bradykinin induces generation of reactive oxygen species in bovine aortic endothelial cells. *Res. Commun. Chem. Pathol. Pharmacol.* **1994**, *84*, 301–314.
19. Woodfin, A.; Hu, D.E.; Sarker, M.; Kurokawa, T.; Fraser, P. Acute NADPH oxidase activation potentiates cerebrovascular permeability response to bradykinin in ischemia-reperfusion. *Free Radic. Biol. Med.* **2011**, *50*, 518–524. [CrossRef]
20. Sarker, M.H.; Hu, D.E.; Fraser, P.A. Acute effects of bradykinin on cerebral microvascular permeability in the anaesthetized rat. *J. Physiol.* **2000**, *528*, 177–187. [CrossRef]
21. Sobey, C.G. Bradykinin B2 receptor antagonism: A new direction for acute stroke therapy? *Br. J. Pharmacol.* **2003**, *139*, 1369–1371. [CrossRef]
22. Touzani, O.; Boutin, H.; Chuquet, J.; Rothwell, N. Potential mechanisms of interleukin-1 involvement in cerebral ischaemia. *J. Neuroimmunol.* **1999**, *100*, 203–215. [CrossRef]
23. Cavalli, G.; De Luca, G.; Campochiaro, C.; Della-Torre, E.; Ripa, M.; Canetti, D.; Oltolini, C.; Castiglioni, B.; Tassan Din, C.; Boffini, N.; et al. Interleukin-1 blockade with high-dose anakinra in patients with COVID-19, acute respiratory distress syndrome, and hyperinflammation: A retrospective cohort study. *Lancet Rheumatol.* **2020**, *2*, e325–e331. [CrossRef]
24. Ucciferri, C.; Auricchio, A.; Di Nicola, M.; Potere, N.; Abbate, A.; Cipollone, F.; Vecchiet, J.; Falasca, K. Canakinumab in a subgroup of patients with COVID-19. *Lancet Rheumatol.* **2020**. [CrossRef]
25. Conti, P.; Ronconi, G.; Caraffa, A.; Gallenga, C.; Ross, R.; Frydas, I.; Kritas, S. Induction of pro-inflammatory cytokines (IL-1 and IL-6) and lung inflammation by Coronavirus-19 (COVI-19 or SARS-CoV-2): Anti-inflammatory strategies. *J. Biol. Regul. Homeost. Agents* **2020**, *34*, 327–331.
26. Stoll, L.L.; McCormick, M.L.; Denning, G.M.; Weintraub, N.L. Antioxidant effects of statins. *Timely Top. Med. Cardiovasc. Dis.* **2005**, *9*, E1. [CrossRef]

27. Freitas, F.; Estato, V.; Reis, P.; Castro-Faria-Neto, H.C.; Carvalho, V.; Torres, R.; Lessa, M.A.; Tibirica, E. Acute simvastatin treatment restores cerebral functional capillary density and attenuates angiotensin II-induced microcirculatory changes in a model of primary hypertension. *Microcirculation* **2017**, *24*. [CrossRef]
28. Margaritis, M.; Sanna, F.; Antoniades, C. Statins and oxidative stress in the cardiovascular system. *Curr. Pharm. Des.* **2017**. [CrossRef]
29. Ali, F.; Zakkar, M.; Karu, K.; Lidington, E.A.; Hamdulay, S.S.; Boyle, J.J.; Zloh, M.; Bauer, A.; Haskard, D.O.; Evans, P.C.; et al. Induction of the cytoprotective enzyme heme oxygenase-1 by statins is enhanced in vascular endothelium exposed to laminar shear stress and impaired by disturbed flow. *J. Biol. Chem.* **2009**, *284*, 18882–18892. [CrossRef]
30. Chartoumpekis, D.; Ziros, P.G.; Psyrogiannis, A.; Kyriazopoulou, V.; Papavassiliou, A.G.; Habeos, I.G. Simvastatin lowers reactive oxygen species level by Nrf2 activation via PI3K/Akt pathway. *Biochem. Biophys. Res. Commun.* **2010**, *396*, 463–466. [CrossRef]
31. Gueler, F.; Park, J.K.; Rong, S.; Kirsch, T.; Lindschau, C.; Zheng, W.; Elger, M.; Fiebeler, A.; Fliser, D.; Luft, F.C.; et al. Statins attenuate ischemia-reperfusion injury by inducing heme oxygenase-1 in infiltrating macrophages. *Am. J. Pathol.* **2007**, *170*, 1192–1199. [CrossRef]
32. Hsu, H.H.; Ko, W.J.; Hsu, J.Y.; Chen, J.S.; Lee, Y.C.; Lai, I.R.; Chen, C.F. Simvastatin ameliorates established pulmonary hypertension through a heme oxygenase-1 dependent pathway in rats. *Respir. Res.* **2009**, *10*, 32. [CrossRef]
33. Makabe, S.; Takahashi, Y.; Watanabe, H.; Murakami, M.; Ohba, T.; Ito, H. Fluvastatin protects vascular smooth muscle cells against oxidative stress through the Nrf2-dependent antioxidant pathway. *Atherosclerosis* **2010**, *213*, 377–384. [CrossRef]
34. Piechota-Polanczyk, A.; Kopacz, A.; Kloska, D.; Zagrapan, B.; Neumayer, C.; Grochot-Przeczek, A.; Huk, I.; Brostjan, C.; Dulak, J.; Jozkowicz, A. Simvastatin Treatment Upregulates HO-1 in Patients with Abdominal Aortic Aneurysm but Independently of Nrf2. *Oxidative Med. Cell. Longev.* **2018**, *2018*, 2028936. [CrossRef]
35. Chambers, D.J.; Fallouh, H.B. Cardioplegia and cardiac surgery: Pharmacological arrest and cardioprotection during global ischemia and reperfusion. *Pharmacol. Ther.* **2010**, *127*, 41–52. [CrossRef]
36. Fraser, P.A.; Dallas, A.D.; Davies, S. Measurement of filtration coefficient in single cerebral microvessels of the frog. *J. Physiol.* **1990**, *423*, 343–361. [CrossRef]
37. Ximenes, V.F.; Kanegae, M.P.; Rissato, S.R.; Galhiane, M.S. The oxidation of apocynin catalyzed by myeloperoxidase: Proposal for NADPH oxidase inhibition. *Arch. Biochem. Biophys.* **2007**, *457*, 134–141. [CrossRef]
38. Kaizu, T.; Tamaki, T.; Tanaka, M.; Uchida, Y.; Tsuchihashi, S.; Kawamura, A.; Kakita, A. Preconditioning with tin-protoporphyrin IX attenuates ischemia/reperfusion injury in the rat kidney. *Kidney Int.* **2003**, *63*, 1393–1403. [CrossRef]
39. Cambridge, H.; Brain, S.D. Mechanism of bradykinin-induced plasma extravasation in the rat knee joint. *Br. J. Pharmacol.* **1995**, *115*, 641–647. [CrossRef]
40. Feletou, M.; Bonnardel, E.; Canet, E. Bradykinin and changes in microvascular permeability in the hamster cheek pouch: Role of nitric oxide. *Br. J. Pharmacol.* **1996**, *118*, 1371–1376. [CrossRef]
41. Holland, J.A.; Pritchard, K.A.; Pappolla, M.A.; Wolin, M.S.; Rogers, N.J.; Stemerman, M.B. Bradykinin induces superoxide anion release from human endothelial cells. *J. Cell. Physiol.* **1990**, *143*, 21–25. [CrossRef] [PubMed]
42. Wambi-Kiesse, C.O.; Katusic, Z.S. Inhibition of copper/zinc superoxide dismutase impairs NO.-mediated endothelium-dependent relaxations. *Am. J. Physiol.* **1999**, *276*, H1043–H1048. [CrossRef] [PubMed]
43. Shigematsu, S.; Ishida, S.; Gute, D.C.; Korthuis, R.J. Bradykinin-induced proinflammatory signaling mechanisms. *Am. J. Physiol. Heart Circ. Physiol.* **2002**, *283*, H2676–H2686. [CrossRef] [PubMed]
44. Furchgott, R.F.; Vanhoutte, P.M. Endothelium-derived relaxing and contracting factors. *FASEB J.* **1989**, *3*, 2007–2018. [CrossRef]
45. Nizamutdinova, I.T.; Maejima, D.; Nagai, T.; Bridenbaugh, E.; Thangaswamy, S.; Chatterjee, V.; Meininger, C.J.; Gashev, A.A. Involvement of histamine in endothelium-dependent relaxation of mesenteric lymphatic vessels. *Microcirculation* **2014**, *21*, 640–648. [CrossRef]
46. Manitsopoulos, N.; Orfanos, S.E.; Kotanidou, A.; Nikitopoulou, I.; Siempos, I.; Magkou, C.; Dimopoulou, I.; Zakynthinos, S.G.; Armaganidis, A.; Maniatis, N.A. Inhibition of HMGCoA reductase by simvastatin protects mice from injurious mechanical ventilation. *Respir. Res.* **2015**, *16*, 24. [CrossRef]

47. Yang, C.H.; Kao, M.C.; Shih, P.C.; Li, K.Y.; Tsai, P.S.; Huang, C.J. Simvastatin attenuates sepsis-induced blood-brain barrier integrity loss. *J. Surg. Res.* **2015**, *194*, 591–598. [CrossRef]
48. Tuuminen, R.; Sahanne, S.; Loukovaara, S. Low intravitreal angiopoietin-2 and VEGF levels in vitrectomized diabetic patients with simvastatin treatment. *Acta Ophthalmol.* **2014**, *92*, 675–681. [CrossRef]
49. Zhang, W.; Yan, H. Simvastatin increases circulating endothelial progenitor cells and reduces the formation and progression of diabetic retinopathy in rats. *Exp. Eye Res.* **2012**, *105*, 1–8. [CrossRef]
50. Caldwell, R.B.; Bartoli, M.; Behzadian, M.A.; El-Remessy, A.E.; Al-Shabrawey, M.; Platt, D.H.; Liou, G.I.; Caldwell, R.W. Vascular endothelial growth factor and diabetic retinopathy: Role of oxidative stress. *Curr. Drug Targets* **2005**, *6*, 511–524. [CrossRef]
51. Dell'Omo, G.; Bandinelli, S.; Penno, G.; Pedrinelli, R.; Mariani, M. Simvastatin, capillary permeability, and acetylcholine-mediated vasomotion in atherosclerotic, hypercholesterolemic men. *Clin. Pharmacol. Ther.* **2000**, *68*, 427–434. [CrossRef] [PubMed]
52. Van Nieuw Amerongen, G.P.; Vermeer, M.A.; Negre-Aminou, P.; Lankelma, J.; Emeis, J.J.; van Hinsbergh, V.W. Simvastatin improves disturbed endothelial barrier function. *Circulation* **2000**, *102*, 2803–2809. [CrossRef] [PubMed]
53. Mooradian, A.D.; Haas, M.J.; Batejko, O.; Hovsepyan, M.; Feman, S.S. Statins ameliorate endothelial barrier permeability changes in the cerebral tissue of streptozotocin-induced diabetic rats. *Diabetes* **2005**, *54*, 2977–2982. [CrossRef] [PubMed]
54. Liesmaa, I.; Kokkonen, J.O.; Kovanen, P.T.; Lindstedt, K.A. Lovastatin induces the expression of bradykinin type 2 receptors in cultured human coronary artery endothelial cells. *J. Mol. Cell. Cardiol.* **2007**, *43*, 593–600. [CrossRef]
55. Blum, C.B. Comparison of properties of four inhibitors of 3-hydroxy-3-methylglutaryl-coenzyme A reductase. *Am. J. Cardiol.* **1994**, *73*, 3d–11d. [CrossRef]
56. Zhou, Q.; Liao, J.K. Pleiotropic effects of statins—Basic research and clinical perspectives. *Circ. J. Off. J. Jpn. Circ. Soc.* **2010**, *74*, 818–826.
57. Schachter, M. Chemical, pharmacokinetic and pharmacodynamic properties of statins: An update. *Fundam. Clin. Pharmacol.* **2005**, *19*, 117–125. [CrossRef]
58. Kunz, M.; Nussberger, J.; Holtmannspötter, M.; Bitterling, H.; Plesnila, N.; Zausinger, S. Bradykinin in blood and cerebrospinal fluid after acute cerebral lesions: Correlations with cerebral edema and intracranial pressure. *J. Neurotrauma* **2013**, *30*, 1638–1644. [CrossRef]
59. Dobrivojevic, M.; Spiranec, K.; Sindic, A. Involvement of bradykinin in brain edema development after ischemic stroke. *Pflug. Arch. Eur. J. Physiol.* **2015**, *467*, 201–212. [CrossRef]
60. Eder, C. Mechanisms of interleukin-1beta release. *Immunobiology* **2009**, *214*, 543–553. [CrossRef]
61. Yang, D.; Elner, S.G.; Bian, Z.M.; Till, G.O.; Petty, H.R.; Elner, V.M. Pro-inflammatory cytokines increase reactive oxygen species through mitochondria and NADPH oxidase in cultured RPE cells. *Exp. Eye Res.* **2007**, *85*, 462–472. [CrossRef] [PubMed]
62. Warboys, C.M.; Toh, H.B.; Fraser, P.A. Role of NADPH oxidase in retinal microvascular permeability increase by RAGE activation. *Investig. Ophthalmol. Vis. Sci.* **2009**, *50*, 1319–1328. [CrossRef] [PubMed]
63. Matsubara, T.; Ziff, M. Increased superoxide anion release from human endothelial cells in response to cytokines. *J. Immunol.* **1986**, *137*, 3295–3298. [PubMed]
64. Wagner, A.H.; Kohler, T.; Ruckschloss, U.; Just, I.; Hecker, M. Improvement of nitric oxide-dependent vasodilatation by HMG-CoA reductase inhibitors through attenuation of endothelial superoxide anion formation. *Arter. Thromb. Vasc. Biol.* **2000**, *20*, 61–69. [CrossRef]
65. Endres, M.; Laufs, U. Effects of statins on endothelium and signaling mechanisms. *Stroke* **2004**, *35*, 2708–2711. [CrossRef]
66. Chen, W.; Pendyala, S.; Natarajan, V.; Garcia, J.G.; Jacobson, J.R. Endothelial cell barrier protection by simvastatin: GTPase regulation and NADPH oxidase inhibition. *Am. J. Physiol. Lung Cell. Mol. Physiol.* **2008**, *295*, L575–L583. [CrossRef]
67. Wassmann, S.; Laufs, U.; Baumer, A.T.; Muller, K.; Konkol, C.; Sauer, H.; Bohm, M.; Nickenig, G. Inhibition of geranylgeranylation reduces angiotensin II-mediated free radical production in vascular smooth muscle cells: Involvement of angiotensin AT1 receptor expression and Rac1 GTPase. *Mol. Pharmacol.* **2001**, *59*, 646–654. [CrossRef]

68. Delbosc, S.; Morena, M.; Djouad, F.; Ledoucen, C.; Descomps, B.; Cristol, J.P. Statins, 3-hydroxy-3-methylglutaryl coenzyme A reductase inhibitors, are able to reduce superoxide anion production by NADPH oxidase in THP-1-derived monocytes. *J. Cardiovasc. Pharmacol.* **2002**, *40*, 611–617. [CrossRef]
69. Hordijk, P.L. Regulation of NADPH oxidases: The role of Rac proteins. *Circ. Res.* **2006**, *98*, 453–462. [CrossRef]
70. Maack, C.; Kartes, T.; Kilter, H.; Schafers, H.J.; Nickenig, G.; Bohm, M.; Laufs, U. Oxygen free radical release in human failing myocardium is associated with increased activity of rac1-GTPase and represents a target for statin treatment. *Circulation* **2003**, *108*, 1567–1574. [CrossRef]
71. Ma, T.; Xue, Y. RhoA-mediated potential regulation of blood-tumor barrier permeability by bradykinin. *J. Mol. Neurosci.* **2010**, *42*, 67–73. [CrossRef] [PubMed]
72. Wojciak-Stothard, B.; Potempa, S.; Eichholtz, T.; Ridley, A.J. Rho and Rac but not Cdc42 regulate endothelial cell permeability. *J. Cell Sci.* **2001**, *114*, 1343–1355. [PubMed]
73. Boland, A.J.; Gangadharan, N.; Kavanagh, P.; Hemeryck, L.; Kieran, J.; Barry, M.; Walsh, P.T.; Lucitt, M. Simvastatin Suppresses Interleukin Iβ Release in Human Peripheral Blood Mononuclear Cells Stimulated With Cholesterol Crystals. *J. Cardiovasc. Pharmacol. Ther.* **2018**, *23*, 509–517. [CrossRef] [PubMed]
74. Yokota, K.; Miyazaki, T.; Hirano, M.; Akiyama, Y.; Mimura, T. Simvastatin inhibits production of interleukin 6 (IL-6) and IL-8 and cell proliferation induced by tumor necrosis factor-alpha in fibroblast-like synoviocytes from patients with rheumatoid arthritis. *J. Rheumatol.* **2006**, *33*, 463–471.

Publisher's Note: MDPI stays neutral with regard to jurisdictional claims in published maps and institutional affiliations.

© 2020 by the authors. Licensee MDPI, Basel, Switzerland. This article is an open access article distributed under the terms and conditions of the Creative Commons Attribution (CC BY) license (http://creativecommons.org/licenses/by/4.0/).

Article

Impairment of PGC-1 Alpha Up-Regulation Enhances Nitrosative Stress in the Liver during Acute Pancreatitis in Obese Mice

Sergio Rius-Pérez [1,†], Isabel Torres-Cuevas [2,†], María Monsalve [3], Francisco J. Miranda [1] and Salvador Pérez [1,*]

1 Department of Physiology, Faculty of Pharmacy, University of Valencia, Avda. Vicente Andres Estelles s/n, 46100 Burjassot, Spain; sergio.rius@uv.es (S.R.-P.); Francisco.J.Miranda@uv.es (F.J.M.)
2 Neonatal Research Group, Health Research Institute La Fe, 46026 Valencia, Spain; Maria.I.Torres@uv.es
3 Instituto de Investigaciones Biomédicas "Alberto Sols" (CSIC-UAM), Arturo Duperier, 4, 28029 Madrid, Spain; mpmonsalve@iib.uam.es
* Correspondence: salvador.perez-garrido@uv.es; Tel.: +34-963-54-3253
† Sergio Rius-Pérez and Isabel Torres-Cuevas are equal first authors.

Received: 31 July 2020; Accepted: 17 September 2020; Published: 19 September 2020

Abstract: Acute pancreatitis is an inflammatory process of the pancreatic tissue that often leads to distant organ dysfunction. Although liver injury is uncommon in acute pancreatitis, obesity is a risk factor for the development of hepatic complications. The aim of this work was to evaluate the role of PGC-1α in inflammatory response regulation in the liver and its contribution to the detrimental effect of obesity on the liver during acute pancreatitis. For this purpose, we induced acute pancreatitis by cerulein in not only wild-type (WT) and PGC-1α knockout (KO) mice, but also in lean and obese mice. PGC-1α levels were up-regulated in the mice livers with pancreatitis. The increased PGC-1α levels were bound to p65 to restrain its transcriptional activity toward *Nos2*. Lack of PGC-1α favored the assembly of the p65/phospho-STAT3 complex, which promoted *Nos2* expression during acute pancreatitis. The increased transcript *Nos2* levels and the pro-oxidant liver status caused by the down-regulated expression of the PGC-1α-dependent antioxidant genes enhanced nitrosative stress and decreased energy charge in the livers of the PGC-1α KO mice with pancreatitis. It is noteworthy that the PGC-1α levels lowered in the obese mice livers, which increased the *Nos2* mRNA expression and protein nitration levels and decreased energy charge during pancreatitis. In conclusion, obesity impairs PGC-1α up-regulation in the liver to cause nitrosative stress during acute pancreatitis.

Keywords: acute pancreatitis; obesity; nitrosative stress; PGC-1α; liver

1. Introduction

Acute pancreatitis (AP) is an inflammatory disorder of the pancreas that often leads to a systemic inflammatory response and organ failure [1]. Currently, AP is the main cause of hospital admission for gastrointestinal problems in the USA [2], with a mortality rate of around 30% in patients who develop organ failure [3]. Several pieces of evidence suggest that oxidative and nitrosative stresses play an essential role in AP pathogenesis [4]. Oxidative stress amplifies the inflammatory process that leads to oxidative damage and contributes to the progression of extrapancreatic complications [5,6]. Nitrosative stress is a well-known feature of AP, with inducible nitric oxide synthase (NOS2) being the main source of nitric oxide (NO) in the pancreas during the course of this pathology [7–10]. In fact, *Nos2*-deficient mice exhibit a low degree of pancreatic inflammation and tissue damage in the pancreas with AP [10].

Obesity is a chronic inflammatory condition that increases the appearance of local and systemic complications and mortality in patients with AP [11–13]. Numerous factors contribute to systemic injury in obese patients with AP, such as the uncontrolled cytokine response, the release of unsaturated fatty acids and damage-associated molecular patterns [13]. The pulmonary, cardiovascular, and renal systems are more frequently affected in AP via these mediators [14]. On the contrary, liver damage is less common, but interestingly its appearance is used as a prognostic value in human AP and its failure invariably leads to death [15]. Remarkably, obesity only triggers hepatic injury during AP in genetically obese fa/fa Zucker rats compared to lean rats [16]. Furthermore, fatty livers can imply much higher rates for local complications, organ failure and mortality during AP [17].

PPARγ co-activator 1α (PGC-1α) is a transcriptional co-activator that is dysregulated in obesity and is important for maintenance of balance in the production of reactive oxygen species (ROS) during inflammatory processes [18]. Indeed, PGC1α modulates the expression of mitochondrial antioxidant defense genes, including manganese superoxide dismutase (*Sod2*), peroxiredoxin (*Prx*) 3, *Prx5* and catalase [19,20]. PGC-1α overexpression is known to suppress the expression of the pro-inflammatory cytokines triggered by tumor necrosis factor-α (TNF-α) in C2C12 muscle cells [21]. Low PGC-1α levels in inflamed tissues increase ROS production and contribute to increased inflammatory response [22].

In the present work, we address the role of PGC-1α in inflammatory response regulation in the liver during acute pancreatitis. Furthermore, we explore the precise contribution of PGC-1α in the liver to the detrimental effect of obesity on acute pancreatitis.

2. Materials and Methods

2.1. Animals

C57BL/6 J PGC-1$\alpha^{-/-}$ mice were originally provided by Dr. Bruce Spiegelman (Dana–Farber Cancer Institute, Harvard Medical School, Boston, MA, USA). Subsequently, a colony was established at the Institute of Biomedical Research "Alberto Sols" (Madrid, Spain) animal facility. The generation and phenotype of PGC-1α knockout (KO) mice have been described previously [23].

Male C57BL/6 J PGC-1$\alpha^{+/+}$ (22.4 ± 1.5 g; n = 12) and C57BL/6 J PGC-1$\alpha^{-/-}$ (21.9 ± 1.7 g; n = 12) mice were used and fed a standard diet. The male C57BL/6 J mice purchased from Jackson Laboratory (Bar Harbor, ME, USA) were used, and fed either standard chow (TD.08485, Envigo, Barcelona, Spain) (lean: 22.9 ± 1.0 g; n = 10) or a high-fat diet with 40% calories from fat (TD.88137, Envigo, Barcelona, Spain) (obese: 29.7 ± 1.8 g; n = 10) for 12 weeks.

All the animals were housed under standard environmental conditions (20–22 °C, 50 ± 10% humidity, 12 h light–dark cycle) with food and water ad libitum. Experiments were conducted in compliance with legislation on the protection of animals used for scientific purposes in Spain (RD 53/2013) and with EU (Directive 2010/63/EU). Protocols were approved by the Ethics Committee of Animal Experimentation and Welfare of the University of Valencia (Ethical Protocol Code A1529666350463, Valencia, Spain). This was approved by the Regional Ministry of Agriculture, Environment, Climate Change and Rural Development of Generalitat Valenciana with Code 2018/VSC/PEA/0190 type 2.

2.2. Experimental Model of Acute Pancreatitis

Acute pancreatitis was induced in 12-week-old mice by seven intraperitoneal cerulein injections (Sigma-Aldrich, St. Louis, MO, USA) (50 µg/kg body weight) at 1-h intervals [24]. Physiological saline (0.9% NaCl) was administered to the control group (sham mice). Animals were sacrificed 1 h after the seventh cerulein injection. Mice were sacrificed by euthanization under anesthesia with isoflurane 3–5% and were then exsanguinated. The pancreas and liver were immediately removed. Sacrifice was confirmed by cervical dislocation.

2.3. RNA Extraction and RT-qPCR Analysis of Gene Expression

Total RNA was isolated using TRIzol reagent (Sigma-Aldrich, St. Louis, MO, USA) following the manufacturer's instructions. The RNA concentration was measured in a NanoDrop Lite spectrophotometer (Thermo Scientific, Waltham, MA, USA), and purity was determined by the optical density (OD) 260/280 ratio. RNA was reverse transcribed to cDNA with the PrimeScript RT Reagent Kit (Perfect Real Time) (Takara Bio Inc., Kusatsu, Japan) following the manufacturer's instructions. The RNA levels of the genes were performed in a thermal cycler (I-Cycler + IQ Multicolor Real-Time OCR Detection System, Biorad, Hercules, CA, USA) by using the SYBR Green PCR Master Mix (Takara Bio Inc., Kusatsu, Japan). The employed specific primers are shown in Table 1.

Table 1. The oligonucleotides used for RT-qPCR.

Target Gene (mm)	Direct/Reverse Oligonucleotide
Ppargc1a (Gene ID: 19017)	F -> TTAAAGTTCATGGGGCAAGC R -> TAGGAATGGCTGAAGGGATG
Sod2 (Gene ID: 20656)	F -> GGCCAAGGGAGATGTTACAA R -> GAACCTTGGACTCCCACAGA
Prx3 (Gene ID: 11757)	F -> CAAGAAAGAATGGTGGTTTGG R -> TGCTTGACGACACCATTAGG
Nos2 (Gene ID: 18126)	F ->GCATCCCAAGTACGAGTGGGT R ->GAAGTCTCGAACTCCAATC
Tbp (Gene ID: 21374)	F ->CAGCCTTCCACCTTATGCTC R -> CCGTAAGGCATCATTGGACT

RT-qPCR was performed by running TaqMan gene expression assays and the TaqMans PCR Master Mix (Applied Biosystems, Life Technologies Corporation, Carlsbad, CA, USA). A list of the analyzed genes and TaqMan probes is presented in Table 2.

Table 2. The TaqMan® probe used for RT-qPCR.

Target Gene (mm)	Sonda TaqMan
Tnfα (Gene ID: 21926)	Mm00443258_g1
Il6 (Gene ID: 16193)	Mm00446190_m1
Tbp (Gene ID: 21374)	Mm01277042_m1

The results were normalized using the TATA binding protein (*Tbp*) as housekeeping. The threshold cycle (CT) was determined and the relative gene expression was expressed as follows: fold change = 2–Δ(ΔCT), where ΔCT = CT target – CT housekeeping, and Δ(ΔCT) = ΔCT treated – ΔCT control.

2.4. Western Blot Analysis

The liver and pancreas tissue samples were frozen at −80 °C until homogenization (Politron Generator FSH-G 5/085 from Thermo Fisher Scientific, Waltham, MA, USA,) in extraction buffer (100 mg/mL) on ice. Lysis buffer (20 mm Tris–HCl, pH 7.5, 1 mM EDTA, 150 mM NaCl, 0.1% SDS, 1% Igepal, 30 mM sodium pyrophosphate, 50 mM sodium fluoride, 1 mM sodium orthovanadate) and a protease inhibitor cocktail (Sigma-Aldrich) at a concentration of 4 μL/mL were employed. Homogenates were centrifuged for 15 min at 15,000 rpm and 4 °C. The concentration of the proteins in each homogenate was measured by the bicinchoninic acid (BCA) protein assay (Thermo Fisher

Scientific, Waltham, MA, USA). Blots were visualized using a chemiluminescence (ECL) detection kit Western blotting substrate (Fisher Scientific, Madrid, Spain). Signals were captured by the ChemiDoc XRS and Imaging System (Bio-rad, Richmond, CA, USA). The density of bands was measured by version 2.0.1 of the Image Lab Software (Bio-rad, Richmond, CA, USA).

The employed antibodies were: anti-β-tubulin (1:1000, ab6046 from Abcam, Cambridge, UK); anti-PGC-1α (1:500, sc-518025 from Santa Cruz Biotechnology, Dallas, TX, USA); anti-p65 (1:1000, #8242 from Cell Signaling Technology, Danvers, MA, USA); anti-phospho-p65 (Ser 536) (1:100, #3033 from Cell Signaling); anti-Nitro-tyrosine (1:1000, #9691 from Cell Signaling); anti-STAT3 (1:1000, #9132 from Cell Signaling); anti-phospho-STAT3 (Tyr705) (1:1000, #9131 from Cell Signaling); anti-GAPDH (1:1000, #2118 from Cell Signaling); anti-NOS2 (1:1000, ab178945 from Abcam); anti-IgG (1:1000, #7076 from Cell Signaling).

2.5. Co-Immunoprecipitation

Protein–protein interactions were analyzed by co-immunoprecipitation experiments. Whole-cell extracts were prepared and subjected to immunoprecipitation with specific antibodies against PGC-1α (sc-518025, Santa Cruz, Dallas, TX, USA) and p65 (1:1000, #8242 from Cell Signalling) as previously described [25]. The presence of both NF-κB and p-STAT3 in immunoprecipitates was evaluated by a Western blot with their corresponding antibodies (p65 and p-STAT3).

2.6. Redox Pairs and Protein Nitration and Chlorination by UPLC-MS/MS Analysis

Redox pairs, namely, oxidized glutathione (GSSG)/reduced glutathione (GSH), γ-glutamilcystine/ γ-glutamilcysteine and cysteine (Cyss)/cysteine (Cys), were analyzed from the frozen liver samples homogenized in phosphate buffered saline (PBS) with 10 mM N-ethylmaleimide. Then, perchloric acid was added to obtain a 4% concentration and centrifuged at 15,000× *g* for 15 min at 4 °C. The concentration of analytes was determined in the supernatants by Ultra Performance Liquid Chromatography—mass spectrometry UPLC-MS/MS. This method was performed following the protocol of Escobar et al. [26]

Protein nitration and chlorination were determined by calculating the ratio 3NO2-Tyrosine/ p-Tyrosine and 3Cl-Tyrosine/p-Tyrosine. The protocol consists of homogenizing frozen liver with lysis buffer (100 mg/mL). Next proteins were precipitated with trichloroacetic acid (TCA) (10%, *v*/*v*), and pellets were resuspended in sodium acetate (50 mmol/L, Ph 7.2) (Sigma-Aldrich, St. Louis, MO, USA). Immediately, the protein digestion from tissue extracts was carried out according to Hensley's method [27]. To finish pronase activity, TCA was used to precipitate it. Then, samples were centrifuged (5000 rpm, 4 °C, 5 min) and the supernatant from each sample was injected into the chromatographic system to be quantified by UPLC-MS/MS according to Torres-Cuevas et al. [28].

Data were acquired and processed with the MassLynx 4.1 software and the QuanLynx 4.1 software (Waters), respectively.

2.7. Energy Charge Determined by UPLC-MS/MS

The energy charge (E.C.) is an index relative to ATP, ADP and AMP concentrations as indicated by the formula E.C. = ((ATP) + 0.5(ADP))/((ATP) + (ADP) + (AMP)) [29].

The determination has been performed in the liver tissue by means of ultra-high-resolution liquid chromatography coupled to a mass spectrometry tandem (UPLC-MS/MS). The system used is Acquity UPLC-Xevo TQD from Waters (Milford, MA, USA). The samples were processed from frozen liver samples (−80 °C), homogenized in water: methanol (1:3) cold at 4 °C (100 mg/mL). In this step, the methanol precipitates the proteins present in the homogenate and, once precipitated, to be eliminated, the samples were centrifuged for 20 min at 15,000 rpm at 4 °C. The supernatant obtained was analyzed by UPLC-MS/MS according to Jiang Y et al. with slight modifications [30]. The precipitate was suspended in ammonium acetate buffer pH = 7 to determine the protein concentration in the sample.

2.8. Biochemical Assays

Lipase, amylase and aspartate aminotransferase (AST) activity were determined in plasma by the LIPASE-LQ, AMYLASE-LQ and GOT (AST)-LQ, respectively, (Spinreact, Girona, Spain). The procedures were performed according to the indications of the kit.

Triglycerides and total lipids were determined in liver tissue by the TRIGLYCERIDES-LQ and TOTAL LIPIDS, respectively (Spinreact, Girona, Spain). The procedures were performed according to the indications of the kit in liver homogenate in PBS (100 mg/mL). Results were normalized by protein concentration.

2.9. Histological Analysis

Pieces of pancreas were rapidly removed, fixed in 4% paraformaldehyde for 24 h and embedded in paraffin, 4 μM sections were prepared using an automatic microtome, and then stained with hematoxylin and eosin for microscopic analysis. Pancreatic sections were assessed at 20× magnification over 10 separate fields for severity of pancreatitis by scoring for edema and inflammatory infiltrate according to Van Laethem et al. [31].

2.10. Statistical Analysis

All the values were expressed as means ± SE. To analyze the significance of the quantitative variables, the statistical treatment one-way analysis of variance (ANOVA), followed by Tukey's post-hoc test, was used. $p < 0.05$ was the limit to accept statistically significant differences. The results were managed with the statistical tool of the GraphPad Prism 8 software (GraphPad Software Inc., La Jolla, CA, USA).

3. Results

3.1. PGC-1α Levels are Up-Regulated in Mice Livers after Inducing Acute Pancreatitis

In order to determine the role of PGC-1α in liver tissue during AP, our first approach was to measure its transcriptional and protein levels. First, we confirmed the appropriate AP induction by a histological analysis of the pancreas and determined plasma amylase and lipase activity (see Figure S1 Supplementary Materials).

Interestingly, *Ppargc1a* mRNA expression was up-regulated in the livers of the mice with AP (Figure 1A). Then, we set out to verify whether this transcriptional up-regulation would result in a rise in protein levels. The Western blot analysis showed that PGC-1α protein expression was higher in the livers of the mice with cerulein-induced pancreatitis than in the control mice (Figure 1B).

3.2. PGC-1α Restrains Nos2 Expression in the Liver after Acute Pancreatitis in Mice

Having determined the induction of PGC-1α in the liver during AP, the inflammatory response of AP in liver tissue was studied using the PGC-1α deficient mice (Figure 2A).

For this purpose, the hepatic transcriptional expression of cytokines *Tnfα* and interleukin-6 (*Il6*), as well as *Nos2*, a marker of cellular stress, was analyzed. The *Tnfα* and *Il6* mRNA levels did not vary after administering cerulein to any experimental group (Figure 2B). *Nos2* gene expression remained unchanged in the wild-type (WT) mice after inducing AP (Figure 2B). However, and unlike the changes observed in the *Tnfα* and *Il6* gene expressions, a marked increase (around 6-fold) was detected in the transcription of *Nos2* in the livers of the PGC-1α deficient mice with AP (Figure 2B). Western blot also showed that NOS2 protein expression was higher in the livers of the KO mice with AP (Figure 2C).

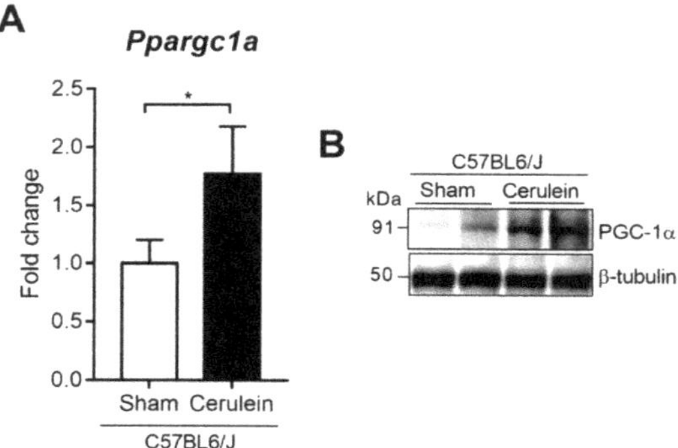

Figure 1. (**A**) mRNA relative expression of *Ppargc1a* versus *Tbp* (TATA-binding protein; housekeeping) in the livers of the control (Sham) and 1 h after the cerulein-induced acute pancreatitis (Cerulein) mice. (**B**) Representative Western blot of PGC-1α in the livers of the control (Sham) and 1 h after the cerulein-induced acute pancreatitis (Cerulein) mice. β-tubulin was used as the loading control. There were six mice per group. Statistical difference is indicated as * $p < 0.05$ vs. Sham.

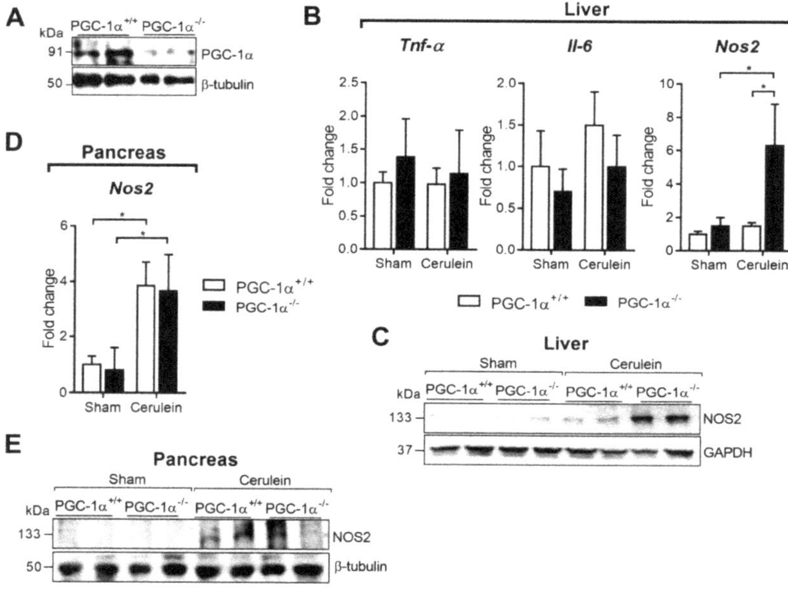

Figure 2. (**A**) Representative Western blot of PGC-1α in the livers of PGC-1$\alpha^{+/+}$ (WT—wild-type) and PGC-1$\alpha^{-/-}$ (KO—knockout) mice. B-tubulin was used as the loading control. (**B**) mRNA relative expression of *Tnfa, Il6* and *Nos2* versus *Tbp* (TATA binding protein; housekeeping) in the livers of the sham PGC-1$\alpha^{+/+}$ (WT) and PGC-1$\alpha^{-/-}$ (KO) mice and at 1 h after cerulein-induced acute pancreatitis (AP) (Cerulein). (**C**) Representative Western blot of NOS2 in the livers of the sham PGC-1$\alpha^{+/+}$ (WT) and PGC-1$\alpha^{-/-}$ (KO) mice and at 1 h after cerulein-induced AP (Cerulein). GAPDH was used as the loading control. (**D**) mRNA relative levels of *Nos2* versus *Tbp* in the pancreas of the sham PGC-1$\alpha^{+/+}$ (WT) and PGC-1$\alpha^{-/-}$ (KO) mice and at 1 h after cerulein-induced AP (Cerulein). (**E**) Representative Western blot of NOS2 in the pancreas of the sham PGC-1$\alpha^{+/+}$ (WT) and PGC-1$\alpha^{-/-}$ (KO) mice and at 1 h after cerulein-induced AP (Cerulein). B-tubulin was used as the loading control. There were six mice per group. Statistical difference is indicated as p * < 0.05 vs. Sham.

We previously reported that *Il6* mRNA levels were selectively up-regulated in the pancreas of the PGC-1α KO mice with AP compared to their WT littermates, conversely to other pro-inflammatory cytokines [32]. Here, we observed that although *Nos2* mRNA expression was up-regulated in the pancreas upon pancreatitis induction, no significant differences appeared between the PGC-1α KO mice and the WT mice (Figure 2D). Similarly, the Western blot also showed NOS2 induction during AP, although there was no change between WT and KO mice with pancreatitis (Figure 2D).

3.3. PGC-1α Avoids the Assembly of the Complex between p65 and Phospho-STAT3 in the Liver during Experimental Acute Pancreatitis

By taking into account our previous studies, in which the plasma levels of IL-6 increased in the PGC-1α KO mice with pancreatitis [32], we considered studying whether NF-κB and STAT3 activation are involved in inducing *Nos2* in the liver upon PGC-1α deficiency.

The Western blot analysis showed a dramatic increase in the phosphorylation of the p65 subunit of NF-κB in the livers of the PGC-1α KO mice after AP induction (Figure 3A). Regarding STAT3 activation, AP triggered an increase in its phosphorylated form in both the WT and KO mice in liver tissue (Figure 3A). Interestingly, the STAT3 phosphorylation levels were higher in the PGC-1α-deficient mice compared to the WT under AP conditions (Figure 3A).

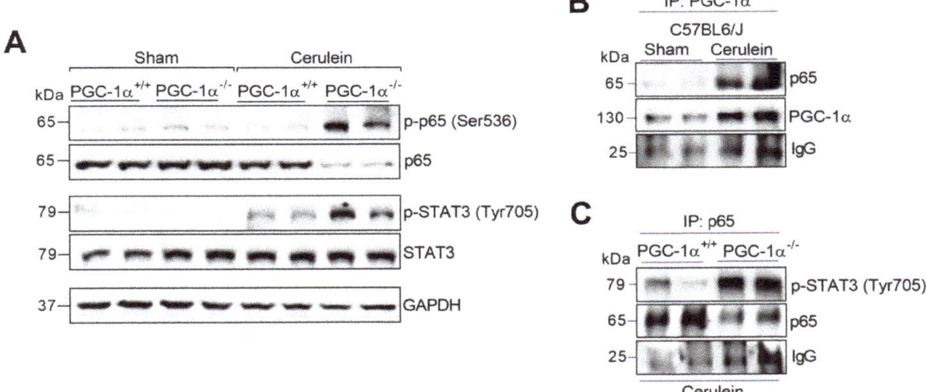

Figure 3. (**A**) Representative Western blot of p-p65 (Ser536), p65, p-STAT3 (Tyr705) and STAT3 in the livers of the sham PGC-1α$^{+/+}$ (WT) and PGC-1α$^{-/-}$ (KO) mice and at 1 h after cerulein-induced AP (Cerulein). GAPDH was used as the loading control. (**B**) Representative Western blot of p65 and PGC-1α in the PGC-1α immunoprecipitate of the livers of the sham PGC-1α$^{+/+}$ (WT) mice and at 1 h after cerulein-induced AP mice. (**C**) Representative Western blot of p-STAT3 (Tyr705) and p65 in the p65 immunoprecipitate of the livers of the PGC-1α$^{+/+}$ (WT) mice and PGC-1α$^{-/-}$ (KO) mice with pancreatitis (Cerulein). IgG was used as the loading control. There were six mice per group.

It has been previously shown that the p65 subunit of NF-κB may complex with p-STAT3 favoring the expression of *Nos2* [33]. Additionally, our research group has shown that PGC-1α forms a protein complex with p65 in the pancreas during AP [32]. Based on this background, we decided to evaluate the role of PGC1alpha in p65-binding to p-STAT3 in the liver during AP. According to our immunoprecipitation studies, the augmented levels of PGC-1α found in the liver after the induction of AP bound to p65 (Figure 3B) as we previously reported in pancreatic tissue. However, the lack of PGC-1α markedly increased the binding of p65 to p-STAT3 in KO mice with pancreatitis (Figure 3C), suggesting that the induction of PGC-1α in the liver during AP may inhibit the formation of the p65/p-STAT3 complex.

3.4. PGC-1α Deficiency Downregulates Antioxidant Gene Expression and Increases Oxidative Stress in the Liver with Acute Pancreatitis

Taking into account the role of PGC-1α in the regulation of antioxidant genes, the mRNA expression of *Sod2* and *Prdx3* was measured in the liver of the PGC-1α KO and WT mice. As expected, PGC-1α deficiency triggered the down-regulation of *Sod2* and *Prdx3* under basal conditions and during pancreatitis (Figure 4A).

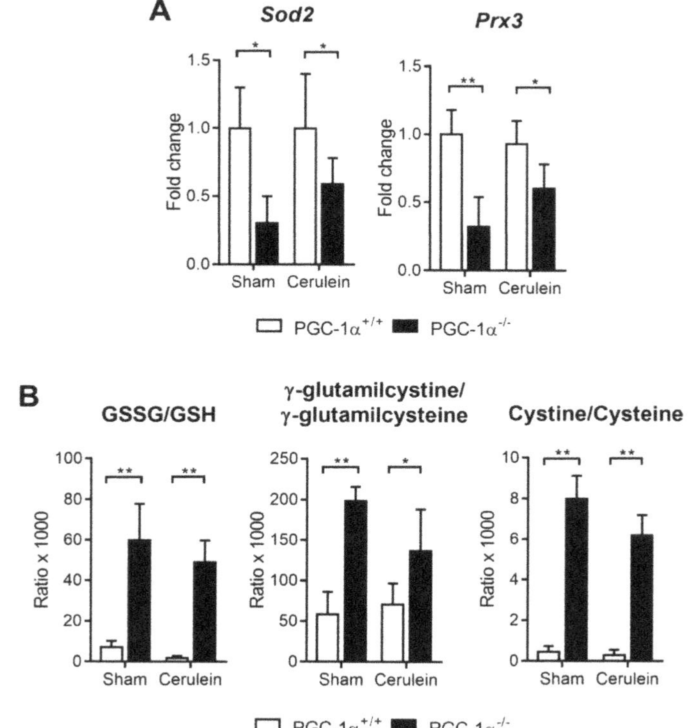

Figure 4. (**A**) mRNA relative expression of *Sod2* and *Prx3* versus the *Tbp* (TATA-binding protein; housekeeping) in the livers of the sham PGC-1$\alpha^{+/+}$ (WT) and PGC-1$\alpha^{-/-}$ (KO) mice and at 1 h after cerulein-induced AP (Cerulein). (**B**) The GSSG/GSH, g-glutamilcystine/g-glutamilcysteine and cystine/cysteine ratios in the livers of the sham PGC-1$\alpha^{+/+}$ (WT) and PGC-1$\alpha^{-/-}$ (KO) mice and at 1 h after cerulein-induced AP (Cerulein). There were six mice per group. The statistical difference is indicated as * $p < 0.05$ and ** $p < 0.001$.

Additionally, the liver redox status was assessed by measuring disulfide pairs, Cyss/Cys, γ-glutamyl cystine/γ-glutamyl cysteine and GSSG/GSH. According to the antioxidant genes expression, the GSSG/GSH, γ-glutamyl cystine/γ-glutamyl cysteine and Cyss/Cys ratios were significantly higher in the livers of the PGC-1α KO mice than in the WT mice (Figure 4B), which supports the relevance of PGC-1α for maintaining redox homeostasis in liver during AP.

3.5. PGC-1α Prevents Protein Nitration in the Liver after Inducing Experimental Acute Pancreatitis

In accordance with the *Nos2* mRNA levels found in the pancreas and liver of the mice with AP, we observed that AP produced increased protein nitration in the pancreas, but not in the liver of the WT mice (Figure 5A). Remarkably, and consistently with the increased *Nos2* mRNA expression observed in the livers of the PGC-1α KO mice with pancreatitis, we detected higher levels of 3-nitrotyrosine, a marker of protein nitration measured by mass spectrometry, in the livers of these mice compared to the other groups (Figure 5B). This increase in 3-nitrotyrosine was confirmed by Western blot in the livers of the PGC-1α KO mice with cerulein-induced pancreatitis (Figure 5C). However, we did not observe changes in 3-chlorotyrosine, parameter associated with inflammation (see Figure S2A in the Supplementary Materials) [34].

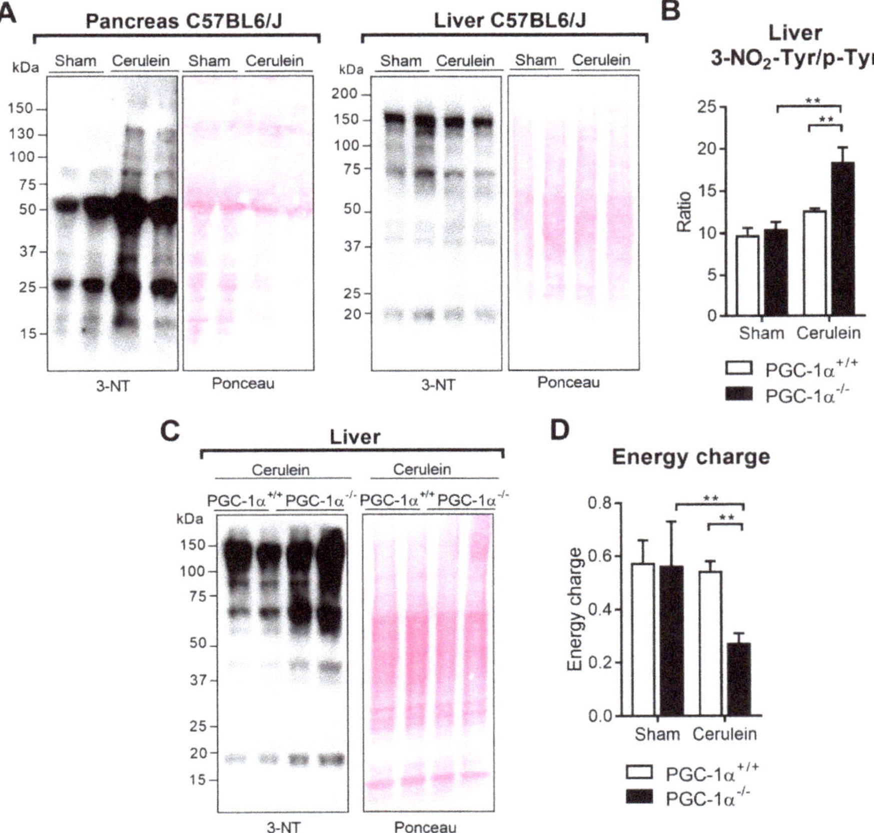

Figure 5. (**A**) Representative Western blot of 3-Nitrotyrosine (NT) in the pancreas and liver of the control (Sham) and 1 h after cerulein-induced acute pancreatitis (Cerulein) mice. Ponceau was used as the loading control. (**B**) Determination of the 3NO2-Tyr/p-Tyr ratio in the livers of the sham PGC-1$\alpha^{+/+}$ (WT) and PGC-1$\alpha^{-/-}$ (KO) mice and at 1 h after cerulein-induced AP (Cerulein). (**C**) Representative Western blot of 3-NT in the liver after inducing acute pancreatitis (Cerulein) in the PGC-1$\alpha^{+/+}$ (WT) and PGC-1$\alpha^{-/-}$ (KO) mice. Ponceau was used as the loading control. (**D**) Energy charge in the livers of the sham PGC-1$\alpha^{+/+}$ (WT) and PGC-1$\alpha^{-/-}$ (KO) mice and at 1 h after cerulein-induced AP (Cerulein). There were six mice per group. The statistical difference is indicated as ** $p < 0.001$.

Considering that mitochondria are the main source of peroxynitrite and are also highly susceptible to the effects of nitrosative stress [35], we set out to calculate the energy charge in liver tissue during AP. The results revealed a significant drop in this parameter in the livers of the KO mice with pancreatitis

(Figure 5D), although these changes did not alter the serum AST levels in AP (see Figure S2B in the Supplementary Materials).

3.6. PGC-1α Levels Lower in the Livers of Obese Mice under Basal Conditions and during Pancreatitis

PGC-1α has been previously reported to be dysregulated in obese animals and patients [18]. As we know that obesity increases the risk of systemic complications in AP [36], we decided to measure the PGC-1α levels in the livers of the lean and obese mice under basal conditions and during AP. First, we corroborated that high-fat diet induced weight gain, hyperglycemia and fatty liver (Table 3).

Table 3. Parameters of high-fat diet-induced obese mice.

	Lean Group	Obese Group
Body weight (g)	22.9 ± 1.1	29.7 ± 1.8 **
Blood glucose (mg/dl)	137.8 ± 25.8	201.6 ± 50.3 *
Liver Triglycerides (mg/g protein)	86.4 ± 22.4	138.8 ± 28.4 **
Liver Total Lipids (mg/g protein)	665.8 ± 208.4	1581.7 ± 437.7 **

The statistical difference is indicated as * $p < 0.05$ and ** $p < 0.001$.

The *Ppargc1a* mRNA levels were markedly down-regulated in the livers of the obese mice under basal conditions (Figure 6A). Interestingly, the obese mice did not exhibit any increased *Ppargc1a* expression, which was found in the lean mice after inducing AP (Figure 6A).

After bearing in mind our findings in the PGC-1α KO mice, we decided to study whether the PGC-1α deficiency detected in the obese mice would impact nitrosative stress in the livers of these mice during AP. We observed that *Nos2* gene expression was up-regulated in the obese mice under basal conditions (Figure 6B). We stress that this up-regulation was higher in the obese mice with pancreatitis (Figure 6B). We also observed a marked increase in the binding of p65 to p-STAT3 in the obese mice with pancreatitis (Figure 6C), which supports the notion that lack of PGC-1α in obese mice livers may be associated with inducing *Nos2* during AP.

The protein nitration levels were higher in the livers of the obese mice than those of the lean mice with pancreatitis (Figure 6D). Interestingly, the obese mice showed a lower hepatic energy charge in relation to the lean mice with pancreatitis (Figure 6E). As we observed in the PGC-1α-deficient mice, this drop in energy charge did not change the AST levels in serum (see Figure S2C in the Supplementary Materials).

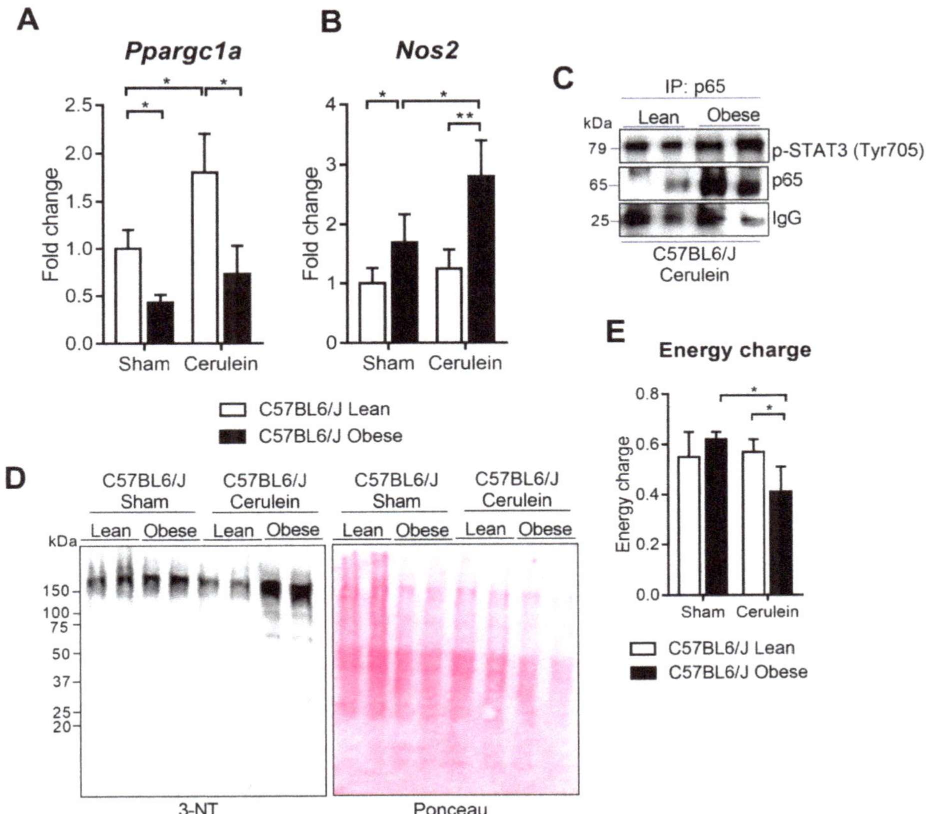

Figure 6. (**A**) mRNA relative expression of *Ppargc1a* versus *Tbp* (TATA-binding protein; housekeeping) in the livers of the sham lean and obese mice and 1 h after cerulein-induced acute pancreatitis (Cerulein). (**B**) mRNA relative expression of *Nos2* versus *Tbp* in the livers of the sham lean and obese mice and 1 h after cerulein-induced acute pancreatitis (Cerulein). (**C**) Representative Western blot of p-STAT3 (Tyr705) and p65 in p65 immunoprecipitate of the sham lean and obese mice and 1 h after cerulein-induced acute pancreatitis (Cerulein). IgG was used as the loading control. (**D**) Representative Western blot of 3-NT in the livers of the sham lean and obese mice and 1 h after cerulein-induced acute pancreatitis (Cerulein). Ponceau was used as the loading control. (**E**) Energy charge in the livers of the sham lean and obese mice and 1 h after cerulein-induced acute pancreatitis (Cerulein). There were five mice per group. Statistical difference is indicated as * $p < 0.05$ and ** $p < 0.001$.

4. Discussion

Of the different systemic complications to occur during AP, the commonest are those that affect the cardiovascular, renal, and pulmonary systems [14]. Although most of the pancreatic enzymes and mediators released by an inflamed pancreas pass through the liver before entering systemic circulation, liver failure is a rare pathophysiological condition in the course of AP with a significant prognostic value for its severity [15,16]. It is noteworthy that obesity, which is a risk factor for AP development, decisively contributes to the appearance of systemic complications, including those that affect liver function [13,16]. The present work highlights the role of PGC-1α in regulating the relation between obesity and liver injury in the course of AP. We particularly demonstrate that obesity impairs hepatic

PGC-1α up-regulation and, thus, enhances *Nos2* transcriptional expression and causes nitrosative stress in the liver during AP.

Our results show that AP induction in mice increases transcriptional and protein PGC-1α expressions in the liver, which is crucial for preventing *Nos2* up-regulation. In pancreatic tissue, our group has recently shown that PGC-1α binds to phospho-p65 to repress *Il6* gene expression during pancreatitis. Consequently, PGC-1α deficiency enhances NF-κB-dependent *Il6* transcription in the pancreas by augmenting circulating IL-6 levels and increasing local and systemic inflammatory responses [32]. In accordance with these findings, here we found higher levels for both NF-κB and STAT3 activation in the livers of the PGC-1α KO mice with pancreatitis. PGC-1α deficiency may be responsible for activating p65, as we have previously reported in pancreatic tissue [32]. It is also accompanied by STAT3 hyperactivation, probably due to the high IL-6 plasma levels achieved in PGC-1α KO mice during pancreatitis. Yet despite the activation of both the NF-κB and STAT3 signaling pathways in the livers of the PGC-1α KO mice, we did not observe any increase in the *Tnfα* or *Il6* expression levels in these mice. Strikingly, and unlike pancreatic tissue, lack of PGC-1α in the liver specifically induced *Nos2* transcription during AP development.

Although *Nos2* expression can be induced by different transcription factors (AP-1, C/EBP, CREB, IRF-1, NF-κB, NF-IL6, Oct-1, SRF, STAT1α) [37], its transcription has also been described to be specifically regulated by the formation of complexes between p65 and STAT3 in the promoter region of *Nos2* [33]. During pancreatitis, our results revealed that p65 bound to PGC-1α in the liver. The formation of this complex has been previously reported in both human cardiac cells and the heart, liver and pancreas of mice [32,38,39]. The present work shows that, as a consequence of PGC-1α deficiency, p65 bound to p-STAT3 in the liver, which could justify the particular increase in the *Nos2* transcription found in the livers of the PGC-1α KO mice.

The present results confirm the presence of nitrosative stress in the pancreas during AP in accordance with augmented pancreatic *Nos2* expression levels [8,9]. Nevertheless, we did not note any increase in the protein nitration levels in the liver after inducing pancreatitis, which is consistent with the PGC-1α-dependent repression of *Nos2* in this tissue. It is well-known that nitrosative stress is implied in the liver pathophysiology [40]. In particular, peroxynitrite accumulation in LPS-treated mice is one of the commonest causes for acute liver injury [41,42]. In fact, hepatocytes respond to LPS treatment with IL-6- and TNF-α-mediated NO production, which induces a hepatic acute phase response [43]. Furthermore, reactive nitrogen species (RNS) production contributes to nitrate critical amino acid residues, which render hepatocytes more susceptible to oxidative damage [40]. In our work, lack of PGC-1α in the liver lowered the antioxidant defense gene expression, which triggered increased oxidative stress. This pro-oxidant environment in the livers of the PGC-1α KO mice, together with specific *Nos2* induction, would explain the more marked nitration pattern observed in the livers of these mice with pancreatitis. Hence, these results demonstrate that PGC-1α is crucial for preventing nitrosative stress in the liver during AP development.

The mitochondrion is the organelle most susceptible to the consequences of nitrosative stress [35]. Numerous studies have shown that mitochondrial dysfunction participates both in the progression of organ failure and in the development of the systemic inflammatory response syndrome (SIRS) [44]. Interestingly, Trumbecakaite et al. showed that AP triggered mitochondrial failure in the pancreas, kidney, and lung, while the liver preserved mitochondrial function during the development of PA [45]. According to this work, our results reveal that the energy charge was unchanged in the liver of WT mice with pancreatitis. However, the induction of PA in mice deficient in PGC-1α caused a decrease in the energy charge in the liver of these mice, confirming the essential role of PGC-1α in the maintenance of mitochondrial homeostasis during inflammatory processes.

According to our results, *Nos2* transcription in the liver during pancreatitis seems to depend on PGC-1α levels. PGC-1α expression lowers in skeletal muscle in both mice with genetic obesity (ob/ob) and fat diet-induced obesity [46]. Furthermore, it has been found that a high-fat diet inhibits PGC-1α expression in mice liver and induces non-alcoholic fatty liver development [39]. Accordingly, our research group recently showed a drop in the protein and transcriptional levels of PGC-1α in the pancreas of obese Zucker rats and in mice with fat diet-induced obesity [32]. Here, we confirm that a high-fat diet in mice down-regulated *Ppargc1a* expression in the liver under basal conditions, and that this decrease was also maintained after inducing AP. Consequently, we observed higher *Nos2* gene expression and protein nitration levels and decreased energy charge in the livers of obese mice compared to the lean mice with pancreatitis.

Obesity is associated with increased *Nos2* expression in insulin-sensitive tissue in rodents and humans [47]. Abdominal obesity increases free fatty acids in the liver, which induces superoxide anion formation and up-regulates *Nos2* gene expression to result in peroxynitrite synthesis and to lead to both mitochondrial dysfunction and liver injury [48]. After taking into account that high-fat diet administration aggravates AP-induced hepatic injury via oxidative stress [49], our herein reported findings provide new insights into the detrimental effect of obesity on liver complications during AP, which highlights the key role of PGC-1α deficiency in this regard.

5. Conclusions

The present work demonstrates the essential role of PGC-1α in inflammatory response and nitrosative stress during AP, particularly in liver tissue. First, pancreatitis leads to marked PGC-1α induction in the liver. Our results also suggest that PGC-1α binds to p65 by acting as a repressor of NF-κB transcriptional activity and, in turn, prevents the formation of a transcriptional complex between p65 and p-STAT3. Therefore, PGC-1α deficiency triggers the increase in the *Nos2* expression mediated by the p65/p-STAT3 complex in the liver, and consequently results in higher protein nitration levels and a decrease in the energy charge. Finally, obesity triggers PGC-1α deficiency in the liver and enhances nitrosative stress during pancreatitis. Therefore, our results highlight the protective role of PGC-1α in the liver for preventing nitrosative stress during AP.

Supplementary Materials: The following are available online at http://www.mdpi.com/2076-3921/9/9/887/s1, Figure S1 and Figure S2 are in the attached supplementary document.

Author Contributions: S.R.-P. and I.T.-C. induced AP in mice. S.R.-P. and I.T.-C. performed most of the assays. F.J.M. did the statistical data analysis. M.M. provided the PGC-1α KO mice and performed the genotyping of the KO and WT mice. S.R.-P., I.T.-C. and F.J.M. prepared all the figures in the manuscript. S.P. designed the study, supervised the experiments and wrote the manuscript. All the authors contributed to update and critically review the literature, wrote the manuscript, and supervised and approved the final version. All authors have read and agreed to the published version of the manuscript.

Funding: This work was supported by Grant GV/2019/153 from the Generalitat Valenciana, Conselleria d'Educació, Investigació, Cultura I Esport.

Conflicts of Interest: The authors declare that there is no conflict of interest.

References

1. Crockett, S.D.; Wani, S.; Gardner, T.B.; Falck-Ytter, Y.; Barkun, A.N. American Gastroenterological Association Institute Clinical Guidelines Committee, American Gastroenterological Association Institute Guideline on Initial Management of Acute Pancreatitis. *Gastroenterology.* **2018**, *154*, 1096–1101. [CrossRef] [PubMed]
2. Lankisch, P.G.; Apte, M.; Banks, P.A. Acute pancreatitis. *Lancet* **2015**, *386*, 85–96. [CrossRef]
3. Schepers, N.J.; Bakker, O.J.; Besselink, M.G.; Ali, U.A.; Bollen, T.L.; Gooszen, H.G.; van Santvoort, H.C.; Bruno, M.J. Impact of characteristics of organ failure and infected necrosis on mortality in necrotising pancreatitis. *Gut* **2019**, *68*, 1044–1051. [CrossRef] [PubMed]
4. Pérez, S.; Pereda, J.; Sabater, L.; Sastre, J. Redox signaling in acute pancreatitis. *Redox Biol.* **2015**, *5*, 1–14. [CrossRef]
5. Tsai, K.; Wang, S.; Chen, T.; Kong, C.; Lee, S.; Lu, F. Oxidative stress: An important phenomenon with pathogenetic significance in the progression of acute pancreatitis. *Gut* **1998**, *42*, 850–855. [CrossRef]
6. Sweiry, J.H.; Mann, G.E. Role of oxidative stress in the pathogenesis of acute pancreatitis. *Scand. J. Gastroenterol.* **1996**, *31*, 10–15. [CrossRef]

7. Escobar, J.; Pereda, J.; Arduini, A.; Sandoval, J.; Moreno, M.L.; Pérez, S.; Sabater, L.; Aparici, L.; Cassinello, N.; Hidalgo, J.; et al. Oxidative and nitrosative stress in acute pancreatitis. *Modul. Pentoxifylline Oxypurinol. Biochem. Pharmacol.* **2011**, *83*, 122–130. [CrossRef]
8. Rius-Pérez, S.; Pérez, S.; Torres-Cuevas, I.; Martí-Andrés, P.; Taléns-Visconti, R.; Paradela, A.; Guerrero, L.; Franco, L.; López-Rodas, G.; Torres, L.; et al. Blockade of the trans-sulfuration pathway in acute pancreatitis due to nitration of cystathionine β-synthase. *Redox Biol.* **2019**, *28*, 101324. [CrossRef]
9. Ang, A.D.; Adhikari, S.; Ng, S.W.; Bhatia, M. Expression of Nitric Oxide Synthase Isoforms and Nitric Oxide Production in Acute Pancreatitis and Associated Lung Injury. *Pancreatology* **2009**, *9*, 150–159. [CrossRef]
10. Cuzzocrea, S.; Mazzon, E.; Dugo, L.; Serrino, I.; Centorrino, E.; Ciccolo, A.; Van de Loo, F.; Britti, D.; Caputi, A.; Thiemermann, C. Inducible nitric oxide synthase-deficient mice exhibit resistance to the acute pancreatitis induced by cerulein. *Shock* **2002**, *17*, 416–422. [CrossRef]
11. Yadav, D.; Lowenfels, A.B. The epidemiology of pancreatitis and pancreatic cancer. *Gastroenterology* **2013**, *144*, 1252–1261. [CrossRef] [PubMed]
12. Emerenziani, S.; Pier Luca-Guarino, M.; Trilo-Asensio, L.M.; Altamore, A.; Ribolsi, M.; Balestrieri, P.; Cicala, M. Role of Overweight and Obesity in Gastrointestinal Disease. *Nutrients* **2020**, *12*, 111. [CrossRef] [PubMed]
13. Khatua, B.; El-Kurdi, B.; Singh, V.P. Obesity and pancreatitis. *Curr. Opin. Gastroenterol.* **2017**, *33*, 374–382. [CrossRef] [PubMed]
14. Garg, P.K.; Singh, V.P. Organ Failure due to Systemic Injury in Acute Pancreatitis. *Gastroenterology* **2019**, *156*, 2008–2023. [CrossRef]
15. Folch-Puy, E. Importance of the liver in systemic complications associated with acute pancreatitis: The role of Kupffer cells. *J. Pathol.* **2007**, *211*, 383–388. [CrossRef]
16. Segersvärd, R.; Tsai, J.A.; Herrington, M.K.; Wang, F. Obesity alters cytokine gene expression and promotes liver injury in rats with acute pancreatitis. *Obesity (Silver Spring)* **2008**, *16*, 23–28. [CrossRef]
17. Yoon, S.B.; Lee, I.S.; Choi, M.H.; Lee, K.; Ham, H.; Oh, H.J.; Park, S.H.; Lim, C.-H.; Choi, M.-G. Impact of Fatty Liver on Acute Pancreatitis Severity. *Gastroenterol. Res. Pract.* **2017**, *2017*, 4532320. [CrossRef]
18. Rius-Pérez, S.; Torres-Cuevas, I.; Millán, I.; Ortega, Á.L.; Pérez, S. PGC-1α, Inflammation, and Oxidative Stress: An Integrative View in Metabolism. *Oxid. Med. Cell Longev.* **2020**, *2020*, 1452696. [CrossRef]
19. Valle, I.; Alvarez-Barrientos, A.; Arza, E.; Lamas, S.; Monsalve, M. PGC-1alpha regulates the mitochondrial antioxidant defense system in vascular endothelial cells. *Cardiovasc. Res.* **2005**, *66*, 562–573. [CrossRef]
20. Olmos, Y.; Sánchez-Gómez, F.J.; Wild, B.; García-Quintans, N.; Cabezudo, S.; Lamas, S.; Monsalve, M. SirT1 regulation of antioxidant genes is dependent on the formation of a FoxO3a/PGC-1α complex. *Antioxid. Redox Signal.* **2013**, *19*, 1507–1521. [CrossRef]
21. Eisele, P.S.; Salatino, S.; Sobek, J.; Hottiger, M.O.; Handschin, C. The peroxisome proliferator-activated receptor γ coactivator 1α/β (PGC-1) coactivators repress the transcriptional activity of NF-κB in skeletal muscle cells. *J. Biol. Chem.* **2013**, *288*, 2246–2260. [CrossRef] [PubMed]
22. Yuan, S.; Liu, X.; Zhu, X.; Qu, Z.; Gong, Z.; Li, J.; Xiao, L.; Yang, Y.; Liu, H.; Sun, L.; et al. The Role of TLR4 on PGC-1α-Mediated Oxidative Stress in Tubular Cell in Diabetic Kidney Disease. *Oxid. Med. Cell Longev.* **2018**, *2018*, 6296802. [CrossRef] [PubMed]
23. Lin, J.; Wu, P.-H.; Tarr, P.T.; Lindenberg, K.S.; St-Pierre, J.; Zhang, C.-Y.; Mootha, V.K.; Jäger, S.; Vianna, C.R.; Reznick, R.M.; et al. Defects in adaptive energy metabolism with CNS-linked hyperactivity in PGC-1alpha null mice. *Cell* **2004**, *119*, 121–135. [CrossRef] [PubMed]
24. Silva-Vaz, P.; Margarida-Abrantes, A.; Castelo-Branco, M.; Gouveia, A.; Botelho, M.F.; Ghiherme-Tralhao, J. Murine Models of Acute Pancreatitis: A Critical Appraisal of Clinical Relevance. *Int. J. Mol. Sci.* **2019**, *20*, 2794. [CrossRef]
25. Monsalve, M.; Wu, Z.; Adelmant, G.; Puigserver, P.; Fan, M.; Spiegelman, B.M. Direct coupling of transcription and mRNA processing through the thermogenic coactivator PGC-1. *Mol. Cell* **2000**, *6*, 307–316. [CrossRef]
26. Escobar, J.; Sánchez-Illana, A.; Kuligowski, J.; Torres-Cuevas, I.; Solberg, R.; Garberg, H.T.; Huun, M.U.; Saugstad, O.D.; Vento, M.; Cháfer-Pericás, C. Development of a reliable method based on ultra-performance liquid chromatography coupled to tandem mass spectrometry to measure thiol-associated oxidative stress in whole blood samples. *J. Pharm. Biomed. Anal.* **2016**, *123*, 104–112. [CrossRef]
27. Henseley, K.; Maidt, M.L.; Pye, Q.N.; Stewart, C.A.; Wack, M.; Tabatabaie, T.; Floyd, R.A. Quantitation of protein-bound 3-nitrotyrosine and 3,4-dihydroxyphenylalanine by high-performance liquid chromatography with electrochemical array detection. *Anal. Biochem.* **1997**, *251*, 187–195. [CrossRef]

28. Torres-Cuevas, I.; Kuligowsky, J.; Cárcel, M.; Cháfer-Pericas, C.; Asensi, M.; Solberg, R.; Cubells, E.; Nuñez, A.; Saugtad, O.D.; Vento, M.; et al. Protein-bound tyrosine oxidation, nitration and chlorination by-products assessed by ultraperformance liquid chromatography coupled to tandem mass spectrometry. *Anal. Chem. Acta* **2016**, *913*, 104–110. [CrossRef]
29. Li, L.; Lu-Nan, Y.; Xiao-Li, C.; Wu-Sheng, L.; Xiao-Dong, X.; Yan-Tao, W. Hepatic adenylate energy charge levels in patients with hepatoma after hepatic artery embolization. *World J. Gastroenteroly* **1998**, *15*, 109–111. [CrossRef]
30. Jiang, Y.; Chengjun, S.; Ding, X.; Yuan, D.; Chen, K.; Gao, B.; Chen, Y.; Sun, A. Simultaneous determination of adenine nucleotides, creatine phosphate and creatine in rat liver by high performance liquid chromatography electrospray ionization-tandem mass spectrometry. *J. Pharm. Biomed. Anal.* **2012**, *66*, 258–263. [CrossRef]
31. Van Laethem, J.L.; Eskinazi, R.; Louis, H.; Rickaert, F.; Robberecht, P.; Deviere, J. Multisystemic production of interleukin 10 limits the severity of acute pancreatitis in mice. *Gut* **1998**, *43*, 408–413. [CrossRef] [PubMed]
32. Pérez, S.; Rius-Pérez, S.; Finamor, I.; Martí-Andrés, P.; Prieto, I.; García, R.; Monsalve, M.; Sastre, J. Obesity causes Pgc-1α deficiency in the pancreas leading to marked Il-6 up-regulation via NF-κB in acute pancreatitis. *J. Pathol.* **2018**, *247*, 48–59. [CrossRef] [PubMed]
33. Ma, J.F.; Sanchez, B.J.; Hall, D.T.; Tremblay, A.K.; Di Marco, S.; Gallouzi, I. STAT3 promotes IFNγ/TNFα-induced muscle wasting in an NF-κB-dependent and IL-6-independent manner. *EMBO Mol. Med.* **2017**, *9*, 622–637. [CrossRef] [PubMed]
34. Nybo, T.; Dieterich, S.; Gamon, L.F.; Chuang, C.Y.; Hammer, A.; Hoefler, G.; Malle, E.; Rogowska-Wrzesinska, A.; Davies, M.J. Chlorination and oxidation of the extracellular matrix protein laminin and basement membrane extracts by hypochlorous acid and myeloperoxidase. *Redox Biol.* **2019**, *20*, 496–513. [CrossRef]
35. Radi, R.; Cassina, A.; Hodara, R. Nitric oxide and peroxynitrite interactions with mitochondria. *Biol. Chem.* **2002**, *383*, 401–409. [CrossRef] [PubMed]
36. Pérez, S.; Finamor, I.; Martí-Andrés, P.; Pereda, J.; Campos, A.; Domingues, R.; Haj, F.; Sabater, L.; de-Madaria, E.; Sastre, J. Role of obesity in the release of extracellular nucleosomes in acute pancreatitis: A clinical and experimental study. *Int. J. Obes. (Lond)* **2018**, *43*, 158–168. [CrossRef] [PubMed]
37. Kleinert, H.; Pautz, A.; Linker, K.; Schwarz, P.M. Regulation of the expression of inducible nitric oxide synthase. *Eur. J. Pharmacol.* **2004**, *500*, 255–266. [CrossRef]
38. Alvarez-Guardia, D.; Palomer, X.; Coll, T.; Davidson, M.M.; Chan, T.O.; Feldman, A.M.; Laguna, J.C.; Vázquez-Carrera, M. The p65 subunit of NF-kappaB binds to PGC-1alpha, linking inflammation and metabolic disturbances in cardiac cells. *Cardiovasc. Res.* **2010**, *87*, 449–458. [CrossRef]
39. Barroso, W.A.; Victorino, V.J.; Jeremias, I.C.; Petroni, R.C.; Ariga, S.K.K.; Salles, T.A.; Barbeiro, D.F.; de Lima, T.M.; de Souza, H.P. High-fat diet inhibits PGC-1α suppressive effect on NFκB signaling in hepatocytes. *Eur. J. Nutr.* **2018**, *57*, 1891–1900. [CrossRef]
40. Reiniers, M.J.; van Golen, R.F.; van Gulik, T.M.; Heger, M. Reactive Oxygen and Nitrogen Species in Steatotic Hepatocytes: A Molecular Perspective on the Pathophysiology of Ischemia-Reperfusion Injury in the Fatty Liver. *Antioxid Redox Signal* **2014**, *21*, 1119–1142. [CrossRef] [PubMed]
41. Islam, M.S.; Yu, H.; Miao, L.; Liu, Z.; He, Y.; Sun, H. Hepatoprotective Effect of the Ethanol Extract of Illicium henryi against Acute Liver Injury in Mice Induced by Lipopolysaccharide. *Antioxidants* **2019**, *8*, 446. [CrossRef] [PubMed]
42. Proniewski, B.; Kij, A.; Sitek, B.; Kelley, E.E.; Chlopicki, S. Multiorgan Development of Oxidative and Nitrosative Stress in LPS-Induced Endotoxemia in C57Bl/6 Mice: DHE-Based In Vivo Approach. *Oxid. Med. Cell Longev.* **2019**, *2019*, 7838406. [CrossRef] [PubMed]
43. Saad, B.; Frei, K.; Scholl, F.A.; Fontana, A.; Maier, P. Hepatocyte-Derived Interleukin-6 and Tumor-Necrosis Factor α Mediate the Lipopolysaccharide-Induced Acute-Phase Response and Nitric Oxide Release by Cultured Rat Hepatocytes. *Eur. J. Biochem.* **1995**, *229*, 349–355. [CrossRef] [PubMed]
44. Dare, A.J.; Phillips, A.R.; Hickey, A.J.; Mittal, A.; Loveday, B.; Thompson, N.; Windsor, J.A. A systematic review of experimental treatments for mitochondrial dysfunction in sepsis and multiple organ dysfunction syndrome. *Free Radic. Biol. Med.* **2009**, *47*, 1517–1525. [CrossRef]
45. Trumbeckaite, S.; Kuliaviene, I.; Deduchovas, O.; Kincius, M.; Baniene, R.; Virketyte, S.; Bukauskas, D.; Jansen, E.; Kupčinskas, L.; Borutaite, V.; et al. Experimental acute pancreatitis induces mitochondrial dysfunction in rat pancreas, kidney and lungs but not in liver. *Pancreatology* **2013**, *13*, 216–224. [CrossRef]

46. Crunkhorn, S.; Dearie, F.; Mantzoros, C.; Gami, H.; da Silva, W.S.; Espinoza, D.; Faucette, R.; Barry, K.; Bianco, A.C.; Patti, M.E. Peroxisome proliferator activator receptor gamma coactivator-1 expression is reduced in obesity: Potential pathogenic role of saturated fatty acids and p38 mitogen-activated protein kinase activation. *J. Biol. Chem.* **2007**, *282*, 15439–15450. [CrossRef]
47. Kaneki, M.; Shimizu, N.; Yamada, D.; Chang, K. Nitrosative Stress and Pathogenesis of Insulin Resistance. *Antioxid. Redox Signal.* **2007**, *9*, 319–329. [CrossRef]
48. García-Ruiz, C.; Rodriguez-Juan, C.; Díaz-Juan, T.; Del Hoyo, P.; Colina, F.; Muñoz-Yagüe, T.; Solís-Herruzo, J.A. Uric acid and anti-TNF antibody improve mitochondrial dysfunction in ob/ob mice. *Hepatology* **2006**, *44*, 581–591. [CrossRef]
49. Fangchao, M.; Jia, Y.; Man, L.; Mingwei, X.; Yupu, H.; Yu, Z.; Yudong, Y.; He, X.; Hongzhong, J.; Weixing, W. Magnesium isoglycyrrhizinate alleviates liver injury in obese rats with acute necrotizing pancreatitis. *Pathol. Res. Pract.* **2019**, *215*, 106–114. [CrossRef]

© 2020 by the authors. Licensee MDPI, Basel, Switzerland. This article is an open access article distributed under the terms and conditions of the Creative Commons Attribution (CC BY) license (http://creativecommons.org/licenses/by/4.0/).

Review

The Role of Dietary Phenolic Compounds in Epigenetic Modulation Involved in Inflammatory Processes

Milan Číž [1,*], Adéla Dvořáková [1,2], Veronika Skočková [1,2] and Lukáš Kubala [1,2,3]

[1] Institute of Biophysics of the Czech Academy of Sciences, 612 65 Brno, Czech Republic; dvorakova.adela@ibp.cz (A.D.); skockova@ibp.cz (V.S.); kubalal@ibp.cz (L.K.)

[2] Department of Experimental Biology, Faculty of Science, Masaryk University, 625 00 Brno, Czech Republic

[3] International Clinical Research Center, St. Anne's University Hospital, 656 91 Brno, Czech Republic

* Correspondence: milanciz@ibp.cz; Tel.: +420-541-517-104

Received: 30 June 2020; Accepted: 20 July 2020; Published: 3 August 2020

Abstract: A better understanding of the interactions between dietary phenolic compounds and the epigenetics of inflammation may impact pathological conditions and their treatment. Phenolic compounds are well-known for their antioxidant, anti-inflammatory, anti-angiogenic, and anti-cancer properties, with potential benefits in the treatment of various human diseases. Emerging studies bring evidence that nutrition may play an essential role in immune system modulation also by altering gene expression. This review discusses epigenetic mechanisms such as DNA methylation, post-translational histone modification, and non-coding microRNA activity that regulate the gene expression of molecules involved in inflammatory processes. Special attention is paid to the molecular basis of NF-κB modulation by dietary phenolic compounds. The regulation of histone acetyltransferase and histone deacetylase activity, which all influence NF-κB signaling, seems to be a crucial mechanism of the epigenetic control of inflammation by phenolic compounds. Moreover, chronic inflammatory processes are reported to be closely connected to the major stages of carcinogenesis and other non-communicable diseases. Therefore, dietary phenolic compounds-targeted epigenetics is becoming an attractive approach for disease prevention and intervention.

Keywords: diseases; immune system; inflammation; NF-κB

1. Introduction

Nutritional phenolic compounds and their impact on human health have attracted great interest in the last three decades. Despite an increasing amount of available data, final conclusions are still diverse and further studies are necessary. Nowadays, dietary phenolic compounds are also studied from the point of view of their possible effects on epigenetic mechanisms. Although epigenetic changes occur in many diseases, only a few studies have reported the effect of natural phenolic compounds on the prevention and treatment of diseases other than cancer. There are a limited number of studies that directly describe the connection between dietary phenolic compounds, epigenetic changes, and inflammatory processes. This review discusses this issue and comments briefly on the consecutive impacts of phenolic compounds on individual diseases. For this purpose, we performed a literature search of topics, including polyphenols, epigenetics, and inflammation using the Web of Science citation indexing service. Studies describing effects of phenolic compounds other than epigenetic ones were excluded.

2. Dietary Phenolic Compounds

Phenolic compounds are a class of plant secondary metabolites widely present in fruits and various plant-derived beverages such as tea or wine [1], though vegetables, legumes, and grains are also non-negligible sources [2]. They generally contribute to plant defense reactions against pathogens, herbivores, or oxidative stress. Over 8000 phenolic compounds have already been described in edible plants [3]. Taken together, phenolic compounds are an important part of the human diet, as they are the most abundant antioxidants consumed by humans [4].

As mentioned above, phenolic compounds are well-known for their antioxidant properties [5–7]; they may protect cells against oxidative damage and, in this way, reduce the risk of various diseases associated with oxidative stress [4]. Anti-inflammatory, anti-angiogenic, and anti-cancer properties have already been described for various phenolic compounds and the positive effects of certain phenolic compounds have been observed in a wide range of non-communicable diseases (neurodegenerative diseases, chronic inflammation, cancer, cardiovascular diseases, type 2 diabetes, and obesity). Some phenolic compounds may support beneficial intestinal microflora, which could also positively impact chronic disease risk [8].

The use of phenolic compounds as a therapeutic measure can encounter certain problems. Phenolic compounds can present a potential health risk when consumed in high concentrations—for example, as food supplements instead as a natural part of the diet. Although phenolic compounds are strong antioxidants, they can also display pro-oxidative effects under certain conditions, as in the presence of transition ions. In the human body, phenolic pro-oxidants can be produced in the gastrointestinal tract when there is an abundance of phenolic compounds and, in high concentrations, could adversely affect the organism [9]. A different kind of problem is represented by isoflavones, which display estrogenic effects. It was observed that soy isoflavones may impact growth and pubertal development in children fed soy-based food in infancy. On the other hand, a potential negative effect of soy isoflavones on women with or at-risk for estrogen-sensitive breast cancer was not proven [8].

Another problem is phenolic compound bioavailability. It must be taken into consideration that while the exposure of humans to dietary phenolic compounds is frequent and long lasting, the fate of dietary phenolic compounds in the body—specifically, their bioavailability and tissue distribution—is not well understood. In particular, there is still a lack of clinical data on the effective dosage of individual phenolic compounds, the metabolic fate of phenolic compounds, their bioavailability and distribution in individual organs, tissues and cell types, the possible synergistic or antagonistic effects of individual compounds contained in various foods, and the possible interactions with gut microflora or lipid domains in cell membranes [10].

Structurally, phenolic compounds range from simple molecules to complex compounds. They are derivatives of phenylalanine or its precursor shikimic acid and are often conjugated with sugar residues, organic acids, or other phenolic molecules [2,3].

The classification of phenolic compounds is based on their chemical structure, or, more precisely, on the number of phenol rings and the way these rings bind together [2]. Phenolic compounds are divided into at least 10 different classes, which include flavonoids, phenolic acids, lignans, stilbenes, and others [11]. The classification of selected phenolic compounds and their main sources are summarized in Table 1. The structure of the cited phenolic families is shown in Figure 1.

Table 1. Phenolic compounds and their epigenetic effects.

Classification		Examples of Phenolic Compounds	Main Sources	Possible Targets	References
Phenolic Acids	Hydroxybenzoic acids	**gallic**, protocatechuic, **ellagic acids**	berries, olive oil, tea	HATs, HDACs	[12]
	Hydroxycinnamic acids	coumaric, ferulic, sinapic, **caffeic acids**	fruits (blueberries, cherries, apples), grains	DNMTs	[13]
Flavonoids	Flavonols	**quercetin**, kaempferol, **fisetin**	vegetables (leek, broccoli), blueberries	HATs, HDACs	[14,15]
	Flavones	**luteolin**, apigenin	parsley, celery	HATs, HDACs	[16]
	Flavanones	naringenin, hesperetin, eriodictyol	citrus fruits	not determined	
	Isoflavones	**genistein**, daidzein, glycitein	legumes (soya)	DNMTs, HATs, SIRT1, miRNA	[17–19]
	Flavanols	catechin, **epicatechin**, gallocatechin, epigallocatechin, **epigallocatechin-3-gallate**	green tea, cocoa	DNMTs, HATs, HDACs, HMTs	[20–22]
	Anthocyanins	cyanidin	fruits (blackcurrants, blackberries)	not determined	
Lignans		secoisolariciresinol, matairesinol	linseed	not determined	
Stilbenes		**resveratrol**, pterostilbene	grapes, red wine, blueberries	SIRT1, DNMTs, miRNA	[23–26]
Phenolic alcohols		tyrosol, hydroxytyrosol	virgin olive oil	not determined	
Curcuminoids		**curcumin**	turmeric	DNMTs, miRNA, HATs, HDACs	[27–30]

The table summarizes classification of phenolic compounds, their main sources, and possible epigenetic targets in inflammation. The bold phenolic compounds are discussed in the text in more details. DNMTs: DNA methyltransferases; HATs: histone acetyltransferases; HDACs: histone deacetylases; HMTs: histone methyltransferases; miRNA: microRNA; SIRT: sirtuin.

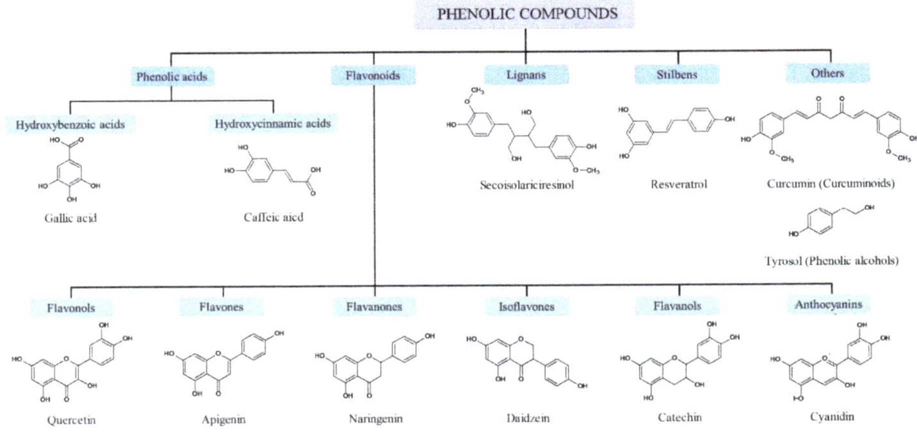

Figure 1. Classification and structure of the cited phenolic families.

2.1. Phenolic Acids

Phenolic acids consist of derivatives of benzoic and cinnamic acids [2].

There are a few edible plants with a higher content of hydroxybenzoic acids; these include certain red fruits (blackberries, raspberries), black radish, onions, olives, and tea. Gallic acid and protocatechuic acid are the main compounds in this group; the main source of gallic acid is tea, while high levels of protocatechuic acid can be found in raspberries and olive oil [2]. Slightly less abundant is ellagic acid, which occurs mostly in berries (raspberries, strawberries), pomegranates, and nuts [31].

Hydroxycinnamic acids are a more common type of phenolic acid. The most abundant are coumaric, caffeic, ferulic, and sinapic acids. Hydroxycinnamic acids are usually found in the form of glycosylated derivatives or esters of quinic, shikimic, and tartaric acids. For example, coumaric and caffeic together with quinic acid form chlorogenic acids, which are common in various types of fruits and coffee beans [32]. Caffeic acid is, overall, the most copious phenolic acid and the most abundant hydroxycinnamic acid in fruits (especially high contents can be found in blueberries, kiwis, plums, cherries, and apples). The most abundant phenolic acid in grains is ferulic acid [2].

2.2. *Flavonoids*

Flavonoids are the largest class of polyphenols [33]. All compounds share a similar structure with two aromatic rings and are divided into subclasses based on the oxidation status of the central pyran ring (the third ring which links the two aromatic rings). Presently, six subclasses of flavonoids have been identified: flavonols, flavones, flavanones, isoflavones, anthocyanins, and flavanols [34].

2.2.1. Flavonols

Flavonols are the most abundant flavonoids in food and include quercetin, kaempferol, and fisetin. Flavonols mostly occur in glycosylated form, often bound to glucose or rhamnose. High amounts of flavonols can be found in some vegetables (onion, curly kale, leeks, broccoli) and blueberries. These flavonoids are usually accumulated in the outer parts of plant as their biosynthesis is stimulated by light [2].

2.2.2. Flavones

Flavones are less common than flavonols and their main representatives are luteolin and apigenin. Similarly to flavonols, they are also found in the form of glycosides. The most important sources of flavones are parsley and celery [2].

2.2.3. Flavanones

High concentrations of flavanones are present only in citrus fruits; such flavanones include naringenin, hesperetin, and eriodictyol, which are found in grapefruit, oranges, and lemons, respectively. Tomatoes and mint are minor sources of flavanones. All flavanones are usually glycosylated by a disaccharide [2].

2.2.4. Isoflavones

Isoflavones, also called phytoestrogens, show structural similarities to estrogens, as the configuration of their hydroxyl groups is analogous to that in the estradiol molecule. Although isoflavones are not steroids, they are able to bind to estrogen receptors, thanks to their similarity to estradiol. The most prevalent isoflavones are genistein, daidzein, and glycitein, which are often glycosylated by glucose [35]. During digestion, consumed isoflavones may be converted by colonic microflora into equol, another isoflavone, which shows even greater phytoestrogenic properties than the original isoflavones [36]. The main sources of isoflavones in the human diet are legumes, of which the most important is soya [35].

2.2.5. Flavanols

Flavanols are found in both monomeric and polymeric form; the monomers are called catechins while the polymers are called proanthocyanidins [37]. Unlike other subclasses of flavonoids, flavanols do not occur in glycosylated form [2].

Catechins are the most common bioactive substances in green tea; another important source is cocoa. Red wine and some types of fruits (mostly apricots and cherries) contain slightly lower but still substantial amounts of catechins. The most abundant catechins in fruits are catechin and epicatechin, while in tea, the most common catechins are gallocatechin, epigallocatechin, and epigallocatechin gallate (EGCG) [37–39].

2.2.6. Anthocyanins

Anthocyanins are pigments located in vacuoles of plant epidermal cells; they are usually glycosylated. The most common anthocyanin is cyanidin, with some of the highest amounts present in blackcurrants and blackberries. Anthocyanins are generally abundant in fruits, but red wine and some cereals and vegetables (eggplant, cabbage, beans, onions, radishes) are also good sources of them. Anthocyanins occur mostly in the peel and their concentration increases during fruit ripening [2].

2.3. Lignans

Lignans are composed of two phenylpropane units and usually occur in free form. The richest source of lignans is linseed, but smaller amounts can also be found in algae, lentils, wheat, certain vegetables (garlic, asparagus, carrots), and fruits (pears, plums). The main lignan in linseed is secoisolariciresinol; matairesinol could be found in lower concentrations. After consumption, lignans are converted to enterodiol and enterolactone by intestinal microflora [40–42].

2.4. Stilbenes

A low level of stilbenes occurs in the human diet [2], but despite this, one stilbene in particular, resveratrol, has been studied intensively because of its possible healthful effects [8]. Resveratrol is produced by plants in response to pathogen attack or stress conditions and occurs mostly in glycosylated form, named piceid [43,44]. Resveratrol and piceid are most abundant in the skins of grapes and subsequently are extracted into red wine; minor sources are peanuts and certain berries (cranberries, blueberries) [45]. Another currently studied stilbene is pterostilbene, which is the methoxylated analogue of resveratrol and occurs mainly in blueberries and grapes [46].

2.5. Phenolic Alcohols

The main phenolic alcohols are tyrosol and hydroxytyrosol and their richest source is virgin olive oil [47]. Tyrosol can also be found in beverages such as wine and beer [48], while hydroxytyrosol is present in red wine and is additionally generated during digestion after red wine consumption [49].

2.6. Curcuminoids

Curcuminoids have a symmetric molecule with two similar aromatic rings. The main curcuminoid, curcumin, can mostly be found in turmeric, which is used as a spice as well as a medicinal herb with a wide range of uses in Indian and Chinese medicine [50].

3. Mechanisms of Epigenetic Regulation

An epigenetic modification is defined as an alteration in gene expression that does not involve a change in the DNA sequence [51]. Epigenetic modifications determine the expression or silencing of genes. The epigenome is very stable in general, but some of these modifications can be affected by the environment (e.g., nutrition, infections, chemical agents, drugs, cytokines, or hormones) [52,53]. The epigenetic modulation of gene expression is crucial for eukaryotic life, as its dysfunction occurs in numerous human diseases [54]. These epigenetic modifications are also responsible for the regulation of the expression of genes related to the immune system; thus, they can modify innate and adaptive immune responses [52]. Epigenetic mechanisms are able to modulate host defense and determine specific immune signaling pathways. The recognition of pathogens by the innate immune system triggers the expression of genes to produce cytokines, chemokines, and transcription factors. The production of these specific inflammatory mediators can also be controlled by epigenetic mechanisms [55].

Among the most recognized epigenetic modifications are DNA methylation, histone modification, and non-coding RNA, which play an important role in cellular differentiation processes [55].

3.1. DNA Methylation

The methylation of DNA is an epigenetic modification in which methyl groups are added to the DNA molecule. It is a reversible covalent modification of DNA, enabled by DNA methyltransferases (DNMTs) [56]. DNMTs catalyze the addition of the methyl group from S-adenosyl-L-methionine to the C5 position of cytosines, mainly in 5′-cytosine-phosphate-guanine-3′ (CpG) islands [57]. A family of DNA methyltransferase enzymes responsible for the methylation of mammalian DNA comprises DNMT1, DNMT3A, DNMT3B, and DNMT3L. DNMT1 is the most abundant and major maintenance methyltransferase, which is responsible for the propagation of existing DNA methylation patterns. DNMT3A and DNMT3B are two related proteins that function mainly as de novo methyltransferases that set up DNA methylation patterns early in development. DNMT3L acts as a cofactor for DNMT3A during embryogenesis, but is catalytically inactive [58].

Generally, a lack of DNA methylation in promoter regions of genes is often associated with transcription activation and gene expression. Conversely, methylation in gene promoters can be responsible for gene silencing [57]. The hypermethylation of DNA CpG islands plays a significant role in the epigenetic mechanism for silencing genes crucial for cell cycle regulation, receptors, DNA repair, and apoptosis. [20]. Therefore, the regulation of DNMT activity by specific inhibitors represents a promising therapeutic target with respect to arresting abnormal cellular processes [58].

3.2. Histone Modification

In the eukaryotic nucleus, 147 bp of DNA associates with histone proteins to form a nucleosome—the basic structural unit of chromatin [59]. The packaging of genomic DNA in chromatin is a process intimately dependent on histone proteins that participates in the regulation of gene expression. Chromatin dynamics are facilitated by two mechanisms: complexes remodeling chromatin and enzymes modifying histone proteins [55]. Post-translational modifications of histone proteins that are essential for the epigenetic control of gene expression take place predominantly at their N-terminal tails, which project out of the nucleosomal core and are rich in lysine and arginine residues [54]. Histone modifications include methylation, acetylation, phosphorylation, citrullination (deimination), and ubiquitination, among others [56], and they work synergistically to promote transcriptional activation or repression [52,60].

3.2.1. Histone Acetylation

Acetylation of lysine residues of histones is catalyzed by histone acetyltransferases (HATs) with acetyl-CoA as the acetyl group donor [61]. To balance the acetylation of histones, histone deacetylases (HDACs) remove acetyl groups from the lysine residues [55]. Histone acetylation neutralizes the positive charge of lysine residues, resulting in the weakening of histone-DNA interactions, which, in turn, leads to the decondensation of chromatin and the increased accessibility of DNA for transcription factors. Thus, it is generally accepted that acetylation is associated with transcription activation, while histone deacetylation is associated with histone condensation and transcription repression [61]. The degree of acetylation has been shown to modulate the regulation of inflammatory genes, DNA repair, and cell proliferation [55].

3.2.2. Other Histone Modifications

There are other modifications of histones with significant effects on gene expression. One of these is the phosphorylation of histone proteins, which is also associated with the activation of transcription. Serine or threonine residues are accessible to phosphorylation by histone kinases, while phosphates are removed by diverse phosphatases. HMTs add methyl groups to arginine to form mono- or dimethylated residues, or to lysine, which can accept one, two, or three methyl groups [56]. The methylation of arginine can be reversed by deiminases, which block methylation of arginine by changing it into citrulline [56,61]. Further, citrullination is the conversion of arginine residues to form the nonstandard amino acid citrulline by removal of the imino group. This reaction is catalyzed by peptidyl arginine deiminases [56]. Histone arginine citrullination is often compared to histone lysine acetylation because of the loss of positive charge resulting in the decondensation of chromatin and the possible activation of transcription [54].

3.3. Non-Coding RNAs

Small non-coding RNAs can also affect the epigenome but these mechanisms of epigenetic control are less well understood than the mechanisms based on DNA and histone modifications [53,56]. Small non-coding RNAs are principally derived from larger RNA precursor molecules by means of cleavage with RNAse III-family enzymes and include microRNAs (miRNAs), short interfering RNAs (siRNAs), PIWI-interacting RNAs (piRNAs), and small nuclear RNAs (snRNAs) [62].

miRNAs are able to negatively control gene expression by binding target mRNA and thus by mediating its degradation and the inhibition of its translation [63]. siRNAs, which are produced by the cleavage of larger double-stranded RNA molecules (dsRNAs), can degrade homologous mRNAs and thus interfere with the translation process [64]. snRNAs control pre-messenger RNA as a part of spliceosome machinery and regulate transcription by RNA polymerase II via regulating the level of active positive transcription elongation factor b in the nucleus [65]. piRNAs associate with encoding regulatory PIWI proteins to configure a piRNA-induced silencing complex. This is connected to the post-transcriptional silencing of genes and epigenetic reprogramming [66].

In addition, miRNAs and siRNAs are both presumably involved in the process of DNA methylation and even histone modification [56,60]. Further, piRNA mediates histone H3K9 trimethylation and silences transposable elements [53].

4. Epigenetic Modifications in Inflammation

Inflammation, as a typical innate immune response, is very complex and relies on the precise control of many functional mechanisms acting at different levels, including also the regulation of gene expression by epigenetic modifications. While acute inflammatory response is mediated mainly by neutrophils, chronic inflammatory response is mediated mostly by monocytes and macrophages producing pro-inflammatory cytokines [67]. Inflammation is closely connected to the release of reactive oxygen species (ROS) and reactive nitrogen species (RNS) by polymorphonuclear and mononuclear phagocytic cells. Increased levels of ROS and RNS lead to the formation of oxidative stress and tissue damage [68]. Deregulation of the extent of inflammatory response is a characteristic feature of chronic inflammatory diseases. Genetic, epigenetic, and environmental factors play a role in this pathological condition [67].

4.1. NF-κB Signaling

Nuclear factor NF-κB is a ubiquitous transcriptional factor in eukaryotic cells and is crucial for inflammatory immune response. It modulates the regulation of many genes (for chemokines, cytokines, cell adhesion molecules, etc.) including various mediators that intensify inflammatory response and worsen the status of inflammatory disease [1,69]. NF-κB consists of different subunits, of which p65 and p50 are the most common and well-studied. In unstimulated cells, NF- κB is often inactive and forms a complex with its inhibitor subunit IκBα. NF-κB activation by different stimuli leads to degradation of the IκBα subunit, and released dimer p65-p50 is then translocated to the nucleus, where it binds to its promoters and induces the expression of inflammatory genes. The p65 subunit is the key transcriptionally active component of NF-κB [70]. Phosphorylation of the p65 subunit is required for its transcriptional regulation [71], while acetylation of the p65 subunit is necessary for its full transcriptional activity and for the regulation of NF-κB signaling. p65 acetylation and deacetylation, which are mediated by HATs and HDACs, represent an important epigenetic target during inflammation. A decreased level of p65 acetylation leads to the inhibition of NF-κB target gene expression, such as the expression of interleukin-6 (IL-6) or TNF-α (tumor necrosis factor alpha) genes, and can even prevent the translocation of p65 to the nucleus. The regulation of NF-κB also affects the expression of COX (cyclooxygenase) and iNOS (inducible nitric oxide synthase). The modulation of p65 acetylation thus represents a promising target in chronic inflammation therapy [1,70]. Additionally, two closely related HATs, p300 and CBP (CREB-binding protein), which are transcriptional cofactors, exhibit a strong synergism with the p65 subunit [12].

The generation of ROS also leads to the activation of various signaling pathways such as extracellular signal-regulated kinases (ERKs). It can further induce NF-κB-dependent immune response and alterations in histone protein acetylation/deacetylation status, which finally results in changed transcription initiation [68]. All these ROS-mediated mechanisms result in augmented pro-inflammatory cytokine levels, which contribute to the chronicity of the inflammatory process.

4.2. Chronic Inflammation

During chronic inflammatory processes, the epigenetic mechanisms involved in the regulation of gene expression are modified. For example, ROS and RNS generated during oxidative stress may cause a loss of DNA methylation after mitosis, since DNMT1 does not recognize oxidatively modified methyl groups. On the other hand, halogen derivatives produced by inflammatory processes imitate cytosine methylation, resulting in increased methylation by DNMT1. The levels of some pro-inflammatory cytokines (e.g., TNF-α) are tightly modulated by epigenetic mechanisms. In contrast, some other cytokines (e.g., IL-6) can provoke histone protein modifications and DNA methylation and thus regulate other gene expressions and processes. Chronic inflammatory processes are generally characterized by epigenetic changes based on the enrichment of hypo-acetylated histones and hypermethylated CpGs. The modulation of both HDACs and DNMT1 activities is thus therapeutically beneficial in chronic inflammatory diseases and both enzymes represent a transcriptional repression complex [67].

5. Epigenetics, Inflammation, and Phenolic Compounds

Phenolic compounds have the ability to modulate gene expression through the regulation of epigenetic mechanisms including either DNA methylation, histone modification, or miRNA expression. In general, several phenolic compounds are able to activate HDACs (e.g., fisetin), activate (e.g., genistein) or inhibit (e.g., EGCG, curcumin) HATs, activate (e.g., resveratrol) sirtuins (SIRT), and inhibit DNMTs (e.g., EGCG) [14,21,23,27,72]. Some phenolic compounds are associated with the regulation of miRNA expression (e.g., genistein, curcumin) [17,28]. Many phenolic compounds exhibit anti-inflammatory properties and have been shown to inhibit the secretion and production of pro-inflammatory cytokines and decrease the production of ROS and nitric oxide (NO) [73].

On the basis of available data, it is generally accepted that phenolic compounds are also able to modify epigenetic mechanisms involved in the modulation of the immune system and diseases connected to inflammation [72]. Many phenolic compounds show the potential to modulate NF-κB activity and chromatin remodeling through the regulation of HDACs and HATs [1]. The epigenetic effects of selected phenolic compounds related to inflammation are summarized in Table 1. Animal and human studies in which the effects of phenolic compounds on epigenetic changes under inflammatory conditions were studied are listed in Tables 2 and 3.

Table 2. Summary of the studies performed on animal models.

Species	Strain Characteristics	Sample Size (per Group)	Pathological Condition	Type of Study	Phenolic Compound Intervention	Reference
Mus musculus	CD45.2 + C57BL/6	n = 9–10	repeated social defeat stress	observational, cross-sectional	dihydrocaffeic acid 5 mg/kg/d for 2 weeks, delivered through drinking water, treatment repeated 4 months later	[13]
Mus musculus	BALB/c 6 weeks old 20–25 g body weight	n = 3		intervention	epigallocatechin-3-gallate 100 mmol/L, 50 mmol/L	[21]
Mus musculus	C57BL/6J males 6–8 weeks old 20–30 g body weight	n = 12	inflammation	intervention	curcumin 200 mg/kg (i.p.)	[74]
Rattus norvegicus	Sprague–Dawley females 250–300 g body weight	n = 6	inflammation	intervention	curcumin 40 mg/kg (final injection volume of 0.2 mL)	[75]
Rattus norvegicus	Wistar females 10 weeks old 180–220 g body weight	n = 10	ovariectomized diabetic rat	intervention	genistein 1 mg/kg/day, subcutaneously, for 8 weeks	[18]
Rattus norvegicus	Sprague–Dawley males two weeks old	n = 19 (normal group), 16 (normal intervention group), 29 (diabetic group), 25 (diabetic intervention group)	diabetes	intervention	resveratrol intraperitoneally 10 mg/kg/day, for 10 weeks	[25]

Table 3. Summary of human studies.

Type of Study	Sample Size	Participant Characteristics	Dietary Intervention	Disease	Reference
intervention	n = 128	age ≥40 years	one capsule/day of resveratrol 500 mg/day (n = 43), 40 mg/day (n = 43) or placebo (n = 42), for 6 months	type 2 diabetes mellitus	[76]
intervention	n = 214	men and women 20 years of age	Treated group (45 men and 65 women) received the cocoa cream product (13 g/unit; 1 g cocoa/unit, 6 units/d; 6 g cocoa/d) for 2 weeks. Control group consisted of 46 men and 58 women.	cardiovascular disease risk	[77]

5.1. Phenolic Acids

Gallic and ellagic acids can affect the activity of HATs and HDACs in monocytic cells with induced inflammation. Kiss and co-workers proved that both acids decreased HAT activity in TNF-α-activated human monocytic (THP-1) cells and that ellagic acid also increased HDAC activity. This led to the attenuation of inflammatory response and improved the survival of cells [12].

Another phenolic acid, dihydrocaffeic acid (DHCA), which is a metabolite of caffeic acid, influences the methylation level of DNA. It is known that stressful conditions can change the methylation level of DNA in humans and animals, which also affects immunity, as the methylation of genes related to immunity leads to less effective immune response [78,79]. Blaze and colleagues proved that treatment with DHCA caused a decrease in DNA methylation level in peripheral leucocytes from mice exposed to stressful conditions [13]. In another experiment, DHCA also decreased the level of DNA methylation in human and mice peripheral leucocytes exposed to lipopolysaccharide (LPS) in vitro, which means DHCA can exhibit an anti-inflammatory effect [13].

5.2. Stilbenes

Resveratrol exhibits anti-cancer, anti-oxidative, and anti-angiogenic properties with potential benefits for human immunity or in the treatment of some diseases including diabetes, rheumatoid arthritis, metabolic disorder, and cardiovascular or neurodegenerative diseases. It has been shown that resveratrol directly targets immune cells, such as macrophages, large lymphocytes, and dendritic cells [58,80–82]. It can also affect several epigenetic targets including DNMTs and HDACs, especially SIRT1 deacetylase from class III HDACs [23,58,82–84]. SIRT1 can deacetylate the p65 subunit of NF-κB and also inhibit NF-κB-mediated inflammation [85]. Schug and co-workers demonstrated that SIRT1 inhibited the transcriptional activity of NF-κB through the deacetylation of the p65 subunit, affecting the expression of inflammatory cytokines by macrophages [86]. TNF-α induced NF-κB p65 expression in human umbilical vein endothelial cells (HUVECs) was diminished by SIRT1 activation triggered by resveratrol treatment [24]. In another experiment, Bo and colleagues found that SIRT1 expression in peripheral blood mononuclear cells was increased in the case of type 2 diabetes mellitus patients receiving resveratrol for six months. The elevation of SIRT1 led to significantly reduced levels of H3K56 acetylation and it is known from previous research that a high acetylation level of H3K56 is related to genes playing an important role in the long-term effect of hyperglycemia on the body [76]. Besides SIRT1, resveratrol can also affect the methylation of DNA. Lou et al. observed that longer resveratrol administration to diabetic rats led to a decline in the levels of pro-inflammatory cytokines (IL-1, IL-6, TNF-α, and interferon-γ) and enhanced the production of anti-inflammatory IL-10 in the cells of arterial intima. At the same time, they detected a higher level of the methylation of CpG islands in these pro-inflammatory genes and a lower level of CpG methylation in the IL-10 gene compared to untreated diabetic individuals. These results indicate that resveratrol can decrease inflammation also through changes in DNA methylation and has a protective effect on the aorta or other arteries under hyperglycemic conditions [25]. The available experimental data also suggest that resveratrol and its analogs are able to modulate miRNA expression in various diseases [26,87].

5.3. Flavonols

Quercetin has been shown to inhibit the production of COX and lipoxygenase, inhibit the molecular level of NF-κB, and block pro-inflammatory cytokine production or inhibit mitogen-activated protein kinase; it thus possesses anti-inflammatory, anti-cancer, and anti-diabetic qualities [73,80]. Quercetin was reported to increase histone H3 acetylation in HL-60 (human promyelocytic leukemia cells); therefore, quercetin shows potential for the activation of HATs or the inhibition of HDACs [88] and probably possesses histone demethylase inhibitor activity [89]. On the other hand, Xiao et al. found that quercetin suppressed p300 activity in human breast cancer and endothelial cells and thus reduced the p300-mediated acetylation of NF-κB [15]. Ruiz and colleagues demonstrated that quercetin inhibited NF-κB binding to the pro-inflammatory interferon-γ-inducible protein (IP-10) and macrophage inflammatory protein 2 (MIP-2) gene promoters in murine intestinal epithelial cells, therefore further blocking cofactor recruitment and HATs activity at the chromatin of these promoters. In addition to reducing HAT activity, quercetin inhibited the TNF-induced acetylation and phosphorylation of histone H3 and NF-κB cofactor CBP/p300 at the IP-10 and MIP-2 gene promoter [90].

Fisetin also exhibits anti-inflammatory properties and may provide potential benefits in the treatment of several diseases. For example, diabetes is closely associated with chronic inflammation, as high glucose levels have been implicated in the activation of histone acetylation and the activation of NF-κB. Kim et al. observed that fisetin suppressed pro-inflammatory cytokine release via the NF-κB signaling pathway in THP-1 cells. The supplementation of cells with fisetin activated HDACs and suppressed HATs, resulting in deacetylation of the p65 subunit of NF-κB and finally in diminished pro-inflammatory cytokine release [14].

5.4. Flavones

Luteolin is reported to have antioxidant, anti-inflammatory, and anti-cancer properties. It is related to the activation of nuclear factor 2 (erythroid-derived 2)-like 2 (Nrf-2), a transcription factor which causes the transactivation of antioxidant genes [91,92]. Luteolin also displays similar effects as fisetin, as it activates HDACs and inhibits HATs in high glucose-treated THP-1 cells. Luteolin inhibits HAT activity and the expression of CBP/p300 protein, which results in deacetylation of the p65 subunit and subsequently in the decreased activity of NF-κB. This leads to the suppression of inflammatory cytokine release, such as IL-6 and TNF-α [16]. Recently, Kim et al. found that luteolin in combination with fisetin not only suppressed HAT and NF-κB activity and inflammatory cytokine release in THP-1 cells, but also significantly inhibited ROS production and activated SIRT1 expression [93].

5.5. Flavanols

The tea polyphenol EGCG exhibits anti-NF-κB activity in several pathological conditions, such as chronic inflammation or cancer. Choi and colleagues proved that EGCG was responsible for HAT inhibition and induced hypoacetylation of the p65 subunit. Furthermore, EGCG reduced the TNF-α-induced expression of IL-6, COX-2 and iNOS in HEK293, THP-1 cells, and even primary peritoneal macrophages [21]. EGCG was also shown to inhibit DNMTs and thus reactivate methylation-silenced genes in cancer cell lines [20]. Recently, it was demonstrated that EGCG mitigates vascular inflammatory response through a repressive epigenetic effect on the NF-κB signaling pathway [94]. The positive effect of EGCG was also observed in the case of regulatory T cells (Tregs), which are negative regulators of inflammation. Tregs function and their number are often decreased in obese individuals compared with lean subjects. The role of EGCG in the modulation of Tregs was studied in obese and lean human subjects in vitro; Tregs were isolated and cultured in the absence or presence of EGCG. It was observed that EGCG treatment enhanced HDAC activity and decreased NF-κB activity in both groups. Furthermore, EGCG also increased the production of anti-inflammatory cytokine IL-10 by suppressing the NF-κB signaling pathway [95].

Cordero-Herrera and colleagues demonstrated that epicatechin, a main cocoa flavanol, prevented the increased acetylation of H3K9 and the dimethylation of H3K4, and also decreased the dimethylation of H3K9 in THP-1 cells cultured under high glucose conditions by affecting HDAC4 levels and HAT activity. They further showed that both the expression level of NF-κB and the release of TNF-α were decreased in these cells [22]. Crescenti and co-workers showed that the consumption of cocoa led to a decreased DNA methylation level in human peripheral mononuclear leucocytes [77]. It was confirmed previously that elevated DNA methylation in these cells is associated with various non-communicable diseases such as cardiovascular disorders and obesity [96] or insulin resistance [97].

5.6. Isoflavones

Genistein has been shown to possess anti-diabetic, anti-inflammatory, and anticancer properties. It is able to affect cancer cell survival and activate tumor suppressor genes by epigenetic changes [18,98]. Various cancers have been associated with the hypermethylation of CpG islands at the regulatory sites of tumor suppressor genes with subsequent gene silencing and genistein has been shown to decrease DNMT activity. Furthermore, genistein is able to increase HAT activity, as histone acetylation is related to a loosened chromatin structure and the induction of tumor suppressor gene expression in various cancer cell types [19,98]. In another study, genistein was able to increase SIRT1 levels in a model of ovariectomized diabetic rats, which led to a subsequent decrease in NF-κB and IL-1β protein levels, this providing evidence of the anti-inflammatory effect of genistein [18]. Recently, Zhang et al. demonstrated that genistein played an important role in the prevention of atherosclerosis through the regulation of miRNA expression. This study indicates that genistein could reverse oxidized low-density lipoprotein-induced inflammation through the regulation of miRNA-155/SOCS1 (suppressor of the cytokine signaling-1), which was accompanied by the inhibition of the NF-κB signaling pathway in HUVECs [84].

5.7. Curcuminoids

Curcumin is well-known for its anti-inflammatory, anti-oxidative, and anti-lipidemic properties and is able to modulate several diseases (e.g., neurological disorders, diabetes) via epigenetic regulation. It suppresses the function of NF-κB by decreasing the acetylation level of the p65 subunit [70], which regulates the anti-inflammatory response of the enzymatic activities of COX and iNOS. Curcumin also downregulates the expression of NF-κB-related gene products (e.g., TNF-α, IL-1, IL-6, IL-8, adhesion molecules) [74,99] and modulates signaling pathways controlling anti-oxidative properties through the regulation of Nrf-2 [29]. Several studies have recently shown that curcumin can affect DNMTs and HDACs and thus reverse the silencing of key genes [29,30,75]. Moreover, curcumin has also been shown to alter the profiles of miRNA expression [28] and to inhibit HATs [27]. HAT inhibition resulted in the significantly reduced acetylation of histone H3 levels in the IL-6 promoter, as well as decreased IL-6 mRNA expression and IL-6 protein release by rheumatoid arthritis synovial fibroblasts [100]. In addition, in a study using THP-1 cells as a model for human monocytes, curcumin was reported to have an effect on histone acetylation and pro-inflammatory cytokine secretion under high-glucose conditions. The results showed that curcumin treatment not only significantly reduced HAT activity and the level of p300 (a co-activator of NF-κB), but also induced HDAC2 expression. The results indicate that curcumin decreases high glucose-induced cytokine production in monocytes via epigenetic changes involving NF-κB [70]. A similar effect of curcumin (increased HDAC and reduced HAT activity) was also observed in TNF-α-activated THP-1 cells [12]. In another study, Yuan and colleagues investigated the role of curcumin-mediated epigenetic modulation of the expression of TREM-1 (triggering receptor expressed on myeloid cells 1) proteins. TREMs constitute a family of immunoglobulin cell surface receptors expressed on macrophages and neutrophils, which are capable of regulating various immunological events in both innate and adaptive immune cells. Yuan and colleagues demonstrated that curcumin inhibited the methylation and acetylation of H3K4 and p300 in the TREM-1 promoter in vitro in bone marrow derived macrophages and in vivo in a septic lung injury model, which led to the reduced binding of p65 to the TREM-1 promoter in response to LPS. Their study also indicated that the inhibition of TREM-1 by curcumin is independent of ROS production [19]. Curcumin also inhibits the phosphorylation and degradation of IκBα and the subsequent translocation of the p65 subunit of NF-κB to the nucleus [101].

6. Conclusions

Many phenolic compounds exhibit anti-inflammatory properties and have been shown to inhibit the secretion and production of pro-inflammatory cytokines and decrease the production of reactive oxygen and nitrogen species. Recent findings suggest that phenolic compounds can regulate epigenetic mechanisms involved in immune system modulation. Phenolic compounds have the ability to modulate gene expression through the regulation of epigenetic mechanisms including DNA methylation, histone modification, and miRNA expression.

In addition, however, the regulation of HATs, HDACs, and SIRT1 activity, which all influence NF-κB signaling, seems to be a crucial mechanism of the epigenetic control of inflammation by phenolic compounds (Figure 2). Furthermore, this mechanism occurs to a greater or lesser extent in many different phenolic classes. The modulation of epigenetic modifications by phenolic compounds is, therefore, of great interest in the context of clinical approaches to immune-mediated diseases and diseases such as rheumatoid arthritis, cancer, cardiovascular disease, atherosclerosis, metabolic disorders, neurodegenerative diseases, obesity, and diabetes.

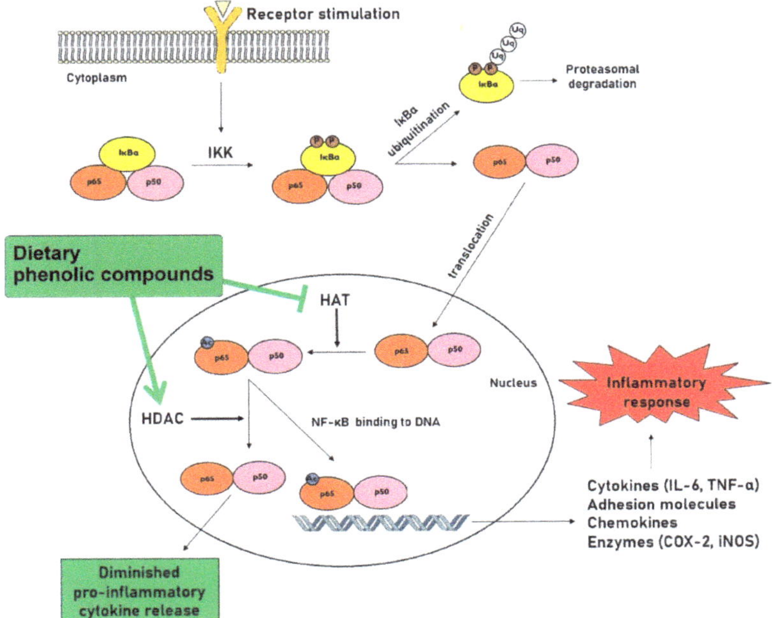

Figure 2. The regulation of inflammatory response by dietary phenolic compounds through HAT inhibition or HDAC activation. NF-κB is a nuclear factor responsible for the transcription of genes connected to inflammatory immune response. It is found in cytoplasm associated with its inhibitory subunit IκBα, which masks the nuclear translocation signal. Upon cell stimulation, the inhibitory subunit is phosphorylated by IκB kinase (IKK), which leads to its ubiquitination and degradation by proteasome. The released NF-κB dimer (consisting of the p65 and p50 subunits) is then translocated into the nucleus and initiates the expression of genes, including genes for various mediators that intensify inflammatory response. HATs and HDACs have been implicated in regulating NF-κB activity. While HATs have a role in NF-κB-mediated transcriptional activation, HDACs have been shown to reverse this process and to repress NF-κB-mediated gene transcription. Many dietary phenolic compounds show a potential to modulate NF-κB-mediated inflammation through the regulation of HAT or HDAC activity. Ac: acetylation; COX-2: cyclooxygenase-2; HAT: histone acetyltransferase; HDAC: histone deacetylase; IKK: IκB kinase; IL-6: interleukin 6; iNOS: inducible nitric oxide synthase; NF-κB: nuclear factor kappa B; P: phosphorylation; TNF-α: tumor necrosis factor alpha; Uq: ubiquitination.

Taking all the above information together, the use of dietary phenolic compounds to induce epigenetic modifications appears to be a promising approach with respect to disease prevention and the development of treatment strategies. Nevertheless, the role of phenolic compounds in the regulation of epigenetic mechanisms is complex and still poorly understood. Thus, it represents an attractive field of research.

Author Contributions: All authors contributed equally to the preparation of the manuscript. All authors have read and agreed to the published version of the manuscript.

Funding: This study was performed with the research support of IBP CAS (No. 68081707) and MEYS CR (National Program of Sustainability II project no. LQ1605 Translational Medicine).

Conflicts of Interest: The authors declare that they have no conflict of interest.

List of Abbreviations

CBP	CREB-binding protein
CpG	5′-cytosine-phosphate-guanine-3′
COX	cyclooxygenase
DHCA	dihydrocaffeic acid
DNMT	DNA methyltransferase
EGCG	epigallocatechin gallate
ERK	extracellular signal-regulated kinase
HAT	histone acetyltransferase
HDAC	histone deacetylase
HMT	histone methyltransferases
HUVEC	umbilical vein endothelial cells
IKK	IκB kinase
IL	interleukin
iNOS	inducible nitric oxide synthase
IP-10	interferon-γ-inducible protein
LPS	lipopolysaccharide
MIP	macrophage inflammatory protein
miRNA	microRNA
NF-κB	nuclear factor kappa B
Nrf-2	nuclear factor 2 (erythroid-derived 2)-like 2
piRNA	PIWI-interacting RNA
RNS	reactive nitrogen species
ROS	reactive oxygen species
siRNA	short interfering RNA
SIRT	sirtuin
snRNA	small nuclear RNA
SOCS	suppressor of the cytokine signaling
TNF-α	tumor necrosis factor alpha
Tregs	regulatory T cells
TREM-1	triggering receptor expressed on myeloid cells 1

References

1. Owona, B.A.; Ebrahimi, A.; Schluesenner, H. Epigenetic effects of natural polyphenols: A focus on SIRT1-mediated mechanisms. *Mol. Nutr. Food Res.* **2013**, *58*, 22–32. [CrossRef]
2. Manach, C.; Scalbert, A.; Morand, C.; Rémésy, C.; Jiménez, L. Polyphenols: Food sources and bioavailability. *Am. J. Clin. Nutr.* **2004**, *79*, 727–747. [CrossRef] [PubMed]
3. Bravo, L. Polyphenols: Chemistry, dietary sources, metabolism, and nutritional significance. *Nutr. Rev.* **1998**, *56*, 317–333. [CrossRef] [PubMed]
4. Scalbert, A.; Manach, C.; Morand, C.; Rémésy, C.; Jimenez, L. Dietary Polyphenols and the Prevention of Diseases. *Crit. Rev. Food Sci. Nutr.* **2005**, *45*, 287–306. [CrossRef] [PubMed]
5. Číž, M.; Čížová, H.; Denev, P.; Kratchanova, M.; Slavov, A.; Lojek, A. Different methods for control and comparison of the antioxidant properties of vegetables. *Food Control.* **2010**, *21*, 518–523. [CrossRef]
6. Denev, P.; Kratchanov, C.G.; Ciz, M.; Lojek, A.; Kratchanova, M.G. Bioavailability and Antioxidant Activity of Black Chokeberry (Aronia melanocarpa) Polyphenols: in vitro and in vivo Evidences and Possible Mechanisms of Action: A Review. *Compr. Rev. Food Sci. Food Saf.* **2012**, *11*, 471–489. [CrossRef]
7. Denev, P.; Kratchanova, M.; Číž, M.; Lojek, A.; Vasicek, O.; Nedelcheva, P.; Blazheva, D.; Toshkova, R.; Gardeva, E.; Yossifova, L.; et al. Biological activities of selected polyphenol-rich fruits related to immunity and gastrointestinal health. *Food Chem.* **2014**, *157*, 37–44. [CrossRef]
8. Cory, H.; Passarelli, S.; Szeto, J.; Tamez, M.; Mattei, J. The Role of Polyphenols in Human Health and Food Systems: A Mini-Review. *Front. Nutr.* **2018**, *5*, 87. [CrossRef]
9. Zhang, H.; Tsao, R. Dietary polyphenols, oxidative stress and antioxidant and anti-inflammatory effects. *Curr. Opin. Food Sci.* **2016**, *8*, 33–42. [CrossRef]

10. Joven, J.; Micol, V.; Segura-Carretero, A.; Alonso-Villaverde, C.; Menendez, J.A.; Platform, B.F.C. Polyphenols and the Modulation of Gene Expression Pathways: Can We Eat Our Way Out of the Danger of Chronic Disease? *Crit. Rev. Food Sci. Nutr.* **2014**, *54*, 985–1001. [CrossRef]
11. Hardy, T.M.; Tollefsbol, T.O. Epigenetic diet: Impact on the epigenome and cancer. *Epigenomics* **2011**, *3*, 503–518. [CrossRef] [PubMed]
12. Kiss, A.K.; Granica, S.; Stolarczyk, M.; Melzig, M.F. Epigenetic modulation of mechanisms involved in inflammation: Influence of selected polyphenolic substances on histone acetylation state. *Food Chem.* **2012**, *131*, 1015–1020. [CrossRef]
13. Blaze, J.; Wang, J.; Ho, L.; Mendelev, N.; Haghighi, F.; Pasinetti, G.M. Polyphenolic Compounds Alter Stress-Induced Patterns of Global DNA Methylation in Brain and Blood. *Mol. Nutr. Food Res.* **2018**, *62*. [CrossRef] [PubMed]
14. Kim, H.J.; Kim, S.H.; Yun, J.-M. Fisetin Inhibits Hyperglycemia-Induced Proinflammatory Cytokine Production by Epigenetic Mechanisms. *Evidence-Based Complement. Altern. Med.* **2012**, *2012*, 1–10. [CrossRef]
15. Xiao, X.; Shi, D.; Liu, L.; Wang, J.; Xie, X.; Kang, T.; Deng, W. Quercetin Suppresses Cyclooxygenase-2 Expression and Angiogenesis through Inactivation of P300 Signaling. *PLoS ONE* **2011**, *6*, e22934. [CrossRef]
16. Kim, H.J.; Lee, W.; Yun, J.-M. Luteolin Inhibits Hyperglycemia-Induced Proinflammatory Cytokine Production and Its Epigenetic Mechanism in Human Monocytes. *Phytotherapy Res.* **2014**, *28*, 1383–1391. [CrossRef]
17. Zhang, H.P.; Zhao, Z.X.; Pang, X.F.; Yang, J.; Yu, H.X.; Zhang, Y.H.; Zhou, H.; Zhao, J.H. Genistein Protects Against Ox-LDL-Induced Inflammation Through MicroRNA-155/SOCS1-Mediated Repression of NF-AB Signaling Pathway in HUVECs. *Inflammation* **2017**, *40*, 1450–1459. [CrossRef]
18. Yousefi, H.; Alihemmati, A.; Karimi, P.; Alipour, M.R.; Habibi, P.; Ahmadiasl, N. Effect of genistein on expression of pancreatic SIRT1, inflammatory cytokines and histological changes in ovariectomized diabetic rat. *Iran. J. Basic Med. Sci.* **2017**, *20*, 423–429.
19. Majid, S.; Dar, A.A.; Shahryari, V.; Hirata, H.; Ahmad, A.; Saini, S.; Tanaka, Y.; Dahiya, A.V.; Dahiya, R. Genistein reverses hypermethylation and induces active histone modifications in tumor suppressor gene B-Cell translocation gene 3 in prostate cancer. *Cancer* **2009**, *116*, 66–76. [CrossRef]
20. Fang, M.Z.; Wang, Y.; Ai, N.; Hou, Z.; Sun, Y.; Lu, H.; Welsh, W.; Yang, C.S. Tea polyphenol (-)-epigallocatechin-3-gallate inhibits DNA methyltransferase and reactivates methylation-silenced genes in cancer cell lines. *Cancer Res.* **2003**, *63*, 7563–7570.
21. Choi, K.C.; Jung, M.G.; Lee, Y.H.; Yoon, J.C.; Kwon, S.H.; Kang, H.B.; Kim, M.J.; Cha, J.H.; Kim, Y.J.; Jun, W.J.; et al. Epigallocatechin-3-Gallate, a Histone Acetyltransferase Inhibitor, Inhibits EBV-Induced B Lymphocyte Transformation via Suppression of RelA Acetylation. *Cancer Res.* **2009**, *69*, 583–592. [CrossRef] [PubMed]
22. Cordero-Herrera, I.; Chen, X.P.; Ramos, S.; Devaraj, S. (-)-Epicatechin attenuates high-glucose-induced inflammation by epigenetic modulation in human monocytes. *Eur. J. Nutr.* **2017**, *56*, 1369–1373. [CrossRef] [PubMed]
23. Borra, M.T.; Smith, B.C.; Denu, J.M. Mechanism of Human SIRT1 Activation by Resveratrol. *J. Boil. Chem.* **2005**, *280*, 17187–17195. [CrossRef] [PubMed]
24. Pan, W.; Yu, H.; Huang, S.; Zhu, P.-L. Resveratrol Protects against TNF-α-Induced Injury in Human Umbilical Endothelial Cells through Promoting Sirtuin-1-Induced Repression of NF-KB and p38 MAPK. *PLoS ONE* **2016**, *11*, e0147034. [CrossRef] [PubMed]
25. Lou, X.-D.; Wang, H.; Xia, S.-J.; Skog, S.; Sun, J. Effects of Resveratrol on the Expression and DNA Methylation of Cytokine Genes in Diabetic Rat Aortas. *Arch. Immunol. Ther. Exp.* **2014**, *62*, 329–340. [CrossRef]
26. Kumar, A.; Rimando, A.M.; Levenson, A.S. Resveratrol and pterostilbene as a microRNA-mediated chemopreventive and therapeutic strategy in prostate cancer. *Ann. NY Acad. Sci.* **2017**, *1403*, 15–26. [CrossRef]
27. Neckers, L.; Trepel, J.; Lee, S.; Chung, E.-J.; Lee, M.-J.; Jung, Y.-J.; Marcu, M.G. Curcumin is an Inhibitor of p300 Histone Acetylatransferase. *Med. Chem.* **2006**, *2*, 169–174. [CrossRef]
28. Sun, M.; Estrov, Z.; Ji, Y.; Coombes, K.R.; Harris, D.H.; Kurzrock, R. Curcumin (diferuloylmethane) alters the expression profiles of microRNAs in human pancreatic cancer cells. *Mol. Cancer Ther.* **2008**, *7*, 464–473. [CrossRef]

29. Guo, Y.; Shu, L.; Zhang, C.; Su, Z.-Y.; Kong, A.-N. Curcumin inhibits anchorage-independent growth of HT29 human colon cancer cells by targeting epigenetic restoration of the tumor suppressor gene DLEC1. *Biochem. Pharmacol.* **2015**, *94*, 69–78. [CrossRef]
30. Maugeri, A.; Mazzone, M.G.; Giuliano, F.; Vinciguerra, M.; Basile, G.; Barchitta, M.; Agodi, A. Curcumin Modulates DNA Methyltransferase Functions in a Cellular Model of Diabetic Retinopathy. *Oxidative Med. Cell. Longev.* **2018**, *2018*, 1–12. [CrossRef]
31. Vattem, D.; Shetty, K. biological functionality of ellagic acid: A review. *J. Food Biochem.* **2005**, *29*, 234–266. [CrossRef]
32. Crupi, P.; Bleve, G.; Tufariello, M.; Corbo, F.; Clodoveo, M.L.; Tarricone, L. Comprehensive identification and quantification of chlorogenic acids in sweet cherry by tandem mass spectrometry techniques. *J. Food Compos. Anal.* **2018**, *73*, 103–111. [CrossRef]
33. Fonayet, J.V.; Millán, S.; Martí, M.P.; Borràs, E.; Arola, L. Advanced separation methods of food anthocyanins, isoflavones and flavanols. *J. Chromatogr. A* **2009**, *1216*, 7143–7172. [CrossRef]
34. Harborne, J.B.; A Williams, C. Advances in flavonoid research since 1992. *Phytochemistry* **2000**, *55*, 481–504. [CrossRef]
35. Coward, L.; Smith, M.; Kirk, M.; Barnes, S. Chemical modification of isoflavones in soyfoods during cooking and processing. *Am. J. Clin. Nutr.* **1998**, *68*, 1486S–1491S. [CrossRef]
36. Setchell, K.D.R.; Brown, N.M.; Lydeking-Olsen, E. The Clinical Importance of the Metabolite Equol—A Clue to the Effectiveness of Soy and Its Isoflavones. *J. Nutr.* **2002**, *132*, 3577–3584. [CrossRef]
37. Lakenbrink, C.; Lapczynski, S.; Maiwald, B.; Engelhardt, U.H. Flavonoids and other polyphenols in consumer brews of tea and other caffeinated beverages. *J. Agric. Food Chem.* **2000**, *48*, 2848–2852. [CrossRef]
38. Graham, H.N. Green tea composition, consumption, and polyphenol chemistry. *Prev. Med.* **1992**, *21*, 334–350. [CrossRef]
39. Guyot, S.; Marnet, N.; Drilleau, J.-F. Thiolysis-HPLC characterization of apple procyanidins covering a large range of polymerization states. *J. Agric. Food Chem.* **2001**, *49*, 14–20. [CrossRef]
40. Thompson, L.U.; Robb, P.; Serraino, M.; Cheung, F. Mammalian lignan production from various foods. *Nutr. Cancer* **1991**, *16*, 43–52. [CrossRef]
41. Adlercreutz, H.; Mazur, W. Phyto-oestrogens and Western diseases. *Ann. Med.* **1997**, *29*, 95–120. [CrossRef] [PubMed]
42. Heinonen, S.; Salonen, J.T.; Liukkonen, K.; Poutanen, K.; Wähälä, K.; Deyama, T.; Nishibe, S.; Adlercreutz, H. In vitro metabolism of plant lignans: New precursors of mammalian lignans enterolactone and enterodiol. *J. Agric. Food Chem.* **2001**, *49*, 3178–3186. [CrossRef] [PubMed]
43. Bavaresco, L. Role of viticultural factors on stilbene concentrations of grapes and wine. *Drugs Under Exp. Clin. Res.* **2003**, *29*, 181–187.
44. Delmas, D.; Lançon, A.; Colin, D.; Jannin, B.; Latruffe, N. Resveratrol as a Chemopreventive Agent: A Promising Molecule for Fighting Cancer. *Curr. Drug Targets* **2006**, *7*, 423–442. [CrossRef]
45. Das, D.K.; Mukherjee, S.; Ray, D. Resveratrol and red wine, healthy heart and longevity. *Hear. Fail. Rev.* **2010**, *15*, 467–477. [CrossRef]
46. Lee, P.-S.; Chiou, Y.-S.; Ho, C.-T.; Pan, M.-H. Chemoprevention by resveratrol and pterostilbene: Targeting on epigenetic regulation. *BioFactors* **2017**, *44*, 26–35. [CrossRef]
47. Cabrini, L.; Barzanti, V.; Cipollone, M.; Fiorentini, D.; Grossi, G.; Tolomelli, B.; Zambonin, L.; Landi, L. Antioxidants and total peroxyl radical-trapping ability of olive and seed oils. *J. Agric. Food Chem.* **2001**, *49*, 6026–6032. [CrossRef]
48. Covas, M.I.; Miró-Casas, E.; Fitó, M.; Farré-Albadalejo, M.; Gimeno, E.; Marrugat, J.; De La Torre, R. Bioavailability of tyrosol, an antioxidant phenolic compound present in wine and olive oil, in humans. *Drugs Under Exp. Clin. Res.* **2003**, *29*, 203–206.
49. De La Torre, R.; Covas, M.I.; Pujadas, M.A.; Fitó, M.; Farré, M. Is dopamine behind the health benefits of red wine? *Eur. J. Nutr.* **2006**, *45*, 307–310. [CrossRef]
50. Amalraj, A.; Pius, A.; Gopi, S.; Gopi, S. Biological activities of curcuminoids, other biomolecules from turmeric and their derivatives—A review. *J. Tradit. Complement. Med.* **2016**, *7*, 205–233. [CrossRef]
51. Pan, M.-H.; Lai, C.-S.; Wu, J.-C.; Ho, C.-T. Epigenetic and disease targets by polyphenols. *Curr. Pharm. Des.* **2013**, *19*, 6156–6185. [CrossRef] [PubMed]

52. Suárez-Álvarez, B.; Raneros, A.B.; Ortega, F.; López-Soto, A. Epigenetic modulation of the immune function: A potential target for tolerance. *Epigenetics* **2013**, *8*, 694–702. [CrossRef]
53. Ospelt, C.; Gay, S.; Klein, K. Epigenetics in the pathogenesis of RA. *Semin. Immunopathol.* **2017**, *39*, 409–419. [CrossRef] [PubMed]
54. Fuhrmann, J.; Thompson, P.R. Protein Arginine Methylation and Citrullination in Epigenetic Regulation. *ACS Chem. Boil.* **2015**, *11*, 654–668. [CrossRef] [PubMed]
55. Morandini, A.C.; Santos, C.F.; Yilmaz, O. Role of epigenetics in modulation of immune response at the junction of host–pathogen interaction and danger molecule signaling. *Pathog. Dis.* **2016**, *74*. [CrossRef] [PubMed]
56. Gibney, E.R.; Nolan, C.M. Epigenetics and gene expression. *Heredity* **2010**, *105*, 4–13. [CrossRef]
57. Castillo-Aguilera, O.; Depreux, P.; Halby, L.; Arimondo, P.B.; Goossens, L. DNA Methylation Targeting: The DNMT/HMT Crosstalk Challenge. *Biomolecules* **2017**, *7*, 3. [CrossRef]
58. Fernandes, G.F.S.; Silva, G.D.B.; Pavan, A.R.; Chiba, D.E.; Chin, C.M.; Dos Santos, J.L. Epigenetic Regulatory Mechanisms Induced by Resveratrol. *Nutrients* **2017**, *9*, 1201. [CrossRef]
59. Leshner, M.; Wang, S.; Lewis, C.; Zheng, H.; Chen, X.A.; Santy, L.; Wang, Y. PAD4 mediated histone hypercitrullination induces heterochromatin decondensation and chromatin unfolding to form neutrophil extracellular trap-like structures. *Front. Immunol.* **2012**, *3*. [CrossRef]
60. Chuang, J.C.; Jones, P.A. Epigenetics and microRNAs. *Pediatr. Res.* **2007**, *61*, 24r–29r. [CrossRef]
61. Xu, Y.-M.; Du, J.-Y.; Lau, A.T.Y. Posttranslational modifications of human histone H3: An update. *Proteomics* **2014**, *14*, 2047–2060. [CrossRef] [PubMed]
62. Frías-Lasserre, D.; Villagra, C.A. The Importance of ncRNAs as Epigenetic Mechanisms in Phenotypic Variation and Organic Evolution. *Front. Microbiol.* **2017**, *8*. [CrossRef] [PubMed]
63. Yekta, S.; Shih, I.-H.; Bartel, B. MicroRNA-Directed Cleavage of HOXB8 mRNA. *Science* **2004**, *304*, 594–596. [CrossRef] [PubMed]
64. Agrawal, N.; Dasaradhi, P.V.N.; Mohmmed, A.; Malhotra, P.; Bhatnagar, R.K.; Mukherjee, S.K. RNA Interference: Biology, Mechanism, and Applications. *Microbiol. Mol. Boil. Rev.* **2003**, *67*, 657–685. [CrossRef] [PubMed]
65. Muniz, L.; Egloff, S.; Kiss, T. RNA elements directing in vivo assembly of the 7SK/MePCE/Larp7 transcriptional regulatory snRNP. *Nucleic Acids Res.* **2013**, *41*, 4686–4698. [CrossRef]
66. Siomi, M.C.; Sato, K.; Pezic, D.; Aravin, A.A. PIWI-interacting small RNAs: The vanguard of genome defence. *Nat. Rev. Mol. Cell Boil.* **2011**, *12*, 246–258. [CrossRef]
67. Milagro, F.I.; Mansego, M.L.; De Miguel, C.; Martínez, J.A. Dietary factors, epigenetic modifications and obesity outcomes: Progresses and perspectives. *Mol. Asp. Med.* **2013**, *34*, 782–812. [CrossRef]
68. Adler, V.; Yin, Z.; Tew, K.D.; Ronai, Z.A. Role of redox potential and reactive oxygen species in stress signaling. *Oncogene* **1999**, *18*, 6104–6111. [CrossRef]
69. Olefsky, J.M. IKK epsilon: A Bridge between Obesity and Inflammation. *Cell* **2009**, *138*, 834–836. [CrossRef]
70. Yun, J.-M.; Jialal, I.; Devaraj, S. Epigenetic regulation of high glucose-induced proinflammatory cytokine production in monocytes by curcumin. *J. Nutr. Biochem.* **2011**, *22*, 450–458. [CrossRef]
71. Shakibaei, M.; Buhrmann, C.; Mobasheri, A. Resveratrol-mediated SIRT-1 Interactions with p300 Modulate Receptor Activator of NF-kappa B Ligand (RANKL) Activation of NF-kappa B Signaling and Inhibit Osteoclastogenesis in Bone-derived Cells. *J. Biol. Chem.* **2011**, *286*, 11492–11505. [CrossRef] [PubMed]
72. Cuevas, A.; Saavedra, N.; Salazar, L.A.; Abdalla, D.S. Modulation of Immune Function by Polyphenols: Possible Contribution of Epigenetic Factors. *Nutrients* **2013**, *5*, 2314–2332. [CrossRef] [PubMed]
73. Chen, S.; Jiang, H.; Wu, X.; Fang, J. Therapeutic Effects of Quercetin on Inflammation, Obesity, and Type 2 Diabetes. *Mediat. Inflamm.* **2016**, *2016*, 1–5. [CrossRef] [PubMed]
74. Yuan, Z.; Syed, M.A.; Panchal, D.; Rogers, D.; Joo, M.; Sadikot, R.T. Curcumin mediated epigenetic modulation inhibits TREM-1 expression in response to lipopolysaccharide. *Int. J. Biochem. Cell Boil.* **2012**, *44*, 2032–2043. [CrossRef]
75. Boyanapalli, S.S.; Huang, Y.; Su, Z.-Y.; Cheng, D.; Zhang, C.; Guo, Y.; Rao, R.; Androulakis, I.P.; Kong, A.-N. Pharmacokinetics and Pharmacodynamics of Curcumin in regulating anti-inflammatory and epigenetic gene expression. *Biopharm. Drug Dispos.* **2018**, *39*, 289–297. [CrossRef]

76. Bo, S.; Togliatto, G.; Gambino, R.; Ponzo, V.; Lombardo, G.; Rosato, R.; Cassader, M.; Brizzi, M.F. Impact of sirtuin-1 expression on H3K56 acetylation and oxidative stress: A double-blind randomized controlled trial with resveratrol supplementation. *Acta Diabetol.* **2018**, *55*, 331–340. [CrossRef]
77. Crescenti, A.; Solà, R.; Valls, R.M.; Caimari, A.; Del Bas, J.M.; Anguera, A.; Anglès, N.; Arola, L. Cocoa Consumption Alters the Global DNA Methylation of Peripheral Leukocytes in Humans with Cardiovascular Disease Risk Factors: A Randomized Controlled Trial. *PLoS ONE* **2013**, *8*, e65744. [CrossRef]
78. Uddin, M.; Aiello, A.E.; Wildman, D.E.; Koenen, K.C.; Pawelec, G.; Santos, R.D.L.; Goldmann, E.; Galea, S. Epigenetic and immune function profiles associated with posttraumatic stress disorder. *Proc. Natl. Acad. Sci. USA* **2010**, *107*, 9470–9475. [CrossRef] [PubMed]
79. Klengel, T.; Mehta, D.; Anacker, C.; Rex-Haffner, M.; Pruessner, J.C.; Pariante, C.M.; Pace, T.W.W.; Mercer, K.B.; Mayberg, H.S.; Bradley, B.; et al. Faculty of 1000 evaluation for Allele-specific FKBP5 DNA demethylation mediates gene-childhood trauma interactions. *Nat. Neurosci.* **2013**, *16*, 33–41. [CrossRef]
80. Gao, Y.; Tollefsbol, T.O. Impact of Epigenetic Dietary Components on Cancer through Histone Modifications. *Curr. Med. Chem.* **2015**, *22*, 2051–2064. [CrossRef]
81. Ding, S.; Jang, H.; Fang, J. Regulation of Immune Function by Polyphenols. *J. Immunol. Res.* **2018**, *2018*, 1–8. [CrossRef] [PubMed]
82. Maugeri, A.; Barchitta, M.; Mazzone, M.G.; Giuliano, F.; Basile, G.; Agodi, A. Resveratrol Modulates SIRT1 and DNMT Functions and Restores LINE-1 Methylation Levels in ARPE-19 Cells under Oxidative Stress and Inflammation. *Int. J. Mol. Sci.* **2018**, *19*, 2118. [CrossRef] [PubMed]
83. Kong, S.; Yeung, P.; Fang, D. The class III histone deacetylase sirtuin 1 in immune suppression and its therapeutic potential in rheumatoid arthritis. *J. Genet. Genom.* **2013**, *40*, 347–354. [CrossRef] [PubMed]
84. Kala, R.; Shah, H.N.; Martin, S.L.; Tollefsbol, T.O. Epigenetic-based combinatorial resveratrol and pterostilbene alters DNA damage response by affecting SIRT1 and DNMT enzyme expression, including SIRT1-dependent γ-H2AX and telomerase regulation in triple-negative breast cancer. *BMC Cancer* **2015**, *15*, 672. [CrossRef]
85. Yang, H.; Zhang, W.; Pan, H.; Feldser, H.G.; Lainez, E.; Miller, C.; Leung, S.; Zhong, Z.; Zhao, H.; Sweitzer, S.; et al. SIRT1 Activators Suppress Inflammatory Responses through Promotion of p65 Deacetylation and Inhibition of NF-κB Activity. *PLoS ONE* **2012**, *7*, e46364. [CrossRef]
86. Schug, T.T.; Xu, Q.; Gao, H.; Peres-Da-Silva, A.; Draper, D.W.; Fessler, M.B.; Purushotham, A.; Li, X. Myeloid Deletion of SIRT1 Induces Inflammatory Signaling in Response to Environmental Stress. *Mol. Cell. Boil.* **2010**, *30*, 4712–4721. [CrossRef]
87. Tili, E.; Michaille, J.-J. Promiscuous Effects of Some Phenolic Natural Products on Inflammation at Least in Part Arise from Their Ability to Modulate the Expression of Global Regulators, Namely microRNAs. *Molecules* **2016**, *21*, 1263. [CrossRef]
88. Tseng, B.; Lee, W.-J.; Chen, Y.-R.; Tseng, T.-H. Quercetin induces FasL-related apoptosis, in part, through promotion of histone H3 acetylation in human leukemia HL-60 cells. *Oncol. Rep.* **2011**, *25*, 583–591. [CrossRef]
89. Abdulla, A.; Zhao, X.; Yang, F. Natural Polyphenols Inhibit Lysine-Specific Demethylase-1 in vitro. *J. Biochem. Pharmacol. Res.* **2013**, *1*, 56–63.
90. Ruiz, P.A.; Braune, A.; Hölzlwimmer, G.; Quintanilla-Fend, L.; Haller, D. Quercetin inhibits TNF-induced NF-kappaB transcription factor recruitment to proinflammatory gene promoters in murine intestinal epithelial cells. *J. Nutr.* **2007**, *137*, 1208–1215. [CrossRef]
91. Reichard, J.F.; Motz, G.T.; Puga, A. Heme oxygenase-1 induction by NRF2 requires inactivation of the transcriptional repressor BACH1. *Nucleic Acids Res.* **2007**, *35*, 7074–7086. [CrossRef] [PubMed]
92. Rangarajan, P.; Karthikeyan, A.; Dheen, S.T. Role of dietary phenols in mitigating microglia-mediated neuroinflammation. *Neuromolecular Med.* **2016**, *18*, 453–464. [CrossRef] [PubMed]
93. Kim, A.; Yun, J.-M. Combination Treatments with Luteolin and Fisetin Enhance Anti-Inflammatory Effects in High Glucose-Treated THP-1 Cells Through Histone Acetyltransferase/Histone Deacetylase Regulation. *J. Med. Food* **2017**, *20*, 782–789. [CrossRef] [PubMed]
94. Liu, D.; Perkins, J.T.; Hennig, B. EGCG prevents PCB-126-induced endothelial cell inflammation via epigenetic modifications of NF-κB target genes in human endothelial cells. *J. Nutr. Biochem.* **2015**, *28*, 164–170. [CrossRef]
95. Yun, J.-M.; Jialal, I.; Devaraj, S. Effects of epigallocatechin gallate on regulatory T cell number and function in obese v. lean volunteers. *Br. J. Nutr.* **2010**, *103*, 1771–1777. [CrossRef]
96. Kim, M.; Long, T.I.; Arakawa, K.; Wang, R.; Yu, M.C.; Laird, P.W. DNA Methylation as a Biomarker for Cardiovascular Disease Risk. *PLoS ONE* **2010**, *5*, e9692. [CrossRef]

97. Zhao, J.Y.; Goldberg, J.; Bremner, J.D.; Vaccarino, V. Global DNA Methylation Is Associated with Insulin Resistance A Monozygotic Twin Study. *Diabetes* **2012**, *61*, 542–546. [CrossRef]
98. Zhang, Y.; Chen, H. Genistein, an epigenome modifier during cancer prevention. *Epigenetics* **2011**, *6*, 888–891. [CrossRef]
99. Boyanapalli, S.S.S.; Kong, A.-N.T. "Curcumin, the King of Spices": Epigenetic Regulatory Mechanisms in the Prevention of Cancer, Neurological, and Inflammatory Diseases. *Curr. Pharmacol. Rep.* **2015**, *1*, 129–139. [CrossRef]
100. Wada, T.T.; Araki, Y.; Sato, K.; Aizaki, Y.; Yokota, K.; Kim, Y.T.; Oda, H.; Kurokawa, R.; Mimura, T. Aberrant histone acetylation contributes to elevated interleukin-6 production in rheumatoid arthritis synovial fibroblasts. *Biochem. Biophys. Res. Commun.* **2014**, *444*, 682–686. [CrossRef]
101. Karunaweera, N.; Raju, R.; Gyengesi, E.; Munch, G. Plant polyphenols as inhibitors of NF-kappa B induced cytokine production a potential anti-inflammatory treatment for Alzheimer's disease? *Front. Mol. Neurosci.* **2015**, *8*. [CrossRef] [PubMed]

© 2020 by the authors. Licensee MDPI, Basel, Switzerland. This article is an open access article distributed under the terms and conditions of the Creative Commons Attribution (CC BY) license (http://creativecommons.org/licenses/by/4.0/).

Article

Cashew (*Anacardium occidentale* L.) Nuts Counteract Oxidative Stress and Inflammation in an Acute Experimental Model of Carrageenan-Induced Paw Edema

Marika Cordaro [1,†], Rosalba Siracusa [2,†], Roberta Fusco [2,†], Ramona D'Amico [2], Alessio Filippo Peritore [2], Enrico Gugliandolo [2], Tiziana Genovese [2], Maria Scuto [3], Rosalia Crupi [4], Giuseppina Mandalari [2], Salvatore Cuzzocrea [2,5,*], Rosanna Di Paola [2,*] and Daniela Impellizzeri [2]

1 Department of Biomedical, Dental and Morphological and Functional Imaging University of Messina, Via Consolare Valeria, 98125 Messina, Italy; cordarom@unime.it

2 Department of Chemical, Biological, Pharmaceutical and Environmental Sciences, University of Messina, 98166 Messina, Italy; rsiracusa@unime.it (R.S.); rfusco@unime.it (R.F.); rdamico@unime.it (R.D.); aperitore@unime.it (A.F.P.); egugliandolo@unime.it (E.G.); tgenovese@unime.it (T.G.); gmandalari@unime.it (G.M.); dimpellizzeri@unime.it (D.I.)

3 Department of Biomedical and Biotechnological Sciences, University of Catania, Via Santa Sofia 97, 95123 Catania, Italy; mary-amir@hotmail.it

4 Department of Veterinary Sciences, University of Messina, 98168 Messina, Italy; rcrupi@unime.it

5 Department of Pharmacological and Physiological Science, Saint Louis University School of Medicine, Saint Louis, MO 63104, USA

* Correspondence: salvator@unime.it (S.C.); dipaolar@unime.it (R.D.P.); Tel.: +39-090-6765-208 (S.C. & R.D.P.)

† The authors equally contributed to the work.

Received: 30 June 2020; Accepted: 20 July 2020; Published: 24 July 2020

Abstract: Background: *Anacardium occidentale* L. is a medicinal plant with powerful anti-oxidative and anti-inflammatory properties. Acute inflammatory events cause tissue alterations, decrease of anti-oxidative endogenous enzymes such as superoxide dismutase, catalase and glutathione, neutrophils infiltration, increase in the activities of myeloperoxidase, malondialdehyde, and pro-inflammatory release. Methods: Paw edema was induced by subplantar injection of carrageenan into the right hind paw in rats, but 30 min before a group of animals were orally treated with 100 mg/kg of cashew nuts to evaluate the anti-inflammatory and anti-oxidative response. Results: In the present work, we found that (1) cashew nuts reduced the development of carrageenan-induced paw edema limiting the formation of edema and pain; (2) cashew nuts ameliorated the diminutions of the anti-oxidative enzymes caused by carrageenan injection; (3) cashew nuts decreased myeloperoxidase malondialdehyde activity induced by carrageenan; and (4) cashew nuts acted by blocking pro-inflammatory cytokines response and nitrate/nitrite formation stimulated by carrageenan injection. Conclusions: The mechanisms of anti-inflammatory and analgesic effects exerted by cashew nuts were relevant to oxygen free radical scavenging, anti-lipid peroxidation, and inhibition of the formation of inflammatory cytokines.

Keywords: paw edema; cashew nuts; antioxidant; inflammation; polyphenols; analgesic

1. Introduction

Inflammation is the first physiological response to tissue injury involving a complex cascade of reactions which can be provoked by numerous agents such as toxic compounds, microbes, etc., [1,2].

The changes that happen during acute inflammatory event have physiologically functions in controlling infection and restoring tissue to its normal state. The acute inflammatory state is generally composed of four sub-events distinctly into: (1) Exudation of fluid that helps deliver plasma proteins to sites of damage; (2) infiltration of neutrophils that leads to remove pathogens and cellular fragments; (3) vasodilation that the delivery of necessary proteins and cells (like exudation) and increasing tissue temperature; (4) pain and loss of function help to enforce rest and lower the risk of further tissue damage [3].

When acute inflammatory response was controlled, the result is the elimination of the infectious agents followed by a resolution and repair phase [3]. However, when the inflammation is uncontrolled it can be harmful to health [4–6]. Main events during the inflammatory state involve nitric oxide (NO) imbalance, lipid peroxidation, cytokines release, and maybe the most important, neutrophil-derived reactive oxygen species (ROS) formation [7].

The instability of free radicals is fundamentally the result of the loss of an electron that leads to an heightened reactivity and to a constantly "steal" electrons from other molecules starting a dangerous chain reaction called "free radical damage" [8]. Principal targets of these "steal" are proteins, lipids, and DNA/RNA, and all these modifications in different and several molecules may increase the chances of mutagenesis. In fact, ROS/RNS overproduction over a prolonged period of time can cause serious injury of the cellular structure and functions. For this reason, it is mandatory to remove it quickly [9]. Free radicals are important mediators that initiate inflammatory processes and, consequently, their neutralization by antioxidants and radical scavengers can attenuate inflammation. In order to minimize the damage caused by free radicals, the organism utilizes several enzyme such as superoxide dismutase (SOD) and catalase (CAT) and cofactor such as glutathione (GSH) [10,11]. Often, however, the physiological endogenous response put in place by the antioxidant enzymes may not be sufficient to limit ROS production [12].

Actually, the primary treatment during acute inflammatory event is the use nonsteroidal anti-inflammatory drugs (NSAIDs), in particular to limit neutrophil migration and oxygen free-radical generation, but several studies demonstrated that long-term use could lead to a lot of side effects, such as cardiovascular and gastrointestinal complications [13,14].

For this reason, it is mandatory to find new molecules to counteract novel drugs for treatment of acute inflammation and pain.

In the past decades, there has been a growing interest in studying and quantifying the antioxidant and anti-inflammatory constituents of vegetables in terms of their potential health functionality through action against inflammatory conditions [15,16]. The use of plant with known anti-inflammatory and/or antioxidant properties can be of great significance in therapeutic anti-inflammatory treatments. In particular, plants or fruit or nuts, rich in phenolic compounds are known for their wide ranges of biological activities, including anticancer, antibacterial, antioxidant, antidiabetic, and anti-inflammatory properties which could constitute an alternative in therapeutics [17].

Numerous studies, has been carried out on nuts demonstrating that a diet enriched with walnuts decreases serum cholesterol levels compared to a standard healthy diet [18]. By definition, tree nuts are dry fruits with one seed in which the ovary wall becomes hard at maturity. One of the most popular edible tree nuts is cashews (*Anacardium occidentale* L.) [18].

Cashew nuts are rich of unsaturated fatty acids (UFAs) such as oleic (ω-9) and linoleic (ω-6) acid, flavonoids, anthocyanins and tannins, fiber, folate and tocopherols [19–23].

The proposal of nuts as cardio-protective foods was supported from both epidemiological observations suggesting a consistent inverse association between nut intake and development of heart disease and numerous short-term clinical trials that showed beneficial effects of nut intake on the lipid profile [24–28]. Additionally, recent studies proved that the use of cashew nuts (*Anacardium occidentale* L.) can modulate the effects of several chronic inflammatory state such as colitis, degenerative joint disease, dyslipidemia, and others [29–34].

Its anti-inflammatory and anti-oxidative activities is probably due to the inhibition of the biosynthesis of inflammatory mediators by blocking the activities of 5-lipoxygenase (5-LOX) or cyclooxygenase 2 (Cox-2) which makes it a promising treatment for different inflammatory diseases [35,36].

However, until today, nobody evaluated its effects during acute inflammatory events.

Carrageenan-induced paw edema is a common murine experimental model used for the study of new compounds during the acute phase of inflammation [37]. As well-known, following injury induced by carrageenan, there is cell infiltration, mainly neutrophils, that contributes to the inflammatory response by producing myeloperoxidase and pro-inflammatory cytokines [4,6,30,38,39]. Moreover, a critical role during the development of the inflammatory state is the lipid peroxidation and the imbalance between ROS production and anti-oxidant enzyme activities [4,6,30,38,39]. With this background in mind, we used this consolidated experimental model to evaluate for the first time the analgesic, anti-inflammatory, and anti-oxidant effects of cashew nuts.

2. Materials and Methods

2.1. Animals

Male rats (Sprague-Dawley (200–230 g, Envigo, Milan, Italy)) were used throughout. The University of Messina Review Board for animal care (OPBA) approved the study. All animal experiments agree with the new Italian regulations (D. Lgs 2014/26), EU regulations (EU Directive 2010/63), and the ARRIVE guidelines.

2.2. Carrageenan-Induced Paw Edema

After anesthesia with 5.0% isoflurane in 100% O_2 rats were subjected to a subplantar injection of CAR (0.1 mL/rat of a 1% suspension in saline) with a 27-gauge needle into the right hind paw, as described previously by Morris and Britti [40,41]. The animals were sacrificed after 6 h post CAR-injection by isoflurane overdose. All analyses were performed in a blinded manner of experimental groups [42].

2.3. Experimental Groups

Rats were randomly divided into the following groups:

(1) CAR + vehicle (saline): rats were subjected to CAR-induced paw edema (n = 10);

(2) CAR + cashew nuts (100 mg/kg): rats were subjected to CAR-induced paw edema and cashew nuts (100 mg/kg) was administered 30 min before CAR (n = 10);

(3) The sham-operated group underwent the same surgical procedures as the CAR group, except that saline or drugs were administered instead of CAR (n = 10 for all experimental groups).

The tested dose was chosen based on previous studies performed in our laboratories [30]. After sacrifice, paw tissue and blood were collected for histological and biochemical analysis.

In another sets of experiments (n = 6 for each group) we analyzed ROS production and 5-LOX/COX pathways.

2.4. Assessment of CAR-Induced Paw Edema

Edema was assessed as previously described [40]. In short, the volume of the paw was measured with a plethysmometer (Ugo Basile, Comerio, Italy) immediately before carrageenan was injected and for 6 h at hourly intervals subsequently. For each animal, edema was expressed as increase in paw volume (mL) after CAR injection relative to pre-injection value.

2.5. Pain-Related Behavioral Analysis in the CAR-Induced Inflammation

To evaluate the analgesic effects of cashew nuts we made a plantar and Von Frey tests. Briefly, during the plantar test we analyzed the hyperalgesic response to heat at different time point using a Basile Plantar Test (Ugo Basile, Varese, Italy) with a cut-off latency of twenty seconds to prevent tissue injury. A mobile unit containing a high-intensity projector bulb was located to carry on a thermal stimulus directly to a single hind paw from beneath the chamber. The withdrawal latency period of injected paws was determined with an electronic clock circuit and thermocouple. Results are expressed as paw withdrawal latencies [41,43]. Additionally, Von Frey test (BIO-EVF4, Bioseb, Vitrolles, France) was made. The device encloses a force transducer furnished with a plastic tip. The force applied was measured when pressure is applied to the tip. The tip was applied to the hind leg's plantar region, and an increasing force was exerted upwards before the paw was extracted. The withdrawal threshold was defined as the force, expressed in grams, at which the rats removed the paw [41,44,45].

2.6. Myeloperoxidase (MPO) and Malonaldehyde (MDA) Activity

As previously described for MPO evaluation, paw tissues were homogenized in 0.5 percent hexadecyltrimethyl-ammonium bromide dissolved in 10 mM potassium phosphate buffer (pH 7.0) and centrifuged at 20,000× *g* at 4 °C for 30 min. A supernatant aliquot had been allowed to react with a 1.6 mM tetramethylbenzidine/0.1 mM H_2O_2 solution. The rate of absorbance shift was measured at 650 nm, using a spectrophotometer. MPO activity was defined as the amount of enzyme degrading 1 mM of peroxide at 37 °C within 1 min, and expressed in units per gram of wet tissue weight [46,47]. Additionally, for MDA analysis, paw tissues, collected at the end of experiment, were homogenized in 1.15% KCl solution. An aliquot of the homogenate was added to a reaction mixture containing sodium dodecyl sulfate (SDS), acetic acid (pH 3.5), thiobarbituric acid, and distilled water. Samples were then boiled and centrifuged. The supernatant's absorbance was measured at 650 nm using spectrophotometry [46–48].

2.7. Determination of Nitrite/Nitrate Concentration in Paw

Levels of nitrite/nitrate production in the paw tissue were determined as previously described by Costantino et al. [47]. Briefly, at the end of experiment, paws were cut and centrifuged to recover a sample of the edematous fluid. Blood was separated from the fluid sample and nitrite + nitrate (NOx) production, an indicator of NO synthesis, was measured [47]. Concentrations of nitrate were determined by comparison of regular sodium nitrate solutions prepared in saline solution at the OD550.

2.8. Evaluation of Cytokines and Antioxidant Enzymes in Blood

TNF-α, IL-6, IL-1β, and IL-10 levels from each sample were measured in duplicate with highly sensitive rat Elisa kit according to manufacturer's instructions (R&D Systems, Minneapolis, MN, USA) [49]. Additionally, also the levels of SOD, GSH, and CAT were assayed in blood according to manufacturer's instructions (Cusabio Biotech Co., Ltd, Wuhan, Hubei, China) [50–53].

2.9. Histological Examination of the CAR-Inflamed Hind Paw

For histological examination hematoxylin/eosin (H/E) was made and observed blinded to the treatment protocol. Briefly, paw tissues were taken at the end of experiment, and were dehydrated, embedded in Paraplast, and cut into sections of 7 μm and observed under microscopy (Leica DM7, Milan, Italy). The gradation of inflammation was estimated according a score based on 5 point: none, mild, mild/moderate, moderate, moderate/severe, and severe inflammation [54,55].

2.10. Cashew Nuts Nutritional Composition

Nutritional composition of Cashew kernel samples (*Anacardium occidentale* L.) obtained from Burkina Faso, a landlocked country in West Africa, was previously detected [30]. Briefly, 100 g of cashew kernel samples containing moisture 4.86 g, protein 21.01 g, lipids (total) 44.70 g, dietary fiber (total) 3.86 g, sugars (total) 32.80 g, ash 2.68 g, and total phenols 69.64 mg.

2.11. Estimation of Oxidant Levels

At the end of the experiment, through the dichlorodihydrofluoresceindiacetate (H_2DCFDA) staining method we measured the intracellular oxidant levels as previously described [56,57]. Briefly, we dissolved H_2DCFDA probes (Invitrogen Corporation; Carlsbad, CA, USA) in a solution of ethanol with a final concentration of 12.5 mM and we kept it at −80 °C in the dark. Before use, the solution was diluted with potassium phosphate buffer with a final concentration of 125 μM. To obtain the fluorescence reactions, 96-well black microplates were loaded with potassium phosphate buffer to a concentration of 152 μM/well. Then 8 μL diluted tissue homogenate and 40 μL (152 μM dye) were added to get a final concentration of 25 μM. The variation in fluorescence intensity was monitored every 5 min for 30 min with excitation and emission wavelengths set at 485 nm and 538 nm [58].

2.12. Western Blots Analysis for 5-LOX and Cox-2

Western blot examination on cytosolic fraction of the paw tissue was prepared as previously described [59]. Membranes were incubated with anti-5-LOX (1:1000) (Santa Cruz Biotechnology, Heidelberg, Germany), anti-Cox-2 (1:1000) (Santa Cruz Biotechnology, Heidelberg, Germany), and β-actin (1:500) (Santa Cruz Biotechnology, Heidelberg, Germany) for the standardization. Signals were identified with enhanced chemiluminescence (ECL) detection system reagent and the relative expression of the protein bands was measured by densitometry with BIORAD ChemiDocTM XRS+software (Bio-rad, Milan, Italy). A representation of blot signals were imported to analysis software (Image Quant TL, v2003).

2.13. Reagents

All other materials were purchased from Sigma-Aldrich Co. Stock solutions were prepared in nonpyrogenic saline (0.9% NaCl, Baxter Healthcare Ltd., Thetford, Norfolk, UK).

2.14. Data Analysis

All values are expressed as mean ± standard error of the mean of N observations. For in vivo experiments, N represents the number of animals. For experiments involving histology, the photos shown are demonstrative at least three experiments performed on different experimental days on tissue sections collected from all animals in each group. The results were analyzed by two-way ANOVA when the effect of the treatment was investigated in time-dependent mode or by one-way ANOVA when the means of two or more samples were analyzed. All analysis were followed by a Bonferroni post-hoc test for multiple comparisons. In all statistical studies GraphPad Software Prism 8 (La Jolla, CA, USA) was used. A p value of less than 0.05 was considered significant. # $p < 0.05$ vs CAR; ## $p < 0.01$ vs CAR; ** $p < 0.01$ vs sham; *** $p < 0.001$ vs sham.

3. Results

3.1. Effect of Cashew Nuts on CAR-Induced Inflammation and Pain

One of the first sign of intraplantar injection of CAR, was the increase in paw volume in a time-dependent way (Figure 1A) measured at 8 different set point from 0 (time when the experiment started) to 6 h (time when experiment ended). The increase in paw volume leads to pain that was assessed by the development of thermal hyperalgesia (Figure 1B) and mechanical allodynia (Figure 1C). In our study, we found that oral treatment with Cashew nuts at the dose of 100 mg/kg given 30 min before CAR, showed a reduction of the volume of rat paw significantly at 6 h post-CAR as well as a decrease in pain showing an inflammatory activity and algesic response.

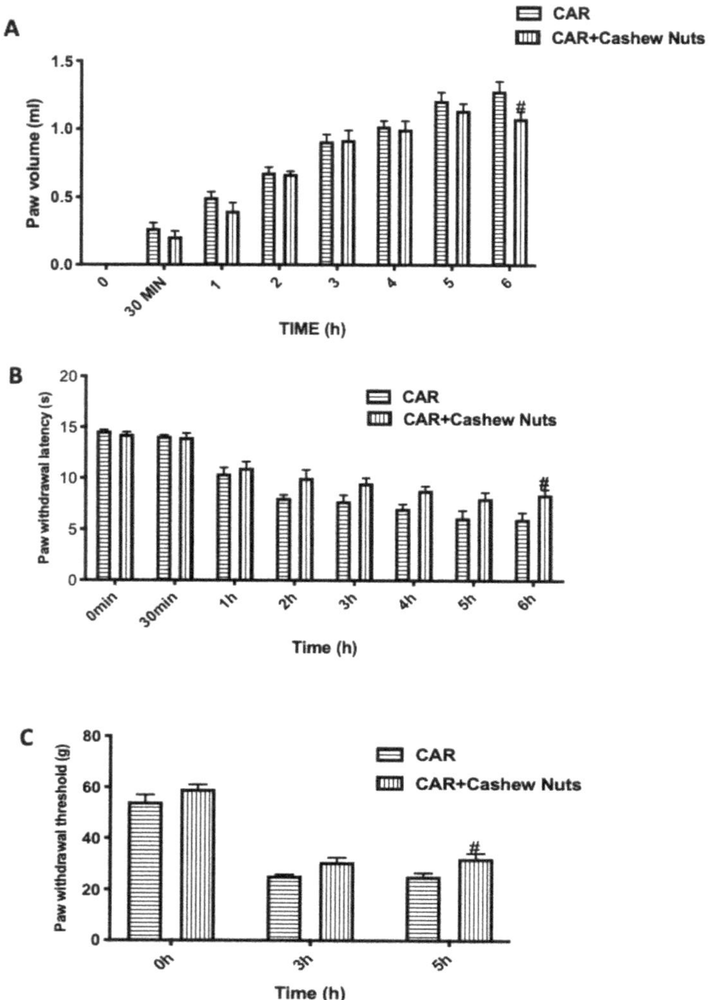

Figure 1. Evaluation of the effects of cashew nuts on carrageenan (CAR)-induced inflammation and pain. Paw edema was induced by subplantar injection of CAR. Paw volume was measured before the subplantar injection and hourly to 6 h. The edema volume is the difference in the paw volume at each time-point and the basal paw volume. Hyperalgesia and mechanical allodynia were assessed at the time points indicated with plantar and Von Frey tests. Cashew nuts administration shows significant improvement in the treatment of inflammation and pain. See materials and methods for further details. Paw volume (**A**); plantar test (**B**); Von Frey test (**C**). # $p < 0.05$ vs. CAR.

3.2. *Effects of Cashew Nuts on Histological Alteration after CAR Injection*

At the end of experiment, a histopathological study was made in paw tissue by H/E examination. A microscopic study of the paw biopsies in CAR group showed edema formation and cellular diffuses infiltration with serious alteration in tissue architecture (Figure 2B and inset B1, see histological score D).

Cashew nuts administration, at the dose of 100 mg/kg, was able to slightly reduce histological injury in paw tissues of rats (Figure 3C, and inset C1, see histological score D) counteracting both cellular infiltration and edema formation. Sham rats showed a normal architecture of paw tissue (Figure 2A and inset A1).

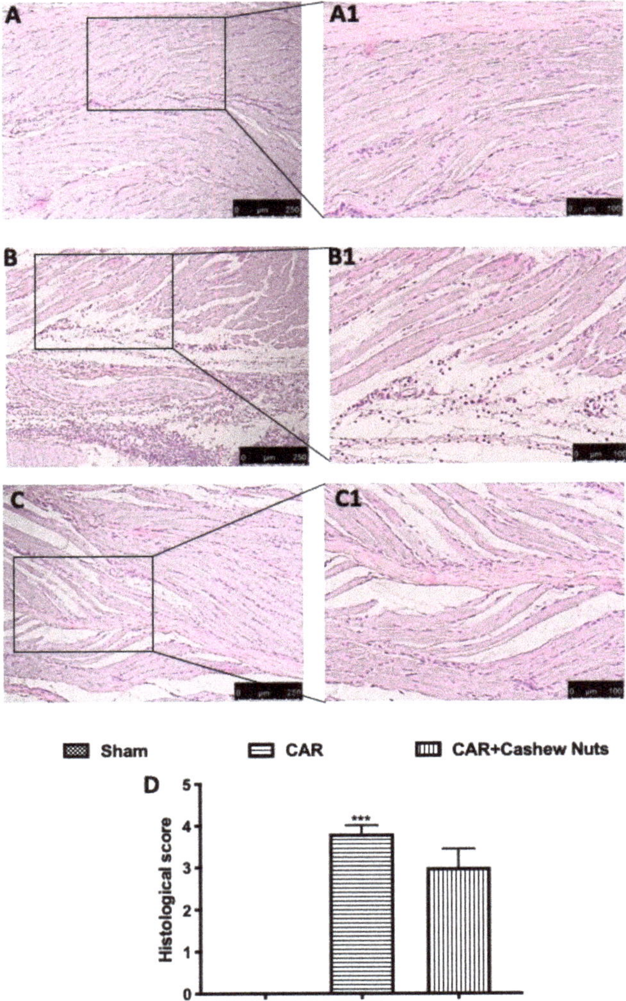

Figure 2. Histological evaluation of paw tissue after cashew nuts treatment following CAR-injection. Investigation of tissue injury of the animals subjected to CAR, was made by H/E staining. CAR group shows a loss of the physiological architecture compared to sham. Cashew nuts administered 30 min before CAR-injection shows reduction in infiltrating cells and edema formation compared to vehicle. Histological score (**D**). See materials and methods for further details. Sham (**A**,**A1**); CAR (**B**,**B1**); cashew nuts (**C**,**C1**). *** $p < 0.001$ vs. sham.

3.3. Effects of Cashew Nuts on Nitrate/Nitrite, MPO, and MDA Activity in CAR-Injured Rats

The development of histological damage was associated with a statistically significant increase in MPO activity (Figure 3A), as an indicator of neutrophil infiltration, and MDA (Figure 3B), marker of lipid peroxidation. In our study, we found that cashew nuts administered 30 min before CAR injection was able to reduce not only MPO activity by inhibiting neutrophil recruitment but also MDA levels.

Additionally, during inflammatory events, NO played a critical role in tissue injury [60]. For this reason, nitrite/nitrate levels were measured in exudate of paw tissues to regulate the expression of nitric oxide (Figure 3C). Oral treatment with cashew nuts at the doses of 100 mg/kg were able to significantly decrease also nitrite/nitrate levels.

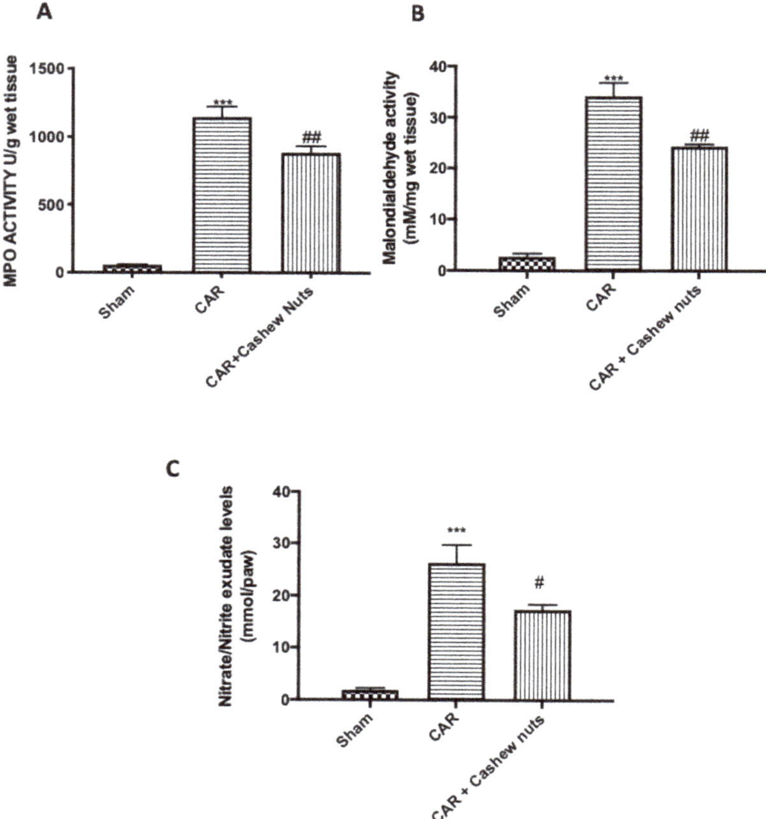

Figure 3. Effects of cashew nuts on nitrate/nitrite, MPO, and MDA activity after CAR-injection. As a consequence of neutrophils infiltration and lipid peroxidation MPO and MDA were assessed. Additionally, considering the key role played by NO during inflammatory events, we analyzed also levels in nitrite/nitrate in paw exudate. Carrageenan induces a significant increase in all parameter taken in consideration. On the other hands, oral treatment with cashew nuts at the dose of 100 mg/kg significantly reduced MPO, MDA, and nitrate/nitrite CAR-induction. MPO (**A**); MDA (**B**); nitrate/nitrite (**C**). See materials and methods for further details. # $p < 0.05$ vs. CAR; ## $p < 0.01$ vs. CAR; *** $p < 0.001$ vs. sham.

3.4. *Effects of Anacardium occidentale L. on Cytokines Production*

As well-know, cytokines exert important effects during inflammatory events, for this reason they can be used as biomarkers in indicating or monitoring inflammation and its progress [61]. In our study, we found a significant increase compared to sham animals in serum pro-inflammatory cytokine levels in the group subjected to CAR (Figure 4A–C) as well as, a significant decrease in IL-10 production was detected (Figure 4D). Cashew nuts administration given 30 min before CAR-injection at the dose of 100 mg/kg was able to significantly decrease pro-inflammatory cytokines production, and on the other hand, significantly increase IL-10 release.

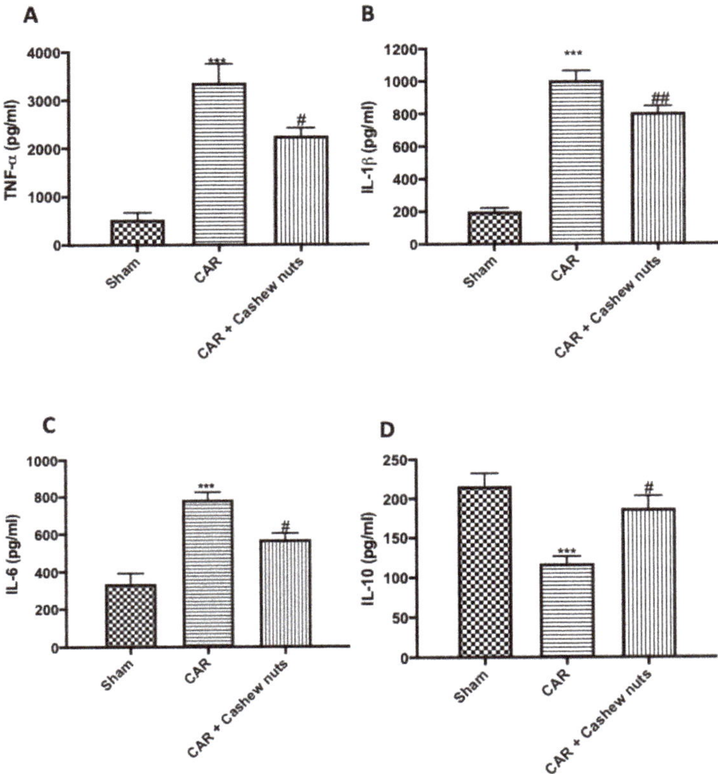

Figure 4. Effects of cashew nuts on cytokines production CAR-induced. Investigation of cytokines is mandatory when talking of inflammatory condition. As predicted, during CAR-caused inflammation, we observed a significant increase in TNF-α, IL-1β, IL-6 levels; vice versa we observed a significant decrease in IL-10 production. Cashew nuts administration 30 min before CAR-injection at the dose of 100 mg/kg significantly reduced pro-inflammatory cytokines expression and, vice versa, increased anti-inflammatory expression of IL-10. TNF-α (**A**), IL-1β (**B**), IL-6 (**C**), and IL-10 (**D**). See materials and methods for further details. # $p < 0.05$ vs. CAR; ## $p < 0.01$ vs. CAR; *** $p < 0.001$ vs. sham.

3.5. *Effect of Cashew Nuts on CAR-Induced Oxidative Stress*

Considering a variety of oxidants and free radicals that are implicated in the pathogenesis of inflammatory process and considering that dietary components also may contribute to the antioxidant defense either by providing redox active compounds that can directly scavenge or neutralize free radicals, we investigated the oxidative stress through H_2DCFDA probes and ELISA kits. As supposed, after CAR-injection, we observed a very important increase of ROS production (Figure 5A) and, on the other hand, a decrease in SOD (Figure 5B), GSH (Figure 5C), and CAT (Figure 5D) activity compared to sham animals. After cashew nuts treatment, decrease in oxidative stress and increase in the activity of SOD, GSH, and CAT were observed.

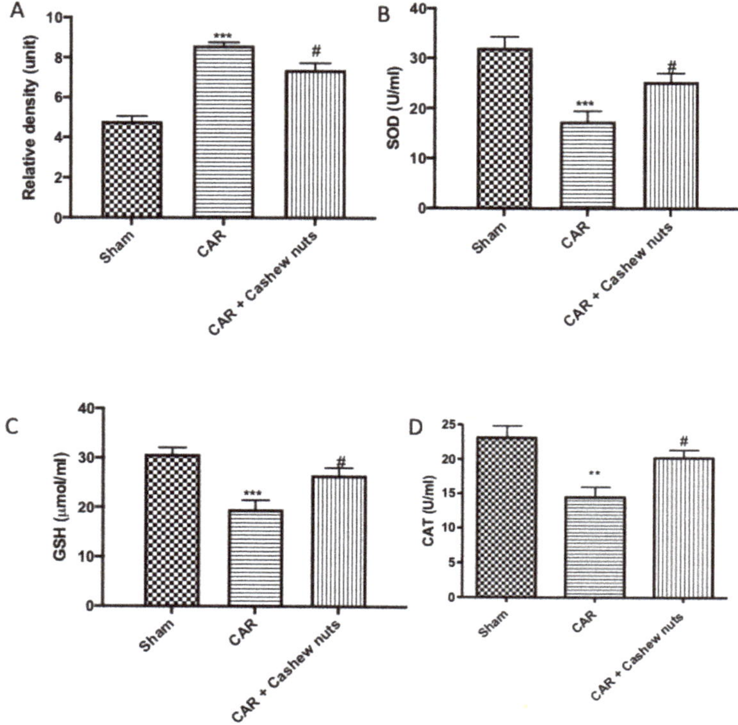

Figure 5. Anti-oxidant effects of cashew nuts after CAR-induction. Oxidative stress is a major component of acute inflammatory condition. First line of defense against free radical production (**A**) is represented by SOD (**B**), CAT (**C**), and GSH (**D**). Administration of cashew nuts at the dose of 100 mg/kg significantly increased the activity of the anti-oxidant enzymes SOD, CAT, and GSH which had been significantly reduced by carrageenan injection. SOD (**A**), CAT (**B**), and GSH (**C**). See materials and methods for further details. # $p < 0.05$ vs. CAR; ** $p < 0.01$ vs. sham; *** $p < 0.001$ vs. sham.

3.6. Effect of Cashew Nuts on CAR-Induced 5-LOX and Cox-2 Expressions

One of the most important role, during inflammatory events, is done by the mediators of the arachidonic acid cascade from COX and LOX pathways, that, as well-known, are modulated by flavonoids [62]. For better understanding the molecular mechanism of cashew nut, we investigate by Western blots, 5-LOX and Cox-2 expressions. As speculated, after CAR injection we found a significant increase in both expression, compared to sham animals. After oral treatment with cashew nuts at the dose of 100 mg/kg, we found a significant decrease in both (Figure 6A,B).

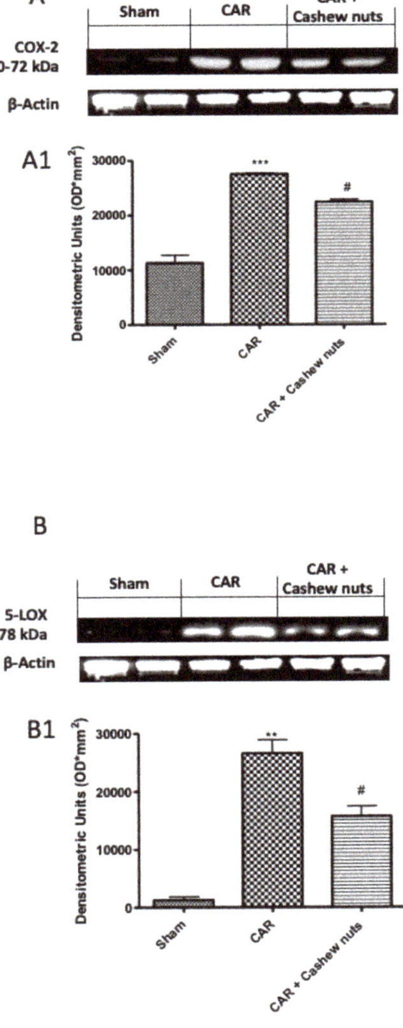

Figure 6. Effects of cashew nuts on 5-LOX and Cox-2 expressions. The inhibition of the biosynthesis of inflammatory mediators through the modulation of the activities of 5-LOX and/or Cox-2 considered as a promising approach to treat inflammatory diseases. For these reason we investigated the effect of cashew nuts treatment by Western blots on both enzymes. We found a significant increase after CAR injection, compared to sham, that was significantly decreased in 5-LOX as well as in Cox-2 expression. Representative western blot for Cox-2 (**A**) and LOX-5 (**B**) and respectively densitometric analysis (**A1**,**B1**). See materials and methods for further details. # $p < 0.05$ vs. CAR; ** $p < 0.01$ vs. sham.

4. Discussion

Inflammatory condition are universally identified as one of the most important causes of co-morbidity across the population [63]. When under control, inflammation is a defensive response of a body against invasion by the foreign bodies [64]. An acute inflammatory response is, for definition, represented by redness, heat, swelling, pain, and the loss of function [65,66]. The protective effects of inflammatory cascade and potential for tissue destruction are usually balanced in normal state, whereas, when uncontrolled, inflammation may arise in numerous diseased states like rheumatoid arthritis, multiple sclerosis, inflammatory bowel disease, and many others [4,67–70].

Considering doubts about the side effects of repeated use of synthetic chemicals, there is growing interest in the medicinal uses of natural chemicals and their derivatives as healthier replacements, such as functional products or as nutraceuticals [71].

Until today, the most advantageous therapies for the management of inflammatory state is based on the use of NSAIDs. Unfortunately, the chronic use of NSAIDs is connected with a broad spectrum of side effects ranging from gastrointestinal problems to kidney toxicity [64]. Toxicity and reappearance of signs is a major problem related to currently available synthetic drugs [64]. For these reasons the development of safer anti-inflammatory agents remains to be a subject of great interest [16]. Improvement of anti-inflammatory drugs derived from natural sources is the rational and productive strategy toward the cure of inflammatory ailments [72,73].

The search for natural molecules with antioxidant and anti-inflammatory activities has increased extremely over the past decades because the natural products are safe, efficacious, biocompatible, and cost-effective alternatives to treat inflammatory diseases [64].

In particular, researchers focused the attention on the possibilities that dietary daily intake of sources of antioxidants may offer a cost-effective approach to treating most linked pathways associated with inflammation: the oxidative stress [62–64].

Nuts would be a promising alternative in reducing oxidative damage, owing to their secondary metabolites richness such as polyphenols, flavonoids, tannins, terpenoids, and anthraquinones [74–78]. The most accredited hypothesis may be that polyphenolic components of dietary plants may increase the endogenous antioxidant defense potential and thus modulate cellular redox state, additionally, and sequentially is apt to consider how polyphenols may modulate the redox system and its components in a cell during normal and pathophysiological conditions [78].

Anacardium occidentale L. is a Brazilian plant that is usually consumed in nature and used in folk medicine with high value edible nut and a source of carbohydrates, proteins, phosphorous, iron, zinc, magnesium, fibers, and fatty acids [79]. Actually, it is officially listed in the National System of Medicinal Plants and Herbal Medicine of the specific Italian health system for medicinal purposes [30]. Also, it is a tree rich in anthocyanins, carotenoids, flavonoids, and other polyphenols as well as mineral components [30]. In recent years, it was used for its antioxidant, antigenotoxic, antimutagenic, antiulcerogenic, anti-inflammatory, antibacterial, antifungal, and larvicides activities [30,80–87]. In another studies made in our laboratory we demonstrated, in a murine model of colitis, that cashew nuts treatment, was able to alleviate the clinical signs of colon damage as well as oxidative stress, inflammation, and iNOS, ICAM-1, and P-selectin expressions [31].

Until today, nobody demonstrated the effect of cashew nuts treatment in an acute inflammatory model.

Carrageenan-induced paw edema is a very sensitive and reproducible test used in the screening of new molecules with anti-inflammatory activities [88]. Carrageenan-induced inflammation causes an acute and local inflammatory response that is advantageous for detecting orally active anti-inflammatory agents; therefore, it has significant prognostic value for anti-inflammatory agents acting through mediators of acute inflammation [88].

First step of acute inflammatory response is characterized by edema often formed because of exudation of fluid and plasma proteins [89]. In our work we found that edema formation was reduced significantly at 6 h post-CAR.

Additionally, carrageenan-induced paw edema leads to sensitization of primary sensory neurons, essentially event to inflammatory pain [90]. In humans, this nociceptor sensitization usually leads to clinical conditions known as hyperalgesia defined as an increased response to a painful stimulus or allodynia described as pain evoked by non-noxious stimuli. In our study, we have proven that oral administration of cashew nuts 30 min before CAR, was in grade to reduce hyperalgesia and allodynia significantly at 6 h post-CAR.

Edema and pain in the hind paw of animals as a result of CAR-induced inflammation usually limit their motility and cause trouble in using their hind paw.

During a CAR-induced acute inflammation event, paw tissue loses normal muscle architecture and shows important amassing of infiltrating inflammatory cells and increased inter-fiber space during microscopic observation [6]. During our study, we found that oral administration of cashew nuts decreased infiltrating inflammatory cell as also demonstrated by the significant decrease of MPO assay.

One of the most dangerous consequences of uncontrolled oxidative is cell injury caused by ROS [91]. Since it is complex measuring the free radicals directly in vivo, it is common in use to carry out the quantification of different molecules which can react with these free radicals, such as for example lipids [92]. Considering that lipid peroxides are very reactive compounds, they appear to quickly degrade in a range of sub-products. MDA is one of the most known secondary products of lipid peroxidation, and it is the most used as marker of cell membrane injury [92]. Additionally, another important biomarker, in the pathogenesis of inflammation, is NO that is produced by inducible nitric oxide synthase during the formation of l-citrulline from l-arginine [93]. In our studies we found that cashew nuts at the dose of 100 mg/kg was in grade to significantly diminished lipid peroxidation and NO formation.

Conversion of arachidonic acid to biologically active leukotrienes, a potent mediators of inflammatory reactions is a key point when looking for new molecules that can inhibit inflammatory events. Plant domain is a valuable source for new 5-LOX and dual 5-LOX/COX-inhibitors, because they are fundamentally rich in flavonoids compound [36]. In past, Wagner and colleagues demonstrate that *Anacardium occidentale* L. strongly inhibited prostaglandin synthase, but nobody evaluated these activities in vivo [94]. Our results demonstrate that cashew nuts decrease inflammation probably across to the modulation of 5-LOX and Cox-2.

Inflammatory cascade activates cells and induce production of inflammatory cytokines, such as IL-1β, IL-6, and TNF-α. These molecules can potentially serve as biomarkers for diseases diagnosis, prognosis, and therapeutic management [95–97]. In these work we demonstrated that cashew nuts could inhibit the production of the cytokines involved in carrageenan-induced paw edema.

To counteract ROS formation, cells have developed a complex antioxidant defense system that consists of several enzyme systems involved in the conversion of ROS to less reactive molecules such as O_2 and water [98,99]. The first line of antioxidants defense was composed by SOD, CAT, and GSH.

SOD, is the most powerful antioxidant enzyme in the cell that catalytically converts superoxide radical or singlet oxygen into hydrogen peroxide and molecular oxygen; CAT catalyzes the degradation or reduction of hydrogen peroxide to water and molecular oxygen, completing the detoxification process initiated by SOD, and GSH is the most abundant intracellular non-protein thiol in cells with the functions of removing potentially toxic electrophiles and metals protecting cells from toxic oxygen products [100,101].

In our studies, carrageenan significantly increased ROS production and reduced GSH, SOD, and CAT levels, but this increase/decrease was counteracted by the oral treatment of cashew nuts, suggesting that the inhibition of carrageenan-induced oxidative stress may also explain the analgesic effect.

5. Conclusions

Inflammation studies have been one of the main hubs of global science study. The inflammation is known to be correlated with oxidative processes, mainly because they share some common pathways. Since oxidative stress is common in several degenerative disease, it has been supposed that dietary antioxidants may explain a very important protective effect. Nuts are a main source of antioxidants in the diets worldwide. Nuts are high in antioxidant, in fiber, and in beneficial unsaturated fats and low in saturated fats. Nuts are usually eaten as a snack or added to food to provide both nutrients and bioactive antioxidants. In conclusion, in our work, we demonstrated for the first time that cashew nuts consumption not only brings benefits in experimental mouse models of chronic inflammation, but also in acute inflammation events. In particular was in grade to significantly counteract edema formation and consequently carrageenan-related pain. In addition, oral treatment with 100 mg/kg of cashew nuts significantly decreased MPO and MDA activity as well as nitrate/nitrite formation. Moreover, in agreement with our previous study, we demonstrated for the first time, that cashew nuts administration was able to significantly improve endogenous antioxidant activity, limiting pro-inflammatory cytokines release. Its beneficial effect is probably due to the high content of phenols that mediate activation of 5-LOX COX pathways. Considering all the benefits brought by cashew nuts, its usual consumption in the diet could be considered in order to reduce the events of cellular oxidative stress. Taken together, our result fits with previous study in which it was demonstrated that cashew nuts possess interesting anti-inflammatory, anti-oxidative, and analgesic activities that will be of interest for further investigation.

Author Contributions: Conceptualization: M.C.; formal analysis: T.G. and R.C.; investigation: A.F.P.; methodology: R.S. and D.I.; project administration: S.C.; supervision: R.D.P.; validation: E.G. and M.S.; writing—original draft, R.F. and R.D.; writing—review and editing: G.M. All authors have read and agreed to the published version of the manuscript.

Funding: This research received no external funding.

Acknowledgments: We would like to thank Salma Seetaroo from Ivorienne de Noix de Cajou S.A. of Cote d'Ivoire for providing the cashew kernel samples from West Africa.

Conflicts of Interest: The authors declare no conflict of interest.

References

1. Chen, L.; Deng, H.; Cui, H.; Fang, J.; Zuo, Z.; Deng, J.; Li, Y.; Wang, X.; Zhao, L. Inflammatory responses and inflammation-associated diseases in organs. *Oncotarget* **2018**, *9*, 7204–7218. [CrossRef] [PubMed]
2. Adegbaju, O.D.; Otunola, G.A.; Afolayan, A.J. Anti-inflammatory and cytotoxic evaluation of extracts from the flowering stage of Celosia argentea. *BMC Complement Med. Ther.* **2020**, *20*, 152. [CrossRef] [PubMed]
3. Medzhitov, R. Origin and physiological roles of inflammation. *Nature* **2008**, *454*, 428–435. [CrossRef] [PubMed]
4. Cordaro, M.; Siracusa, R.; Impellizzeri, D.; R, D.A.; Peritore, A.F.; Crupi, R.; Gugliandolo, E.; Fusco, R.; Di Paola, R.; Schievano, C.; et al. Safety and efficacy of a new micronized formulation of the ALIAmide palmitoylglucosamine in preclinical models of inflammation and osteoarthritis pain. *Arthritis Res. Ther.* **2019**, *21*, 254. [CrossRef] [PubMed]
5. Peritore, A.F.; Siracusa, R.; Crupi, R.; Cuzzocrea, S. Therapeutic efficacy of palmitoylethanolamide and its new formulations in synergy with different antioxidant molecules present in diets. *Nutrients* **2019**, *11*, 2175. [CrossRef]
6. Paterniti, I.; Impellizzeri, D.; Cordaro, M.; Siracusa, R.; Bisignano, C.; Gugliandolo, E.; Carughi, A.; Esposito, E.; Mandalari, G.; Cuzzocrea, S. The Anti-inflammatory and antioxidant potential of pistachios (*Pistacia vera* L.) in vitro and in vivo. *Nutrients* **2017**, *9*, 915. [CrossRef]
7. Ahmad, T.; Shinkafi, T.S.; Routray, I.; Mahmood, A.; Ali, S. Aqueous extract of dried flower buds of *Syzygium aromaticum* inhibits inflammation and oxidative stress. *J. Basic Clin. Pharm.* **2012**, *3*, 323–327. [CrossRef]
8. Rao, B.G.; Kiran, P.M.; Raju, A.D.V. Investigation of antioxidant and anti-inflammatory activity of leaves of *Dalbergia paniculata* (Roxb). *Asian Pac. J. Trop. Med.* **2012**, *5*, 455–458. [CrossRef]
9. Hussain, T.; Tan, B.; Yin, Y.; Blachier, F.; Tossou, M.C.; Rahu, N. Oxidative stress and inflammation: What polyphenols can do for us? *Oxid. Med. Cell Longev.* **2016**, *2016*, 7432797. [CrossRef]
10. Lubrano, V.; Balzan, S. Enzymatic antioxidant system in vascular inflammation and coronary artery disease. *World J. Exp. Med.* **2015**, *5*, 218–224. [CrossRef]

11. Lopez-Torres, M.; Perez-Campo, R.; Rojas, C.; Cadenas, S.; Barja, G. Maximum life span in vertebrates: Relationship with liver antioxidant enzymes, glutathione system, ascorbate, urate, sensitivity to peroxidation, true malondialdehyde, in vivo H_2O_2, and basal and maximum aerobic capacity. *Mech. Ageing Dev.* **1993**, *70*, 177–199. [CrossRef]
12. Nathan, C. Points of control in inflammation. *Nature* **2002**, *420*, 846–852. [CrossRef] [PubMed]
13. Wu, C.X.; Liu, Y.; Zhang, J.C. Chronic intermittent hypoxia and hypertension: A review of systemic inflammation and Chinese medicine. *Chin. J. Integr. Med.* **2013**, *19*, 394–400. [CrossRef] [PubMed]
14. Sun, K.; Song, X.; Jia, R.; Yin, Z.; Zou, Y.; Li, L.; Yin, L.; He, C.; Liang, X.; Yue, G.; et al. Evaluation of analgesic and anti-inflammatory activities of water extract of *Galla chinensis* in vivo models. *Evid. Based Complement Alternat. Med.* **2018**, *2018*, 6784032. [CrossRef]
15. Menichini, F.; Tundis, R.; Bonesi, M.; Loizzo, M.R.; Conforti, F.; Statti, G.; De Cindio, B.; Houghton, P.J.; Menichini, F. The influence of fruit ripening on the phytochemical content and biological activity of *Capsicum chinense* Jacq. cv habanero. *Food Chem.* **2009**, *114*, 553–560. [CrossRef]
16. Mueller, M.; Hobiger, S.; Jungbauer, A. Anti-inflammatory activity of extracts from fruits, herbs and spices. *Food Chem.* **2010**, *122*, 987–996. [CrossRef]
17. Boutennoun, H.; Boussouf, L.; Kebieche, M.; Al-Qaoud, K.; Madani, K. In vivo analgesic, anti-inflammatory and antioxidant potentials of *Achillea odorata* from north Algeria. *S. Afr. J. Bot.* **2017**, *112*, 307–313. [CrossRef]
18. Ros, E. Health benefits of nut consumption. *Nutrients* **2010**, *2*, 652–682. [CrossRef]
19. Agila, A.; Barringer, S.A. Volatile profile of cashews (*Anacardium occidentale* L.) from different geographical origins during roasting. *J. Food Sci.* **2011**, *76*, C768–C774. [CrossRef]
20. de Melo, M.; Pereira, D.E.; Sousa, M.M.; Medeiros, D.M.F.; Lemos, L.T.M.; Madruga, M.S.; Santos, N.M.; de Oliveira, M.E.G.; de Menezes, C.C.; Soares, J.K.B. Maternal intake of cashew nuts accelerates reflex maturation and facilitates memory in the offspring. *Int. J. Dev. Neurosci.* **2017**, *61*, 58–67. [CrossRef]
21. Baptista, A.; Goncalves, R.V.; Bressan, J.; Peluzio, M. Antioxidant and antimicrobial activities of crude extracts and fractions of cashew (*Anacardium occidentale* L.), cajui (*Anacardium microcarpum*), and pequi (*Caryocar brasiliense* C.): A systematic review. *Oxid. Med. Cell Longev.* **2018**, *2018*, 3753562. [CrossRef] [PubMed]
22. Alexiadou, K.; Katsilambros, N. Nuts: Anti-atherogenic food? *Eur. J. Intern. Med.* **2011**, *22*, 141–146. [CrossRef]
23. Gomez-Caravaca, A.M.; Verardo, V.; Caboni, M.F. Chromatographic techniques for the determination of alkyl-phenols, tocopherols and other minor polar compounds in raw and roasted cold pressed cashew nut oils. *J. Chromatogr. A* **2010**, *1217*, 7411–7417. [CrossRef] [PubMed]
24. Stewart, R.A.H. Primary prevention of cardiovascular disease with a Mediterranean diet supplemented with extra-virgin olive oil or nuts. *N. Engl. J. Med.* **2018**, *379*, 1388. [CrossRef] [PubMed]
25. What you eat significantly impacts your heart health. A low-fat diet, plus more fruits, grains, nuts, fish and poultry instead of red meat, yields cardiovascular health benefits. *Duke Med. Health News* **2011**, *17*, 4–5.
26. Nash, S.D.; Nash, D.T. Nuts as part of a healthy cardiovascular diet. *Curr. Atheroscler. Rep.* **2008**, *10*, 529–535. [CrossRef]
27. Mukuddem-Petersen, J.; Oosthuizen, W.; Jerling, J.C. A systematic review of the effects of nuts on blood lipid profiles in humans. *J. Nutr.* **2005**, *135*, 2082–2089. [CrossRef]
28. Ros, E. Nuts and novel biomarkers of cardiovascular disease. *Am. J. Clin. Nutr.* **2009**, *89*, 1649S–1656S. [CrossRef]
29. Liu, C.M.; Peng, Q.; Zhong, J.Z.; Liu, W.; Zhong, Y.J.; Wang, F. Molecular and functional properties of protein fractions and isolate from cashew nut (*Anacardium occidentale* L.). *Molecules* **2018**, *23*, 393. [CrossRef]
30. Siracusa, R.; Fusco, R.; Peritore, A.F.; Cordaro, M.; D'Amico, R.; Genovese, T.; Gugliandolo, E.; Crupi, R.; Smeriglio, A.; Mandalari, G.; et al. The antioxidant and anti-inflammatory properties of *Anacardium occidentale* L. cashew nuts in a mouse model of colitis. *Nutrients* **2020**, *12*, 834. [CrossRef]
31. Fusco, R.; Siracusa, R.; Peritore, A.F.; Gugliandolo, E.; Genovese, T.; D'Amico, R.; Cordaro, M.; Crupi, R.; Mandalari, G.; Impellizzeri, D.; et al. The role of cashew (*Anacardium occidentale* L.) nuts on an experimental model of painful degenerative joint disease. *Antioxidants* **2020**, *9*, 511. [CrossRef] [PubMed]

32. Batista, K.S.; Alves, A.F.; Lima, M.D.S.; da Silva, L.A.; Lins, P.P.; de Sousa Gomes, J.A.; Silva, A.S.; Toscano, L.T.; de Albuquerque Meireles, B.R.L.; de Magalhaes Cordeiro, A.M.T.; et al. Beneficial effects of consumption of acerola, cashew or guava processing by-products on intestinal health and lipid metabolism in dyslipidaemic female Wistar rats. *Br. J. Nutr.* **2018**, *119*, 30–41. [CrossRef] [PubMed]
33. Dias, C.C.Q.; Madruga, M.S.; Pintado, M.M.E.; Almeida, G.H.O.; Alves, A.P.V.; Dantas, F.A.; Bezerra, J.K.G.; de Melo, M.; Viera, V.B.; Soares, J.K.B. Cashew nuts (*Anacardium occidentale* L.) decrease visceral fat, yet augment glucose in dyslipidemic rats. *PLoS ONE* **2019**, *14*, e0225736. [CrossRef] [PubMed]
34. Oliveira, A.S.; Nascimento, J.R.; Trovao, L.O.; Alves, P.C.S.; Maciel, M.C.G.; Silva, L.D.M.; Marques, A.A.; Santos, A.; Silva, L.A.; Nascimento, F.R.F.; et al. The anti-inflammatory activity of *Anacardium occidentale* L. increases the lifespan of diabetic mice with lethal sepsis. *J. Ethnopharmacol.* **2019**, *236*, 345–353. [CrossRef] [PubMed]
35. Schneider, I.; Bucar, F. Lipoxygenase inhibitors from natural plant sources. Part 2: Medicinal plants with inhibitory activity on arachidonate 12-lipoxygenase, 15-lipoxygenase and leukotriene receptor antagonists. *Phytother. Res.* **2005**, *19*, 263–272. [CrossRef] [PubMed]
36. Schneider, I.; Bucar, F. Lipoxygenase inhibitors from natural plant sources. Part 1: Medicinal plants with inhibitory activity on arachidonate 5-lipoxygenase and 5-lipoxygenase[sol]cyclooxygenase. *Phytother. Res.* **2005**, *19*, 81–102. [CrossRef]
37. Halici, Z.; Dengiz, G.O.; Odabasoglu, F.; Suleyman, H.; Cadirci, E.; Halici, M. Amiodarone has anti-inflammatory and anti-oxidative properties: An experimental study in rats with carrageenan-induced paw edema. *Eur. J. Pharmacol.* **2007**, *566*, 215–221. [CrossRef]
38. Gugliandolo, E.; D'Amico, R.; Cordaro, M.; Fusco, R.; Siracusa, R.; Crupi, R.; Impellizzeri, D.; Cuzzocrea, S.; Di Paola, R. Effect of PEA-OXA on neuropathic pain and functional recovery after sciatic nerve crush. *J. Neuroinflamm.* **2018**, *15*, 264. [CrossRef]
39. Petrosino, S.; Cordaro, M.; Verde, R.; Schiano Moriello, A.; Marcolongo, G.; Schievano, C.; Siracusa, R.; Piscitelli, F.; Peritore, A.F.; Crupi, R.; et al. Oral Ultramicronized palmitoylethanolamide: Plasma and tissue levels and spinal anti-hyperalgesic effect. *Front. Pharmacol.* **2018**, *9*, 249. [CrossRef]
40. Morris, C.J. Carrageenan-induced paw edema in the rat and mouse. *Methods Mol. Biol.* **2003**, *225*, 115–121. [CrossRef]
41. Britti, D.; Crupi, R.; Impellizzeri, D.; Gugliandolo, E.; Fusco, R.; Schievano, C.; Morittu, V.M.; Evangelista, M.; Di Paola, R.; Cuzzocrea, S. A novel composite formulation of palmitoylethanolamide and quercetin decreases inflammation and relieves pain in inflammatory and osteoarthritic pain models. *BMC Vet. Res.* **2017**, *13*, 229. [CrossRef] [PubMed]
42. Salvemini, D.; Wang, Z.Q.; Wyatt, P.S.; Bourdon, D.M.; Marino, M.H.; Manning, P.T.; Currie, M.G. Nitric oxide: A key mediator in the early and late phase of carrageenan-induced rat paw inflammation. *Br. J. Pharmacol.* **1996**, *118*, 829–838. [CrossRef] [PubMed]
43. Hargreaves, K.; Dubner, R.; Brown, F.; Flores, C.; Joris, J. A new and sensitive method for measuring thermal nociception in cutaneous hyperalgesia. *Pain* **1988**, *32*, 77–88. [CrossRef]
44. Peritore, A.F.; Siracusa, R.; Fusco, R.; Gugliandolo, E.; D'Amico, R.; Cordaro, M.; Crupi, R.; Genovese, T.; Impellizzeri, D.; Cuzzocrea, S.; et al. Ultramicronized palmitoylethanolamide and paracetamol, a new association to relieve hyperalgesia and pain in a sciatic nerve injury model in rat. *Int. J. Mol. Sci.* **2020**, *21*, 509. [CrossRef]
45. Fusco, R.; Siracusa, R.; D'Amico, R.; Peritore, A.F.; Cordaro, M.; Gugliandolo, E.; Crupi, R.; Impellizzeri, D.; Cuzzocrea, S.; Di Paola, R. Melatonin plus folic acid treatment ameliorates reserpine-induced fibromyalgia: An evaluation of pain, oxidative stress, and inflammation. *Antioxidants* **2019**, *8*, 628. [CrossRef]
46. Cuzzocrea, S.; Mazzon, E.; Esposito, E.; Muia, C.; Abdelrahman, M.; Di Paola, R.; Crisafulli, C.; Bramanti, P.; Thiemermann, C. Glycogen synthase kinase-3beta inhibition attenuates the development of ischaemia/reperfusion injury of the gut. *Intensive Care Med.* **2007**, *33*, 880–893. [CrossRef] [PubMed]
47. Costantino, G.; Cuzzocrea, S.; Mazzon, E.; Caputi, A.P. Protective effects of melatonin in zymosan-activated plasma-induced paw inflammation. *Eur. J. Pharmacol.* **1998**, *363*, 57–63. [CrossRef]
48. Impellizzeri, D.; Esposito, E.; Di Paola, R.; Ahmad, A.; Campolo, M.; Peli, A.; Morittu, V.M.; Britti, D.; Cuzzocrea, S. Palmitoylethanolamide and luteolin ameliorate development of arthritis caused by injection of collagen type II in mice. *Arthritis Res. Ther.* **2013**, *15*, R192. [CrossRef] [PubMed]

49. Nithya, S. Anti-inflammatory effect of Elettaria cardamom oil on carrageenan-induced paw edema using rats based on tumor necrosis factor α, interleukin 6, and interleukin 1 levels in serum. *Asian J. Pharm. Clin. Res.* **2018**, *11*, 207. [CrossRef]
50. Shukla, A.; Shukla, R.; Rai, G. Preliminary investigations on *Lens culinaris* Med. Seeds for anti-inflammatory and antioxidant properties. *Asian J. Pharm. Pharmacol.* **2017**, *3*, 29–31.
51. Parlar, A.; Arslan, S.O.; Dogan, M.F.; Cam, S.A.; Yalcin, A.; Elibol, E.; Ozer, M.K.; Uckardes, F.; Kara, H. The exogenous administration of CB2 specific agonist, GW405833, inhibits inflammation by reducing cytokine production and oxidative stress. *Exp. Ther. Med.* **2018**, *16*, 4900–4908. [CrossRef] [PubMed]
52. El-Sayed el, S.M.; Mansour, A.M.; Nady, M.E. Protective effects of pterostilbene against acetaminophen-induced hepatotoxicity in rats. *J. Biochem. Mol. Toxicol.* **2015**, *29*, 35–42. [CrossRef] [PubMed]
53. Vajic, U.J.; Grujic-Milanovic, J.; Miloradovic, Z.; Jovovic, D.; Ivanov, M.; Karanovic, D.; Savikin, K.; Bugarski, B.; Mihailovic-Stanojevic, N. *Urtica dioica* L. leaf extract modulates blood pressure and oxidative stress in spontaneously hypertensive rats. *Phytomedicine* **2018**, *46*, 39–45. [CrossRef] [PubMed]
54. Bang, J.S.; Oh da, H.; Choi, H.M.; Sur, B.J.; Lim, S.J.; Kim, J.Y.; Yang, H.I.; Yoo, M.C.; Hahm, D.H.; Kim, K.S. Anti-inflammatory and antiarthritic effects of piperine in human interleukin 1beta-stimulated fibroblast-like synoviocytes and in rat arthritis models. *Arthritis Res. Ther.* **2009**, *11*, R49. [CrossRef]
55. Coura, C.O.; Souza, R.B.; Rodrigues, J.A.; Vanderlei Ede, S.; de Araujo, I.W.; Ribeiro, N.A.; Frota, A.F.; Ribeiro, K.A.; Chaves, H.V.; Pereira, K.M.; et al. Mechanisms involved in the anti-inflammatory action of a polysulfated fraction from *Gracilaria cornea* in rats. *PLoS ONE* **2015**, *10*, e0119319. [CrossRef] [PubMed]
56. Kalyanaraman, B.; Darley-Usmar, V.; Davies, K.J.A.; Dennery, P.A.; Forman, H.J.; Grisham, M.B.; Mann, G.E.; Moore, K.; Roberts, L.J.; Ischiropoulos, H. Measuring reactive oxygen and nitrogen species with fluorescent probes: Challenges and limitations. *Free Rad. Biol. Med.* **2012**, *52*, 1–6. [CrossRef]
57. Radak, Z.; Naito, H.; Kaneko, T.; Tahara, S.; Nakamoto, H.; Takahashi, R.; Cardozo-Pelaez, F.; Goto, S. Exercise training decreases DNA damage and increases DNA repair and resistance against oxidative stress of proteins in aged rat skeletal muscle. *Pflugers Arch. Eur. J. Physiol.* **2002**, *445*, 273–278. [CrossRef]
58. Ziaaldini, M.M.; Koltai, E.; Csende, Z.; Goto, S.; Boldogh, I.; Taylor, A.; Radak, Z. Exercise training increases anabolic and attenuates catabolic and apoptotic processes in aged skeletal muscle of male rats. *Exp. Gerontol.* **2015**, *67*, 9–14. [CrossRef]
59. Gugliandolo, E.; Fusco, R.; D'Amico, R.; Militi, A.; Oteri, G.; Wallace, J.L.; Di Paola, R.; Cuzzocrea, S. Anti-inflammatory effect of ATB-352, a H2S-releasing ketoprofen derivative, on lipopolysaccharide-induced periodontitis in rats. *Pharmacol. Res.* **2018**, *132*, 220–231. [CrossRef]
60. Waltz, P.; Escobar, D.; Botero, A.M.; Zuckerbraun, B.S. Nitrate/nitrite as critical mediators to limit oxidative injury and inflammation. *Antioxid. Redox Signal.* **2015**, *23*, 328–339. [CrossRef]
61. Kany, S.; Vollrath, J.T.; Relja, B. Cytokines in inflammatory disease. *Int. J. Mol. Sci.* **2019**, *20*, 6008. [CrossRef] [PubMed]
62. Omoboyowa, D.A.; Nwodo, O.F.C.; Joshua, P.E. Anti-diarrhoeal activity of chloroform-ethanol extracts of cashew (*Anacardium occidentale*) kernel. *J. Nat. Prod.* **2013**, *6*, 109–117.
63. Dewanjee, S.; Dua, T.K.; Sahu, R. Potential anti-inflammatory effect of *Leea macrophylla* Roxb. leaves: A wild edible plant. *Food Chem. Toxicol.* **2013**, *59*, 514–520. [CrossRef] [PubMed]
64. Patil, K.R.; Mahajan, U.B.; Unger, B.S.; Goyal, S.N.; Belemkar, S.; Surana, S.J.; Ojha, S.; Patil, C.R. Animal models of inflammation for screening of anti-inflammatory drugs: Implications for the discovery and development of phytopharmaceuticals. *Int. J. Mol. Sci.* **2019**, *20*, 4367. [CrossRef] [PubMed]
65. Calixto, J.B.; Otuki, M.F.; Santos, A.R. Anti-inflammatory compounds of plant origin. Part I. Action on arachidonic acid pathway, nitric oxide and nuclear factor kappa B (NF-kappaB). *Planta Med.* **2003**, *69*, 973–983. [CrossRef] [PubMed]
66. Chung, H.J.; Lee, H.S.; Shin, J.S.; Lee, S.H.; Park, B.M.; Youn, Y.S.; Lee, S.K. Modulation of acute and chronic inflammatory processes by a traditional medicine preparation GCSB-5 both in vitro and in vivo animal models. *J. Ethnopharmacol.* **2010**, *130*, 450–459. [CrossRef]
67. Simmons, D.L. What makes a good anti-inflammatory drug target? *Drug Discov. Today* **2006**, *11*, 210–219. [CrossRef]

68. Debnath, S.; Ghosh, S.; Hazra, B. Inhibitory effect of *Nymphaea pubescens* Willd. flower extract on carrageenan-induced inflammation and CCl(4)-induced hepatotoxicity in rats. *Food Chem. Toxicol.* **2013**, *59*, 485–491. [CrossRef]
69. Fangkrathok, N.; Junlatat, J.; Sripanidkulchai, B. In vivo and in vitro anti-inflammatory activity of *Lentinus polychrous* extract. *J. Ethnopharmacol.* **2013**, *147*, 631–637. [CrossRef]
70. Gualillo, O.; Eiras, S.; Lago, F.; Dieguez, C.; Casanueva, F.F. Elevated serum leptin concentrations induced by experimental acute inflammation. *Life Sci.* **2000**, *67*, 2433–2441. [CrossRef]
71. Ben Khedir, S.; Mzid, M.; Bardaa, S.; Moalla, D.; Sahnoun, Z.; Rebai, T. In vivo evaluation of the anti-inflammatory effect of *Pistacia lentiscus* fruit oil and its effects on oxidative stress. *Evid. Based Complement Alternat. Med.* **2016**, *2016*, 6108203. [CrossRef] [PubMed]
72. Sofidiya, M.O.; Imeh, E.; Ezeani, C.; Aigbe, F.R.; Akindele, A.J. Antinociceptive and anti-inflammatory activities of ethanolic extract of *Alafia barteri*. *Rev. Bras. Farmacogn.* **2014**, *24*, 348–354. [CrossRef]
73. Uddin, G.; Rauf, A.; Siddiqui, B.S.; Muhammad, N.; Khan, A.; Shah, S.U.A. Anti-nociceptive, anti-inflammatory and sedative activities of the extracts and chemical constituents of *Diospyros lotus* L. *Phytomedicine* **2014**, *21*, 954–959. [CrossRef] [PubMed]
74. Meagher, E.A.; Barry, O.P.; Lawson, J.A.; Rokach, J.; FitzGerald, G.A. Effects of vitamin E on lipid peroxidation in healthy persons. *JAMA* **2001**, *285*, 1178–1182. [CrossRef]
75. Vincent, H.K.; Bourguignon, C.M.; Vincent, K.R.; Weltman, A.L.; Bryant, M.; Taylor, A.G. Antioxidant supplementation lowers exercise-induced oxidative stress in young overweight adults. *Obesity* **2006**, *14*, 2224–2235. [CrossRef]
76. Hogan, S.; Canning, C.; Sun, S.; Sun, X.; Zhou, K. Effects of grape pomace antioxidant extract on oxidative stress and inflammation in diet induced obese mice. *J. Agric. Food Chem.* **2010**, *58*, 11250–11256. [CrossRef]
77. Sacchet, C.; Mocelin, R.; Sachett, A.; Bevilaqua, F.; Chitolina, R.; Kuhn, F.; Boligon, A.A.; Athayde, M.L.; Roman Junior, W.A.; Rosemberg, D.B.; et al. Antidepressant-like and antioxidant effects of *Plinia trunciflora* in mice. *Evid. Based Complement. Alternat. Med.* **2015**, *2015*, 601503. [CrossRef]
78. Rahman, I.; Biswas, S.K.; Kirkham, P.A. Regulation of inflammation and redox signaling by dietary polyphenols. *Biochem. Pharmacol.* **2006**, *72*, 1439–1452. [CrossRef]
79. Pereira de Jesus Costa, A.C.; Kelly Dos Santos Silva, M.; Batista de Oliveira, S.; Silva, L.L.; Silva, A.C.; Barroso, R.B.; Macedo Costa, J.R.; Lima Hunaldo, V.K.; Neto, M.S.; Pascoal, L.M.; et al. Effects of cashew nut (*Anacardium occidentale* L.) seed flour in moderately malnourished children: Randomized clinical trial. *J. Nutr. Metab.* **2020**, *2020*, 6980754. [CrossRef]
80. Melo-Cavalcante, A.A.; Dantas, S.M.; Leite Ade, S.; Matos, L.A.; e Sousa, J.M.; Picada, J.N.; da Silva, J. In vivo antigenotoxic and anticlastogenic effects of fresh and processed cashew (*Anacardium occidentale*) apple juices. *J. Med. Food* **2011**, *14*, 792–798. [CrossRef]
81. Melo Cavalcante, A.A.; Rubensam, G.; Picada, J.N.; Gomes da Silva, E.; Fonseca Moreira, J.C.; Henriques, J.A. Mutagenicity, antioxidant potential, and antimutagenic activity against hydrogen peroxide of cashew (*Anacardium occidentale*) apple juice and cajuina. *Environ. Mol. Mutagen.* **2003**, *41*, 360–369. [CrossRef] [PubMed]
82. Behravan, E.; Heidari, M.R.; Heidari, M.; Fatemi, G.; Etemad, L.; Taghipour, G.; Abbasifard, M. Comparison of gastric ulcerogenicity of percolated extract of *Anacardium occidentale* (cashew nut) with indomethacin in rats. *Pak. J. Pharm. Sci.* **2012**, *25*, 111–115. [PubMed]
83. Olajide, O.A.; Aderogba, M.A.; Adedapo, A.D.; Makinde, J.M. Effects of Anacardium occidentale stem bark extract on in vivo inflammatory models. *J. Ethnopharmacol.* **2004**, *95*, 139–142. [CrossRef]
84. Carvalho, N.S.; Silva, M.M.; Silva, R.O.; Nicolau, L.A.; Sousa, F.B.; Damasceno, S.R.; Silva, D.A.; Barbosa, A.L.; Leite, J.R.; Medeiros, J.V. Gastroprotective properties of cashew gum, a complex heteropolysaccharide of Anacardium occidentale, in naproxen-induced gastrointestinal damage in rats. *Drug Dev. Res.* **2015**, *76*, 143–151. [CrossRef] [PubMed]
85. Vilar, M.S.; de Souza, G.L.; Vilar Dde, A.; Leite, J.A.; Raffin, F.N.; Barbosa-Filho, J.M.; Nogueira, F.H.; Rodrigues-Mascarenhas, S.; Moura, T.F. Assessment of phenolic compounds and anti-inflammatory activity of ethyl acetate phase of *Anacardium occidentale* L. bark. *Molecules* **2016**, *21*, 1087. [CrossRef]

86. da Silveira Vasconcelos, M.; Gomes-Rochette, N.F.; de Oliveira, M.L.; Nunes-Pinheiro, D.C.; Tome, A.R.; Maia de Sousa, F.Y.; Pinheiro, F.G.; Moura, C.F.; Miranda, M.R.; Mota, E.F.; et al. Anti-inflammatory and wound healing potential of cashew apple juice (*Anacardium occidentale* L.) in mice. *Exp. Biol. Med.* **2015**, *240*, 1648–1655. [CrossRef]
87. Khan, H.B.; Vinayagam, K.S.; Moorthy, B.T.; Palanivelu, S.; Panchanatham, S. Anti-inflammatory and anti-hyperlipidemic effect of *Semecarpus anacardium* in a high fat diet: STZ-induced type 2 diabetic rat model. *Inflammopharmacology* **2013**, *21*, 37–46. [CrossRef]
88. Dzoyem, J.; McGaw, L.; Kuete, V.; Bakowsky, U. Anti-inflammatory and anti-nociceptive activities of African medicinal spices and vegetables. In *Medicinal Spices and Vegetables from Africa*; Academic Press: Cambridge, MA, USA, 2017; pp. 239–270.
89. Mansouri, M.T.; Hemmati, A.A.; Naghizadeh, B.; Mard, S.A.; Rezaie, A.; Ghorbanzadeh, B. A study of the mechanisms underlying the anti-inflammatory effect of ellagic acid in carrageenan-induced paw edema in rats. *Indian J. Pharmacol.* **2015**, *47*, 292–298. [CrossRef]
90. Gill, N.; Bijjem, K.R.; Sharma, P.L. Anti-inflammatory and anti-hyperalgesic effect of all-trans retinoic acid in carrageenan-induced paw edema in Wistar rats: Involvement of peroxisome proliferator-activated receptor-beta/delta receptors. *Indian J. Pharmacol.* **2013**, *45*, 278–282. [CrossRef]
91. Ayala, A.; Munoz, M.F.; Arguelles, S. Lipid peroxidation: Production, metabolism, and signaling mechanisms of malondialdehyde and 4-hydroxy-2-nonenal. *Oxid. Med. Cell Longev.* **2014**, *2014*, 360438. [CrossRef]
92. Grotto, D.; Maria, L.S.; Valentini, J.; Paniz, C.; Schmitt, G.; Garcia, S.C.; Pomblum, V.J.; Rocha, J.B.T.; Farina, M. Importance of the lipid peroxidation biomarkers and methodological aspects FOR malondialdehyde quantification. *Química Nova* **2009**, *32*, 169–174. [CrossRef]
93. Tsai, D.S.; Huang, M.H.; Tsai, J.C.; Chang, Y.S.; Chiu, Y.J.; Lin, Y.C.; Wu, L.Y.; Peng, W.H. Analgesic and anti-inflammatory activities of *Rosa taiwanensis* nakai in mice. *J. Med. Food* **2015**, *18*, 592–600. [CrossRef] [PubMed]
94. Wagner, H.; Breu, W.; Willer, F.; Wierer, M.; Remiger, P.; Schwenker, G. In vitro inhibition of arachidonate metabolism by some alkamides and prenylated phenols. *Planta Med.* **1989**, *55*, 566–567. [CrossRef] [PubMed]
95. Goldstein, B.I.; Kemp, D.E.; Soczynska, J.K.; McIntyre, R.S. Inflammation and the phenomenology, pathophysiology, comorbidity, and treatment of bipolar disorder: A systematic review of the literature. *J. Clin. Psychiatry* **2009**, *70*, 1078–1090. [CrossRef] [PubMed]
96. Turner, M.D.; Nedjai, B.; Hurst, T.; Pennington, D.J. Cytokines and chemokines: At the crossroads of cell signalling and inflammatory disease. *Biochim. Biophys. Acta* **2014**, *1843*, 2563–2582. [CrossRef] [PubMed]
97. Di Rosa, M.; Giroud, J.P.; Willoughby, D.A. Studies on the mediators of the acute inflammatory response induced in rats in different sites by carrageenan and turpentine. *J. Pathol.* **1971**, *104*, 15–29. [CrossRef]
98. Valerio, D.A.; Cunha, T.M.; Arakawa, N.S.; Lemos, H.P.; Da Costa, F.B.; Parada, C.A.; Ferreira, S.H.; Cunha, F.Q.; Verri, W.A., Jr. Anti-inflammatory and analgesic effects of the sesquiterpene lactone budlein A in mice: Inhibition of cytokine production-dependent mechanism. *Eur. J. Pharmacol.* **2007**, *562*, 155–163. [CrossRef]
99. Uttara, B.; Singh, A.V.; Zamboni, P.; Mahajan, R.T. Oxidative stress and neurodegenerative diseases: A review of upstream and downstream antioxidant therapeutic options. *Curr. Neuropharmacol.* **2009**, *7*, 65–74. [CrossRef]
100. Ighodaro, O.; Akinloye, O. First line defence antioxidants-superoxide dismutase (SOD), catalase (CAT) and glutathione peroxidase (GPX): Their fundamental role in the entire antioxidant defence grid. *Alexandria J. Med.* **2018**, *54*, 287–293. [CrossRef]
101. Du, Z.X.; Zhang, H.Y.; Meng, X.; Guan, Y.; Wang, H.Q. Role of oxidative stress and intracellular glutathione in the sensitivity to apoptosis induced by proteasome inhibitor in thyroid cancer cells. *BMC Cancer* **2009**, *9*, 56. [CrossRef]

© 2020 by the authors. Licensee MDPI, Basel, Switzerland. This article is an open access article distributed under the terms and conditions of the Creative Commons Attribution (CC BY) license (http://creativecommons.org/licenses/by/4.0/).

Review

The Antioxidant Role of Soy and Soy Foods in Human Health

Gianluca Rizzo

Independent Researcher, Via Venezuela 66, 98121 Messina, Italy; gianlucarizzo@email.it; Tel.: +39-3208-976-687

Received: 26 June 2020; Accepted: 16 July 2020; Published: 18 July 2020

Abstract: Oxidative stress seems to play a role in many chronic diseases, such as cardiovascular diseases, diabetes, and some cancers. Research is always looking for effective approaches in the prevention and treatment of these pathologies with safe strategies. Given the central role of nutrition, the identification of beneficial healthy foods can be the best key to having a safe and at the same time effective approach. Soy has always aroused great scientific interest but often this attention is galvanized by the interaction with estrogen receptors and related consequences on health. However, soy, soy foods, and soy bioactive substances seem to have antioxidant properties, suggesting their role in quenching reactive oxygen species, although it was frequently mentioned but not studied in depth. The purpose of this review is to summarize the scientific evidence of the antioxidant properties of soy by identifying the human clinical trials available in the literature. A total of 58 manuscripts were individuated through the literature search for the final synthesis. Soy bioactive substances involved in redox processes appear to be multiple and their use seems promising. Other larger clinical trials with adequate standardization and adequate choice of biomarkers will fill the gap currently existing on the suggestive role of soy in antioxidant mechanisms.

Keywords: soy; soybeans; soy foods; antioxidants; reactive oxygen species; cardiovascular diseases; diabetes; cancer

---

1. Introduction

Soy (*Glycine max* L.) is a very important food in human nutrition. It represents a popular food in South-East Asian cuisine and its characteristics have attracted the interest of the food industry [1]. Its nutritional properties and biotechnological characteristics are attractive and explain the widespread use in the production of processed foods, especially as analogs and substitutes of meat intended for vegetarian and self-conscious users [2]. The lower incidence of some chronic diseases among populations that traditionally use soy and soy foods worldwide has suggested their role in the prevention of some pathologies [3]. However, the evidence from the literature suggests a discrepancy in the effect between Asian and Western populations whose consumption by the latter is very limited and only recently adopted. For example, there is a limited amount of prospective studies that have assessed the association between soybean and type 2 diabetes mellitus (T2DM) in Western populations [4]. In 2016, Ding and colleagues conducted a prospective pooled analysis based on three large North American cohort studies (Health Professionals Follow-Up Study, First and Second Nurses' Health Study) about the influence of soy, soy foods, and soy isoflavones consumption on T2DM risk [5]. The analysis included 163,457 subjects with 9185 T2DM documented incident cases. After multivariate adjustments, there was no association between soy food consumption and T2DM risk but there was a significant inverse association between soy isoflavone and risk of T2DM (HR: 0.89, with 0.83 to 0.96 95% CI; $p = 0.009$ for the trend) for highest compared to the lowest quintile of consumption. A significant inverse correlation persists down to the third quintile of soy isoflavone consumption. Other studies in

China, Singapore, and Hawaii showed an inverse correlation between soy and T2DM risk [6–8] but this was not confirmed in a Japanese study [9].

In 1999, the US Food and Drug Administration (FDA) released a cholesterol-lowering claim about soy protein foods [10], but recently, based on new evidence, a revocation was proposed [11]. In light of this latter hypothesis, Jenkins at al. performed a cumulative meta-analysis including the before-1999 studies used for the original claim and some additional recent trials, finally selecting 46 studies and 2607 participants for the re-analysis [12]. A minimum reduction of 4.0 mg/dL (with −6.7 to −1.3 mg/dL 95% CI; $p = 0.004$) and a maximum reduction of 7.7 mg/dL (with −11.2 to −4.3 mg/dL 95% CI; $p < 0.00001$) for total cholesterol (TC) was observed with a median soy protein intake of 25 g/d. Moreover, a minimum reduction of 4.2 mg/dL (with −6.6 to −1.8 mg/dL 95% CI; $p = 0.0006$) and a maximum reduction of 6.7 mg/dL (with −10.2 to −3.2 mg/dL 95% CI; $p < 0.0002$) for low density lipoprotein cholesterol (LDL-C) was observed. From the previous trials regarding the first claim of the FDA until 2013, the significance never fell below $p = 0.002$. This modest soy effect, if adopted with concomitant synergic dietary interventions, is comparable with a statin-like reduction effect [13]. Likewise, Health Canada has recently approved a soy cholesterol-lowering effect claim [14].

In a dose-response meta-analysis of 23 prospective cohort studies and 330,826 participants, soy consumption was inversely associated with cancer mortality (RR: 0.88, with 0.79 to 0.99 95% CI; $p = 0.03$; I^2: 47.1%) and CVD mortality (RR: 0.85, with 0.72 to 0.99 95% CI; $p = 0.04$; I^2: 50.0%) [15]. Besides, soy isoflavone consumption was also associated with all-cause mortality (RR: 0.90, with 0.82 to 0.98 95% CI; $p = 0.02$; I^2: 0.0%). Soy protein intake was inversely associated with breast cancer mortality (RR: 0.73, with 0.55 to 0.96 95% CI; $p = 0.02$; I^2: 63.5%) but no association emerged for all-cause or CVD mortality. In the linear dose-response analysis, a 10 mg/d increase in soy isoflavone was associated with lowering the breast cancer mortality risk by 9%, whereas a 5 g/d increase in soy protein intake was associated with a lowering of 12%. In a recent cohort study of 52,795 US women from the Adventist church, the substitution of dairy milk with soy milk was associated with reduced breast cancer risk (HR: 0.68, with 0.55 to 0.85 95% CI; $p = 0.001$) [16]. In a systematic review and meta-analysis of 30 studies and 266,699 multi-ethnic participants, an inverse association between soy consumption and the prostate risk was highlighted (RR: 0.70, with 0.58 to 0.85 95% CI; $p < 0.001$; I^2: 68.9%) [17]. Unfermented soy foods (but not fermented foods) were also significantly associated with reduced prostate cancer risk (RR: 0.65, with 0.52 to 0.83 95% CI; $p < 0.001$; I^2: 60.3%). Likewise, genistein and daidzein but not total isoflavone intakes were inversely associated with cancer risk, which implies a wider mechanism of action not limited to isoflavone bioavailability. Moreover, the genistein and daidzein circulating concentrations were not associated with prostate cancer risk.

It has been hypothesized that the effect of soy on chronic diseases depends on specific bioactive substances and primary isoflavones, and that the protective action involves anti-inflammatory, antineoplastic, antiplatelet mechanisms, and, at least in part, antioxidant mechanisms [18–20]. The effect of isoflavones on cardiometabolic risk factors could mainly imply mechanisms unrelated to the interaction with the estrogen receptor [21]. Moreover, oxidative stress seems to influence glycemic tolerance [22], CVD [23], and the genesis of neoplastic pathologies [24]. The implication of oxidative stress in non-communicable diseases suggests the possibility of an effective and safe approach for prevention. The use of soy, soy foods, and soy bioactive substances, widely consumed in the world, could provide an economic and easily available solution in a sustainable nutritional paradigm, already widely discussed for preventive purposes. However, there is a need for comprehensive data collection that defines the effectiveness of soy. The purpose of this review is to clarify the antioxidant properties involved in the beneficial effects of soy, soy foods, and soy bioactive substances on health through the identification of clinical trials on humans.

2. Methods

Consultation of the databases was carried out through a systematic search method using the PubMed search engine [25] and a Trials Cochrane Library Central search [26], to identify the relevant works based on the specific inclusion criteria listed below.

The following string for the query was used: "(ROS [Title/Abstract] OR antioxidant [Title/Abstract] OR antioxidants [Title/Abstract] OR oxidative [Title/Abstract]) AND (soy [Title/Abstract] OR soyfood [Title/Abstract] OR soybeans [Title/Abstract] soyfoods [Title/Abstract] OR soybean [Title/Abstract] OR "soy food" [Title/Abstract] OR "soy beans" [Title/Abstract] OR "soy foods" [Title/Abstract] OR "soy bean" [Title/Abstract] OR tofu [Title/Abstract] OR tempeh [Title/Abstract] OR natto [Title/Abstract]) ".

The formatting has been adapted to the search engine used by including the MeSH terms of interest, where available.

The results of the query were limited to articles in English, including human clinical trials investigating the effect of soy, soy foods, or soy bioactive substances on oxidative stress markers as a primary or secondary outcome. No time restrictions have been applied, specifically from inception of the databases up to 19 May 2020. Interventions that used non-soy isoflavones (e.g., from red clover or *Genista tridentata* L.) were excluded. Case reports, case series, and animal studies were excluded. Besides, the reference lists of the manuscripts identified were checked for additional entries.

3. Human Clinical Studies: The Effect of Soy on Oxidative Biomarkers

A total of 2360 entries were obtained from PubMed and, after limiting the results to articles in English and on humans, 398 entries were selected. A total of 310 results were obtained from the Cochrane library. Following the removal of 62 duplicates, 646 articles were obtained. After a selection by title and abstract, 51 articles were obtained according to the eligibility criteria. Eleven additional articles were extracted from reference lists after reading the full-text already selected. The full-text of 2 articles were unavailable and 2 articles were duplicates of trials already present among the selected publications for a total of 58 articles used for the final synthesis that are summarized in Table 1.

Table 1. Human clinical studies of the antioxidant effect of soy.

Reference	Design	Intervention	Biomarkers	Subjects	Nationality	Outcomes
Ahn-Jarvis et al. 2012 [27]	Cross-over RCT	Soy bread (20 g protein99 mg aglycone isoflavones) or soy beverage (93 mg aglycone isoflavones) per day for 3 weeks each.	Ox-LDL	20 adult subjects with hypercholesterolemia	North-American	Lower LDL oxidation in beverage intervention. Lag time after bread intervention was not significant.
Ashton et al. 2000 [28]	Cross-over RCT	290 g/day tofu (119.8 mg isoflavones) vs. 150 g lean meat for 1 month each.	Ox-LDL	42 healthy men	Australian	Lower LDL oxidation.
Azadbakht et al. 2007 [29]	Cross-over RCT	30 g/day of soy protein (84 mg phytoestrogen), soy nut (102 mg phytoestrogen) or DASH (control) for 8 weeks each.	TBARS, FRAP	42 postmenopausal women with metabolic syndrome	Iranian	Improvement of oxidative biomarkers in soy phases.
Bakhtiari et al. 2019, 2011 [30,31]	Single-blind RCT	35 g/day soy nut (117.2 mg isoflavones), or 35 g/day soy protein (96.2 mg isoflavones) vs. control for 12 weeks.	FRAP, TBARS	25 per arm for old women with metabolic syndrome	Iranian	Improved oxidative biomarkers compared to control and before the intervention. No significant differences between soy interventions.

Table 1. *Cont.*

Reference	Design	Intervention	Biomarkers	Subjects	Nationality	Outcomes
Beavers et al. 2009 [32,33]	Single-blind RCT	3 servings/day of Vanilla soy beverage (90 mg isoflavones) vs. dairy (control) for 4 weeks (with or without exercise bout in different publications).	SOD, GPx, COX-2	Healthy postmenopausal women: 16 treated, 15 control	North-American	Not significant.
Brandão et al. 2019 [34]	Double-Blind RCT	80 mg/day isoflavones vs. placebo for 4 months.	SOD, CAT, GPx, TBARS	Postmenopausal women with insomnia: 19 treated, 19 placebo	South-American	Not significant.
Brown et al. 2004 [35]	Double-Blind RCT	3 soy bars/day (33 g protein) vs. whey bars (placebo) or control for 9-weeks strength training protocol.	ABTS, MPO	Young men experienced weightlifters: 9 soy, 9 placebo, 9 control	North-American	Preserved antioxidant capacity.
Campbell et al. 2006 [36]	Cross-over RCT	39 g soy protein shake single intake (85 mg aglycone isoflavones) vs. 40 g milk protein (placebo) with 6h post-prandial evaluation each.	Ox-LDL	50 healthy men	North-American	Not significant.
Celec et al. 2013 [37]	Single-arm, open-label trial	2 g/kg BW/day soybeans for 1 week.	TBARS, AOPP, ABTS	88 young healthy adults	Slovakian	Improvement of oxidative biomarkers with gender discrepancies (AOPP not significant in men).
Cha et al. 2014 [38]	Double-blind RCT	9.8 g/day of Doenjang supplement vs. placebo pills for 12 weeks.	ORAC, CAT, TRAP	Overweight subjects: 26 treated, 25 placebo	Korean	Increased antioxidant biomarkers.
Chen et al. 2004 [39]	Double-blind RCT	150 mg/day isoflavones vs. placebo for 4 weeks before two sessions of exercise bout at 80% VO2max.	SOD, CAT, GPx	30 healthy men: 15 treatment, 15 placebo	North-American	Higher pre-exercise SOD attenuated post-exercise CAT and GPx exercise-induced increment.
Cicero et al. 2004 [40]	Single-arm, open-label trial	8 g/day soy protein with 2g/day β-sitosterol for 40 days.	Ox-LDL Ab	36 moderately hypercholesterolemic subjects	Italian	Not significant.
Clerici et al. 2011 [41]	Cross-over RCT	80 g/day soy germ enriched pasta (31–33 mg/day aglycone isoflavones) vs. conventional pasta (placebo) for 8 weeks each.	PGF2α, FRAP, GSH, Ox-LDL	26 T2DM patients	Italian	All markers improved in soy phase.
Costa e Silvia et al. 2020 [42]	Double-blind RCT	10 mL/day Brazil nut oil vs soy oil (control) for 30 days.	TBARS, ABTS	31 adult patients with metabolic syndrome (15 BNO and 16 SO)	South-American	MDA decreased in BNO group while TEAC increased in SO group
Djuric et al. 2001 [43]	Single-arm, open-label trial	50 mg/day isoflavones (women), 100 mg/day isoflavones (men) for 3 weeks.	5-OHmdU, PGF2α	20 healthy subjects	North-American	Reduced DNA oxidative damage without significant lipid oxidation changes.
Engelman et al. 2005 [44]	Double-blind RCT	40 g/day of soy protein isolate with or without isoflavones, with or without phytate for 6 weeks.	Ox-LDL, plasma protein carbonyls, PGF2α	55 postmenopausal women (14 w/o I and F; 13 w/o I and w F; 14 w I and w/o F; 14 w I and F)	North-American	Not significant.
Fei et al. 2014 [45]	RCT	10 g/day Soybean oligosaccharides vs. control for 8 weeks.	SOD, GPx, CAT, TBARS	Pregnant women with GDM: 46 treatment, 51 control	Chinese	Improvement of oxidative biomarkers.
Fritz et al. 2003 [46]	Cross-over RCT	2.01 (high), 1.01 (low), or 0.15 (control) mg/Kg BW/day isoflavones for 12 weeks each.	Polar and non-polar LARC	10 healthy premenopausal women	North-American	Urinary non-polar LARC levels were lower in isoflavone diets than control. Polar compounds were not significantly different.

Table 1. *Cont.*

Reference	Design	Intervention	Biomarkers	Subjects	Nationality	Outcomes
Gardner-Thorpe et al. 2003 [47]	Cross-over RCT	3 scones/day (soy flour-based, 120 mg isoflavones) or wheat-based (placebo) for 6 weeks each.	FOX 1, Ox-LDL (Cu or MPO)	20 healthy men	Irish	Improvement of MPO Lag time and FOX 1 without any changes in Ox-LDL with Cu.
Hanachi et al. 2007,2008 [48,49]	Open-label RCT	Soy milk (12.5 g protein, 17.13 mg isoflavones) or soy milk + 1-h walking/day vs. control for 3 months.	FRAP	Postmenopausal women: 15 soy, 12 soy + walking, 10 control	Iranian	Improved antioxidant activity in soy arms.
Hanwell et al. 2009 [50]	Cross-over RCT	Single meal fish oil and/or 96 mg soy isoflavone vs placebo with 6h post-prandial evaluation each.	LOOH, Ox-LDL, ABTS	10 overweight or obese men	North-American	Not significant.
Hariri et al. 2015 [51]	Double-Blind RCT	200 mL/day probiotic soy milk or soy milk (control) for 8 weeks.	SOD, 8-OHdG	40 patients with T2DM (20 Intervention and 20 control)	Iranian	Higher SOD and lower 8-OHdG in the probiotic group but not in the control group.
Hematdar et al. 2018 [52]	Single-blind RCT	1 cup of soybeans, three days/week vs. legume or control groups for 8 weeks.	TBARS	Patients with T2DM: 21 soy, 20 legume, 23 control	Iranian	Not significant.
Heneman et al. 2007 [53]	Cross-over RCT	Single shake intake containing milk protein (placebo), 25 g soy protein with 107 mg isoflavones or soy with <4 mg isoflavones with 8h post-prandial evaluation each.	FRAP, ORAC, PCA-ORAC, Ox-LDL, Whole plasma oxidation	16 healthy adults	North-American	Not significant.
Hill et al. 2004 [54]	Double-blind RCT	39 g/day of soy protein (88 mg aglycone isoflavones) vs. 39 g/day whey protein (placebo), 4 weeks before moderate-intensity weight resistance exercise.	LOOH	Young adult men, recreationally trained: 9 treatment, 9 placebo	North-American	Lower exercise-induced oxidative stress in soy arm.
Jamilian et al. 2015 [55]	Double-Blind RCT	30 g/day textured soy protein (75 mg phytoestrogen) or animal protein (control) for 6 weeks.	FRAP, GPx, TBARS	Women with GDM: 34 soy, 34 control	Iranian	Improvement of biomarkers. After adjustment for baseline values, FRAP and GPx lost significance.
Jamilian et al. 2016 [56]	Double-Blind RCT	50 mg/day soy isoflavones vs. placebo for 12 weeks.	TBARS, GSH, FRAP	70 women with PCOS (35 intervention, 35 placebo)	Iranian	MDA and GSH improved in soy arm, FRAP values were not significantly different after adjustment for baseline.
Jenkins et al. 2000 [57]	Cross-over RCT	93% of animal protein replaced with soy and other vegetable protein (86 mg total isoflavone per 2000 Kcal) vs. LOV low-fat diet (no isoflavones, control) for 1-month each.	Ox-LDL	31 hyperlipidemic subjects	Canadian	Lower LDL oxidation.
Jenkins et al. 2000 [58]	Cross-over RCT	Soy flour added cereal/day (36 g protein, 168 mg isoflavones) vs. cereal control for 3 weeks each.	Ox-LDL	25 hyperlipidemic subjects	North-American	Lower LDL oxidation.

Table 1. *Cont.*

Reference	Design	Intervention	Biomarkers	Subjects	Nationality	Outcomes
Jenkins et al. 2002 [59]	Cross-over RCT	50 g/day soy protein (73 mg isoflavones, treatment) vs. 52 g soy protein (10 mg isoflavones, placebo) vs low-fat dairy for 1 month (control).	Ox-LDL	37 hyperlipidemic men and postmenopausal women	North-American	Not significant. CONTRAST test showed significant negative percentage change of Ox-LDL in soy phase compared to control.
Karamali et al. 2018 [60]	RCT	30 g/day textured soy protein (75 mg phytoestrogen) vs. control for 8 weeks.	FRAP, GPx, TBARS	Patients with PCOS: 30 soy, 30 control	Iranian	Improvement of GPx (per time and group), and MDA (per time). FRAP was not significant. The correlations did not change after metabolic profile adjustments.
Kwak et al. 2013 [61]	Double-Blind RCT	4.5 g/day black soy peptides or casein (placebo) for 8 weeks.	PGF2α, TBARS, SOD	Patients with pre-hypertension: 45 soy, 46 placebo	Korean	Improved antioxidant capacity. A significant correlation between ACE and SOD.
Lee et al. 2006 [62]	Single-blind, Cross-over RCT	30 mL dark soy sauce or colorant (placebo) + 200g boiled rice with post-prandial 4 h evaluation each.	Serum and U PGF2α, U 8-OHdG	24 healthy adults	Singaporean	Plasma 8-isoprostane levels were lower to a greater extent in the soy phase. DNA damage and U 8-isoprostane were not significantly different.
Li et al. 2017 [63]	Double-blind RCT	80 mg/day isoflavones vs. placebo for 4 weeks.	SOD, PGF2α, TBARS, Nrf2	Patients with IHD: 100 treated, 100 placebo	Chinese	Improvement of oxidative biomarkers.
Miraghajani et al. 2012 [64]	Cross-over RCT	240 mL/day of soy milk vs cow's milk for 4 weeks each.	TBARS	25 patients with T2DM and nephropathy	Iranian	Not significant.
Miraghajani et al. 2017 [65]	Double-blind RCT	200 mL/day probiotic soy milk or soy milk (control) for 8 weeks.	PGF2α, TBARS, GPx, GSH, GSR, GSSG, FRAP	Patients with T2DMKD: 20 probiotic, 20 control	Iranian	Higher values for GSH, lower GSSG, in the probiotic arm. The two arms provoked increased activity of GPx (higher for control) and GSR (higher for treatment). Other markers were not significant.
Mitchell et al. 1999 [66]	Open-label RCT	1 L/day soy milk, rice milk or cow's milk for 4 weeks.	Comet Assay, FRAP	Healthy men: 4 soy, 3 rice, 3 cow's	Scottish	Reduced DNA oxidative damage. No effect on FRAP or H2O2-induced DNA damage protection.
Mittal et al. 2014 [67]	Double-blind RCT	75 mg/day isoflavones vs. placebo for 12 weeks.	TBARS, SOD, GPx, CAT	Oophorectomized women: 17 treatment, 17 placebo	Indian	Not significant.
Nestel et al. 1997 [68]	Cross-over RCT	80 mg/day isoflavones or placebo for 5 weeks each	Ox-LDL	21 peri- and postmenopausal women	Australian	Not significant.
Nhan et al. 2005 [69]	Non-randomized Cross-over, CT	1 L soy milk with 113-207 mg/day isoflavones vs low isoflavone soy milk (<4.5 mg/day) for one menstrual cycle each.	U PGF2α	8 premenopausal women	North-American	No significant differences between phases. Age-adjusted Correlation between urinary 8-isoprostane change and isoflavone intake and excretion.

Table 1. *Cont.*

Reference	Design	Intervention	Biomarkers	Subjects	Nationality	Outcomes
Oh et al. 2005 [70]	Single-arm, open-label trial	2 g/day of genistein combined polysaccharides (120 mg genistein, 57 mg daidzein) for 12 weeks.	GPx, CAT, PON	12 postmenopausal women with diabetic retinopathy	Korean	Increased GPx activity. Unchanged CAT and PON.
Önning et al. 1998 [71]	RCT	0.75 (M) 1 (W) L/day of soy or cow milk for 4 weeks.	ABTS	23 Healthy subjects (12 soy, 11 cow)	Swedish	Not significant.
Pusparini et al. 2013 [72]	Double-blind RCT	100 mg/day isoflavones vs. placebo for 12 months	TBARS	Postmenopausal women: 90 treated, 92 placebo	Indonesian	Reduced oxidative marker.
Reverri et al. 2015 [73]	Cross-over RCT	70 g soy nuts/day vs. control snack for 4 weeks each.	Ox-LDL	17 adults with cardiometabolic risk	North-American	Not significant.
Rossi et al. 2000 [74]	Single-Blind RCT	2 soy beverage/day (40 g protein, 44 mg genistein) vs. whey beverage (placebo) for 3 weeks and before and after a strenuous aerobic exercise bout.	ABTS, MPO, Uric acid	10 young males	North-American	Increased TAS in soy arm, reduced MPO and increased uric acid after exercise bout in soy arm.
Ryan-Borchers et al. 2006 [75]	Double-Blind RCT	706 mL/day Soy milk (71.6 mg isoflavones), 706 mL cow's milk + 70 mg isoflavones or cow's milk (placebo) for 16 weeks	PGF2α, 8-OHdG	Postmenopausal women: 18 Soy, 15 cow's milk + isoflavones, 19 control	North-American	Similar reduced DNA damage marker in interventional arms. Not significant isoprostane variation.
Sedaghat et al. 2019 [76]	Single-blind RCT	60 g soy nut/day vs. control for 8 weeks.	FRAP	Patients with T2DM: 34 intervention, 34 control	Iranian	Improved antioxidant capacity.
Sen et al. 2012 [77]	Cross-over RCT	2 soy food servings/day (soy milk, soy nuts or tofu) with 50 mg aglycone isoflavones vs control for 6 months each.	U PGF2α	82 premenopausal women	Hawaiian	Oxidative markers had a little worsened in soy phase.
Serrano et al. 2017 [78]	Cross-over RCT	Postprandial response at 60 min after single intake each of 500 mL soy beverage vs. 500 mL glucose solution (placebo) or 500 mL water (control).	FRAP	29 healthy, non-smokers, young adults	Spanish	Higher antioxidant response.
Steinberg et al. 2003 [79]	Cross-over RCT	25 g/day of soy protein isolate with or without 107 mg of isoflavones or phytate for 6 weeks each.	Ox-LDL	28 healthy postmenopausal women	North-American	Not significant.
Swain et al. 2002 [80]	Double-Blind RCT	40 g SPI high-isoflavone (treatment) vs SPI low-isoflavone (placebo) and 40 g whey protein isolate (control) for 24 weeks.	ABTS	24 perimenopausal women: 4 treatment, 24 placebo, 21 control.	North-American	Not significant time-treatment differences. A positive correlation between TAS and soy intake in multiple regression analysis at 12 weeks.
Taniguchi-Fukatsu et al. 2012 [81]	Cross-over RCT	200 g/day Natto vs. non-viscous boiled soybeans (placebo) for two weeks each.	MDA-LDL	11 overweight adults with IGT	Japanese	Reduced oxidative biomarker.
Tikkanen et al. 1998 [82]	Single-arm, open-label trial	Two soy bar/day with a total of 38 mg isoflavones for 2 weeks.	Ox-LDL	6 healthy subjects	Finnish	Improved antioxidant effect.
Utarwuthipong et al. 2009 [83]	Cross-over RCT	20% of energy intake from soybean oil, rice bran oil, palm oil or 3:1 mixture of RBO/PO for 10 weeks each.	Ox-LDL	16 hypercholesterolemic women	Thai	Increased oxidation lag time in PO, RBO and RBO/PO but reduced in SBO.

Table 1. *Cont.*

Reference	Design	Intervention	Biomarkers	Subjects	Nationality	Outcomes
Vega-López et al. 2005 [84]	Cross-over RCT	25 g/1000 kcal/day animal or soy protein (with or without 40 mg isoflavone/1000 kcal/day) for 42 days each.	TAP, Ox-LDL, U PGF2α, TBARS, Plasma protein carbonyls (on native and oxidized plasma)	42 hypercholesterolemic old adults	North-American	Only TAP and protein carbonyls (oxidized plasma) show amelioration in soy protein phases. MDA was higher in isoflavones diet.
Wiseman et al. 2000 [85]	Cross-over RCT	One textured soy protein burger/day with 15 g protein and 56 mg isoflavones vs. burger without isoflavones (alcohol extracted) as a placebo for 17 days each.	PGF2α, Ox-LDL, TBARS	24 healthy adults	British	8-isoprostane and Ox-LDL were improved in intervention, TBARS was not significant.
Zemel et al. 2010 [86]	Cross-over RCT	3 smoothies/day diary-based or soy-based (control) for 28 days each.	TBARS, PGF2α	20 healthy adults (including 10 overweight)	North-American	Improved antioxidant ability in the diary phase. No significant differences for soy treatment.

Abbreviations: 2,2′-azino-di-(3-ethylbenzthiazoline sulfonate) (ABTS); 5-hydroxymethyl-2′deoxyuridine (5-OHmdU); 8-hydroxy-2′-deoxyguanosine (8-OHdG); 8-iso-prostaglandin F2α (PGF2α); Advanced oxidation protein products (AOPP); Auto-antibodies for oxidized LDL (Ox-LDL Ab); Catalase (CAT); Copper (Cu); Cyclooxygenase-2 (COX-2); Ferric-reducing antioxidant power (FRAP); Ferrous oxidation xylenol orange, version 1 (FOX 1); Glutathione peroxidase (GPx); Glutathione reductase (GSR); Lipid hydroperoxides (LOOH); Lipophilic aldehydes and related carbonyl compounds (LARC); Malondialdehyde-modified low density lipoprotein (MDA-LDL); Myeloperoxidase (MPO); Nuclear factor erythroid 2-related factor 2 (Nrf2); Oxidized glutathione (GSSG); Oxidized LDL (Ox-LDL) ; Oxygen radical absorbance capacity (ORAC); Paraoxonase (PON) ; Perchloric-acid-treated oxygen radical absorbance capacity (PCA-ORAC); Plasma total antioxidant performance (TAP); Reduced glutathione (GSH); Superoxide dismutase (SOD); Thiobarbituric acid reactive substances (TBARS); Total peroxyl radical-tapping anti-oxidant potential (TRAP); Urinary 8-hydroxy-2′-deoxyguanosine (U 8-OHdG); Urinary 8-iso-prostaglandin F2α (U PGF2α).

The number of subjects in trials ranged from 6 to 200 individuals. A total of 22 manuscripts employed North American populations, 11 Iranian, 1 Canadian, 1 Hawaiian, 2 South American, 5 North European, 1 Slovakian, 3 South European, 2 Australian, 1 Indian, and 9 South-East Asian. The design of the trials was heterogeneous: 28 parallel trials, 25 cross-over trials, and 5 single-arm trials. Of these, 40 double-blind RCTs, 7 single-blind RCTs, 7 open-label trials, 1 non-randomized trial, and 3 that were not well-defined RCTs regarding the blinding.

Twenty articles were based on healthy individuals and 20 articles were based on metabolic pathologies, such as diabetes, dyslipidemia, hypercholesterolemia, hypertension or obesity, and simple overweight. Moreover, 2 articles were based on gestational diabetes, 2 on PCOS, 1 on nephropathy, 2 on neuropathic T2DM, and 1 on diabetic retinopathy. Ten articles were based on menopausal healthy populations. The intervention time ranged from 60 post-prandial minutes after single soy ingestion to 12 months of intervention. Selected trials were carried out using soy and soy foods (including traditional, fermented and non-fermented foods, as well as texturized soy) or isolated components, such as isoflavones, oils, sterols, proteins, or oligosaccharides. Two studies used soybeans while 18 manuscripts used soy foods, such as soy drink, tofu, and soy flour, with one manuscript using various soy foods, including soy nuts. Moreover, soy nuts were exclusively used in 4 trials. Three trials used fermented foods, i.e., Doenjang [39], natto [81], and dark soy sauce [62]. Eleven manuscripts used soy protein isolates with associated isoflavones. One manuscript used a mix of proteins and sitosterol [40], 1 black soy peptides [61], and 1 used pasta enriched with soy germ [41]. Eleven manuscripts evaluated the effect of isolated soy isoflavones on oxidative stress markers. One manuscript used a mix of isoflavones and oligosaccharides [70] and 1 oligosaccharide only [45]. Five manuscripts evaluated the use of texturized soybean bars or burgers. Finally, 2 articles assessed the effect of soybean oil [42,83].

Taking into account interventions where concentrations of isoflavones were declared, the daily intake varied from 17.13 mg/day [48] to 207 mg/day [69]. The intake of isoflavones in interventions with soybean oil and oligosaccharides was negligible [42,45,61,83]. The intake of soy protein in selected trials ranged from 8 g/day [40] to 52 g/day [59].

Regarding the oxidative biomarkers, different techniques and molecules have been used which are summarized in Table 2.

Table 2. Biomarkers for antioxidant capacity and oxidative status.

Name	Description
2,2′-azino-di-(3-ethylbenzthiazoline sulfonate) (ABTS)	Total Antioxidant Status (TAS) determined with a colorimetric method based on the inhibition of ABTS (2,2′-azino-di-(3-ethylbenzthiazoline sulfonate)) oxidation by a peroxidase. Values are expressed in Trolox Equivalent antioxidant capacity (TEAC) [87].
5-hydroxymethyl-2′deoxyuridine (5-OHmdU)	It is a modified DNA by-product generated by thymine oxidation, that represents a marker of oxidative damage. After isolation and purification of nucleic acids, 5-OHmdU is detected through GC-MS.
8-hydroxy-2′-deoxyguanosine (8-OHdG)	It is a common marker of oxidative DNA damage involved in cancer genesis. It can be measured in various biological samples with several methods including HPLC, GC-MS, or ELISA [88].
8-iso-prostaglandin F2α (PGF2α)	The most abundant arachidonic acid oxidation by-product. It can be quantified in blood or urine samples by an enzyme immunoassay method (EIA) or GC-MS [89].
Advanced oxidation protein products (AOPP)	Oxidative by-products of plasma proteins, usually in uremic patients, detected with spectrophotometric assay [90].
Catalase (CAT)	An antioxidant enzyme that catalyzes the conversion of hydrogen peroxide into water and oxygen. The enzyme activity is detected spectrophotometrically.
Comet assay	It is a single-cell microgel electrophoresis (SCGE) assay to detect the frequency of DNA breaks or DNA oxidized bases (with a modification of the original assay using an endonuclease to break DNA at sites of oxidized bases) [91].
Cyclooxygenase-2 (COX-2)	It is an enzyme involved in prostanoids metabolism by the conversion of arachidonic acid. Its oxidative action is of interest in oxidative stress and inflammation. It is usually detected using commercial ELISA kits.
Ferric-reducing antioxidant power (FRAP)	Total antioxidant capacity (TAC) determined by a colorimetric assay using ferric-reducing antioxidant power [92].
Ferrous oxidation xylenol orange, version 1 (FOX1)	An assay to detect plasma lipid hydroperoxides through ferrous ion oxidation with the ferric ion indicator xylenol orange reagent [93].
Glutathione peroxidase (GPx)	An antioxidant enzyme that catalyzes the reduction of lipid hydroperoxides and hydrogen peroxide into corresponding alcohols and water, respectively. Its activity is detected spectrophotometrically.
Glutathione reductase (GSR)	An enzyme that catalyzes the reconversion of GSSG into GSH using NADPH. Its activity is detected spectrophotometrically.
Lipophilic aldehydes and related carbonyl compounds (LARC) and lipid hydroperoxides (LOOH)	Aldehydes e carbonyl compounds are intermediate and secondary products of oxidation in lipid peroxidation pathways, respectively [94]. Extracted compounds from bodily fluids are detected by commercial spectrophotometric assays or HPLC separation after 2,4-Dinitrophenylhydrazine (DNPH) derivatives compound formation.
Malondialdehyde-modified LDL (MDA-LDL)	A by-product of oxidized LDL [95]. Compounds are detected by ELISA.
Myeloperoxidase (MPO)	A prooxidant enzyme expressed primarily by neutrophil granulocytes as an antimicrobial defense strategy. Its action may also cause oxidative damage to the host. It is usually detected by ELISA.
Nuclear factor erythroid 2-related factor 2 (Nrf2)	It is an antioxidant related molecule acting as a transcription factor regulating the cellular resistance to oxidative stress [96]. It is detected with a semi-quantitative colorimetric assay kit using a nuclear extract.
Oxidized glutathione (GSSG)	It is the oxidized, disulfide form of glutathione generated after the reaction of GSH with reactive oxygen species (ROS). In healthy conditions, only a minor part of glutathione is in the GSSG form. A lower GSH/GSSG ratio show the oxidative stress status. It is usually detected using commercial ELISA kits.
Oxidized LDL (Ox-LDL) and auto-antibodies for oxidized LDL (Ox-LDL Ab)	An ex-vivo method based on the formation of conjugated dienes catalyzed by incubation with Cu^{2+} or myeloperoxidase as prooxidants, on isolated LDL from plasma [97]. Ox-LDL can be detected also dosing Ox-LDL auto-antibodies.
Oxygen radical absorbance capacity (ORAC)	Measurement of the overall plasma capacity to scavenge oxygen radicals using a peroxyl generator, such as 2,2′-Azobis(2-amidino-propane) dihydrochloride (AAPH), and a fluorescent probe [98].
Paraoxonase (PON)	Antioxidant enzyme preventing lipoprotein oxidation by lactonase and esterase action. The enzyme activity in the solution is measured spectrophotometrically.
Plasma protein carbonyls	Protein oxidation by-products generally detected by a colorimetric ELISA assay.
Plasma total antioxidant performance (TAP)	Fluorometric assay performed in a multilabel counter using dichlorodihydrofluorescein (DCFH), a probe for oxidative activity [99].
Reduced glutathione (GSH)	It is a molecule that reacts with free radicals and peroxides quenching oxidative stress generated by these molecules. It is a cofactor for GPx. It is usually detected using commercial ELISA kits.
Superoxide dismutase (SOD)	An antioxidant enzyme that catalyzes the dismutation of superoxide radicals into hydrogen peroxide and molecular oxygen. It is usually detected using commercial ELISA kits.
Thiobarbituric acid reactive substances (TBARS)	Lipid peroxidation quantification based on a spectrophotometrical or HPLC assay for the quantification malondialdehyde (MDA) with thiobarbituric acid reactive substances through the formation of MDA-TBA adduct [100].
Total peroxyl radical-tapping anti-oxidant potential (TRAP)	Measurement of the radical-trapping capacity of whole plasma using a free-radical probe [101].
Uric acid	It is a metabolite of purine nucleotides oxidative breakdown. Its role in antioxidant capacity has been speculated. It is detected by quantitative colorimetric or fluorometric enzymatic kits.
Whole plasma oxidation	Assy based on conjugated dienes formation in whole plasma by Cu^{2-} and detection by a spectrophotometer [102].

A total of 44 articles used markers based on lipid oxidation (TBARS, PGF2α, LARC, LOOH, and Ox-LDL); 21 articles used indicators of total antioxidant capacity (FRAP, ABTS, TAP, AOPP, ORAC, and whole plasma oxidation); 19 articles assessed the levels of endogenous oxidation biomarkers, such as molecules and enzymes involved in antioxidant defenses or response to oxidative stress (SOD, GPx, GSH, GSSG, GSR, CAT, MPO, COX-2, PON, Nrf-2, and uric acid); 5 articles used nucleic acid oxidation markers (8-OHdG, 5-OHmdU, and the comet assay); and 2 articles used protein carbonyl oxidation markers. About articles with multiple markers usage, no manuscript evaluated at least one of each class of markers listed above (nucleic acids, proteins, fats, endogenous oxidation biomarkers, and total antioxidant capacity) at the same time.

To better summarize the results that emerged from the selected articles, a subdivision was made based on some features of the interventions, evaluating the results based on the type of biomarkers adopted.

3.1. The Effect of Soy, Soy Foods, and Soy Bioactive Molecules on Oxidative Stress

3.1.1. Soy Foods

Of the 18 trials that used soy foods, mixed results were highlighted. The DNA oxidative damage markers showed an improvement effect on the intervention [75], partial effect [66], or no statistically significant effect [51]. Regarding fat oxidation, an improvement effect has been detected [27,28,58], as well as partial [47,69], slightly worsening [69,77], and no statistically significant effect [64,65,75,86]. About endogenous oxidation biomarkers, 1 work showed improvement of markers [74], two works showed mixed results [47,65], and three works did not detect significant variations [32,33,51]. Regarding the markers of total antioxidant capacity, 3 papers did not show a significant effect after the consumption of soy foods [65,66,71], while 3 papers showed improvement in the antioxidant capacity [48,74,78]. No protein oxidation markers were used by the selected trials.

The two trials by Beavers and colleagues [32,33] have been conducted with the same protocol on the same subjects. The two works differed from the presence or not of an exercise bout in the trial protocol. However, the results of the oxidation markers were not significant in both cases. In the work of Miraghajani and colleagues [65], while a partial response on endogenous oxidation biomarkers in both arms was noted, the trial was designed to detect the effect of taking soy milk with probiotic characteristics compared to normal soy milk. An improvement has been achieved compared to non-probiotic soy control, but with the lack of a soy-free placebo arm. A similar design was used in the clinical trial of Hariri and colleagues [51] but with improvements only with probiotic milk and not attributable to soy, because it was not significant in the control group with unfermented soy (effect over time compared to baseline). Zemel and colleagues conducted a clinical trial in which the soy phase was considered as a control [86]. In this circumstance, an improvement effect of the antioxidant capacities was detected but there was no significant variation during the soy phase. As in other works cited above [51,65], Ahn-Jarvis and colleagues did not use soy-free control, being the two phases of intervention based on soy bread or soy drinks with a similar intake of isoflavones [27]. The ameliorative effect on oxidative stress was obtained with the soy drink alone, demonstrating the role of the food matrix on oxidative stress.

Overall, the effect of soy foods seems unconvincing but it is unlikely that the oxidation worsened after soy intake.

Rayan-Brochers and colleagues conducted a clinical trial with three parallel arms that included an intervention with soy milk, one with cow's milk (as placebo) and one with cow's milk added with 70 mg of isoflavones, the same amount obtained from the intervention with soy milk [75]. The outcomes of the two arms of intervention on the oxidation markers were similar, suggesting that the antioxidant effect was borne by isoflavones and not by other components of soy milk. Similarly, in the work of Nhan and colleagues, the control phase of the trial consisted of a soy drink with a low isoflavone content and, although there were no significant differences between the phases of the trial, a correlation emerged between the variations of the markers of oxidative stress and the daily dose of isoflavones [69]. The presence of a biphasic curve has been hypothesized, based on reduced urinary excretion of PGF2α for the intakes of daidzein <60 mg/d but with an increase in the excretion of PGF2α for daily intakes >72 mg.

3.1.2. Pure Isoflavones

Although isoflavones are likely to play a role in the antioxidant properties of soy, the effect of soy foods could depend only in part by the isoflavones' action. Eleven manuscripts used isolated soy isoflavones. Regarding the possible damage of nucleic acids, 2 manuscripts have shown improvement through the intake of isoflavones [43,75]. About the oxidative stress on fatty acids, 3 manuscripts showed an improvement [56,63,72], while 6 manuscripts showed no significant variations [34,43,50,67,68,75], with 1 manuscript showing a partial effect [46]. About endogenous oxidation biomarkers, an improvement was found after the intake of isoflavones in some cases [39,56,63] but not significant in other cases [34,67]. Regarding the effect on total antioxidant capacity, no clinical trial has shown improvement of the markers used [50,56]. No manuscript has evaluated the oxidation status of the proteins following the intake of soy isoflavones.

Overall, as in the case of the soy foods previously evaluated, there does not seem to be a clear protective effect of isolated isoflavones on oxidative stress. However, in Li and colleagues' work, based on a 4-week double-blind RCT on 100 patients with IHD versus 100 placeboes (the largest population sample among selected trials), all four oxidation markers (endogenous and lipid biomarkers) improved after the intervention with soy isoflavones [63].

In the work of Oh and colleagues, 2 g/day of soy polysaccharides combined with isoflavones (about 155 mg/day) caused an increase in GPx but did not influence the CAT and PON levels [70]. In this context, it was not possible to discriminate between the effect of the isoflavones and that of the soy polysaccharides.

3.1.3. Proteins

Results on lipid oxidation markers following intervention with isolated soy proteins showed improvement [29,30,54,57], as well as partial results [59], pejorative [84], or no significant results [36,44,53,79,84]. About the oxidation of proteins, 1 work found no significant effects [44] and 1 showed partial effects [84]. The effects on total antioxidant capacity were ameliorative [29,30,84], partial [80], or not significant [53]. No intervention trials with isolated soy proteins used markers for the oxidation of nucleic acids or the evaluation of endogenous oxidation biomarkers. In the clinical trial of Campbell and colleagues, there was not any change in biomarkers, although it was a post-prandial evaluation after 6 h after a single-meal soy ingestion [36]. Similarly, Heneman and colleagues evaluated oxidative biomarkers after 8 h post-prandial of a single intake of soy protein, without any significant effect in presence or not of isoflavones with soy protein [53]. Furthermore, in the work of Engelman and colleagues [44] and the work of Steinberg and colleagues [79], the effect of soy protein, with or without isoflavones, showed no significant differences in biomarkers. In the work of Swain and colleagues, the effect of isolated soy proteins containing isoflavones did not show any significant differences compared to soy proteins without isoflavones, with an antioxidant effect correlated to soy intake and not to isoflavones [80]. Similarly, in the clinical trial of Jenkins and colleagues, the antioxidant effect seems to refer to the soy supply and not to the isoflavone content [59]. In the work of Vega-López and colleagues [84], the antioxidant effect was independent of the content of the isoflavones but in the case of the evaluation of MDA levels, the pejorative effect had been observed only in the arm with isoflavones.

The antioxidant effect of soy in the form of protein isolates does not suggest a clear protective outcome. From the works comparing a soy protein isolate with or without isoflavone, it seems that the latter is not decisive for obtaining the antioxidant effect. The only negative effect that emerged seems to be related to isoflavones and not to soybeans [84].

Black soy is a variant of soy known to have particularly high levels of phytochemicals with promising applications on human health [103]. In the work of Kwak and coworkers, black soybean peptides were used against a placebo (casein) to evaluate their antioxidant ability [61]. The improvement in lipid oxidation and the increase in SOD levels were significant in the intervention arm.

In Cicero and colleagues' work, 8 g per day of soy protein with beta-sitosterol showed no improvement in the lipid oxidation status compared to the baseline in an open-label single-arm trial [40]. Clerici and colleagues showed the improvement of lipid oxidation markers, total antioxidant capacity, and increased GSH concentration using soy-germ enriched pasta containing 31–33 mg/day of aglycone isoflavones [41]. It was not possible to discriminate between the effect of proteins or other phytochemicals contained in the soy-germ compared to the isoflavones.

3.1.4. Nuts

Among the 5 manuscripts that used soy nuts, 3 trials showed an improvement on lipid oxidation markers [29,30,73] and 3 trials showed an improvement effect on the total oxidation capacity [29,30,76]. No selected manuscript showed negative effects on any marker, mixed-effects, or no significant effects. None of the articles evaluated the DNA oxidation, protein oxidation, or endogenous oxidation biomarkers. The use of soy nuts is very common in the world and represents one of the foods with the highest concentration of isoflavones [73].

The work of Azabakht and colleagues employed two different soy-based interventions with isolated proteins or soy nuts [29]. The improvement in total antioxidant capacity was significant compared to the control for both soybean interventions but higher in the soy nuts group. While, regarding lipid oxidation, there was no difference between the two interventions with soy. Even if the intake of phytoestrogens in the soy nuts arm was slightly greater, a beneficial effect of the food matrix is not excluded. Even in the work of Bakhtiari and colleagues [30], there were no differences in the effect on the oxidation markers between the soy protein arm and the soy nut arm, although the latter had higher isoflavones levels, as highlighted in the work of Azabakht and colleagues [29]. In the work of Sen and colleagues already cited, various soy-based foods, including soy nuts, were used, but the choice of the types of foods was a participants' option, with a total daily intake of 50 mg of isoflavones [77]. There was, therefore, no sub-analysis for the specific effect concerning a single food used.

The limited number of trials using soy nuts that emerged from the selection does not allow solid conclusions to be drawn. However, albeit limited, an improvement in lipid oxidation and total antioxidant capacity were shown. The use of soy nuts is very promising for its widespread use in the world, the high content of isoflavones, and the minimum amount of transformation to obtain the final product. The comparison between soy proteins containing phytoestrogens and soy nuts seems to show an overlapping or slightly unbalanced effect towards soy nuts and this could be further explored.

3.1.5. Snack and Texturized Soy Foods

Opposed to traditional soy foods, there are foods obtained through transformation techniques that lead to the change of the texture and taste of soy through high-pressure and high-temperature procedures. These lead to the formation of soy matter useful for the production of snacks, burgers, and other meat substitutes. Generally, the soy from these processes undergoes a profound transformation that is thought to have a negative effect on the bioavailability of isoflavones [1]. Five trials using snacks or textured soy proteins were selected. Three manuscripts showed an improvement in lipid markers [55,60,82] while 1 manuscript showed mixed results [85]. Two articles showed improvement of endogenous oxidation biomarkers [35,60] while 1 showed no significant variation [55]. One manuscript showed improvement in total antioxidant capacity [35] while 2 manuscripts did not reach significance [55,60]. No manuscript adopted markers for the oxidation of proteins or nucleic acids. No manuscript showed worsening outcomes after soy interventions. Overall, the results suggest an ameliorative effect on oxidation by snack or soy-textured products, although the number of trials is still limited, and not all of them have shown significant results. Since this type of highly processed food has often a reduced concentration of bioavailable isoflavones, the antioxidant effect could partly depend on other components. No trial indicated isoflavone content in aglycone equivalents. However, in 1 trial the control was obtained through the alcoholic extraction of isoflavones, still indicating a role for these bioactive substances in the antioxidant effect [85]. Nevertheless, there was no evaluation of the effect from the baseline; so, it was not possible to evaluate the contribution of other soy components.

3.1.6. Fermented Soy Foods

Fermentation seems to favor the breakdown of the glycosidic bond of the isoflavones and therefore the transformation into the aglycone form with consequent increases in the bioavailability of the isoflavones thanks to the greater diffusion capacity of these compounds through the intestinal mucosal barrier [104]. Among the 3 trials identified, 1 trial showed an improvement in lipid oxidation [81], 1 showed mixed results [62], 1 trial showed enhanced total antioxidant capacity and endogenous oxidation biomarkers [38], while 1 work did not show significant effects on DNA oxidation [62]. No selected manuscripts used markers of protein oxidation. Even if none of the works showed negative results, the number of trials that emerged does not allow definitive conclusions to be drawn. Fermented soybeans are considered highly bioavailable foods rich in isoflavones and other bioactive compounds thanks to the action of microorganisms. Other works will help to understand if this aspect represents an advantage for the antioxidant effect of fermented soy foods. In the trial of Taniguchi-Fukatsu and colleagues, although the content of isoflavones was not declared, the placebo phase consisted of boiled, unfermented soybeans [81]. The improvement of the oxidation biomarkers indicated that the effect depended on the modification of soy during fermentation. It needs to be clarified if it was a matrix effect, a greater bioavailability of soy phytochemicals after transformation, or depending on other molecules or substances of microbial origin released by the starter microorganisms.

3.1.7. Soybeans

Only 2 papers used soybeans in the intervention arm of the selected clinical trials. Hamatdar and colleagues did not show any significant differences in lipid oxidation status using three cups of soybeans per week [52]. However, the intake could be insufficient to trigger an effect. Celec and colleagues showed improvement in lipid oxidation and total antioxidant capacity with 2 g of soybeans per kg of body weight per day [37]. Some discrepancies have been identified in the gender sub-analysis, with the loss of significance on total antioxidant capacity among men.

3.1.8. Soybean Oil

From the selected manuscripts, two trials used soybean oil. Soybean oil does not contain isoflavones due to the extractive procedures during production that does not allow the recovery of these substances [2]. In the clinical trial of Costa and Silvia et al., the total antioxidant capacity improved in the soybean oil arm (control) compared to the intervention with Brazilian walnut oil [42]. However, a worsening effect was shown on lipid oxidation. The worsening of the lipid oxidation biomarkers found would depend on the greater concentration of polyunsaturated fatty acids in soybean oil, easily prone to peroxidation. Conversely, the Brazilian walnut oil had a lower concentration of polyunsaturated fatty acids and involved a greater supply of saturated fatty acids, less prone to peroxidation. Similarly, in the work of Utarwuthipong and colleagues, the soybean oil arm showed a reduced lag time of LDL oxidation compared to soybean oil and/or rice oil, demonstrating the abovementioned hypothesis of greater susceptibility [83].

3.2. *The Effect of Soy on Exercise-Born Oxidative stress*

Five articles assessed the effect of soy on oxidation markers in conjunction with an exercise session [33,35,39,48,54,74]. The exercise-born oxidative stress is a phenomenon known and studied also for the possible implications in skeletal muscle adaptation [105]. The soybean intake seems to improve the total antioxidant capacity [35,48,74] and the oxidation status of the fatty acids, although derived from just one trial [54]. Endogenous oxidation biomarkers also seem to improve [35,39,74], although one work showed no significant results [33]. None of the identified works showed outcomes on markers of protein or nucleic acid oxidation. None of the selected works showed worsening effects following soy intake. It is very difficult to give a conclusive statement, based on the great heterogeneity of the experimental setting, such as time, type, and intensity of exercise.

3.3. Soy Postprandial Antioxidant Effect

Five articles assessed the effect of soy from 1 to 8 h in the postprandial period [36,50,53,62,78]. Among the trials selected, only 1 article showed partially reduced lipid oxidation after soy intake [62] and another one showed improvement in total antioxidant capacity [78]. In the remaining trials, there was a lack of significance for the antioxidant effect on lipids, the total antioxidant capacity, and oxidation of nucleic acids. Maybe, the limited time elapsed between the intake and the assessment was too short to have a clear effect, but currently, it does not seem that soy improves the redox status in the postprandial period after its ingestion.

3.4. The Soy Antioxidant Effect in Menopause

Eleven articles recruited healthy individuals in the perimenopausal phase [32–34,44,48,67,68,72,75,79,80]. Menopause is a period of hormonal changes in women in which the risk of metabolic pathologies increases [106]. Among the selected trials, the work of Swain et al. showed a correlation between total antioxidant capacity and soy intake, independent of the isoflavone content [80], while Pusparini and colleagues showed an improvement in the oxidative status of lipids following the intake of soy isoflavones [72]. In the double-blind, parallel RCT of Ryan-Borchers et al., the reduction of nucleic acid oxidation was similar in the arm with soy milk compared with cow's milk added with soy isoflavones [75]. All the other works showed no significant improvement in the oxidation markers after soy intake.

3.5. Bioavailability and Metabolism Od Soy Isoflavones

Although it is not possible to draw definitive conclusions, some clinical trials suggest a role for isoflavones in the soy antioxidant effect. For this purpose, it is useful to understand the absorption of isoflavones through the evaluation of serum or urinary levels of both isoflavones and metabolites, such as equol, *O*-desmethylangolensin (*O*-DMA), 6-hydroxy-*O*-desmethylangolensin (6-OH-*O*-DMA), dihydrodaidzein (DHD), and dihydrogenistein. Among the manuscripts selected, 19 trials evaluated the levels of isoflavones and related metabolites [27–29,37,39,41,43,53,57,59,66,69,73,75,77,79,82,84,85]. In general, although with a wide individual variability, the soybean intake lead to an increase in blood [28,29,37,39,41,43,53,66,73,75,79,82,84,85] or urinary isoflavones [57,59], but not in all trials [69]. The levels of metabolites after soy treatment were also higher in plasma [79,85] and urine [59]. In the clinical trial of Steinberg and colleagues, although the number of equol-producers (individuals capable of metabolizing daidzein into equol) was not sufficient to assess a statistical significance, the inclusion of only these subjects in the sub-group analysis (10 out of 28 individuals) did not change the results of oxidative biomarkers that remained not significant [79]. In the cross-over trial of Sen et al., the equol-producers were equally distributed between the groups at baseline [77]. A significant positive correlation was shown between the urinary concentrations of isoflavones and the level of urinary isoprostane; it was persistent after the stratification by the equol-producer but lost significance for the subjects who were not equol-producers. In the manuscript of Mitchell and colleagues, 1 out of the 4 participants in the intervention group was an equol-producer [66]. However, no sub-analysis was possible for obvious numerical reasons. A gender-specific variability in gut microbial competence for isoflavones metabolism was proposed [107]; however, Celec and colleagues, although they have found differences in oxidative response between men and women, did not investigate isoflavone metabolism [37]. Ahn-Jarvis and colleagues highlighted that their cohort was represented by 30%

equol-producers and all but three subjects were O-DMA and DHD producers. They showed a gender difference in the excretion of isoflavone metabolites, greater in women than in men after taking soy bread [27]. Despite these differences, no gender variability in the excretion of isoflavones was highlighted following the intake of soy drink, despite a comparable intake of aglycone isoflavones in the two interventions. Increased production of equol, O-DMA, and DHD was shown in the soy bread arm compared to soy beverage with concomitant lower urine concentrations of daidzein, but only in women. Similarly, Ryan-Borchers and colleagues showed differences in the absorption and metabolization of isoflavones when taking soy drink or isolated isoflavones added to cow's milk [75]. While the circulating genistein levels were more elevated in the intervention arm with isolated isoflavones than in the soy drink, equol concentrations increased in the soy drink arm while they did not show significant changes in the arm with isolated isoflavones. Furthermore, urinary genistein and daidzein were higher in the arm with isoflavones but the equol excretion was greater in the arm with soy drink. Despite these differences, there were no different outcomes between the two interventions. This suggests an effect of the matrix or other components on the metabolizing capacity of soy isoflavones, which is not necessarily complementary to the absorption capacity. In the trial of Reverri and colleagues, 47% of the participants were equol-producers while 71% were O-DMA producers [73]. However, the ability to produce isoflavone metabolites was not related to oxidative stress or other metabolic markers in the linear mixed model. Nhan et al. found a statistically significant positive association between the combined urinary excretion of daidzein, genistein, and equol, as well as change in urinary PGF2α [69]. However, only 2 out of 8 participants were equol-producers so it was not possible to highlight a clear effect based on the ability to metabolize isoflavones.

4. Oxidative Stress Implication in Human Health: Pathophysiological and Cellular Mechanisms

Traditionally, oxidative stress has been implicated in mechanisms of aging through the production of free radicals with potentially harmful effects on cellular structures [108]. The oxidation of proteins, nucleic acids, and lipids leads to an increase in the risk of various pathologies, including cancer [109]. Similarly, oxidative stress is recognized as a major contributor to the initiation and progression of cardiovascular dysfunctions [110,111] and atherosclerosis [112]. Among the various mechanisms involved, ROS play a role in angiogenesis, causing cell apoptosis of endothelial cells and increasing the adhesive capacity of monocytes [23]. Moreover, the increase in visceral fat promotes the release of inflammatory cytokines, which, in turn, are responsible for the generation of the oxidative stress implicated in thrombogenesis and atherogenesis [113]. Chronic hyperglycemia has been proposed as a cause of mitochondrial and extramitochondrial oxidative stress [114]. In particular, mitochondria cope with the oxidizing action of oxygen during electron leakage [115]. Oxidative stress appears to affect glucose tolerance, as hypothesized following cell culture studies, suggesting that oxidative stress impairs insulin-mediated translocation of GLUT4 [116] and suppressing the gene expression of insulin in pancreatic β-cells [117]. ROS generated from glucose metabolism act as signals for insulin secretion [118], so there is an equilibrium between the formation and removal of intracellular ROS acting on glycolysis and the Krebs pathways by uncoupling the stimulus-secretion mechanisms that impair insulin secretion [119]. The pathogenesis of diabetes mellitus has been linked to oxidative and inflammatory mechanisms that appear to be responsible for the long-term consequences of diabetes [120].

Oxidative stress represents a possible mechanism in dyslipidemia and hyperglycemia through the reduction of the volume of beta cells of the pancreatic Langerhans' islets [121]. The same vascular effect mentioned above could explain, at least in part, the micro and macrovascular complications typical of diabetes [122]. For these reasons, potential approaches to diabetes have been proposed through the use of antioxidants [123–125]. For example, the synergic treatment of rosiglitazone and metformin improves vascular function through the reduction of inflammation and oxidative stress [126]. Even in secondary prevention, mitigating oxidative stress was proposed as an approach for diabetes, as well as for associated cardiovascular diseases [127,128]. Oxidative stress can also induce modification of nucleic acids, such as breaking of the scaffold of DNA and other epigenetic modifications, e.g., increased methylation, with consequent effects on gene expression and therefore acting on the pathogenesis of cancer [129,130].

Over the years, the literature on soy and its potential health effects have focused primarily on soy proteins and the associated isoflavones. Phytoestrogens could manifest their antioxidant effect thanks to their ability to donate hydrogen atoms to free radicals, manifesting a quenching action on oxidative stress [131]. However, genistein and daidzein likely have a lower antioxidant capacity than quercetin, due to the presence of only one hydroxyl group compared to the two hydroxyl groups in the B-ring of the latter [132]. In cell cultures, the action of soy isoflavones increases the expression of the metallothioneins and antioxidant enzymes, thus also showing an indirect antioxidant effect [133,134]. Another indirect effect on oxidative stress could be represented by the ability of soy isoflavones to modulate the cellular management of iron by removing molecules that can trigger Fenton reactions in the cytosol [135]. Absorption and metabolization seem to be crucial aspects for the efficacy of isoflavones and explain the frequent discrepancies between in vitro and in vivo studies [136,137]. Despite the great interest, the biological activity of soy is not limited to proteins and phytoestrogens but also other functional components, such as peptides, saponins, phytosterols, protease inhibitors, and phytic acid [138,139]. Recently, research has shown great interest in bioactive soybean peptides, such as lunasin, BowmanBirk inhibitor, and soymorphins, especially for their ACE-inhibitor-like efficacy as an effective approach in the fight against hypertension and cardiovascular diseases [138]. Bioactive peptides are 2- to 20-amino-acid-long molecules that are activated only after being detached from the parent protein by food transformation or intestinal digestion processes [140]. Lunasin and soy protein seem to play a ROS scavenger action [141,142]. Many amino acids seem to contribute to the antioxidant effect of peptides [143]. Since the fermentation of foods seems to cut the protein components of soy, the antioxidant effect also increases in comparison with unfermented foods [144]. Another class of bioactive compounds, of which soy is the main source, are saponins [145]. The antioxidant capacity of these compounds is far greater than that of SOD [146]. However, the effectiveness of these compounds depends on their bioavailability. These compounds in food are found in the glycoside form and therefore not bioavailable until the oligosaccharide moieties are cut to obtain the aglycone forms, called soyasapogenols [139]. As in the case of isoflavones, saponins are not naturally found in the aglycone form but can be obtained following the processing of soy foods [139]. Although saponins are generally considered to be poorly bioavailable, gut bacteria can hydrolyze oligosaccharides moieties [147]; however, saponins or their metabolites were not recovered in the urine of healthy women after intake [148]. Other components present in soybeans, although in limited quantities, are phytosterols and triterpenes. These compounds are mainly studied for their cholesterol-lowering action, which does not seem to require intestinal absorption to exert their action, being generally poorly absorbed and not requiring the simultaneous presence of cholesterol in the intestinal lumen [139]. The mechanism of action is still unclear but an antineoplastic effect of phytosterols has been proposed through an antioxidant effect [149]. Other phytoestrogens different from isoflavones are present in soy, mainly characterized in other plant foods such as lignans [150,151]. As with other bioactive compounds discussed, the absorption of lignans is closely linked to the ability to metabolize the glycoside form by the intestinal microbiota, capable of transforming them into enterolignans [152]. Although not specifically shown for soy lignans, an antioxidant and cytoprotective effect has been shown, based on the ability of lignans to scavenge free radicals [153–157]. Phytic acid, a molecule responsible for the accumulation of phosphorus in the plant, is also capable of exerting an antioxidant effect through its ability to chelate metal ions [156,157]. However, phytic acid may not be effectively absorbed from the intestinal lumen [158]. In addition to isoflavones and lignans, other polyphenols can be found in soybeans, including phenolic acids, anthocyanins, tannins, and stilbenes, with strong antioxidant power. These components are often associated with external layers of soybean and with a wide variability, depending on the cultivars and processing [103].

5. Remarks and Future Perspectives

Literature shows an evident gap regarding the antioxidant mechanisms of soy in health. This discrepancy of the literature is particularly relevant taking into consideration that the antioxidant effect of soy is frequently cited but not described in depth. The causes of this phenomenon are manifold: (i) The interest is largely focused on the interaction between isoflavones and the estrogen receptors. Even if this effect is the most evident, the currently available evidence suggests that it is not sufficient to justify the effects of soy on health. Furthermore, isoflavones itself can act with multiple mechanisms, probably concerted with non-isoflavone soy phytochemicals. (ii) There are many markers of oxidative stress but there is no consensus of methods and cutoffs adequate to detect the antioxidant ability of a substance in vivo. (iii) The presence of numerous confounding factors, especially dietary ones, raises critical issues in detecting the real effects of soy against oxidation. This has frequently led to the use of animal models whose data are difficult to transfer to humans for physiological differences. For example, rodent and non-human primate differently metabolize isoflavones. Moreover, there are discrepancies between experimental animal conditions and human real life. (iv) The detection of a physiological effect could be greatly different by the type of food used (soybeans, traditional soy foods, and soy-based meat analogs and supplements) and by the population reference (healthy individuals, patients with metabolic diseases, pre or postmenopausal, athletes, etc.).

6. Conclusions

Overall, the antioxidant effect seems to play an important role in the beneficial effects of soy and its bioactive substances in human health. More in-depth studies are needed solve inconsistencies in the literature. There is a need for clinical trials that simultaneously evaluate the antioxidant effect of soy on the main groups of markers available (i.e., on nucleic acids, proteins, fats, total antioxidant capacity, and endogenous enzymes). It is necessary to evaluate these aspects with adequate time laps to avoid interpreting transitory effects. Furthermore, the use of large populations would allow a correct assessment of the presence of equol-compliant individuals that could, at least in part, clarify some of the discordant data present in the literature.

Funding: This research received no external funding.

Conflicts of Interest: The author declares no conflict of interest.

References

1. Zaheer, K.; Akhtar, M.H. An updated review of dietary isoflavones: Nutrition, processing, bioavailability and impacts on human health. *Crit. Rev. Food Sci. Nutr.* **2017**, *57*, 1280–1293. [CrossRef] [PubMed]
2. Rizzo, G.; Baroni, L. Soy, Soy Foods and Their Role in Vegetarian Diets. *Nutrients* **2018**, *10*, 43. [CrossRef] [PubMed]
3. Messina, M. Modern applications for an ancient bean: Soybeans and the prevention and treatment of chronic disease. *J. Nutr.* **1995**, *125*, 567S–569S. [PubMed]
4. Zamora-Ros, R.; Forouhi, N.G.; Sharp, S.J.; González, C.A.; Buijsse, B.; Guevara, M.; Van Der Schouw, Y.T.; Amiano, P.; Boeing, H.; Bredsdorff, L.; et al. The Association between Dietary Flavonoid and Lignan Intakes and Incident Type 2 Diabetes in European Populations: The EPIC-InterAct study. *Diabetes Care* **2013**, *36*, 3961–3970. [CrossRef]
5. Ding, M.; Pan, A.; Manson, J.E.; Willett, W.C.; Malik, V.; Rosner, B.; Giovannucci, E.; Hu, F.B.; Sun, Q. Consumption of soy foods and isoflavones and risk of type 2 diabetes: A pooled analysis of three US cohorts. *Eur. J. Clin. Nutr.* **2016**, *70*, 1381–1387. [CrossRef]
6. Villegas, R.; Gao, Y.-T.; Yang, G.; Li, H.-L.; Elasy, T.A.; Zheng, W.; Shu, X.-O. Legume and soy food intake and the incidence of type 2 diabetes in the Shanghai Women's Health Study. *Am. J. Clin. Nutr.* **2008**, *87*, 162–167. [CrossRef]
7. Misso, M.L.; Murata, Y.; Boon, W.C.; Jones, M.; Britt, K.; Simpson, E.R. Cellular and Molecular Characterization of the Adipose Phenotype of the Aromatase-Deficient Mouse. *Endocrinology* **2003**, *144*, 1474–1480. [CrossRef]

8. Morimoto, Y.; Steinbrecher, A.; Kolonel, L.N.; Maskarinec, G. Soy consumption is not protective against diabetes in Hawaii: The Multiethnic Cohort. *Eur. J. Clin. Nutr.* **2011**, *65*, 279–282. [CrossRef]
9. Nanri, A.; Mizoue, T.; Takahashi, Y.; Kirii, K.; Inoue, M.; Noda, M.; Tsugane, S. Soy Product and Isoflavone Intakes Are Associated with a Lower Risk of Type 2 Diabetes in Overweight Japanese Women. *J. Nutr.* **2010**, *140*, 580–586. [CrossRef]
10. Food labeling: Health claims; soy protein and coronary heart disease. Food and Drug Administration, HHS. Final rule. *Fed. Regist.* **1999**, *64*, 57700–57733.
11. Food Labeling: Health Claims; Soy Protein and Coronary Heart Disease. Available online: https://www.federalregister.gov/documents/2017/10/31/2017-23629/food-labeling-health-claims-soy-protein-and-coronary-heart-disease (accessed on 15 June 2020).
12. Jenkins, D.J.A.; Mejia, S.B.; Chiavaroli, L.; Viguiliouk, E.; Li, S.S.; Kendall, C.W.C.; Vuksan, V.; Sievenpiper, J.L. Cumulative Meta-Analysis of the Soy Effect Over Time. *J. Am. Heart Assoc.* **2019**, *8*, e012458. [CrossRef] [PubMed]
13. Jenkins, D.J.A.; Jenkins, D.J.; Marchie, A.; Faulkner, D.A.; Wong, J.M.W.; De Souza, R.J.; Emam, A.; Parker, T.L.; Vidgen, E.; Lapsley, K.G.; et al. Effects of a Dietary Portfolio of Cholesterol-Lowering Foods vs Lovastatin on Serum Lipids and C-Reactive Protein. *J. Am. Med. Assoc.* **2003**, *290*, 502–510. [CrossRef] [PubMed]
14. Canada, H. Summary of Health Canada's Assessment of a Health Claim about Soy Protein and Cholesterol Lowering. Available online: https://www.canada.ca/en/health-canada/services/food-nutrition/food-labelling/health-claims/assessments/summary-assessment-health-claim-about-protein-cholesterol-lowering.html (accessed on 15 June 2020).
15. Nachvak, S.M.; Moradi, S.; Anjom-Shoae, J.; Rahmani, J.; Nasiri, M.; Maleki, V.; Sadeghi, O. Soy, Soy Isoflavones, and Protein Intake in Relation to Mortality from All Causes, Cancers, and Cardiovascular Diseases: A Systematic Review and Dose-Response Meta-Analysis of Prospective Cohort Studies. *J. Acad. Nutr. Diet.* **2019**, *119*, 1483–1500.e17. [CrossRef] [PubMed]
16. Fraser, G.E.; Jaceldo-Siegl, K.; Orlich, M.; Mashchak, A.; Sirirat, R.; Knutsen, S. Dairy, soy, and risk of breast cancer: Those confounded milks. *Int. J. Epidemiol.* **2020**. [CrossRef]
17. Applegate, C.C.; Rowles, J.L.; Ranard, K.M.; Jeon, S.; Erdman, J.W. Soy Consumption and the Risk of Prostate Cancer: An Updated Systematic Review and Meta-Analysis. *Nutrients* **2018**, *10*, 40. [CrossRef]
18. Kulling, S.E.; Honig, D.M.; Metzler, M. Oxidative metabolism of the soy isoflavones daidzein and genistein in humans in vitro and in vivo. *J. Agric. Food Chem.* **2001**, *49*, 3024–3033. [CrossRef]
19. Kulling, S.E.; Honig, D.M.; Simat, T.J.; Metzler, M. Oxidative in vitro metabolism of the soy phytoestrogens daidzein and genistein. *J. Agric. Food Chem.* **2000**, *48*, 4963–4972. [CrossRef]
20. Durazzo, A.; Gabrielli, P.; Manzi, P. Qualitative Study of Functional Groups and Antioxidant Properties of Soy-Based Beverages Compared to Cow Milk. *Antioxidants* **2015**, *4*, 523–532. [CrossRef]
21. Van Der Velpen, V.; Geelen, A.; Schouten, E.G.; Hollman, P.; Afman, L.A.; Veer, P.V. 'T Estrogen Receptor–Mediated Effects of Isoflavone Supplementation Were Not Observed in Whole-Genome Gene Expression Profiles of Peripheral Blood Mononuclear Cells in Postmenopausal, Equol-Producing Women. *J. Nutr.* **2013**, *143*, 774–780. [CrossRef]
22. Monnier, L.; Mas, E.; Ginet, C.; Michel, F.; Villon, L.; Cristol, J.-P.; Colette, C. Activation of Oxidative Stress by Acute Glucose Fluctuations Compared with Sustained Chronic Hyperglycemia in Patients with Type 2 Diabetes. *J. Am. Med. Assoc.* **2006**, *295*, 1681–1687. [CrossRef]
23. Taniyama, Y.; Griendling, K.K. Reactive Oxygen Species in the Vasculature. *Hypertension* **2003**, *42*, 1075–1081. [CrossRef] [PubMed]
24. Marnett, L.J. Oxyradicals and DNA damage. *Carcinogensis* **2000**, *21*, 361–370. [CrossRef] [PubMed]
25. PubMed. Available online: https://pubmed.ncbi.nlm.nih.gov/ (accessed on 19 May 2020).
26. Trials Search | Cochrane Library. Available online: https://www.cochranelibrary.com/central (accessed on 19 May 2020).
27. Ahn-Jarvis, J.; Clinton, S.K.; Riedl, K.M.; Vodovotz, Y.; Schwartz, S.J. Impact of food matrix on isoflavone metabolism and cardiovascular biomarkers in adults with hypercholesterolemia. *Food Funct.* **2012**, *3*, 1051–1058. [CrossRef] [PubMed]
28. Ashton, E.L.; Dalais, F.S.; Ball, M.J. Effect of meat replacement by tofu on CHD risk factors including copper induced LDL oxidation. *J. Am. Coll. Nutr.* **2000**, *19*, 761–767. [CrossRef] [PubMed]

29. Azadbakht, L.; Kimiagar, M.; Mehrabi, Y.; Esmaillzadeh, A.; Hu, F.B.; Willett, W.C. Dietary soya intake alters plasma antioxidant status and lipid peroxidation in postmenopausal women with the metabolic syndrome. *Br. J. Nutr.* **2007**, *98*, 807–813. [CrossRef]
30. Bakhtiari, A.; Hajian-Tilaki, K.; Omidvar, S.; Nasiri-Amiri, F. Clinical and metabolic response to soy administration in older women with metabolic syndrome: A randomized controlled trial. *Diabetol. Metab. Syndr.* **2019**, *11*, 1–12. [CrossRef]
31. Bakhtiary, A.; Yassin, Z.; Hanachi, P.; Rahmat, A.; Ahmad, Z.; Halalkhor, S.; Firouzjahi, A. Evaluation of the oxidative stress and glycemic control status in response to soy in older women with the metabolic syndrome. *Irani. Red Crescent Med. J.* **2011**, *13*, 795–804.
32. Beavers, K.M.; Serra, M.C.; Beavers, D.P.; Cooke, M.; Willoughby, D.S. Soymilk supplementation does not alter plasma markers of inflammation and oxidative stress in postmenopausal women. *Nutr. Res.* **2009**, *29*, 616–622. [CrossRef]
33. Beavers, K.M.; Serra, M.C.; Beavers, D.P.; Cooke, M.B.; Willoughby, D.S. Soy and the exercise-induced inflammatory response in postmenopausal women. *Appl. Physiol. Nutr. Metab.* **2010**, *35*, 261–269. [CrossRef]
34. Brandão, L.C.; Hachul, H.; Bittencourt, L.R.A.; Baracat, E.C.; Tufik, S.; D'Almeida, V. Effects of isoflavone on oxidative stress parameters and homocysteine in postmenopausal women complaining of insomnia. *Boil. Res.* **2009**, *42*, 281–287. [CrossRef]
35. Brown, E.C.; DiSilvestro, R.A.; Babaknia, A.; Devor, S.T. Soy versus whey protein bars: Effects on exercise training impact on lean body mass and antioxidant status. *Nutr. J.* **2004**, *3*, 22. [CrossRef]
36. Campbell, C.G.; Brown, B.D.; Dufner, D.; Thorland, W.G. Effects of soy or milk protein durign a high-fat feeding challenge on oxidative stress, inflammation, and lipids in healthy men. *Lipids* **2006**, *41*, 257–265. [CrossRef] [PubMed]
37. Celec, P.; Hodosy, J.; Palffy, R.; Gardlík, R.; Halčák, L.; Ostatníková, D. The Short-Term Effects of Soybean Intake on Oxidative and Carbonyl Stress in Men and Women. *Molecules* **2013**, *18*, 5190–5200. [CrossRef] [PubMed]
38. Cha, Y.-S.; Park, Y.; Lee, M.; Chae, S.-W.; Park, K.; Kim, Y.; Lee, H.-S. Doenjang, a Korean Fermented Soy Food, Exerts Antiobesity and Antioxidative Activities in Overweight Subjects with the PPAR-γ2 C1431T Polymorphism: 12-Week, Double-Blind Randomized Clinical Trial. *J. Med. Food* **2014**, *17*, 119–127. [CrossRef] [PubMed]
39. Chen, C.-Y.; Bakhiet, R.M.; Hart, V.; Holtzman, G. Isoflavones restore altered redox homeostasis of antioxidant enzymes in healthy young men undergoing 80% peak oxygen consumption (VO2pk) exercise. *Nutr. Res.* **2004**, *24*, 347–359. [CrossRef]
40. Cicero, A.F.G.; Minardi, M.; Mirembe, S.; Pedro, E.; Gaddi, A. Effects of a new low dose soy protein/?-sitosterol association on plasma lipid levels and oxidation. *Eur. J. Nutr.* **2004**, *43*, 319–322. [CrossRef]
41. Clerici, C.; Nardi, E.; Battezzati, P.M.; Asciutti, S.; Castellani, D.; Corazzi, N.; Giuliano, V.; Gizzi, S.; Perriello, G.; Di Matteo, G.; et al. Novel Soy Germ Pasta Improves Endothelial Function, Blood Pressure, and Oxidative Stress in Patients with Type 2 Diabetes: Figure 1. *Diabetes Care* **2011**, *34*, 1946–1948. [CrossRef]
42. Silva, L.M.C.E.; De Melo, M.L.P.; Reis, F.V.F.; Monteiro, M.C.; Dos Santos, S.M.; Gomes, B.A.Q.; Da Silva, L.H.M. Comparison of the Effects of Brazil Nut Oil and Soybean Oil on the Cardiometabolic Parameters of Patients with Metabolic Syndrome: A Randomized Trial. *Nutrients* **2020**, *12*, 46. [CrossRef]
43. Djuric, Z.; Chen, G.; Doerge, D.R.; Heilbrun, L.K.; Kucuk, O. Effect of soy isoflavone supplementation on markers of oxidative stress in men and women. *Cancer Lett.* **2001**, *172*, 1–6. [CrossRef]
44. Engelman, H.M.; Alekel, D.L.; Hanson, L.N.; Kanthasamy, A.G.; Reddy, M.B. Blood lipid and oxidative stress responses to soy protein with isoflavones and phytic acid in postmenopausal women. *Am. J. Clin. Nutr.* **2005**, *81*, 590–596. [CrossRef]
45. Fei, B.-B.; Ling, L.; Hua, C.; Ren, S.-Y. Effects of soybean oligosaccharides on antioxidant enzyme activities and insulin resistance in pregnant women with gestational diabetes mellitus. *Food Chem.* **2014**, *158*, 429–432. [CrossRef] [PubMed]
46. Fritz, K.L.; Seppanen, C.; Kurzer, M.S.; Csallany, A.S. The in vivo antioxidant activity of soybean isoflavones in human subjects. *Nutr. Res.* **2003**, *23*, 479–487. [CrossRef]
47. Gardner-Thorpe, D.; O'Hagen, C.; Young, I.S.; Lewis, S. Dietary supplements of soya flour lower serum testosterone concentrations and improve markers of oxidative stress in men. *Eur. J. Clin. Nutr.* **2003**, *57*, 100–106. [CrossRef] [PubMed]

48. Hanachi, P.; Golkho, S.; Ahmadi, A.; Barantalab, F. The effect of soymilk on alkaline phosphatase, total antioxidant levels, and vasomotor symptoms in menopause women. *Iran. J. Basic Med. Sci.* **2007**, *10*, 162–168. [CrossRef]
49. Hanachi, P.; Golkho, S. Assessment of Soy Phytoestrogens and Exercise on Lipid Profiles and Menopause Symptoms in Menopausal Women. *J. Boil. Sci.* **2008**, *8*, 789–793. [CrossRef]
50. Hanwell, H.; Kay, C.D.; Lampe, J.W.; Holub, B.J.; Duncan, A.M. Acute Fish Oil and Soy Isoflavone Supplementation Increase Postprandial Serum (n-3) Polyunsaturated Fatty Acids and Isoflavones but Do Not Affect Triacylglycerols or Biomarkers of Oxidative Stress in Overweight and Obese Hypertriglyceridemic Men. *J. Nutr.* **2009**, *139*, 1128–1134. [CrossRef]
51. Hariri, M.; Salehi, R.; Feizi, A.; Mirlohi, M.; Ghiasvand, R.; Habibi, N. A randomized, double-blind, placebo-controlled, clinical trial on probiotic soy milk and soy milk: Effects on epigenetics and oxidative stress in patients with type II diabetes. *Genes Nutr.* **2015**, *10*, 1–8. [CrossRef]
52. Hematdar, Z.; Ghasemifard, N.; Phishdad, G.; Faghih, S. Substitution of red meat with soybean but not non-soy legumes improves inflammation in patients with type 2 diabetes; a randomized clinical trial. *J. Diabetes Metab. Disord.* **2018**, *17*, 111–116. [CrossRef]
53. Heneman, K.M.; Chang, H.C.; Prior, R.L.; Steinberg, F.M. Soy protein with and without isoflavones fails to substantially increase postprandial antioxidant capacity. *J. Nutr. Biochem.* **2007**, *18*, 46–53. [CrossRef]
54. Hill, S.; Box, W.; DiSilvestro, R.A. Moderate intensity resistance exercise, plus or minus soy intake: Effects on serum lipid peroxides in young adult males. *Int. J. Sport Nutr. Exerc. Metab.* **2004**, *14*, 125–132. [CrossRef]
55. Jamilian, M.; Asemi, Z. The Effect of Soy Intake on Metabolic Profiles of Women with Gestational Diabetes Mellitus. *J. Clin. Endocrinol. Metab.* **2015**, *100*, 4654–4661. [CrossRef] [PubMed]
56. Jamilian, M.; Asemi, Z. The Effects of Soy Isoflavones on Metabolic Status of Patients with Polycystic Ovary Syndrome. *J. Clin. Endocrinol. Metab.* **2016**, *101*, 3386–3394. [CrossRef] [PubMed]
57. Jenkins, D.J.; Kendall, C.W.; Garsetti, M.; Rosenberg-Zand, R.S.; Jackson, C.-J.; Agarwal, S.; Rao, A.V.; Diamandis, E.; Parker, T.; Faulkner, D.; et al. Effect of soy protein foods on low-density lipoprotein oxidation and ex vivo sex hormone receptor activity—A controlled crossover trial. *Metabolism* **2000**, *49*, 537–543. [CrossRef]
58. Jenkins, D.J.A.; Kendall, C.W.; Vidgen, E.; Vuksan, V.; Jackson, C.-J.; Augustin, L.; Lee, B.; Garsetti, M.; Agarwal, S.; Rao, A.; et al. Effect of soy-based breakfast cereal on blood lipids and oxidized low-density lipoprotein. *Metabolism* **2000**, *49*, 1496–1500. [CrossRef] [PubMed]
59. Jenkins, D.J.A.; Jenkins, D.J.; Jackson, C.-J.C.; Connelly, P.; Parker, T.; Faulkner, D.; Vidgen, E.; Cunnane, S.C.; Leiter, L.A.; Josse, R.G. Effects of high- and low-isoflavone soyfoods on blood lipids, oxidized LDL, homocysteine, and blood pressure in hyperlipidemic men and women. *Am. J. Clin. Nutr.* **2002**, *76*, 365–372. [CrossRef]
60. Karamali, M.; Kashanian, M.; Alaeinasab, S.; Asemi, Z. The effect of dietary soy intake on weight loss, glycaemic control, lipid profiles and biomarkers of inflammation and oxidative stress in women with polycystic ovary syndrome: A randomised clinical trial. *J. Hum. Nutr. Diet.* **2018**, *31*, 533–543. [CrossRef]
61. Kwak, J.H.; Kim, M.; Lee, E.; Lee, S.-H.; Ahn, C.-W.; Lee, J.H. Effects of black soy peptide supplementation on blood pressure and oxidative stress: A randomized controlled trial. *Hypertens. Res.* **2013**, *36*, 1060–1066. [CrossRef]
62. Lee, C.-Y.J.; Isaac, H.B.; Wang, H.; Huang, S.H.; Long, L.-H.; Jenner, A.M.; Kelly, R.P.; Halliwell, B.; Lee, J.C.-Y. Cautions in the use of biomarkers of oxidative damage; the vascular and antioxidant effects of dark soy sauce in humans. *Biochem. Biophys. Res. Commun.* **2006**, *344*, 906–911. [CrossRef]
63. Li, Y.; Zhang, H. Soybean isoflavones ameliorate ischemic cardiomyopathy by activating Nrf2-mediated antioxidant responses. *Food Funct.* **2017**, *8*, 2935–2944. [CrossRef]
64. Miraghajani, M.S.; Esmaillzadeh, A.; Najafabadi, M.M.; Mirlohi, M.; Azadbakht, L. Soy Milk Consumption, Inflammation, Coagulation, and Oxidative Stress among Type 2 Diabetic Patients with Nephropathy. *Diabetes Care* **2012**, *35*, 1981–1985. [CrossRef]
65. Miraghajani, M.; Zaghian, N.; Mirlohi, M.; Feizi, A.; Ghiasvand, R. The Impact of Probiotic Soy Milk Consumption on Oxidative Stress among Type 2 Diabetic Kidney Disease Patients: A Randomized Controlled Clinical Trial. *J. Ren. Nutr.* **2017**, *27*, 317–324. [CrossRef] [PubMed]
66. Mitchell, J.; Collins, A. Effects of a soy milk supplement on plasma cholesterol levels and oxidative DNA damage in men–a pilot study. *Eur. J. Nutr.* **1999**, *38*, 138–143. [CrossRef] [PubMed]

67. Mittal, R.; Mittal, N.; Hota, D.; Suri, V.; Aggarwal, N.; Chakrabarti, A. Antioxidant effect of isoflavones: A randomized, double-blind, placebo controlled study in oophorectomized women. *Int. J. Appl. Basic Med. Res.* **2014**, *4*, 28–33. [CrossRef] [PubMed]
68. Nestel, P.J.; Yamashita, T.; Sasahara, T.; Pomeroy, S.; Dart, A.; Komesaroff, P.A.; Owen, A.; Abbey, M. Soy Isoflavones Improve Systemic Arterial Compliance but Not Plasma Lipids in Menopausal and Perimenopausal Women. *Arterioscler. Thromb. Vasc. Boil.* **1997**, *17*, 3392–3398. [CrossRef]
69. Nhan, S.; Anderson, K.; Nagamani, M.; Grady, J.J.; Lu, L.-J.W. Effect of a Soymilk Supplement Containing Isoflavones on Urinary F2 Isoprostane Levels in Premenopausal Women. *Nutr. Cancer* **2005**, *53*, 73–81. [CrossRef] [PubMed]
70. Oh, H.Y.; Kim, S.S.; Chung, H.-Y.; Yoon, S. Isoflavone Supplements Exert Hormonal and Antioxidant Effects in Postmenopausal Korean Women with Diabetic Retinopathy. *J. Med. Food* **2005**, *8*, 1–7. [CrossRef]
71. Önning, G.; Åkesson, B.; Öste, R.; Lundquist, I. Effects of consumption of oat milk, soya milk, or cow's milk on plasma lipids and antioxidative capacity in healthy subjects. *Ann. Nutr. Metab.* **1998**, *42*, 211–220. [CrossRef]
72. Pusparini; Dharma, R.; Suyatna, F.D.; Mansyur, M.; Hidajat, A. Effect of soy isoflavone supplementation on vascular endothelial function and oxidative stress in postmenopausal women: A community randomized controlled trial. *Asia Pac. J. Clin. Nutr.* **2013**, *22*, 357–364.
73. Reverri, E.J.; LaSalle, C.D.; Franke, A.A.; Steinberg, F.M. Soy provides modest benefits on endothelial function without affecting inflammatory biomarkers in adults at cardiometabolic risk. *Mol. Nutr. Food Res.* **2014**, *59*, 323–333. [CrossRef]
74. Rossi, A.L.; Blostein-Fujii, A.; DiSilvestro, R.A. Soy Beverage Consumption by Young Men. *J. Nutraceuticals, Funct. Med Foods* **2001**, *3*, 33–44. [CrossRef]
75. Ryan-Borchers, T.A.; Park, J.S.; Chew, B.P.; McGuire, M.K.; Fournier, L.R.; Beerman, K.A. Soy isoflavones modulate immune function in healthy postmenopausal women. *Am. J. Clin. Nutr.* **2006**, *83*, 1118–1125. [CrossRef] [PubMed]
76. Sedaghat, A.; Shahbazian, H.; Rezazadeh, A.; Haidari, F.; Jahanshahi, A.; Latifi, S.M.; Shirbeigi, E. The effect of soy nut on serum total antioxidant, endothelial function and cardiovascular risk factors in patients with type 2 diabetes. *Diabetes Metab. Syndr. Clin. Res. Rev.* **2019**, *13*, 1387–1391. [CrossRef] [PubMed]
77. Sen, C.; Morimoto, Y.; Heak, S.; Cooney, R.V.; Franke, A.A.; Maskarinec, G. Soy foods and urinary isoprostanes: Results from a randomized study in premenopausal women. *Food Funct.* **2012**, *3*, 517–521. [CrossRef] [PubMed]
78. Serrano, J.C.; Martín-Gari, M.; Cassanye, A.; Granado-Serrano, A.B.; Portero-Otín, M. Characterization of the post-prandial insulinemic response and low glycaemic index of a soy beverage. *PLoS ONE* **2017**, *12*, e0182762. [CrossRef] [PubMed]
79. Steinberg, F.M.; Guthrie, N.L.; Villablanca, A.C.; Kumar, K.; Murray, M.J. Soy protein with isoflavones has favorable effects on endothelial function that are independent of lipid and antioxidant effects in healthy postmenopausal women. *Am. J. Clin. Nutr.* **2003**, *78*, 123–130. [CrossRef] [PubMed]
80. Swain, J.H.; Alekel, D.L.; Dent, S.B.; Peterson, C.T.; Reddy, M.B. Iron indexes and total antioxidant status in response to soy protein intake in perimenopausal women. *Am. J. Clin. Nutr.* **2002**, *76*, 165–171. [CrossRef] [PubMed]
81. Taniguchi-Fukatsu, A.; Yamanaka-Okumura, H.; Naniwa-Kuroki, Y.; Nishida, Y.; Yamamoto, H.; Taketani, Y.; Takeda, E. Natto and viscous vegetables in a Japanese-style breakfast improved insulin sensitivity, lipid metabolism and oxidative stress in overweight subjects with impaired glucose tolerance. *Br. J. Nutr.* **2011**, *107*, 1184–1191. [CrossRef]
82. Tikkanen, M.J.; Wähälä, K.; Ojala, S.; Vihma, V.; Adlercreutz, H. Effect of soybean phytoestrogen intake on low density lipoprotein oxidation resistance. *Proc. Natl. Acad. Sci. USA* **1998**, *95*, 3106–3110. [CrossRef]
83. Utarwuthipong, T.; Komindr, S.; Pakpeankitvatana, V.; Songchitsomboon, S.; Thongmuang, N. Small Dense Low-Density Lipoprotein Concentration and Oxidative Susceptibility Changes after Consumption of Soybean Oil, Rice Bran Oil, Palm Oil and Mixed Rice Bran/Palm Oil in Hypercholesterolaemic Women. *J. Int. Med Res.* **2009**, *37*, 96–104. [CrossRef]
84. Vega-López, S.; Yeum, K.-J.; Lecker, J.L.; Ausman, L.M.; Johnson, E.J.; Devaraj, S.; Jialal, I.; Lichtenstein, A.H. Plasma antioxidant capacity in response to diets high in soy or animal protein with or without isoflavones. *Am. J. Clin. Nutr.* **2005**, *81*, 43–49. [CrossRef]

85. Wiseman, H.; O'Reilly, J.D.; Adlercreutz, H.; Mallet, A.I.; Bowey, E.A.; Rowland, I.R.; Ab Sanders, T. Isoflavone phytoestrogens consumed in soy decrease F2-isoprostane concentrations and increase resistance of low-density lipoprotein to oxidation in humans. *Am. J. Clin. Nutr.* **2000**, *72*, 395–400. [CrossRef] [PubMed]
86. Zemel, M.; Sun, X.; Sobhani, T.; Wilson, B. Effects of dairy compared with soy on oxidative and inflammatory stress in overweight and obese subjects. *Am. J. Clin. Nutr.* **2009**, *91*, 16–22. [CrossRef]
87. Miller, N.J.; Rice-Evans, C.A. Factors Influencing the Antioxidant Activity Determined by the $ABTS^{\bullet+}$ Radical Cation Assay. *Free. Radic. Res.* **1997**, *26*, 195–199. [CrossRef] [PubMed]
88. Santella, R.M. Immunological methods for detection of carcinogen-DNA damage in humans. *Cancer Epidemiol. Biomark. Prev.* **1999**, *8*, 733–739.
89. Ii, L.R.; Morrow, J.D. The generation and actions of isoprostanes. *Biochim. Biophys. Acta* **1997**, *1345*, 121–135. [CrossRef]
90. Witko-Sarsat, V.; Friedlander, M.; Capeillère-Blandin, C.; Nguyen-Khoa, T.; Nguyen, A.T.; Zingraff, J.; Jungers, P.; Descamps-Latscha, B. Advanced oxidation protein products as a novel marker of oxidative stress in uremia. *Kidney Int.* **1996**, *49*, 1304–1313. [CrossRef] [PubMed]
91. Collins, A.; Duthie, S.; Dobson, V.L. Direct enzymic detection of endogenous oxidative base damage in human lymphocyte DNA. *Carcinogenesis* **1993**, *14*, 1733–1735. [CrossRef] [PubMed]
92. Benzie, I.; Strain, J. The Ferric Reducing Ability of Plasma (FRAP) as a Measure of "Antioxidant Power": The FRAP Assay. *Anal. Biochem.* **1996**, *239*, 70–76. [CrossRef] [PubMed]
93. Wolff, S.P. [18] Ferrous ion oxidation in presence of ferric ion indicator xylenol orange for measurement of hydroperoxides. In *Methods in Enzymology*; Academic Press: San Diego, CA, USA, 1994; Volume 233, pp. 182–189. [CrossRef]
94. Kim, S.-S.; Gallaher, D.; Csallany, A.S. Lipophilic aldehydes and related carbonyl compounds in rat and human urine. *Lipids* **1999**, *34*, 489–496. [CrossRef]
95. Amaki, T.; Suzuki, T.; Nakamura, F.; Hayashi, D.; Imai, Y.; Morita, H.; Fukino, K.; Nojiri, T.; Kitano, S.; Hibi, N.; et al. Circulating malondialdehyde modified LDL is a biochemical risk marker for coronary artery disease. *Heart* **2004**, *90*, 1211–1213. [CrossRef]
96. Ma, Q. Role of nrf2 in oxidative stress and toxicity. *Annu. Rev. Pharmacol. Toxicol.* **2013**, *53*, 401–426. [CrossRef]
97. Esterbauer, H.; Striegl, G.; Puhl, H.; Rotheneder, M. Continuous Monitoring ofin VztroOxidation of Human Low Density Lipoprotein. *Free. Radic. Res. Commun.* **1989**, *6*, 67–75. [CrossRef] [PubMed]
98. Cao, G.; Alessio, H.M.; Cutler, R.G. Oxygen-radical absorbance capacity assay for antioxidants. *Free. Radic. Boil. Med.* **1993**, *14*, 303–311. [CrossRef]
99. Aldini, G.; Yeum, K.-J.; Russell, R.M.; Krinsky, N.I. A method to measure the oxidizability of both the aqueous and lipid compartments of plasma. *Free. Radic. Boil. Med.* **2001**, *31*, 1043–1050. [CrossRef]
100. Janero, D.R. Malondialdehyde and thiobarbituric acid-reactivity as diagnostic indices of lipid peroxidation and peroxidative tissue injury. *Free. Radic. Boil. Med.* **1990**, *9*, 515–540. [CrossRef]
101. Wayner, D.; Burton, G.; Ingold, K.; Locke, S. Quantitative measurement of the total, peroxyl radical-trapping antioxidant capability of human blood plasma by controlled peroxidation. *FEBS Lett.* **1985**, *187*, 33–37. [CrossRef]
102. Schnitzer, E.; Pinchuk, I.; Bor, A.; Fainaru, M.; Samuni, A.M.; Lichtenberg, D. Lipid oxidation in unfractionated serum and plasma. *Chem. Phys. Lipids* **1998**, *92*, 151–170. [CrossRef]
103. Ganesan, K.; Xu, B. A Critical Review on Polyphenols and Health Benefits of Black Soybeans. *Nutrients* **2017**, *9*, 455. [CrossRef]
104. Messina, M. Soy and Health Update: Evaluation of the Clinical and Epidemiologic Literature. *Nutrients* **2016**, *8*, 754. [CrossRef]
105. Peake, J.M.; Suzuki, K.; Wilson, G.; Hordern, M.; Nosaka, K.; MacKinnon, L.; Coombes, J.S. Exercise-Induced Muscle Damage, Plasma Cytokines, and Markers of Neutrophil Activation. *Med. Sci. Sports Exerc.* **2005**, *37*, 737–745. [CrossRef]
106. Stangl, V.; Baumann, G.; Stangl, K. Coronary atherogenic risk factors in women. *Eur. Heart J.* **2002**, *23*, 1738–1752. [CrossRef] [PubMed]

107. Wiseman, H.; Casey, K.; Bowey, E.A.; Duffy, R.; Davies, M.; Rowland, I.R.; Lloyd, A.S.; Murray, A.; Thompson, R.; Clarke, D.B. Influence of 10 wk of soy consumption on plasma concentrations and excretion of isoflavonoids and on gut microflora metabolism in healthy adults. *Am. J. Clin. Nutr.* **2004**, *80*, 692–699. [CrossRef] [PubMed]
108. Harman, D. Aging: A Theory Based on Free Radical and Radiation Chemistry. *J. Gerontol.* **1956**, *11*, 298–300. [CrossRef] [PubMed]
109. Hansel, B.; Giral, P.; Nobecourt, E.; Chantepie, S.; Bruckert, E.; Chapman, M.J.; Kontush, A. Metabolic Syndrome Is Associated with Elevated Oxidative Stress and Dysfunctional Dense High-Density Lipoprotein Particles Displaying Impaired Antioxidative Activity. *J. Clin. Endocrinol. Metab.* **2004**, *89*, 4963–4971. [CrossRef]
110. Dröge, W. Free Radicals in the Physiological Control of Cell Function. *Physiol. Rev.* **2002**, *82*, 47–95. [CrossRef]
111. Harrison, D.G.; Góngora, M.C. Oxidative Stress and Hypertension. *Med Clin. North Am.* **2009**, *93*, 621–635. [CrossRef]
112. Young, I.S.; McEneny, J. Lipoprotein oxidation and atherosclerosis. *Biochem. Soc. Trans.* **2001**, *29*, 358–362. [CrossRef]
113. Grundy, S.M.; Information, P.E.K.F.C. Metabolic syndrome update. *Trends Cardiovasc. Med.* **2016**, *26*, 364–373. [CrossRef]
114. Newsholme, P.; Haber, E.P.; Hirabara, S.M.; Rebelato, E.L.O.; Procopio, J.; Morgan, D.; Oliveira-Emilio, H.C.; Carpinelli, A.R.; Curi, R. Diabetes associated cell stress and dysfunction: Role of mitochondrial and non-mitochondrial ROS production and activity. *J. Physiol.* **2007**, *583*, 9–24. [CrossRef]
115. Banerjee, M.; Vats, P. Reactive metabolites and antioxidant gene polymorphisms in Type 2 diabetes mellitus. *Redox Boil.* **2013**, *2*, 170–177. [CrossRef]
116. Rudich, A.; Tirosh, A.; Potashnik, R.; Hemi, R.; Kanety, H.; Bashan, N. Prolonged oxidative stress impairs insulin-induced GLUT4 translocation in 3T3-L1 adipocytes. *Diabetes* **1998**, *47*, 1562–1569. [CrossRef] [PubMed]
117. Vasu, S.; McClenaghan, N.H.; McCluskey, J.T.; Flatt, P. Cellular responses of novel human pancreatic β-cell line, 1.1B4 to hyperglycemia. *Islets* **2013**, *5*, 170–177. [CrossRef] [PubMed]
118. Morgan, D.; Rebelato, E.; Abdulkader, F.; Graciano, M.F.R.; Oliveira-Emilio, H.R.; Hirata, A.E.; Rocha, M.S.; Bordin, S.A.; Curi, R.; Carpinelli, A.R. Association of NAD(P)H Oxidase with Glucose-Induced Insulin Secretion by Pancreatic β-Cells. *Endocrinology* **2009**, *150*, 2197–2201. [CrossRef] [PubMed]
119. Newsholme, P.; Morgan, D.; Rebelato, E.; Oliveira-Emilio, H.C.; Procopio, J.; Curi, R.; Carpinelli, A. Insights into the critical role of NADPH oxidase(s) in the normal and dysregulated pancreatic beta cell. *Diabetologia* **2009**, *52*, 2489–2498. [CrossRef]
120. Evans, J.; Goldfine, I.D.; Maddux, B.A.; Grodsky, G.M. Oxidative Stress and Stress-Activated Signaling Pathways: A Unifying Hypothesis of Type 2 Diabetes. *Endocr. Rev.* **2002**, *23*, 599–622. [CrossRef] [PubMed]
121. Matsumoto, K.; Sera, Y.; Abe, Y.; Tominaga, T.; Horikami, K.; Hirao, K.; Ueki, Y.; Miyake, S. High serum concentrations of soluble E-selectin correlate with obesity but not fat distribution in patients with type 2 diabetes mellitus. *Metab. Clin. Exp.* **2002**, *51*, 932–934. [CrossRef] [PubMed]
122. Donato, A.J.; Eskurza, I.; Silver, A.E.; Levy, A.S.; Pierce, G.L.; Gates, P.E.; Seals, D.R. Direct Evidence of Endothelial Oxidative Stress With Aging in Humans. *Circ. Res.* **2007**, *100*, 1659–1666. [CrossRef]
123. Giacco, F.; Brownlee, M. Oxidative stress and diabetic complications. *Circ. Res.* **2010**, *107*, 1058–1070. [CrossRef]
124. Badawi, A.; Garcia-Bailo, B.; El-Sohemy, A.; Haddad, P.S.; Arora, P.; Benzaied, F.; Karmali, M. Vitamins D, C, and E in the prevention of type 2 diabetes mellitus: Modulation of inflammation and oxidative stress. *Boil. Targets Ther.* **2011**, *5*, 7–19. [CrossRef]
125. Tabatabaei-Malazy, O.; Larijani, B.; Abdollahi, M. A novel management of diabetes by means of strong antioxidants' combination. *J. Med Hypotheses Ideas* **2013**, *7*, 25–30. [CrossRef]
126. Wang, T.-D.; Chen, W.-J.; Cheng, W.-C.; Lin, J.-W.; Chen, M.; Lee, Y.-T. Relation of Improvement in Endothelium-Dependent Flow-Mediated Vasodilation after Rosiglitazone to Changes in Asymmetric Dimethylarginine, Endothelin-1, and C-Reactive Protein in Nondiabetic Patients with the Metabolic Syndrome. *Am. J. Cardiol.* **2006**, *98*, 1057–1062. [CrossRef] [PubMed]
127. American Diabetes Association Standards of Medical Care in Diabetes–2010. *Diabetes Care* **2010**, *33*, S11–S61. [CrossRef] [PubMed]

128. Jørgensen, C.H.; Gislason, G.H.; Ahlehoff, O.; Andersson, C.; Torp-Pedersen, C.; Hansen, P.R. Use of secondary prevention pharmacotherapy after first myocardial infarction in patients with diabetes mellitus. *BMC Cardiovasc. Disord.* **2014**, *14*, 4. [CrossRef]
129. Federico, A.; Morgillo, F.; Tuccillo, C.; Ciardiello, F.; Loguercio, C.; Loguercio, C. Chronic inflammation and oxidative stress in human carcinogenesis. *Int. J. Cancer* **2007**, *121*, 2381–2386. [CrossRef] [PubMed]
130. Ziech, D.; Franco, R.; Pappa, A.; Panagiotidis, M. Reactive Oxygen Species (ROS)—-Induced genetic and epigenetic alterations in human carcinogenesis. *Mutat. Res. Mol. Mech. Mutagen.* **2011**, *711*, 167–173. [CrossRef] [PubMed]
131. Mitchell, J.H.; Gardner, P.T.; McPhail, D.B.; Morrice, P.C.; Collins, A.R.; Duthie, G.G. Antioxidant Efficacy of Phytoestrogens in Chemical and Biological Model Systems. *Arch. Biochem. Biophys.* **1998**, *360*, 142–148. [CrossRef]
132. Cao, G.; Sofic, E.; Prior, R.L. Antioxidant and Prooxidant Behavior of Flavonoids: Structure-Activity Relationships. *Free. Radic. Boil. Med.* **1997**, *22*, 749–760. [CrossRef]
133. Raschke, M.; Rowland, I.R.; Magee, P.J.; Pool-Zobel, B.L. Genistein protects prostate cells against hydrogen peroxide-induced DNA damage and induces expression of genes involved in the defence against oxidative stress. *Carcinogenesis* **2006**, *27*, 2322–2330. [CrossRef]
134. Suzuki, K.; Koike, H.; Matsui, H.; Ono, Y.; Hasumi, M.; Nakazato, H.; Okugi, H.; Sekine, Y.; Oki, K.; Ito, K.; et al. Genistein, a soy isoflavone, induces glutathione peroxidase in the human prostate cancer cell lines LNCaP and PC-3. *Int. J. Cancer* **2002**, *99*, 846–852. [CrossRef]
135. Persichini, T.; Maio, N.; Di Patti, M.C.B.; Rizzo, G.; Colasanti, M.; Musci, G. Genistein up-regulates the iron efflux system in glial cells. *Neurosci. Lett.* **2010**, *470*, 145–149. [CrossRef]
136. Axelson, M.; Kirk, D.N.; Farrant, R.D.; Cooley, G.; Lawson, A.M.; Setchell, K.D. The identification of the weak oestrogen equol [7-hydroxy-3-(4′-hydroxyphenyl) chroman] in human urine. *Biochem. J.* **1982**, *201*, 353–357. [CrossRef] [PubMed]
137. Hwang, J.; Wang, J.; Morazzoni, P.; Hodis, H.N.; Sevanian, A. The phytoestrogen equol increases nitric oxide availability by inhibiting superoxide production: An antioxidant mechanism for cell-mediated LDL modification. *Free. Radic. Boil. Med.* **2003**, *34*, 1271–1282. [CrossRef]
138. Chatterjee, C.; Gleddie, S.; Xiao, C.-W. Soybean Bioactive Peptides and Their Functional Properties. *Nutrients* **2018**, *10*, 1211. [CrossRef] [PubMed]
139. Kang, J.; Badger, T.M.; Ronis, M.J.J.; Wu, X. Non-isoflavone Phytochemicals in Soy and Their Health Effects. *J. Agric. Food Chem.* **2010**, *58*, 8119–8133. [CrossRef] [PubMed]
140. Singh, B.P.; Vij, S.; Hati, S. Functional significance of bioactive peptides derived from soybean. *Peptides* **2014**, *54*, 171–179. [CrossRef] [PubMed]
141. Lule, V.K.; Garg, S.; Pophaly, S.D.; Hitesh; Tomar, S.K. Potential Health Benefits of Lunasin: A Multifaceted Soy-Derived Bioactive Peptide. *J. Food Sci.* **2015**, *80*, R485–R494. [CrossRef]
142. Draganidis, D.; Karagounis, L.; Athanailidis, I.; Chatzinikolaou, A.; Jamurtas, A.Z.; Fatouros, I.G.; Athanailiidis, I. Inflammaging and Skeletal Muscle: Can Protein Intake Make a Difference? *J. Nutr.* **2016**, *146*, 1940–1952. [CrossRef]
143. Ranamukhaarachchi, S.; Meissner, L.; Moresoli, C.; Regestein, L. Production of antioxidant soy protein hydrolysates by sequential ultrafiltration and nanofiltration. *J. Membr. Sci.* **2013**, *429*, 81–87. [CrossRef]
144. Xiao, Y.; Wang, L.; Rui, X.; Li, W.; Chen, X.; Jiang, M.; Dong, M. Enhancement of the antioxidant capacity of soy whey by fermentation with Lactobacillus plantarum B1–6. *J. Funct. Foods* **2015**, *12*, 33–44. [CrossRef]
145. Oakenfull, D. Saponins in food—A review. *Food Chem.* **1981**, *7*, 19–40. [CrossRef]
146. Yoshiki, Y.; Okubo, K. Active Oxygen Scavenging Activity of DDMP (2,3-Dihydro-2,5-dihydroxy-6-methyl-4 H -pyran-4-one) Saponin in Soybean Seed. *Biosci. Biotechnol. Biochem.* **1995**, *59*, 1556–1557. [CrossRef]
147. Hu, J.; Zheng, Y.L.; Hyde, W.; Hendrich, S.; Murphy, P.A. Human Fecal Metabolism of Soyasaponin I. *J. Agric. Food Chem.* **2004**, *52*, 2689–2696. [CrossRef]
148. Hu, J.; Reddy, M.B.; Hendrich, S.; Murphy, P.A. Soyasaponin I and Sapongenol B Have Limited Absorption by Caco-2 Intestinal Cells and Limited Bioavailability in Women. *J. Nutr.* **2004**, *134*, 1867–1873. [CrossRef]
149. Woyengo, T.A.; Ramprasath, V.R.; Jones, P.J.H. Anticancer effects of phytosterols. *Eur. J. Clin. Nutr.* **2009**, *63*, 813–820. [CrossRef] [PubMed]

150. Mazur, W.M.; Duke, J.A.; Wähälä, K.; Rasku, S.; Adlercreutz, H. Isoflavonoids and Lignans in Legumes: Nutritional and Health Aspects in Humans 11The method development and synthesis of the standards and deuterium-labelled compounds was supported by National Institutes of Health Grants No. 1 R01 CA56289-01 and No. 2 R01 CA56289-04, and analytical work by the EU research contract FAIR-CT95-0894. *J. Nutr. Biochem.* **1998**, *9*, 193–200. [CrossRef]
151. Clavel, T.; Doré, J.; Blaut, M. Bioavailability of lignans in human subjects. *Nutr. Res. Rev.* **2006**, *19*, 187–196. [CrossRef] [PubMed]
152. Heinonen, S.; Salonen, J.T.; Liukkonen, K.; Poutanen, K.; Wähälä, K.; Deyama, T.; Nishibe, S.; Adlercreutz, H. In vitro metabolism of plant lignans: New precursors of mammalian lignans enterolactone and enterodiol. *J. Agric. Food Chem.* **2001**, *49*, 3178–3186. [CrossRef]
153. Chin, Y.-W.; Chai, H.-B.; Keller, W.J.; Kinghorn, A.D. Lignans and Other Constituents of the Fruits ofEuterpe oleracea (Açai) with Antioxidant and Cytoprotective Activities. *J. Agric. Food Chem.* **2008**, *56*, 7759–7764. [CrossRef]
154. Willför, S.; Ahotupa, M.O.; Hemming, J.E.; Reunanen, M.H.T.; Eklund, P.C.; Sjöholm, R.E.; Eckerman, C.S.E.; Pohjamo, S.P.; Holmbom, B.R. Antioxidant Activity of Knotwood Extractives and Phenolic Compounds of Selected Tree Species. *J. Agric. Food Chem.* **2003**, *51*, 7600–7606. [CrossRef]
155. Sutthivaiyakit, S.; Na Nakorn, N.; Kraus, W.; Sutthivaiyakit, P. A novel 29-nor-3,4-seco-friedelane triterpene and a new guaiane sesquiterpene from the roots of Phyllanthus oxyphyllus. *Tetrahedron* **2003**, *59*, 9991–9995. [CrossRef]
156. Liu, J.; Li, Y.; Mei, C.; Ning, X.; Pang, J.; Gu, L.; Wu, L. Phytic acid exerts protective effects in cerebral ischemia-reperfusion injury by activating the anti-oxidative protein sestrin2. *Biosci. Biotechnol. Biochem.* **2020**, *84*, 1401–1408. [CrossRef] [PubMed]
157. Porres, J.M.; Stahl, C.H.; Cheng, W.-H.; Fu, Y.; Roneker, K.R.; Pond, W.G.; Lei, X.G. Dietary Intrinsic Phytate Protects Colon from Lipid Peroxidation in Pigs with a Moderately High Dietary Iron Intake. *Proc. Soc. Exp. Boil. Med.* **1999**, *221*, 80–86. [CrossRef]
158. Grases, F.; Simonet, B.M.; Vucenik, I.; Prieto, R.M.; March, J.G.; Shamsuddin, A.M.; Costa-Bauzá, A. Absorption and excretion of orally administered inositol hexaphosphate (IP6or phytate) in humans. *BioFactors* **2001**, *15*, 53–61. [CrossRef] [PubMed]

© 2020 by the author. Licensee MDPI, Basel, Switzerland. This article is an open access article distributed under the terms and conditions of the Creative Commons Attribution (CC BY) license (http://creativecommons.org/licenses/by/4.0/).

Article

In Vitro and In Vivo Nutraceutical Characterization of Two Chickpea Accessions: Differential Effects on Hepatic Lipid Over-Accumulation

Mariangela Centrone [1], Patrizia Gena [1], Marianna Ranieri [1], Annarita Di Mise [1], Mariagrazia D'Agostino [1], Maria Mastrodonato [2], Maria Venneri [1], Davide De Angelis [3], Stefano Pavan [3], Antonella Pasqualone [3], Carmine Summo [3], Valentina Fanelli [3], Giovanna Valenti [1,4], Giuseppe Calamita [1,4] and Grazia Tamma [1,4,*]

[1] Department of Biosciences Biotechnologies and Biopharmaceutics, University of Bari Aldo Moro, 70125 Bari, Italy; mariangela.centrone@uniba.it (M.C.); annapatrizia.gena@uniba.it (P.G.); marianna.ranieri@uniba.it (M.R.); annarita.dimise@uniba.it (A.D.M.); mariagrazia.dagostino@uniba.it (M.D.); maria.venneri@uniba.it (M.V.); giovanna.valenti@uniba.it (G.V.); giuseppe.calamita@uniba.it (G.C.)

[2] Department of Biology, University of Bari Aldo Moro, 70125 Bari, Italy; maria.mastrodonato@uniba.it

[3] Department of Soil, Plant and Food Sciences, University of Bari Aldo Moro, 70125 Bari, Italy; davide.deangelis@uniba.it (D.D.A.); stefano.pavan@uniba.it (S.P.); antonella.pasqualone@uniba.it (A.P.); carmine.summo@uniba.it (C.S.); valentina.fanelli@uniba.it (V.F.)

[4] Istituto Nazionale di Biostrutture e Biosistemi, (I.N.B.B.), 00136 Rome, Italy

[*] Correspondence: grazia.tamma@uniba.it; Tel.: +39-080-544-2388

Received: 28 February 2020; Accepted: 22 March 2020; Published: 24 March 2020

Abstract: Dietary habits are crucially important to prevent the development of lifestyle-associated diseases. Diets supplemented with chickpeas have numerous benefits and are known to improve body fat composition. The present study was undertaken to characterize two genetically and phenotypically distinct accessions, MG_13 and PI358934, selected from a global chickpea collection. Rat hepatoma FaO cells treated with a mixture of free fatty acids (FFAs) (O/P) were used as an in vitro model of hepatic steatosis. In parallel, a high-fat diet (HFD) animal model was also established. In vitro and in vivo studies revealed that both chickpea accessions showed a significant antioxidant ability. However, only MG_13 reduced the lipid over-accumulation in steatotic FaO cells and in the liver of HFD fed mice. Moreover, mice fed with HFD + MG_13 displayed a lower level of glycemia and aspartate aminotransferase (AST) than HFD mice. Interestingly, exposure to MG_13 prevented the phosphorylation of the inflammatory nuclear factor kappa beta (NF-kB) which is upregulated during HFD and known to be linked to obesity. To conclude, the comparison of the two distinct chickpea accessions revealed a beneficial effect only for the MG_13. These findings highlight the importance of studies addressing the functional characterization of chickpea biodiversity and nutraceutical properties.

Keywords: legumes; liver; hepatic steatosis; lipid dyshomeostasis; ROS

1. Introduction

Lifestyle and diet patterns play a critical role in disease development. Several studies demonstrate that healthy aging is often associated with a plant-based diet [1]. In contrast, the so-called Western diet, which is enriched in red meat and fats, combined with sedentary a lifestyle, significantly increases the risk of developing metabolic and energy balance disorders including insulin resistance, hyperlipidemia, hepatic steatosis, and obesity, that are hallmarks of the metabolic syndrome [2]. Nowadays, unhealthy lifestyle and diet-associated diseases can be considered a worldwide rising issue. Therefore, there is a growing interest in characterizing selective food groups for their physiological benefits, which could be useful to prevent chronic disease risk [3].

Legumes are a fundamental component of the diet of several human populations and chickpea (*Cicer arietinum* L.) represents the second most widely cultivated grain legume worldwide [4]. Chickpeas are characterized by having a high content of carbohydrates, proteins, fats, vitamins, and fibers [5] that reduce the risk of incidence of chronic diseases [6]. Numerous studies have demonstrated that chickpeas and ethanol extract of chickpeas can alter lipid metabolism by reducing the formation of lipid droplets during 3T3-L1 adipocytes differentiation [7]. Moreover, the consumption of chickpeas slows down the glycemic response [8] and weakens insulin resistance, type-2 diabetes, and dyslipidemia associated with a high-fat diet (HFD) [9].

A HFD is associated with the overproduction of reactive oxygen species (ROS). Imbalances between oxidants and antioxidant endogenous reserve play a role in the onset of several disorders [10]. Indeed, a HFD increases the expression level of nicotinamide adenine dinucleotide phosphate (NADPH) oxidase while reduces antioxidant enzymes [11,12]. Cellular ROS enhance lipid peroxidation, thereby possibly altering the expression and the activity of the oxidative-sensitive nuclear factor kappa beta (NF-kB) and other chemokines [12]. NF-kB is a crucial player in controlling various cellular processes such as inflammation, cell survival, and proliferation. NF-kB is normally sequestered in the cytosol by binding the inhibitory proteins named IkBs [13]. Two signaling pathways can lead to NF-kB activation [14]. The main mechanism for the classical stimulation of the NF-kB pathway includes phosphorylation of IkB by the IκB kinase (IKK), its ubiquitylation and degradation. These molecular events promote the activation and nuclear translocation of NF-kB [14]. Phosphorylation of NF-kB at serine-536, by multiple kinases, provides an additional level of regulation of NF-kB that affects cell proliferation by controlling the expression of numerous cell cycle regulatory proteins [15,16] such as cyclin D1 and proliferating cell nuclear antigen (PCNA) [17]. The canonical pathway is switched on by various cytokines, such as TNFα that increases significantly secondary to a HFD [18]. Numerous studies have strongly indicated that food habits play a key role in the development of several disorders. Medical plants and a balanced diet have been reported to slow down chronic diseases. However, evidence for the consumption of chickpeas in dyslipidemia, insulin resistance, and type 2 diabetes are few and sometimes discordant [19,20].

Recent studies on global chickpea germplasm collections reveal a high level of genetic, phenotypic, and compositional diversity [4,21,22]. The present study was carried out to investigate the physiological effect of two genetically and phenotypically diverse chickpea accessions in a cellular and animal model of nonalcoholic fatty liver disease.

2. Materials and Methods

2.1. Chemicals and Antibodies

All chemicals were obtained from Merck (Merck KGaA, Darmstadt, Germany). Monoclonal antibodies against pNF-kB p65 (27.Ser 536), NF-kB p65 (F-6), and PCNA (PC10) were purchased from Santa Cruz Biotechnology (Tebu Bio, Milan, Italy). Secondary anti-mouse conjugated to horseradish peroxidase (HRP) was obtained from Bio-Rad (Bio-Rad Laboratories, Inc., Hercules, CA, USA). The tert-butyl hydroperoxide (tBHP) was a kind gift from A. Signorile (University of Bari, Aldo Moro, Bari. Italy). The RayBio® Mouse Insulin ELISA Kit was from RayBiotech (Norcross, GA, USA).

2.2. Chickpea Selection and Preparation of Extracts

Two chickpea accessions, namely MG_13 and PI358934, were selected from the germplasm collections previously characterized, as they belong to distinct genetic clusters and display markedly different phenotypic features [22,23]. For each accession, 40 plants were grown at the experimental farm "P. Martucci" of the University of Bari (41°01′22.1″ N and 16°54′21.0″ E) from 2013 to 2014. At crop maturity, seeds were bulked separately for each accession and were cleaned to remove broken seeds, dust, and other undesirable matter. To obtain the wholemeal flour with uniform particle size, the seeds were ground by an electric mill (Model ETA, Vercella Giuseppe, Mercenasco, Italy) equipped with a sieve of 0.6 mm. The chemical and nutritional composition of the two chickpea accessions is reported in the Supplementary Material of a study by Summo et al. [4]. MG_13 is characterized by higher protein and lower lipid content than PI358934, as well as by higher oleic acid and lower linoleic acid content. For the preparation of the extracts, 2 g of chickpea flour were mixed with 20 mL of a 30/70 (*v/v*) ethanol/water solution in centrifuge tubes and stirred for 2 h in the dark. Then, the extracts were centrifuged at 12,000× *g* for 10 min. The supernatant was recovered and filtered through nylon filters (pore size 0.45 μm). Finally, the extracts were freeze-dried by a laboratory lyophilizer (Analytical Control De Mori s.r.l., Milano, Italy) and stored at −20 °C until the analysis.

2.3. Cell Culture, Treatments, Crystal Violet Assay, and ROS Detection

Rat hepatoma FaO cells were grown in Coon's modified Ham's F12 (Euroclone, Milan, Italy) supplemented with 10% fetal bovine serum (FBS) (Thermo Fisher Scientific, Waltham, MA, USA), 100 I. u./mL penicillin, 100 μg/mL streptomycin (Euroclone, Milan, Italy) at 37 °C in 5% CO_2 and used at 70% to 80% confluence. For treatments, cells were incubated for 24 h in high glucose medium, without FBS, supplemented with 0.25% bovine serum albumin (BSA) and exposed to chickpea extracts obtained from the accessions MG_13 and PI358934, respectively. Cells were left under basal condition or stimulated with a mixture of free fatty acids (FFAs) (0.5 mM oleate and 0.25 mM palmitate) for 3 h [24,25].

Crystal violet assay was performed as previously described [26]. Cells viability was detected by measurement of the optical density at 595 nm (DO595) with a microplate reader (Bio-Rad Laboratories, Inc., Hercules, CA, USA).

ROS were detected as previously shown [26,27] and using the dihydrorhodamine-123 (10 μM). The fluorescence emission signal was recorded using a fluorimeter (RF-5301PC, Shimadzu Corporation, Kyoto, Japan) at excitation and emission wavelengths of 508 and 529 nm, respectively.

2.4. Lipid Inclusions Staining

FaO cells were fixed with 4% paraformaldehyde (PFA) in phosphate buffer saline (PBS) for 20 min. After washing in 50% isopropanol, cells were stained with 0.12% Oil Red O solution and counterstained with hematoxylin and eosin. Alternatively, livers were fixed quickly in 4% PFA, dehydrated through a graded ethanol series and embedded in paraffin wax. Lipid inclusions analysis was performed as previously described [26,28]. In paraffin sections stained with periodic acid-Schiff (PAS), lipid droplets were characterized by circular shape and chromatic uniformity (white areas, because the lipids have been removed from the droplets by the solvent used for dehydration). The obtained data were analyzed using an Image J–Particles Analyzer (https://imagej.net). Statistical analysis was performed with GraphPad Prism (GraphPad Software, San Diego, CA, USA) by one-way ANOVA followed by Dunnett's Multiple Comparisons test with * $p < 0.05$ considered statistically different.

2.5. Cell and Liver Lysates

FaO cells were seeded onto 60 mm diameter Petri dishes. After treatments, cells were lysed as previously described [29]. Alternatively, 30 to 50 mg of frozen liver tissue per mouse sample was homogenized with a mini-potter in an ice-cold RIPA buffer, in the presence of proteases (1 mM PMSF, 2 mg/mL leupeptin, and 2 mg/mL pepstatin A) and phosphatases (10 mM NaF and 1 mM sodium orthovanadate) inhibitors. Suspensions were then centrifuged at 12,000× *g* for 10 min at 4 °C. Lysates were used for SDS-PAGE and Western blotting analysis.

2.6. Gel Electrophoresis and Western Blotting

Proteins were separated using 10% or 12% stain-free polyacrylamide gels (Bio-Rad Laboratories, Inc., Hercules, CA, USA) under reducing conditions as previously described [26]. Protein bands were electrophoretically transferred onto Immobilon-P membranes (Merck KGaA, Darmstadt, Germany) for Western blot analysis, blocked in TBS-Tween-20 containing 3% bovine serum albumin (BSA) and incubated with primary antibodies overnight. Immunoreactive bands were detected with secondary goat anti-mouse horseradish peroxidase-coupled antibodies. Membranes were incubated with Super Signal West Pico Chemiluminescent Substrates (Thermo Fisher Scientific, Waltham, MA, USA), and the signals were visualized with the ChemiDoc System gels (Bio-Rad Laboratories, Inc., Hercules, CA, USA). Obtained bands were normalized to total protein using the stain-free technology gels (Bio-Rad Laboratories, Inc., Hercules, CA, USA). Densitometry analysis was performed using Image Lab gels (Bio-Rad Laboratories, Inc., Hercules, CA, USA). Data were analyzed using GraphPad Prism (GraphPad Software, San Diego, CA, USA).

2.7. Animals and Experimental Protocol

Forty-five C57BL/6J male mice (Envigo RMS S.R.L., San Pietro al Natisone, Udine, Italy), 3 weeks old and about 18 g of initial body weight were housed in a temperature-controlled room (25 °C) on a 12 h light-dark cycle. Mice were fed with a control diet (CTR) for one week. Mice had free access to water and diet according to the assigned dietary formulation for 16 weeks. The four kinds of diets (Altromin Rieper SpA, Vandoies, Italy) were the following: (1) a standard control diet (CTR) containing 10% fat, 24% proteins, and 66% of carbohydrates (metabolized energy 3.514 kcal/kg diet); (2) a high-fat diet (HFD) made of 45% fat, 19% proteins, and 36% of carbohydrates (metabolized energy 4.496 kcal/kg diet) (3) a high-fat plus chickpea diet (accession number: MG_13) (HFD + MG_13) having the same metabolized energy of the HFD diet, as well as the same composition except that 10% of MG_13 raw crushed chickpea seed flour replaced crude fibers and ashes of HFD; (4) a high-fat plus chickpea diet (accession number: PI358934) (HFD + PI358934) similar to the previous one except that 10% of PI358934 raw crushed chickpea seed flour replaced crude fibers and ashes of HFD. The experimental diets were stored at −20 °C to avoid rancidity.

Before killing by cervical dislocation, mice were deprived of food for 12 h. Blood glucose levels were measured from the tail vein using a freestyle glucometer (Accu-Check Aviva, Roche, Basel, Switzerland). Livers were removed quickly, weighed, and fixed for histochemical analysis or frozen using liquid N_2. The serum was separated from blood samples and promptly stored at −80 °C.

Animal experiments were carried out following the Directive 2010/63/UE, enforced by Italian D.L. 26/2014, and approved by the animal care and the Committee of the University of Bari (OPBA), Bari, Italy and the Italian Ministry of Health, Rome, Italy (authorization n.326/2018-PR).

2.8. Serum Insulin Measurement

Serum insulin secretion was determined using an ELISA kit, following the manufacturer's instructions. Briefly, each sample was incubated for 2.5 h at room temperature with gentle shaking. In parallel, increasing concentrations (400, 200, 100, 50, 25, 12.5, and 6.25 μlU/mL) of a standard protein reproducing mouse insulin were incubated as an internal standard. Wells were washed with the appropriate washing solution and incubated with biotinylated antibodies for 1 h at room temperature. Then, wells were washed and incubated with streptavidin solution for 45 min at room temperature. After the last four washes, TMB substrate reagent was added to each well and incubated for 30 min in the dark at room temperature. Optical density was measured at 450 nm by a microplate reader (iMark, Bio-Rad Laboratories, Inc., Hercules, CA, USA).

2.9. Statistical Analysis

All values are reported as means ± S.D. Statistical analysis was performed by one-way ANOVA followed by Dunnett's multiple comparisons test with * $p < 0.05$ considered statistically different.

3. Results

3.1. Characterization of the Chickpeas Extracts in FaO Cells

Chickpeas have been repeatedly reported to have a beneficial effect in preventing chronic diseases and reducing adiposity. However, a clear demonstration of the health benefits of chickpeas is still missing. In this study, chickpea extracts obtained from two different accessions (MG_13 and PI358934) of a global chickpea collection [4,22] have been characterized. The potential effects of the extracts on cell viability were evaluated by performing the crystal violet assay. FaO cells were exposed for 24 h to increasing concentrations of the chickpea extracts derived from the two accessions (0.1 mg/mL, 1 mg/mL, 2 mg/mL, and 10 mg/mL). Compared to the untreated condition (CTR), treatment with the extract isolated from the accession MG_13 did not alter cell viability at the concentrations used in this study (Figure 1A). In contrast, treatment with the extract obtained from the accession PI358934 reduced cell viability at the highest concentration (10 mg/mL) (**** $p < 0.0001$ vs. CTR). No relevant cytotoxic effect was observed at 0.1 mg/mL, 1 mg/mL, and 2 mg/mL doses (Figure 1B). As with other pulses, chickpea seeds contain several bioactive compounds including anthocyanins, carotenoids, and phenols [4] which are important to counteract oxidative stress. Reactive species can potentially promote inflammation and fibrosis during the development of hepatic steatosis. Thus, to investigate the functional effects of chickpea extracts and their correlation with oxidative signals and inflammation, a liver steatosis cellular model was established by exposing FaO cells to 0.75 mmol/L oleate/palmitate (O/P) for 3 h [24,25,30]. Compared to the untreated cells (CTR), exposure to O/P led to a significant increase of the reactive oxygen species (ROS) as assessed by a fluorimetric method (* $p < 0.05$ vs. CTR). As an internal and additional positive control, cells were treated with the oxidant tert-butyl hydroperoxide showing a relevant increase of ROS content as compared with that measured in CTR cells (**** $p < 0.0001$ vs. CTR). Exposure to both chickpea extracts at 0.1 mg/mL and 10 mg/mL displayed a similar ability to decrease intracellular ROS induced by the O/P treatment (Figure 2). Hence, the lowest (0.1 mg/mL) concentration of the extracts was used in the following experiments.

3.2. Effects of Chickpea Extracts on Lipid Accumulation in FaO Cells

Lipid accumulation was visualized using Oil Red O staining. This staining can detect neutral triglycerides and glycolipids in the control (CTR) and steatotic hepatocytes (O/P) exposed or not to the chickpea extracts (0.1 mg/mL) for 24 h (Figure 3). Treatment with O/P promoted lipid droplet over-accumulation that was not observed in the cells pretreated with the extract obtained from the MG_13 accession (MG_13 + O/P). In contrast, exposure to the extract isolated from the PI358934 accession did not hamper the over-accumulation of fat induced by the O/P mixture (PI358934 + O/P). Compared to the untreated cells, the O/P treatment resulted in a significant increase in lipid accumulation (**** $p < 0.0001$ vs. CTR). Similar to CTR cells, incubation with each one of the extracts from the two accessions did not lead to an over-accumulation of lipids within the cells.

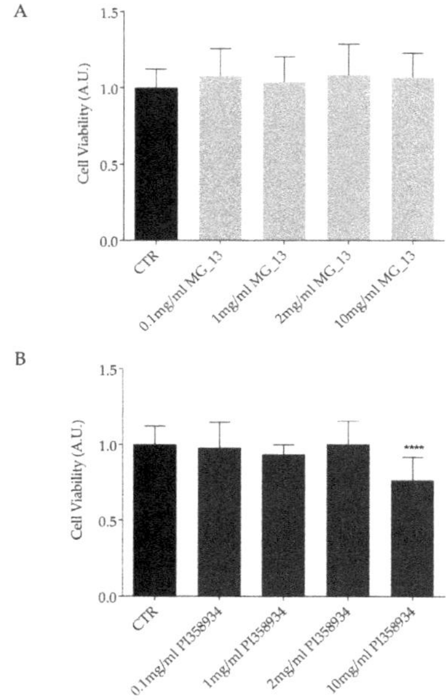

Figure 1. Cell viability of FaO cells exposed to the chickpea extracts. FaO cells were left under basal condition (CTR) or exposed to different concentrations (0.1 mg/mL, 1 mg/mL, 2 mg/mL, or 10 mg/mL) of chickpea extract MG_13. (**A**) and PI358934; (**B**) as described in the Methods Section. Cells were stained with crystal violet solution. Data are presented as means ± S.D. of 3 independent experiments and analyzed by one-way ANOVA followed by Dunnett's multiple comparisons test. (**** $p < 0.0001$ vs. control diet (CTR)).

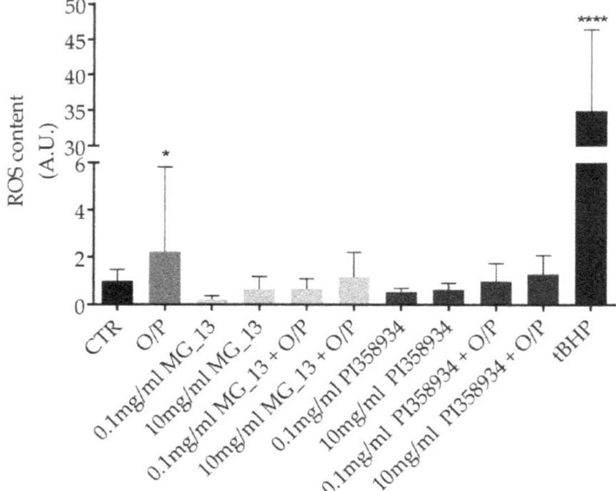

Figure 2. Reactive oxygen species (ROS) content in FaO cells. ROS content was measured using dihydrorhodamine-123 fluorescence in FaO cells treated as described in the Methods Section. As a positive control, cells were treated with tert-butyl hydroperoxide (tBHP). Data are shown as means ± S.D. of 3 independent experiments and analyzed by one-way ANOVA followed by Dunnett's multiple comparisons test. (* $p < 0.05$ vs. CTR and **** $p < 0.0001$ vs. CTR).

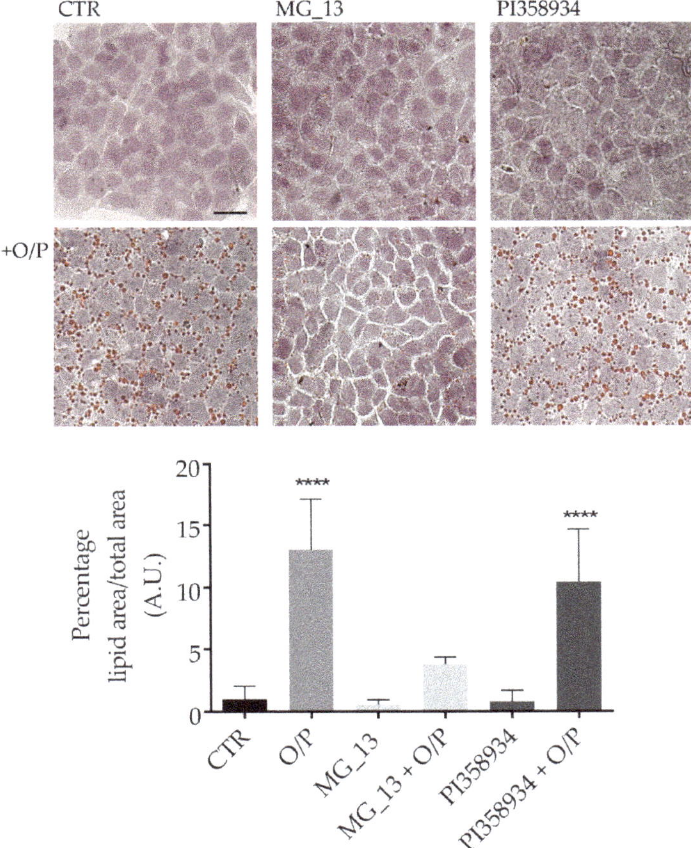

Figure 3. Detection of intracellular lipid droplets. FaO cells were left under basal condition (CTR) or treated with MG_13 (0.1 mg/mL), PI358934 (0.1 mg/mL) or with chickpea extract in the presence of a mixture of free fatty acids (FFAs) (oleate/palmitate). Oil Red O staining revealed lipid droplet formation. Lipid droplets are stained dark red and were analyzed using Image-J software (Bar 25 μm). Data are shown as a percentage of lipid area/total area and are presented as means ± S.D. of 3 independent experiments and analyzed by one-way ANOVA followed by Dunnett's multiple comparisons test. (**** $p < 0.0001$ vs. CTR).

3.3. *Effects of Chickpea Extracts on NF-kB Phosphorylation in FaO Cells*

Western blotting analysis revealed that in steatotic hepatocytes (O/P), phosphorylation of NF-kB at serine-536 significantly increased as compared with untreated cells (** $p < 0.01$ vs. CTR) (Figure 4). Preincubation with the extract obtained from the MG_13 accession impaired the increase of NF-kB phosphorylation at S536 induced by O/P (MG_13 + O/P). In contrast, phosphorylation of NF-kB significantly increased (* $p < 0.05$ vs. CTR) in steatotic hepatocytes pretreated with the extract PI358934 (PI358934 + O/P) at a similar level to that observed in O/P treated cells. Treatment with only the extracts, in the absence of O/P, did not alter NF-kB phosphorylation (Figure 4).

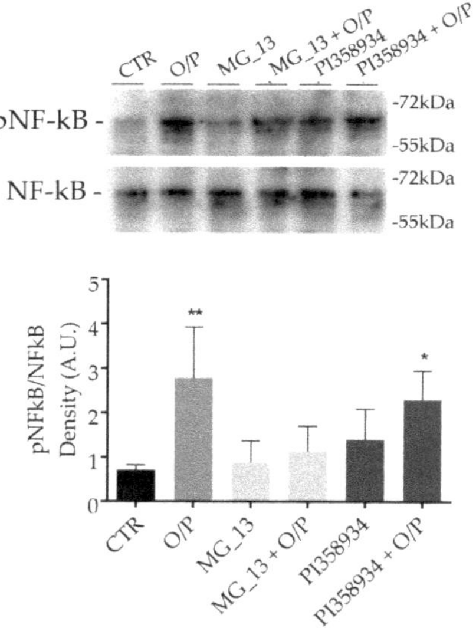

Figure 4. Effects of the chickpea extracts on nuclear factor kappa beta (NF-kB) phosphorylation. FaO cells were treated as described in the Methods Section. An equal amount of proteins (30 μg/lane) was separated by gel electrophoresis and immunoblotted for evaluation of pNF-kB and total NF-kB levels. Densitometric analysis revealed that the O/P treatment caused a significant increase in pNF-kB levels as compared with cells under basal condition (CTR). Treatment with MG_13 (0.1 mg/mL) prevented the O/P-dependent increase of pNF-kB. In contrast, treatment with PI358934 (0.1 mg/mL) did not impair the O/P-action. No significant alterations in pNF-kB levels were observed by the sole treatment with chickpea extract. Data are expressed as means ± S.D. and were analyzed by one-way ANOVA followed by Dunnett's multiple comparisons test. (** $p < 0.01$ vs. CTR and * $p < 0.05$ vs. CTR, $n = 4$).

3.4. Animal Studies: Body Weight, Food Intake, Food Efficiency Ratio

To further investigate the biological actions of the two chickpea accessions, an animal model of nonalcoholic fatty liver disease (NAFLD) was established by feeding mice with a high-fat diet. As already described in the Methods Section, four different diets were randomly assigned to C57BL/6J adult male mice for 16 weeks. The initial and final body weights, the body weight gain, the food intake, and the feed efficiency ratios (FER) of the mice are indicated in Table 1. Compared to CTR mice, the final body weight and weight gain were significantly higher in the HFD groups (**** $p < 0.0001$). The addition of the flour obtained from the accession PI358934 (HFD + PI358934) significantly increased the final body weight and the FER with respect to those measured in HFD mice (#### $p < 0.0001$).

Table 1. Body weight, food intake, and food efficiency ratio of CTR and high-fat diet (HFD) mice.

Parameters	CTR (12)	HFD (12)	HFD + MG_13 (12)	HFD + PI358934 (9)
Initial body weight (g)	22.78 ± 1.30	23.63 ± 1.16	22.68 ± 1.02	24.19 ± 0.69 *
Final body weight (g)	31.04 ± 1.04 ####	40.1 ± 3.00 ****	43.69 ± 6.08 ****	49.91 ± 4.04 **** ####
Weight gain (g/d)	0.069 ± 0.009 ####	0.1384 ± 0.019 ****	0.176 ± 0.036 **** ##	0.216 ± 0.024 **** ####
Food intake (g/d)	3.176 ± 0.286	3.036 ± 0.115	3.787 ± 0.698 **** ####	2.911 ± 0.226
FER (%)	2.185 ± 0.365 ####	4.557 ± 0.667 ****	4.661 ± 1.293 ****	7.423 ± 1.025 **** ####

Values represent the mean ± S.D., the number of mice is indicated in the brackets. * Significantly different versus CTR (**** $p < 0.0001$; * $p < 0.05$). # Significantly different versus HFD (#### $p < 0.0001$ and ## $p < 0.01$). Food efficiency ratio (FER)

3.5. Serum Parameters and Liver Function

Selected serum parameters are listed in Table 2 showing that glycemia was higher in the HFD groups as compared with that measured in the control mice. However, the serum glucose measured in the HFD + MG_13 group was significantly lower than the glycemia measured in the HFD mice ($^{\#}$ $p < 0.05$ vs. HFD). Serum insulin levels were slightly changed among groups although these differences were not statistically significant ($p > 0.05$). Total cholesterol was increased in mice fed with HFD in the presence of chickpea flours more than in the group kept on the HFD diet (HFD + MG_13 vs. CTR, ** $p < 0.01$ and HFD + PI358934 vs. CTR, ** $p < 0.01$). However, the increase measured in the HFD + PI358934 mice was higher than the one measured in the HFD mice (HFD + PI358934 vs. HFD, $^{\#}$ $p < 0.05$). Serum triglycerides were increased in the HFD mice that received the flours. To evaluate the liver function, the levels of alkaline phosphatases and the aspartate aminotransferase (AST) were determined. Interestingly, the alkaline phosphatases level did not change among groups. In contrast, a significant increase of the AST was found in the HFD and HFD + PI358934 groups as compared with the control mice (HFD vs. CTR, * $p < 0.05$ and HFD + PI358934 vs. CTR, ** $p < 0.01$). No significant alteration was measured in the HFD + MG_13 animals as compared with the CTR and HFD groups. Moreover, the liver weight normalized to the total body weight significantly increased only in mice fed with HFD + PI358934 (HFD + PI358934, 0.0394 ± 0.0031 vs. CTR: 0.0303 ± 0.0005, ** $p < 0.01$).

Table 2. Biochemical analysis of serum parameters of CTR and HFD mice.

Parameters	CTR (12)	HFD (12)	HFD + MG_13 (12)	HFD + PI358934 (9)
Glycemia (mg/dL)	87.42 ± 21.46 ####	214.6 ± 32.84 ****	184.6 ± 12.89 **** #	209.4 ± 41.92 ****
Insulin (µIU/mL)	15.75 ± 4.55	26.51 ± 8.82	23.68 ± 13.41	16.05 ± 2.32
Total cholesterol (mg/dL)	63.00 ± 27.40	107.0 ± 9.64	186.3 ± 58.07 **	221.0 ± 27.87 ** #
Triglycerides (mg/dL)	76.00 ± 39.40	128.0 ± 22.72	165.3 ± 11.37 *	181.7 ± 56.36 *
Alkaline phosphatases (U/l)	66.00 ± 10.54	65.33 ± 28.31	48.00 ± 15.13	64.00 ± 7.00
AST (U/l)	279.0 ± 130.9 #	582.7 ± 130.6 *	486.3 ± 89.39	703.7 ± 49.92 **

Values represent the mean ± S.D. The number of mice is indicated in the brackets. * significantly different versus CTR (**** $p < 0.0001$; ** $p < 0.01$ and * $p < 0.05$). # significantly different versus HFD (#### $p < 0.0001$ and # $p < 0.05$). Aspartate aminotransferase (AST).

3.6. Effects of Chickpea Diets on Lipid Accumulation and NF-kB Phosphorylation in Liver

Lipid inclusions analysis in the liver revealed that HFD feeding promotes the over-accumulation of lipid droplets (**** $p < 0.0001$ vs. CTR) (Figure 5). A similar increase in lipid accumulation was found in liver sections prepared from the HFD + PI358934 fed mice, likely suggesting that this accession has no beneficial effect on hepatic lipid over-accumulation (**** $p < 0.0001$ vs. CTR). In contrast, a significant reduction in lipid droplets over-accumulation as compared with the HFD samples was detected in liver sections of the HFD + MG_13 fed group (#### $p < 0.0001$ vs. HFD), likely indicating that this accession prevented liver adiposity. Indeed, an increase of the cytoplasmic glycogen concentration was found in the HFD + MG_13 mice as assessed by PAS staining. NF-kB can be considered a general biomarker of inflammation and is often activated in patients with dyslipidemia. The Western blotting analysis (Figure 6) of liver lysates revealed that NF-kB phosphorylation at S536 increased in the HFD and HFD + PI358934 fed mice (* $p < 0.05$ vs. CTR). In contrast, the degree of NF-kB phosphorylation in the HFD + MG_13 fed mice was similar to that obtained in the liver lysates of the CTR mice.

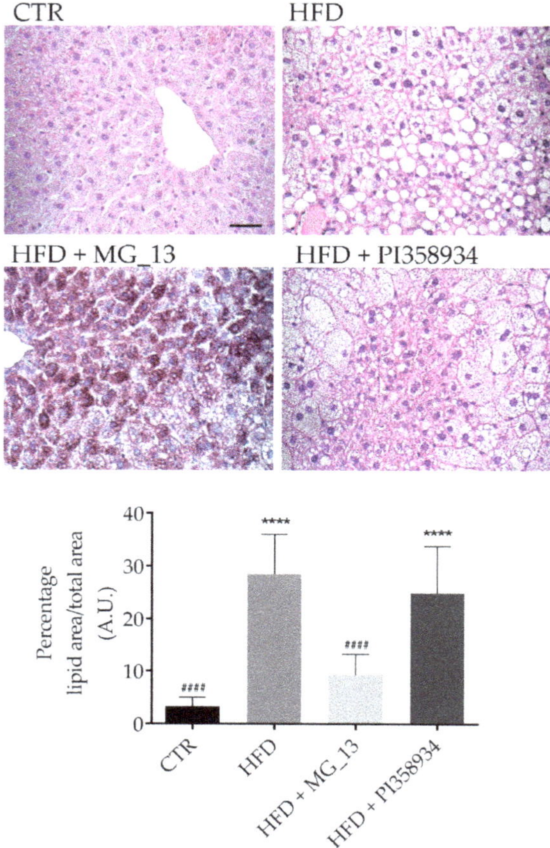

Figure 5. Histochemical features of liver parenchyma of mice. Paraffin sections of mice livers were stained with PAS-hematoxylin and eosin. PAS reaction demonstrated glycogen stores (stained red) in the hepatocytes. The lipid droplets being approximately circular in the section were counted with Image-J software (Bar 50 µm). Data are shown as a percentage of lipid area/total area and are presented as means ± S.D. of 3 independent experiments and analyzed by one-way ANOVA followed by Dunnett's multiple comparisons test. (**** $p < 0.0001$ vs. CTR and #### $p < 0.0001$ vs. HFD).

3.7. Effects of Chickpea Diets on ROS Production and Proliferating Cell Nuclear Antigen (PCNA) Expression in Liver

Fresh liver sections were subjected to a fluorimetric assay to detect ROS content as described in the Methods Section (Figure 7). Compared to the CTR group, the HFD groups displayed a significant increase in ROS content (* $p < 0.05$ vs. CTR). In contrast, a relevant decrease in ROS production was found in the livers of the HFD mice that received the chickpea flours, likely indicating that the consumption of these flours exerts an antioxidant effect. The Western blotting analysis of liver lysates (Figure 8) revealed that the expression of PCNA, a marker of cell proliferation, increased in the HFD fed mice as compared with the CTR group. Interestingly, the expression level of PCNA in liver lysates of the HFD + MG_13 and HFD + PI358934 fed mice was significantly lower as compared with lysates from the HFD fed animals, and thereby similar to that detected in lysates of CTR animals (HFD vs. CTR, * $p < 0.05$)

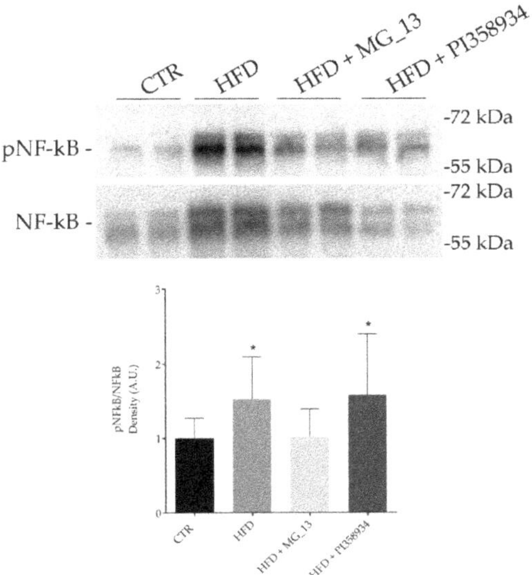

Figure 6. Effects of chickpea diets on NF-kB phosphorylation. Liver lysates were prepared as described in the Methods Section. An equal amount of proteins (30 µg/lane) was separated by gel electrophoresis and immunoblotted for the evaluation of pNF-kB and total NF-kB levels. Densitometric analysis revealed that the HFD diet (HFD, $n = 12$) caused a significant increase in pNF-kB levels as compared with the mice fed with a normal diet (CTR, $n = 12$). The diet with HFD + MG_13 ($n = 12$) prevented the HFD dependent increase of pNF-kB. Data are expressed as means ± S.D. and were analyzed by one-way ANOVA followed by Dunnett's multiple comparisons test (* $p < 0.05$ vs. CTR).

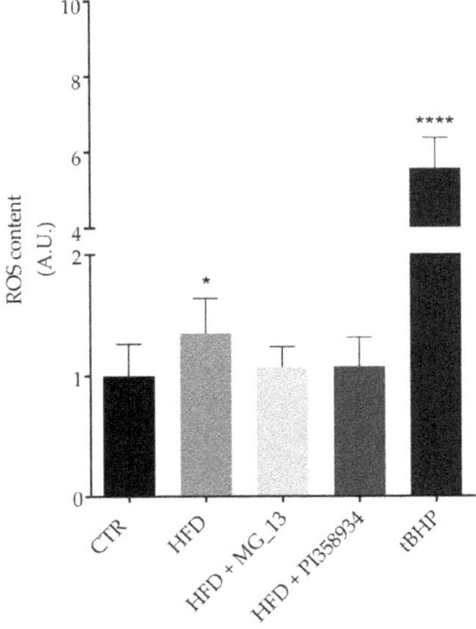

Figure 7. ROS content in liver tissue. ROS content was measured using dihydrorhodamine-123 fluorescence in a liver slice as described in the Methods Section. As a positive control, the liver sections were treated with tBHP. Data are expressed as means ± S.D. and analyzed by one-way ANOVA followed by Dunnett's multiple comparisons tests (**** $p < 0.0001$ vs. CTR and * $p < 0.05$ vs. CTR).

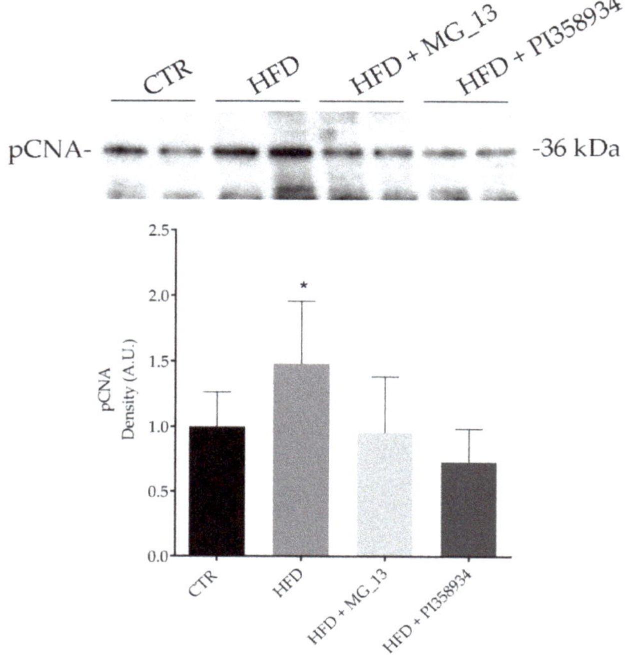

Figure 8. Effects of chickpea diets proliferating cell nuclear antigen (PCNA) expression. Liver lysates were prepared as described in the Methods Section. An equal amount of proteins (30 µg/lane) was separated by gel electrophoresis and immunoblotted for evaluation of the PCNA levels. Densitometric analysis revealed that the HFD diet (HFD) caused a significant increase in the PCNA levels as compared with mice fed a normal diet (CTR). No significant alterations in PCNA levels were observed in mice fed the HFD + MG_13 diet or HFD + PI358934 diet. Data are expressed as means ± S.D. and analyzed by one-way ANOVA followed by Dunnett's multiple comparisons test (* $p < 0.05$ vs. CTR).

4. Discussion

Overweight and obesity associated with a high-fat diet have become a global health concern having a tremendous impact on clinical care [31]. High-fat diets increase the risk of the incidence of several clinical disorders such as dyslipidemia, hypertension, type-2 diabetes, and liver diseases [32,33]. It has been reported, that prolonged high-fat feeding promoted nonalcoholic fatty liver disease (NAFLD) that can lead to nonalcoholic steatohepatitis (NASH) and cirrhosis [34,35]. Therefore, the identification of new dietary supplements or the characterization of specific foods that reduce hepatic fat over-accumulation and improve the systemic lipid profile is an unmet need. The present study was undertaken to investigate the possible beneficial effects of dietary consumption of chickpeas. Legumes are traditionally recommended to reduce lipid accumulation and to prevent insulin resistance, possibly because they provide several nutrients including low glycemic index carbohydrates and dietary fibers [36]. However, clear evidence is missing for the recommendation of chickpeas in controlled diets with the aim of improving health conditions. It is not clear at what amount beneficial effects can be gained. Short- and long-term consumption leads to differential responses of glycemia, insulin, and other serum parameters [19]. Another crucial aspect to be investigated is whether genetic diversity and a different composition in cultivated material is an important variable for determining the effects of chickpea consumption on human health [19,20,36]. Here, both in vitro and in vivo studies were performed to compare the effects of two genetically and phenotypically diverse chickpea accessions on lipid accumulation, ROS generation, and the activation of the inflammatory factor NF-kB. Exposure of hepatic FaO cells to free fatty acids (FFAs) mimicked cellular steatosis [30]. Accordingly, in this report, Oil Red O staining revealed the presence of numerous intracellular lipid droplets in O/P treated FaO cells. Furthermore, abnormal intrahepatic fat accumulation led to a considerable upregulation of the oxidative signals in NAFLD patients [34]. Fluorimeter measurements showed that the O/P treatment promoted ROS generation as compared with the untreated cells. Interestingly, both chickpea accessions reduced the ROS production induced by the O/P treatment. The significant antioxidant effect could be related to the presence, in these chickpea accessions, of bioactive molecules such as carotenoids, anthocyanins, and phenols [23]. Oxidative stress and inflammation are involved in the onset and progression of numerous diseases. Therefore, phytocoumpounds, such as phenols with proven antioxidant ability, are considered health-promoting natural antioxidants [37,38]. Higher production of reactive species can occur secondary to intracellular altered lipid metabolism and accumulation [34]. However, only the accession MG_13 prevented lipid over-accumulation induced in steatotic hepatocytes by the O/P treatment. It is well established that HFD feeding causes an increase of intrahepatic fat, abnormal functions of lysosomes and mitochondria, leading to the activation of NF-kB [39,40]. The nuclear factor NF-kB is a master regulator of inflammatory and oxidative signals. Numerous reports have shown that metabolic diseases associated with obesity can be accompanied by abnormal NF-kB expression and activity [30,41]. In 12-weeks feeding HFD mice, a significant upregulation of NF-kB signaling was also demonstrated [41]. Therefore, to further investigate the differential effects of the two chickpea accessions, the activity of NF-kB was evaluated. Steatotic FaO cells displayed a significant increase of NF-kB phosphorylation at serine 536, consistent with an activation of the NF-kB pathway. To confirm the in vitro findings, an HFD mouse model was established. As predicted, HFD feeding increased total body weight. However, mice receiving the flour of the accession PI358934 displayed a higher body weight as compared with CTR and HFD mice. A similar trend was also found for the food efficiency ratio (FER). These observations suggested that animals fed with the PI358934 accession gained more weight as compared with the HFD + MG_13 fed mice. Moreover, compared to CTR mice, HFD feeding increased blood glucose. Interestingly, MG_13 supplementation was associated with a significant reduction of the hyperglycemia induced by HFD, likely indicating that the consumption of this accession could have a beneficial effect in decreasing blood glucose. The supplementation of the accession PI358934 did not exert any beneficial effect on glycemia. Indeed, mice receiving the HFD + MG_13 diet displayed a higher liver glycogen deposition than the HFD and HFD + PI358934 fed mice. Accordingly, rats fed a diet enriched with chickpea were characterized by having an increased level of glucokinase activity that correlated to a major liver glycogen deposition as compared with the HFD fed rats. These findings proposed that the consumption of chickpea could ameliorate the abnormal lipid profile [20]. Moreover, the serum level of the AST, which is a known biomarker of liver injury, was significantly higher in the HFD and HFD + PI358934 groups. An evaluation of the intrahepatic fat infiltration revealed that the lipid area in the liver tissue was significantly higher in the HFD and HFD + PI358934 fed mice as compared with the CTR mice. In contrast, previous reports have demonstrated that dietary chickpeas ameliorated the lipid profile in rats fed an HFD diet containing lard at 20% [9]. Whereas, in the present study, the high-fat diet included a higher fat level. However, similar to the in vitro data, the liver ROS content induced by the HFD diet was significantly reduced in the mice receiving the chickpea supplemented diets even though the phosphorylation level of NF-kB was reduced only in the liver of the HFD + MG_13 fed mice as compared with the HFD fed animals. Indeed, the activity of NF-kB in the liver of the HFD + PI358934 fed mice was similar to that measured in the HFD mice. Together, these findings indicate that the MG_13 accession exerted a stronger beneficial effect as compared with the PI358934 accession that contained a higher level of linoleic acid than the MG_13 accession [4]. It has been proposed that a high intake of linoleic acid can activate inflammation pathways through the accumulation of arachidonic acid and the release of several oxidative and inflammatory molecules that can play an important role in the onset of chronic diseases [42–44]. Nevertheless, the differential level of linoleic acid in the chickpea accessions partially explain the present data. Overall, it is plausible to hypothesize that other composition factors that differentiate these accessions are important for activating specific antioxidant and anti-inflammatory pathways that bring health benefits.

5. Conclusions

In conclusion, this study compares, for the first time, the effects of two different chickpea accessions revealing that only the MG_13 has beneficial responses to significantly attenuate the hepatic steatosis induced by a high-fat diet. The obtained results indicated that the beneficial actions of the single chickpea accession could be related to the presence of specific bioactive compounds or the combination of selective molecules. Hence, the identification of accession-specific phytocompounds should be an important step to explain the difference between the accessions MG_13 and PI358934. Overall, these findings highlight the importance of the genotypic, phenotypic, and nutraceutical characterization of chickpea cultivars of global food interest.

Author Contributions: Conceptualization, M.C., G.T., G.C., and G.V.; methodology, S.P., P.G., M.C., M.D., M.R., A.D.M., M.V., D.D.A. and V.F.; investigation, G.T., M.C., M.D., M.M., A.D.M., C.S., P.G., and A.P.; data curation, G.T., M.R., G.C., S.P., M.C., M.D., A.P., C.S., P.G., and A.D.M.; writing, G.T. and M.C.; project administration, G.T., and C.S.; funding acquisition, G.T., and C.S. All authors have read and agreed to the published version of the manuscript.

Funding: This research has been performed within the project "LegumeGEneticREsources as a tool for the development of innovative and sustainable food TEchnological system" supported under the "Thought for Food" Initiative by the Agropolis Foundation (through the"Investissementsd'avenir" programme with reference number ANR-10-LABX-0001-01), the Fondazione Cariplo, and the Daniel & Nina Carasso Foundation.

Acknowledgments: We acknowledge the farm "Iannone Anna" for multiplication of the seeds of the accession MG_13 used in field experiments. In addition, we acknowledge Gaetano Devito for animal caring.

Conflicts of Interest: The authors declare no conflict of interest.

References

1. Pejcic, T.; Tosti, T.; Dzamic, Z.; Gasic, U.; Vuksanovic, A.; Dolicanin, Z.; Tesic, Z. The Polyphenols as Potential Agents in Prevention and Therapy of Prostate Diseases. *Molecules* **2019**, *24*, 3982. [CrossRef] [PubMed]
2. Ismaiel, A.; Dumitrascu, D.L. Cardiovascular Risk in Fatty Liver Disease: The Liver-Heart Axis-Literature Review. *Front. Med.* **2019**, *6*, 202. [CrossRef] [PubMed]
3. Portincasa, P.; Calamita, G. Phytocompounds modulating Aquaporins: Clinical benefits are anticipated. *Food Chem.* **2019**, *274*, 642–650. [CrossRef] [PubMed]
4. Summo, C.; De Angelis, D.; Ricciardi, L.; Caponio, F.; Lotti, C.; Pavan, S.; Pasqualone, A. Data on the chemical composition, bioactive compounds, fatty acid composition, physico-chemical and functional properties of a global chickpea collection. *Data Brief* **2019**, *27*, 104612. [CrossRef] [PubMed]
5. Rachwa-Rosiak, D.; Nebesny, E.; Budryn, G. Chickpeas-composition, nutritional value, health benefits, application to bread and snacks: A review. *Crit. Rev. Food Sci. Nutr.* **2015**, *55*, 1137–1145. [CrossRef] [PubMed]
6. Mirmiran, P.; Yuzbashian, E.; Asghari, G.; Sarverzadeh, S.; Azizi, F. Dietary fibre intake in relation to the risk of incident chronic kidney disease. *Br. J. Nutr.* **2018**, *119*, 479–485. [CrossRef]
7. Shinohara, S.; Gu, Y.; Yang, Y.; Furuta, Y.; Tanaka, M.; Yue, X.; Wang, W.; Kitano, M.; Kimura, H. Ethanol extracts of chickpeas alter the total lipid content and expression levels of genes related to fatty acid metabolism in mouse 3T3-L1 adipocytes. *Int. J. Mol. Med.* **2016**, *38*, 574–584. [CrossRef]
8. Valentín-Gamazo, I.G. Chickpea flour ingredient slow glycemic response to pasta in healthy volunteers. *Food Chem.* **2003**, *81*, 5. [CrossRef]
9. Yang, Y.; Zhou, L.; Gu, Y.; Zhang, Y.; Tang, J.; Li, F.; Shang, W.; Jiang, B.; Yue, X.; Chen, M. Dietary chickpeas reverse visceral adiposity, dyslipidaemia and insulin resistance in rats induced by a chronic high-fat diet. *Br. J. Nutr.* **2007**, *98*, 720–726. [CrossRef]
10. Matsuzawa-Nagata, N.; Takamura, T.; Ando, H.; Nakamura, S.; Kurita, S.; Misu, H.; Ota, T.; Yokoyama, M.; Honda, M.; Miyamoto, K.; et al. Increased oxidative stress precedes the onset of high-fat diet-induced insulin resistance and obesity. *Metab. Clin. Exp.* **2008**, *57*, 1071–1077. [CrossRef]

11. Charradi, K.; Elkahoui, S.; Limam, F.; Aouani, E. High-fat diet induced an oxidative stress in white adipose tissue and disturbed plasma transition metals in rat: Prevention by grape seed and skin extract. *J. Physiol. Sci. Jps* **2013**, *63*, 445–455. [CrossRef] [PubMed]
12. Furukawa, S.; Fujita, T.; Shimabukuro, M.; Iwaki, M.; Yamada, Y.; Nakajima, Y.; Nakayama, O.; Makishima, M.; Matsuda, M.; Shimomura, I. Increased oxidative stress in obesity and its impact on metabolic syndrome. *J. Clin. Investig.* **2004**, *114*, 1752–1761. [CrossRef] [PubMed]
13. Wan, F.; Lenardo, M.J. The nuclear signaling of NF-kappaB: Current knowledge, new insights, and future future perspectives. *Cell Res.* **2010**, *20*, 24–33. [CrossRef] [PubMed]
14. Liu, T.; Zhang, L.; Joo, D.; Sun, S.C. NF-kappaB signaling in inflammation. *Signal Transduct. Target. Ther.* **2017**, *2*. [CrossRef] [PubMed]
15. Buss, H.; Dorrie, A.; Schmitz, M.L.; Hoffmann, E.; Resch, K.; Kracht, M. Constitutive and interleukin-1-inducible phosphorylation of p65 NF-{kappa}B at serine 536 is mediated by multiple protein kinases including I{kappa}B kinase (IKK)-{alpha}, IKK{beta}, IKK{epsilon}, TRAF family member-associated (TANK)-binding kinase 1 (TBK1), and an unknown kinase and couples p65 to TATA-binding protein-associated factor II31-mediated interleukin-8 transcription. *J. Biol. Chem.* **2004**, *279*, 55633–55643. [CrossRef]
16. Buss, H.; Handschick, K.; Jurrmann, N.; Pekkonen, P.; Beuerlein, K.; Muller, H.; Wait, R.; Saklatvala, J.; Ojala, P.M.; Schmitz, M.L.; et al. Cyclin-dependent kinase 6 phosphorylates NF-kappaB P65 at serine 536 and contributes to the regulation of inflammatory gene expression. *PLoS ONE* **2012**, *7*. [CrossRef]
17. Zhu, Q.C.; Gao, R.Y.; Wu, W.; Guo, B.M.; Peng, J.Y.; Qin, H.L. Effect of a high-fat diet in development of colonic adenoma in an animal model. *World J. Gastroenterol.* **2014**, *20*, 8119–8129. [CrossRef]
18. Da Costa, R.M.; Neves, K.B.; Mestriner, F.L.; Louzada-Junior, P.; Bruder-Nascimento, T.; Tostes, R.C. TNF-alpha induces vascular insulin resistance via positive modulation of PTEN and decreased Akt/eNOS/NO signaling in high fat diet-fed mice. *Cardiovasc. Diabetol.* **2016**, *15*, 119. [CrossRef]
19. Nestel, P.; Cehun, M.; Chronopoulos, A. Effects of long-term consumption and single meals of chickpeas on plasma glucose, insulin, and triacylglycerol concentrations. *Am. J. Clin. Nutr.* **2004**, *79*, 390–395. [CrossRef]
20. Zulet, M.A.; Macarulla, M.T.; Portillo, M.P.; Noel-Suberville, C.; Higueret, P.; Martinez, J.A. Lipid and glucose utilization in hypercholesterolemic rats fed a diet containing heated chickpea (Cicer aretinum L.): A potential functional food. *Int. J. Vitam. Nutr. Res. Int. Z. Fur Vitam.—Und Ernahr. J. Int. De Vitaminol. Et De Nutr.* **1999**, *69*, 403–411. [CrossRef]
21. De Giovanni, C.; Pavan, S.; Taranto, F.; Di Rienzo, V.; Miazzi, M.M.; Marcotrigiano, A.R.; Mangini, G.; Montemurro, C.; Ricciardi, L.; Lotti, C. Genetic variation of a global germplasm collection of chickpea (Cicer arietinum L.) including Italian accessions at risk of genetic erosion. *Physiol. Mol. Biol. Plants Int. J. Funct. Plant Biol.* **2017**, *23*, 197–205. [CrossRef] [PubMed]
22. Pavan, S.; Lotti, C.; Marcotrigiano, A.R.; Mazzeo, R.; Bardaro, N.; Bracuto, V.; Ricciardi, F.; Taranto, F.; D'Agostino, N.; Schiavulli, A.; et al. A Distinct Genetic Cluster in Cultivated Chickpea as Revealed by Genome-wide Marker Discovery and Genotyping. *Plant Genome* **2017**, *10*. [CrossRef] [PubMed]
23. Summo, C.; De Angelis, D.; Ricciardi, L.; Caponio, F.; Lotti, C.; Pavan, S.; Pasqualone, A. Nutritional, physico-chemical and functional characterization of a global chickpea collection. *J. Food Compos. Anal.* **2019**, *84*. [CrossRef]
24. Baldini, F.; Portincasa, P.; Grasselli, E.; Damonte, G.; Salis, A.; Bonomo, M.; Florio, M.; Serale, N.; Voci, A.; Gena, P.; et al. Aquaporin-9 is involved in the lipid-lowering activity of the nutraceutical silybin on hepatocytes through modulation of autophagy and lipid droplets composition. *Biochim.Biophys. Acta Mol. Cell Biol. Lipids* **2020**, *1865*, 158586. [CrossRef]
25. Grasselli, E.; Cortese, K.; Voci, A.; Vergani, L.; Fabbri, R.; Barmo, C.; Gallo, G.; Canesi, L. Direct effects of Bisphenol A on lipid homeostasis in rat hepatoma cells. *Chemosphere* **2013**, *91*, 1123–1129. [CrossRef]
26. Ranieri, M.; Di Mise, A.; Difonzo, G.; Centrone, M.; Venneri, M.; Pellegrino, T.; Russo, A.; Mastrodonato, M.; Caponio, F.; Valenti, G.; et al. Green olive leaf extract (OLE) provides cytoprotection in renal cells exposed to low doses of cadmium. *PLoS ONE* **2019**, *14*, e0214159. [CrossRef]
27. Tamma, G.; Ranieri, M.; Di Mise, A.; Centrone, M.; Svelto, M.; Valenti, G. Glutathionylation of the aquaporin-2 water channel: A novel post-translational modification modulated by the oxidative stress. *J. Biol. Chem.* **2014**, *289*, 27807–27813. [CrossRef]

28. Liquori, G.E.; Calamita, G.; Cascella, D.; Mastrodonato, M.; Portincasa, P.; Ferri, D. An innovative methodology for the automated morphometric and quantitative estimation of liver steatosis. *Histol. Histopathol.* **2009**, *24*, 49–60. [CrossRef]
29. Ranieri, M.; Tamma, G.; Di Mise, A.; Russo, A.; Centrone, M.; Svelto, M.; Calamita, G.; Valenti, G. Negative feedback from CaSR signaling to aquaporin-2 sensitizes vasopressin to extracellular Ca2. *J. Cell Sci.* **2015**, *128*, 2350–2360. [CrossRef]
30. Vecchione, G.; Grasselli, E.; Voci, A.; Baldini, F.; Grattagliano, I.; Wang, D.Q.; Portincasa, P.; Vergani, L. Silybin counteracts lipid excess and oxidative stress in cultured steatotic hepatic cells. *World J. Gastroenterol.* **2016**, *22*, 6016–6026. [CrossRef]
31. Yumuk, V.; Tsigos, C.; Fried, M.; Schindler, K.; Busetto, L.; Micic, D.; Toplak, H. Obesity Management Task Force of the European Association for the Study of, O. European Guidelines for Obesity Management in Adults. *Obes. Facts* **2015**, *8*, 402–424. [CrossRef] [PubMed]
32. Maric, C.; Hall, J.E. Obesity, metabolic syndrome and diabetic nephropathy. *Contrib. Nephrol.* **2011**, *170*, 2835. [CrossRef]
33. Oosterveer, M.H.; van Dijk, T.H.; Tietge, U.J.; Boer, T.; Havinga, R.; Stellaard, F.; Groen, A.K.; Kuipers, F.; Reijngoud, D.J. High fat feeding induces hepatic fatty acid elongation in mice. *PLoS ONE* **2009**, *4*, e6066. [CrossRef]
34. Vanni, E.; Bugianesi, E.; Kotronen, A.; De Minicis, S.; Yki-Jarvinen, H.; Svegliati-Baroni, G. From the metabolic syndrome to NAFLD or vice versa? *Dig. Liver Dis. Off. J. Ital. Soc. Gastroenterol. Ital. Assoc. Study Liver* **2010**, *42*, 320330. [CrossRef] [PubMed]
35. Velazquez, K.T.; Enos, R.T.; Bader, J.E.; Sougiannis, A.T.; Carson, M.S.; Chatzistamou, I.; Carson, J.A.; Nagarkatti, P.S.; Nagarkatti, M.; Murphy, E.A. Prolonged high-fat-diet feeding promotes non-alcoholic fatty liver disease and alters gut microbiota in mice. *World J. Hepatol.* **2019**, *11*, 619637. [CrossRef] [PubMed]
36. Ferreira, H.; Vasconcelos, M.; Gil, A.M.; Pinto, E. Benefits of pulse consumption on metabolism and health: A systematic review of randomized controlled trials. *Crit. Rev. Food Sci. Nutr.* **2020**, 1–12. [CrossRef]
37. Bungau, S.; Abdel-Daim, M.M.; Tit, D.M.; Ghanem, E.; Sato, S.; Maruyama-Inoue, M.; Yamane, S.; Kadonosono, K. Health Benefits of Polyphenols and Carotenoids in Age-Related Eye Diseases. *Oxid. Med. Cell. Longev.* **2019**, *2019*, 9783429. [CrossRef]
38. Fascella, G.; D'Angiolillo, F.; Mammano, M.M.; Amenta, M.; Romeo, F.V.; Rapisarda, P.; Ballistreri, G. Bioactive compounds and antioxidant activity of four rose hip species from spontaneous Sicilian flora. *Food Chem.* **2019**, *289*, 56–64. [CrossRef]
39. Feldstein, A.E.; Werneburg, N.W.; Canbay, A.; Guicciardi, M.E.; Bronk, S.F.; Rydzewski, R.; Burgart, L.J.; Gores, G.J. Free fatty acids promote hepatic lipotoxicity by stimulating TNF-alpha expression via a lysosomal pathway. *Hepatology* **2004**, *40*, 185–194. [CrossRef]
40. Li, Z.; Berk, M.; McIntyre, T.M.; Gores, G.J.; Feldstein, A.E. The lysosomal-mitochondrial axis in free fatty acid-induced hepatic lipotoxicity. *Hepatology* **2008**, *47*, 1495–1503. [CrossRef]
41. Carlsen, H.; Haugen, F.; Zadelaar, S.; Kleemann, R.; Kooistra, T.; Drevon, C.A.; Blomhoff, R. Diet-induced obesity increases NF-kappaB signaling in reporter mice. *Genes Nutr.* **2009**, *4*, 215–222. [CrossRef] [PubMed]
42. Calder, P.C.; Grimble, R.F. Polyunsaturated fatty acids, inflammation and immunity. *Eur. J. Clin. Nutr.* **2002**, *56* (Suppl. 3), S14–S19. [CrossRef] [PubMed]
43. Fritsche, K.L. Too much linoleic acid promotes inflammation-doesn't it? *Prostaglandins Leukot. Essent. Fat. Acids* **2008**, *79*, 173–175. [CrossRef] [PubMed]
44. Simopoulos, A.P. The importance of the omega-6/omega-3 fatty acid ratio in cardiovascular disease and other chronic diseases. *Exp. Biol. Med.* **2008**, *233*, 674–688. [CrossRef]

© 2020 by the authors. Licensee MDPI, Basel, Switzerland. This article is an open access article distributed under the terms and conditions of the Creative Commons Attribution (CC BY) license (http://creativecommons.org/licenses/by/4.0/).

Article

Concentration Dependence of the Antioxidant and Prooxidant Activity of Trolox in HeLa Cells: Involvement in the Induction of Apoptotic Volume Decrease

Maria Elena Giordano, Roberto Caricato and Maria Giulia Lionetto *

Department of Biological and Environmental Sciences and Technologies (DiSTeBA), University of Salento, Via prov.le Lecce-Monteroni, 73100 Lecce, Italy; elena.giordano@unisalento.it (M.E.G.); roberto.caricato@unisalento.it (R.C.)

* Correspondence: giulia.lionetto@unisalento.it; Tel.: +39-0832-298-668; Fax: +39-0832-298-626

Received: 3 October 2020; Accepted: 26 October 2020; Published: 29 October 2020 

Abstract: Trolox (6-hydroxy-2,5,7,8-tetramethylchroman-2-carboxylic acid), a hydrophilic analog of vitamin E, is known for its strong antioxidant activity, being a high radical scavenger of peroxyl and alkoxyl radicals. Under particular conditions, Trolox may also exhibit prooxidant properties. The present work aimed at studying the dual antioxidant/prooxidant behavior of Trolox over a wide range of concentrations (from 2.5 to 160 μM) in HeLa cells. In particular, the study addressed the dose-dependent effects of Trolox on the oxidative cell status and vitality of HeLa cells, focusing on the potential role of the vitamin E analog in the induction of one of the first steps of the apoptotic process, Apoptotic Volume Decrease (AVD). In HeLa cells, Trolox showed significant antioxidant activity, expressed as the ability to reduce the endogenous ROS production detected by the ROS-sensitive probe 5-(and-6)-chloromethyl-2′,7′-dichlorodihydrofluorescein diacetate (CM-H_2DCFDA), at low concentrations (range: 2.5–15 μM), but exerted a dose-dependent prooxidant effect at higher concentrations after 24 h exposure. The prooxidant effect was paralleled by the reduction in cell viability due to the induction of the apoptotic process. The dual behavior, antioxidant at lower concentrations and prooxidant at higher concentrations, was evident also earlier after 2 h incubation, and it was paralleled by the isotonic shrinkage of the cells, ascribed to AVD. The use of SITS, known Cl^- channel blocker, was able to completely inhibit the Trolox-induced isotonic cell shrinkage, demonstrating the involvement of the vitamin E analog in the alteration of cell volume homeostasis and, in turn, in the AVD induction. In conclusion, the study shed light on the concentration dependence of the Trolox antioxidant/prooxidant activity in HeLa cells and revealed its role in the induction of one of the first events of apoptosis, AVD, at high concentrations.

Keywords: Trolox; HeLa; antioxidant; prooxidant; AVD; apoptosis

---

1. Introduction

Trolox (6-hydroxy-2,5,7,8-tetramethylchroman-2-carboxylic acid) is a hydrophilic analog of α-tocopherol, the most active and the most common form of tocopherols (vitamin E) in the human body. α-tocopherol is the major lipid antioxidant of biomembranes; it prevents membrane oxidative damage through inhibition of polyunsaturated fatty acids peroxidation by scavenging lipid peroxyl radicals [1,2]. Besides, it quenches and reacts with singlet oxygen and slowly reacts with superoxide anions [3]. Trolox shows the same antioxidant activity of α-tocopherol, but compared to α-tocopherol, which is lipid soluble, it lacks the phytyl tail, it is more hydrosoluble, and it has the advantage to reach both the water and the lipid compartments of cells.

Trolox is known for its high radical scavenging activity of peroxyl and alkoxyl radicals [4] and as such, it is often used as a reference in several biochemical assays, in which the radical scavenging activity of studied compounds is expressed as Trolox equivalents. It has been demonstrated to act as a strong antioxidant in several cellular model systems. In human skin fibroblasts, it was found to lower intracellular ROS levels and lipid peroxidation and to induce a less oxidized mitochondrial thiol redox state [5]. In IPEC-J2 cells, Trolox reduced intracellular oxidative stress, it improved wound-healing capacity and paracellular permeability following exposure to exogenous prooxidant compounds [6]. In thymocytes, it was able to prevent peroxynitrite-mediated oxidative stress and apoptosis [7]. Trolox contributed to the protection of human and murine primary alveolar type II cells from injury by cigarette smoke-induced oxidative stress [8] and prevented the generation of oxidative stress in primary adult rat optic nerve head astrocytes showing induced reactive astrocytosis [9]. Moreover, Trolox was demonstrated to prevent lipid peroxidation induced by CYP2E1 in Hep G2 cells [10], to attenuate impaired proliferation of oxidatively stressed myoblasts overexpressing parkin interacting substrate (PARIS/ZNF746) [11], to inhibit DNA damage formation induced by singlet oxygen in human lymphoblast WTK-1 cells27 [12], and to protect erythrocytes during photodynamic treatment [13].

Trolox has been reported to prevent oxidative stress-induced apoptosis in several cell types, such as mouse thymocytes [14], renal NRK-52e cells (normal rat kidney 52e) [15], rat myocardial H9c2 cells [16], renal proximal tubular epithelial LLC-PK1 cells [17].

Although Trolox possesses strong antioxidant activity, it might also exhibit prooxidant properties in particular conditions, as also observed for other antioxidant compounds, including α-tocopherol [18–20]. Like α-tocopherol, the Trolox antioxidant activity arises from its ability to donate hydrogen from the hydroxyl group of the chromanol ring [21] to reactive species. This, in turn, drives the formation of phenoxyl radical (PhO·), which can oxidize ascorbate and other biomolecules to radicals [21]. Phenoxyl radical can extract a hydrogen atom from the bisallylic methylene groups of polyunsaturated fatty acids, inducing lipid peroxidation, although the rate constant for this reaction is slow, being about $10^{-1} \pm 0.05$ M^1s^1 [22]. Several studies demonstrated Trolox to exhibit prooxidant properties in the presence of free metal ions [23]. For example, in astrocytes, Trolox increased the Cu^{2+}-induced ROS generation and cytotoxicity [24], while in erythrocytes Trolox stimulated ferric ion-catalyzed ascorbate oxidation [25].

The synergic prooxidant action of Trolox has been described also for nonmetallic compounds. It has been reported to induce a synergistic prooxidant effect with superoxide generating selenium compounds in mouse keratinocytes [26], to potentiate As_2O_3-induced oxidative stress in APL cell line, myeloma, and breast cancer cells [27], and to enhance curcumin's cytotoxicity in A2780 cells [28].

At higher concentrations, Trolox has been described to induce lipid peroxidation, GSH oxidation, and cytotoxicity [29,30].

The dual antioxidant/prooxidant behavior of Trolox is a multifaceted phenomenon, showing cell-type specificity and being influenced by the experimental conditions used. Concentration is one of the most determinant factors; however, most of the studies exploring the cellular effects of Trolox have been carried out on single concentrations of the vitamin E analog.

The present study aimed at investigating the dual antioxidant/prooxidant behavior of Trolox over a wide range of concentrations (from 2.5 to 160 μM) on a model cell line, HeLa cells. In particular, we addressed if there are dose-dependent effects of Trolox on the oxidative cell status of HeLa cells and if the vitamin E analog is involved in the induction of the apoptotic process focusing on one of the first step of the apoptotic process, Apoptotic Volume Decrease (AVD). One of the main hallmarks of apoptosis is represented by cell shrinkage, which occurs in two distinct stages: the early shrinkage, named, AVD, starting in the first hours, represented by an isotonic shrinkage of the cells [31–33], and the second phase related to cell fragmentation or formation of the apoptotic body [34]. Recent studies have pointed out the importance of AVD in the onset and progression of apoptosis, suggesting its critical role in the regulation of apoptotic nucleases and the activation of caspases. AVD arises from a net loss of K^+, Cl^-, and organic osmolytes from the cell [31,35,36], which, in turn, is followed by an osmotic loss of water and consequent cell shrinkage. In particular, in the present work, we addressed if Trolox can induce AVD thought its effect on Cl^- channels.

In this study, HeLa cells were chosen as the experimental cell model for addressing the objectives of the work. HeLa cells are one of the most studied cellular models, widely utilized in several fields from cancer research, to drug development, gene expression, and cell death pathways. Moreover, they are particularly useful for the objectives of this study because their AVD response is well known and characterized [37].

2. Methods

2.1. Materials

All chemicals were reagent grade. Cell culture materials were acquired from EuroClone (Paignton-Devon, UK). The cell-permeant probe 5-(and-6)-chloromethyl-2′,7′-dichlorodihydrofluorescein diacetate (CM-H_2DCFDA) and Alexa Fluor® 488 annexin V were purchased from Life Technologies-Molecular Probes (Waltham, MA, USA). All the other reagents were purchased from Sigma Aldrich (St. Louis, MO, USA). HeLa cells were purchased from ATCC (Manassas, VA, USA).

2.2. Intracellular Oxidative Stress Detection

The intracellular oxidative stress was assessed using a ROS-sensitive probe, 5-(and-6-)-chloromethyl -2′,7′-dichlorodihydrofluorescein diacetate, acetyl ester (CM-H_2DCFDA) (Ex/Em: 492–495/517–527 nm) (Thermo Fisher Scientific, Waltham, MA, USA). HeLa cells were grown as a monolayer in Dulbecco's Modified Eagle's Medium with 4500 mg glucose/L (DMEM) supplemented with 10% FBS, 40 IU/mL penicillin G, 2 mM L-glutamine, and 100 μg/mL streptomycin under a 95% air/5% CO_2 atmosphere. Cells were plated (1×10^5 per mL) into 96-well black plate and incubated for 24 h to allow the cells to attach to the bottom of the plate. Afterwards, the cells were incubated with increasing concentrations of Trolox (respectively, 2.5, 5, 10, 15, 20, 40, 80, and 160 mM) for 2 or 24 h. Then, they were charged with the cell-permeant fluorescent probe CM-H_2DCFDA 5 μM. Subsequent oxidation, this probe yields a fluorescent adduct that is trapped inside the cell. Fluorescence was then measured by the Synergy™ (BioTek Instruments, Inc., Winooski, VT, USA) multi-mode microplate reader.

2.3. Cell Viability Assessment by MTT Assay and Propidium Iodide

Cell viability was assessed by the MTT test according to Latronico et al. [38]. The MTT assay was carried out after plating HeLa cells (1×10^5 per mL) into 96-well plate. After 24 h incubation, the cells were preincubated with Trolox at the final concentrations of 2.5, 5, 10, 15, 20, 40, 80, and 160 μM for 24 h. Then, 20 μL MTT (0.5 mg/mL in PBS) was put on each well. After an incubation of 4 h at 37 °C, the medium was discharged and DMSO (100 μL) was added. Absorbance was measured at 570 nm with a spectrophotometer (EON BioTek Instruments, Winooski, VT, USA).

For propidium iodide cell viability assessment, HeLa cells (1×10^5 per mL) were plated into 6-well plate and incubated for 24 h. Then, the cells were incubated with Trolox (2.5, 5, 10, 15, 20, 40, 80, and 160 μM) for 24 h. Following incubation, they were washed with PBS, detached by gentle trypsinization, and then incubated with propidium iodide (final concentration of 50 μg/mL) for 10 min. Then, cells were washed again and spectrofluorimetrically analyzed by Synergy™ (BioTek Instruments, Inc., Winooski, VT, USA) multi-mode microplate reader (λ_{ex}: 535 nm and λ_{em}: 617 nm).

2.4. Estimation of Changes in Cell Volume

Cell volume changes were assessed by the measurement of changes in cell size through morphometric analysis of the cell area according to Lionetto et al. [32]. Cultured cells were observed by an inverted microscope in bright field (NIKON TE300 Eclipse E600, Nikon, Tokyo, Japan), and the 2-dimensional images obtained from a video camera (TK-C1381, JVC, Yokohama, Japan) were digitalized and analyzed using the LUCIA images analysis software (Nikon, Tokyo, Japan). At least a minimum of 300 cells/condition was analyzed.

2.5. Detection of Apoptosis

HeLa cells (1×10^5 per mL) were plated into 6-well plate and incubated for 24 h. Then, the cells were preincubated with Trolox at the final concentrations of 2.5, 40, 80, and 160 μM for 14 h. Then, they were washed with PBS, detached by gentle trypsinization, and then resuspended in annexin V binding buffer (HEPES 10 mM, pH 7.4, NaCl 150 mM, $CaCl_2$ 2.5 mM in PBS) and incubated with Alexa Fluor® 488 annexin V (final concentration 2.5 μg/mL) and propidium iodide (final concentration 50 μg/mL) for 10 min. Then, the cells were washed and transferred in a 96-well black plate. Fluorescence was measured by the Synergy™ (Biotek) multi-mode microplate reader (annexin V: λ_{ex}: 488 nm and λ_{em}: 530 nm; propidium iodide: λ_{ex}: 535 nm and λ_{em}: 617 nm).

2.6. Statistics

Values are given as the mean ± S.E.M. The statistical significance of data was analyzed by one-way ANOVA, Dunnett's test, and Student's *t*-test. Percentage values were arcsin transformed before the analysis. Data are expressed as mean ± SEM.

3. Results

3.1. Dose-Dependent Effect of Trolox on Basal ROS Production and Cell Viability after 24 h Exposure

HeLa cells were treated for 24 h with increasing concentrations of Trolox in the range of 2.5–160 μM, and then they were charged with the cell-permeable ROS-sensitive probe CM-H_2DCFDA to investigate the effect of the vitamin E analog on the intracellular oxidative status of the cells. The incubation time of 24 h was established based on previous works, demonstrating 24 h incubation as an appropriate time period for Trolox-induced biological responses to be evoked in several cell types [39–41]. Figure 1 shows the percentage variation of the fluorescence emitted by the cells calculated compared to control cells (not exposed to Trolox). Trolox was able to exert an antioxidant activity at lower concentrations (2.5–15 μM), as indicated by the negative percentage variation of the CM-H_2DCFDA fluorescence, which is indicative of a decrease in basal ROS production. The antioxidant effect size was about 20% in the concentration range of 2.5–10 μM. A decrease in the antioxidant effect was recorded at 15 μM Trolox. The antioxidant activity disappeared at 20 μM. On the other hand, at higher concentrations (from 40 to 160 μM) a dose-dependent prooxidant effect was recorded, as indicated by the positive increase in the percentage variation of the CM-H_2DCFDA fluorescence.

These results demonstrate a dose-dependent dual behavior of Trolox on the basal ROS production of HeLa cells following 24 h exposure.

In parallel to the assessment of the oxidative status of the cells, cell viability was analyzed by both MTT test and propidium iodide staining in HeLa cells following 24 h exposure (Figure 2A,B).

The measurement of the metabolic activity of the cells by the MTT assay (Figure 2A) showed a dose-dependent reduction in cell viability with a maximum effect observed at 160 μM Trolox.

Moreover, as shown in Figure 2B, HeLa cells exposed for 24 h to Trolox at concentrations of 40, 80, and 160 μM showed a significant increase in the fluorescence of PI compared to control cells, while no significant change was observed at lower Trolox concentrations (Figure 2B). Propidium iodide cannot pass through a viable cell membrane, but it binds to DNA intercalating with the double helix in dead cells. Therefore, the results obtained with propidium iodide confirmed the viability results obtained with the MTT test.

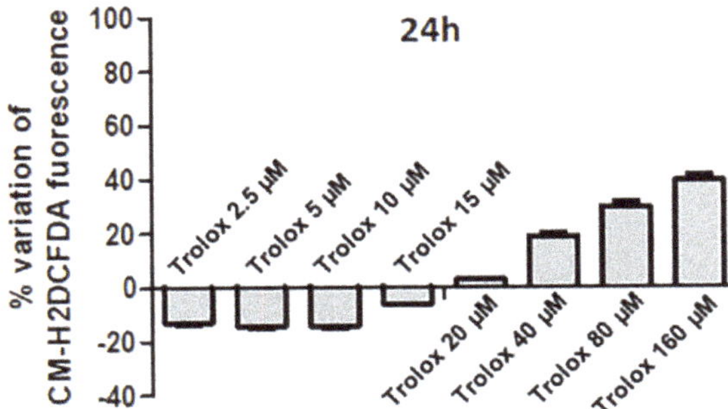

Figure 1. Percentage variation of fluorescence intensity in HeLa cells exposed for 24 h to increasing concentrations of Trolox (from 2.5 to 160 µM) and then charged with 5-(and-6)-chloromethyl-2′,7′-dichlorodihydrofluorescein diacetate (CM-H_2DCFDA). The ordinates indicate the percentage variation of the probe fluorescence intensity, which was calculated as follows: (fluorescence of control cells − fluorescence of treated cells)/(fluorescence of control cells) × 100. Data are expressed as mean ± SEM of 3 independent experiments.

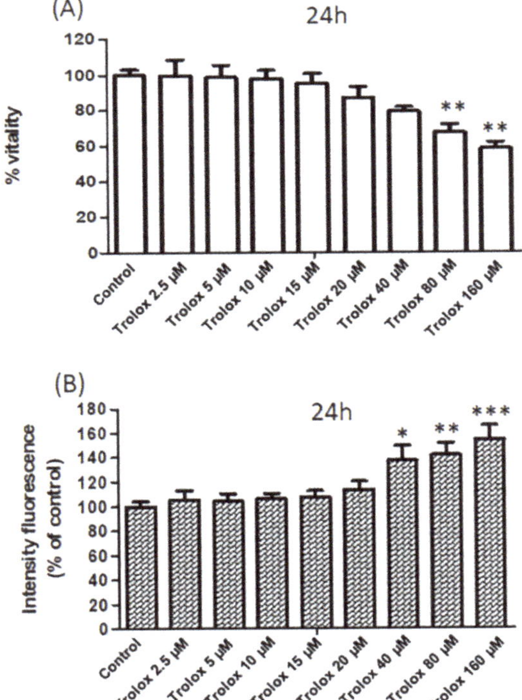

Figure 2. (**A**) The vitality of HeLa cells exposed for 24 h to increasing concentrations of Trolox (from 2.5 to 160 µM) assessed by MTT test. Data are expressed as percentage vs. control. (**B**) The vitality of HeLa cells exposed for 24 h to increasing concentrations of Trolox (from 2.5 to 160 µM) assessed by the fluorescence intensity (% vs. control) of propidium iodide. Data are expressed as mean ± SEM of 3 independent experiments. The statistical significance of data was analyzed by one-way ANOVA and Dunnett test. * = $p < 0.05$; ** = $p < 0.01$; *** = $p < 0.001$.

3.2. Apoptosis Detection

To deepen the knowledge about the reduced cell viability caused by the exposure to high Trolox concentrations in HeLa cells, the hypothesis of apoptosis induction was tested. One of the hallmarks of apoptosis is represented by the translocation of phosphatidylserine from the inner to the outer leaflet of the plasma membrane, thus exposing phosphatidylserine to the external cellular environment [42]. HeLa cells were exposed to 40, 80, and 160 μM Trolox, respectively, which were the concentrations found to induce oxidative stress and to decrease cell viability dose dependently, and to Trolox 2.5 μM for comparison. After 14 h exposure, the cells were double-stained with annexin V and propidium iodide, and the relative fluorescence was recorded with a multiplate reader. A shorter incubation time of 14 h was chosen in this case, after preliminary tests, in order to detect, if any, the early translocation of phosphatidylserine through annexin V binding before the occurrence of cell death. It is known that the kinetic profile of phosphatidylserine exposure in the outer layer of plasma membrane precedes the loss of membrane integrity. As previously assessed in other cell types, 14 h after the exposure to an apoptotic stimulus phosphatidylserine translocation is detectable, while the loss of membrane integrity is not yet [43]. As reported in Figure 3, the percentage variation of the annexin V fluorescence appeared strongly increased in cells exposed to 40, 80, and 160 μM Trolox for 14 h, but not in cells exposed to 2.5 μM Trolox.

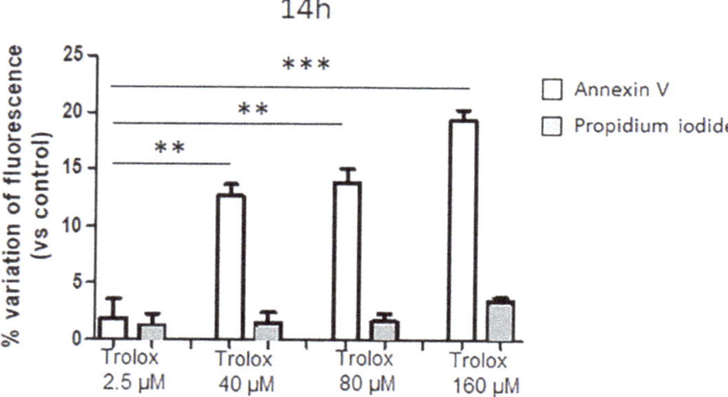

Figure 3. Effect of Trolox after 14 h exposure on annexin V and propidium iodide fluorescence (expressed as percentage variations of control) in HeLa cells. Details as Figure 1. Data are expressed as mean ± SEM of 3 independent experiments. Statistical significance of differences was assessed by one-way ANOVA and Dunnett test. ** = $p < 0.01$; *** = $p < 0.001$.

3.3. Dose-Dependent Effect of Trolox on Basal ROS Production and Cell Volume after 2 h Exposure

After having demonstrated that Trolox can induce apoptosis at high concentrations, we tested the hypothesis of the possible involvement of the vitamin E analog in the induction of one of the first events of apoptosis, the Apoptotic Volume Decrease, which is known to occur in the first 1–2 h [31,44]. Therefore, the investigation of the dose-dependent effects of Trolox on HeLa cells was deepened at 2 h exposure.

HeLa cells exposed for 2 h to different concentrations of Trolox (range: 2.5–160 μM) and then charged with CM-H_2DCFDA showed the same dual behavior observed at 24 h. A decrease in the basal ROS level was evident at lower concentrations (2.5–20 μM), as indicated by the reduced fluorescence of the intracellularly trapped probe, while an increase in the basal ROS levels was detected at higher concentrations (40–160 μM) (Figure 4A).

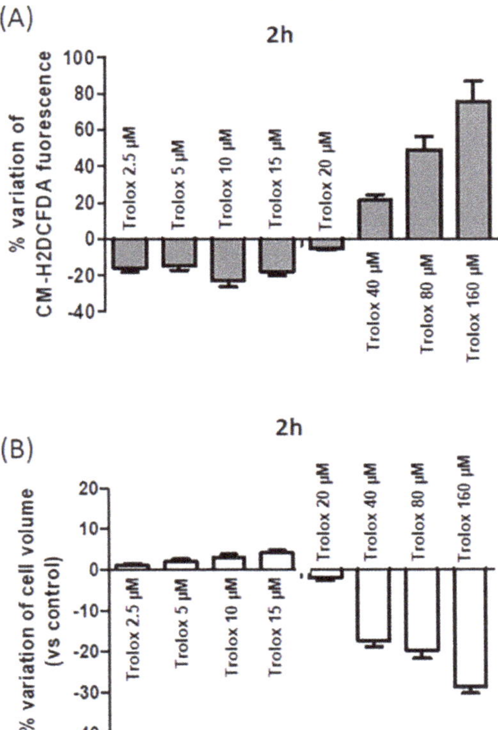

Figure 4. (**A**) Percentage variation of fluorescence intensity in HeLa cells exposed for 2 h to increasing concentrations of Trolox (from 2.5 to 160 µM) and then charged with CM-H_2DCFDA. Details as Figure 1. (**B**) Effect of increasing concentrations of Trolox on cell volume after 2 h incubation. Data are expressed as cell volume percentage variation calculated as follows: (cell size of control cells − cell size of treated cells/cell size of control cells) × 100. Data are expressed as mean ± SEM of 3 independent experiments.

Comparing the percentage variation of CM-H_2DCFDA fluorescence at the highest concentrations tested after 2 and 24 h, an increased percentage variation was observed after 2 h exposure. For example, at 160 µM, the percentage variation of CM-H_2DCFDA fluorescence was about 75% after 2 h incubation (Figure 4A), while it was about 40% after 24 h (Figure 1). It cannot be excluded that after 24 h exposure, the occurrence of cell death could cause a possible leakage of the intracellularly trapped probe from the cells, which could contribute to some extent to decrease the fluorescence signal.

In parallel, the cell size of the cells exposed to different Trolox concentrations was measured. A decrease in cell dimension at 40, 80, and 160 µM after 2 h was evident (Figure 4B), and it was indicative of isotonic cell shrinkage, namely, AVD.

3.4. *Effect of SITS on the Trolox-Induced AVD*

AVD is known to be due to the loss of K^+ and Cl^- from the cells [45]. As reported in Figure 5, when HeLa cells were preincubated with the 0.5 mM disulfonic stilbene derivative SITS (4-Acetamido-4′-isothiocyanato-stilbene-2,2′-disulfonic acid), a known inhibitor of Cl^- channels [46], and then exposed to Trolox 80 and 160 µM for 2 h, the Trolox-induced isotonic shrinkage was completely prevented. On the other hand, SITS alone was not able to produce any significant alteration of cell size.

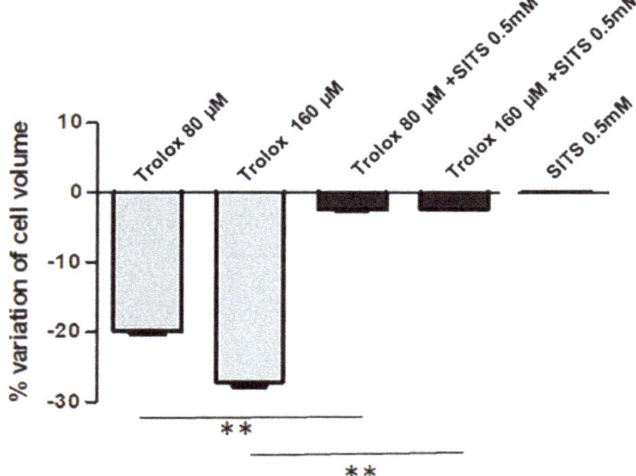

Figure 5. Effect of SITS (4-acetamido-4′-isothiocyanostilbene) (0.5 mM) on the Trolox-induced cell volume decrease. Details as Figure 4B. Data are reported as mean ± S.E.M of 3 independent experiments. Statistical significance was assessed by one-way ANOVA and Dunnett test. ** = $p < 0.01$.

4. Discussion

Oxidative stress, which occurs when the production and accumulation of reactive species in cells and tissues are not balanced by the antioxidant defenses of a biological system, has been recognized as an important factor in the genesis of several diseases, including chronic and degenerative diseases [47]. This accounts for the great interest devoted in the last years to antioxidant compounds and to their capacity in protecting cells from oxidative damage. However, some studies have pointed out that antioxidants can exert prooxidant effects under particular conditions, depending on their concentration and the nature of surrounding molecules [48]. This study is aimed to analyze the dose-dependent effect of Trolox, a widely utilized synthetic analog of vitamin E, on the basal ROS production and viability of HeLa cells, focusing on the ability of the compound to induce apoptosis at high concentrations.

In HeLa cells, Trolox showed significant antioxidant activity, expressed as the ability to reduce the endogenous ROS production detected by the ROS-sensitive probe CM-H_2DCFDA. The antioxidant activity was expressed at concentrations included in the range of 2.5–15 µM. At 20 µM, the Trolox antioxidant activity appeared reduced compared to the previous concentrations, as observed after 2 h incubation, or disappeared, as observed after 24 h incubation. At 40 µM, Trolox activity was reversed becoming prooxidant. The prooxidant effect showed a marked concentration-dependence with the highest effect observed at the highest concentration tested (160 µM). The prooxidant effect was accompanied by a reduction in cell viability as assessed by MTT test and propidium iodide staining.

These results are in agreement with results obtained on HUVEC cells exposed for 24 h to a wide range of Trolox concentrations [39]. However, in the case of HeLa cells, the prooxidant effects appeared at lower concentrations compared to HUVEC cells, where the prooxidant behavior, accompanied by a reduction in cell viability, was detected at much higher concentrations (500–1000 µg/mL) [39]. This points out the cell-type specificity in the dual behavior of Trolox and in the threshold concentration for the inversion of its antioxidant behavior in prooxidant behavior. The cell-type specificity could be influenced by the oxidative status of the cell and by the intracellular antioxidant defenses. The appearance of the prooxidant behavior as a consequence of the accumulations of phenoxyl radical (PhO·), arising, in turn, from the antioxidant activity of Trolox, can depend on the intracellular concentrations of antioxidants, such as GSH or ascorbic acid, able to neutralize the radical forms of Trolox. In our experimental model, 40 µM seems to be the threshold concentration for the inversion.

In our study, the reduction in the cell viability was ascribed to apoptosis, as assessed by the early detection (after 14 h incubation) of phosphatidylserine translocation by annexin V binding.

The study on the prooxidant effect of Trolox in HeLa cells was deepened by the analysis of the early effects of the vitamin E analog after 2 h incubation. The dual behavior, antioxidant at lower concentrations and prooxidant at higher concentrations, was evident already after 2 h incubation, with a pronounced increase in the intracellular oxidative stress at 80 and 160 μM. At this early stage, the prooxidant effect of Trolox was paralleled by the isotonic shrinkage of the cells, ascribed to AVD. This represents one of the first events of apoptosis, known to start within 0.5–2 h after apoptosis induction [37]. AVD is mainly due to KCl release from the cells, followed by osmotic loss of water. KCl occurs thanks to the activation of K^+ and Cl^- channels in several cell types [31–33,49]. The Cl^- channel involved in this process is known to be the volume-sensitive outwardly rectifying (VSOR) anion channel in different cell types [31,50,51]. They are known to be activated by osmotic cell swelling and to be involved in cell volume regulation and apoptotic cell death. VSOR channels are expressed in several cell types including HeLa cells [52,53].

In our study, we tested the hypothesis of possible induction of these channels by exposure to high concentrations of Trolox, 80 and 160 μM, respectively, which are the concentrations able to exert a marked prooxidant effect in HeLa cells. The use of SITS, known to block Cl^- channels including VSOR channel [54], was able to completely inhibit the Trolox-induced isotonic cell shrinkage, demonstrating the role of the vitamin E analog in the induction of volume-sensitive Cl^- channels, and, in turn, in AVD induction. On the other hand, the only treatment with SITS did not produce any alterations in cell volume, suggesting that these channels are normally closed in basal conditions.

As regards the mechanism through which Trolox can activate volume-sensitive Cl^- channels in HeLa cells, VSOR have been demonstrated to be sensitive to ROS in HeLa cells [48]. The obtained results demonstrate that Trolox at high concentrations can increase the intracellular ROS basal level. This, in turn, could activate volume-sensitive Cl^- channels, which play a key role in the apoptotic process. As demonstrated in HeLa, U937, and PC12 cells [31,45,53], the activation of AVD-inducing VSOR Cl^- channel is an early requisite event of apoptosis, because its inhibition can prevent further apoptotic events such as caspase-3 activation, cytochrome *c* release, and DNA laddering. Therefore, this study, for the first time, disclosed the role of Trolox at high concentrations in AVD and VSOR Cl^- channel activation, although future studies will be needed to analyze the effect of high Trolox concentrations on AVD induction in comparison with other oxidant agents and to deepen the mechanisms and signaling events underlying the Trolox effect.

5. Conclusions

In conclusion, this study clarified the concentration dependence of the antioxidant and prooxidant activity of Trolox in HeLa cells over a wide range of concentrations, disclosing the threshold inversion concentration of its antioxidant/prooxidant behavior. Although this study has been carried out on a single cellular type, the obtained results can be used as a model for further studies on other antioxidants in other cell types. In fact, this study underlines the need to evaluate in advance the dose dependence of the effects of the substance under investigation, its possible prooxidant behavior, and the antioxidant/prooxidant inversion threshold before carrying out any evaluation on its biological effect on any experimental system.

Moreover, this study disclosed for the first time the role of the vitamin E analog at high concentrations in the induction of one of the first events of apoptosis, AVD, through the induction of oxidative stress and VSOR Cl^- channel activation.

Author Contributions: Conceptualization, M.E.G. and M.G.L.; data curation, R.C.; investigation, M.E.G.; methodology, R.C.; supervision, M.G.L.; writing—original draft, M.G.L.; writing—review and editing, M.E.G. All authors have read and agreed to the published version of the manuscript.

Funding: This research received no external funding.

Acknowledgments: This study was supported by the MIUR (Italian Ministry of Education, University and Research).

Conflicts of Interest: The authors declared that there are no conflict of interest.

References

1. Massey, K.D.; Burton, K.P. Free radical damage in neonatal rat cardiac myocyte cultures: Effects of alpha-tocopherol, Trolox and phytol. *Free Radic. Biol. Med.* **1990**, *8*, 449. [CrossRef]
2. Lúcio, M.; Nunes, C.; Gaspar, D.; Ferreira, H.; Lima, J.L.F.C.; Reis, S. Antioxidant Activity of Vitamin E and Trolox: Understanding of the factors that govern lipid peroxidation studies in vitro. *Food Biophys.* **2009**, *4*, 312–320. [CrossRef]
3. Birgelius-Flohe, R.; Traber, M.G. Vitamin E Function and metabolism. *FASEB J.* **1999**, *13*, 1145–1155. [CrossRef]
4. Huang, D.; Ou, B.; Prior, R.L. The chemistry behind antioxidant capacity assays. *J. Agric. Food Chem.* **2005**, *53*, 1841–1856. [CrossRef]
5. Distelmaier, F.; Valsecchi, F.; Forkink, M.; van Emst-de Vries, S.; Swarts, H.G.; Rodenburg, R.J.T.; Verwiel, E.T.P.; Smeitink, J.A.M. Trolox-sensitive reactive oxygen species regulate mitochondrial morphology, oxidative phosphorylation. *Antioxid. Redox. Sign.* **2012**, *17*, 1652–1669.
6. Vergauwen, H.; Tambuyzer, B.; Jennes, K.; Degroote, J.; Wang, W.; De Smet, S.; Michiels, J.; Van Ginneken, C. Trolox and Ascorbic Acid Reduce Direct and Indirect Oxidative Stress in the IPEC-J2 Cells, an In Vitro Model for the Porcine Gastrointestinal Tract. *PLoS ONE* **2015**, *10*, e0120485. [CrossRef] [PubMed]
7. Salgo, M.G.; Pryor, W.A. Trolox inhibits peroxynitrite-mediated oxidative stress and apoptosis in rat thymocytes. *Arch. Biochem. Biophys.* **1996**, *333*, 482–488. [CrossRef]
8. Messier, E.M.; Bahmed, K.; Tuder, R.M.; Chu, H.W.; Bowler, R.P.; Kosmider, B. Trolox contributes to Nrf2-mediated protection of human and murine primary alveolar type II cells from injury by cigarette smoke. *Cell Death Dis.* **2013**, *4*, e573. [CrossRef]
9. Ghosh, A.K.; Rao, V.R.; Stubbs, E.B.; Kajga, S. Antioxidants protect against reactive astrocytosis-induced sensitization to oxidative stress. *Invest. Ophth Vis. Sci.* **2019**, *60*, 3791.
10. Caro, A.A.; Thompson, S.; Tackett, J. Increased oxidative stress and cytotoxicity by hydrogen sulfide in HepG2 cells overexpressing cytochrome P450 2E1. *Cell Biol. Toxicol.* **2011**, *27*, 439–453. [CrossRef]
11. Bae, J.H.; Jeong, H.J.; Kim, H.; Leem, Y.E.; Ryu, D.; Park, S.C.; Lee, Y.I.; Cho, S.C.; Kan, J.S. ZNF746/PARIS overexpression induces cellular senescence through FoxO1/p21 axis activation in myoblasts. *Cell Death Dis.* **2020**, *11*, 359. [CrossRef] [PubMed]
12. Ouedraogo, G.D.; Redmond, R.W. Secondary reactive oxygen species extend the range of photosensitization effects in cells: DNA damage produced via initial membrane photosensitization. *Photochem. Photobiol.* **2003**, *77*, 192–203. [CrossRef]
13. Besselink, G.A.; van Engelenburg, F.A.; Ebbing, I.G.; Hilarius, P.M.; de Korte, D.; Verhoeven, A.J. Additive effects of dipyridamole and Trolox in protecting human red cells during photodynamic treatment. *Vox Sang.* **2003**, *85*, 25–30. [CrossRef] [PubMed]
14. Forrest, V.J.; David, Y.H.; McClain, E.; Robinson, D.H.; Ramakrishnan, N. Oxidative stress-induced apoptosis prevented by trolox. *Free Radic. Biol. Med.* **1994**, *16*, 675–684. [CrossRef]
15. Guo, C.; He, Z.; Wen, L.; Zhu, L.; Lu, Y.; Deng, S.; Yang, Y.; Wei, Q.; Yuan, H. Cytoprotective effect of trolox against oxidative damage and apoptosis in the NRK-52e cells induced by melamine. *Cell Biol. Int.* **2012**, *36*, 183–188. [CrossRef] [PubMed]
16. Davargaon, R.S.; Sambe, A.D.; Muthangi, S. Trolox prevents high glucose-induced apoptosis in rat myocardial H9c2 cells by regulating GLUT-4 and antioxidant defense mechanism. *IUBMB Life* **2019**, *71*, 1876–1895. [CrossRef]
17. Xiao, T.; Choudhary, S.; Zhang, W.; Ansari, N.H.; Salahudeen, A. Possible involvement of oxidative stress in cisplatin-induced apoptosis in LLC-PK1 cells. *J. Toxicol. Environ. Health A* **2003**, *66*, 469–479. [CrossRef]
18. Singh, P.P.; Chandra, A.; Mahdi, F.; Ray, A.; Sharma, P. Reconvene and reconnect the antioxidant hypothesis in Human health and disease. *Ind. J. Clin. Biochem.* **2010**, *25*, 225–243. [CrossRef]
19. Castañeda-Arriaga, R.; Pérez-González, A.; Reina, M.; Alvarez-Idaboy, J.R.; Galano, A. Comprehensive investigation of the antioxidant and pro-oxidant effects of phenolic Compounds: A Double-Edged Sword in the Context of Oxidative Stress? *J. Phys. Chem.* **2018**, *122*, 6198–6214. [CrossRef]

20. Albertini, R.; Abuja, P.M. Prooxidant and antioxidant properties of Trolox C, analog of vitamin E, in oxidation of low-density lipoprotein. *Free Radic. Res.* **1999**, *30*, 181–188. [CrossRef]
21. Sharma, M.K.; Buettner, G.R. Interaction of vitamin C and vitamin E during free radical stress in plasma: An ESR study. *Free Radic. Biol. Med.* **1993**, *14*, 649–653. [CrossRef]
22. Ingold, K.U.; Bowry, V.W.; Stocker, R.; Walling, C. Autoxidation of lipids and antioxidation by alpha-tocopherol and ubiquinol in homogeneous solution and in aqueous dispersions of lipids: Unrecognized consequences of lipid particle size as exemplified by oxidation of human low density lipoprotein. *Proc. Natl. Acad. Sci. USA* **1993**, *90*, 45–49. [CrossRef] [PubMed]
23. Poljsak, B.; Raspor, P. The antioxidant and pro-oxidant activity of vitamin C and trolox in vitro: A comparative study. *Appl. Toxicol.* **2008**, *28*, 183–188. [CrossRef] [PubMed]
24. Gyulkhandanyan, A.V.; Feeney, C.J.; Pennefather, P.S. Modulation of mitochondrial membrane potential and reactive oxygen species production by copper in astrocytes. *J. Neurochem.* **2003**, *87*, 448–460. [CrossRef]
25. Ko, K.M.; Yick, P.K.; Poon, M.K.T.; Ip, S.P. Prooxidant and antioxidant effects of trolox on ferric ion-induced oxidation of erythrocyte membrane lipids. *Mol. Cell Biochem.* **1994**, *141*, 65–70. [CrossRef]
26. Stewart, M.S.; Spallholz, J.E.; Neldner, K.H.; Pence, B.C. Selenium compounds have disparate abilities to impose oxidative stress and induce apoptosis. *Free Radic. Biol. Med.* **1999**, *26*, 42–48. [CrossRef]
27. Diaz, Z.; Colombo, M.; Mann, K.K.; Su, H.; Smith, K.N.; Scott Bohle, D.; Schipper, H.M.; Miller, W.H., Jr. Trolox selectively enhances arsenic-mediated oxidative stress and apoptosis in APL and other malignant cell lines. *Bood* **2005**, *105*, 1237–1245. [CrossRef]
28. Zheng, J.; Payne, K.; Jori, E.; Taggart, H.J.; Stuart, E.; Ding, L.W.Q. Trolox Enhances Curcumin's Cytotoxicity through Induction of Oxidative Stress. *Cell Physiol. Biochem.* **2012**, *29*, 353–360. [CrossRef]
29. Bowry, V.W.; Stocker, R. Tocopherol-mediated peroxidation. The pro-oxidant effect of vitamin E on the radicalinitiated oxidation of human low-density lioprotein. *J. Am. Chem. Soc.* **1993**, *115*, 6029–6044. [CrossRef]
30. Tafazoli, S.; Wright, J.S.; O'Brien, P.J. Prooxidant and antioxidant activity of vitamin E analogues and troglitazone. *Chem. Res. Toxicol.* **2005**, *18*, 1567–1574. [CrossRef]
31. Maeno, E.; Ishizaki, Y.; Kanaseki, T.; Hazama, A.; Okada, Y. Normotonic cell shrinkage because of disordered volume regulation is an early prerequisite to apoptosis. *Proc. Natl. Acad. Sci. USA* **2000**, *97*, 9487–9492. [CrossRef] [PubMed]
32. Lionetto, M.G.; Giordano, M.E.; Calisi, A.; Caricato, R.; Hoffmann, E.K.; Schettino, T. Role of BK channels in the Apoptotic Volume Decrease in native eel intestinal cells. *Cell Physiol. Biochem.* **2010**, *25*, 733–744. [CrossRef] [PubMed]
33. Antico, S.; Lionetto, M.G.; Giordano, M.E.; Caricato, R.; Schettino, T. Cell Volume Regulation and Apoptotic Volume Decrease in rat distal colon superficial enterocytes. *Cell Physiol. Biochem.* **2013**, *32*, 1551–1565. [CrossRef] [PubMed]
34. Burg, E.D.; Remillard, C.V.; Yuan, J.X.J. K^+ channels in apoptosis. *J. Membr. Biol.* **2006**, *209*, 3–20. [CrossRef]
35. Poulsen, K.A.; Andersen, E.C.; Klausen, T.K.; Hougaard, C.; Lambert, I.H.; Hoffmann, E.K. Deregulation of Apoptotic volume decrease and ionic movements in Multidrug Resistant Tumour cells: Role of the chloride permeabililty. *Am. J. Physiol.* **2009**, *298*, C14–C25. [CrossRef]
36. Bortner, C.D.; Cidlowski, J.A. Apoptotic volume decrease and the incredible shrinking cell. *Cell Death Differ.* **2002**, *9*, 1307–1310. [CrossRef]
37. Maeno, E.; Tsubata, T.; Okada, Y. Apoptotic Volume Decrease (AVD) is independent of mitochondrial dysfunction and initiator caspase activation. *Cells* **2012**, *1*, 1156–1167. [CrossRef]
38. Latronico, S.; Giordano, M.E.; Urso, E.; Lionetto, M.G.; Schettino, T. Effect of the flame retardant Tris (1,3-dichloro-2-propyl) Phosphate (TDCPP) on Na^+-K^+-ATPase and Cl^- transport in HeLa cells. *Toxicol. Mech. Method* **2018**, *28*, 599–606. [CrossRef]
39. Wattamwar, P.P.; Hardas, S.S.; Butterfield, D.A.; Anderson, K.W.; Dziubla, T.D. Tuning of the pro-oxidant and antioxidant activity of trolox through the controlled release from biodegradable poly(trolox ester) polymers. *J. Biomed. Mater. Res. Part A* **2011**, *99*, 184–191. [CrossRef]
40. O'Gara, B.A.; Murray, P.M.; Hoyt, E.M.; Leigh-Logan, T.; Smeaton, M.B. The Vitamin E analog Trolox reduces copper toxicity in the annelid Lumbriculus variegatus but is also toxic on its own. *Neurotoxicology* **2006**, *27*, 604–614. [CrossRef]

41. Morabito, C.; Guarnieri, S.; Cucina, A.; Bizzarri, M.; Mariggiò, M.A. Antioxidant Strategy to Prevent Simulated Microgravity-Induced Effects on Bone Osteoblasts. *Int. J. Mol. Sci.* **2020**, *21*, 3638. [CrossRef]
42. Leventis, P.A.; Grinstein, S. The distribution and function of phosphatidylserine in cellular membranes. *Annu. Rev. Biophys.* **2010**, *39*, 407–427. [CrossRef] [PubMed]
43. Kupcho, K.; Shultz, J.; Hurst, R.; Hartnett, J.; Zhou, W.; Machleidt, T.; Grailer, J.; Worzella, T.; Riss, T.; Lazar, D.; et al. A real-time, bioluminescent annexin V assay for the assessment of apoptosis. *Apoptosis* **2019**, *24*, 184–197. [CrossRef] [PubMed]
44. Chang, S.H.; Phelps, P.C.; Berezesky, I.K.; Ebersberger, M.L.; Trump, B.F., Jr. Studies on the Mechanisms and Kinetics of Apoptosis Induced by Microinjection of Cytochrome c in Rat Kidney Tubule Epithelial Cells (NRK-52E). *Am. J. Pathol.* **2000**, *156*, 637–649. [CrossRef]
45. Yu, S.P.; Choi, D.W. Ions, cell volume, and apoptosis. *Proc. Natl. Acad. Sci. USA* **2000**, *97*, 9360–9362. [CrossRef] [PubMed]
46. Kokubun, S.; Saigusa, A.; Tamura, T. Blockade of Cl channels by organic and inorganic blockers in vascular smooth muscle cells. *Pflfigers Arch.* **1991**, *418*, 204–213. [CrossRef] [PubMed]
47. Pizzino, G.; Irrera, N.; Cucinotta, M.; Pallio, G.; Mannino, F.; Arcoraci, V.; Squadrito, F.; Altavilla, D.; Bitto, A. Oxidative Stress: Harms and Benefits for Human Health. *Oxid. Med. Cell. Longev.* **2017**, 8416763. [CrossRef]
48. Villanueva, C.; Kross, R.D. Antioxidant-induced stress. *Int. J. Mol. Sci.* **2012**, *13*, 2091–2099. [CrossRef]
49. Hasegawa, Y.; Shimizu, T.; Takahashi, N.; Okada, Y. The apoptotic volume decrease is an upstream event of MAP kinase activation during staurosporine-induced apoptosis in HeLa cells. *Int. J. Mol. Sci.* **2012**, *13*, 9363–9379. [CrossRef] [PubMed]
50. D'Anglemont de Tassigny, A.; Souktani, R.; Henry, P.; Ghaleh, B.; Berdeaux, A. Volume-sensitive chloride channels (ICl, vol) mediate doxorubicin-induced apoptosis through apoptotic volume decrease in cardiomyocytes. *Fundam. Clin. Pharm.* **2004**, *18*, 531–538. [CrossRef]
51. Lee, E.L.; Shimizu, T.; Ise, T.; Numata, T.; Kohno, K.; Okada, Y. Impaired activity of volume-sensitive Cl^- channel is involved in cisplatin resistance of cancer cells. *J. Cell. Physiol.* **2007**, *211*, 513–521. [CrossRef] [PubMed]
52. Shimizu, T.; Numata, T.; Okada, Y. A role of reactive oxygen species in apoptotic activation of volume sensitive Cl(-) channel. *Proc. Natl. Acad. Sci. USA* **2004**, *101*, 6770–6773. [CrossRef] [PubMed]
53. Sato-Numata, K.; Numata, T.; Inoue, R.; Okada, Y. Distinct pharmacological and molecular properties of the acid-sensitive outwardly rectifying (ASOR) anion channel from those of the volume-sensitive outwardly rectifying (VSOR) anion channel. *Pflug. Arch.* **2016**, *468*, 795–803. [CrossRef] [PubMed]
54. Arreola, J.; Melvin, J.E.; Begenisich, T. Volume-activated chloride channels in rat parotid acinar cells. *J. Physiol.* **1995**, *484*, 677–687. [CrossRef]

Publisher's Note: MDPI stays neutral with regard to jurisdictional claims in published maps and institutional affiliations.

© 2020 by the authors. Licensee MDPI, Basel, Switzerland. This article is an open access article distributed under the terms and conditions of the Creative Commons Attribution (CC BY) license (http://creativecommons.org/licenses/by/4.0/).

MDPI
St. Alban-Anlage 66
4052 Basel
Switzerland
Tel. +41 61 683 77 34
Fax +41 61 302 89 18
www.mdpi.com

Antioxidants Editorial Office
E-mail: antioxidants@mdpi.com
www.mdpi.com/journal/antioxidants

www.ingramcontent.com/pod-product-compliance
Lightning Source LLC
LaVergne TN
LVHW071628170726
843515LV00009B/2504

* 9 7 8 3 0 3 6 5 1 6 4 4 8 *